Aquatic Entomology

The Jones and Bartlett Series in Biology

AIDS: The Biological Basis, Third Edition
Alcamo

AIDS: Science and Society, Fourth Edition
Fan / Conner / Villarreal

Aquatic Entomology
McCafferty / Provonsha

Botany, Third Edition
Mauseth

The Cancer Book
Cooper

Cell Biology: Organelle Structure and Function
Sadava

Creative Evolution?!
Campbell / Schopf

Defending Evolution: A Guide to the Evolution/Creation Controversy
Alters

Early Life, Second Edition
Margulis / Dolan

Electron Microscopy, Second Edition
Bozzola / Russell

Elements of Human Cancer
Cooper

Encounters in Microbiology
Alcamo

Essential Genetics: A Genomic Perspective, Third Edition
Hartl / Jones

Essentials of Molecular Biology, Fourth Edition
Malacinski

Evolution, Third Edition
Strickberger

Experimental Techniques in Bacterial Genetics
Maloy

Exploring the Way Life Works: The Science of Biology
Hoagland / Dodson / Hauck

Fundamentals of Microbiology, Sixth Edition
Alcamo

Genetics, Fifth Edition
Hartl / Jones

Genetics of Populations, Second Edition
Hedrick

Genomic and Molecular Neuro-Oncology
Zhang / Fuller

Grant Application Writer's Handbook
Reif-Lehrer

How Pathogenic Viruses Work
Sompayrac

Human Anatomy and Physiology Coloring Workbook and Study Guide
Anderson

Human Biology: Health, Homeostasis and the Environment, Fourth Edition
Chiras

Human Embryonic Stem Cells: An Introduction to the Science and Therapeutic Potential
Kiessling / Anderson

Human Genetics: The Molecular Revolution
McConkey

The Illustrated Glossary of Protoctista
Margulis / McKhann / Olendzenski

Introduction to the Biology of Marine Life, Eighth Edition
Sumich

Laboratory Research Notebooks
Jones and Bartlett Publishers

A Laboratory Textbook of Anatomy and Physiology, Seventh Edition
Donnersberger / Lesak

Major Events in the History of Life
Schopf

Medical Biochemistry
Bhagavan

Microbes and Society
Alcamo

Microbial Genetics, Second Edition
Maloy / Cronan / Freifelder

Missing Links: Evolutionary Concepts and Transitions Through Time
Martin

Oncogenes, Second Edition
Cooper

100 Years Exploring Life, 1888–1988, The Marine Biological Laboratory at Woods Hole
Maienschein

The Origin and Evolution of Humans and Humanness
Rasmussen

Origin and Evolution Of Intelligence
Schopf

Plant Cell Biology: Structure and Function
Gunning / Steer

Plants, Genes, and Crop Biotechnology, Second Edition
Chrispeels / Sadava

Population Biology
Hedrick

Protein Microarrays
Schena

Statistics: Concepts and Applications for Science
LeBlanc

Vertebrates: A Laboratory Text
Wessels

Aquatic Entomology

The Fishermen's and Ecologists' Illustrated Guide to Insects and Their Relatives

W. PATRICK MCCAFFERTY

WITH ILLUSTRATIONS BY

ARWIN V. PROVONSHA

JONES AND BARTLETT PUBLISHERS

Sudbury, Massachusetts

BOSTON TORONTO LONDON SINGAPORE

World Headquarters
Jones and Bartlett Publishers
40 Tall Pine Drive
Sudbury, MA 01776
978-443-5000
info@jbpub.com
www.jbpub.com

Jones and Bartlett Publishers Canada
P.O. Box 19020
Toronto, ON M5S 1X1
CANADA

Jones and Bartlett Publishers International
Barb House, Barb Mews
London, W6 7PA
UK

The following are gratefully acknowledged for the use of quoted material appearing as epigraphs. A & C Black Limited (G.E.M. Skues); Mrs. Norma Millay Ellis (*The Harp-Weaver and Other Poems* by Edna St. Vincent Millay, 1923); Harcourt Brace Jovanovich, Inc. (*Afternoon of a Pawnbroker and Other Poems* by Kenneth Fearing, 1943); Mrs. James Thurber (*The White Deer* by James Thurber, 1945); and Liveright Publishing Corp. (*The Bridge* by Hart Crane, 1930).

Library of Congress Cataloging in Publication Data
McCafferty, W. Patrick.
Aquatic Entomology.
Includes bibliography and index.
1. Insects, Aquatic—North America—Identification. 2. Insects—Identification. 3. Insects—North America—Identification. I. Title
QL473.M35 595.7097 83 -13577

ISBN: 0-86720-017-0

ABOUT THE COVER: Copyright © 1981 by Arwin V. Provonsha. A freshly hatched subimago of *Stenonema vicarium*, a large and variable flatheaded mayfly that develops in eastern and north-central streams. The appearance of this dun, known to fisherman as the Ginger Quill or March Brown, can signal some excellent fly fishing throughout the day in early summer.

PUBLISHER Arthur C. Bartlett
MANUSCRIPT EDITOR Dick Johnson
BOOK AND JACKET DESIGN Paula Schlosser
PRODUCTION Hal Lockwood, Bookman Productions
COMPOSITION Interactive Composition Corp.
COLOR SEPARATIONS Colorscan

Printed in the United States of America
03 10 9

IN MEMORIAL TO HERBERT H. ROSS,
a guiding light in entomological excellence

In Nature's infinite book of secrecy
A little I can read

William Shakespeare

PREFACE

AQUATIC BIOLOGY AND NATURAL HISTORY are subjects of interest to many, whether they be related to one's vocational education, one's avocation, or purely to one's appreciation for the living order of the world. This book has been written not only for entomologists, ecologists, and students of aquatic entomology, but also for sports fishermen, naturalists, and environmental assessment specialists. For those who may not have some vested interest in the field of aquatic entomology per se but do have a general interest in nature and ecology, this book will provide a pictorial introduction to some of the most fascinating life forms on earth and, I hope, "wet" the appetite for understanding the aquatic insects, their environment, and their relationship to human life.

Although originally conceived as an illustrated field guide to the aquatic insects of North America, the book evolved to embody much more during its writing. It covers many of the approaches to introducing principles of aquatic entomology that have been developed in a course I have taught over a ten-year span, and it incorporates a considerable amount of general biological and ecological information generated by contemporary researchers. In fact, it is largely due to the tremendous growth of aquatic entomology and the discoveries of the past 15 years or so that a book of this kind is possible. For the fly fisherman, in particular, a good deal of applied information was added. To my knowledge, the book is the first comprehensive work to provide a sound scientific introduction to insects for the fly fisherman, including identification, up-to-date association of fishermen's names and correct scientific names, and complete discussions of insect biology applicable to scientific fishing.

Although peripheral to aquatic entomology, special chapters dealing with macroinvertebrates, semiaquatic insects, crustaceans, arachnids, and collembolans are included to round out the book. These chapters will be especially appealing for teachers and students who, even if primarily concerned with freshwater insects, cannot study them at the exclusion of forms to which they are related either taxonomically or environmentally.

Because of the broad audience for which the book is intended, technical jargon and multisyllabic scientific terms are used sparingly and avoided whenever possible. Common names of aquatic insects are emphasized and used liberally in hopes of standardizing the vernacular of North American aquatic insect groups and bringing the book within the realm of the nonspecialist and general reader.

The unique working relationship between the illustrator and myself has been the most important factor in the compilation of this book. The artwork, including the more than 1,000 original illustrations, should help advance the general awareness of aquatic insects and raise the subject to a level of interest demonstrated by nature-enthusiasts of the more popular animal groups. The descriptions and illustrations should enable users of this book to recognize the vast majority of aquatic insects they will encounter.

In the study of insects, more so perhaps than for any other major group of familiar organisms, emphasis is placed on the family level of classification, because families of insects are often the smallest group conveniently learned by the general student. Whereas individual species of birds, fishes, or flowering plants may be competently learned by students of those groups, such a task would normally be quite impossible for the student of insects (perhaps 8,000 or more aquatic species occur north of Mexico). Families of aquatic insects, therefore, are of primary concern in this book, though important genera and species have not been neglected. Useful references are indicated at the ends of pertinent chapters for those readers who may be interested in generic and species identification.

I am indebted to many individuals for their critiques, information, suggestions, and specimens they have loaned for illustration. The following are gratefully acknowledged for reading parts of the manuscript or chapters within their area of expertise: Donald Alstad, William M. Beck, Jr., Mark Deyrup, George F. Edmunds, Jr., Benjamin A. Foote, Todd L. Harris, William L. Hilsenhoff, John Keltner, B. Elwood Montgomery, Andrew P. Nimmo, Arwin Provonsha, Carl. W. Schaefer, Paul J. Spangler, Kenneth W. Stewart, and Rebecca Surdick-Pifer. I, alone, am responsible for any errors or failure to communicate effectively with readers.

The following additional individuals are gratefully acknowledged for providing specific information or specimens for illustration: S. W. T. Batra, Richard W. Baumann, Andrew F. Bednarik, Daniel Bloodgood, Warren U. Brigham, Marion Buegler, Oliver S. Flint, Jr., Wills Flowers, George Godfrey, K. G. Andrew Hamilton, Donald G. Huggins, Mike Lawson, Marc Minno, Charles W. O'Brien, William L. Peters, John T. Polhemus, Annelle R. Soponis, Stephen Tessler, John D. Unzicker, Robert Waltz, and Donald W. Webb.

To George F. Edmunds, Jr., I am especially thankful for his inspiration and support in this endeavor. To Nadine McCafferty and Judy Provonsha, without whose patience and understanding over the past few years this book could never have been completed, I offer my sincerest gratitude. And finally I wish to express my gratitude to Arthur Bartlett and Ric Davern for their keen interest in the book, to Dick Johnson, Paula Schlosser, and Hal Lockwood for their editorial, design, and production assistance, and to Nadine McCafferty for typing the many drafts.

W. *Patrick McCafferty*
March 1981

CONTENTS

LIST OF INSECTS DETAILED BY WHOLE SPECIMEN, BLACK-AND-WHITE ILLUSTRATIONS

(SEE COLOR PLATES FOR ADDITIONAL SPECIES REPRESENTED IN COLOR)

PART I

BACKGROUND

CHAPTER 1

Introduction and Organization

CONCERN WITH INSECTS is assuredly as old as the human race itself. Pictorial representations of insects date from some of the earliest cave sketchings and entombments known to archeologists, and observations of insects are found in the earliest known writings of scholars, even predating Aristotle by as many as 1,000 years or more. In particular, aquatic insects must always have preoccupied humankind. The spectacularly massive seasonal emergences of the short-lived adult forms of aquatic insects from the rivers and lakes of the world could hardly have gone unnoticed, and the bites of mosquitoes and certain other aquatic flies must have seriously plagued our earliest ancestors.

There are many historical and cultural examples of human awareness of aquatic insects. Aelianus, in *De Animalum Natura*, which dates from the second century A.D., wrote of caddisflies flying over the waters of the Astraeus in Macedonia and of the use of artificial flies for catching trout and graylings. Swarms of mosquitoes are depicted on thirteenth-century American Indian pottery from New Mexico. References to the short-lived adults of riverine mayflies are contained in Hungarian folksongs that have probably been passed down through centuries; and poets throughout the world have always been enamored with the symbolic relationship between human existence and the ephemeral nature of certain aquatic insects.

Indians of western North America are known to have gathered the great masses of brine flies that washed ashore along coastal marshes and

When in the spring the Tisza blossoms,
Thousands of mayflies play above its running ripples.
None of them lives until I count hundred:
The Tisza turns to a grave-yard when it blossoms.

Our love was like the life of a mayfly.
It ended before it really began to blossom.
But with tears in my eyes I am longing
For the Tisza when it blossoms!

Hungarian folksong

salt lakes and to have used them for food. To the Aztecs of Mexico, these morsels were known as "puxi," and to the Paiutes to the north they were known as "koo-cha-bee." Even today, Papuan tribes of New Guinea relish the large mayflies that seasonally emerge from their rivers; the people of China and other parts of Asia eat giant water bugs after boiling them in salt water; and the eggs of certain water bugs are prepared for sale and human consumption in Mexico, where they are known as "ahualte."

Historical Perspective

Following the authoritarian writings about biology contributed by the ancient Greek philosophers, it was not until the sixteenth and seventeenth centuries that scientific observations of insects began to permeate the works of European anatomists and naturalists. Much of this upsurge paralleled the early use of the microscope and the accumulation of specimens that were being gathered from explorations throughout the world.

The 1675 work *Ephemeri Vita*, by the Dutch anatomist Jan Swammerdam, stands as a scientific landmark in the study of aquatic insects. Swammerdam extensively studied the burrowing mayfly, *Palingenia longicauda*. Besides fully detailing the anatomy of this insect, along with its eggs, burrows, and its transformation from an aquatic to a terrestrial form, Swammerdam distinguished the gills as being respiratory organs and recognized the importance of certain internal structures as well as sexual characters of the adults.

About the same time, the Englishman Charles Cotton, who many recognize as the father of modern fly fishing, contributed to the 1676 edition of Izaak Walton's *The Compleat Angler*. In this edition, some of the first, albeit prosaic, correlations between aquatic insect biology and fly fishing appeared, although some noteworthy observations had already appeared in John Taverner's *Certaine Experiments Concerning Fish and Fruite,* published in 1600. These works were the rudiments of a long line of treatises on angling and aquatic insects in the British Isles that were to appear over the next three centuries.

Important advances in biological thinking during the eighteenth century opened the door to a more intensive study of insects worldwide. The implementation of a utilitarian system of classifying organisms is particularly noteworthy; the 10th edition of *Systema Naturae* (1758), by the Swedish biologist Carolus Linnaeus, marks the beginning of modern classification. Linnaeus catalogued many insects, including aquatic species, but it was his Danish student, Fabricius, who is considered to have been the first outstanding insect taxonomist. During this century, the only real progress in aquatic entomology from the fly fisherman's point of view was contained in Charles Bowlker's *The Art of Angling*, published in 1780. Bowlker introduced elaborate fly-tying methods and designs intended to imitate the detailed structures of aquatic insects.

Significant contributions to knowledge of North American insects began in the nineteenth century. Two notable entomologists who described many aquatic insect species were Thomas Say and Benjamin Walsh. Say's *American Entomology, or Descriptions of the Insects of North America* appeared in three volumes between 1817 and 1828. Say's interests as a naturalist were broad, but he seems to have been particularly fond of the fauna of rivers, and he described many mussels as well as insects from these environments. Walsh amassed a large collection of aquatic insects, mainly from the vicinity of Rock Island, Illinois, near the Mississippi River. His descriptions were published between the years 1862 and 1864. Also during this period, the American ecologist and limnologist Stephen Forbes had come to realize the various and important roles that aquatic insects played in an aquatic ecosystem. His paper "The Lake as a Microcosm" was read to the Peoria Scientific Association in 1887, and today stands as an early classic of aquatic ecology.

British anglers had continued to sophisticate their sport during the nineteenth century and had begun a scientific application of the knowledge of forms and habits of aquatic insects to fly fishing. Alfred Ronald's *Fly Fisher's Entomology*, which appeared in 1836, emphasized theories of imitation and set a new standard for discussing and illustrating aquatic insects. These trends were followed by a number of authors; other important works published during this period include Michael Theakston's *List of Natural Flies* (1853) and Frederick Halford's *Dry Fly Entomology* (1897).

Fly fishing in North America was greatly influenced by European writing, and to some extent still is today. Early literary contributions of American authors to angling entomology had a rather little noted beginning with George Washington Bethune's by-lines to the first American edition of Walton's ever-charming *The Compleat Angler*, which appeared in 1847. Later in the nineteenth century, Theodore Gordon, who came to be known as the master of American fly fishing, set the standard for modern American fly tying.

Advances in understanding aquatic insects were being made on another front during the late nineteenth century and were to revolutionize strategies for controlling some of the world's most dreaded diseases. In 1878, Patrick Manson, working in China, discovered that mosquitoes transmit filariasis. Soon after, discoveries were made that certain mosquitoes were also the vectors of malaria and yellow fever. Ronald Ross, in India, proved the association between anopheline mosquitoes and malaria in 1897, and Walter Reed and his co-workers in the United States Army Yellow Fever Commission, working in Cuba, demonstrated in 1900 the association between yellow fever and the mosquito *Aedes aegypti*.

It is impossible to review in this brief introduction the many important aquatic entomologists and outstanding discoveries about aquatic insects that they made in the twentieth century. Nevertheless, the American James G. Needham (1868–1957) probably should be credited as much as any other one worker with formalizing aquatic entomology into a distinctive discipline in the first half of this century. Needham, together with his students and colleagues, published on nearly every major order of aquatic insects and seeded much of the present interest of researchers.

The study of the taxonomy, biology, and ecology of aquatic insects has flourished in the twentieth century and has received the attention of many pre-eminent biologists throughout the world. Not only have entomologists come to specialize in the study of certain groups of aquatic insects, thus immensely advancing our knowledge of diversity, relationships, and natural histories of many of the previously poorly known forms, but the study of freshwater ecology has become a vastly important field, at least partly because of the realization that water is a vital resource to be protected and preserved. Educational institutions have met the challenge of a growing need for entomologists, ecologists, fisheries biologists, and technicians with special knowledge of aquatic insects by rapidly incorporating aquatic entomology courses into curricula at the college and university level.

Every traveller is a self-taught entomologist.
Oliver Wendell Holmes

As a result of the multiple-use pressures that are placed on streams, rivers, and lakes by the ever-increasing human population, sports fishermen have witnessed the dwindling productivity of some of their favorite haunts, which have become more challenging to fish. Today's fisherman is more appreciative than ever before of the delicate balance of life in these habitats. Great strides have been made in bridging the gap between biology and fishing, and a more precise knowledge of aquatic insects is demanded of the modern fly fisherman.

Applications of Aquatic Entomology

Aquatic entomology is simply the study of aquatic insects. Although it is a discipline that can stand alone, it often interacts with other disciplines or forms an integral part of them. For example, many aquatic flies are of importance as disease vectors, hence aquatic entomology interacts with medical entomology and parasitology; moreover, although aquatic insects contribute significantly to a freshwater ecosystem, they are just one of many groups of organisms that, together, must be considered in the study of aquatic ecology. Some of the applications of aquatic entomology are briefly reviewed below.

Environmental Protection

It would be difficult to overestimate the importance of aquatic insects as food items for other animals, particularly in the food webs associated with wetland environments. Many fishes, amphibians, shorebirds, waterfowl, and other animals forage heavily on both the aquatic and terrestrial stages of aquatic insects, which are essential to their survival.

It is imperative that society fully understand the potential consequences of the uses and alterations of the earth's natural environment. Environmental scientists have come to realize that uses or changes of waterways and lakes may have dire short-term or even irreversible long-term effects not only on the quality of water itself but also on aquatic ecosytems. As a result, research has been employed to ascertain the effects of perturbations on our water resources and to increase our knowledge about the makeup of natural communities of aquatic organisms and their relationships with natural environments.

To this end, aquatic organisms that may be affected by an impending alteration or activity are often surveyed as part of an environmental assessment or impact study. This should be done, for example, before impounding a river to create a reservoir; before dredging a stream; before constructing power-plant cooling facilities, sewage treatment plants, or factories that may deliver potentially toxic effluents into a river or lake; before mining or deforestation activities; or before spraying chemicals over wetlands and forests. Measurements of the richness and diversity of aquatic insect species in relationship to the chemical and physical characters of their environment provide very usable indices for such baseline studies and are commonly made for these purposes.

Aquatic insects and other bottom-dwelling organisms in freshwater systems are also monitored in order to gauge subtle and profound effects that changes in water quality have on aquatic life. Changes in the composition of bottom-dwelling communities, as measured both qualitatively and quantitatively, will reflect, to various degrees, either nonoptimum or intolerable quality shifts that result from the addition of pollutants to the water. Sources of such pollutants may be continual, intermittent, or accidental, and they may originate either from precise points or from large areas.

Biomonitoring of this sort has some decided advantages over chemical and other types of water analyses. Bottom-dwelling insects and other invertebrates, by their very nature, maintain a relatively stable position in the aquatic environment. Thus the community composition can reflect either previous or long-range shifts in water quality. Analyses of water chemistry or of highly mobile aquatic animals such as fishes tend to reflect the quality of water only at the times that samples are taken.

Ecology is destined to become the lore of Round River, a belated attempt to convert our collective knowledge of biotic materials into a collective wisdom of biotic navigation. This, in the last analysis, is conservation.

Aldo Leopold

Aquatic insects are also used in bioassay work. Laboratory bioassays are performed to determine the toxicological effects of pesticides on aquatic insects. The pesticides may be those that are intentionally used in aquatic or adjacent environments for the control of mosquitoes and other pests or they may be those that inadvertently find their way into aquatic ecosystems—for example, via agricultural runoff. Such bioassays are commonly used to ascertain effects on nontarget aquatic insect species, and thus have a direct relationship to environmental protection. In addition, several aquatic insects are studied in the laboratory in order to determine the mode of action of pesticides and the relative tolerances that these insects have for pollutants, such as heavy metals, organic enrichment, and heated waters.

Aquatic insects for which sufficient toxicological data are available have some potential for being used in field bioassays. That is, caged stock populations could be implanted at various points in a stream—for example, at the source (or supposed source) of a pollutant, upstream from the source, and downstream from the source. The relative mortality of these implanted populations could then be used as pollution indicators. Certain fish species have been commonly used in such field bioassays.

Fisheries Resources

Aquatic insects and other aquatic invertebrates are staples in the diets of many freshwater fishes. Because aquatic insects are, to a large degree, responsible for converting plant material into animal tissue in freshwater ecosystems, they can be of immense importance in the food chains leading to fish production. It follows that the degree to which an aquatic environment is able to support fish populations is, in part, directly related to the relative abundance of certain aquatic insects, and that efficient fish management is dependent on integrated studies that include aquatic entomology.

Fisheries biologists are cognizant of the fundamental and complex relationship between primary aquatic productivity, the abundance and diversity of fish food organisms, and the quality of fish populations. Fisheries resource research must include not only the study of insect ecology but also the study of food preferences and feeding habits of fish species. A simple yet important index of the interaction between a fish and its environment is a so-called forage ratio. This ratio takes into account the percentage of a food item, such as an aquatic insect, in the stomach contents of a fish as well as the percentage of that same food item among all the available food in the fish's environment. A combination of methods for studying food habits are used by fisheries biologists to appraise the kinds and volume of food eaten by fishes. Ideals in fisheries management are approached when this information is correlated with food availability, preferences and requirements, nutritional values, and the efficiency with which different food organisms are assimilated.

Certain aquatic insects are of concern in fisheries management because they are predators. Many dragonflies, water bugs, dobsonflies, water beetles, and a few others will readily attack small fishes and fry. Although such predation is probably of minor consequence in most instances, it is a factor to be considered by the fisheries biologist. These predatory insects are of particular concern when small ponds are stocked with small fry, as in aquacultural enterprises; however, it should be realized that the majority of aquatic insects are highly desirable in pond fish cultures. Some species have a potential for being propagated as feed.

Aquatic Insect Pests

Many of the most important pests of humans and livestock are aquatic insects. Endeavors to control these insects continue as a prolonged war, with some battles won and others seemingly hopeless. The inestimable amount of human and material resources that have been devoted to this war are certainly justifiable if one considers the consequences of such diseases as malaria, encephalitis, yellow fever, and several others that are insect borne. There are many other aquatic insects, besides the disease carriers, that are also considered to be direct or indirect pests, and an overview of aquatic insects as pests follows. Many aquatic insect pests warrant control measures, and others do not; details of control strategies, which fall under the subject of pest management, are, however, beyond the scope of this book.

The most notorious aquatic insect pests are those that, as terrestrial adults, bite people. Black flies, sand flies, biting midges, mosquitoes, horse flies, and deer flies all fall into this category. Some are carriers or potential carriers of human diseases and diseases of livestock, poultry, and waterfowl. All of these pests are blood feeders, and their bites can be painful and sometimes cause allergic reactions. Large numbers of such pests can render an area virtually intolerable for purposes of habitation or recreation and can cause severe decreases in livestock production.

Mosquito and gnat control has been and continues to be carried out intensively throughout the world. Insecticides are used for killing both the aquatic and the terrestrial stages; certain breeding habitats are regularly eliminated; and predators (including some carnivorous aquatic insects) and pathogens are used to varying degrees for control. Abatement organizations, which have been established to carry out regular surveillance and control, are common in those areas where outbreaks of these pests pose serious problems. The World Health Organization has been particularly concerned with these pests in less developed countries. Experience has shown that whatever control and eradication practices are employed, their long-range success and the maintenance of natural environments depend on thorough and sound biological and ecological study of not only the pest species but also other species with which they are associated.

Some aquatic insects are pestiferous merely because of their large numbers. Certain mayflies, brine flies, shore flies, midges, and caddisflies emerge in great masses from aquatic habitats, especially large rivers and lakes. Some flies also breed in large numbers in sewage treatment facilities. When mass emergences occur, the swarming individuals are sometimes a nuisance in populated or recreational areas. Many of these insects are attracted to lights, and this may accentuate the problem considerably. Burrowing mayflies have been known to pile up several feet deep on shores of lakes and rivers. On river bridges in some parts of the Midwest, they have actually had to be removed with snow plows in order to clear the way for traffic. Furthermore, some mayflies and caddisflies have been implicated as allergens; fine particles of the decayed remains of such massive populations can enter the atmosphere to be inhaled and cause allergic reactions in generally susceptible people.

For the sake of conservation, aquatic insects that are merely an occasional nuisance should perhaps be tolerated whenever possible: Their overall benefit to the ecology of a natural environment would seem to outweigh any occasional unpleasantries to humans.

A few aquatic insect species are serious pests of rice crops. The constantly flooded fields in which many varieties of rice are cultivated provide an excellent environment for aquatic insects. Species that feed directly on rice plants, either by boring into the stems, mining the leaves, or feeding externally, include the Rice Water Weevil; some aquatic flies, such as the Rice Seed Midge and the Rice Leaf Miner; and the Rice Borer, which is a semiaquatic caterpillar. Damage to rice crops is often sufficient to warrant integrated pest-management practices to control these pests. Even insects that do not feed on rice may cause damage; the burrowing activities of large water scavenger beetles have been known to uproot entire plants.

A number of aquatic insects that cause no medical or agricultural problems or are not a general nuisance may have some occasional importance as pests, as the following examples show. Small aquatic insects, such as phantom midges, have been known to gain entrance to drinking-water systems through filters. Biting water bugs and other undesirable insects commonly enter swimming pools. Insects in large numbers, such as certain caddisflies, may clog irrigation tunnels. Certain burrowing mayflies burrow into and damage dock pilings on large rivers in Asia and Africa. Adult mayflies, midges, and caddisflies may cause damage to automobile finishes, particularly where the automobiles are parked in large, lighted lots in the vicinity of rivers and streams; upon being attracted by the reflective surfaces and alighting there, the insects may become stuck and die or even lay sticky egg masses. As mentioned previously, large numbers of adult burrowing mayflies sometimes obstruct bridges and roads.

Biological Control

Many aquatic weeds are undesirable when they occur in large numbers in ponds and streams. They may clog waterways and water treatment facilities, hinder fishing and access to the water, cause silting of shellfish beds, and enhance habitats that harbor mosquitoes and other pests. In pond fish cultures, weeds use up fertilizers without significantly increasing fish food and interfere with the activities of desirable fishes, such as bass, which by their feeding help to control overproduction of plankton and certain less desirable fishes.

A number of herbicides can be used to control weeds, and a few weed-eating fishes, such as the grass carp, have been used to some extent for control. Several insects feed on aquatic weeds. Although their effect in controlling weed populations is probably only minimal in natural situations, some hold the potential for use as control agents if introduced into environments in sufficient numbers. Among those that could possibly be used in this way are some of the aquatic and semiaquatic caterpillars, water weevils, aquatic leaf beetles, and even some Old World burrowing mayflies. The use of a nonaquatic leaf beetle for the biological control of alligator weed has already met with some success.

Freshwater snails serve as intermediate hosts for several human and domestic animal diseases, particularly in tropical and subtropical regions of the world. Marsh fly larvae, which naturally feed on freshwater snails, have received considerable attention from researchers as possible biological agents for controlling these hosts and thus the disease pathogens they harbor.

The larvae of certain water beetles and other predatory aquatic insects readily feed on mosquitoes, and thus have the potential of being managed for mosquito control. The predatory larvae of certain water scavenger beetles hold some promise for controlling pests of rice, such as the rice seed midges.

The beneficial role that certain aquatic insects play in sewage treatment should be mentioned as being at least indirectly related to biological control. One method of sewage treatment employs a series of oxidation ponds in which sewage is broken down by the activity of bacteria. Algae and decomposing sludge are eventual by-products in such a system. Midges contribute significantly to the efficacy of the system by feeding on the algae and sediments that, if left unchecked, would fill or clog the system. Because of their burrowing activities, the midges also increase the exchange of nutrients and oxygen between the water and the bottom mud.

Fluid sewage is commonly treated in trickling filter systems. The sewage is passed through a filter bed of rocks or objects that provide a large amount of surface area, and the bacterial and algal film that forms on the filtering surfaces promotes the breakdown of the sewage. The feeding activities of certain aquatic flies, such as moth flies, prevent the surface film of ooze from becoming excessive and clogging the filtering system. If the flies become too numerous, however, they can adversely scour the filter, and they may emerge massively to form pest swarms.

Sports Fishing

Since aquatic insects are the "meat and potatoes" of many freshwater game fishes, the importance of aquatic entomology to the sports fisherman and in particular the fly fisherman is as paramount as knowledge of the fishes themselves. That such a statement is no exaggeration can be seen merely by browsing through the fishing literature, wherein the subject of insects and flies has received voluminous attention. Many very recent fishing treatises, such as *Selective Trout* by D. Swisher and C. Richards, *Nymphs* by E. Schwiebert, *Hatches* by A. Caucci and B. Nastasi, and *Meeting and Fishing the Hatches* by C. R. Meck, to name just a few, have been devoted almost entirely to aquatic insects, their use, and imitations. A lengthy vocabulary of fly fishing terms and insect names has originated from fishermen, and further exemplifies the importance of entomology to the sport. The fishermen's names of mayflies in North America, alone, are remarkably extensive (see Appendix).

Numerous kinds of live aquatic insects are used by bait fishermen. The most commonly used species are the larger forms, such as giant stoneflies, hellgrammites, burrowing mayflies, dragonflies, and crane flies. Many of these are commercially available in bait shops in various regions throughout North America. For example, burrowing mayflies are stocked in certain bait shops in parts of the Midwest and Canada because of their popularity with ice fishermen in these regions; giant stoneflies (salmonflies) are sold in regions of the West because of their popular use on trout streams; and rattailed maggots (mousies) are often available in bait shops in the Midwest and South because of their popularity in fishing for bass and other warmwater fishes. Of course, nonaquatic insects, such as crickets, are also very popular with bait fishermen, although the bait species often sold as "water crickets" are actually common stonefly larvae.

Fly fishing is a multimillion dollar sports industry that has its basis in the fact that insect-feeding fishes (mainly trout) can be caught with artificial lures fashioned after insects and, in some instances, other invertebrates or minnows. Flies, as the lures have long been known, are usually made from parts of feathers, animal hairs, and thread that are tied onto a fish hook. Many fishermen pride themselves on their skill in fly tying, and many books have been devoted to the various methods of fly tying and fly patterns. Large selections of flies are also available from fishing supply houses and bait shops.

Some flies are based on the winged stages of such groups as mayflies, caddisflies, stoneflies, and midges, and are fished as dry flies; that is, they are fished primarily on the water surface with a water repellent line, and fishes rise to them. Other flies are fished wet (below the surface), and may resemble either the winged stages or the immature aquatic stages, in which case the term "nymph fishing" is applied. Some flies, known as attractors, do not necessarily resemble any particular insect.

The feeding habits of trout can vary considerably with the variety of trout, time of day and season, region, and local habitat. The search images that are formed by trout, and to which they will respond, apparently depend to varying degrees on the form, activity, time of appearance, abundance, and possibly even the color of the insects present in their habitat. The great challenge to the sophisticated fly fisherman is to utilize all of this information in the choice of fishing tactics and selection of flies in order to outsmart the fish or trick it into taking a fly. To this end, experience, knowledge of fish habits, and an ability to recognize the aquatic insects that occur naturally in an environment are all essential.

Among fly fishermen, there are two general schools of thought. The first is the imitationist school, which includes those fishermen who believe that their catches are attracted by the naturalness of the flies used. In other words, the closer the fly imitates the exact insect species that the fish is

When artful flies the Angler would prepare,
This task of all deserves his utmost care:
Nor verse nor prose can ever teach him well
What masters only know, and practice tell;
Yet thus at large I venture to support,
Nature best followed best secures the sport:
Of flies the kinds; their seasons, and the breed,
Their shapes, their hue, with nice observance heed:
Which most the Trout admires, and where obtain'd,
Experience will teach, or perchance some friend.

Moses Browne

attuned to at the time, the better the fishing. The second school is the presentationist school, which includes those fishermen who believe that their catches are a result of their fishing skills and who give relatively lesser regard to the naturalness of flies they use. A review of the fly fishing literature, past and present, and the logic propounded by experienced fishermen will not resolve these two views. Matching the fly to the exact prey of the fish is obviously of greater importance to the imitationists (who seem to form the majority); however, regardless to which school a fly fisherman belongs, the one with an understanding of aquatic insects and their behavior will have a definite advantage in knowing how, when, and where to fish. Fly fishing strategies, as they may be enhanced by a knowledge of aquatic insect taxonomy, biology, and ecology, are touched upon in those chapters of this book that cover the relevant groups of insects.

Scope and Organization of the Book

This book is divided into four major parts. Part I (Chapters 1–5) deals with the background information necessary for a thorough understanding of the descriptions and discussions in the remainder of the book. This part of the book will be of most value to those readers who have little or no previous knowledge of entomology or aquatic biology. The introductions to morphology, life cycles, classification, habitats, ecology, aquatic adaptations, field and laboratory methods, and freshwater invertebrates encompass, to a large degree, an explanation of the technical terminology that is indispensable to the student of aquatic entomology.

Part II includes an order by order coverage of the aquatic insects of North America north of Mexico (Chapters 7–16) and a picture key to the orders (Chapter 6). The approximate number of species in those orders is given in Table 1.1. Each ordinal chapter contains a consistently comparative treatment of several subjects. Following a short introduction to the order, the sequence of topics includes a morphological description of postembryonic life stages, a discussion of similar groups with which the order may sometimes be confused, an overview of the life history (including habits of nonaquatic stages), the aquatic habitats in which the group is encountered, a general discussion of the adaptations and behavior of the aquatic stages, and an introduction to the higher classification and taxonomic characters relevant to family classification. The bulk of each ordinal chapter is devoted to the aquatic families within the order, including their identification, North American distribution, primary ecology, and behavioral adaptations.

TABLE 1.1 APPROXIMATE NUMBERS OF AQUATIC INSECT SPECIES OCCURRING IN NORTH AMERICA NORTH OF MEXICO

Order	Number
Ephemeroptera (mayflies)	700
Odonata (damselflies, dragonflies)	450
Plecoptera (stoneflies)	500
Hemiptera (water bugs)	400*
Megaloptera (fishflies, dobsonflies, alderflies)	50
Neuroptera (spongillaflies)	6
Coleoptera (water beetles)	1,000*
Trichoptera (caddisflies)	1,200
Lepidoptera (aquatic caterpillars)	50
Diptera (midges, mosquitoes, aquatic gnats and flies)	3,500*

*Includes semiaquatic species.

North American distribution is given by general regions, which are indicated in Figure 1.1. In certain respects, these geographic regions also constitute broad natural units, but they are established primarily as a convenience to readers who may want to know whether a group is present or absent or common or uncommon in a particular region of interest. It should be noted that the term "North America," as used throughout the book, does not include Mexico. The states of Florida, Texas, and Alaska are biogeographically distinct enough with regard to their aquatic fauna that they have warranted separate status as regions. If a group is noted as being "widespread," it should be interpreted as being found in all regions, unless otherwise indicated.

The families are covered in a more-or-less systematic sequence within each chapter and are listed by both their common and their scientific names. The discussions of familial habitat preferences, adaptations, and behavioral traits vary in extent according to the size and importance of the family, and most examples are drawn from the larger or more common genera. Some members of the families are mentioned in relation to their interest to fly fishermen, their use as indicators of water pollution, and their application to other aspects of aquatic entomology. For those families that can be divided into subfamilies, the identifying traits and distributions of the subfamilies are given.

Accompanying the family treatments are labeled illustrations for identifying aquatic insects and emphasizing important characteristics. Each major ordinal chapter includes a pictorial key to the aquatic stages of families in the order covered. A set of color plates depicting many representative and common species appears after Part II, following page 335.

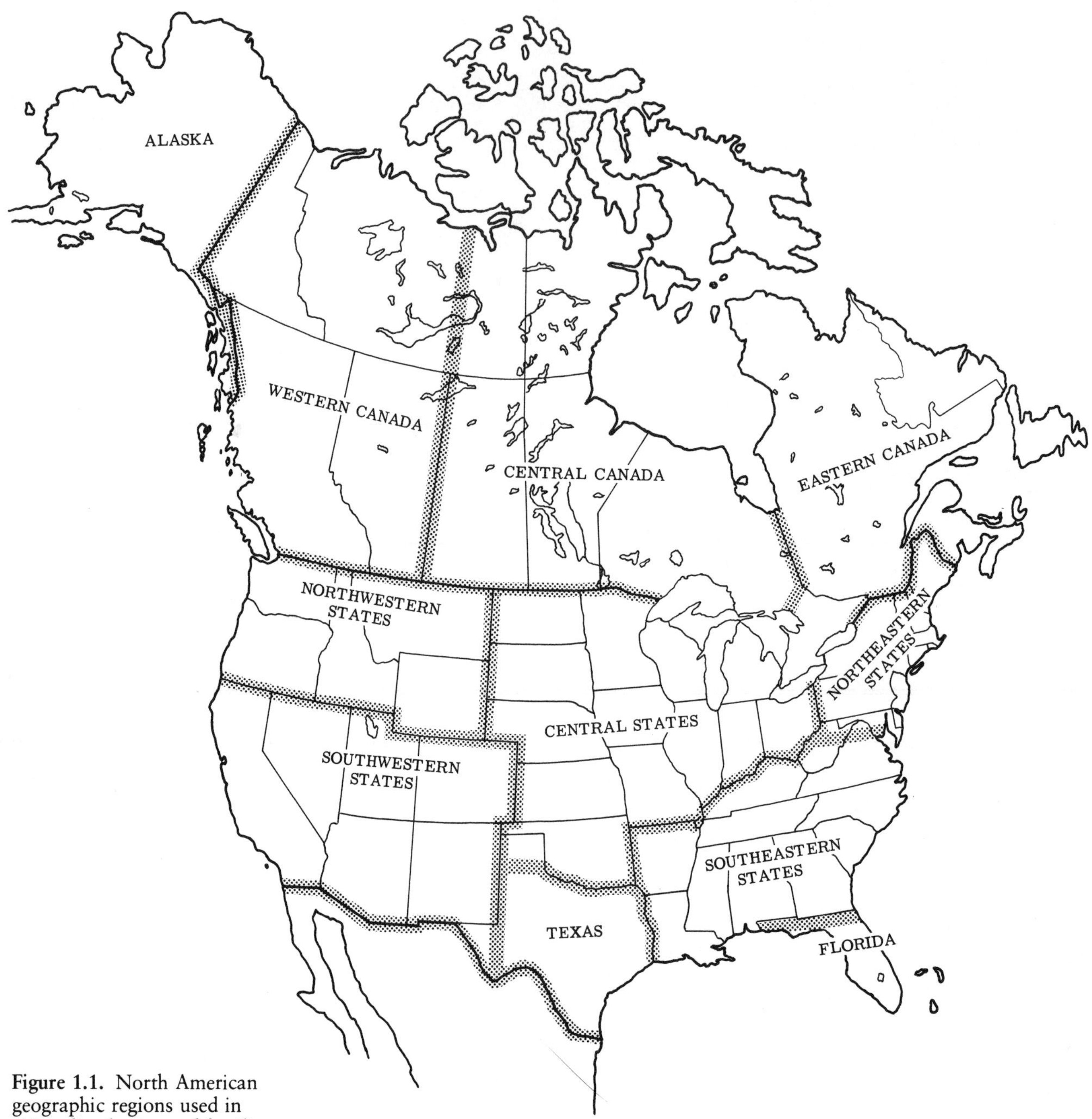

Figure 1.1. North American geographic regions used in citing distributions of families and subfamilies.

References to these color illustrations are given in the diagnostic sections that cover individual families in each order.

Part III of the book (Chapters 17–20) covers groups of insects that are only indirectly or peripherally associated with aquatic environments or are found in atypical wet habitats. Many of these insects are those that have been termed semiaquatic. Thus there are chapters on shore-dwelling insects (including many of the marine as well as inland species), on insects associated with emergent vegetation, on insects that inhabit plant cups and tree holes, and on diving wasps. Many orders covered in Part II are also included in Part III, and others are introduced there for the first time. Family coverage within Part III follows the basic format as described for Part II.

Since several families of insects that contain species considered to be strictly aquatic (and covered in Part II) also contain species that fall into the various environmental categories covered in Part III, chapters in these two parts of the book are completely cross-referenced. Those families that contain at least some aquatic species are treated in detail within Part II; however, if some species of the family are shore-dwelling or tree-hole inhabiting, for example, then that family is also listed in the appropriate chapters of Part III, wherein the reader is referred back to the chapter that includes its detailed discussion. On the other hand, if a family contains species that are shore-dwelling, for example, but does not contain any strictly aquatic species, then its detailed discussion appears in the chapter on shore-dwelling insects, but it is also listed and cross-referenced in the appropriate ordinal chapter of Part II. Thus a reader should readily be able to identify a specimen to family either when the order is known but the exact habitat is in question (in which case Part II should usually be consulted first) or when the exact habitat is known but the order is not (in which case either Parts II or III could be consulted first).

Part IV of the book (Chapters 21–23) deals with the close relatives of insects (other arthropods) that occur in freshwater environments of North America. These forms, although not insects, are often treated within the purview of aquatic entomology, since they are not only taxonomically related but many have close ecological relationships with insects and are commonly encountered in the field along with aquatic insects. Again, the basic format of treating these groups is the same as for the insects but with a lesser degree of detail. The marine forms of these arthropods, particularly of the crustaceans, are so extensive, however, that they are beyond a reasonable coverage in this book.

References used in the compilation of information are listed at the ends of most chapters. For readers seeking sources of detailed information on a particular order, the main list of references is preceded in the ordinal chapters by a brief selection of annotated general references.

A comprehensive glossary appears at the end of the book. Although many technical morphological, functional, and ecological terms are defined at their first usage, the glossary should make the book highly usable for beginning students, fishermen, and others who may use certain parts of the book independently of other parts and who may be unfamiliar with terminology.

References

Ali, A. 1980. Nuisance chironomids and their control: A review. *Bull. Entomol. Soc. Amer.* 26:3–16.

Batra, S. W. T. 1977. Bionomics of the aquatic moth, *Acentropus niveus* (Olivier), a potential biological control agent for Eurasian watermilfoil and Hydrilla. *J. N. Y. Entomol. Soc.* 85:143–152.

Gaufin, A. R. 1973. Use of aquatic invertebrates in the assessment of water quality, pp. 96–116. In *Biological methods for the assessment of water quality* (J. Cairns, Jr. and K. L. Dickson, eds.). Amer. Soc. Test. Mat., Philadelphia.

Gingrich, A. 1974. *The fishing in print.* Winchester Press, New York. 344 pp.

Harwood, R. F., and M. T. James. 1979. *Entomology in human and animal health* (7th ed.). Macmillan, New York. 548 pp.

Henson, E. B. 1966. Aquatic insects as inhalant allergens: A review of American literature. *Ohio J. Sci.* 66:529–539.
Hilsenhoff, W. L. 1977. *Use of arthropods to evaluate water quality of streams.* Wisc. Dept. Nat. Res. Tech. Bull. 100. 15 pp.
Hitchcock, S. W. 1965. The seasonal fluctuation of limnetic *Chaoborus punctipennis* and its role as a pest in drinking water. *J. Econ. Entomol.* 58:902–904.
Lagler, K. F. 1956. *Freshwater fishery biology* (2nd ed.). Wm. C. Brown, Dubuque. 421 pp.
Petersen, A. 1956. *Fishing with natural insects.* Spahr & Glenn, Columbus. 176 pp.
Ross, H. H. 1965. *A textbook of entomology* (3rd ed.). John Wiley & Sons, New York. 539 pp.
Swisher, D., and C. Richards. 1971. *Selective trout.* Crown, New York. 184 pp.
Szent-Ivany, J. J. H., and I. V. Ujházy. 1973. Ephemeroptera in the regimen of some New Guinea people and in Hungarian folksongs. *Eatonia* 17:1–6.
Usinger, R. L. (ed.). 1956. *Aquatic insects of California.* Univ. Calif. Press, Berkeley. 508 pp.
Williams, D. D. 1980. Applied aspects of mayfly biology, pp. 1–17. In *Advances in Ephemeroptera biology* (J. F. Flannagan and K. E. Marshall, eds.). Plenum, New York.
Wirth, W. W. 1971. The brine flies of the genus *Ephydra* in North America (Diptera: Ephydridae). *Ann. Entomol. Soc. Amer.* 64:357–377.
Zalom, F. G., and A. A. Grigarick. 1980. Predation by *Hydrophilus triangularis* and *Tropisternus lateralis* in California rice fields. *Ann. Entomol. Soc. Amer.* 73:167–171.

CHAPTER 2

Identifying Aquatic Insects

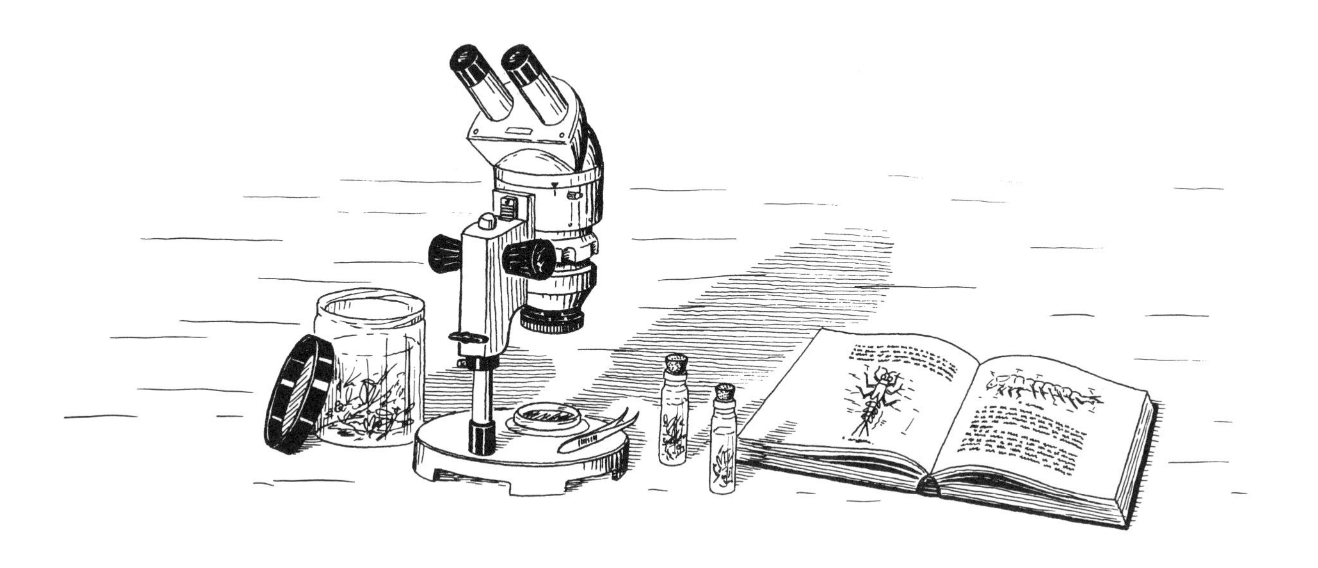

THE ABILITY TO DISTINGUISH among the many kinds of aquatic insects—to call them by their correct names and thereby be able to learn about them or discuss them intelligently—is based foremost on a knowledge of their forms and characters and the means by which they are classified and named. The subjects of insect morphology, metamorphosis, and classification are therefore introduced in this chapter. The depth of coverage of these subjects is such that the remainder of the book can be used efficiently by those who have little or no background in entomology. The various ways in which the book can be used for identification purposes are also discussed.

External Morphology

Much of understanding morphology and morphological descriptions is a matter of developing a basic vocabulary. This chapter therefore covers the fundamental terminology of body parts, including their shapes and relative positions. More technical terminology is treated in other chapters where it is specifically required and is to a large degree self-explained in labeled illustrations.

Orientation and Shapes

In order to locate important diagnostic characters, it is necessary to know the terms used to indicate relative positions on the insect body and body parts. The more commonly used terms are defined below; also see Figures 2.1–2.4.

Anterior: Refers to the head end of the body or that part of a structure closest to the head of the body.

Posterior: Refers to the tail end of the body or that part of a structure closest to the tail of the body.

Dorsal: Refers to the upper or top part of the body or structure.

Ventral: Refers to the lower or bottom part of the body or structure.

Lateral: Refers to the side of the body or structure.

Medial: Refers to the longitudinal midline of the body or structure.

Basal: Refers to the origin of a structure, generally closest to the point of attachment to the body.

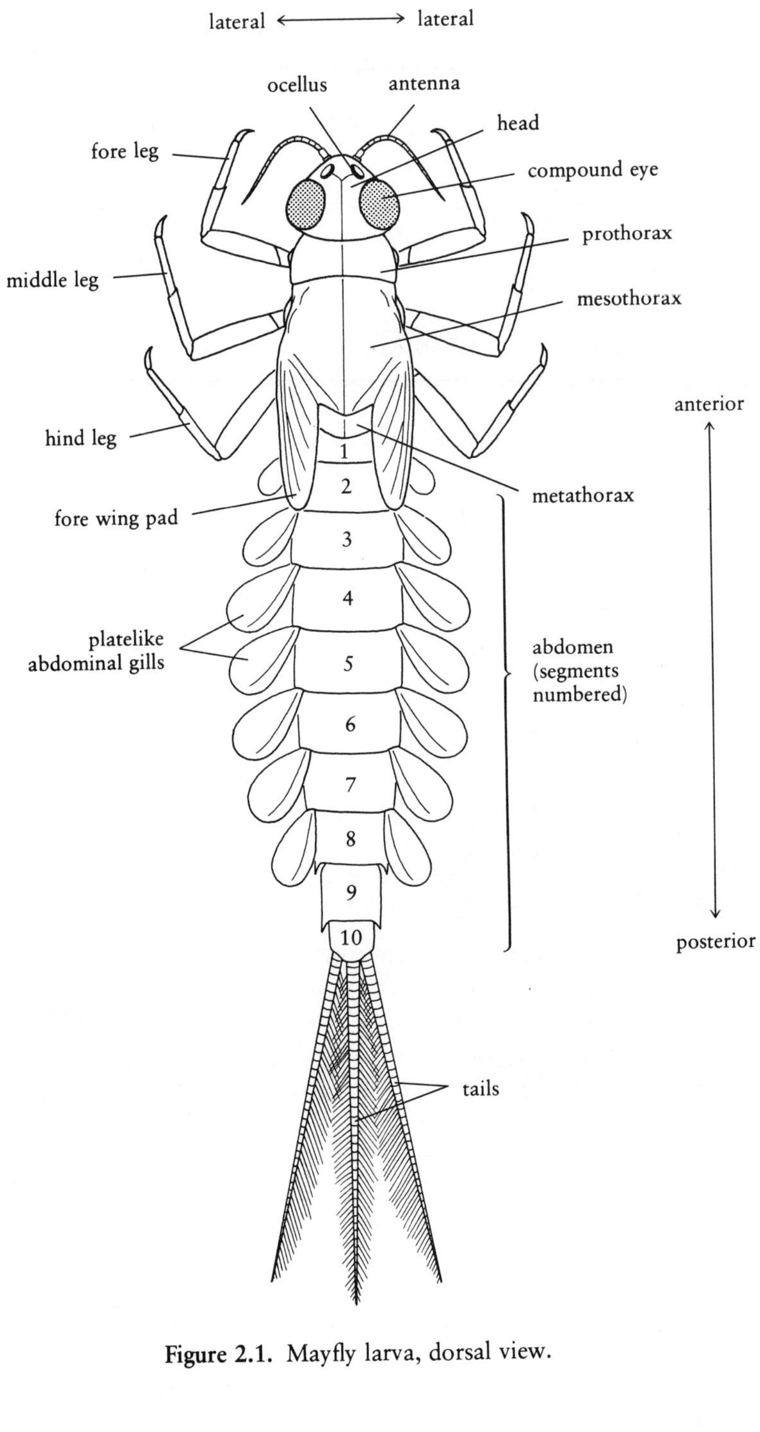

Figure 2.1. Mayfly larva, dorsal view.

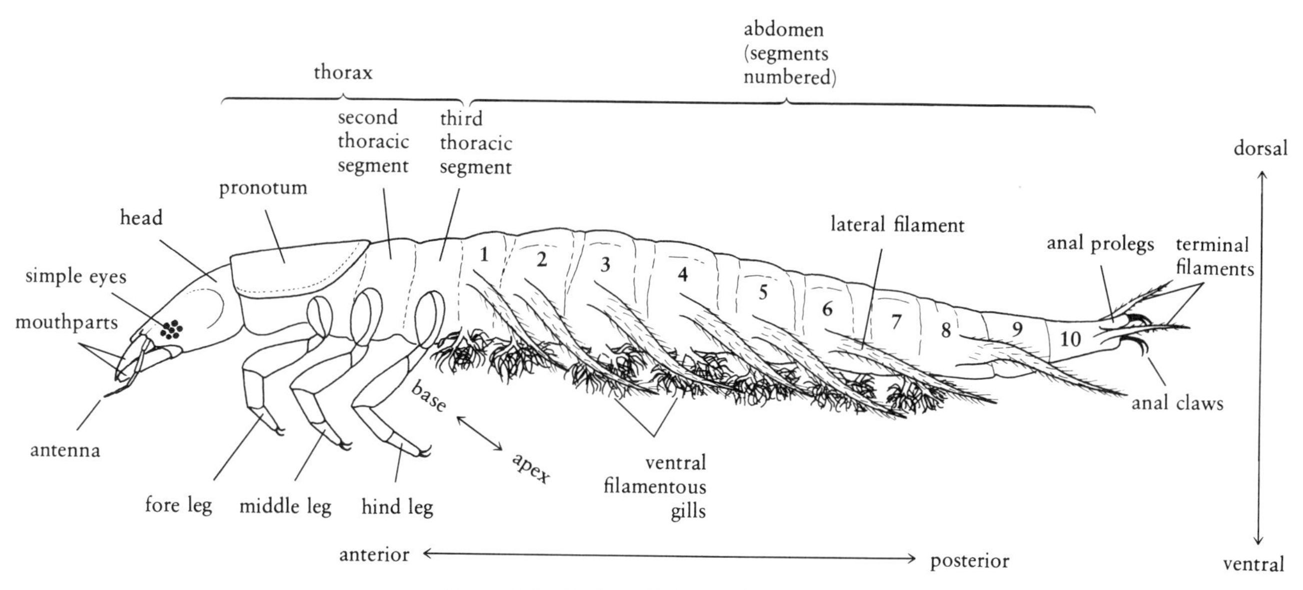

Figure 2.2. Dobsonfly larva, lateral view.

Distal: Refers to that part of a structure furthermost from its point of attachment to the body. The **apex** is the distal end or tip of a structure. The term **apical** is generally synonymous with *distal.*

Shapes of bodies or body parts are often important in describing insects. A **dorsoventrally flattened** body or structure is flattened or depressed from top to bottom and is generally wider (side to side) than it is thick (top to bottom). A **laterally flattened** body or structure is flattened or compressed from the sides and is generally thicker (top to bottom) than it is wide (side to side).

Other important terms refer to the shape of a margin or end of a structure. A **rounded** or **convex** structure is more or less evenly curved outwardly. An **emarginate** or **concave** structure is notched or hollowed out or has an inwardly curved margin. A **truncate** structure is more or less square-shaped or has a blunt apex. An **acute** structure has a more or less pointed apex.

Body Divisions

The generalized insect body (Figs. 2.1, 2.2) is divided into many ringlike **segments** and three major **regions**. The first region is the **head**, which appears to be a single segment but is actually composed of six or seven fused segments. The head is usually capsulelike and contains the feeding apparatus and many of the sensory organs.

The second region, or **thorax**, is posterior to the head and is composed of three segments: the **prothorax** (anterior or first thoracic segment), the **mesothorax** (second or middle segment), and the **metathorax** (posterior or third segment). Appendages used for moving the body are located on the thorax.

Posterior to the thorax is the third body region, or **abdomen.** The abdomen is generally the longest region and is composed of several segments (often from 8 to 11). Various appendages and external structures are located on the abdomen of some insects.

Body regions are not distinguishable in some insects. When the thorax and abdomen are undifferentiated, they are together known as the **trunk.**

The Body Wall

The outer skin of insects is actually an external skeleton known as the **exoskeleton.** The exoskeleton can be relatively hard (**sclerotized** or **platelike**) or thin and soft (**membranous** or **fleshy**), depending on the kind of insect or part of the body. The texture of the exoskeleton and the presence or absence of various kinds of outgrowths are often diagnostic.

SCLERITES

Any piece of body wall is a **sclerite.** A sclerite is bounded by membranous areas or lines known as **sutures. Tergites** are dorsal sclerites (often platelike areas) of a body segment. **Sternites** are the ventral sclerites of a segment. Some insects also have sclerites on the sides of the body, or **pleural** area.

OUTGROWTHS OF THE BODY WALL

Socketed outgrowths are **setae**, and these are referred to in this book by the descriptive terms **hairs**, **bristles**, **spurs**, or **scales**, depending on their thickness and shape. **Spines** and **processes** are robust, often acute, unsocketed outgrowths or extensions of the exoskeleton. The location and density of any of the above outgrowths can be important in identifying insects. The presence and location of punctures, furrows, and sutures can also be important.

Aquatic insects sometimes have additional outgrowths, such as **gills**, **filaments**, and **papillae**. These are often thin-walled and of various shapes and are most often located on the abdomen, but sometimes on the thorax, head, or appendages. Gills may be platelike (Fig. 2.1) or filamentous (Fig. 2.2) and branched or unbranched. Papillae are sometimes eversible.

Structures of the Head

The insect head possesses the eyes, antennae, and mouthparts. These structures vary considerably and thus are important for identification.

EYES

A pair of **compound eyes** (multifaceted) is present laterally on the heads of many insects (Fig. 2.1). Compound eyes are reduced to a single facet or group of single facets in many other insects (Fig. 2.2). In addition to the eyes, many insects possess two or three **ocelli** (single-faceted organs), which appear as spots on the dorsal or anterior part of the head.

ANTENNAE

Antennae are paired, segmented appendages of the head that usually arise between the eyes. They are highly variable in size and shape, often being filamentous (Figs. 2.1, 2.2) or variously expanded or clubbed.

MOUTHPARTS

Because mouthparts are highly variable, depending on the feeding adaptations of the insect, only a generalized plan is discussed here (Fig. 2.3). The **labrum** is a platelike "upper lip" that is attached to the front margin of the head and is the most anterior mouthpart. Posterior to the labrum and at the sides of the mouth is a pair of **mandibles**, followed by a pair of **maxillae.** Each maxilla possesses a lateral armlike appendage known as the **maxillary palp.** At the posterior of the mouth is the "lower lip," or **labium.** The labium can generally be divided into the body of the labium and the **labial palps**, which are lateral, segmented, armlike appendages. An additional

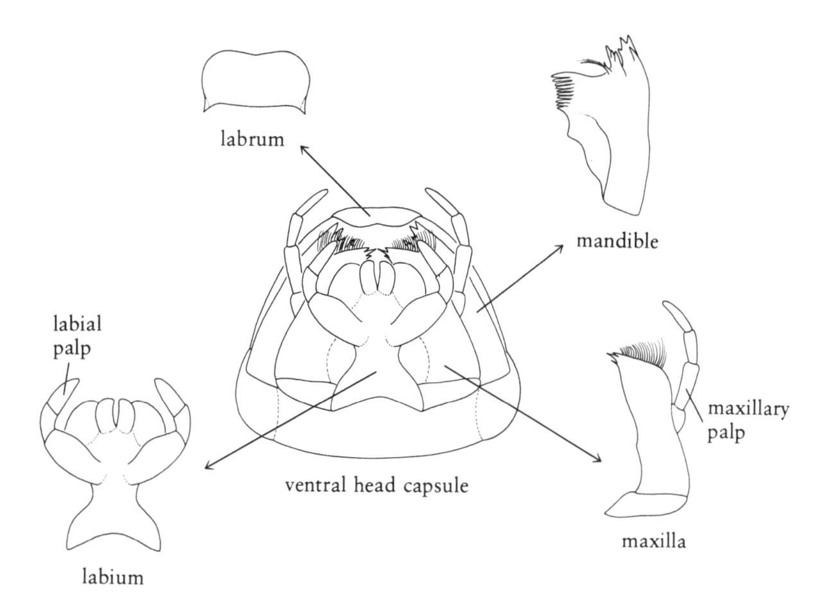

Figure 2.3. Generalized mouthparts, mayfly larva.

mouthpart present in some insects is the **hypopharynx**, a medial structure closely associated with the base of the labium and generally located between the maxillae.

Structures of the Thorax

Segments of the thorax are often relatively robust, since they encase muscles used to move the wings and legs. The terms **notum** or **notal** refer to the dorsal portion of a segment; the size, shape, and degree of sclerotization of the **pronotum** (dorsal portion of the first thoracic segment) and other nota are often diagnostically important.

LEGS

In the generalized insect (Figs. 2.1, 2.2), each thoracic segment possesses a pair of segmented legs. The basal leg segment, the **coxa**, is followed distally by the **trochanter**, **femur**, **tibia**, and **tarsus** (Fig. 2.4A). The tarsus is often subdivided into tarsal segments (segment 1 is always the most basal). A **claw** or pair of claws is usually present at the end of the leg. The pairs of legs are referred to as the **fore**, **middle**, and **hind legs**. Filamentous gills are present at the base of one or more pairs of legs in certain aquatic insects.

Many immature insects do not possess thoracic legs, but some possess prolegs. **Prolegs** are fleshy, usually unsegmented, leglike (usually short) outgrowths. Thoracic prolegs are present most often as a single ventral pair on the first thoracic segment.

WINGS

One or two pairs of wings are present in most adult insects. The generalized wing is a membranous, flaplike structure that articulates with the side of the thorax. The **fore wings**, or first pair of wings, are located on the second thoracic segment, and the **hind wings**, or second pair of wings, are located on the third thoracic segment. When only one pair of wings is present, the wings are located on the second thoracic segment.

Rigidity of membranous wings is maintained in part by the **wing veins** (Fig. 2.4B); **longitudinal veins** run from the base lengthwise, and **crossveins** connect longitudinal veins. The anterior margin of the wing is known as the

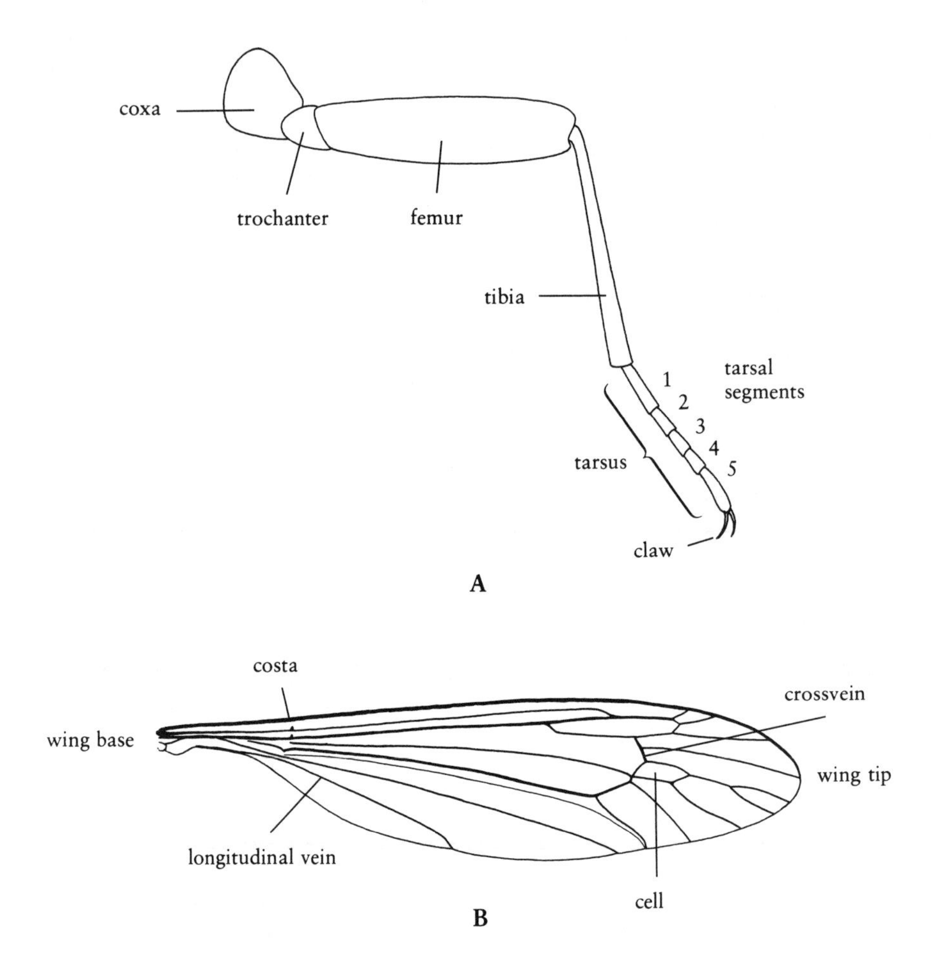

Figure 2.4. A. Beetle leg. **B.** Crane fly wing.

costa. Cells are areas of the wing enclosed by veins. The complex terminology of wing veins differs somewhat from group to group; veins are illustrated and labeled in parts of this book where they must be used.

Fore wings are sometimes variously modified into structures that, when folded over the back of the insect, protect the hind wings. Wings of certain species of aquatic insects never become fully developed. Immature insects may or may not have externally developing wings. These developing wings, including their encasements when present, are referred to as **wing pads.**

Structures of the Abdomen

The number of abdominal segments and an indication of the exact segments on which certain structures occur are often important for identification. Segments are consistently numbered, beginning with the most anterior segment as segment 1. Abdominal segments may appear to be subdivided in some insects.

Paired structures that sometimes occur on the abdomen include lateral filaments, gills of various kinds, and prolegs (Figs. 2.1, 2.2). These occur only on certain segments or serially on several segments. Those that occur at the end of the abdomen are often referred to as **terminal** or **anal** (e.g., **anal prolegs**).

Diagnostically important structures often occur at the end of the abdomen. These include **breathing tubes** or **siphons, caudal lamellae,** anal prolegs, various lobes, gills, filaments, and papillae. Many aquatic insects possess segmented, tail-like structures at the end of the abdomen, and although these are known to specialists by various technical terms, they are herein simply referred to as **tails.** **Genitalia** (the primary external reproductive organs) are located on the abdomen of adult insects, but they are not generally used for the level of identification covered in this book.

Thus the homely and often repulsive grubs, maggots, and caterpillars, which are all the larval forms of the beetles, flies and butterflies, respectively, enter the third stage as worm-like, crawling creatures, and emerge from it as beautiful winged forms, sometimes glistening and gleaming with all the colors of the rainbow.

W. S. Blatchley

Metamorphosis and Life Stages

During its life cycle (Fig. 2.5) every aquatic insect goes through a series of changes in form (the process of metamorphosis) as it develops from egg to adult. These different forms of the insect are known as life **stages**, and they make up two distinct phases of development. The first phase—embryonic development—occurs in the **egg** stage. The second phase—postembryonic development—includes all stages between the time the egg hatches and the time the insect becomes reproductively mature. Reproductive maturity is reached in the **adult** stage (sometimes also called the **imago**), which possesses functional sexual organs and is able to reproduce. The ecological requirements and associated morphological and behavioral adaptations of the different stages of an aquatic insect may be quite different; usually one or more stages are terrestrial. A knowledge of life cycles and stages is therefore very important to the student of aquatic entomology.

Postembryonic Development

Postembryonic development includes both growth and maturation of the individual. **Growth** is generally considered to be increase in size, whereas **maturation** is the progressive morphological and physiological attainment of adult attributes (wings, genitalia, etc.), which are fully developed at adulthood.

Since insects have an exoskeleton, it must be periodically shed, and a new, larger one reformed to accommodate the growing individual. The process of shedding the exoskeleton is known as **molting**, and the shed exoskeleton is commonly referred to as a **cast skin.** The period between any two molts is known as a stadium, and the individual during this interval is referred to as an **instar.** The number of developmental instars may be constant or variable, depending on the species or group of insects. The number of instars during postembryonic development may range from 4 to 40 or more.

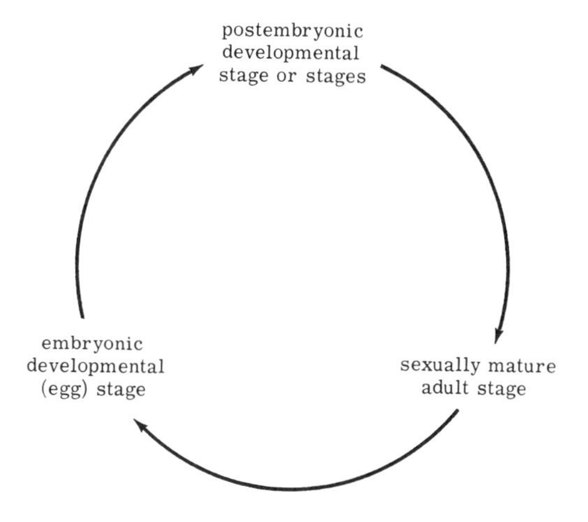

Figure 2.5. Insect life cycle.

During the postembryonic phase of development, insects can generally be referred to as **immatures**, even though there may be either one or two major stages during this phase, depending on the kind of insect and the kind of metamorphosis it undergoes. These postembryonic stages are discussed below.

LARVA

The **larval stage** occurs in all aquatic insects. It follows the egg stage and is followed by either the adult stage or a transitional stage that precedes the adult. Growth (and maturation to varying degrees) occur during the larval stage, and there are three or more instars (larval instars). The relative terms "young larvae" and "mature larvae" refer to early instar larvae and late instar larvae, respectively. Larval instars may vary considerably from one to another in many aquatic insects and are sometimes given special names.

SUBIMAGO

The **subimago** (or subimaginal stage) is a unique transitional stage found only in mayflies. It follows the larval stage and precedes the adult stage. Unlike other pre-adult stages known among insects, it is a fully winged form. The subimago is a maturation stage and includes only one instar.

PUPA

The **pupal stage** is a transitional stage that occurs in more advanced groups of insects. It is primarily a maturation stage and includes only one instar. Depending on the kind of insect, the pupa may be quiescent or active. It also may be either free-living or encased in a cocoon or puparium. A **puparium** is found in some advanced groups of Diptera and is made from the modified skin of the last larval instar.

Kinds of Metamorphosis

Two basic kinds of metamorphosis occur among aquatic insects (see Table 2.1). The kinds of postembryonic stages involved in development determine the kind of metamorphosis. Larvae conveniently reflect these different kinds of metamorphosis by their possession of certain morphological traits.

INCOMPLETE METAMORPHOSIS

Aquatic insects that do not possess a pupal stage have an incomplete kind of metamorphosis, which is generally regarded to be a more primitive kind of development (see Fig. 2.6A). The larvae of these insects undergo both growth and maturation. The larvae (except for very young larvae)

TABLE 2.1 KINDS OF METAMORPHOSIS FOUND AMONG AQUATIC INSECTS

Incomplete Metamorphosis	Complete Metamorphosis
Ephemeroptera	Megaloptera
Odonata	Neuroptera
Plecoptera	Coleoptera
Hemiptera	Trichoptera
	Lepidoptera
	Diptera

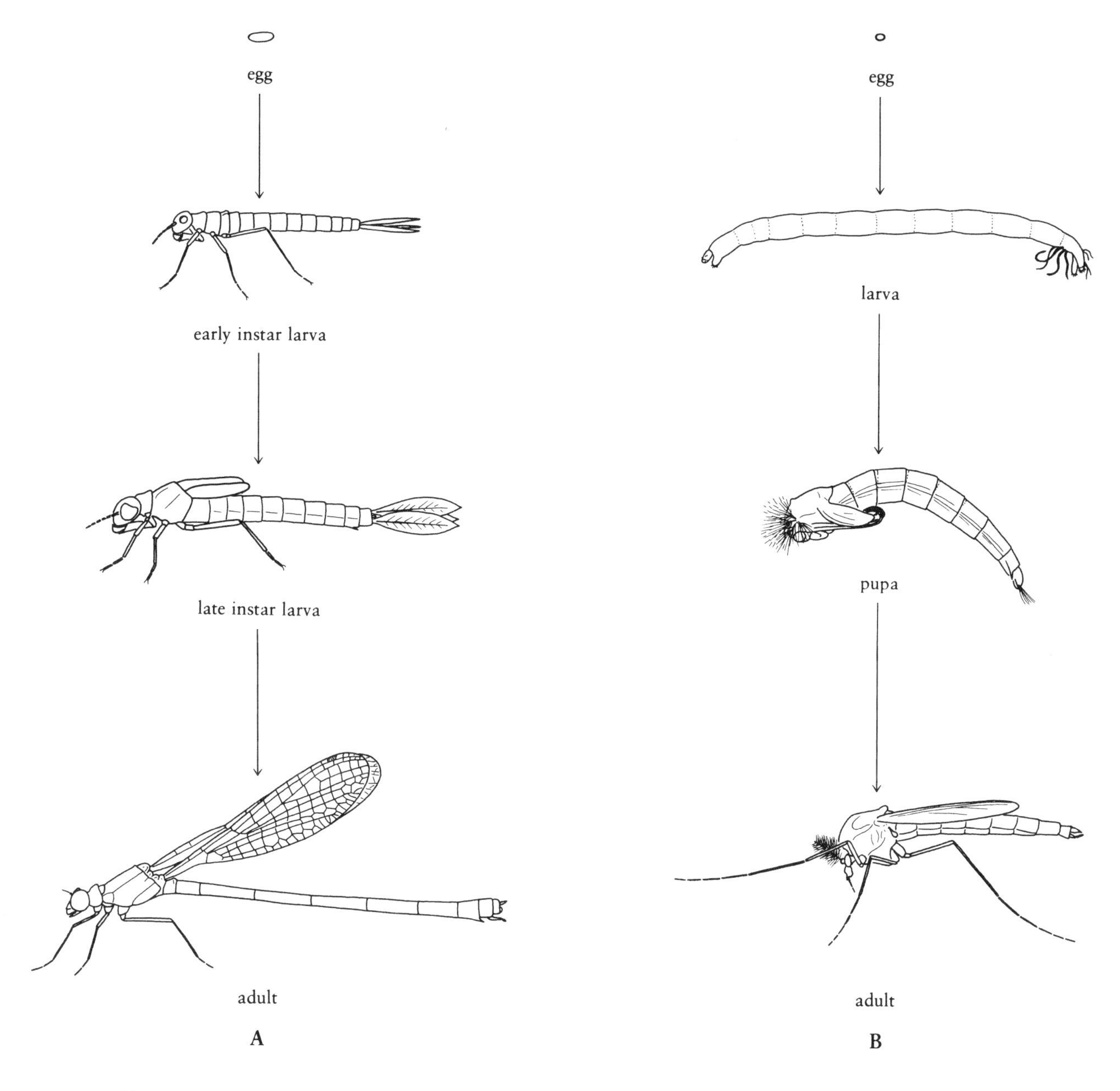

Figure 2.6. Metamorphosis. A. Incomplete metamorphosis, damselfly. B. Complete metamorphosis, midge.

possess well-developed compound eyes and wing pads (in species that have wings as adults). The degree of morphological and ecological difference between larvae and adults varies considerably from one insect group to the next.

The terms **nymph** and **naiad** are sometimes used for the larval stage of insects with incomplete metamorphosis. The term *larva*, however, is more universally accepted in the zoological sciences.

COMPLETE METAMORPHOSIS

Aquatic insects that possess a pupal stage have a complete kind of metamorphosis, which is generally regarded to be a more advanced kind of development (see Fig. 2.6B). In these insects growth takes place mainly in the larval stages, and maturation takes place mainly in the pupal stage. Larvae lack wing pads and compound eyes; their eyes are either highly reduced or composed of a single facet or group of single facets. Differences between larvae and adults are usually extreme.

Classification

In order to communicate about aquatic insects, species and groups of species must have names. These names should be commonly acceptable as far as possible, and must be formulated according to a common set of rules. A brief, pertinent introduction to zoological classification is presented below.

Taxonomic Ranking

All animals are arranged into groups at various levels known as **taxonomic ranks** or **categories**. These ranks form a hierarchy of classification, so that animals are grouped into progressively more inclusive groupings. The major taxonomic ranks are listed below, proceeding from the higher, more inclusive ranks to the lower, less inclusive ones. Infracategories not listed are sometimes used to further subdivide these ranks but are usually of less general importance.

Phylum
Class
Order
Suborder
Family
Subfamily
Genus
Species

This book deals with arthropods, which are animals recognized as a group at the phylum rank. Within the phylum Arthropoda, one group of arthropods, the insects, is of primary concern. Insects are recognized as a group at the class rank (within the class Hexapoda). Within each of ordinal Chapters 7–16, groups recognized at the familial rank, which is of primary importance in this book, are described and discussed. When the family can be subdivided into groups recognized at the subfamilial rank, these groups are described. Finally, examples of behavior, adaptations, or use to the fisherman are discussed for groups recognized as genera or species.

The **species** is the most basic unit of animal classification, since organisms recognized as members of an individual species are usually similar and able to interbreed in nature. Pigeonholing groups of species into genera, groups of genera into families, etc., is a method that not only provides a means of classification, but can also be meaningful in expressing the natural relationships of species. Taxonomists are not always in agreement, however, as to the best way to group species or subdivide higher groups.

Names of Aquatic Insects

A **taxon** is a group of naturally related organisms designated by a name, and the taxon or taxonomic name may be used for a group at any taxonomic rank. For example, the beetles, or Coleoptera, constitute a taxon recognized as an order, and the Salmonfly, or *Pteronarcys californica*, is a taxon recognized as a species.

All taxa must have internationally acceptable scientific names based on rules of nomenclature. These names commonly have Latin or Greek derivatives. Many taxa are also known by nonscientific (common or vernacular) names, but since there are no rules governing these names, they vary greatly among languages and in regional usage. It should be noted that in the formation of the many English names of insects that include the word *fly*, this word stands apart if the name is that of a true fly (member of the order Diptera)—for example, "black fly" or "shore fly." In other orders, however, *fly* is incorporated into the name as a suffix—for example, "caddisfly" or "dragonfly."

Scientific names of taxa classified at certain higher ranks have consistent endings. For example, the taxonomic names of animal families always end with the suffix *-idae*, as in Chironomidae, and the names of subfamilies always end in *-inae*, as in Diamesinae. All taxonomic names of animals at the rank of genus or above are capitalized.

The scientific name of a species is composed of the name of the genus to which it belongs, followed by its specific name. It is therefore a compound or **binomial name**, such as *Pteronarcys californica.* Note that the generic name is capitalized and that the specific name is not, and also that the entire name is italicized. The binomial must be unique for each species. Other species of *Pteronarcys* have the same generic name, but their specific name is different (e.g., *Pteronarcys dorsata*); species of different genera may have the same specific name, but their generic names are different (e.g., *Calineuria californica*).

Genius detects through the fly, through the caterpillar, through the grub, through the egg, the constant individual; through countless individuals the fixed species, through many species the genus, through all genera the steadfast type; through all kingdoms of organized life the eternal unity.

Ralph Waldo Emerson

Names of species occasionally have to be changed, usually when the species is found to belong to a genus different from the one in which it was originally placed or it has been discovered that two different names have been used for the same species. According to the law of priority, the earliest used name usually takes precedence for usage, and the subsequently used name becomes a **synonym** no longer available for usage.

Using This Book

This book is designed so that aquatic insects can be identified to the correct order and family (and other categories when applicable) by any of several methods. One method for field identification of common groups is simply to match the illustrations with the specimens in hand. Those insects illustrated in color were especially chosen to aid in this method. Once a color figure is found that resembles the insect, the reader can refer to the appropriate chapter for further information to verify or modify the identification. If the order is known, a quick look through the black-and-white figures in the pertinent chapter will in many instances lead to a correct family identification.

A second method is to use the picture keys that appear in many chapters. If the order is unknown or in doubt, consult the key to orders in Chapter 6. Once the order is known, use the key in the chapter on that order to identify the family.

A third method is to read through the comparative diagnoses of families in the appropriate chapter. This is most advantageous, of course, when the order is known and the diagnoses are used in conjunction with the illustrations. Characters that differentiate easily confused families are stressed in the discussions of the families.

No matter what method of identification is used initially, it is always advisable to cross-check it with the alternative methods. The information on distribution, habitat, and behavior included in the discussion of each taxonomic group should also prove very helpful in verifying identifications.

CHAPTER 3

Living in Water

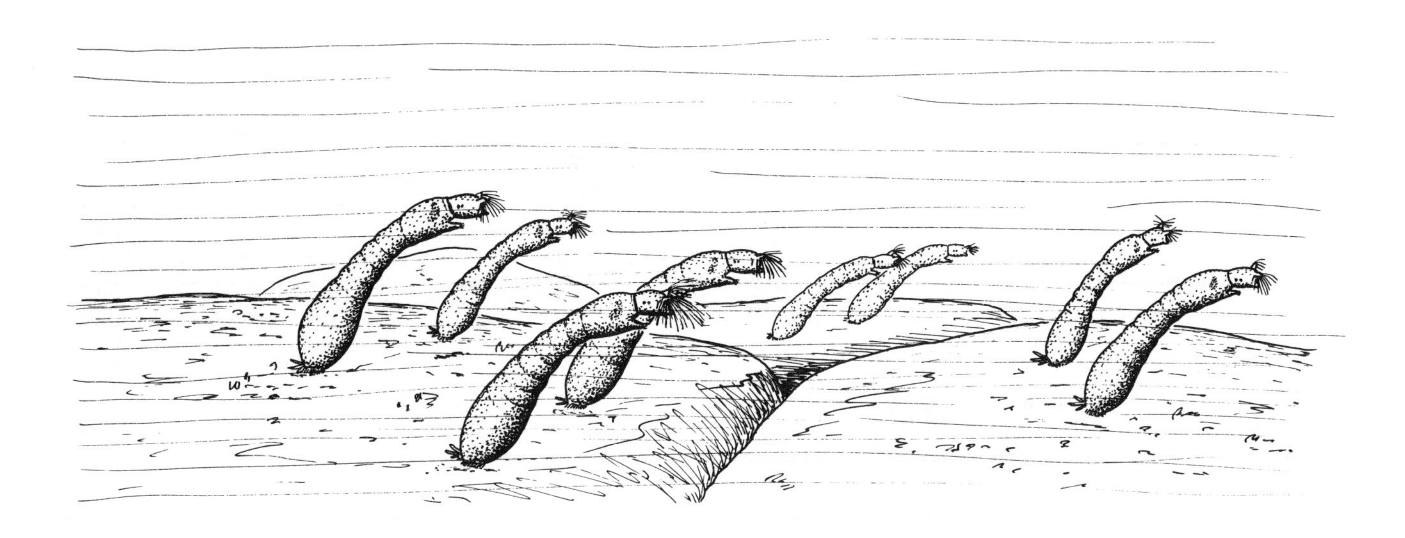

INSECTS CONSTITUTE a large and predominantly terrestrial group of organisms, and only about 10 percent or less of their species are aquatic. The aquatic environment presents many obstacles to insect life that differ markedly from those of a terrestrial environment. Nonetheless, many diverse aquatic adaptations have evolved in insects, reflecting their successful invasion of water—a phenomenon that has probably occurred independently several times in their history. Special respiratory systems, behavioral traits, and mechanisms for maintaining proper internal salt concentrations have evolved, as have adaptations for moving about in water, residing in unique aquatic microhabitats, and using unique food resources. Since nearly all aquatic insects have retained their primitive ties to land and air by having certain terrestrial life stages, they also possess special adaptations for making the transition from one environment to the other.

The primary intent of this chapter is to define and categorize the major aquatic adaptations found among insects. These adaptations can be just as important as descriptors of aquatic insect groups as morphological traits. They are usually highly correlated with morphology and are often indicative of the kind of aquatic habitat in which an insect lives.

Aquatic Habitats

In order to appreciate the adaptations of aquatic insects and their preference for particular habitats, it is important to have an acquaintance with the major kinds of freshwater environments.

Lotic Habitats

Lotic habitats are flowing-water habitats, such as rivers and streams. A lotic system (drainage system) includes everything from trickling headwaters or seepage springs to the larger rivers that empty into lakes and oceans. The turbulence of flowing waters provides a natural means of aerating the water, thus making oxygen readily available to animal life.

The bundle of sticks crawling about in the water, green worms under stones in the stream, swarms of "flies" around the lights along river and lake. . . . They are but a few isolated phenomena, however, in a picture of life histories and interrelationships varied in pattern and interesting in detail.

Herbert H. Ross

Streams in which water flows continually throughout the year are known as **perennial streams,** and streams in which the flow discontinues during part of the year are known as **intermittent streams**. Intermittent streams may become totally dry or retain water temporarily only in small pools.

Terms such as "river," "stream," "brook," and "creek" are not always precise. Unfortunately, what may be considered a creek by some is considered a river by others. There is, however, a natural progression in size from the smallest tributaries (headwaters) to the largest rivers to which they contribute, and streams can be conveniently categorized by their **stream order** (Fig. 3.1) as follows: first-order streams have no tributaries, but when two such streams join, the resultant stream is a second-order stream; and when two second-order streams join, the resultant stream is a third-order stream; etc. There is also a general overall progression in many drainage systems from shallower (and more turbulent in mountainous regions), relatively nutrient-poor environments to deeper, sometimes slower, nutrient-rich environments (such as found in large rivers).

Throughout the course of a stream there is usually an alternation of so-called zones reflected by the relative current velocity, as explained below (see also Fig. 3.1).

EROSIONAL ZONE

An **erosional** zone is an area of a stream in which the water velocity is fast enough to carry small particles in suspension. This zone is often typified by riffles, and the stream bottom is devoid of silt (except perhaps in small pockets). The stream bottom generally consists of stones, gravel, and sometimes sand, depending on the swiftness of the water (even gravel and stones will be carried by rapid water). With all other factors being ideal, riffle areas may often be expected to have a relatively diverse insect fauna.

DEPOSITIONAL ZONE

A **depositional** zone is an area of a stream in which the current is relatively slow and small particles fall out of suspension and become deposited as silt on the stream bottom. Depositional areas often become predominant in larger streams. Stream pools, slow reaches, backwaters, and slow edgewaters are typically depositional in nature. With all other factors being ideal, these slower waters may be expected to have fewer bottom-dwelling insect species than riffle areas, but sometimes they have large numbers of individuals of a few dominant species.

You will not hear it as the sea; even stone
Is not more hushed by gravity. . . . But slow,
As loth to take more tribute—sliding prone
Like one whose eyes were buried long ago.
The River, spreading, flows—and spends your dream.

Hart Crane

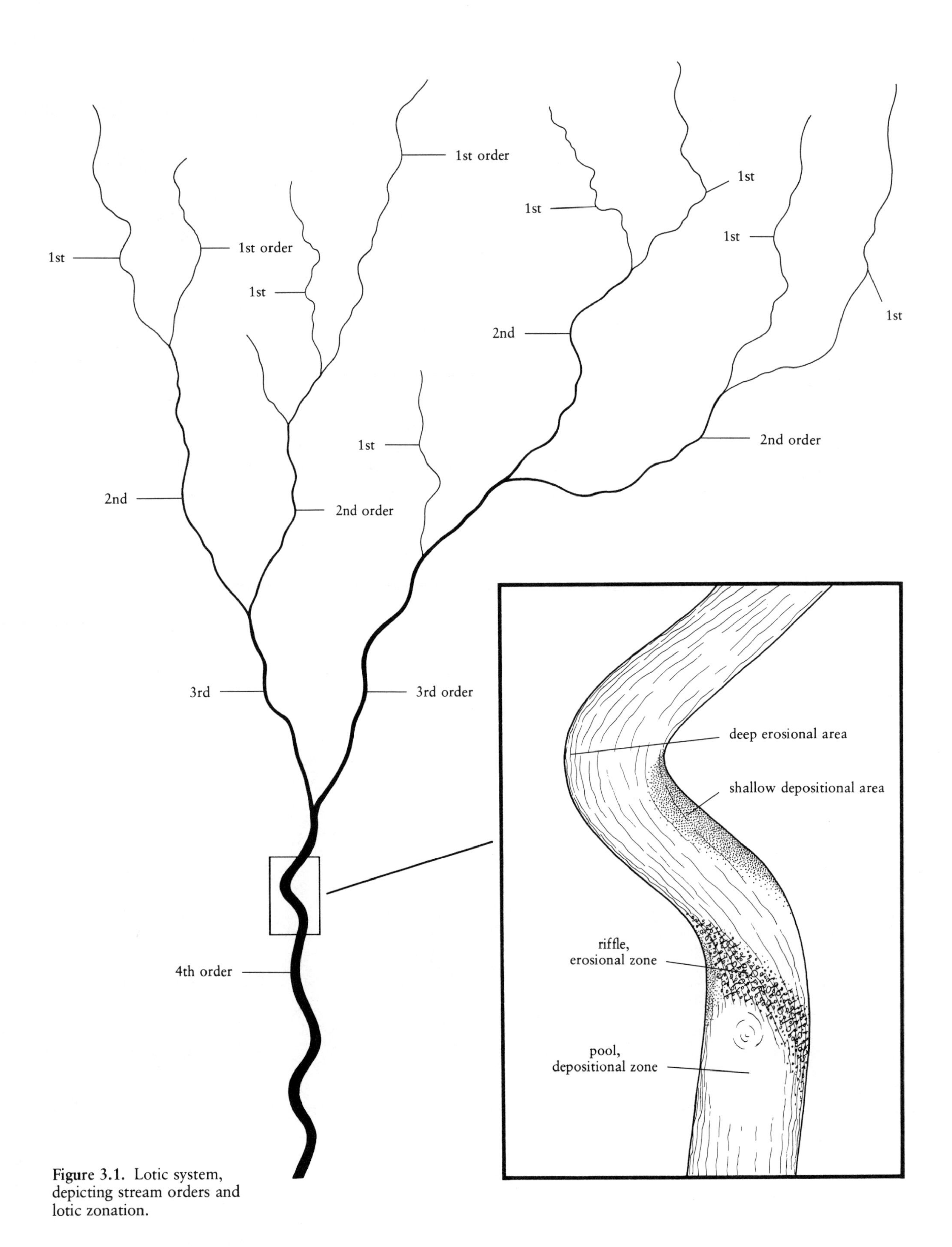

Figure 3.1. Lotic system, depicting stream orders and lotic zonation.

Lentic Habitats

Lentic habitats are standing-water habitats, such as lakes, ponds, pools, and swamps. Basins may receive water from drainage streams and may also have some points of outflow, so that there is at least some turnover of water. Salts become concentrated in evaporation basins with no outflow points, eventually resulting in salt lakes. Lentic systems are primarily depositional in nature; however, shore areas of lakes may be subject to continual wave action and thus be essentially erosional. Lentic habitats and depositional zones of streams often have several kinds of similarly adapted insects in common; erosional zones of streams and wave-swept shores of lentic habitats also often have several kinds of similarly adapted insects in common.

As with the terms "creek" and "river," the terms "pond" and "lake" are often relative and subject to regional interpretation when referring to small bodies of water. Lentic systems are categorized in several ways, sometimes according to the relative degree of nutrients that they contain. For example, **oligotrophic** lakes have a low level of nutrients and biological productivity and a high level of dissolved oxygen at all times; **eutrophic** lakes are rich in nutrients and biological productivity and have periodic low levels of dissolved oxygen. Most lentic systems are something between these two extremes, and highest numbers of aquatic insects are usually found in habitats that are at least somewhat eutrophic.

Identifiable zones within lentic systems are helpful in describing the habitats of lentic insects (Fig. 3.2).

LITTORAL ZONE

The shallow shore area that extends to the limit of rooted aquatic plants is known as the **littoral** zone. Swamps, some ponds, and very shallow lakes are entirely or almost entirely littoral in nature. Numbers of lentic insect species tend to be highest in littoral areas.

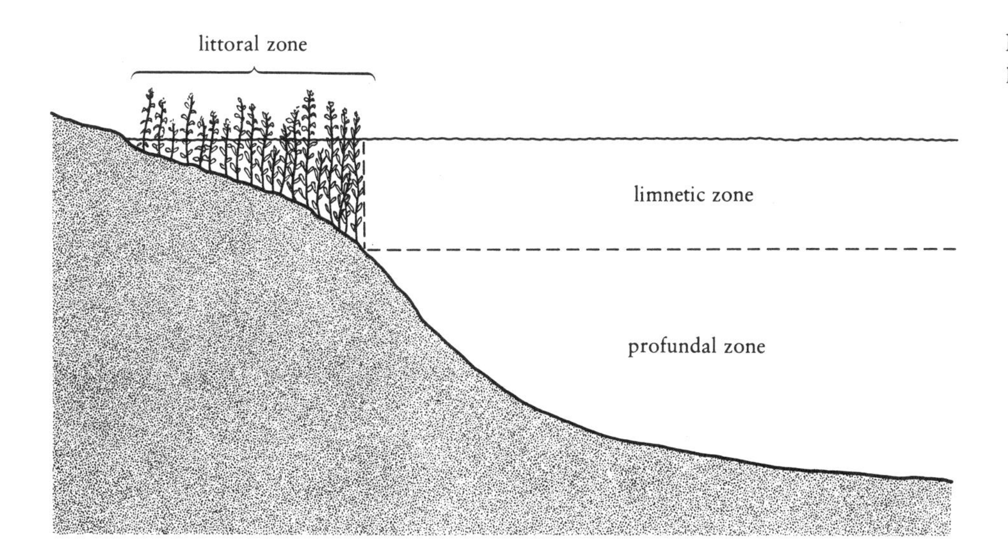

Figure 3.2. Cross section of a pond, depicting lentic zonation.

TABLE 3.1 ENVIRONMENTAL DISTRIBUTION OF THE LIFE STAGES OF AQUATIC INSECTS (APPLICABLE ONLY TO AQUATIC SPECIES OF THE ORDERS)

Orders	Submergent Adults	Submergent Eggs	Submergent Larvae	Submergent Pupae
Ephemeroptera[1]	N	E	E	
Odonata[1]	N	S[2]	E	
Plecoptera[1]	N[3]	P	E	
Hemiptera	P	S	P	
Megaloptera[1]	N	N	E	N
Neuroptera	N	N	E	N
Coleoptera	P	S	P	S[4]
Trichoptera[1]	N	P	E[5]	E[5]
Lepidoptera	N[6]	P[7]	E[7]	E[7]
Diptera	N	P	E	P

E = Exclusive representation (with possibly rare exceptions).
P = Primary representation.
S = Some representation.
N = No representation (with possibly rare exceptions).
[1]Entirely aquatic orders.
[2]Many eggs are deposited in vegetation; many are also submergent.
[3]One species has permanently submerged adults in North America.
[4]Those few submergent pupae in North America occur in air-filled cocoons.
[5]A few species in North America are terrestrial at least in part.
[6]One species is submergent as a short-winged adult in North America.
[7]Some semiaquatic forms occur above or below the waterline in emergent vegetation.

LIMNETIC ZONE

The open water area to the depth of light penetration is known as the **limnetic** zone. Generally, the limnetic zone is differentiated from the littoral zone as being beyond the littoral zone of deeper ponds and lakes. Relatively few floating or surface-dwelling insect species occur in the limnetic zone, although microscopic plant and animal life may be abundant.

PROFUNDAL ZONE

The open, deep-water area below the level of light penetration (including the bottom) is known as the **profundal** zone. There are generally few insects found in this zone, but some bottom-dwelling insects occur in abundance at considerable depths.

Adapting the Life Cycle to Living in Water

The life stages of insects were introduced in Chapter 2. Depending on the aquatic insect, one or more of these stages are aquatic, whereas other stages are terrestrial at least in part (with the exception of one flightless stonefly and some flightless water boatmen). A summary of the environmental distribution of the primary life stages among aquatic insect orders is presented in Table 3.1.

Transition Between Environments

The actual transition from an aquatic existence to a terrestrial one, and vice versa, present special problems for insects and require special adaptations.

FROM AQUATIC TO TERRESTRIAL

Most aquatic insects leave the water in association with a major metamorphic phase—that is, a transformation to the subimago (mayflies), to the pupa (e.g., beetles, Megaloptera, and some Diptera), or to the adult (dragonflies and damselflies, stoneflies, caddisflies, and many Diptera). Many insects simply swim or crawl out of the water onto some exposed shoreline or object just prior to transforming to a terrestrial stage. Larvae of many insects, however, swim or float (sometimes aided by a bubble of gas) to the surface of the water and, after breaking the surface film, transform to a winged stage at the surface. The breaking of the surface film is sometimes aided by an unwettable part of the exoskeleton, and newly formed winged stages often rest, at least momentarily, on the cast larval or pupal skin while wings dry and become ready for flight. Other insects transform to the terrestrial stage under water and immediately swim to the surface (often aided by their wings) or float to the surface (sometimes enveloped in a bubble).

In the parlance of aquatic entomology, the transformation from an aquatic stage to a terrestrial one is often termed **hatching,** and this term is especially applied by fishermen when insects fly, or **hatch**, from the water (e.g., subimagos of many mayflies and adults of many caddisflies and midges). The term **emergence** is best applied generally to the transformation to adult, regardless of the environment involved or the stage that precedes it.

We have our submarines, in which hydraulic ingenuity displays its highest resources. The Caddis-worms have theirs, which emerge, float on the surface, dip down and even stop at mid-depth by releasing gradually their surplus air. And this apparatus, so perfectly balanced, so skillful, requires no knowledge on the part of its constructor. It comes into being of itself, in accordance with the plans of the universal harmony of things.

J. Henri Fabre

FROM TERRESTRIAL TO AQUATIC

Many insects enter the water for the first time in their life cycles either as eggs or young larvae. Females of many aquatic insects **oviposit** (the act of laying or depositing eggs) in the water by dropping eggs on or into the water while in flight or after coming to rest on the water surface. Some females enter the water and dive or crawl to an adequate underwater surface for ovipositing their eggs (e.g., some mayflies, stoneflies, caddisflies, and aquatic moths). Females of some species oviposit in or on emergent or floating vegetation (e.g., damselflies and many dragonflies, some water bugs, water beetles, and aquatic moths). Those larvae that come from eggs oviposited above the waterline in aquatic vegetation usually immediately crawl or fall into the water upon **eclosion** (hatching from the egg).

Eggs of some species are oviposited in or on moist shoreline areas or dry portions of stream beds (e.g., some stoneflies, water bugs, water beetles, midges, and a few caddisflies). Either the young larvae enter the water after eclosion or the eggs eclose only after becoming submerged. Megalopterans, spongillaflies, watersnipe flies, some damselflies and soldier flies, and a few caddisflies oviposit on vegetation or structures that overhang water; their eggs or young larvae eventually fall into the water. Eggs of some giant water bugs are oviposited on the male's back.

Leaving and entering the water is not always restricted to a transformation phase of the life cycle. For example, adult water bugs and water beetles may leave and re-enter water either to escape adverse conditions or to disperse to other streams or ponds. Some aquatic larvae may also leave and re-enter the water under certain conditions.

Synchronizing the Life Cycle

Prevailing climatic conditions usually govern the duration of a particular stage of an aquatic insect, the length of the entire life cycle, and the times at which such events as egg eclosion, larval growth, pupation, emergence, and oviposition take place. The above events are coordinated to take advantage of seasonally favorable conditions for growth and reproduction in the species' environment.

DEVELOPMENTAL TIMES AND ENVIRONMENTS

A **generation** is a complete life cycle of an insect (e.g., from egg to egg or adult to adult), and generation time is the length of time required to complete a single life cycle. Some aquatic insects have one generation per year and are termed **univoltine**. Some have multiple generations in one year and are termed **polyvoltine** or **multivoltine**. Others require more than one year to complete a generation. When multiple generations per year exist or generation time is longer than a year, there are sometimes overlapping generations. This, of course, means that certain individuals of these species occur in different phases of their life cycle concurrently. These different generations can be distinguished as **broods** or **cohorts**.

Extremes of generation times in North American aquatic insects are exemplified by some mosquitoes and some other Diptera that require as little as one week or less to complete a generation, and thus potentially have several generations during warm periods of the year, and by an arctic midge and a few dragonflies that possibly require as much as five or seven years to complete a generation. Extremely short generation time is often an adaptation that allows the insects to take advantage of short and sometimes unpredictable periods of optimal aquatic conditions for development (e.g., in temporary pools or intermittent streams). Relatively long generation time may be necessary to allow for adequate growth of large species, or it may reflect a tendency to increase the proportionate time spent in the aquatic environment, since it is the aquatic larval stage that is long-lived.

In North American temperate environments, where warm periods are seasonal and become even shorter at higher latitudes and altitudes, the active terrestrial stages of aquatic insects are generally relatively short-lived, with extreme examples being common in arctic and high-altitude environments. Winter stoneflies and few other aquatic insects, however, have become peculiarly adapted to emerging in winter months.

Some aquatic insect populations emerge over an extended period, and their larvae will be found to coexist as slightly different size classes. Development of the individuals of others may be closely synchronized; that is, emergence of all individuals occurs within a relatively very short period. Synchronous emergence is common in species that have short-lived adults or small populations: a large proportion of the adult individuals must be present simultaneously to increase the chances of mating and insure adequate reproduction. Adult **swarming** (congregating of flying adults to facilitate mating) also is common in many forms with relatively synchronized emergences.

Most of the active or developmental time during the life cycle of aquatic insects is spent in the aquatic environment (there are some exceptions). Furthermore, some mayflies, caddisflies, and midges are typical of insects in which the terrestrial adult stage is very short-lived in comparison to the aquatic stages. Insects that spend long periods in the terrestrial environment usually do so in an inactive stage (e.g., overwintering adults of some water bugs and water beetles, overwintering last instar larvae of some semiaquatic moths and water beetles, or eggs oviposited in dry areas).

DORMANCY

Many aquatic insects are inactive, or dormant, during parts of their life cycle. **Dormancy** can be defined as a state or period in the insect's life when there is no active mobility or development and a very low or minimal level of life-sustaining functions. It is usually an adaptation for passing through periods of adverse environmental conditions and is relatively common in regions that have considerable seasonal fluctuations of temperature and available water.

Dormancy is sometimes necessary for the completion of development. For example, the eggs of a few mayflies must be subjected to low temperatures and undergo dormancy before they are able to resume development and eclose. This dormancy is sometimes known as **obligate diapause**.

Overwintering may or may not take place in a dormant state. Overwintering dormancy occurs as a response to lowering temperature or shortening daylength and allows the insect to live through periods too cold for normal activity or development. It may involve the aquatic larvae, the aquatic or terrestrial adults, or the eggs (whether eggs are submerged or not).

Aestivation in aquatic insects is a type of dormancy that occurs as a response to drought conditions or warm temperatures. Eggs oviposited in dry areas or aquatic habitats that subsequently become dry may undergo a period of dormancy until aquatic conditions become favorable for eclosion and subsequent larval development. Larval aestivation occurs among some aquatic Diptera, and some stoneflies and fishflies may burrow into moist substrate or remain under rocks of dry intermittent stream beds and become relatively dormant until water is available. Some stonefly larvae become dormant if water becomes too warm.

Another type of dormancy is found among adults of a few caddisfly species that inhabit intermittent aquatic environments. These adults undergo extended periods of inactivity during dry summer months prior to mating and oviposition, so that oviposition corresponds to wetter seasonal periods. And finally, some larvae of aquatic insects undergo dormancy not as a result of temperature or water fluctuation but during periods when food sources are not available. For example, this phenomenon is known among a few caddisflies that feed on decaying leaves and evidently respond to changing daylengths as a triggering mechanism for dormancy or reactivity. After a period of summer dormancy, they resume activity when leaves fall at the end of the summer.

Residing in the Aquatic Environment

Aquatic insects possess certain attributes that are specifically fitted for residing in a particular kind or segment of a habitat (microhabitat). Some of these adaptations are directly related to the physical space (habitat niche) that the insect occupies. Most aquatic insects can be conveniently classified on the basis of their physical habitat and the means by which they move about or maintain themselves. Members of each of these categories may have certain adaptive traits in common regardless of their taxonomic relationships. A few examples of the categories are illustrated in Figure 3.3.

Benthos

Benthic insects (also collectively referred to as **benthos**) include those insects that are bottom dwellers. This definition is commonly extended in aquatic entomology to include any insects that reside on or in any substrate within the aquatic habitat. **Substrates** with which benthic insects may be associated include not only bottom surfaces but also any fixed or floating inorganic or organic object (e.g., stems of aquatic plants, driftwood, rock outcroppings). The majority of aquatic insects are benthic. In using the following benthic subgroupings, keep in mind that a few species may not clearly fit any of them, and some species may fit more than one.

CLINGERS

Benthic insects that cling steadfastly to substrates in strong flowing waters or wave-beaten littoral areas of lakes can be termed **clingers**. Some are equipped with well-developed, grasping tarsal claws (e.g., some adult riffle beetles and longtoed beetles and some larval minnowlike mayflies) or with well-developed anal claws or hooks at the end of the abdomen (e.g., larvae of a few netspinning caddisflies, midges, dobsonflies, and some beetles). Some highly specialized clingers possess attachment discs along the ventral portion of their bodies that allow them to grip rocky substrates in rapids (e.g., those flatheaded mayfly larvae that have an abdominal disc formed by modified gills or thick brushes of hairs). Larvae of netwinged midges possess a series of attachment discs along their ventral margin, and, in essence, the entire body of a water penny larva is an attaching disclike structure. Some other larval insects possess attachment discs at the end of the abdomen (e.g., larval black flies). Some case-making benthic insects may be considered clingers, since they have various ways of maintaining their position—for example, by cementing their cases to the substrate or providing them with ballast.

Many **rheophilic** insects (those occurring in strong flowing water) have a low profile or highly streamlined body that minimizes the frictional force of water. Prime examples are the larvae of minnowlike and brushlegged mayflies. Rheophilic insects also reduce the frictional force of water by the way in which they orient their bodies in relation to the current.

SPRAWLERS

Benthic insects that crawl about on various surfaces of such substrates as rocks, fine sediments, woody debris, or leaf packs can be termed **sprawlers**. They occur in running and still waters, and many commonly reside on the undersides of rocks (e.g., larvae of some flatheaded mayflies, stoneflies, and case-making caddisflies) or in porous areas of rocks or debris (e.g., larvae of some midges, stoneflies, and spiny crawling mayflies). Some predaceous sprawlers (e.g., larvae of fishflies and horse flies and some dragonflies, damselflies, and freeliving caddisflies) are relatively active forms. Some sand-dwelling sprawlers (e.g., a few larval clubtail dragonflies, flatheaded mayflies, midges, and case-making caddisflies) and some other sprawlers, such as the larvae of little stout crawling mayflies and some crane flies and midges, often become partially covered with sediment.

CLIMBERS

Benthic insects that commonly reside on aquatic plant stems, root systems along banks, filamentous algae, or mosses can be termed **climbers**. They usually inhabit the slower reaches or edgewaters of streams and rivers and are very common among the littoral vegetation of ponds, lakes, and swamps. Most are adapted for climbing, but some occasionally swim from one substrate to another, and others are relatively stationary. A few examples of insects that fit in this category include larvae of many dragonflies

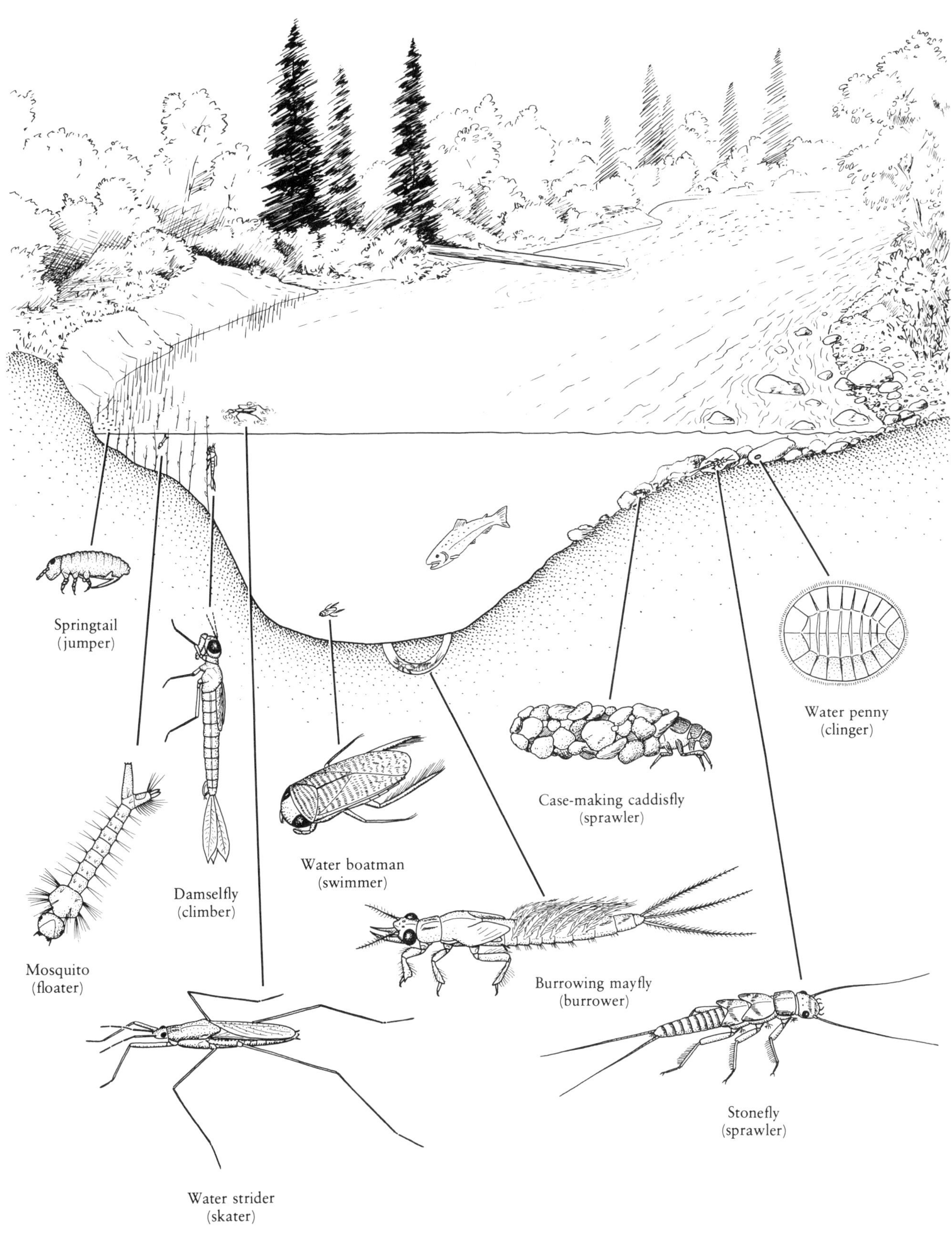

Figure 3.3. Cross section of a stream, with diagrammatic examples of habitat orientation among aquatic insects.

and damselflies, some aquatic caterpillars, a few riffle beetle adults, and larvae of some small squaregilled mayflies. Water scorpions and a few other water bugs are climbers, but often must maintain partial contact with the water surface. Some shore fly larvae (and pupae) may also be included in this category, for although they may not actively climb about, some attach their posterior breathing apparatus to underwater stems, and many occur in algal mats.

BURROWERS

Benthic insects that burrow into soft bottom substrates and live in this so-called **interstitial** habitat are termed **burrowers**. The substrate is usually silt, clay, or silt-sand; however, a few insects live interstitially at some depths in coarse sand or gravelly substrates. Burrowers generally inhabit ponds and lakes as well as the pools, slower reaches, and bank areas of rivers and streams. Examples of burrowers include the larvae of burrowing mayflies, some dragonflies, many midges, a few caddisflies, and some beetles. Some burrowing mayflies, caddisflies, and midges form well-defined burrows or tubes. Furthermore, most burrowing mayflies possess structural adaptations that aid in digging (e.g., broadened fore legs, shovel-like processes of the head, and tusks).

Many very young larvae of aquatic insects that are otherwise not considered burrowers live in the interstitial habitats of stream beds, and some aquatic insects burrow as a response to drying conditions. Larvae of some insects (e.g., a few stoneflies and crane flies) can exist for long periods in deep underground waters. Aquatic insects that are borers or miners in plants may also be considered burrowers (e.g., larvae of a few midges, crane flies, longlegged flies, and shore flies; and some caterpillars).

Plankton/Nekton

Aquatic insects that occur freely in the water fit into this grouping. **Plankton** refers to those that passively float or are freely suspended, and **nekton** refers more specifically to those that swim. Nekton and plankton are considered together, since many insects float part of the time and swim or actively move at other times.

FLOATERS

Some floating insects, especially those that must maintain contact with the air-water interface for respiratory purposes, live at or near the water surface (e.g., many fly pupae, such as those of some shore flies and moth flies; and larvae of some flies, such as mosquitoes, dixid midges, and many marsh flies). Floating larvae, however, are active part of the time, and some, such as mosquitoes, may dive when disturbed. Adult whirligig beetles, which rest on the surface much of the time, are best categorized as swimmers. The eggs of some mosquitoes are laid as rafts that float on the water surface. Other floating insects live at considerable depths in ponds and lakes. The larvae of phantom midges are capable of vertical migration and may occupy various depths, depending on the time of day. Some floaters are equipped with hydrostatic organs, and some others swallow air bubbles that increase their buoyancy.

SWIMMERS

Most of the water beetles and water bugs that periodically surface for air are highly adapted for swimming. They have streamlined bodies and often possess legs that are oarlike and equipped with swimming hairs. Some insects that are otherwise benthic are capable of efficient swimming from one resting place to another (e.g., some larval minnowlike and pronggill mayflies). Some midges and case-making caddisflies are also proficient swimmers, and the adult whirligig beetles, which swim at the surface, are perhaps the most proficient swimmers of all.

DRIFT

Lotic benthic insects that temporarily become suspended in the water and are carried downstream by the current are known collectively as **drift.** Drifting may result from catastrophic events, such as spates, severe lowering of the water level, or pollution. Many benthic insects, however, drift in a periodic manner under normal conditions. This kind of drifting is commonly known as **behavioral drift.**

Behavioral drifting can take place at any time, although it usually occurs at high rates during certain regular periods within a 24-hour cycle. Drifting occurs for only short distances (a few meters or sometimes much less) at any one time; however, it may be repeated in a stepwise fashion. Commonly, behavioral drift occurs at a relatively high rate immediately following sundown, followed by a slight rate of drift for most of the night, and finally another somewhat lower peak preceding sunup. This is known as a **bigeminus pattern** of drift. Another pattern, known as an **alternans pattern**, involves a low peak of drift after sundown and a high peak preceding sunup.

The causes and adaptive value of behavioral drift are not clearly understood for most species. Drifting may be related to the density of individuals or the optimum carrying capacity of the aquatic habitat, to periodic predation pressure, or to downstream migration. Periodic drift may be due to an endogenous mechanism, or biological clock, of the insect; it may be a response to the degree of light or darkness or to temperature; or it may result from some combination of these factors.

Neuston

Aquatic insects that live on the water surface are referred to as **neuston**, or neustonic insects. These insects do not normally break the surface film. They either walk, skate, or jump on the water surface. The claws or parts of the legs that come into contact with the water are unwettable by virtue of a waxy layer of the exoskeleton or hairs, and they are often structured so that they bend rather than break the surface film. Insects that commonly walk or rest on surface film include water treaders and marsh treaders.

SKATERS

Typical water skaters include the water striders and shortlegged striders. These two groups possess preapical claws; that is, the claws are not at the end of the leg but somewhat above the level of the tip, and they do not break the surface film. Skaters usually inhabit quiet waters, but some are found on riffles.

JUMPERS

Some species of springtails are found on the water surface and regularly jump off the water using a springlike structure on the abdomen. Pygmy mole crickets also have the ability to jump off the water surface, using specialized hind legs to do so.

TABLE 3.2 MAJOR FEEDING HABITS FOUND AMONG AQUATIC INSECTS (SEE TEXT FOR EXPLANATION OF CATEGORIES)

Order	Feeding Category				
	MACROVORES			MICROVORES	
	Herbivores	*Carnivores*	*Detritivores*	*Suspension Feeders*	*Bottom Feeders*
Ephemeroptera	S	S	S	S	C
Odonata		E			
Plecoptera	S	C	C		S
Hemiptera	S	C	C[1]		S
Megaloptera		E			
Neuroptera		E			
Coleoptera	C	C	S		C
Trichoptera	S	C	S	C	C
Lepidoptera	C				S
Diptera	C	C	S	C	C

E = Exclusive representation.

C = Considerable representation.

S = Some representation.

[1]Scavengers on animal detritus.

Obtaining Food

Many adaptations for feeding (trophic adaptations) are found among aquatic insects. Some species are specific in their food preferences, and others are more general feeders. Food may be generated within the aquatic ecosystem (**autochthonous**) or originate in an adjacent terrestrial ecosystem (**allochthonous**) and subsequently fall or be blown into the water—for example, falling leaves and branches of overhanging trees as well as terrestrial insects. Food consists of either living material or dead and decomposing animal and plant material. Decomposing material is commonly referred to as **detritus.** Food items may be obtained either from the benthic substrate or from the **seston** (the materials suspended in the water).

Some food items are microscopic organisms, such as unicellular algae, bacteria, some fungi, and some **zooplankton** (planktonic animals), or microscopic bits of detritus. Macroscopic food items (generally more than 10 cubic microns) include most aquatic insects, many other aquatic invertebrates, filamentous algae, aquatic vascular plants, leaf detritus, and dead insects or other animals. Insects that feed on microscopic food items can be termed **microvores,** and those that feed on macroscopic food items can be termed **macrovores.**

Systems for classifying animals into feeding groups or trophic categories variously take into account the dominant kinds and location of the food, the method of feeding, and size of the food. It is difficult to incorporate all of these criteria into consistent feeding categories for aquatic insects. The following system is relatively simple but is adequate for the level of treatment in this book. Table 3.2 summarizes the major feeding habits of aquatic insects. Keep in mind that some aquatic insects change their feeding habits and food preferences as they grow or as seasons change. For example, many of the early instar larvae, simply by virtue of their size, are microvores, and only become macrovores when they grow larger. The examples and data in Table 3.2 are applicable to relatively mature larvae and aquatic adults.

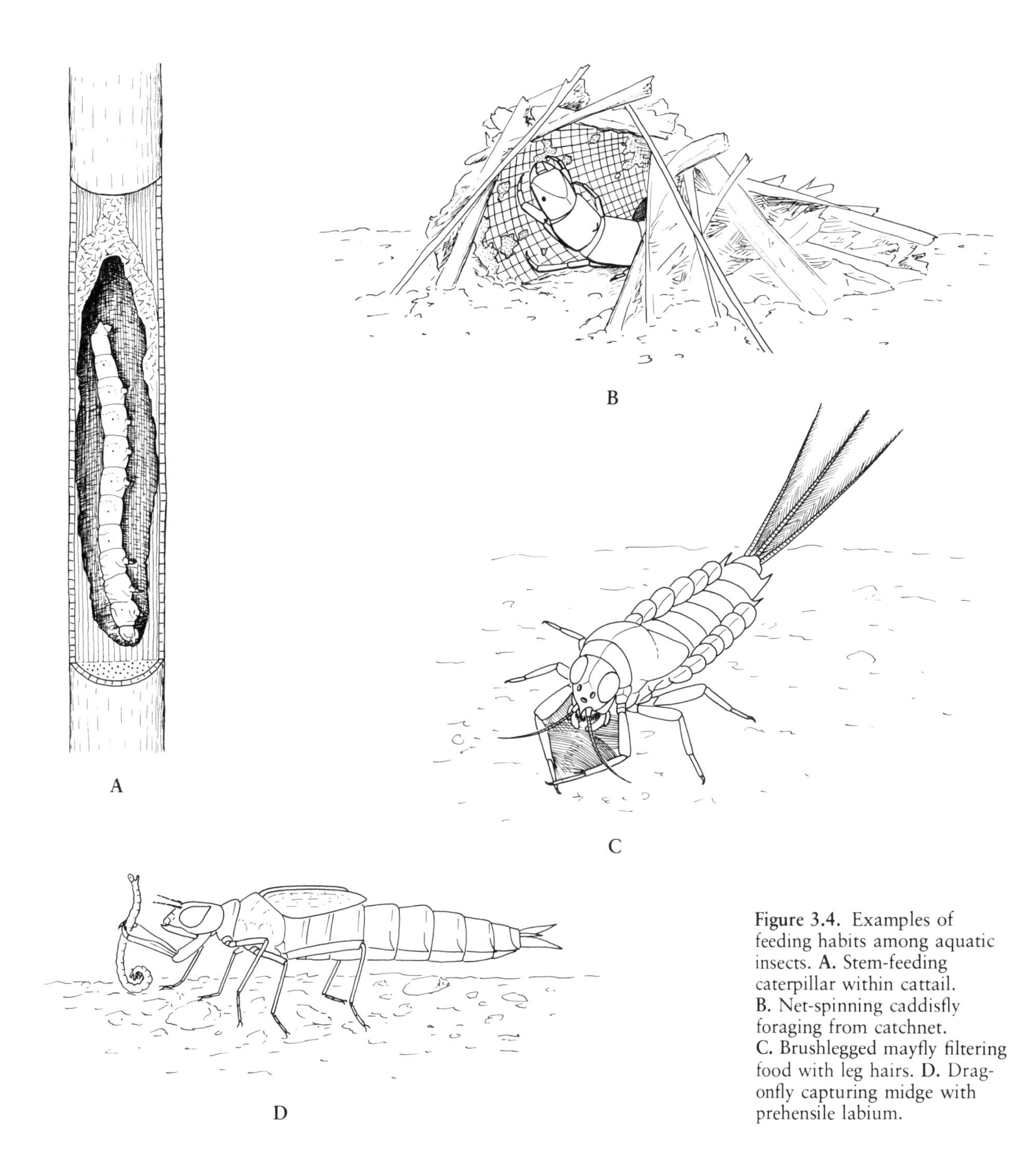

Figure 3.4. Examples of feeding habits among aquatic insects. **A.** Stem-feeding caterpillar within cattail. **B.** Net-spinning caddisfly foraging from catchnet. **C.** Brushlegged mayfly filtering food with leg hairs. **D.** Dragonfly capturing midge with prehensile labium.

Herbivores

Herbivores, or herbivorous aquatic insects, feed on living plant material. Some species are obligate herbivores; i.e., their diet consists exclusively of plant material. Many herbivores will also feed to various degrees on animal material or detritus. Those that feed on both plants and animals are called **omnivores**. Those that feed on both living and decomposing plant and animal material can be called **omnivore-detritivores**.

MACROVORES

Herbivorous macrovores feed on aquatic vascular plants or some filamentous algae. Most of them chew solid plant tissue and are obligate (e.g., larvae of some case-making caddisflies; many aquatic and semiaquatic caterpillars (Fig. 3.4A); the adults and some larvae of water beetles, such as some water scavenger beetles, aquatic leaf beetles, and water weevils; and many aquatic fly larvae). Some caterpillars, midges, and other flies bore into the plant on which they feed. Semiaquatic Homoptera, such as some leaf hoppers and plant hoppers, feed by sucking the juices of emergent plants.

MICROVORES

Many of these herbivores are much more apt to be general in their food preferences and often select food items more on the basis of size and availability than on whether it is plant, animal, or detrital in nature. Many feed on seston and are known as **suspension feeders**. Diatoms make up a significant portion of the diets of many suspension feeders. Prime examples of such feeders include the netspinning caddisflies and a few midge larvae that construct elaborate nets to capture seston floating downstream (Fig. 3.4B), and black fly larvae, which use their mouth brushes for the same purpose. Brushlegged mayfly larvae, which are omnivorous, have well-developed brushes of hairs on their fore legs that filter seston (large and small) in stream currents (Fig. 3.4C).

Many others in this category bottom feed on encrusted algae, including diatoms (usually along with microscopic detrital material). Bottom feeders include a large number of aquatic insects, such as larvae of many mayflies (Fig. 3.5C), larvae of many case-making caddisflies, a few water boatmen, bottom-dwelling caterpillars, water penny larvae, adults and larvae of many riffle beetles, and larvae of many midges as well as other aquatic flies. Many shore fly larvae forage extensively on blue-green algae.

Carnivores

Carnivores, or carnivorous aquatic insects, feed on living animal material. Many are obligate, but some are omnivorous. In addition to attacking live prey, many will scavenge on dead animal material, and thus are technically also detritivores. Those carnivores that attack and immediately kill their prey are known as **predators**, whereas those that feed on their prey and reside with it for a prolonged period before or without killing it are known as **parasites**.

MACROVORES

Predatory aquatic insects that attack other insects, small crustaceans, worms, and even small fishes are found in many orders. Many are partly **cannibalistic**; that is, they will attack members of their own species. Some predators are equipped with strong, opposable mouthparts for biting and chewing their prey. Other predators pierce their prey and suck body fluids with tubelike mouthparts.

The larvae of all dragonflies and damselflies are predators; some stalk their prey, and others rest secretively in ambush; all are equipped with an elaborate labium that is quickly extended to capture and hold prey (Fig. 3.4D). The larvae of all Megaloptera are predators with well-developed mandibles, and most water bugs are predators or predator-scavengers with well-developed piercing and sucking mouthparts. Many water bugs (e.g., giant water bugs (Fig. 3.5A), creeping water bugs, and water scorpions) have **raptorial** fore legs adapted for grasping prey in a pincerlike manner. Examples of other major aquatic insect groups with various predatory habits include the larvae of many stoneflies, a few mayflies, some caddisflies (including some suspension feeders), and many Diptera, and the larvae and adults of many water beetles.

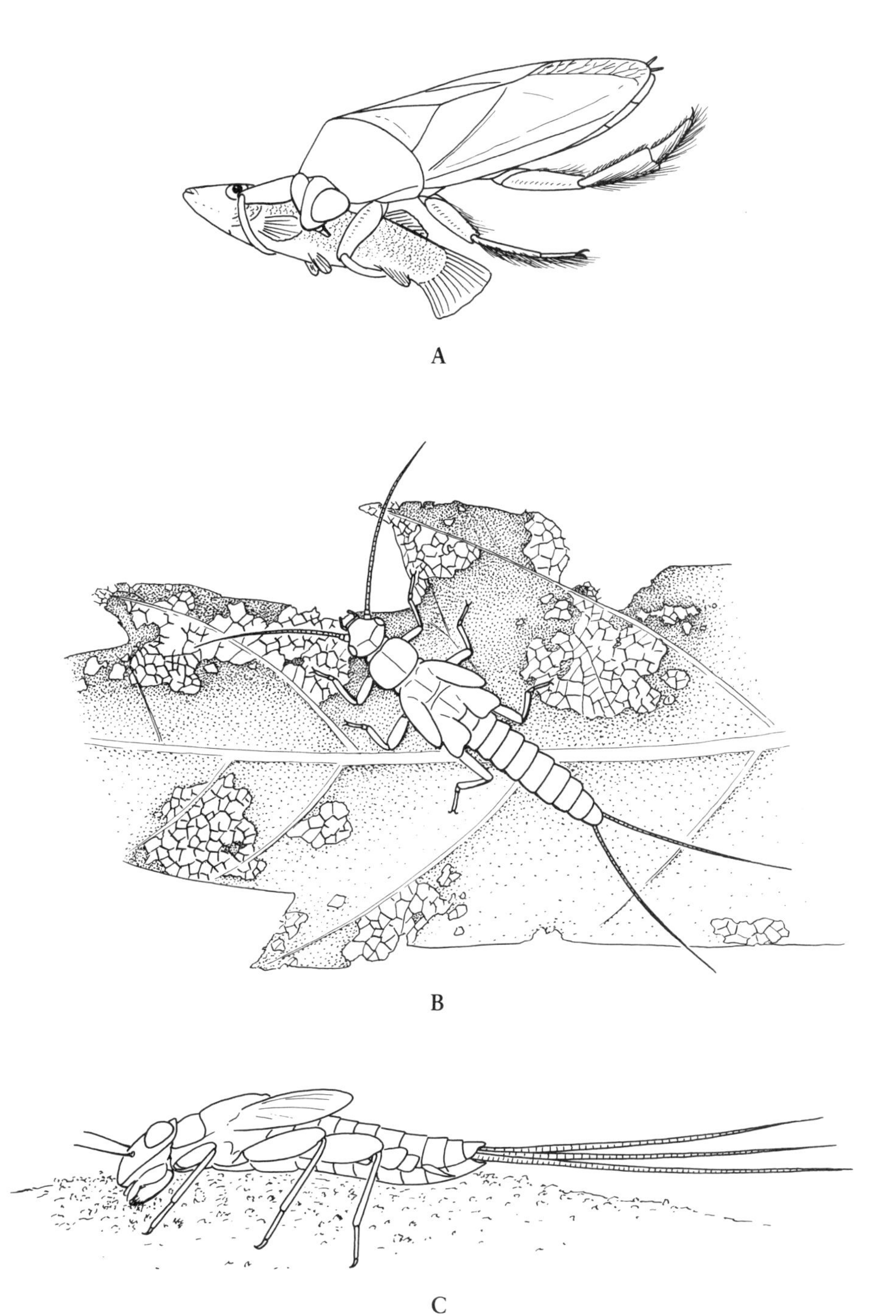

Figure 3.5. Examples of feeding habits among aquatic insects. **A.** Giant water bug impaling small fish with piercing mouthparts. **B.** Winter stonefly feeding on leaf detritus. **C.** Flatheaded mayfly feeding from substrate.

Some aquatic insects are parasites. Spongillafly larvae suck juices and cellular material from freshwater sponges. The larvae of many marsh flies are, to a degree, parasitic on snails, and a few midge larvae are thought to be parasitic on some stoneflies and mayflies. In addition, the larvae of a number of diving wasps are internal parasites of eggs, larvae, and pupae of several aquatic insects.

MICROVORES

Most carnivores in this category are general feeders and can be regarded as omnivore-detritivores. They include many of the suspension feeders that filter microscopic food items from the seston (e.g., larvae of many caddisflies, some mayflies, and a number of Diptera). They also include species that do not filter but actively prey on small suspended zooplankton (e.g., larvae of phantom midges and some midges and mosquitoes).

Detritivores

Detritivores feed on detritus. The preferred detritus may be autochthonous or allochthonous, either plant or animal material. In most cases, the primary nutrient value of feeding on plant detritus is derived from the microorganisms (bacteria and fungi) that reside and grow on the detritus.

MACROVORES

Leaf detritus is the most common food item for the majority of detritivores in this category. Some aquatic insects that feed on leaf detritus use other foods during periods when leaf detritus is not available. Examples of leaf detritivores are the larvae of many stoneflies (Fig. 3.5B) and some case-making caddisflies. A few crane flies and riffle beetles, among others, feed on woody detritus and may be termed **xylophagous.**

A number of the surface-dwelling water bugs, in addition to preying on live insects, readily scavenge on dead insects that float on the surface film. There have even been reports of a few aquatic insects feeding on dead fishes.

MICROVORES

Small detrital particles are important sources of food for many species. In general, the suspension feeders (e.g., many caddisfly larvae) can be included in this category, as can bottom feeders, such as the larvae of most mayflies (Fig. 3.5C) and many caddisflies and midges. Many are herbivore-detritivores or omnivore-detritivores, as mentioned previously. Some, especially among burrowers and rattailed maggot larvae, ingest fine sedimentary material and derive nutrients from fine detritus contained there; these forms are sometimes known as **deposit feeders.**

When simple curiosity passes into the love of knowledge as such, and the gratification of the aesthetic sense of the beauty of completeness and accuracy seems more desirable than the easy indolance of ignorance; when the finding out of the causes of things becomes a source of joy, and he is counted happy who is successful in the search, common knowledge of Nature passes into what our forefathers called Natural History. . . .

Thomas Henry Huxley

Breathing in the Aquatic Environment

Oxygen is necessary for the vital functions of life. The source of oxygen for terrestrial organisms is the atmosphere. Oxygen can also exist in water in a dissolved form and thereby be available to submerged organisms.

Insects, in general, possess a respiratory system that consists of a network of tubes through which gaseous oxygen is distributed to parts of the body. This is known as a **tracheal system.** The largest respiratory tubes, or **trachea,** of the generalized insect originate as external openings along the body wall; these openings are known as **spiracles.** In the generalized insect pattern, spiracles are located on the sides of thoracic and abdominal segments (one pair per segment).

The generalized external respiratory system of insects is well suited for a terrestrial life and is essentially the system found in surface-dwelling aquatic insects and many semiaquatic insects. Among submergent aquatic insects, however, either the system or the behavior associated with obtaining oxygen has been modified.

The respiratory adaptations of submergent aquatic insects, whether structural, behavioral, or physiological, are extremely varied. The general respiratory categories are discussed below (see also Table 3.3).

TABLE 3.3 RESPIRATORY ADAPTATIONS OF SUBMERGED AQUATIC INSECTS (SEE TEXT FOR EXPLANATION OF CATEGORIES)

Order	Major Respiratory Category			
	AEROPNEUSTIC		HYDROPNEUSTIC	
	Surface Breathers	*Endophytic Breathers*	*Cutaneous Respiration*[1]	*Spiracular Respiration*[2]
Ephemeroptera			E	
Odonata			E	
Plecoptera			E	
Hemiptera	C		S(?)	S
Megaloptera	C		C	
Neuroptera			E	
Coleoptera	C	S	C	C
Trichoptera			E	
Lepidoptera	S	S	C	S(?)
Diptera	C	S	C	S

E = Exclusive representation.

C = Considerable representation.

S = Some representation.

[1]Includes those with and without gills.

[2]With or without spiracular gills.

Aeropneustic Insects

Aeropneustic aquatic insects include submerged or partially submerged forms that use atmospheric oxygen (in the air). They possess functional spiracles, although the number and location of spiracles are often modified. Since aeropneustic insects are not limited by the amount of dissolved oxygen present in water, they may be found in a wide variety of aquatic habitats.

SURFACE AIR BREATHERS

These aquatic insects use atmospheric oxygen, even though they may be completely submerged or at least partially submerged most of the time. **Continual-contact breathers**, or **tube breathers**, maintain some bodily connection with the air-water interface. The point or points of air contact are either spiracles at the ends of tubes or open ends of tubes that lead to spiracles. The thoracic **respiratory horns** (Fig. 3.6A) of many aquatic fly pupae are good examples of such structures, as are the terminal **breathing tubes** or **siphons** of water scorpions, mosquito larvae, many shore fly larvae, and rattailed maggots (Fig. 3.6B). The tips of such breathing tubes are unwettable or have unwettable hairs that help maintain them at the surface.

Periodic-contact breathers, or **air-storage breathers**, are aeropneustic insects that live submerged but must swim, crawl, or float to the surface periodically to obtain air. These insects are able to carry an air supply under water, thus allowing them to continue respiration in a submerged state until the oxygen in their air store has been depleted. Air may be captured and stored in several ways. For example, many adult water beetles (Fig. 3.6C) and water bugs, which constitute the large majority of periodic-contact breathers, capture a bubble in an **underwing chamber** (beneath the folded wings) or on other parts of the body, and a few beetle larvae and caterpil-

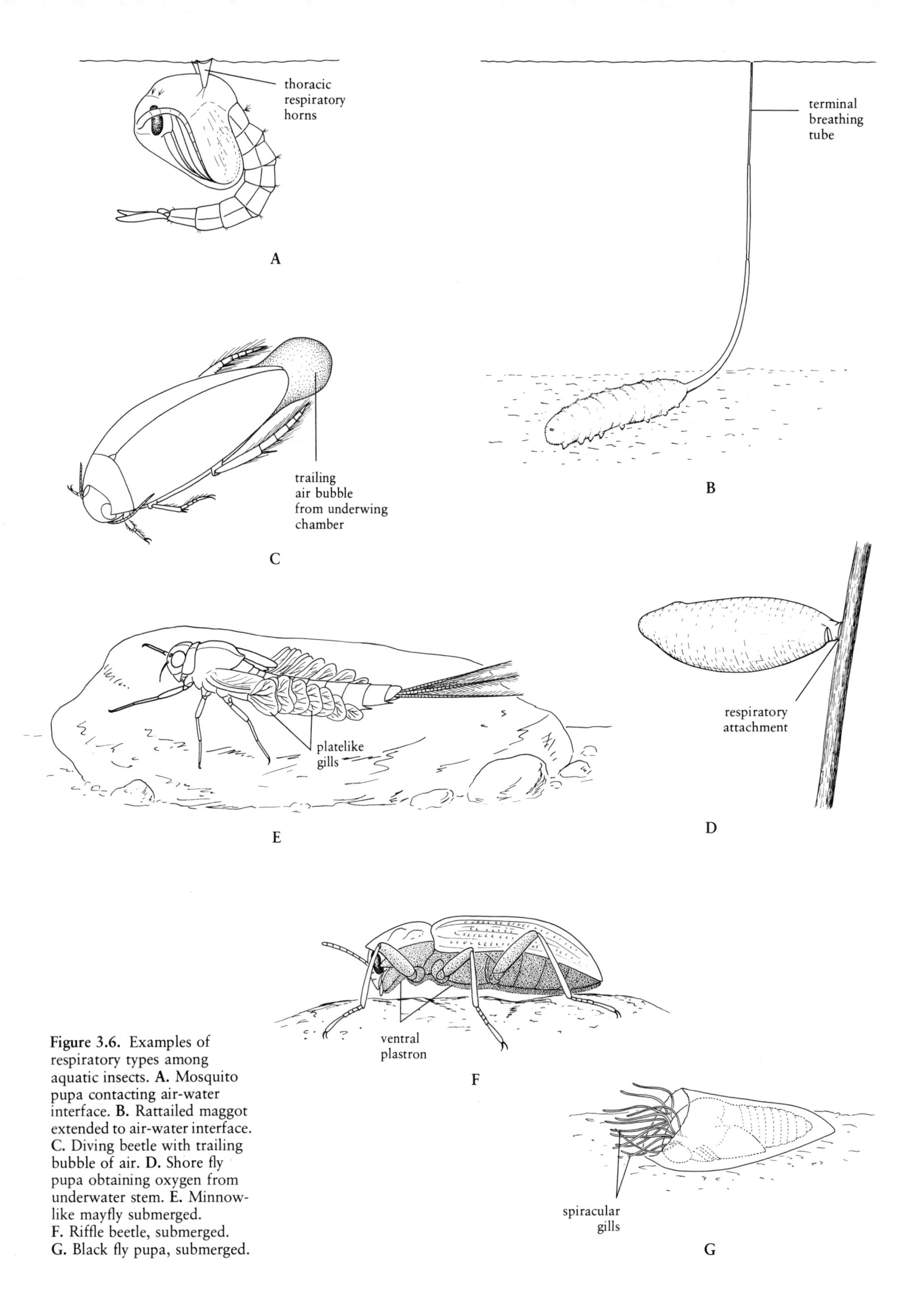

Figure 3.6. Examples of respiratory types among aquatic insects. **A.** Mosquito pupa contacting air-water interface. **B.** Rattailed maggot extended to air-water interface. **C.** Diving beetle with trailing bubble of air. **D.** Shore fly pupa obtaining oxygen from underwater stem. **E.** Minnow-like mayfly submerged. **F.** Riffle beetle, submerged. **G.** Black fly pupa, submerged.

lars obtain air with spiracles upon surfacing and store a surplus internally in enlarged trachea. (A few fly larvae also carry a bubble of air under water.)

Dense coverings of microscopic, unwettable hairs or scales hold the bubble or layer of air to the insect's body while submerged. The densely covered areas may be extensive but at least include those regions of the body where spiracles are located (e.g., the abdomen of bugs and adult beetles). The film of air held in such a manner is called a **plastron.** The plastron keeps the spiracles dry and allows aeropneustic respiration to take place under water.

An important factor related to underwater air storage contributes to the ability of most periodic-contact breathers to remain submerged for extended times: the air bubble, or plastron, acts as a **physical gill.** That is, oxygen present in a dissolved form in the water will to some extent diffuse into and replenish the bubble as long as nitrogen, which is also contained in the air bubble, has not dissipated. Thus the length of submergence of periodic-contact breathers is to some degree a function of the availability of dissolved oxygen, and they may not be entirely aeropneustic.

ENDOPHYTIC BREATHERS

These aeropneustic insects pierce vascular plants with modified spiracles and obtain oxygen available within underwater stems or roots. Examples include the larvae of aquatic leaf beetles and larvae and pupae of two mosquitoes and some shore flies (Fig. 3.6D).

Hydropneustic Insects

Hydropneustic aquatic insects are submerged forms that use dissolved oxygen. Most do not have functional spiracles, but many do. Generally, they are limited in their habitat distribution to some extent by the amount of dissolved oxygen available in the water, and as a result are apt to be more common in lotic and well-aerated lentic habitats (with some exceptions). A few aquatic insects are capable of both hydropneustic and aeropneustic respiration.

CUTANEOUS UPTAKE

Most hydropneustic insects obtain oxygen directly through the body wall and are not reliant on functional spiracles. This so-called **cutaneous respiration** may occur generally over the softer, or membranous, areas of the body. It is found in the permanently and totally submergent stages of most aquatic insects—that is, the larvae of mayflies, damselflies and dragonflies, stoneflies, a few water bugs, spongillaflies, alderflies, some water beetles, caddisflies, a few aquatic moths, and many aquatic flies. Some of the tube-breathing aeropneustic insects are also capable of cutaneous respiration in oxygenated water (e.g., soft-bodied aquatic fly larvae and fishflies).

Many hydropneustic insects possess membranous outgrowths on their bodies that add to the total amount of surface area available for cutaneous uptake and, in some larvae, serve as the primary sites for external respiration. Such outgrowths include most structures known as **gills** that occur on the larvae of mayflies (Fig. 3.6E), many stoneflies, dobsonflies, some caterpillars, many caddisflies, some beetles, and some aquatic flies. Other membranous appendages that have been termed *filaments* or *lamellae* (e.g., the caudal lamellae of some damselfly larvae) may also be respiratory sites.

Gills are basically of two types: (1) flat, platelike structures that are commonly supplied with tracheal branches, and (2) filamentous structures that are fleshy and tubelike but may be highly branched and may or may not be direct outgrowths of the tracheal system. The term *gill* infers a respiratory organ; it perhaps has been too liberally applied to structures found on aquatic insects, since some of the so-called gills (especially some platelike gills) have been shown to have other or additional functions, such as protection, swimming, clinging, stabilization, osmoregulation, or burrow maintenance.

Many hydropneustic insects ventilate parts of their bodies in order to increase the amount of renewed water available to them and thus aid respiration. Examples include the rhythmic beating of the gills of many mayfly larvae, the undulating of the abdomen of caddisfly larvae within tubecases, and the "push-ups" of some stonefly larvae. The repeated intake and expulsion of water from the rectal chamber of dragonfly larvae is another example of ventilation. The rate of ventilation often increases as a response to decreased dissolved oxygen concentrations.

SPIRACULAR UPTAKE

Some hydropneustic insects use functional spiracles while continually submerged. Open spiracles are covered by a layer of air and never come in direct contact with the water, since the insects would otherwise be subject to drowning. The layer of air, or plastron, mediates the uptake of dissolved oxygen by acting as a physical gill (as discussed above under periodic-contact, aeropneustic insects). Adult riffle beetles (Fig. 3.6F) and longtoed beetles, as well as some creeping water bugs and possibly a few other aquatic insects, are capable of maintaining a plastron indefinitely under considerable water pressure. These insects can thus remain submerged in well-oxygenated water without having to resurface. Creeping water bugs ventilate their ventral plastron by rowing the hind legs beneath the abdomen.

A few insects (e.g., black fly (Fig. 3.6G) and some midge and crane fly pupae) possess branched outgrowths of the thoracic spiracles that are known as **spiracular gills**. Oxygen is obtained from the water via a plastron that is associated with the spiracular gills (in some midge pupae, body fluid associated with the spiracular gill, rather than a plastron, mediates the transfer of oxygen from the water to the insect). Certain spiracular-gilled pupae are capable of aeropneustic respiration if they happen to become exposed to air during drying conditions. Some dance fly pupae possess a series of unbranched spiracular gills along the sides of the body.

Maintaining Proper Internal Environments

The maintenance of specific concentrations of internal salts or ions necessary for vital biochemical functions in animals is a phenomenon known as **osmoregulation**. Internal ions and water used metabolically and lost by excretion must be replaced. The availability of replacement ions and water depends partly on the specific kind of external environment in which the animal lives. As a result, osmoregulation is basically a matter of regulating the rate of intake and absorption of water or ions from the external environment and regulating their rate of excretion from the internal environment.

Among terrestrial insects, ions and water lost via metabolic use or excretion are replaced by oral intake and absorption in the gut. The rectum of these insects is very important in controlling, by reabsorption, the amount of water and ions lost by excretion. For example, in arid environments, where water conservation is critical to survival, insects osmoregulate by rectal reabsorption of water and the excretion of concentrated urine.

Osmoregulation is also critical in aquatic environments, both for the majority of aquatic insects that live in freshwaters, where the concentration of ions is lower than that of the insect's internal fluids, and for those insects that live in salt or brackish waters, where the concentration of ions is higher than that of the insect's internal fluids. The degree of tolerance for various concentrations of salinity in water and the relative ability of the insect to osmoregulate are important factors influencing the habitat distribution of aquatic species. The various osmoregulatory adaptations known among aquatic insects are introduced below.

Hyperosmotic Regulation

In freshwater environments, a gradient between the internal fluids of the insect body and the external aquatic medium (known as an osmotic gradient) favors a flux of water into the insect body and a flux of ions from the body. The insect must therefore regulate the removal of surplus water and the intake and retention of ions. This type of osmoregulation is known as **hyperosmotic regulation.**

The removal of excess water is accomplished by the excretion of dilute urine, and ions are retained to some degree by reabsorption in the rectum or other parts of the gut. These functions cannot entirely compensate for ion loss, which is inevitable, hence freshwater insects possess additional osmoregulatory adaptations for active ion absorption. The exact means of ion absorption is still unknown for many aquatic insects. The categories of ion absorption given below, however, are based on known mechanisms. Generally, freshwater insects are able to adjust well to decreases in the salt concentrations of water, but most are unable to cope with more than slight increases of salt concentration and are, for the most part, intolerant of salt water.

CHLORIDE CELLS

The larvae of mayflies and stoneflies and the adults and larvae of submergent water bugs possess scattered, specialized cells, known as **chloride cells,** on various surfaces of their bodies. These cells function in the uptake of ions from the water. Among mayfly larvae, chloride cells occur on different parts of the body, depending on the species, but they are usually concentrated on the gills and sides of the abdomen (Fig. 3.7A). The larvae are able to ventilate these areas when in still waters. Among stoneflies, chloride cells commonly occur on the lateral and ventral abdomen, intersegmental areas, and gills. Among water bugs, chloride cells are distributed on any parts of the body except those covered by the folded wings or plastron.

The rate of ion absorption can be regulated as follows. The number or concentration of chloride cells will increase with a decrease in the salt concentration of the water. In addition, changes in the chloride cell membrane, such as convoluting, may occur as a response to external decreases in salt concentration. Increases in salt concentration of the water can result in the smoothing of the cell membrane and degeneration of chloride cells.

CHLORIDE EPITHELIA

The larvae of damselflies and dragonflies and many caddisflies and aquatic Diptera possess various patches of integument that are rich in cells that function in the uptake of ions. These patches are known as **chloride epithelia.** The chloride epithelia of caddisflies vary in number and are located on the ventral and sometimes dorsal parts of the abdomen (Fig. 3.7B). These areas are actively ventilated. The chloride epithelia of aquatic flies are situated at the sides of the anal opening. The chloride epithelia of dragonflies (Fig. 3.7C) and damselflies are located internally in the rectal chamber. There are three such rectal chloride epithelia in damselflies; the number varies among dragonflies. Rectal chloride epithelia are ventilated by the rectal chamber.

Insects possessing chloride epithelia compensate for slight shifts in the salt concentration of the water by changing the rate of ventilation or by increasing or decreasing the size of the chloride epithelia. For example, the rate of rectal ventilation in dragonflies will increase as a response to decreased salt concentration in the water, and the size of the chloride epithelia of aquatic flies and caddisflies may increase as a response to a similar environmental shift.

PAPILLAE

Thin-walled, somewhat balloonlike outgrowths of the body that are located posteriorly on the abdomen of many aquatic fly and caddisfly larvae are known as **papillae**. Such papillae may occur as more or less permanent appendages posterior to the anus (e.g., anal papillae of midge and mosquito larvae; Fig. 3.7D) or as evaginations of the rectal wall that protrude from the anus (e.g., preanal papillae of black fly larvae; Fig. 3.7E). Papillae range in number from 2 to 12 and are occasionally branched.

The walls, or epithelium, of the papillae are capable of ion uptake. Insects can adjust to shifts in salt concentrations by an increase or decrease in the size of the papillae. For example, the papillae will increase in size (a process sometimes known as functional hypertrophy) in response to decreasing salt concentrations in the water, and will decrease in size in response to increasing salt concentrations.

DRINKING

Ion absorption takes place in the gut of some predaceous diving beetle larvae and alderfly larvae. These insects must drink water to facilitate this process. It also follows that they must then excrete a considerable amount of dilute urine. In the beetle larvae (Fig. 3.7F), the ion absorption takes place in a part of the gut known as the ileum, which is relatively long and lined with a highly developed ion-absorbing epithelium. The rate of ion absorption may, in part, be regulated by the rate of drinking.

Hypoosmotic Regulation

Relatively few species of aquatic insects live submerged in salt or brackish waters (primarily the larvae of some Diptera, such as midges, mosquitoes, and shore flies). The osmotic gradient between the internal fluids of these insects and their aquatic environment favors the influx of ions and the outflux of water. Osmoregulation to conserve water and at the same time eliminate excess ions is a process known as **hypoosmotic regulation.** Most hypoosmotic-regulating insects drink the concentrated external medium

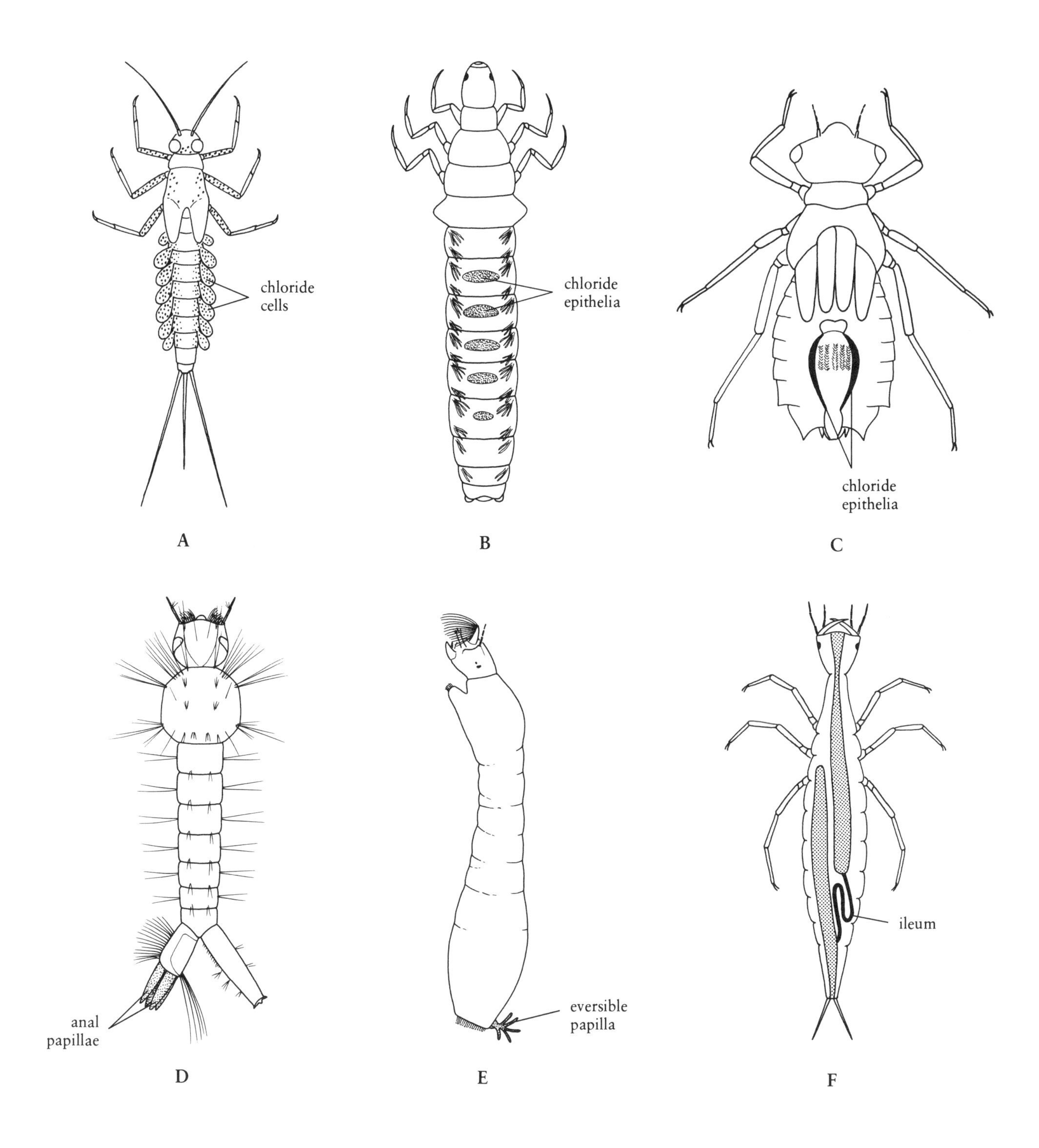

and then during the process of excretion reabsorb water and excrete highly concentrated urine.

Some species that can live in either salt water or freshwater have a part of their rectum (or in some, the ileum) adapted for water reabsorption and another part adapted for ion reabsorption. The former region becomes functional in salt water, and the latter region becomes functional in freshwater. Only in one species (a saltwater mosquito) have capabilities of ion excretion as well as ion absorption been demonstrated. The anal papillae of these mosquitoes absorb ions in freshwater and excrete ions in salt water.

Figure 3.7. Examples of structural osmoregulatory adaptations. **A.** Mayfly larva. **B.** Caddisfly larva. C. Dragonfly larva showing internal rectal chamber. D. Mosquito larva. **E.** Black fly larva. **F.** Predaceous diving beetle larva showing internal gut. [Adapted from Komnick (1977).]

References

Corbet, P. S. 1963. *A biology of dragonflies*. Quadrangle Books, Chicago. 247 pp.

Cummins, K. W. 1973. Trophic relations of aquatic insects. *Annu. Rev. Entomol.* 18:183–206.

Cummins, K. W., C. A. Tryon, and R. T. Hartman (eds.). 1966. *Organism-substrate relationships in streams*. Spec. Publ. 4, Pymatuning Lab. Ecol., Univ. Pittsburgh. 145 pp.

DeMarch, B. G. E. 1976. Spatial and temporal patterns in macrobenthic stream diversity. *J. Fish. Res. Brd. Canada* 33:1261–1270.

Edmunds, G. F., Jr., S. L. Jensen, and L. Berner. 1976. *The mayflies of North and Central America*. Univ. Minn. Press, Minneapolis. 330 pp.

Hinton, H. E. 1968. Spiracular gills. *Adv. Ins. Physiol.* 5:65–162.

Hynes, H. B. N. 1970. *The ecology of running waters*. Univ. Toronto Press, Toronto. 555 pp.

Hynes, H. B. N. 1974. Further studies on the distribution of stream animals within the substratum. *Limnol. Oceanogr.* 19:92–99.

Jander, R. 1975. Ecological aspects of spatial orientation. *Annu. Rev. Ecol. Systematics* 6:171–188.

Komnick, H. 1977. Chloride cells and chloride epithelia of aquatic insects. *Inter. Rev. Cyt.* 49:285–329.

Mackay, R. J., and G. B. Wiggins. 1979. Ecological diversity in Trichoptera. *Annu. Rev. Entomol.* 24:185–208.

Merritt, R. W., and K. W. Cummins (eds.). 1978. *An introduction to the aquatic insects of North America*. Kendall/Hunt, Dubuque. 441 pp.

Oliver, D. R. 1971. Life history of the Chironomidae. *Annu. Rev. Entomol.* 16:211–230.

Rabeni, C. F., and G. W. Minshall. 1977. Factors affecting microdistribution of stream benthic insects. *Oikos* 29:33–43.

Ruttner, F. 1953. *Fundamentals of limnology*. Univ. Toronto Press, Toronto. 295 pp.

Smith, R. L. 1966. *Ecology and field biology*. Harper and Row, New York. 686 pp.

Thorpe, W. H. 1950. Plastron respiration in aquatic insects. *Biol. Rev.* 25: 344–390.

Usinger, R. L. (ed.). 1956. *Aquatic insects of California*. Univ. Calif. Press, Berkeley. 508 pp.

Wallace, J. B., and R. W. Merritt. 1980. Filter-feeding ecology of aquatic insects. *Annu. Rev. Entomol.* 25:103–132.

Waters, T. F. 1972. The drift of stream insects. *Annu. Rev. Entomol.* 17:253–272.

Wesenberg-Lund, C. 1943. *Biologie der süsswasserinsekten*. Springer, Berlin. 682 pp.

Wiggins, G. B. 1977. *Larvae of the North American caddisfly genera*. Univ. Toronto Press, Toronto. 401 pp.

Williams, D. D. 1979. Aquatic habitats of Canada and their insects, pp. 211–234. In *Canada and its insect fauna* (H. V. Danks, ed.). Mem. Entomol. Soc. Canada. 108.

Wootton, R. J. 1972. The evolution of insects in fresh water ecosystems, pp. 69–82. In *Essays in hydrobiology* (R. B. Clark and R. J. Wootton, eds.). Univ. Exeter Press, Exeter.

CHAPTER 4

From Field to Laboratory

TO DETERMINE the various aquatic insects of a habitat and to understand the dynamics of an aquatic insect community, it is important to employ efficient sampling methods. The choice of methods depends on the kind of habitat being investigated and the purpose of the investigation—for example, general or specific collecting; qualitative or quantitative analysis. Whether one is a fly fisherman ascertaining the species present at a favorite fishing site at a specific time, a student making a collection of aquatic insects, or an ecologist studying the aquatic ecosystem, a familiarity with sampling techniques is fundamental.

Once aquatic insects are collected, additional methods for handling the samples must be used; again, these depend on the nature of the investigation. It may be necessary to keep the insects alive, to rear one stage to another, to make a preserved collection, or to examine traits for identification or tying flies. Some generally important field and laboratory methods of sampling and handling aquatic insects are reviewed below.

Sampling Aquatic Insects

In recent years sampling technology has been vastly improved, especially for studying benthic organisms, and many of the new elaborate devices as well as standard equipment are now commercially available. It is beyond the scope of this book to delve into all of this technology and its applications; an introduction to basic methods should prove adequate for most readers.

The Aquatic Environment

The major aquatic environments and the methods used to sample them can be conveniently categorized as follows.

SUBSTRATES OF SHALLOW FLOWING WATERS

Aquatic insect life is usually very productive and diverse on the bottoms of shallow streams. Perhaps the most basic sampling device for this habitat is the **hand screen** (sometimes called a **kick screen**). It is simply a rectangular piece of screen or netting fastened along two of its edges to vertically held, long handles (Fig. 4.1A). The bottom edge of the screen is held flush against the substrate while the substrate and plant material upstream is kicked or stirred up so as to dislodge insects that the current can carry into the screen.

If the handles are long enough, one person can kick while at the same time holding the hand screen. Two persons can sample larger bottom areas because the tasks of kicking and holding can be separated. In slow or slack water, the kicker should help generate the flow of water; and when the substrate is soft, it should be kicked into deeply to dislodge burrowing insects.

Variation in design of devices for intercepting dislodged upstream insects is limited only by one's imagination. Sacklike nets with sturdily framed mouths are more effective in entrapping fast-swimming, elusive benthos and preventing loss of insects when removed from the water. The finer the mesh, the more minute the benthos that will be retained; however, the extra sediment retained will require more sorting.

The above devices offer a good method of qualitative sampling. The standard **surber sampler** or a modification of it can be employed for more quantitatively oriented studies, since only a prescribed area of substrate is sampled. The surber sampler (Fig. 4.1B) consists of a sacklike net, framed at the mouth (usually square or rectangular), and having an additional frame of the same dimensions extending in front of the mouth at a right angle. The substrate enclosed by the bottom frame is agitated with the hands so as to dislodge insects, which are then swept into the net. Although the size of the framed area (usually a square foot) will be consistent from sample to sample, the actual bottom contact area will vary depending on the composition and heterogeneity of the substrate.

Elaborations of quantitative bottom samplers usually involve variations in the size of the frame, the type of bottom enclosure (e.g., a **box sampler** prevents insects outside the area from being inadvertently sampled), and the method of dislodgement (e. g., a vacuum device). Modifications of soil **core samplers** can be used to sample insects within the substrate and determine their vertical distribution.

A **drift net** (Fig. 4.1C) is a sacklike net, framed at the mouth, and secured in the water so that it can remain for prolonged, unattended periods. Such nets collect drifting insects and are most effective during peak behavioral drift periods—usually overnight or at least immediately after sundown or before sunup. Drift netting can also be effective in sampling insects that rise to the surface prior to emergence or hatching; and when nets are located at the surface, floating cast skins can also be collected.

Drift nets require a bag deep enough to prevent insects from escaping or being washed out. It is advisable to empty the net often if there is much detritus in the water, otherwise the netting will become clogged and cause a

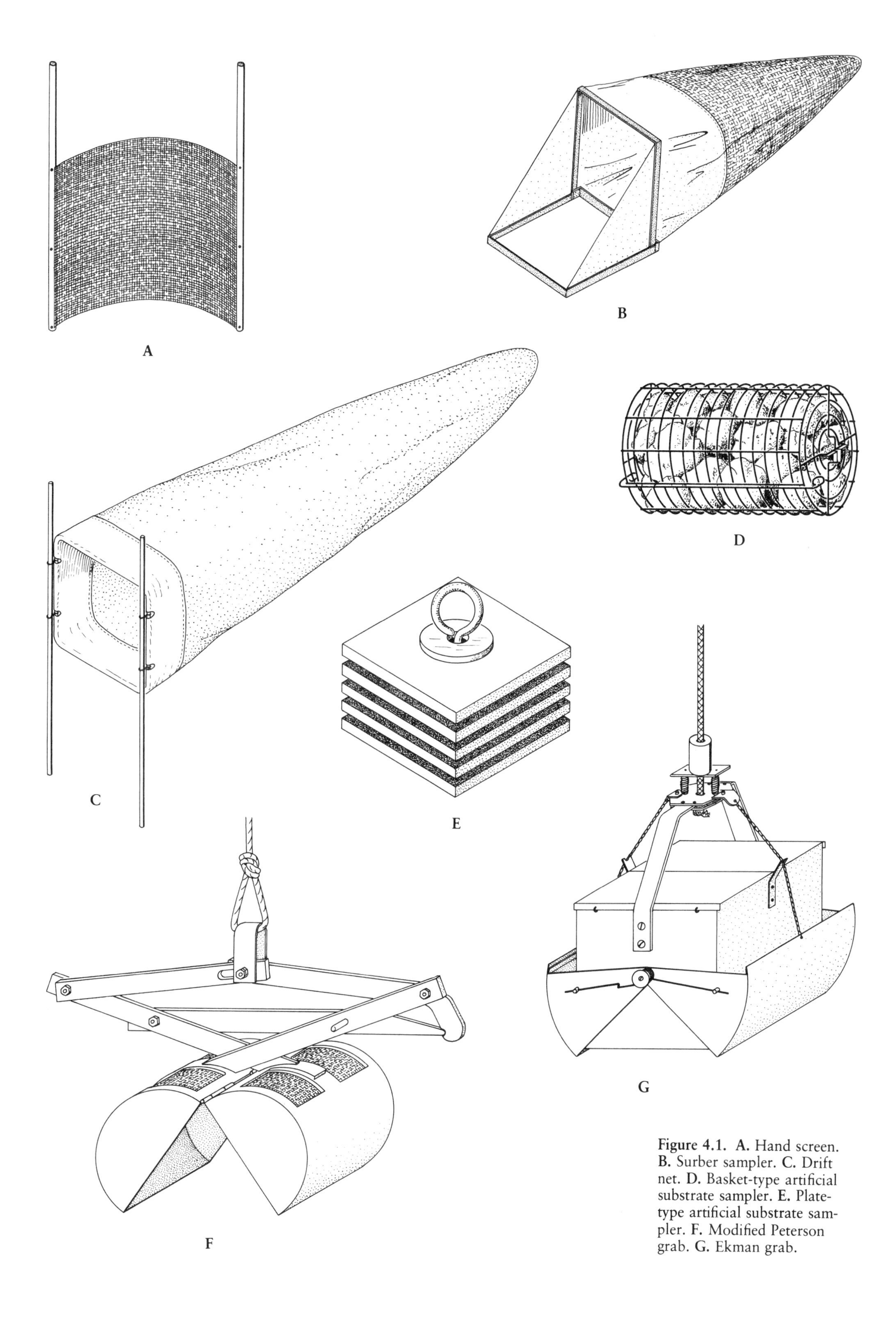

Figure 4.1. **A.** Hand screen. **B.** Surber sampler. **C.** Drift net. **D.** Basket-type artificial substrate sampler. **E.** Plate-type artificial substrate sampler. **F.** Modified Peterson grab. **G.** Ekman grab.

backwash. Drift nets may be used as quantitative samplers, since it is possible to calculate the amount of water sampled for any period.

Another basic technique used for rocky or debris-strewn substrate involves lifting out rocks and driftwood and **picking** the clinging and sprawling insects. The many species that seek out darker microhabitats are often located on the undersides of these objects. Secretive species are sometimes found by picking through porous or algae-covered areas of rocks and the interstices of decaying wood.

An **artificial substrate sampler** simulates the naturally available substrate and can be secured on the bottom or suspended in the water so that it may be colonized by benthic insects. These samplers commonly consist either of wire mesh baskets (Fig. 4.1D) filled with rocks or other objects or of a series of rigidly fastened and spaced plates (Fig. 4.1E).

Displacement of the substrate sampler can be prevented by anchoring it with stakes or weights or by tying it to an immovable object in the water or on the shore. The amount of time required for optimum colonization will vary, and although it generally takes about four weeks for a community to stabilize, some early succession colonization may occur almost immediately. Since the sampler is usually left in place for long periods of time, it is sometimes necessary to mark its location. In order to prevent insect loss, the sampler should be carefully retrieved by placing it in a bag or fine mesh net held under water. The insects can then be washed from the sampler and the bag or net.

SUBSTRATES OF DEEP FLOWING OR STILL WATERS

This habitat can be difficult to sample because of its remoteness, and although the fauna may not be as diverse as in some other habitats, there are important and interesting insects to be found. Several kinds of **grabs** (Fig. 4.1F, G) are commonly used for sampling. Generally, grabs have an opening and closing mouth that engulfs a portion of substrate either automatically when they hit bottom or when they are triggered.

Grabs are lowered and retrieved by a cable or long handle, usually from boats, steep banks, bridges, or docks. Many are heavy and require the use of a winch. Samples usually have to be sieved or sifted in order to find all insects. Grabs provide quantitative samples insofar as the surface area sampled is consistent with each sample.

Heavy duty nets can be used to drag along the substrate. The use of such **dredges** (Fig. 4.2A) often requires a powerful boat. Dredges can be fitted with rakes for stirring substrate in front of them and with runners to stabilize them while being towed. Dredging may be particularly advantageous in studying the bottom fauna of large rivers.

Artificial substrate samplers, discussed above for shallow waters, are also applicable to deep-water habitats, but anchoring, marking, and retrieving are generally more complicated. Drift nets can also be used in deep flowing waters, but they usually must be suspended from banks or boats. Trolling with drift nets suspended from boats can be used to sample large or tidal rivers.

BANKS AND MARGINAL VEGETATION OF STILL OR FLOWING WATERS

Many insect species can be found in marginal or littoral habitats. **Dip nets** (shallow or spoonlike reinforced nets with long handles) can be passed through the water and marginal vegetation to collect swimming or climbing

Figure 4.2. **A.** Dredge. **B.** Apron net. **C.** Triangular dip net. **D.** Plankton tow. **E.** Mosquito dipper. **F.** Enamel pan.

insects. **Apron nets** (Fig. 4.2B)—dip nets that have a large mesh screen placed over part of the mouth—are particularly useful in thickly vegetated areas.

Some dip nets have frames with straight sides (Fig. 4.2C) so that one side can be effectively pulled along the substrate to collect bottom dwellers. Different sized tea strainers or facsimiles thereof are also useful in sampling shallow waters. Hand screens, which were discussed for sampling shallow flowing waters, can be used, but are sometimes not preferred because they may be easily torn in thick vegetation or debris and often must be passed through the water if there is little current.

Sieving and picking through algae and vascular plant material may also be productive. Many secretive insects dwell among filamentous algae, and many species are associated with stems, roots, leaves, and culms of emergent vegetation.

OPEN WATERS AND WATER SURFACES

Many insects live on or near the surface of water, a few remain suspended in deep open water, and many swim in shallower waters. Dip nets are useful for sampling shallow open waters, but **plankton tows** (Fig. 4.2D) must be used in deeper waters. A plankton tow is a fine mesh net that can be pulled through water at various depths; a boat is often required for towing.

For insects that live near the surface, a **dipper** (Fig. 4.2E) is a useful sampler. Dippers, such as the mosquito dipper, are simply small cups with handles of various sizes.

Jumping surface dwellers are often elusive and difficult to collect with nets. An alternative method of collecting them involves using a **pan trap**, a shallow pan (Fig. 4.2F) with a little detergent water placed in it. The pan trap can be passed along the water surface or pushed through vegetation, allowing insects to jump into it. Jumpers are usually immobilized, since a surface film is not maintained in the detergent water.

The Terrestrial Environment

Sampling terrestrial stages of aquatic insects requires methods very different from those used in aquatic environments. Collecting terrestrial stages is important for correlating the different stages of many aquatic insects; and for the fly fisherman, the successful use of fly imitations often depends on which species are emerging or flying at the time.

BASIC INSECT NETS

Dragonflies often patrol above the water in search of prey; many mayflies, caddisflies, midges, and others swarm over or near water; and many species fly near the water to oviposit. **Aerial nets** are a basic tool for capturing these flying insects.

The successful use of aerial nets obviously depends on the user's ability to see and reach the insects. The difficulty in seeing the many forms that fly in twilight can be overcome somewhat by learning to focus at different distances to perceive individuals as silhouettes against the source of visible light. The difficulty of collecting forms that fly out of arm's reach can be overcome sometimes by using aerial nets with extendible handles (Fig. 4.3A).

There is hardly a better way of obtaining the smaller and less active species than the old-fashioned entomologists' method of beating the bushes over an inverted umbrella, or sweeping the streamside vegetation with a heavy beating net.

James G. Needham and P. W. Claassen

A **sweep net** (Fig. 4.3B) differs from a standard aerial net by having a more durable frame and the fore part of the net reinforced with material resistant to tearing. Sweep nets are used to sweep through marginal and riparian vegetation where insects may often be found resting, particularly during the daytime.

Collection of flying forms need not be restricted to the immediate vicinity of water. Many species swarm over meadows, roads, or trails, and a few species are known to disperse overland.

TRAPS AND LIGHTS

The major limitations of netting flying insects (seeing and reaching) are overcome to a great extent by luring and trapping. These methods, however, generally require more complex equipment.

Tentlike or boxlike nets suspended or floated so that insects will enter them upon hatching or crawl up the netting and there transform to the flying stage are generally known as **emergent traps** (Fig. 4.3C). A collecting jar may or may not be connected to the emergent trap. Many variations of these traps are used in flowing and stillwater environments.

Many flying aquatic insects are attracted to lights, and many light-sampling devices have been designed for luring these insects. The **light trap** (Fig. 4.3D) is one such device. Although design varies, light traps generally consist of a light source that can be multidirectionally oriented or, if reflectors are used, oriented in a specific direction, such as toward the adjacent aquatic environment. Many light traps have a funnel that facilitates entrapment and leads from the open lighted area down to a collecting container. A fluid killing agent is sometimes used in the collecting container.

Light traps can be operated unattended for long periods; however, they are nonselective, and the larger, more active, or less easily killed forms may destroy fragile forms collected with them. Subimagos of mayflies taken in this manner obviously have no opportunity to transform to the sometimes preferable adult stage.

Nocturnal species may advantageously be looked for in spiders' webs, and on lamps, adjacent to rivers; and wherever such lamps happen to be close to white walls or placarded hoardings, numbers of specimens are apt to be attracted by the illuminated surfaces.

Rev. A. E. Eaton

Other light-sampling systems that do not have an entrapment or killing component are often used because the user can select and handle insects in an optimal fashion, depending on the kinds of insects being sought. These devices can be collectively referred to as **picking lights**.

Picking lights (Fig. 4.3E) consist of a light source placed on the ground or hung from a tripod, wire, or branch. The picking light is enhanced by some type of reflector or background, such as a white sheet suspended vertically or laid on the ground. The sheet adds to the effectiveness and orientation of the light and also provides a working area on which the insects will land, thereby facilitating selective sampling.

Power sources for light traps and picking lights can be batteries or battery packs that are separate or incorporated into the lighting unit or gas-powered electrical generators. Portability, weight, noise, and power output and longevity are all factors to be considered in choosing a power source.

Black lights and mercury vapor lamps are often preferred over white incandescent lamps, since they are apparently more attractive to many insects. The attractiveness of different wavelengths of light is not thoroughly understood and may vary depending on the kind of insect. Some users combine different lights, and some also have more success on warm or overcast nights. Since the nocturnal activity of many species is periodic, they are more apt to be collected only during certain periods of the night. Late-night and pre-dawn sampling may yield some of the more uncommonly taken species.

Adults can be collected from lighted areas near water (e.g., yard or street lights, car lots, store front windows), especially in more rural areas or smaller towns, often with amazing results. Long-handled aerial nets or **sticky poles** (Fig. 4.3F) greatly aid collecting in these situations.

A relatively new method of sampling flying aquatic insects that involves the use of **sticky traps** has great potential. Sticky traps (Fig. 4.3G) consist of solid pieces of clear plastic or similar material (a single piece or several small ones interconnected) that are suspended over a flyway (e.g., stretched horizontally over a stream or hung vertically from a bridge). The pieces are coated with a clear sticky substance, and the trap works on the principle that insects will inadvertently fly into the trap and be stuck there. These insects are removed with a solvent. The construction, weight, suspension, and general operation of these traps unfortunately prevents their use by many collectors.

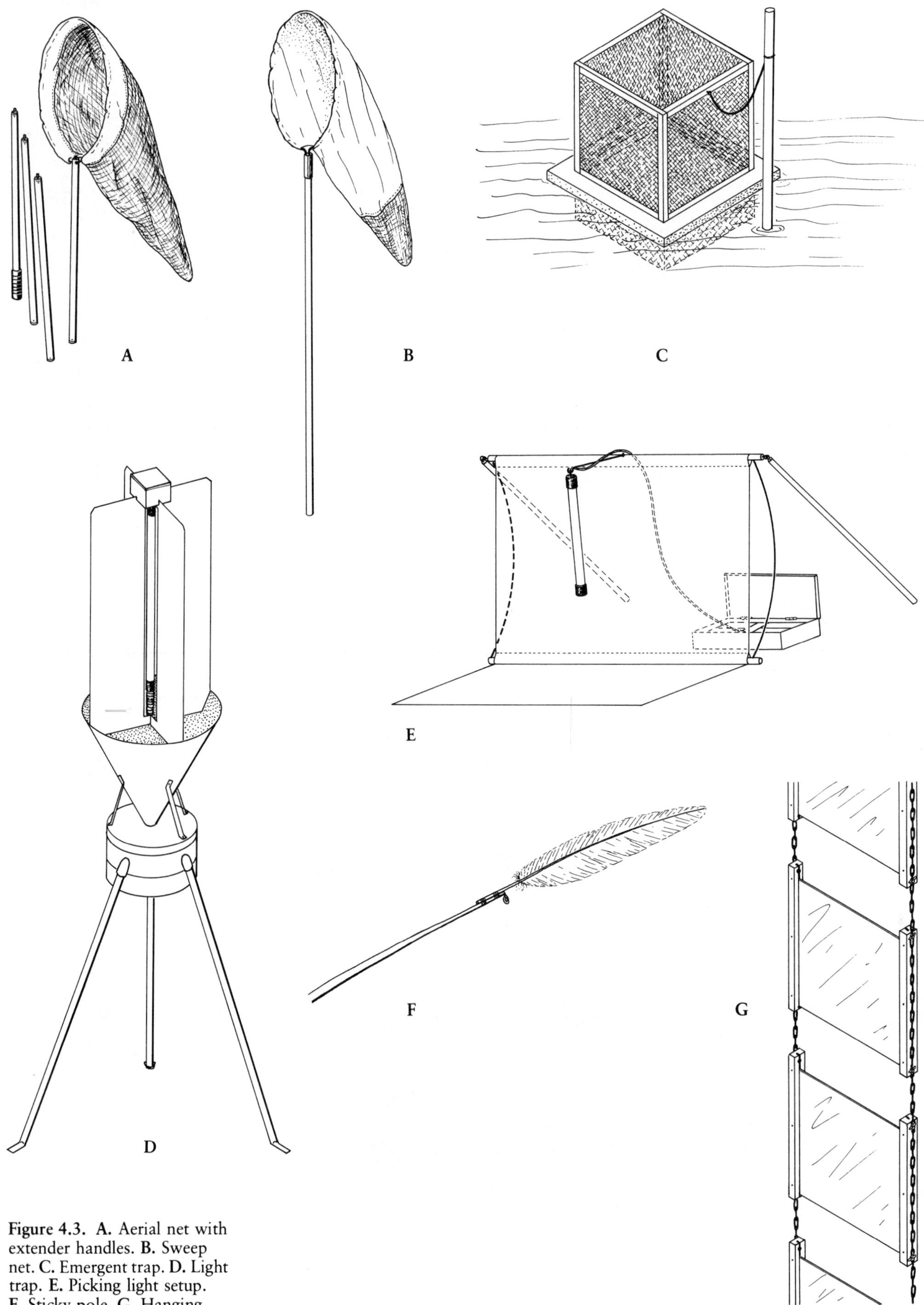

Figure 4.3. **A.** Aerial net with extender handles. **B.** Sweep net. **C.** Emergent trap. **D.** Light trap. **E.** Picking light setup. **F.** Sticky pole. **G.** Hanging sticky trap.

A naturally occurring sticky trap is a spider's orb web. A perusal of webs among marginal vegetation or on bridges can give clues as to what aquatic species have recently been flying in the area.

GROUND COLLECTING

Many aquatic insects pupate in soil or under debris or rocks adjacent to the water. No special equipment is required for collecting these pupae except perhaps a shovel, bucket, and a set of sieves for sifting through soil samples.

Some collecting may be done by searching marginal areas. Adults of some aquatic insects commonly crawl about or rest on the ground or shoreline rocks and driftwood. The cast skins of certain larvae that crawl out of the water to transform can also be found in these habitats.

Rearing Aquatic Insects

Rearing can refer to raising or culturing live aquatic insects or, more specifically, to raising an insect from one life stage to another. It is sometimes necessary to associate larvae, pupae, or subimagos with adults for identification purposes, since descriptions of many species have historically been based on adults and their immature stages remain poorly known. Bait propagation as well as many scientific studies require that aquatic insects be kept alive in artificial environments for prolonged periods.

Field Rearing

Field rearing is usually undertaken for the purposes of associating adult and immature stages of insects. Generally, insects to be field reared should be in an advanced state of development—that is, ready to transform. Mature larvae that possess external wing pads often have darkened pads or convoluted wings evident within the pads. It is usually more difficult to recognize the mature state of development in other larvae. If the insects undergo a pupal stage prior to adulthood, it is usually advisable to transport them to the laboratory or otherwise be able to maintain them for prolonged periods. In caddisflies and certain other insects, the pupa exhibits enough adult characteristics so that the rearing of larva to pupa is adequate for association.

Field **rearing chambers** are generally small containers that hold one or more live insects. They may be enclosed and retain water (any cup, pan, or bucket) or may be made entirely of screen or possess cut-out screened areas so that water can flow through them (Fig. 4.4A, B). When such chambers are placed in the natural aquatic environment, either anchored on the bottom or placed in floats (Fig. 4.4A), fresh water is available to the insects, and maintaining optimal water temperature is generally not a problem. These factors have to be given special consideration when rearing lotic and hydropneustic species.

When terrestrial stages are being sought, there should be adequate dry space at the top of the chamber. This area needs to be enclosed to prevent escape and at the same time screened to prevent suffocation. The top screen also provides a surface to which the insects can cling. It is often desirable to place emergent sticks or rocks in the chamber since some forms crawl out of the water to undergo transformation.

Keep rearing chambers, particularly nonsubmerged chambers, out of direct sunlight because water temperatures may rise quickly to a lethal level.

Even if the chamber is in the shade, an ice bath or cooler may be required for nonsubmerged chambers on warm days. Keep in mind, however, that overly depressed temperatures may thwart emergence. If lotic and hydropneustic insects are being reared in nonsubmerged chambers, portable aquarium aerators or minnow pumps can be used to oxygenate and help agitate the water.

Mass rearing involves placing many individuals presumed to be the same species in the same chamber. This is the easiest method of obtaining large numbers of reared individuals, but if more than one species is present, exact associations may be difficult to obtain. Individual rearing, although requiring more equipment and work, ensures exact associations.

Cast skins should always be retained with reared material. These skins, although fragile, usually exhibit important external characteristics. It is also important that reared forms (usually adults) be allowed sufficient time to complete drying, coloring, and stretching before they are preserved, since only after this part of the maturation is complete do certain characteristics become apparent in the individuals.

It is often not sufficient to rear mayfly larvae merely to the subimaginal stage. Exceptions are those subimagos that are well known to fly fishermen and, in fact, may be the most important stage from the fisherman's point of view. Otherwise, subimagos should be reared to adults by placing them in terrestrial rearing chambers or so-called **sub chambers**. Sub chambers can be anything from small cups (Fig. 4.4C) and jars to larger box chambers (Fig. 4.4D).

Subimagos usually require one or two days before molting to adults. Proper temperature and humidity are required for success. Chambers should be kept out of sunlight; a little wetted paper toweling may help maintain the desired humidity. Subimagos should be handled delicately, since too much pressure applied to the wings or body can result in incomplete molting.

If young larvae of aquatic insects are desired, eggs can sometimes be obtained from gravid female adults. Eggs of many species can be squeezed or teased from the female, placed in water, and taken to the laboratory. Knowledge of the biology of the species and the incubation requirements of its eggs is often necessary for this procedure to be successful.

Transporting Live Specimens

Transporting live specimens from their natural environment to an artificial one for rearing or study purposes presents little problem for some species, whereas for other, sensitive species (e.g., certain lotic and hydropneustic forms), transport must be undertaken carefully while controlling certain conditions.

Generally, insects should be transported in water from their natural environment that is maintained slightly below field temperatures. This can be accomplished with coolers or boxes provided with ice or cool packs placed around small water and specimen containers (a system is shown in Fig. 4.4E). Slight decreases in temperature usually do not have adverse effects, but increases may cause death directly by heat shock or indirectly by depleting dissolved oxygen. Rising water temperatures may also cause premature emergence. Oxygen can be maintained in transport containers by portable aerators, and several air stones can usually be run off a single aerator.

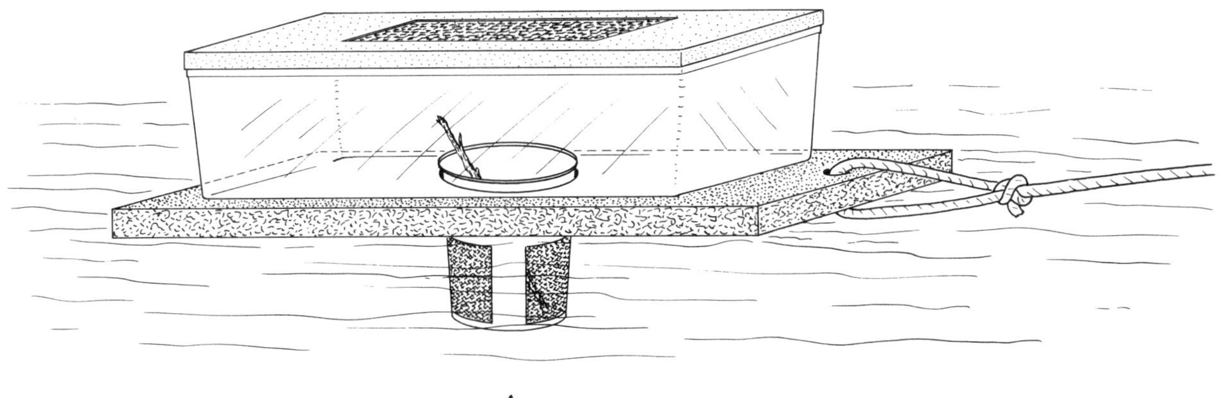

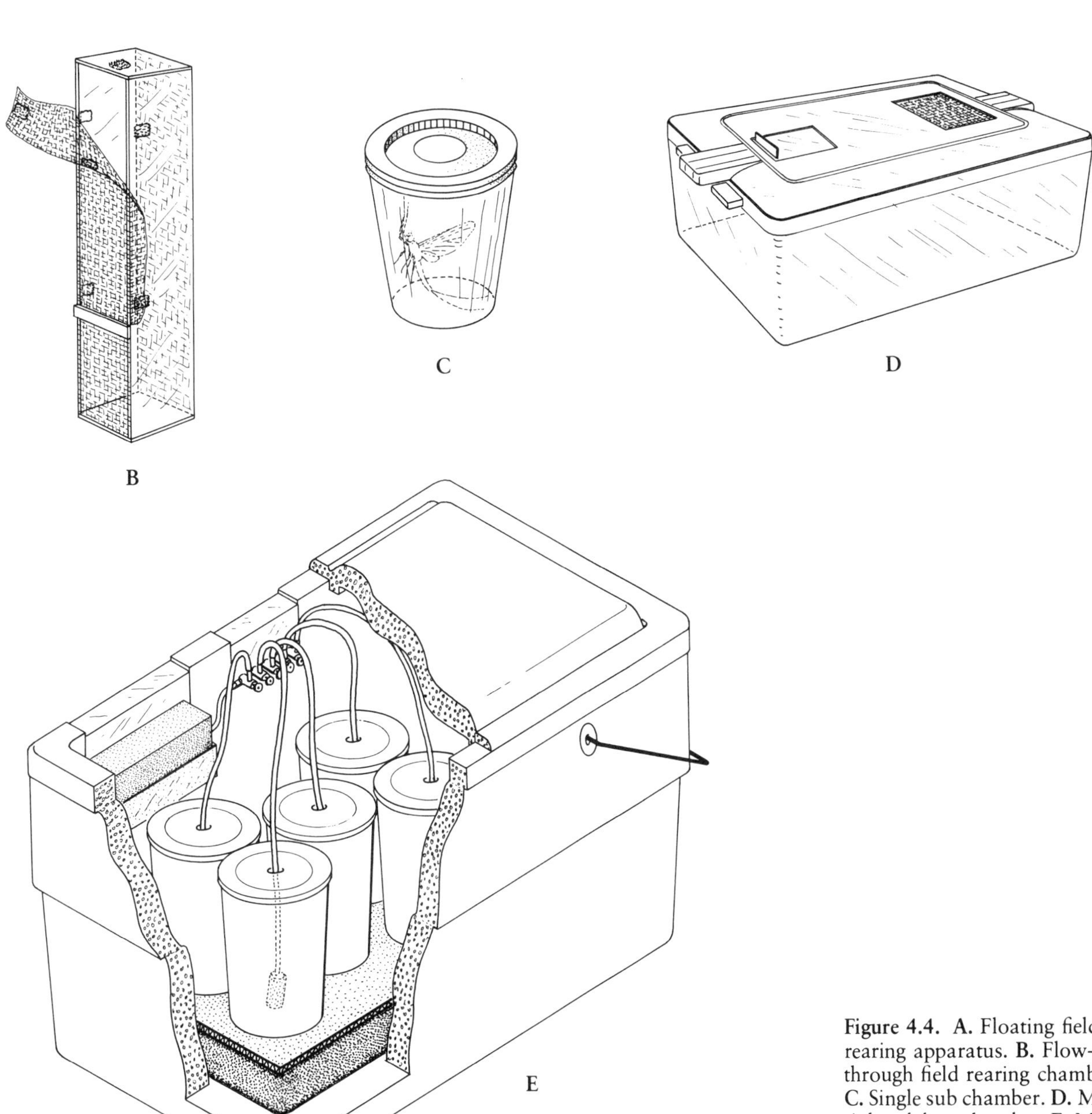

Figure 4.4. **A.** Floating field rearing apparatus. **B.** Flow-through field rearing chamber. C. Single sub chamber. **D.** Multiple sub box chamber. **E.** Multiple rearing transport container.

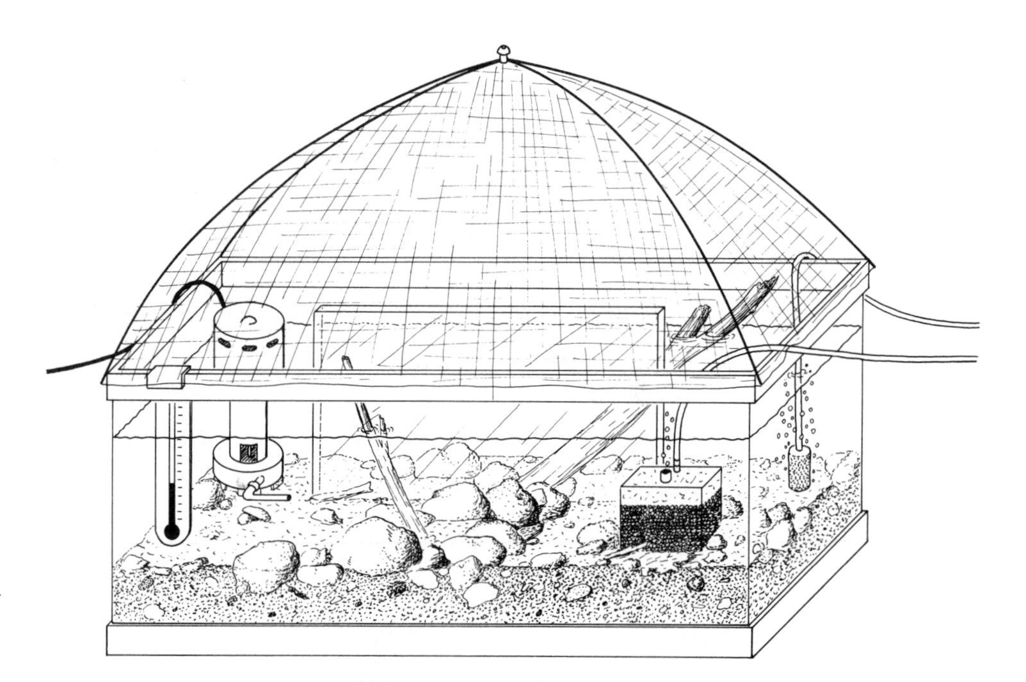

Figure 4.5. Composite laboratory rearing and culture aquarium.

The provision of some natural substrate in the containers can also be important when transporting benthic forms. The insects are not as easily injured if there is material on which to cling or rest. Individuals will tend to use each other as substrate or use the air stones as substrate if natural substrate is not provided.

Laboratory Rearing

Many of the suggestions given above for field rearing also apply to laboratory rearing. Nevertheless, additional factors must be considered when culturing aquatic insects. Laboratory rearing is basically a matter of simulating necessary field conditions; water source, water flow, temperature, turbidity, substrate, light, food, and oxygenation can all be important, depending on the individual requirements of the species.

Various tanks or **aquaria** can be used to maintain aquatic insects. These containers can be covered with netting if there is a possibility of flying stages escaping. A composite aquarium fitted with many of the possible additions discussed below is shown in Figure 4.5. Elaborate **artificial streams** are available in some laboratories, but their design and operation are beyond the scope of this book.

Water from the natural environment can be used in aquaria, and it can be replaced if necessary with tap water that has been dechlorinated by letting it stand. The water can be cooled as desired by several methods, including room-temperature modification, circulation through a cooling reservoir, or by employing refrigeration units or ice baths. If water cannot be cooled to field temperatures, then its temperature must be allowed to rise very slowly so as to have the least possible adverse effects on the insects.

Water should be oxygenated by bubbling air into the aquarium from individual aerators or from a central air compressor. Air from a compressor should not contain any impurities, such as oil droplets. Aeration usually also agitates the water somewhat, and this may be important for lotic insects and for some small insects that are unable to break the surface film when hatching from the water unless it is agitated. For some lotic insects, such as suspension feeders, the aquarium environment may be improved by adding a small water pump. The water flow can be directed in a large aquarium by an appropriately placed vertical plate.

Natural substrate should be added to the aquarium. For short-term cultures of detritivores, there is often sufficient food provided with the substrate. A variety of nutrient sources can be added, such as grain, decaying leaves from an aquatic environment, and powdered fish food. For cultures of carnivorous insects, it is necessary to add an adequate supply of prey organisms, such as midge or mosquito larvae, tubificid worms, small mayflies, or crustaceans; remember that many carnivorous aquatic insects are cannibalistic under crowded conditions. Many insects have very specific food requirements that must be considered.

For aquaria that have silt substrates (for burrowing insects) or that are subject to algal or fungal blooms, a filtering system should be used at least periodically to reduce turbidity in the water. Most standard aquarium filters are adequate.

Emergent objects should be placed in aquaria if insects that crawl out of the water to transform are being reared. When rearing larvae of species that pupate on or in dry or moist soil (e.g., fishflies, some crane flies, and many water beetles), an artificial shoreline should be provided within the aquarium.

Handling and Preserving Aquatic Insect Specimens

Preserved specimens of aquatic insects and information on their habitats are very valuable taxonomic and ecological data. Identified collections serve as helpful reference sources, as baseline data for particular environments, as repositories for voucher specimens used in taxonomic and ecological research, and as part of the preserved natural resources of our environment.

Handling Specimens in the Field

The proper handling of specimens begins at the sample site. Correct procedures will greatly increase the chances of securing a collection that can be studied adequately and preserved for the future.

FIELD PRESERVATION

Specimens taken from the water can be placed directly into any of several fluid **killing agents** or **preservatives**, depending on the needs of the user. The most commonly used fluid is a 90 to 95% solution of **ethyl alcohol**. Strong alcohol is easy to use, will immediately kill most insects, and even if somewhat diluted will initially preserve the insects. The volume of specimens in a container of alcohol should not exceed about 25% of the fluid volume. Other alcohols, such as isopropyl, are substituted by some users.

Fluids other than alcohol are sometimes preferred in the field because they may be less subject to dilution, retain specimen color better, prevent soft parts from swelling, harden specimens more, or penetrate specimens more effectively. **Carnoy fluid** is a commonly used alternative fixative for soft-bodied insects. It consists of 10% glacial acetic acid, 60% of 95% ethyl alcohol, and 30% chloroform.

Fixatives that help prevent breakage and dilution and have strong penetrative powers are desirable when there is considerable extraneous material mixed with samples. **Kahle's fluid** (11% formalin, 28% of 95% ethyl alcohol, 2% glacial acetic acid, and 59% water) and the similar **Pampel's fluid** (10% formalin, 30% of 95% ethyl alcohol, 7% glacial acetic acid, and 53% water) are two such fixatives. Both have the added advantages of toughening tissue yet keeping specimens relatively soft, fixing color somewhat, and having high water contents, which can be added in the field.

Reasonable color retention can be accomplished for some soft-bodied specimens by fixing them directly in boiling water. Larger larvae should be boiled about one minute or left in hot water for about three minutes. Smaller larvae may require as little as 10 seconds in boiling or hot water. Many larvae will float to the top when protein fixation is attained. These specimens should then be transferred to the alcohol preservative. Simply heating specimens in their alcohol preservative immediately after they have been killed will work for some forms.

Field sample containers, such as vials or jars (Fig. 4.6A), should not be overly crowded with specimens, since specimen breakage and dilution are more apt to occur. Certain large and active insects, such as Megaloptera larvae, should initially be placed in separate containers, since they usually die slowly and can tear up other specimens in the meantime; large beetles, giant water bugs, and others present a similar problem.

If field samples are to be moved about, field containers should be completely filled with fluid to evacuate any air. This will prevent unnecessary specimen breakage. If specimens cannot be sorted for a day or more, it is usually a good idea to replace the collecting fluid with fresh alcoholic preservative.

Aquatic entomologists usually place soft-bodied terrestrial adults directly into 75 to 80% ethyl alcohol. Although all insects can be killed in this manner, certain adults, such as dragonflies, moths, and hard-bodied forms, are often placed into a dry **killing jar** (Fig. 4.6B). This is a closeable container in which toxic fumes are produced from such agents as cyanide crystals covered by a layer of plaster of paris, ethyl acetate soaked into plaster of paris, or a piece of pest strip. Specimens should then be placed in containers where they will not be broken as they become dry (e.g., in boxes between layers of paper toweling or in glassine envelopes).

FIELD DATA

Information about the habitat and circumstances in which specimens were found is basic to the usefulness of specimens. These data contribute greatly to what must be learned of the distribution, ecology, and biology of each species.

The minimum amount of field data to be recorded consists of geographic location, date, and name of collector or study project. Locale usually includes the state or province, the county or equivalent subunit, the water system (name of stream, river, lake, etc.), and the precise site on the system, stated in such a way that it can be relocated (e.g., by reference to a specific road, bridge, altitude, etc.).

Additional field data commonly recorded include water and air temperatures, flow regime of the water (e.g., riffles or pools), substrate type if a benthic sample, associated plant material, method of sampling, and other

A

B

IN: Harrison Co.
Mill Cr. at Hwy. 25, 3 mi. S.
New Middletown VII-6-1979
J. W. Smith
S-130

Wormaldia
shawnee (Ross)
det: J.W. Smith 1979

C

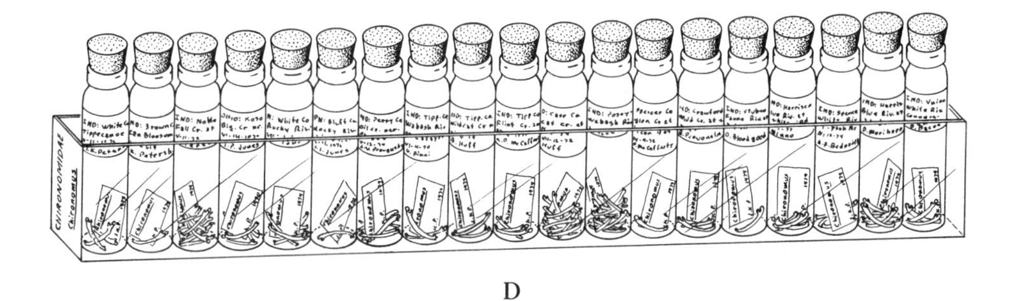

D

E

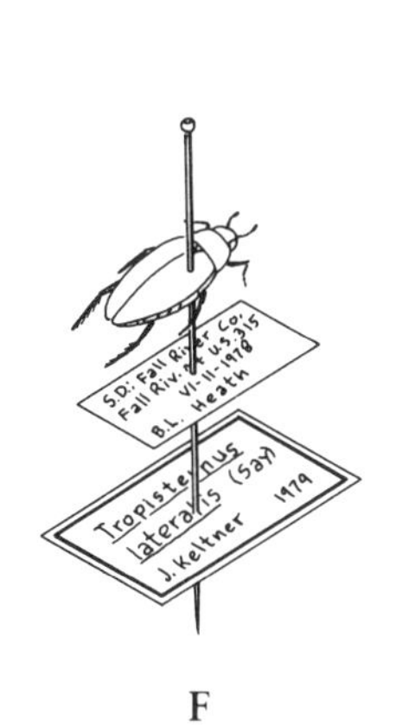

F

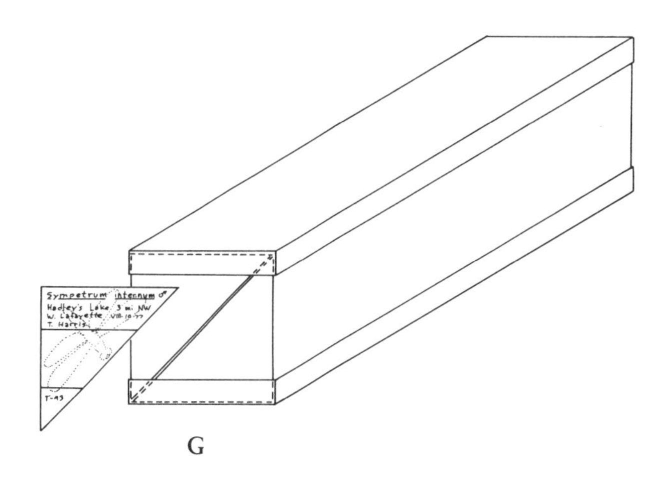

G

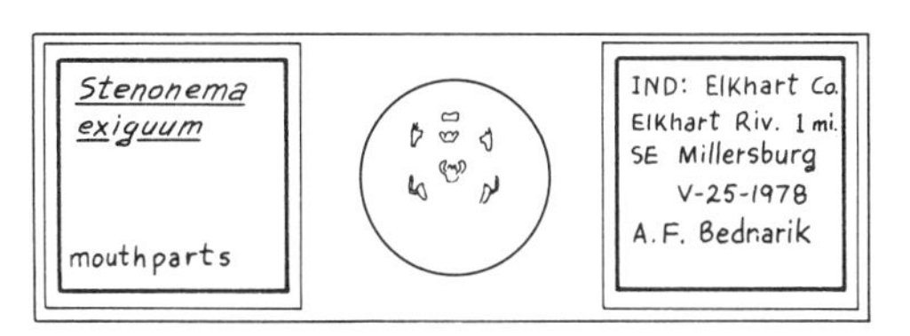

H

Figure 4.6. **A.** Field collecting container. **B.** Killing jar. **C.** (upper) Collecting data label. (lower) Identification label. **D.** Vial rack with stoppered specimen vials. **E.** Storage jar with vials. **F.** Pinned insect with label orientation. **G.** Papered insect with storage box. **H.** Slide mount.

general observations. Certain studies require detailed quantitative measurements of the environment and water quality.

It may be difficult to keep all data with each sample. An alternative is to place a code number with each sample that refers to all information recorded about it in the field. This information can be conveniently kept in a separate log or **field book**. Code numbers should remain with specimens selected from the sample at each stage of sorting.

Maintaining a Collection

Many museums and educational institutions maintain permanent collections of aquatic insects as reference and learning tools, and individuals often maintain collections for various purposes. It is also important that representative specimens upon which environmental assessments and ecological studies have been based are housed in permanent collections and available for examination.

PERMANENT PRESERVATION

Most aquatic insect specimens are stored permanently in 70 to 80% ethyl alcohol as **wet collections**. Specimens may become brittle in higher concentrations or deteriorate in lower concentrations. Sometimes 1% ionol or a small amount of glycerine is added to the alcohol to help keep the specimens wetted for a period in case the alcohol is accidentally allowed to evaporate. Ionol, which is an antioxidant, has the added advantage of lessening bleaching of the specimens, which may occur over time.

It is sometimes preferable to store certain adult aquatic insects in a dry state. These specimens are most easily handled or mounted before they become completely dry and brittle. If specimens were initially placed in alcohol, they can simply be removed to dry, or they can be placed on paper towels soaked with acetone and then covered with a jar to prevent acetone evaporation. Specimens allowed to dry in this manner retain inflation of soft body parts and are not as fragile or brittle as they would be otherwise.

Previously dried specimens can be relaxed by humidifying them in a closed container having a layer of wet sand or sawdust. The addition of a few carbolic acid crystals to the relaxing chamber will help prevent the growth of molds under these conditions.

Adult insects can be mounted on standard insect pins (Fig. 4.6F) as **pinned collections**. Large insects with spread wings that take up much space may alternatively be placed in small, individual folded papers or envelopes (Fig. 4.6G) for permanent storage as **papered collections**.

Minute insects or insect parts can be mounted on microscope slides (Fig. 4.6H) as **slide collections**. Many slide-mounting media are temporary, and for permanent storage, a stable, nonevaporative medium must be used. Oil-soluble resins, such as **balsam** or dammar, last longest. Some water-soluble media, such as **Hoyer's** medium, are also relatively long-lived but often require that the perimeter of the coverslip be sealed (or ringed) to prevent eventual deterioration.

Before making permanent slide mounts (or before being able to examine some specimens in detail), it is sometimes necessary to clear dark pigments or internal organs that obscure important exoskeletal features. This is done by placing specimens in a **clearing agent**, such as strong **KOH** (potassium hydroxide) in water. Clearing time is greatly shortened if the KOH is heated, but too much clearing will debilitate specimens.

One popular method of permanent slide mounting involves placing a drop or two of balsam (with a little cellusolve added) onto a slide, orienting the clean specimen properly in the medium, and then placing a coverslip on the mount so as to displace any air bubbles. **Cellusolve** (ethylene glycol monoethyl ether) dilutes the balsam to a workable consistency and allows the specimen to be placed in the balsam directly from the alcohol preservative without having to dehydrate it by rinsing in absolute alcohol. It also gives the medium some slight clearing properties. The slide should be dried before storing, and this can be facilitated by placing the slide flat in a drying cabinet heated by a light bulb for a few weeks. Special mounting techniques, including staining, are required in certain cases, and should be explored by the serious student.

CURATING A COLLECTION

Usable collections must be properly labeled, and at least two kinds of labels should remain with the specimens. The first is the **collecting data label** (Fig. 4.6C, upper), which contains the minimum field data (discussed above). Such data are generally not subject to change. Additional information can be catalogued somewhere else and indexed by a code number on the label.

The second label is the **identification label** (Fig. 4.6C, lower), which contains the taxonomic name of the specimen. This label is subject to change, since the taxonomic name may have to be changed in the future, and it is for this reason that this label should be separate from the collecting data label. The name of the identifier and date of identification are sometimes included on the identification label to aid taxonomists.

Additional information labels sometimes also accompany specimens. They may be used to indicate whether a specimen has been reared or dissected, where other parts or associated stages are located, or if the specimen has been treated or stained in a special manner.

Labels for wet specimens should always be put inside the container with the specimens, preferably so that they may be easily read from the outside. One method is shown in Figure 4.6D. The ink used for such labels should obviously not be soluble in water or alcohol. Labels for pinned insects should be as small as possible and placed on the pin below the specimen; the lowest label should be the identification label (Fig. 4.6F). Labels for slides should be affixed to the slide (Fig. 4.6H).

A variety of storage systems can be used for housing insect collections. Wet collections are commonly stored in specimen vials placed in racks or jars. Vials kept in racks (Fig. 4.6D) are especially susceptible to evaporation and must be periodically curated to replace alcohol. These vials should be stoppered with locking screw caps or with neoprene or other hard stoppers that will not swell; corks should not be used except for temporary purposes. Vials that are completely filled, stoppered with cotton, and submerged in alcohol-filled jars (Fig. 4.6E) require less curation but are more cumbersome if specimens are being regularly used. Long-fibered surgical cotton is preferred because specimens' claws and spines are less apt to get stuck in this kind of cotton.

A pinned collection can be kept in boxes or drawers that can be securely closed. The pinning bottom (e.g., polyethylene foam) should be somewhat soft and resistant to fumigants. To prevent injury from insect pests, a small amount of napthalene (moth ball), paradichlorobenzene, or pest strip should be placed in the container and periodically replaced.

A papered collection can be stored like files in boxes or special trays (Fig. 4.6G) but must be regularly fumigated. Slide collections can be kept in slide boxes or on trays. They should always be stored flat, with the coverslip side up, and should be occasionally checked for deterioration of the mounting medium.

Collections are usually arranged and catalogued either systematically or alphabetically. In the latter system, insects are arranged alphabetically by the names of orders, families, genera, and species. In the systematic method, taxonomic groups and species are arranged according to their natural relationships. This system is convenient for comparing closely related species but often requires a separate index to names. Some collections are additionally arranged by geographic region or study project.

SHIPPING SPECIMENS

It has become a common practice for people who study aquatic insects to seek identifications or verifications from taxonomic specialists. This usually requires shipping specimens, and because many aquatic insect specimens are exceedingly fragile and subject to damage, this can be a risky undertaking. Much of the destruction of samples can be avoided if a few simple recommendations are heeded.

Pinned, soft-bodied aquatic insects should not be shipped, but hand carried whenever possible. Shipping invariably leads to damage of brittle specimens no matter what precautions are taken in packing.

Vials that must be shipped should be completely filled and voided of any air bubbles, since bubbles have the same effect on specimens that small rocks would have when the vials are shaken. This requires that vials with stoppers be overfilled and "burped" with a teasing needle or piece of wire. A cotton plug placed in the vial is also helpful. Care should be taken that the bottoms of the vials will not break out under this pressure, and the stoppers or screw caps should be secured with tape. Each vial should then be individually wrapped with toweling or placed in spaced cutouts in a styrofoam block. The vials should then be placed in a shipping box or mailing tube with plenty of shock-absorbing packing material placed between the vials and sides of the container.

References

Beak, T. W., T. C. Griffing, and A. G. Appleby. 1973. Use of artificial substrate samplers to assess water pollution, pp. 227–241. In *Biological methods for the assessment of water quality* (J. Cairns and K. L. Dickson, eds.). Amer. Soc. Test. Mat., Philadelphia.

Borror, D. J., D. M. DeLong, and C. A. Triplehorn. 1976. *An introduction to the study of insects* (4th ed.). Holt, Rinehart and Winston, New York. 852 pp.

Campbell, J. et al. 1973. *A preliminary compilation of literature pertaining to the culture of aquatic invertebrates and macrophytes.* Fish. Res. Brd. Canada Tech. Rep. 227. 25 pp.

DeMarch, B. et al. 1975. *A compilation of literature pertaining to the culture of aquatic invertebrates, algae, and macrophytes*, Vol. II. Fish. Mar. Serv. Tech. Rep. 576. 32 pp.

Edmonds, W. T., Jr. 1976. *Collecting and preserving Kansas invertebrates.* St. Biol. Serv. Kans. Tech. Bull. 3. 73 pp.

Edmunds, G. F., Jr., S. L. Jensen, and L. Berner. 1976. *The mayflies of North and Central America*. Univ. Minn. Press, Minneapolis. 330 pp.

Edmunds, G. F., Jr., and W. P. McCafferty. 1978. A new J. G. Needham collecting device for collecting adult mayflies (and other out-of-reach insects). *Entomol. News* 89:193–194.

Harris, T. L., and W. P. McCafferty. 1977. Assessing aquatic insect flight behavior with sticky traps. *Gr. Lakes Entomol.* 10:233–239.

Lehmkuhl, D. M. 1979. *How to know the aquatic insects*. Wm. C Brown, Dubuque. 168 pp.

McCafferty, W. P. 1974. An economical and efficient vial rack for the storage of small, fluid-preserved specimens. *Ann. Entomol. Soc. Amer.* 67:996–997.

Merritt, R. W., and K. W. Cummins (eds.). 1978. *An introduction to the aquatic insects of North America*. Kendall/Hunt, Dubuque. 441 pp.

Milne, M. J. 1938. The "metamorphotype method" in Trichoptera. *J. N.Y. Entomol. Soc.* 46:435–437.

Montgomery, B. E. 1959. A new type of tray for specimens of Odonata. *Proc. N.C.B. Entomol. Soc. Amer.* 14:15–16.

Provonsha, A. V., and W. P. McCafferty. 1975. New techniques for associating the stages of aquatic insects. *Gr. Lakes Entomol.* 8:105–109.

Weber, C. I. (ed.). 1973. *Biological field and laboratory methods for measuring the quality of surface waters and effluents*. NERC/EPA, Cincinnati. 176 pp.

CHAPTER 5

Arthropods and Other Common Freshwater Macroinvertebrates

ALTHOUGH AQUATIC INSECTS (members of the invertebrate phylum Arthropoda) are the major topic of this book, they form only part of the invertebrate communities of most freshwater habitats. At a minimum, students of aquatic entomology should have a casual familiarity with other freshwater macroinvertebrates, since these will often be sampled along with aquatic insects. More importantly, from an ecological point of view, these invertebrates are often intimately associated with the insects as food, predators, parasites, or competitors for available resources or space.

Macroinvertebrates are conveniently described as those animals without backbones that are generally visible to the unaided eye (i.e., usually larger than 0.5 mm at their greatest dimension). This definition obviously excludes microscopic freshwater animals, such as protozoans, gastrotrichs, most rotifers, and water bears; although these organisms are also very important to the ecology of freshwater, they are beyond the scope of this book, as are the freshwater plants and vertebrate animals.

The following treatment of macroinvertebrate groups is not intended to be comprehensive but merely a convenient review of the major forms for those interested primarily in aquatic insects. Some very rare or minor groups of freshwater macroinvertebrates are not treated. The Arthropoda are introduced here and diagnostically compared with other freshwater macroinvertebrates. In addition, the three classes of freshwater arthropods (crustaceans, arachnids, and insects) are diagnostically compared. The crustaceans and arachnids are more fully treated in Chapters 22 and 23, respectively.

SPONGES
(Porifera)

DIAGNOSIS: (Fig. 5.1) These are soft, simply organized organisms that are generally amorphous (some with numerous fingerlike projections) and usually range in thickness from 4 to 60 mm. They may be green, yellow, brown, or, occasionally, gray or reddish. They are highly porous, lack organs, and are always found attached to various substrates.

DISCUSSION: This is primarily a marine group, but is represented in freshwater by one family (Spongillidae). The group is widespread in North America, where there are some 25 species. Freshwater sponges are most common in clear streams, rivers, lakes, and ponds, and they are most abundant in shallow water. They are sessile and attach themselves to hard substrates, such as woody debris, rocks, and mussel shells.

Some aquatic insects are commonly associated with freshwater sponges, and some rely on the sponge as a food source. These insects include spongillaflies as well as some caddisflies and midges.

Yes, a stagnant pool, though but a few feet wide, hatched by the sun, is an immense world, an inexhaustible mine of observation to the studious man and a marvel to the child who, tired of his paper boat, diverts his eyes and thoughts a little with what is happening in the water.

J. Henri Fabre

HYDRAZOANS
(Coelenterata)

DIAGNOSIS: (Fig. 5.2) These are cylindrical, radially symmetrical forms that usually range from 2 to 25 mm in length. Body is somewhat transparent in some species and may be yellow, brown, gray, or rarely green. Five to 12 tentacles are present, surrounding a single body opening that functions as both mouth and anus.

DISCUSSION: The common members of this group are hydras, of which there are about 14 species in North America. A few other freshwater coelenterates, including the so-called freshwater jellyfish (*Craspedacusta*), are also found in North America but are much more uncommon.

Hydras live in shallow, clean waters of lakes, ponds, streams, and rivers, rarely in deep waters. They are primarily benthic organisms, attached at one end of their cylindrical body, with the tentacled end free. They are able to move slowly along the substrate and sometimes will release and float. All

Figure 5.1. Sponge

Figure 5.2. Hydra

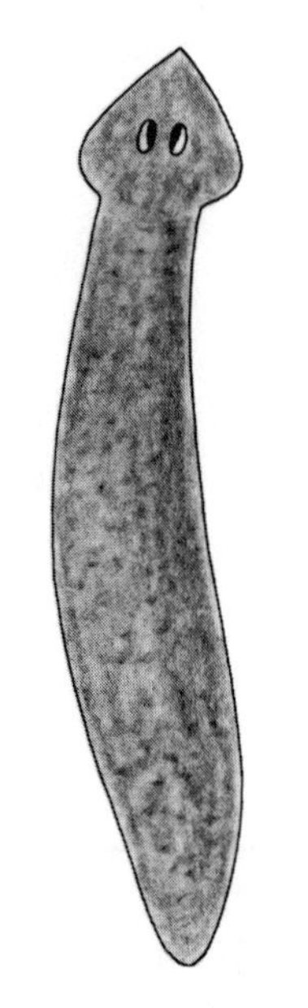

Figure 5.3. Planarian

of the hydrazoans are carnivorous and kill such prey organisms as aquatic insects, worms, and small crustaceans by paralyzing them with poisonous nematocysts released from the body or tentacles. Small fish fry may also on occasion fall victim to the larger hydras.

PLANARIANS
(Platyhelminthes, Turbellaria)

DIAGNOSIS: (Fig. 5.3) These are soft-bodied, elongate, wormlike forms, usually dorsoventrally flattened or at least flattened ventrally. They are generally less than 1 mm in length, but some range to 30 mm. Most are dark colored, and many are mottled. Head area is commonly arrowhead-shaped. A pair of dorsal eyespots is usually present. Mouth and anus are combined into a single ventral opening usually at about midlength along the body.

DISCUSSION: The phylum Platyhelminthes includes the so-called flatworms, many of which are parasitic or marine. Most of the free-living, freshwater forms are planarians, and a few of these are large enough to be considered macroorganisms.

Planarians are usually associated with the substrate of shallow waters. They are often found on the underside of rocks and detritus. Most are carnivores and scavengers that feed on a variety of soft invertebrates.

AQUATIC NEMATODES
(Nematoda)

DIAGNOSIS: (Fig. 5.4) These are slender, cylindrical worms. Freshwater species usually range in length from 0.5 to 2 mm. They may be variously dull colored and are often translucent. Anterior end of body is variously rounded or truncate, and the posterior end is conspicuously tapered. Mouth is terminal at the anterior end, and anus is subterminal at the posterior end. Body is not segmented.

DISCUSSION: Although it is very difficult to estimate, probably as many as 1,000 or more species of free-living freshwater nematodes may eventually be found in North America. Many of the genera and species of freshwater nematodes have widespread distributions, commonly encompassing several continents. In addition, many freshwater nematodes are also known either from soil or marine environments and are thus very adaptable.

Aquatic nematodes can be found in an array of habitats, including clear streams, rivers, ponds, and lakes (at all depths). They are also known from arctic pools, swamps, and trickling filters of sewage treatment facilities. Most are very active benthic forms, but a few species are accomplished swimmers. Some are detritivores, some are primarily herbivores, some are carnivores, and some can be considered generalists. Food substances are variously chewed or sucked, depending on the species of nematode.

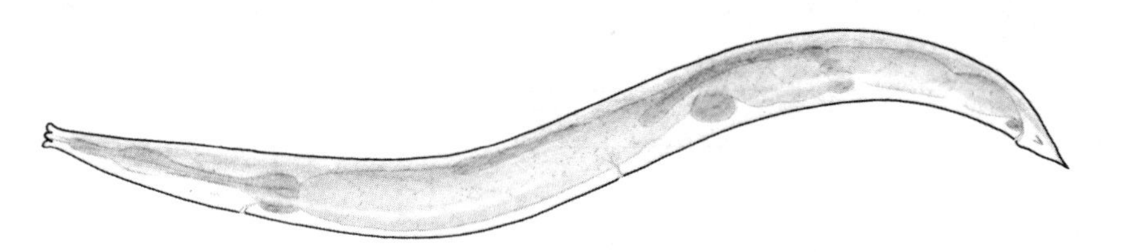

Figure 5.4. Nematode

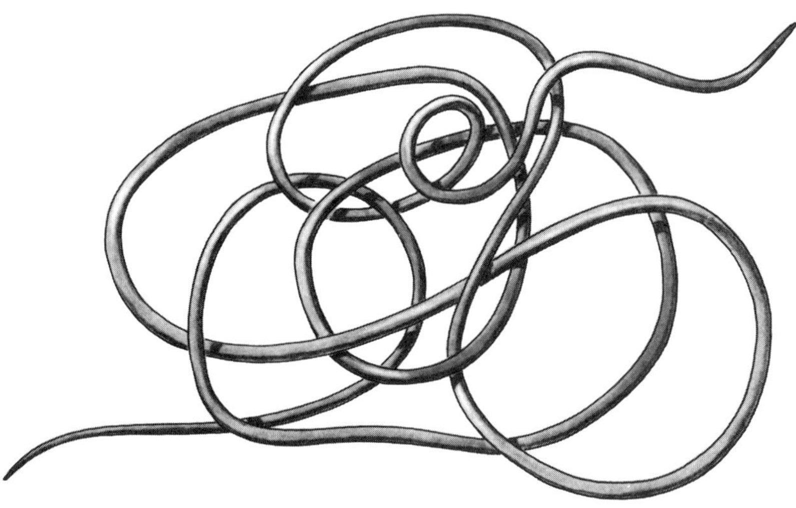

Figure 5.5. Horsehair worm

HORSEHAIR WORMS
(Nematomorpha)

DIAGNOSIS: (Fig. 5.5) These are slender, cylindrical forms as adults, ranging in length from 100 to 700 mm. One or more individuals are often found in a tangled mass. They are generally dark in color. Body is not segmented.

DISCUSSION: Adult horsehair worms are free-living in freshwater or marginal freshwater habitats in North America, where there are about eight described genera and probably somewhat less than 100 species. They may occur on the substrate or in the open water of both flowing and standing waters but are most commonly found in ponds and lakes. Their resemblance to horsehairs is thought to have given rise to the old superstition that horsehairs, when placed in water, became worms. They are also known as "gordian worms."

The larvae of horsehair worms are internal parasites primarily of crickets, grasshoppers, and terrestrial and aquatic beetles. After larvae hatch from the egg, they evidently swim to marginal areas and become encysted on weeds and other objects where they may then be ingested by host insects. The adults bore out of the host body and enter the water. Adult horsehair worms do not ingest food.

AQUATIC EARTHWORMS
(Annelida, Oligochaeta)

DIAGNOSIS: (Figs. 5.6, 5.7) These are generally elongate, cylindrical worms that are usually 1–30 mm but sometimes well over 100 mm in length. Body is segmented, has 7–500 segments, and typically bears a few short bristles or hairs. Color is variable.

DISCUSSION: Of the segmented worms (phylum Annelida), the earthworms (class Oligochaeta) are one of the groups that are well represented in freshwater environments. Although most are poorly known, there may be 200 species or more in North America. Most aquatic earthworms may be found in silty substrates and among the debris and detritus of ponds, lakes, pools, streams, and rivers. Some are associated with algal mats, some are amphibious, and still others may be generally confined to marginal wet environments.

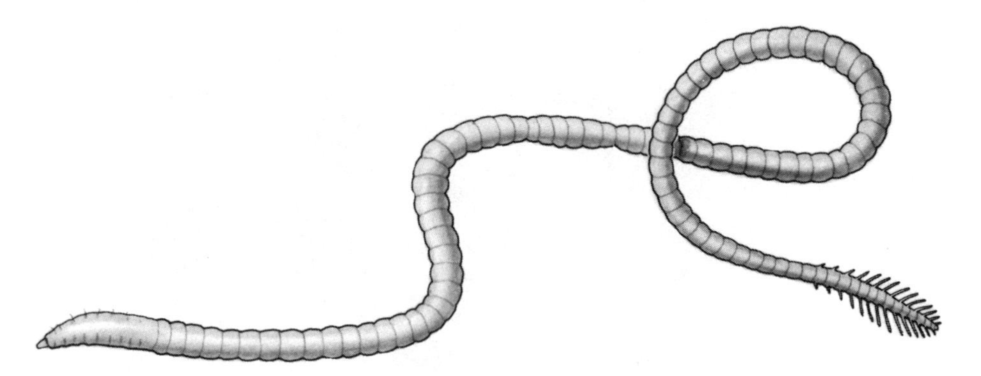

Figure 5.6. Tubificid worm

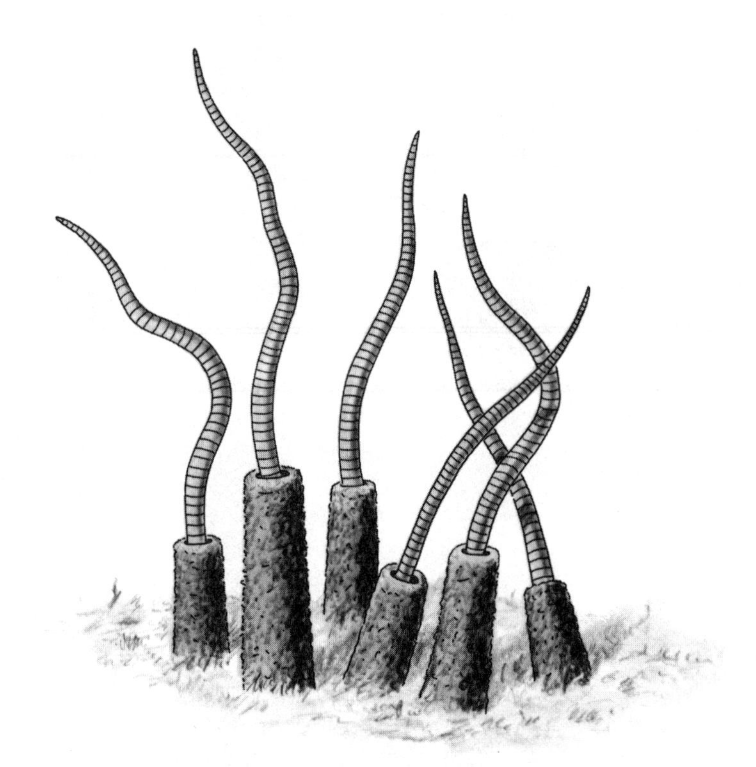

Figure 5.7. Tubificid worms

Many species are typical of terrestrial earthworms in their habits of deposit-feeding on soft substrates and detritus and utilizing the organic fraction for their nutrient source. Some, however, feed primarily on periphyton or detritus, and a few are carnivorous. Most of the tubificid worms build vertical tubes (Fig. 5.7) from which the posterior end of the body may protrude and be waved about.

Many aquatic oligochaetes tolerate low dissolved oxygen concentrations. Dense populations of tubificids (e.g., *Tubifex tubifex*) can often be found in organically polluted streams and rivers, and others can exist in oxygen-poor environments of deep lakes.

LEECHES
(Annelida, Hirudinea)

DIAGNOSIS: (Fig. 5.8) These are dorsoventrally flattened, segmented worms that range in length from 5 to over 400 mm. Many are patterned and brightly colored. They possess both anterior and posterior ventral suckers.

DISCUSSION: There are 63 or more species of freshwater leeches in North America. Leeches (class Hirudinea) constitute one group of segmented worms that comprises primarily freshwater species, although some are marine or terrestrial. Freshwater forms may be found on the substrate or attached to host animals in a wide variety of aquatic habitats.

The anterior and posterior suckers of leeches are used variously for attachment, feeding, or locomotion. Although all leeches may be reputed to be bloodsuckers, only some of the species are actually blood-feeding ectoparasites of vertebrate animals, such as fishes and amphibians. The food habits of other species include scavenging and preying on other invertebrates, such as aquatic insects, mollusks, and other freshwater worms (including leeches).

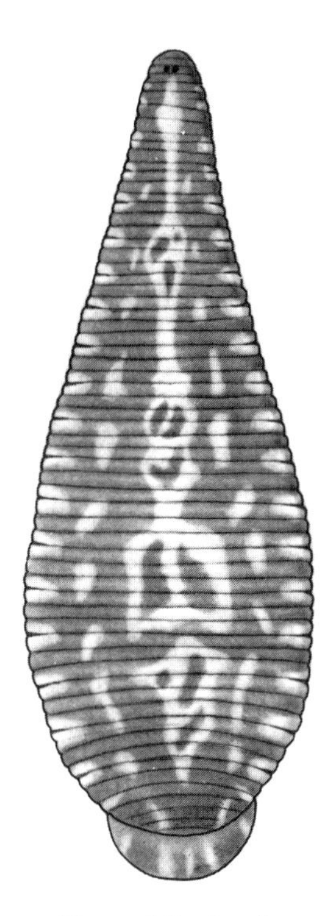

Figure 5.8. Leech

MOSS ANIMALS
(Bryozoa)

DIAGNOSIS: (Fig. 5.9) These are colonial animals that form stalklike, branching colonies, matlike or crustlike colonies, or jellylike colonies. Each organism (or zooid) of a colony exposes its distal crown of tentacles to the water while feeding, but usually retracts it when disturbed. Colonies may reach several centimeters in thickness and may consist of thousands of individuals.

DISCUSSION: This is primarily a marine group of invertebrates, but there are about 20 species of freshwater moss animals known from North America. With some exceptions, colonies are found attached to the undersides of woody debris or rocks of clear, unpolluted ponds or slow reaches of streams. They feed primarily on fine detritus and microorganisms.

Figure 5.9. Part of stalklike moss animal

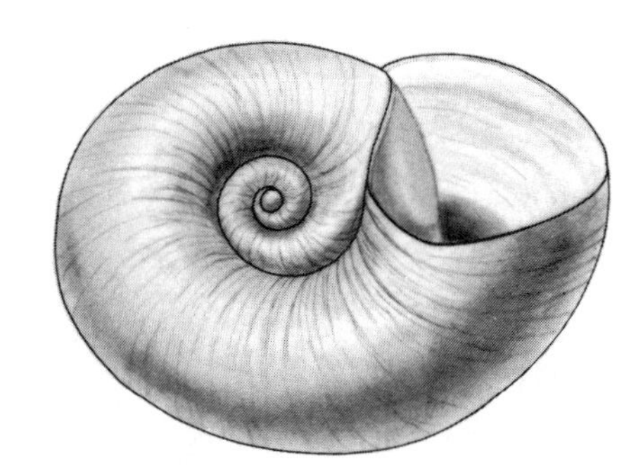

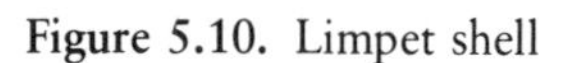

Figure 5.10. Limpet shell

Figure 5.11. Snail shell

SNAILS
(Mollusca, Gastropoda)

DIAGNOSIS: (Figs. 5.10, 5.11) These possess a single (univalve), usually drab-colored shell that is either spiraled or coiled or low and conelike. They generally range in size from 2 to 70 mm. Part of the body protrudes from the aperture of the shell and bears a head with a pair of tentacles.

DISCUSSION: The gastropods are well represented in marine, freshwater, and terrestrial environments. Several hundred species of freshwater snails occur in North America. They are benthic organisms that slowly move about on the substrate of almost all shallow freshwater habitats. Some are known to burrow into soft substrates or detritus during periods of drying in vernal habitats or when shallow habitats become frozen solid.

Calcium carbonate is used in the production of the shell, and it is for this reason that many freshwater snails are more common in hard-water habitats, although some do well in soft water. Many feed on the encrusted growths of algae over which they creep. Others are detritivores or omnivores. Certain freshwater fishes feed extensively on snails, and most marsh fly larvae are predators and parasites of snails.

MUSSELS, CLAMS
(Mollusca, Pelecypoda)

DIAGNOSIS: (Figs. 5.12, 5.13) These possess a two-piece (bivalve) shell, with the two opposing valves connected by a hinge. Shells are variously shaped, commonly oval, and possess concentric growth lines. Shell coloration ranges from black to white, with some species yellow, green, or brown. Size ranges from 2 to 250 mm. Protruding body parts lack head, eyes, and tentacles.

DISCUSSION: The Pelecypoda comprises an entirely aquatic group of the phylum Mollusca. The group is represented in North American freshwater environments by about 300 or more species, often particularly abundant and diverse in medium- to large-sized rivers. The two large groups of freshwater Pelecypoda are the sphaeriacean, or fingernail, clams, which are generally small and thin-shelled, and the generally large and thick-shelled mussels of the family Unionidae.

Mussels occur on or in the substrate. They generally feed by filtering planktonic microorganisms out of the water, although burrowing forms feed on organic detritus strained from the substrate. Their soft body parts are eaten by a number of fishes as well as some other animals, such as muskrats. Some 40 species of mussels in North America were once used for

Figure 5.12. Fingernail clam

Figure 5.13. Unionid mussel

their pearls or shell products. Although mussels once flourished in the rivers of the Midwest, previous harvesting activities, along with impounding and dredging activities, have devastated many populations, so that some species are now extinct and over 100 are considered endangered.

ARTHROPODS
(Arthropoda)

DIAGNOSIS: These are bilaterally symmetrical forms with a segmented body, although segmentation is highly reduced or not apparent in some. They generally range in length from less than 1 mm to 150 mm. Coloration is extremely variable. The body is covered by an exoskeleton that ranges from hard and thick to soft and thin. An anterior head or headlike region possesses various appendages (sometimes internal or not apparent) for feeding. Segmented, paired appendages are often present along the body (these are not present in some immature insects).

DISCUSSION: The freshwater arthropods are the primary subject of this book and therefore are treated in detail in the following chapters. The three classes of the phylum Arthropoda that are of importance in freshwater ecosystems are comparatively diagnosed below.

CLASS CRUSTACEA: See Chapter 22. The freshwater crustaceans (shrimps, scuds, sowbugs, crayfishes, copepods, cladocerans, and related forms) are cylindrical, dorsoventrally flattened, or laterally flattened forms measuring from 1 to 150 mm. They are generally hard-bodied, the exoskeleton being well developed. Body segmentation is usually well developed. Distinct body regions may or may not be recognizable. Body regions include a head and trunk or a head, thorax, and abdomen. Head and thorax are often fused to form a cephalothorax. Body sometimes possesses a shieldlike or valvelike carapace. Head almost always bears two pairs of antennae. Three to 71 pairs of legs are present.

CLASS ARACHNIDA: See Chapter 23. These are usually somewhat oval forms. Water mites are generally less than 4 mm in length, and fishing spiders may be 50 mm or more. Exoskeleton may be soft or hard. Head is generally inconspicuous, either being more or less fused with the remaining body or fused with the thorax to form a cephalothorax. There are four pairs of legs except for larval mites, which possess three pairs of legs.

CLASS HEXAPODA: See Chapters 6–21. The hexapods, which include the insects, generally range in size from 1 to 100 mm. Exoskeleton may be hard or soft. Body segmentation is usually apparent but not always. Body is also usually divided into head, thorax, and abdomen, although these regions are not distinct in many. When antennae are present, there is only a single pair. Three pairs of segmented legs are usually present on the thorax. Fleshy leglike appendages may or may not be present on various thoracic and abdominal segments. Some are wormlike in appearance, with few if any leglike appendages. One or two pairs of wings (sometimes highly modified) or wing pads are present on the second and third thoracic segments of many.

References

Barnes, R. D. 1968. *Invertebrate zoology* (2nd ed.). W. B. Saunders, Philadelphia. 743 pp.

Brinkhurst, R. O., and B. G. M. Jamieson. 1971. *Aquatic Oligochaeta of the world.* Univ. Toronto Press, Toronto. 860 pp.

Burch, J. B. 1972. *Freshwater sphaeriacean clams (Mollusca: Pelecypoda) of North America.* Biota of freshwater ecosystems, U.S.E.P.A., Ident. Man. No. 3. Wash., D.C. 31 pp.

Burch, J. B. 1973. *Freshwater unionacean clams (Mollusca: Pelecypoda) of North America.* Biota of freshwater ecosystems, U.S.E.P.A., Ident. Man. No. 11. Wash., D.C. 176 pp.

Clarke, A. H. 1973. *The freshwater mollusks of the Canadian interior basin.* Malacologia 13. 509 pp.

Eddy, S., and A. C. Hodson. 1961. *Taxonomic keys to the common animals of the north central states* (3rd ed.). Burgess, Minneapolis. 162 pp.

Edmondson, W. T. (ed.). 1959. *Freshwater biology* (2nd ed.). John Wiley & Sons, New York. 1248 pp.

Ferris, V. R., J. M. Ferris, and J. P. Tjepkema. 1973. *Genera of freshwater nematodes (Nematoda) of eastern North America.* Biota of freshwater ecosystems, U.S.E.P.A. Ident. Man. No. 10. Wash., D.C. 38 pp.

Forrest, H. 1963. Taxonomic studies on the hydras of North America. VIII. Description of two new species, with records and a key to North American hydras. *Trans. Amer. Micr. Soc.* 82:6–17.

Kenk, R. 1972. *Freshwater planarians (Turbellaria) of North America.* Biota of freshwater ecosystems, U.S.E.P.A. Ident. Man. No. 1. Wash., D.C. 81 pp.

Klemm, D. J. 1972. *Freshwater leeches (Annelida: Hirudinea) of North America.* Biota of freshwater ecosystems, U.S.E.P.A. Ident. Man. No. 8. Wash., D.C. 53 pp.

Pennak, R. W. 1978. *Fresh-water invertebrates of the United States* (2nd ed.). John Wiley & Sons, New York. 803 pp.

PART II

AQUATIC INSECTS

CHAPTER 6

Key to the Orders and Stages of Aquatic Insects

FOR THE BEGINNING STUDENT, naturalist, or fisherman, ordinal recognition is a basic first step in learning about aquatic insects. The following picture key (Fig. 6.1) is a guide to the orders and life stages (excluding eggs) of those insects most apt to be found in, on, or near water in North America. The key therefore takes into account aquatic and semiaquatic forms as well as terrestrial forms of species that are aquatic or semiaquatic in at least one stage (springtails are included as a matter of convenience, since they are hexapods). Upon determination of the order and stage at hand, the key directs the user to chapters of this book that deal with the appropriate taxonomic and ecological groupings.

Use of the key is based simply on making a series of choices that correctly match observable traits of the insects. The user is always given two alternatives. Choose the alternative that applies to the insect under study and then proceed as directed to the next set of alternatives or until an end point is reached. At each end point, the order and stage is indicated.

If it is known whether the insect stage being studied is aquatic or terrestrial, this could be of some help, since this information is incorporated into the key. Remember that terrestrial insects sometimes fall into the water; however, if an insect is taken on land or in the air it usually can be safely considered a terrestrial stage (with the exception that some aquatic larvae leave the water to molt or overwinter, and either they or their shed skins are sometimes found). In any case, the key is functional without knowledge of exact habitat, and if doubts exist whether an insect stage is aquatic or terrestrial, correct identification and further use of the book will easily resolve this.

Orders will generally key out to more than one end point in the key. This is to accommodate the various stages and the large degree of variability that is encountered in some groups, as well as to allow the use of simple rather than complex or unclear taxonomic characters. The illustrations within the key by no means include all of the types of insects that can be keyed, but were chosen as those that adequately exemplify a particular trait or are representative of larger groups. For most of the insects, a perusal of the illustrations of the chapter or chapters to which the user is directed and of the appropriate color plates will serve as a quick check on the correctness of the ordinal identification. Further use of the book might also be necessary to determine whether a completely terrestrial insect that inadvertently found its way into an aquatic or semiaquatic habitat has been keyed out as an aquatic insect (most of these insects will obviously not fit through the key).

On many accounts these insects are very eligible subjects for scientific research; but so long as they are ill known, and their exact identification a matter difficult of accomplishment their employment in any branch of zoological learning is surrounded with disadvantages too patent to need indication.

Rev. A. E. Eaton

Figure 6.1. INSECTS

THORAX POSSESSES THREE PAIRS OF SEGMENTED LEGS (LEGS ARE SOMETIMES MINUTE OR FUSED TO THE BODY)

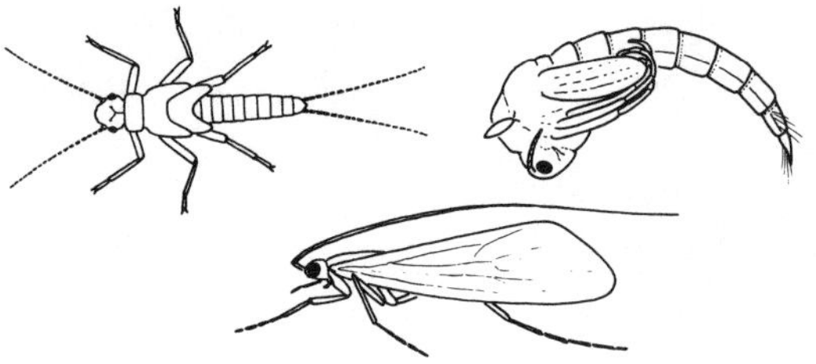

THORAX LACKS THREE PAIRS OF SEGMENTED LEGS (PROLEGS OR PROTUBERANCES MAY SOMETIMES BE PRESENT ON THE THORAX)

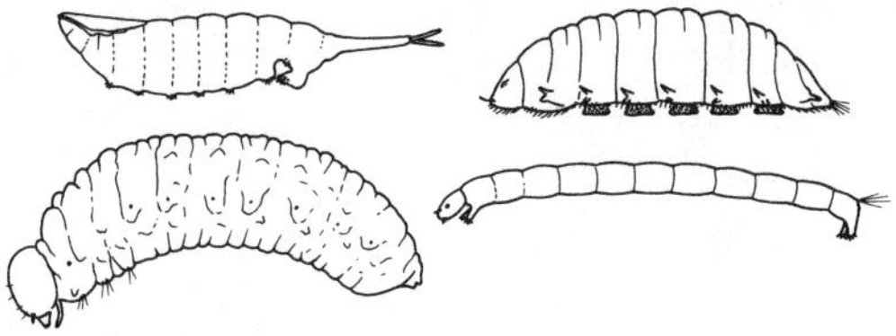

EXTERNAL WINGS OR WING PADS ARE COMPLETELY ABSENT cont. p. 89, top

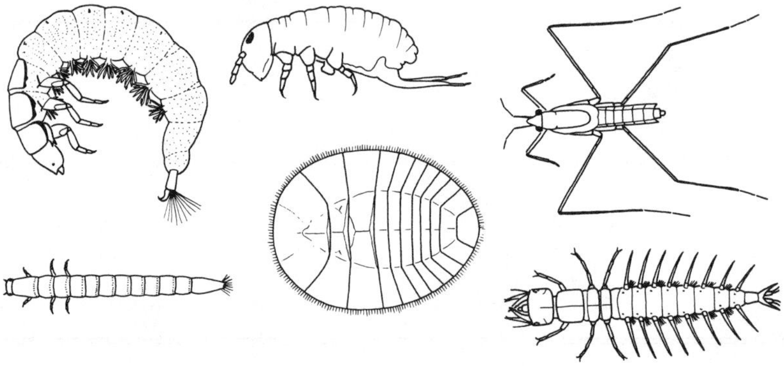

WINGS OR WING PADS ARE PRESENT (FORE WINGS MAY BE FULLY DEVELOPED, RUDIMENTARY, OR VESTIGIAL; FULLY MEMBRANOUS, WITH OR WITHOUT SCALES OR HAIRS, PARTLY MEMBRANOUS, LEATHERY, OR HARDENED AND PLATELIKE; HELD FREE FROM THE BODY OR VARIOUSLY APPRESSED TO IT)

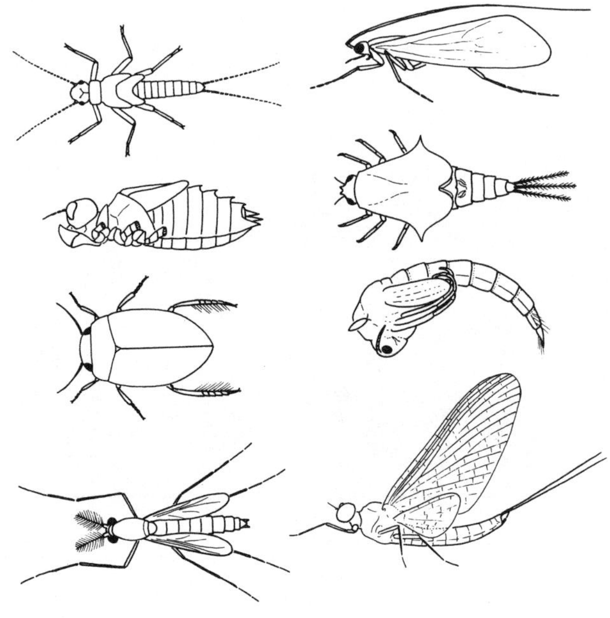

THORACIC LEGS ARE NOT FUSED TO THE BODY

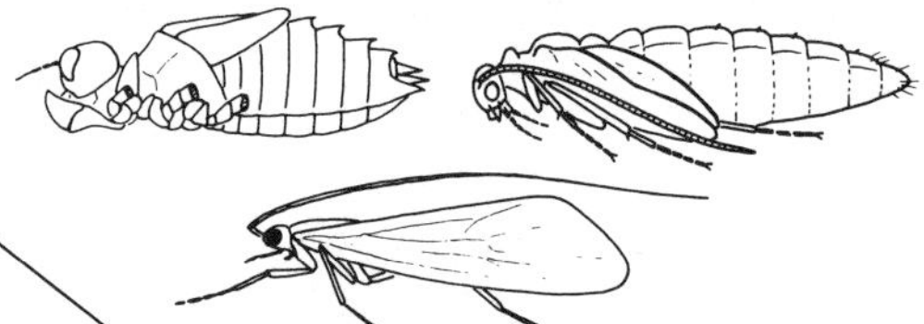

THORACIC LEGS ARE FUSED TO THE BODY

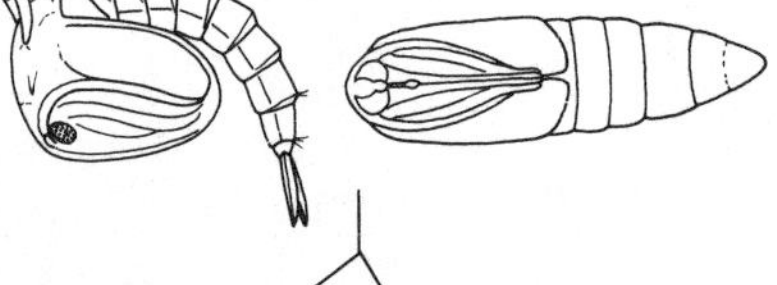

ONLY ONE PAIR OF DEVELOPING WINGS IS PRESENT

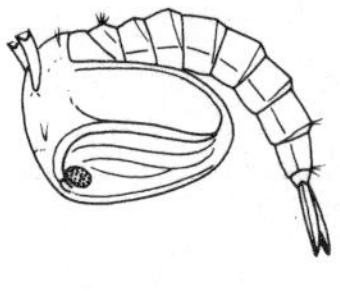

Diptera

Some Aquatic Fly Pupae, Chap. 16. See also: Shore-Dwelling Insects, Chap. 17; Emergent Vegetation Associates, Chap. 18

TWO PAIRS OF DEVELOPING WINGS ARE PRESENT

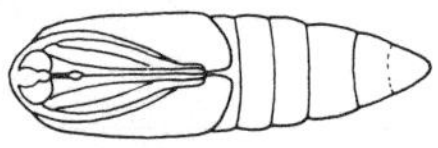

Lepidoptera

All Aquatic Moth Pupae, Chap. 15. See also: Emergent Vegetation Associates, Chap. 18

HEAD IS INDISTINCT, ENTIRELY
OR PARTLY LACKING

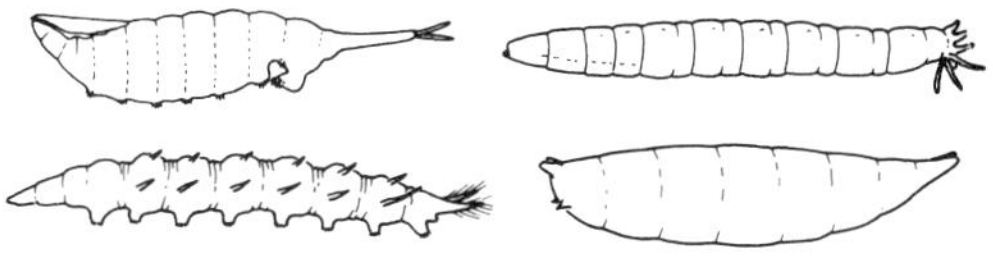

Diptera

Some Aquatic Fly Larvae and Puparia, Chap. 16.
See also: Shore-Dwelling Insects, Chap. 17;
Emergent Vegetation Associates, Chap. 18;
Tree Hole and Plant Cup Residents, Chap. 19

HEAD IS DISTINCT AND
WELL DEVELOPED

PROLEGS ARE OFTEN PRESENT ON EITHER THORAX OR END OF ABDOMEN OR BOTH; IF PROLEGS ARE ABSENT, THEN CONSPICUOUS HAIRS AND/OR DISTINCT PROCESSES ARE PRESENT AT END OF ABDOMEN, AND BODY IS EITHER VERY SLENDER, HEAVILY SCLEROTIZED, OR DIVIDED INTO SEVERAL DISTINCTIVE REGIONS

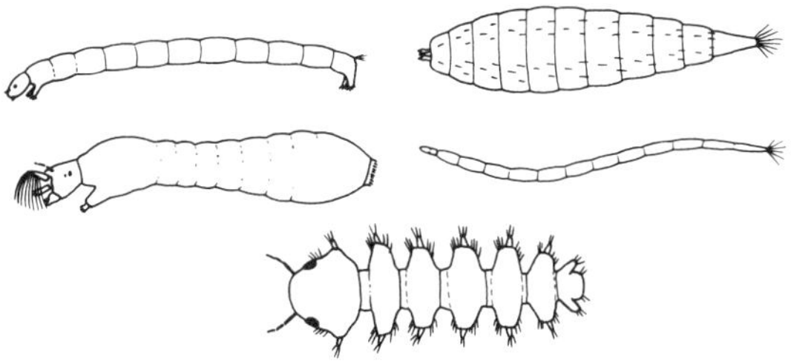

Diptera

Some Aquatic Fly Larvae, Chap. 16.
See also: Shore-Dwelling Insects, Chap. 17;
Tree Hole and Plant Cup Residents, Chap. 19

PROLEGS ARE NOT PRESENT; BODY IS EITHER THICK, FLESHY, AND C-SHAPED OR POSSESSES DORSAL SCLEROTIZED PLATES ON THORACIC SEGMENTS AND ABDOMINAL SEGMENT 8

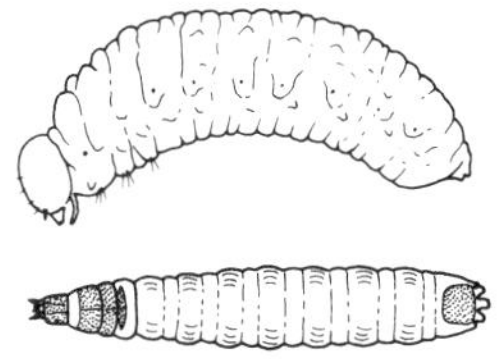

Coleoptera

Some Water Beetle Larvae, Chap. 13.
See also: Shore-Dwelling Insects, Chap. 17

ABDOMEN ENDS IN
THREE TAILS

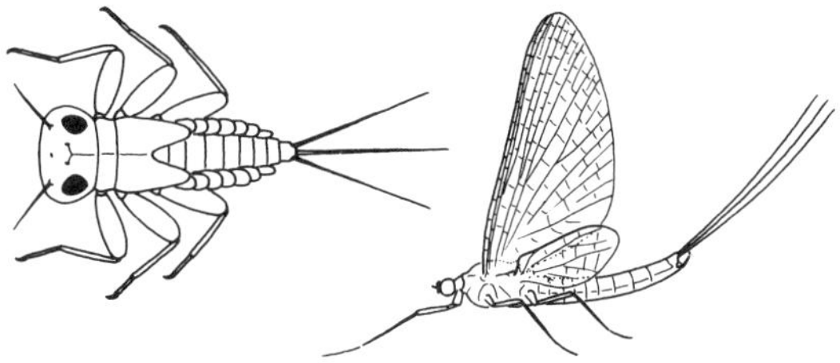

Ephemeroptera

AQUATIC FORMS THAT HAVE
WING PADS AND GILLS

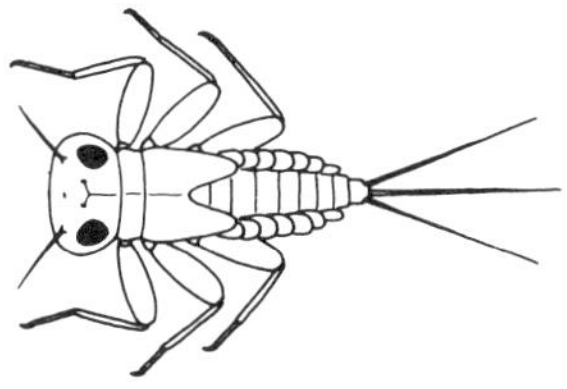

Most Mayfly Larvae, Chap. 7

GENERALLY TERRESTRIAL FORMS
THAT HAVE FULLY DEVELOPED WINGS

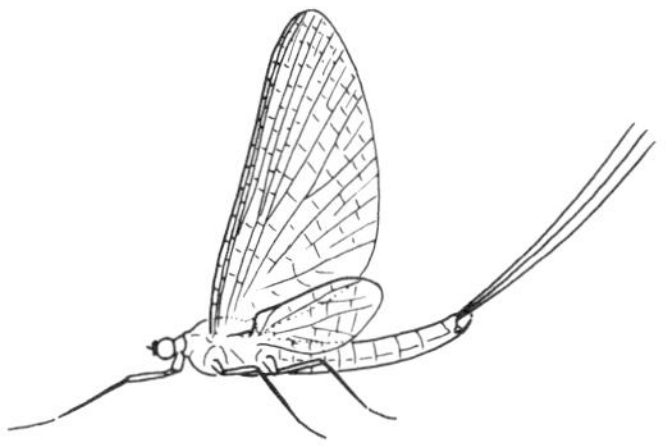

Some Mayfly Adults and Subimagos, Chap. 7

ABDOMEN ENDS IN ONE,
TWO, OR NO TAILS

MOUTH IS NOT IN THE FORM
OF AN ELONGATE BEAK OR
CONELIKE STRUCTURE

cont. p. 86, top

MOUTH IS IN THE FORM
OF AN ELONGATE BEAK OR
CONELIKE STRUCTURE

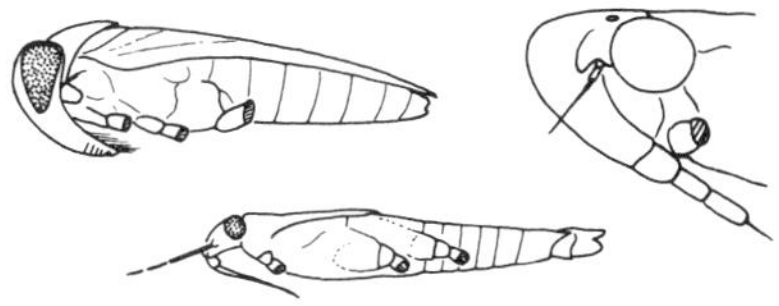

Hemiptera

Some Underwater and Surface Bugs, Chap. 10.
See also: Shore-Dwelling Insects, Chap. 17;
Emergent Vegetation Associates, Chap. 18

cont. from p. 85

LABIUM (LOWER LIP) IS MODIFIED INTO A LARGE MASKLIKE STRUCTURE THAT WHEN AT REST COVERS THE OTHER MOUTHPARTS FROM BELOW (AQUATIC)

Odonata

All Dragonfly and Damselfly Larvae, Chap. 8

LABIUM (LOWER LIP) IS NOT MODIFIED INTO A LARGE MASKLIKE STRUCTURE (AQUATIC AND TERRESTRIAL STAGES)

HIND LEGS ARE ENLARGED AND MODIFIED FOR JUMPING (FORE LEGS ARE MODIFIED FOR DIGGING IN SOME)

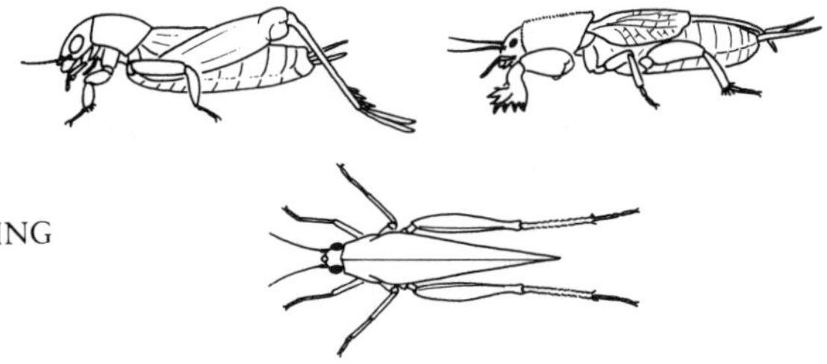

Orthoptera

Some Shore-Dwelling Insects, Chap. 17.
See also: Emergent Vegetation Associates, Chap. 18

HIND LEGS ARE NOT ENLARGED FOR JUMPING

WINGS ARE FULLY DEVELOPED, AND FORE WINGS ARE ENTIRELY MEMBRANOUS, ALTHOUGH THEY MAY BE VARIOUSLY CLOTHED WITH HAIRS OR SCALES (MOST FLYING, GENERALLY TERRESTRIAL ADULTS)

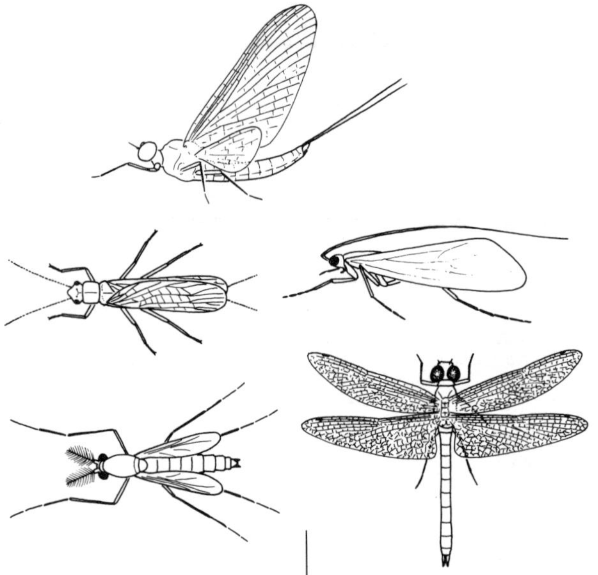

FORE WINGS ARE EITHER REDUCED, SHEATHED IN CASES, PRESENT ONLY AS PADS OR BUDS, OR MODIFIED INTO PROTECTIVE PLATELIKE COVERINGS

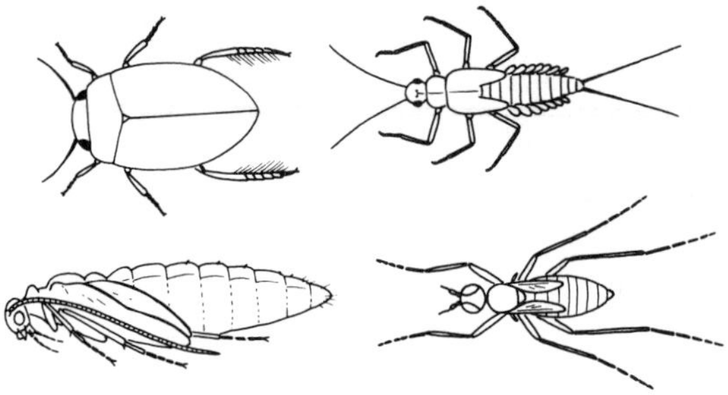

TWO PAIRS OF WINGS ARE PRESENT (HIND WINGS ARE SOMETIMES MUCH SMALLER THAN FORE WINGS)

cont. p. 88, top

ONLY ONE PAIR OF WINGS IS PRESENT

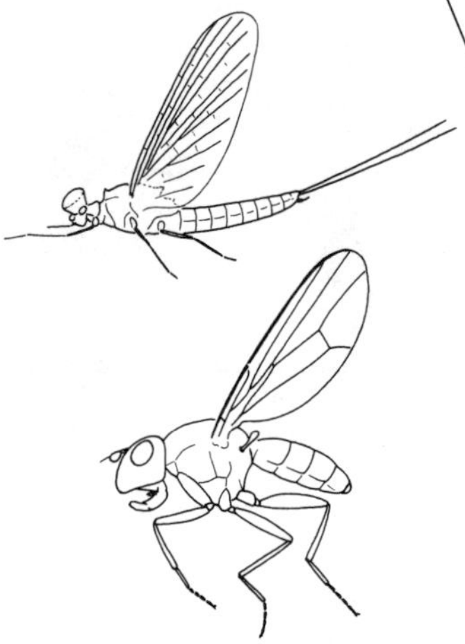

THIRD THORACIC SEGMENT POSSESSES A PAIR OF SMALL PEGLIKE OR KNOBLIKE STRUCTURES

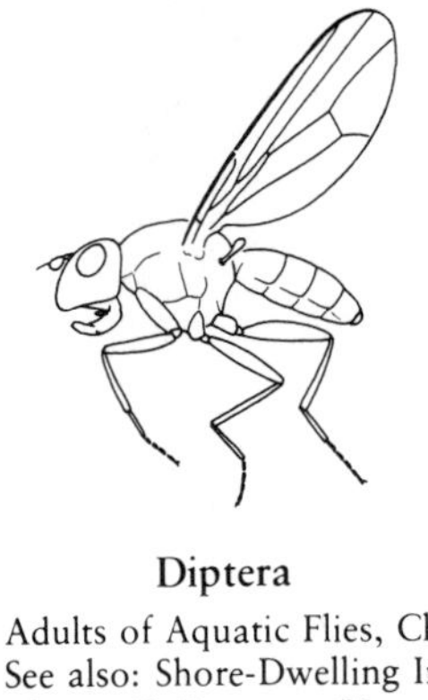

Diptera

Adults of Aquatic Flies, Chap. 16.
See also: Shore-Dwelling Insects, Chap. 17; Emergent Vegetation Associates, Chap. 18

THIRD THORACIC SEGMENT LACKS A PAIR OF PEGLIKE OR KNOBLIKE STRUCTURES (ABDOMEN ENDS IN LONG TAILS)

Ephemeroptera

Some Mayfly Adults and Subimagos, Chap. 7

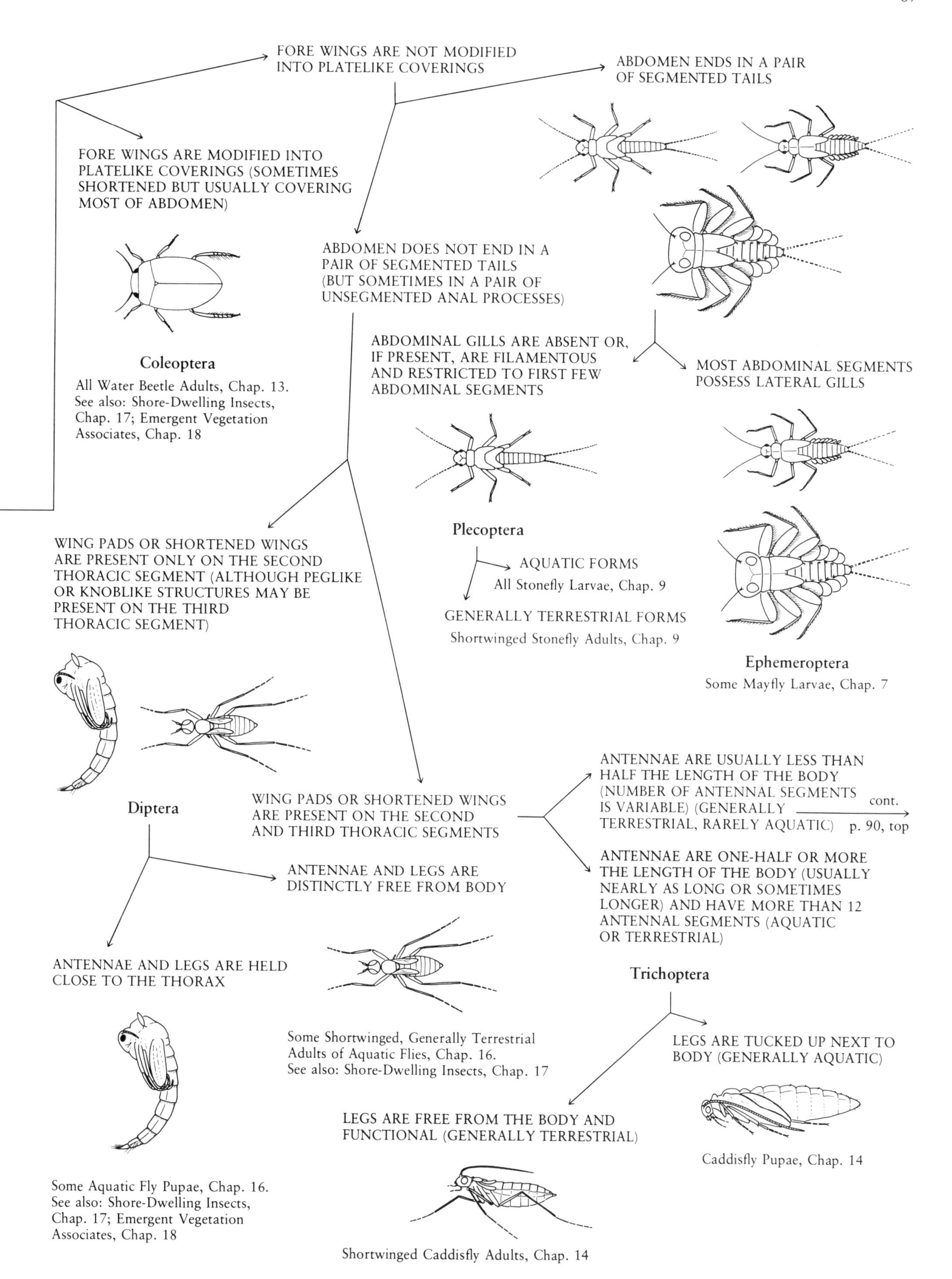
FORE WINGS ARE NOT MODIFIED INTO PLATELIKE COVERINGS
FORE WINGS ARE MODIFIED INTO PLATELIKE COVERINGS (SOMETIMES SHORTENED BUT USUALLY COVERING MOST OF ABDOMEN)
Coleoptera
All Water Beetle Adults, Chap. 13. See also: Shore-Dwelling Insects, Chap. 17; Emergent Vegetation Associates, Chap. 18
ABDOMEN ENDS IN A PAIR OF SEGMENTED TAILS
ABDOMEN DOES NOT END IN A PAIR OF SEGMENTED TAILS (BUT SOMETIMES IN A PAIR OF UNSEGMENTED ANAL PROCESSES)
ABDOMINAL GILLS ARE ABSENT OR, IF PRESENT, ARE FILAMENTOUS AND RESTRICTED TO FIRST FEW ABDOMINAL SEGMENTS
MOST ABDOMINAL SEGMENTS POSSESS LATERAL GILLS
Plecoptera
AQUATIC FORMS
All Stonefly Larvae, Chap. 9
GENERALLY TERRESTRIAL FORMS
Shortwinged Stonefly Adults, Chap. 9
Ephemeroptera
Some Mayfly Larvae, Chap. 7
WING PADS OR SHORTENED WINGS ARE PRESENT ONLY ON THE SECOND THORACIC SEGMENT (ALTHOUGH PEGLIKE OR KNOBLIKE STRUCTURES MAY BE PRESENT ON THE THIRD THORACIC SEGMENT)
Diptera
WING PADS OR SHORTENED WINGS ARE PRESENT ON THE SECOND AND THIRD THORACIC SEGMENTS
ANTENNAE ARE USUALLY LESS THAN HALF THE LENGTH OF THE BODY (NUMBER OF ANTENNAL SEGMENTS IS VARIABLE) (GENERALLY TERRESTRIAL, RARELY AQUATIC)
cont.
p. 90, top
ANTENNAE ARE ONE-HALF OR MORE THE LENGTH OF THE BODY (USUALLY NEARLY AS LONG OR SOMETIMES LONGER) AND HAVE MORE THAN 12 ANTENNAL SEGMENTS (AQUATIC OR TERRESTRIAL)
Trichoptera
ANTENNAE AND LEGS ARE DISTINCTLY FREE FROM BODY
ANTENNAE AND LEGS ARE HELD CLOSE TO THE THORAX
Some Shortwinged, Generally Terrestrial Adults of Aquatic Flies, Chap. 16. See also: Shore-Dwelling Insects, Chap. 17
LEGS ARE TUCKED UP NEXT TO BODY (GENERALLY AQUATIC)
Caddisfly Pupae, Chap. 14
LEGS ARE FREE FROM THE BODY AND FUNCTIONAL (GENERALLY TERRESTRIAL)
Some Aquatic Fly Pupae, Chap. 16. See also: Shore-Dwelling Insects, Chap. 17; Emergent Vegetation Associates, Chap. 18
Shortwinged Caddisfly Adults, Chap. 14

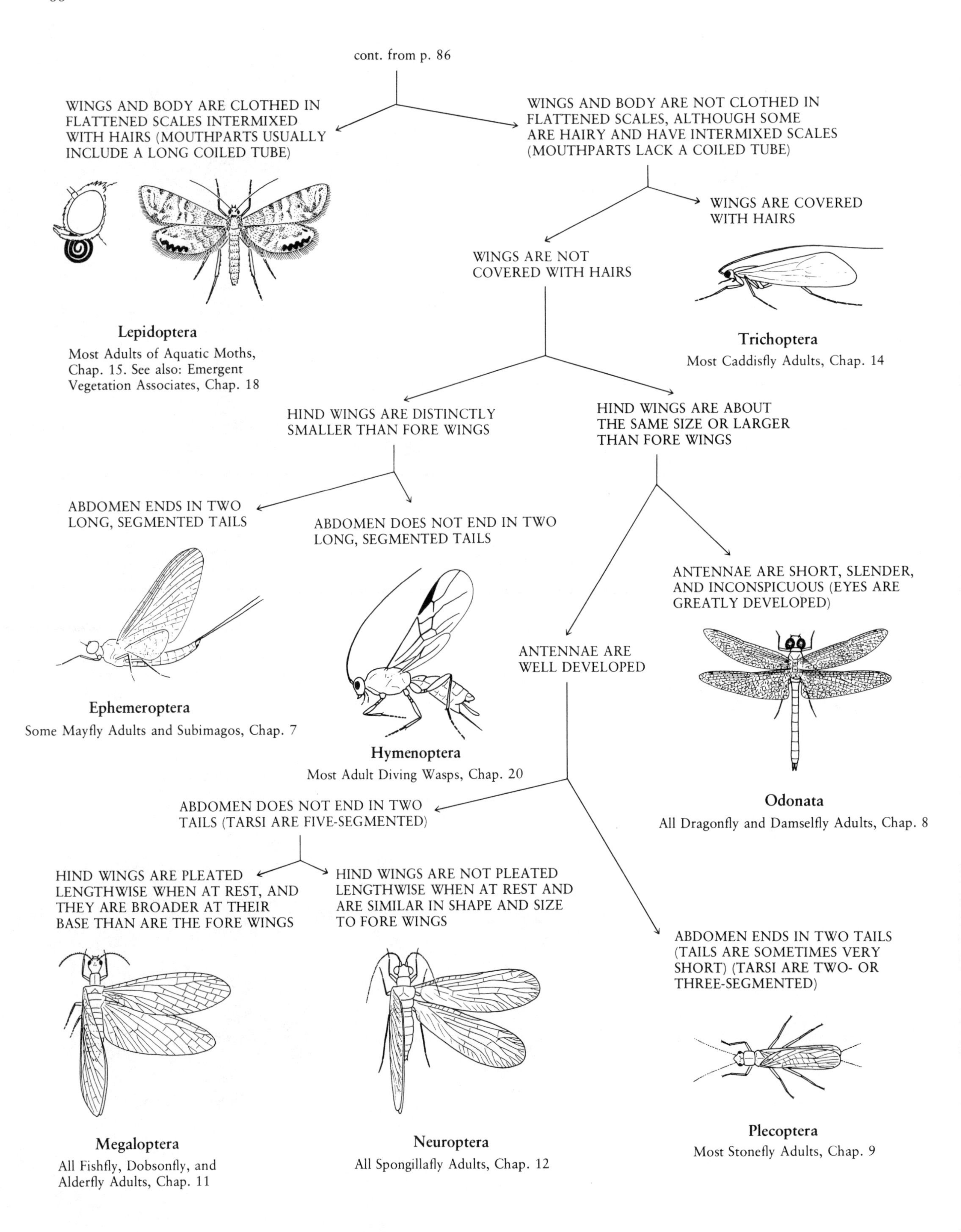

cont. from p. 86
WINGS AND BODY ARE CLOTHED IN FLATTENED SCALES INTERMIXED WITH HAIRS (MOUTHPARTS USUALLY INCLUDE A LONG COILED TUBE)
WINGS AND BODY ARE NOT CLOTHED IN FLATTENED SCALES, ALTHOUGH SOME ARE HAIRY AND HAVE INTERMIXED SCALES (MOUTHPARTS LACK A COILED TUBE)
Lepidoptera
Most Adults of Aquatic Moths, Chap. 15. See also: Emergent Vegetation Associates, Chap. 18
WINGS ARE COVERED WITH HAIRS
WINGS ARE NOT COVERED WITH HAIRS
Trichoptera
Most Caddisfly Adults, Chap. 14
HIND WINGS ARE DISTINCTLY SMALLER THAN FORE WINGS
HIND WINGS ARE ABOUT THE SAME SIZE OR LARGER THAN FORE WINGS
ABDOMEN ENDS IN TWO LONG, SEGMENTED TAILS
ABDOMEN DOES NOT END IN TWO LONG, SEGMENTED TAILS
ANTENNAE ARE SHORT, SLENDER, AND INCONSPICUOUS (EYES ARE GREATLY DEVELOPED)
Ephemeroptera
Some Mayfly Adults and Subimagos, Chap. 7
Hymenoptera
Most Adult Diving Wasps, Chap. 20
ANTENNAE ARE WELL DEVELOPED
Odonata
All Dragonfly and Damselfly Adults, Chap. 8
ABDOMEN DOES NOT END IN TWO TAILS (TARSI ARE FIVE-SEGMENTED)
HIND WINGS ARE PLEATED LENGTHWISE WHEN AT REST, AND THEY ARE BROADER AT THEIR BASE THAN ARE THE FORE WINGS
HIND WINGS ARE NOT PLEATED LENGTHWISE WHEN AT REST AND ARE SIMILAR IN SHAPE AND SIZE TO FORE WINGS
ABDOMEN ENDS IN TWO TAILS (TAILS ARE SOMETIMES VERY SHORT) (TARSI ARE TWO- OR THREE-SEGMENTED)
Megaloptera
All Fishfly, Dobsonfly, and Alderfly Adults, Chap. 11
Neuroptera
All Spongillafly Adults, Chap. 12
Plecoptera
Most Stonefly Adults, Chap. 9

cont. from p. 84

MOUTH IS NOT IN THE FORM OF A BEAK, IF MOUTHPARTS LONG AND SLENDER, THEN THEY PROJECT FORWARD; EYES ARE ABSENT OR CONSIST OF SINGLE OR GROUPS OF SINGLE SPOTS OR FACETS

MOUTH IS IN THE FORM OF A BEAK PROJECTING BELOW THE HEAD; EYES ARE COMPOUND

Hemiptera

Some Surface Bugs, Chap. 10.
See also: Emergent Vegetation Associates, Chap. 18

ABDOMEN USUALLY POSSESSES A FORKED JUMPING ORGAN THAT ORIGINATES ON ITS UNDERSIDE (MINUTE SURFACE-DWELLING ARTHROPODS)

Collembola

Springtails, Chap. 21

ABDOMEN LACKS A JUMPING ORGAN

MOUTH NOT MADE UP OF LONG SLENDER RODS (AQUATIC OR TERRESTRIAL)

MOUTH INCLUDES LONG SLENDER RODS THAT PROJECT FORWARD; WHEN HELD TOGETHER, THEY FORM A SUCKING TUBE (SUBMERGENT)

Neuroptera

All Spongillafly Larvae, Chap. 12

ABDOMEN POSSESSES PAIRS OF SHORT, FLESHY, LEGLIKE STRUCTURES ON THE UNDERSIDE; THESE STRUCTURES END IN A SERIES OF TINY HOOKS

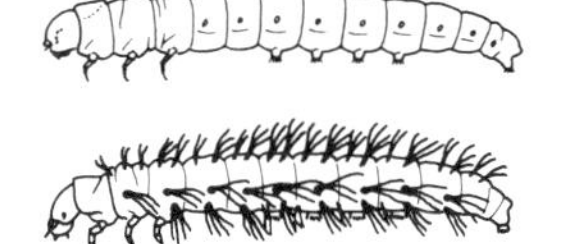

Lepidoptera

Aquatic Caterpillars, Chap. 15.
See also: Emergent Vegetation Associates, Chap. 18

ABDOMEN LACKS PAIRS OF FLESHY LEGS ON THE UNDERSIDE THAT END IN SERIES OF TINY HOOKS

ABDOMEN ENDS VARIOUSLY BUT NEVER IN A PAIR OF PROLEGS EACH HAVING A SINGLE HOOK (IF TERMINAL HOOKS ARE PRESENT, THEN THERE IS A TOTAL OF FOUR HOOKS)

ABDOMEN ENDS IN A PAIR OF SHORT TO LONG PROLEGS (SOMETIMES FUSED TOGETHER) THAT END IN A SINGLE HOOK

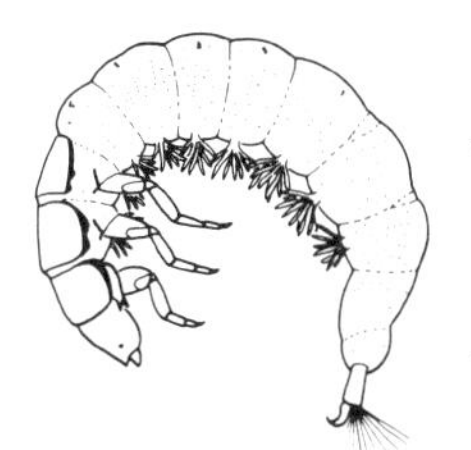

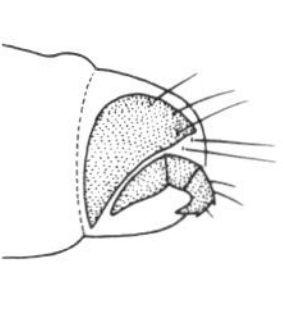

Trichoptera

All Caddisfly Larvae, Chap. 14

ABDOMEN POSSESSES WELL-DEVELOPED LATERAL FILAMENTS

ABDOMEN LACKS WELL-DEVELOPED LATERAL FILAMENTS

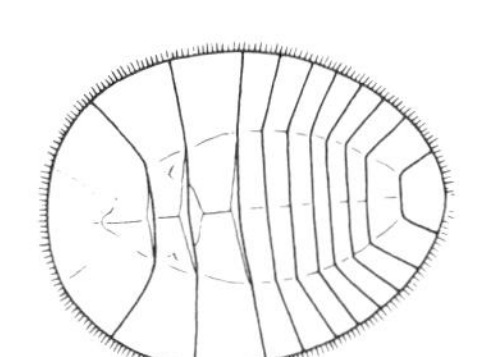

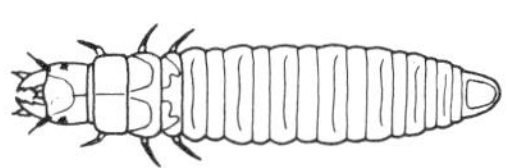

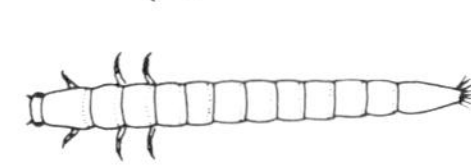

Coleoptera

Some Water Beetle Larvae, Chap. 13.
See also: Shore-Dwelling Insects, Chap. 17; Emergent Vegetation Associates, Chap. 18; Tree Hole and Plant Cup Residents, Chap. 19

ABDOMEN ENDS VARIOUSLY BUT NEVER WITH A SINGLE, UNFORKED, ELONGATE FILAMENT AND NEVER WITH A PAIR OF PROLEGS EACH HAVING A PAIR OF HOOKS (IF FOUR TERMINAL HOOKS ARE PRESENT, THEN TWO PAIRS OF TERMINAL FILAMENTS ARE ALSO PRESENT)

Coleoptera

Some Water Beetle Larvae, Chap. 13

ABDOMEN ENDS EITHER IN A SINGLE, UNFORKED, ELONGATE FILAMENT OR IN A PAIR OF PROLEGS, EACH POSSESSING A PAIR OF HOOKS

Megaloptera

All Fishfly, Dobsonfly, and Alderfly Larvae, Chap. 11

cont. from p. 87

FORE WINGS ARE THICKENED, AND ANTENNAE USUALLY HAVE 11 OR FEWER SEGMENTS

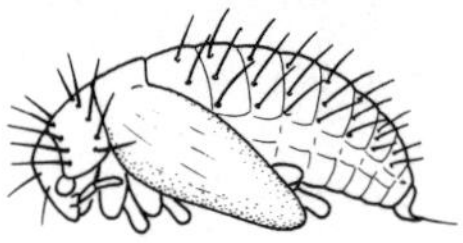

Coleoptera

Generally Terrestrial Water Beetle Pupae, Chap. 13.
See also: Shore-Dwelling Insects, Chap. 17

FORE WINGS ARE NOT THICKENED, AND ANTENNAE USUALLY HAVE 12 OR MORE SEGMENTS

MIDDLE AND HIND LEGS LACK SWIMMING HAIRS (MANDIBLES ARE WELL DEVELOPED) (GENERALLY TERRESTRIAL)

BODY MEASURES OVER 12 MM

Megaloptera

Fishfly, Dobsonfly, and Alderfly Pupae, Chap. 11

BODY MEASURES LESS THAN 10 MM

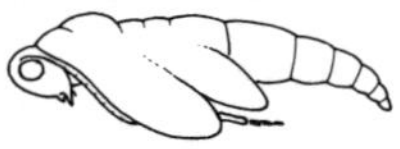

Neuroptera

Spongillafly Pupae, Chap. 12

MIDDLE AND HIND LEGS POSSESS A FRINGE OF SWIMMING HAIRS (MANDIBLES ARE POORLY DEVELOPED) (AQUATIC)

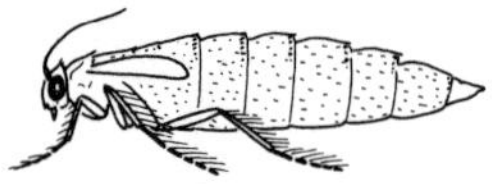

Lepidoptera

Shortwinged Aquatic Moth Adults, Chap. 15

CHAPTER 7

Mayflies

(ORDER EPHEMEROPTERA)

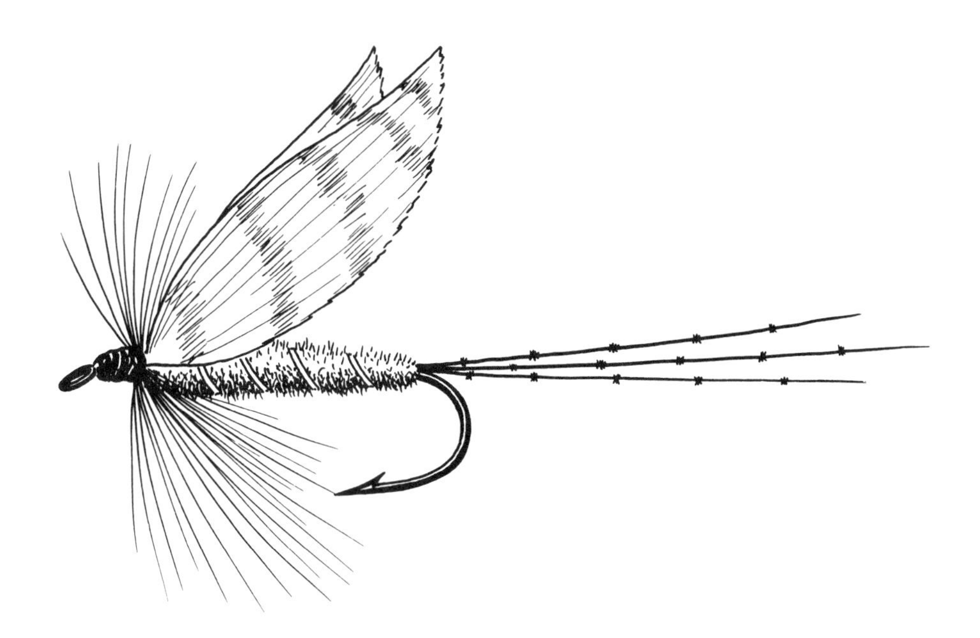

THE MAYFLIES are relatively primitive insects, possessing a number of traits that are thought to have been present in the earliest winged insects, such as tails and an inability to fold the wings flat over the body. All species are aquatic as larvae, and the fragile, terrestrial adults are generally short-lived, hence the allusion to this ephemeral winged stage in their scientific name. Mayflies are unique among the insects in that a fully winged terrestrial life stage, known as the subimago, precedes the sexually mature adult stage. Probably over 700 species occur in North America.

As a group, mayflies are one of the most common and important members of the bottom-dwelling freshwater community. Because most species are detritivores and/or herbivores and are themselves a preferred food of many freshwater carnivores, including other insects and fishes, they form a fundamental link in the freshwater food chain. Many species are highly susceptible to water pollution or occur in very predictable kinds of environments. It is for this reason that mayflies have proved very useful in the analysis or biomonitoring of water quality. Several species emerge in mass numbers, and these mass emergences are among the most spectacular in the insect world. In North America, mayflies may also be known locally by such names as willowflies, shadflies, drakes, duns, spinners, fishflies, and Canadian soldiers.

Mayflies constitute the primary basis for the sport and technique of fly fishing and fly tying. The various stages and species of mayflies serve as models for most of the delicately tied imitations used to lure trout. In North America alone, well over 150 mayfly species commonly occurring

TABLE 7.1 MAYFLIES USED FOR TIED FLY PATTERNS IN NORTH AMERICA

Genus	Numbers of western species upon which specific imitations have been based	Numbers of eastern and/or central species upon which specific imitations have been based	Family
Ameletus	1	0	Siphlonuridae, p. 100
Attenella[1]	0	1	Ephemerellidae, p. 118
Baetis[5]	4	6	Baetidae, p. 102
Baetisca	1	0	Baetiscidae, p. 122
Brachycercus	0	2	Caenidae, p. 121
Caenis	1	5	Caenidae, p. 121
Callibaetis	5	3	Baetidae, p. 102
Cinygma	1	0	Heptageniidae, p. 106
Cinygmula	3	0	Heptageniidae, p. 106
Cloeon	2	3	Baetidae, p. 102
Dannella[1]	0	1	Ephemerellidae, p. 118
Drunella[2]	4	6	Ephemerellidae, p. 118
Epeorus	6	6	Heptageniidae, p. 106
Ephemera	1	3	Ephemeridae, p. 115
Ephemerella	4	6	Ephemerellidae, p. 118
Ephoron	1	2	Polymitarcyidae, p. 113
Eurylophella[1]	0	2	Ephemerellidae, p. 118
Heptagenia	2	1	Heptageniidae, p. 106
Hexagenia	1	5	Ephemeridae, p. 115
Ironodes	1	0	Heptageniidae, p. 106
Isonychia	2	3	Oligoneuriidae, p. 105
Leptophlebia	3	3	Leptophlebiidae, p. 110
Leucrocuta[2]	0	4	Heptageniidae, p. 106
Litobrancha[3]	0	1	Ephemeridae, p. 115
Nixe[2]	1	0	Heptageniidae, p. 106
Paraleptophlebia	8	6	Leptophlebiidae, p. 110
Potamanthus	0	3	Potamanthidae, p. 112
Pseudocloeon	2	2	Baetidae, p. 102
Rhithrogena[5]	3	3	Heptageniidae, p. 106
Serratella[1]	1	2	Ephemerellidae, p. 118
Siphlonurus	3	5	Siphlonuridae, p. 100
Siphloplecton	0	1	Metretopodidae, p. 101
Stenacron[4,5]	0	3	Heptageniidae, p. 106
Stenonema[5]	0	8	Heptageniidae, p. 106
Timpanoga[1]	1	0	Ephemerellidae, p. 118
Tricorythodes	1	4	Tricorythidae, p. 119

[1]Previously included in *Ephemerella*.
[2]Previously included in *Heptagenia*.
[3]Previously included in *Hexagenia*.
[4]Previously included in *Stenonema*.
[5]A number of species names have recently been found to be invalid, so that actual numbers of species are smaller than have been treated in the fishing literature.

in trout waters have been used for fashioning such imitations (see Table 7.1 for a summary of mayfly usage for fly fishing in North America). Fishermen have also given mayflies their own common names. A dictionary of such names in North America is given in the Appendix.

A knowledge of the habitat, distribution, seasonality, morphology, and behavior of mayflies can be of extreme importance to the fly fisherman, since successful fly fishing relies, to a large degree, on precisely matching the predominant insects that happen to be in a given environment at a given time. One general philosophy is that the closer one can come to imitating the exact look, location, and action of the mayfly to which feeding trout are

Around the steel no tortur'd worm shall twine,
No blood of living insect stain my line;
Let me, less cruel, cast feather'd hook,
With pliant rod athwart the pebbled brook,
Silent along the mazy margin stray,
And with fur-wrought fly delude the prey.

John Gay

conditioned at the time, the better one's chances are for successful fishing. It is this challenge to duplicate nature in such a way as to master the game fish at its own "game" that separates fly fishing from other forms of angling and demonstrates that much of fly fishing ability is directly related to the fisherman's knowledge of mayfly biology and behavior.

Larval Diagnosis

These are elongate, cylindrical to flattened forms that, when mature, range in length (not including tails) from 3 to 20 mm, rarely to 30 mm or more. Head possesses well-developed eyes, slender antennae, and chewing mouthparts. They generally have well-developed legs that each have a single claw or rarely no claw. Developing fore wing pads are present; hind wing pads may be present or absent. Arising from the sides of the abdomen is a series of variously directed gills; in some species, most gills are obscured by a pair of larger fore gills (operculate gills) that cover and protect them or by a thoracic shield. The body ends in two or three elongate tails.

Adult Diagnosis

These are delicate insects that possess two or four more-or-less triangular-shaped wings, usually with numerous veins. The wings are held straight up above the body when at rest. Head possesses generally well-developed eyes (more so in males), small antennae, and nonfunctional remnants of larval mouthparts. Legs are of various lengths, the fore legs usually longest. Abdomen is elongate and ends in two or three long, slender tails.

Similar Orders

The larvae of some mayflies may superficially resemble those of stoneflies. A distinction between the two groups, however, can always be made by the presence in mayflies of single claws, rows of abdominal gills, or, if present, a middle tail. Young damselfly larvae may also be mistaken for those of mayflies because their developing caudal lamellae may resemble the tails of mayflies. Mouthparts and claws will distinguish the two groups even when very young.

Adult mayflies are structurally unique and do not closely resemble other winged insects. Nevertheless, all but the experienced observer may sometimes confuse swarms of flying mayflies (especially at dusk) with those of caddisflies or some of the midges. The trailing legs of midges or other flies may superficially resemble the tails of mayflies in flight.

Fishermen often distinguish the various adult insects familiar to them by the way in which the wings are held or appear when at rest. For example, the mayflies are referred to as the "upwings." All others are "downwings," and even more specifically the caddisflies are referred to as "tentwings," the stoneflies as "flatwings," and midges as "glassywings."

Life History

Over the wavelets there is a twitching, a hissing and a rustling, a moving mosaic of bodies, wings, and empty skins. A cloud, like whirling snowflakes, rises and grows denser and denser, now ascending like a mighty column of smoke toward the heavens, then wafted abroad by the gentle evening zephyrs. In this unforgettable drama of nature are the mayflies. . . . those images of the transitory that long age were celebrated in song by the ancient Greeks.

Walter Linsenmaier

Metamorphosis is incomplete, but adults and larvae (also commonly referred to as nymphs or naiads) are very different in form and habit. Most species have one or two generations per year, and larval development time ranges anywhere from a few weeks to as much as two years, depending on the species and local climate. The number of larval instars varies considerably among mayflies, even within the same species, depending on the local environmental conditions. Sometimes eggs or larvae have seasonal periods of nondevelopment, especially at more northern latitudes or higher altitudes.

Once larvae have reached maturity, they are able to transform to subimagos (also known as "duns" or "subs"). This transformation from the aquatic larval stage to the terrestrial, winged subimaginal stage is often referred to as a *hatch*. The hatch can be highly synchronized, with large numbers of individuals hatching during the same period, or it may take place in piecemeal fashion over a period of a week or several weeks. Actual triggering mechanisms for synchronized hatches are not clearly understood but may be related to temperature, daylength, or both.

Before the hatch period, nymph fishing with imitations of those larvae that are abundant and in the mature condition at the location can be productive. The actual hatch is extremely important to the fisherman because it often signals extensive feeding activity by fishes. Depending on the species, larvae may float, swim, or crawl to the water surface to molt; crawl out onto emergent objects to molt; or molt to the subimaginal stage under water, with the subimago having to swim immediately to the surface. During this period of movement from the aquatic to the terrestrial environment, the mayflies can be highly attractive to trout and vulnerable to predation. At these times, the fisherman can have great success with tied flies that are fished wet (submerged), especially the so-called emerger patterns.

As far as the fly fisherman is concerned, one of the most important aspects of the biology of many mayfly species that hatch off the water is that, at the time of transition from larva to subimago, the newly hatched subimago requires a short period of time (from a few seconds to minutes) for the wings to clear the larval wing cases and be readied for flight. The individual thus floats somewhat helplessly on the water. This is perhaps the most vulnerable period of the mayfly's life, and trout will readily rise to the surface to feed on these "sitting duck" individuals. The imitations of the duns of the particular species involved, when fished at the surface or slightly below, are highly productive during these hatches.

Subimagos possess relatively cloudy wings with fringes of minute hairs. Mayflies usually exist as subimagos for one or two days, resting on marginal vegetation much of the time before adult emergence. Changes from the subimago that are apparent in the adult include the clearing of wings, so that the membrane becomes transparent, and increases in the intensity of coloration. Other changes, especially in the male, include the full development of eyes, legs, and genitalia.

Adults (also known as imagos or "spinners") generally do not live for more than a month, usually much less. They are nonfeeding, and their activities consist basically of reproductive behavior (swarming, mating, and oviposition), and in some species, short upstream migration flights.

Although there are some exceptions, swarming usually consists of groups of males rhythmically flying up and down and females flying through this swarm. A female is seized upon from below by a male, and mating takes place in flight. The highly developed eyes of the males of most species are apparently an adaptation for visual mate recognition during swarming. Species that occur in low densities of adults may rely heavily on synchronized emergence times and swarm markers that orient the swarm to certain flight regions in order to ensure successful reproduction. Swarming activity usually occurs only at certain times during a 24-hour period, with dusk and dawn perhaps being the most common times. Little is known about those that swarm at night.

Eggs are laid in three basic ways. (1) Females of some species drop their eggs while flying low over the water. (2) Others lay eggs directly on the water either by intermittently touching the surface or by coming to rest on the surface. (3) Still others submerge themselves and oviposit on the underwater substrate.

Those females that come to rest on the surface for oviposition usually also expire in this position; the presence of ovipositing or spent females will induce surface feeding in fishes. The spent females lie flat on the water surface with their wings spread laterally. Special spreadwing patterns have been designed by fishermen to imitate this look. This behavior has been referred to as a "spinner fall" by fly fishermen. Spinner falls and hatch periods signal the best times for dry fly fishing.

Besides the top-feeding fishes, other predators of the winged stages of mayflies include birds, spiders, dragonflies, damselflies, and insectivorous biting midges. The examination of spider webs on bridges or marginal vegetation is one of the best ways of determining what species of mayflies have recently been flying in a certain locale. Interestingly, a few biting midges with long trailing glands apparently resemble female mayflies closely enough that, after entering a swarm of mayflies, male mayflies are visually enticed into approaching them, at which time they become easy prey for the biting midge.

In some species of North American mayflies, hatching, mating, oviposition, and death take place within a few hours.

To me, after all my eager pursuits, no solid pleasures now remain, but the reflection of a long life spent in meaning well, the sensible conversation of a few good lady ephemerae, and now and then a kind smile. . . .

Benjamin Franklin
from *Soliloquy of a Venerable Ephemera Who Had Lived Four Hundred and Twenty Minutes*

Aquatic Habitats

Mayflies occur in almost all natural and many artificial bodies of freshwater that have adequate supplies of dissolved oxygen. They are common benthos of flowing waters, ponds, and shallow areas of lakes. Some larvae spend much of their time burrowed into substrates, and possibly a great many, besides those known as burrowing mayflies, occur interstitially within soft substrates, especially as young larvae. Although highly atypical of mayflies generally, at least one species has occasionally been found in tide pools, and some species occur in waters with very low dissolved-oxygen levels.

Aquatic Adaptations and Behavior

Mayfly larval respiration is hydropneustic, and gills evidently contribute significantly to the amount of total thin body wall available for oxygen uptake. The gills of some species have other major functions. These include (1) creating a current through a burrow by pulsating movements and thereby maintaining the burrow and providing a flow of food particles and oxygenated water over the body; (2) acting together as an attachment structure and anchoring the individual to rocky substrates in rapid zones of streams and rivers; and (3) providing primary sites for chloride cells, which function in ion uptake necessary for maintaining a proper internal salt balance (osmoregulation). The anterior gills of some mayflies are modified into operculate structures that cover and protect the other gills.

Hiding under stones and driftwood, well aware, no doubt, what enticing morsels they are to a great variety of fishes, we find a number of species of ephemerid larvae. . . .

Stephen A. Forbes

With few exceptions in North America, mayfly larvae are detritivores and/or herbivores that feed on microscopic algae and small bits of organic matter. Food may be suspended in the water, accumulated on substrate surfaces, or intermixed within soft substrate. To a great degree, they are nonselective feeders. Only a few species are primarily carnivorous.

The mayfly suborder Schistonota exhibits a diverse array of adaptations for mobility and habitat orientation. Some larvae are swimmers or riffle dwellers with sleek streamlined bodies and almost finlike tails (e.g., many minnowlike mayflies). Some are sprawlers with extremely flattened bodies and flattened outspread legs (e.g., many flatheaded mayflies). Some are steadfast clingers with strong claws and/or other attachment or friction structures (e.g., some of the flatheaded mayflies). Some are burrowers with robust digging legs, cylindrical bodies, and feathery gills.

Members of the suborder Pannota are generally relatively slow and awkward in movement. Their bodies tend to be stout, and gills are usually protected in one way or another. They are often secretive in habit.

Mayflies contribute significantly to the drift fauna of streams and rivers. The periodicity of drift among mayflies is generally very specific. Since trout may actively feed on drifting larvae, a knowledge of mayfly drift in certain habitats could greatly enhance nymph fishing at times.

Classification and Characters

The mayflies of North America constitute three diverse superfamilies and 12 families of schistonote mayflies (the splitbacks) and 3 superfamilies and 5 families of pannote mayflies (the fusedbacks).

Gill structure is one of the important characters for identifying larvae to higher groups. Species identification of larvae usually requires examination of mouthpart structure, hairs and spination, and to some degree abdominal color patterns. In using the picture key (Fig. 7.5), discussions, or illustrations to identify larvae, the reader should be warned that various structures may change radically with age and that reliable results will be obtained only with mature or nearly mature specimens whose wing pads have darkened or are beginning to darken. By way of example, young larvae of both splitback and fusedback mayflies have wing pads that appear to be fused over most of their length because the wing pads are not fully developed. The fundamental distinction between the groups becomes apparent only when larvae are more mature, and pads are more developed.

Middle tails of larvae are often broken off at the base. A blunt stub rather than a tapered one will, however, indicate this damaged condition. This same rule applies for adults; a tail is considered present in adults only if it is at least as long as the ninth and tenth abdominal segments combined. The presence or absence and the shapes of such structures as **operculate gills** (gills that cover succeeding pairs of gills) and **tusks** (anteriorly projecting outgrowths of the mandibles, usually associated with burrowing behavior) are important to the identification of larvae.

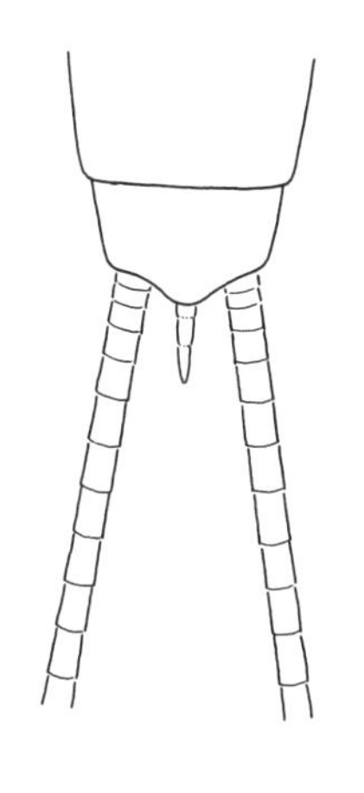

Figure 7.1. Adult, end of abdomen (two tailed)

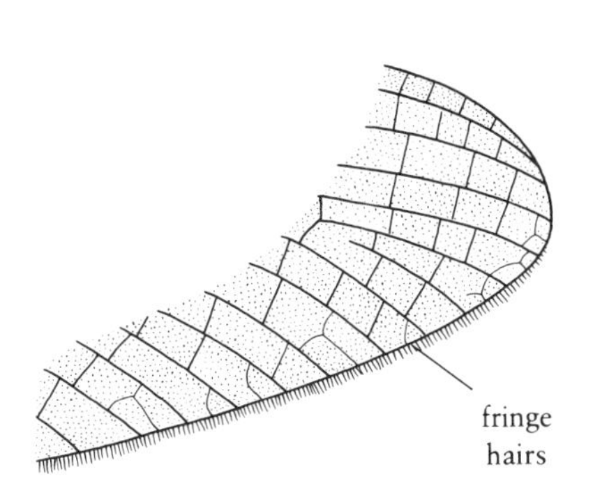

Figure 7.2. Subimago, wing tip

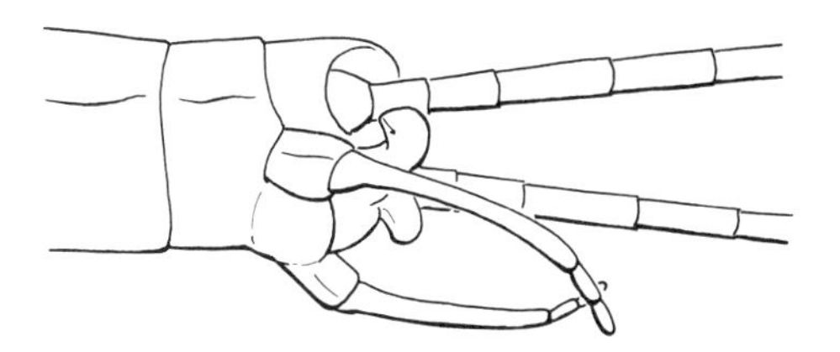

Figure 7.3. Adult ♂, end of abdomen

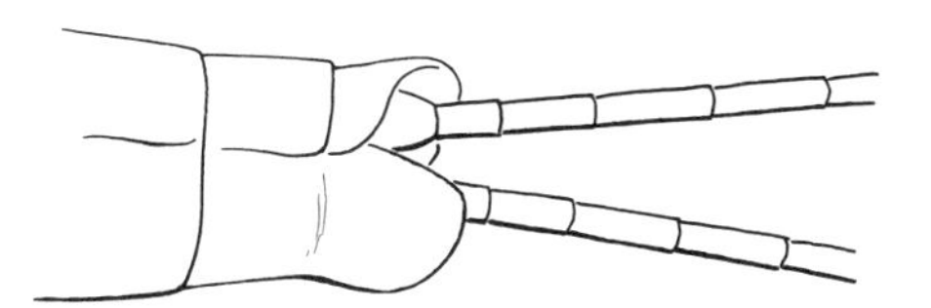

Figure 7.4. Adult ♀, end of abdomen

Wing venation is extremely important for identifying adult mayflies. Genitalic structure and color patterns are also of some importance. In general, only wing venation characters are reliable for subimagos, and this stage is difficult if not impossible to identify otherwise. Nonvenational traits given for adult diagnoses should not be used when examining subimagos. Although wing venation can be complicated, the pertinent wing veins and their abbreviations are indicated clearly in wing drawings.

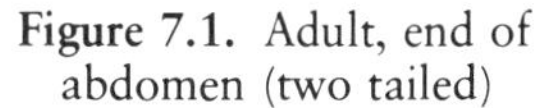

Many mayfly species in North America are known regionally to fishermen by common names. A great profusion of such names has recently appeared in the American fishing literature. Some of these names are borrowed from British mayflies and the earlier fishing literature. Some others allude to the colors that should be used in tying their imitations.

Fishermen's names of mayflies are often extremely confusing, and it becomes difficult to determine exactly what species is being referred to by a particular name. For example, the name Pale Evening Dun has been used for at least seven different species; furthermore, it usually has a different application in western North America than it does in the East, and although it usually refers only to the subimago, it has been referred to both the subimago and the adult of four species. This is an extreme example, but illustrates the confusion that has arisen. An example of the opposite extreme is that of *Ephemera guttulata*, a large and popular mayfly in eastern North America that is known by at least six different fishermen's names. In order to lessen the confusion, a guide to the names that fishermen apply to mayflies is given in the Appendix. This guide serves as a dictionary of North American names and also updates the scientific nomenclature of the pertinent mayfly species.

Fly-fishing certainly partakes more of science than bottom-fishing, and, of course, requires much time, study, and practice, before the Angler can become anything like an adept at making or casting a fly. . . .

Thomas Salter

Figure 7.5. MATURE EPHEMEROPTERA LARVAE

THORAX ROBUST WITH NOTUM FUSED BETWEEN FORE WING PADS FOR AT LEAST HALF LENGTH OF PADS; GILLS ON ABDOMINAL SEGMENT 2 EITHER ABSENT, CONCEALED, OR OPERCULATE

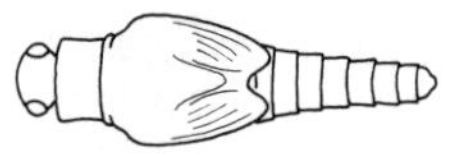

FORE WING PADS FREE, SEPARATE FOR HALF THEIR LENGTH OR MORE; GILLS ON ABDOMINAL SEGMENT 2 PRESENT, VARIABLE, BUT NEVER OPERCULATE

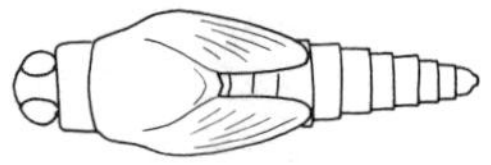

THORACIC NOTUM CARAPACELIKE AND COVERING MUCH OF ABDOMEN

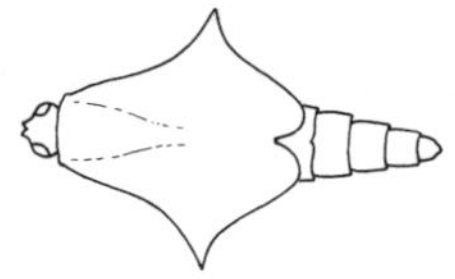

THORACIC NOTUM NOT COVERING MOST OF ABDOMEN

ABDOMINAL GILLS 2–7 DOUBLE, ELONGATE, WITH FRINGED MARGINS

Baetiscidae
(p. 122)

GILLS ON ABDOMINAL SEGMENT 2 ABSENT

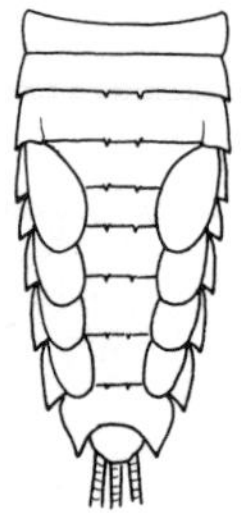

ABDOMINAL GILLS 2–7 NEVER DOUBLE, ELONGATE, AND FRINGED IN COMBINATION

GILLS ON ABDOMINAL SEGMENT 2 PRESENT AND OPERCULATE

GILLS ON ABDOMINAL SEGMENT 2 ROUNDED OR TRIANGULAR

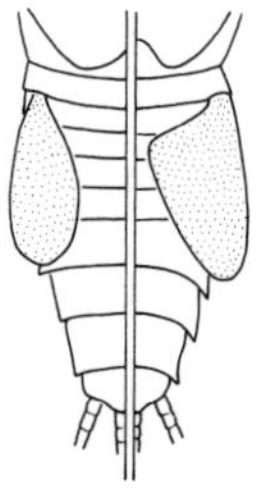

HEAD WITHOUT TUSKS

Ephemerellidae
(p. 118)

GILLS ON ABDOMINAL SEGMENT 2 QUADRATE

HEAD WITH ANTERIORLY PROJECTING TUSKS

Behningiidae
(p. 112)

Tricorythidae
(p. 119)

GILLS ON ABDOMINAL SEGMENT 2 FUSED ALONG MIDLINE

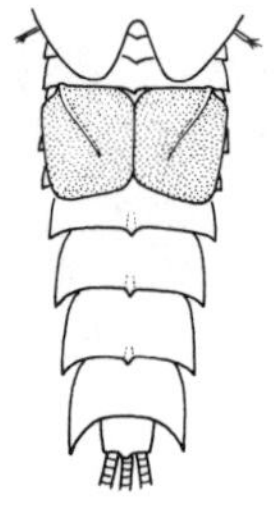

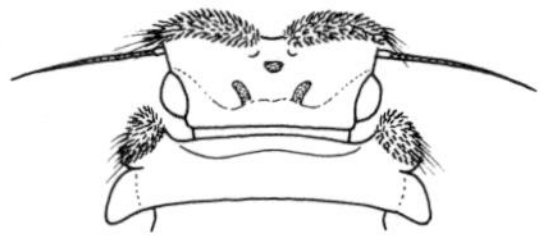

GILLS ON ABDOMINAL SEGMENT 2 NOT FUSED BUT OVERLAPPING

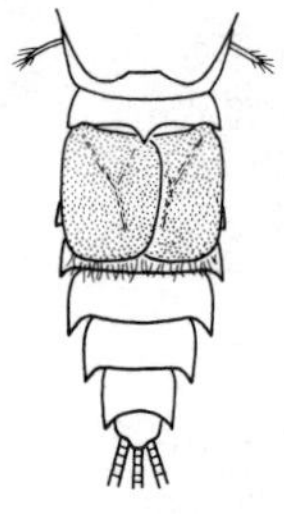

TUSK WITH NO SPINES IN DISTAL HALF

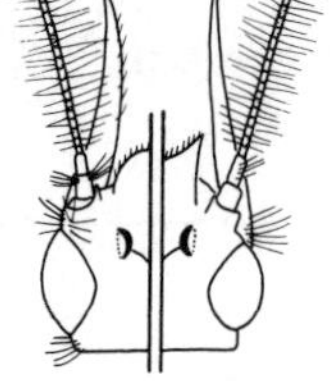

TUSKS WITH DISTAL SPINES

Ephemeridae
(p. 115)

Neoephemeridae
(p. 121)

Caenidae
(p. 121)

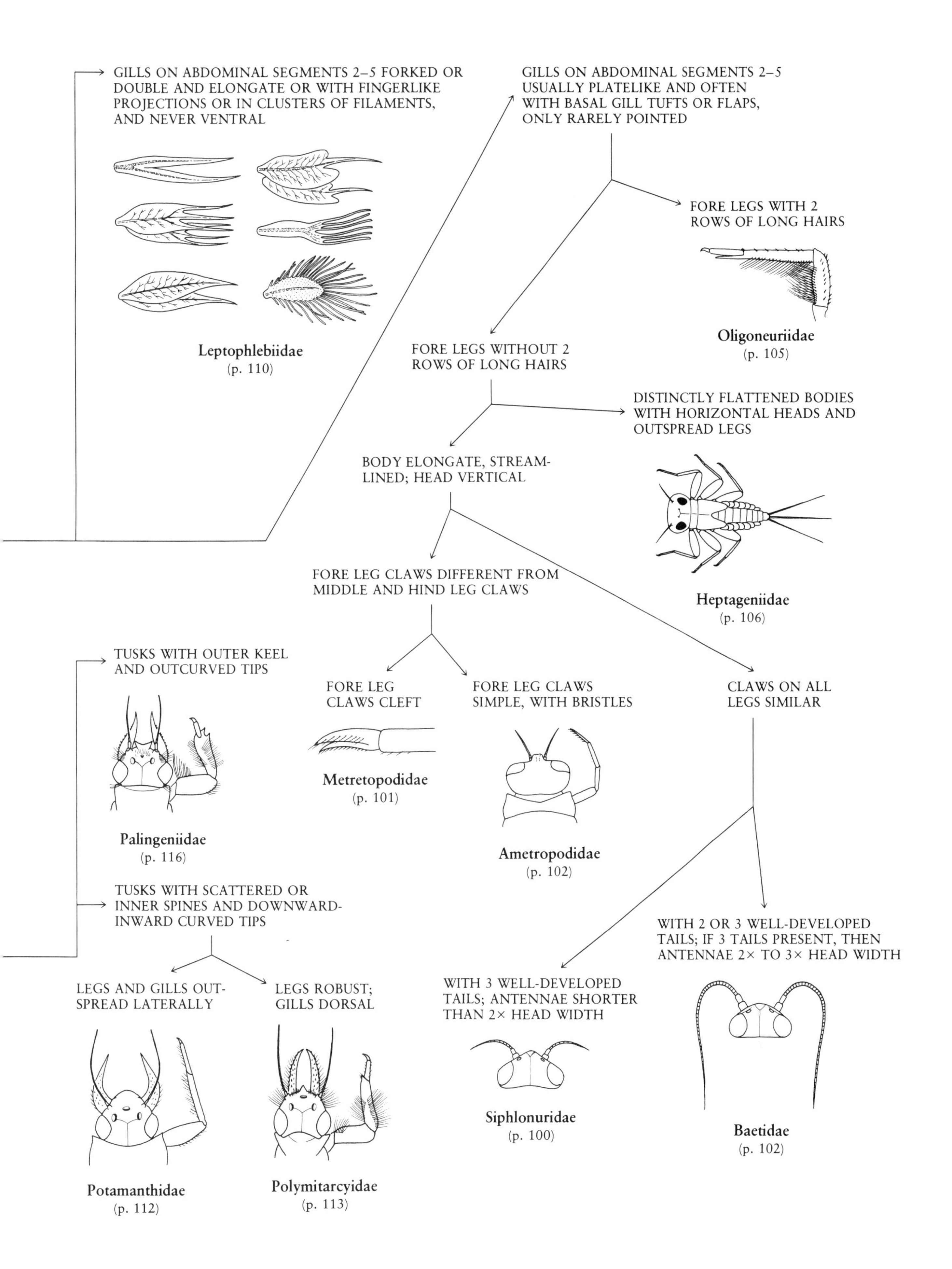

GILLS ON ABDOMINAL SEGMENTS 2–5 FORKED OR DOUBLE AND ELONGATE OR WITH FINGERLIKE PROJECTIONS OR IN CLUSTERS OF FILAMENTS, AND NEVER VENTRAL
Leptophlebiidae
(p. 110)
GILLS ON ABDOMINAL SEGMENTS 2–5 USUALLY PLATELIKE AND OFTEN WITH BASAL GILL TUFTS OR FLAPS, ONLY RARELY POINTED
FORE LEGS WITH 2 ROWS OF LONG HAIRS
Oligoneuriidae
(p. 105)
FORE LEGS WITHOUT 2 ROWS OF LONG HAIRS
DISTINCTLY FLATTENED BODIES WITH HORIZONTAL HEADS AND OUTSPREAD LEGS
Heptageniidae
(p. 106)
BODY ELONGATE, STREAMLINED; HEAD VERTICAL
FORE LEG CLAWS DIFFERENT FROM MIDDLE AND HIND LEG CLAWS
FORE LEG CLAWS CLEFT
Metretopodidae
(p. 101)
FORE LEG CLAWS SIMPLE, WITH BRISTLES
Ametropodidae
(p. 102)
CLAWS ON ALL LEGS SIMILAR
WITH 3 WELL-DEVELOPED TAILS; ANTENNAE SHORTER THAN 2× HEAD WIDTH
Siphlonuridae
(p. 100)
WITH 2 OR 3 WELL-DEVELOPED TAILS; IF 3 TAILS PRESENT, THEN ANTENNAE 2× TO 3× HEAD WIDTH
Baetidae
(p. 102)
TUSKS WITH OUTER KEEL AND OUTCURVED TIPS
Palingeniidae
(p. 116)
TUSKS WITH SCATTERED OR INNER SPINES AND DOWNWARD-INWARD CURVED TIPS
LEGS AND GILLS OUTSPREAD LATERALLY
Potamanthidae
(p. 112)
LEGS ROBUST; GILLS DORSAL
Polymitarcyidae
(p. 113)

Splitback Mayflies
(Suborder Schistonota)

Larvae have fore wing pads that are free for at least half of the pad length. The thorax generally does not appear wide and robust. Gill series are usually well developed along the sides of the abdomen. These are relatively active and diverse mayflies.

MINNOWLIKE AND FLATHEADED MAYFLIES
(Superfamily Baetoidea)

PRIMITIVE MINNOW MAYFLIES
(Family Siphlonuridae)

LARVAL DIAGNOSIS: (Fig. 7.6) These are streamlined forms that, when mature, measure 6–20 mm and more excluding tails. Long axis of the head is more or less vertical. Antennae are no longer than twice the width of the head. Fore legs do not possess two rows of long hairs. Platelike gills are oriented dorsally on abdominal segments 1–7. Posterior abdominal segments have sharp posterolateral spines. Three tails are present.

ADULT DIAGNOSIS: (Figs. 7.7, 7.8) Wing vein CuA of the fore wings is attached to the hind margin of the wing by a series of veinlets. Vein MP of hind wings is forked near its base or unbranched. Two or rarely three tails are present.

DISCUSSION: The larvae of primitive minnow mayflies can be confused with those of the relatively common and similar small minnow mayflies (Figs. 7.13, 7.14). Small minnow mayflies, however, typically have longer antennae, and the posterolateral corners of abdominal segments 8 and 9 are not as well developed as is typical of primitive minnow mayflies. Brush-legged mayflies of the genus *Isonychia* are also similar in body shape to the primitive minnow mayflies, but the former group possesses distinct rows of long filtering hairs on the fore legs (Fig. 7.16).

About 50 species in two common and widespread genera, *Siphlonurus* and *Ameletus*, and five other less common genera occur in North America.

Aquatic habitats include various littoral areas of lentic waters, small streams and springs, and sandy-bottomed rivers. Larvae, although benthic, are generally excellent swimmers and sometimes may be seen swimming intermittently in clear waters. Many species opportunistically feed on other tiny aquatic insects when they are available.

Mature larvae leave the water to hatch to the subimago, and their cast skins can sometimes be found on rocks or debris a few inches above the waterline. Males of some species are not known, and these species are presumably parthenogenetic.

A few species, especially of *Siphlonurus*, are valuable to the fly fisherman. In the West, imitations of the Gray Drake, which usually refers to *Siphlonurus occidentalis*, can be productive in the early fall on quiet edgewaters. This species is also known by the names Brown Quill Spinner and Great Summer Drake. Other species of *Siphlonurus* in both western and eastern regions are of lesser importance. Interestingly, *Siphlonurus* larvae usually occur in lentic habitats or quiet edgewaters of streams; however, oviposition takes place in flowing waters, and hence the adults may be seen flying over such habitats.

SUBFAMILY SIPHLONURINAE: **Larvae** (Fig. 7.6) have their fore legs directed forward. The group is widespread in North America.

PRIMITIVE MINNOW MAYFLIES
(Siphlonuridae)

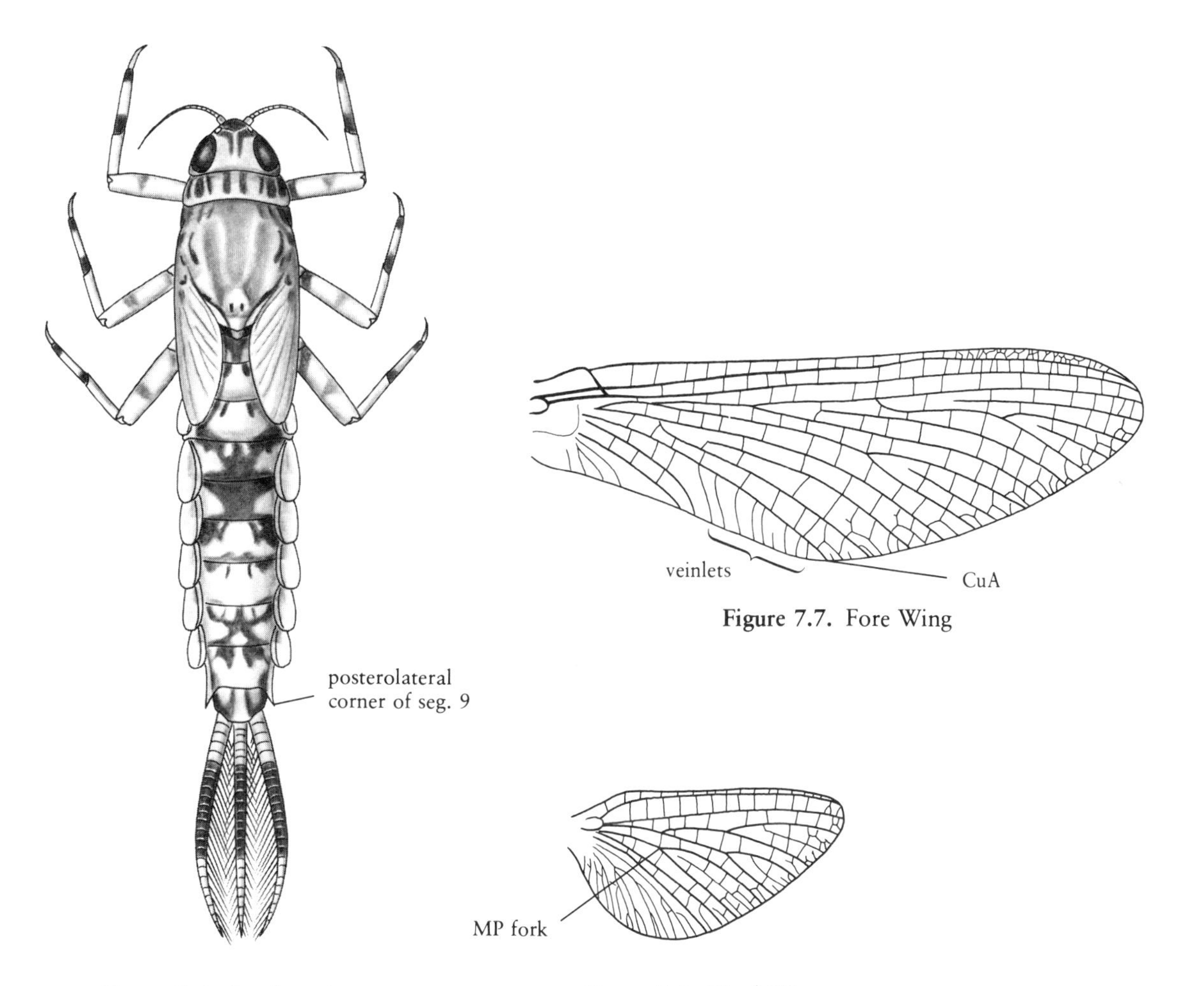

Figure 7.6. *Ameletus* larva

Figure 7.7. Fore Wing

Figure 7.8. Hind Wing

SUBFAMILY ACANTHAMETROPODINAE: **Larvae** have their fore legs directed posteriorly. Body is partially flattened dorsoventrally. These are rare sand dwellers, known only from central Canada and western, central, and southeastern states excluding Florida.

CLEFTFOOTED MINNOW MAYFLIES
(Family Metretopodidae)

LARVAL DIAGNOSIS: (Figs. 7.9–7.11) These are generally similar to the primitive minnow mayflies. When mature they measure 9–16 mm excluding tails. Claw of fore legs is cleft, and claws of middle and hind legs are longer than the tibiae of those legs.

ADULT DIAGNOSIS: (Fig. 7.12) These are generally similar to primitive minnow mayflies. In the fore wings, the MP_2 vein at its base is arched somewhat from the MP_1 vein, and two to four cubital intercalary veins are present. Two tails are present.

DISCUSSION: The unique claws (Figs. 7.9–7.11) of the larvae of cleftfooted minnow mayflies will readily distinguish them from other minnowlike mayflies. This small group consists of two little-known genera and

CLEFTFOOTED MINNOW MAYFLIES
(Metretopodidae)

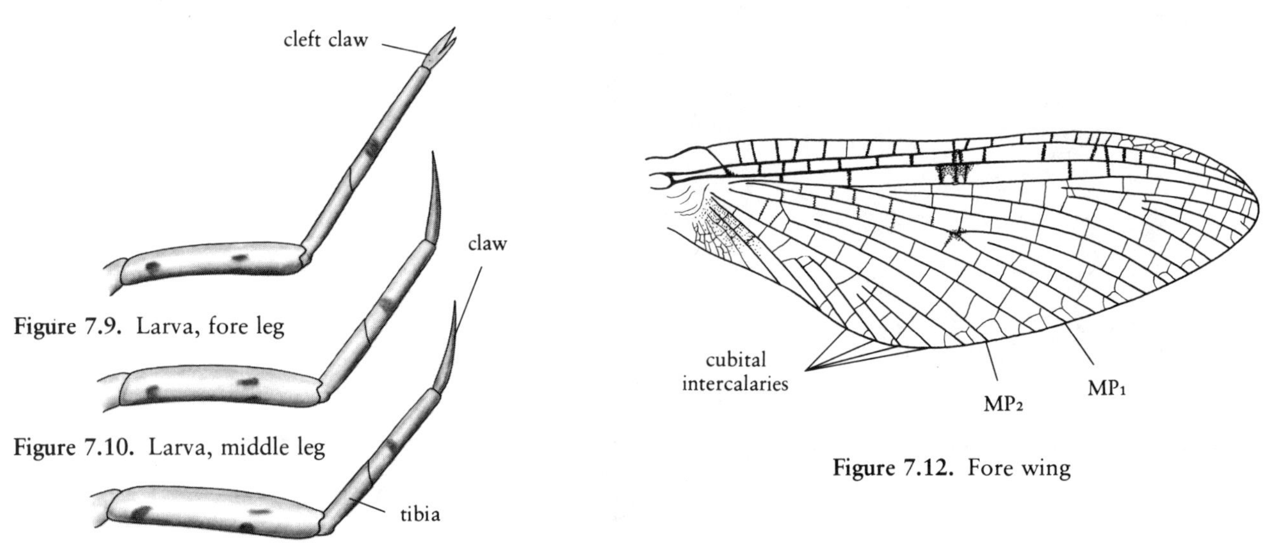

Figure 7.9. Larva, fore leg

Figure 7.10. Larva, middle leg

Figure 7.11. Larva, hind leg

Figure 7.12. Fore wing

about eight species in North America. Cleftfooted minnow mayflies are distributed across Canada and Alaska and also occur in the eastern half of the United States.

The larvae dwell primarily among vegetation in slow, shallow edges or swollen banks of large rivers and northern lakes. Adults are generally early season emergers at lower latitudes.

The Great Speckled Lake Olive (*Siphloplecton basale*) provides both nymphal and dun patterns for fishing lake waters of north-central and northeastern states and eastern Canada.

SAND MINNOW MAYFLIES
(Family Ametropodidae)

LARVAL DIAGNOSIS: (see key Fig. 7.5) These are generally similar to primitive minnow mayflies. The body, 14–18 mm when mature, is somewhat flattened. Fore legs are relatively small. Gills are fringed with short hairs. Three tails are present.

ADULT DIAGNOSIS: These are generally similar to primitive minnow mayflies. Vein A_1 of the fore wings is attached to the hind margin by several veinlets. Hind wings have a sharp costal projection.

DISCUSSION: Although the claws of the middle and hind legs of the larvae are long and slender as in the cleftfooted minnow mayflies, the claws on the fore legs are not cleft. The genus *Ametropus* is known locally from western states, central and western Canada, and rarely from northernmost areas of upper central states. Only about three species occur in North America.

The larvae are adapted to living in the clean, shifting sands of rivers, remaining partially buried when at rest.

SMALL MINNOW MAYFLIES
(Family Baetidae)

LARVAL DIAGNOSIS: (Figs. 7.13, 7.14; Plate I, Fig. 1) These usually have very streamlined bodies that measure 3–12 mm when mature. Head is vertically oriented, and antennae are usually longer than twice the head's

SMALL MINNOW MAYFLIES
(Baetidae)

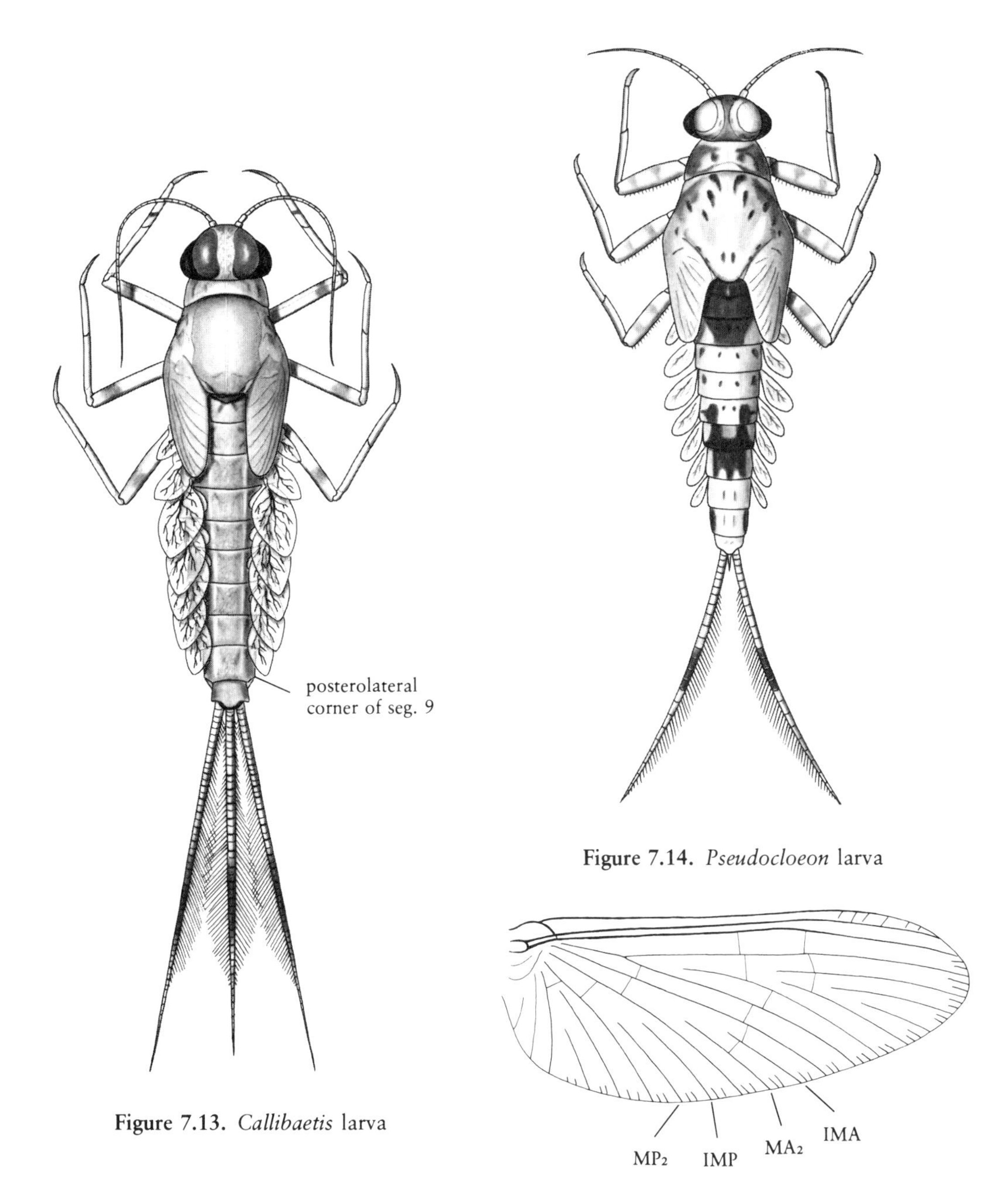

Figure 7.13. *Callibaetis* larva

Figure 7.14. *Pseudocloeon* larva

Figure 7.15. Fore wing

width. Hind wing pads are sometimes absent or minute. Platelike gills are present on abdominal segments 1–7, 1–5, or 2–7. Posterior abdominal segments usually lack posterolateral spines but rarely have moderately developed spines. Two or three tails are present.

ADULT DIAGNOSIS: (Fig. 7.15; Plate I, Figs. 2–4) Eyes of male have a unique turbinate portion. Veins IMA, MA_2, IMP, and MP_2 of the fore wings are not attached basally. Hind wings are reduced, veinless, or absent. Two tails are present.

DISCUSSION: Small minnow mayflies are most similar to some of the primitive minnow mayflies as larvae, although usually smaller; the length of antennae and the posterolateral corners of abdominal segments 8 and 9 should therefore be examined closely. Well over 100, mostly widespread species in 10 genera occur in North America.

Larvae live in a wide variety of lentic and lotic habitats, often constituting a major component of the bottom-dwelling community. Habitats of the very common genus *Baetis* include torrential mountain streams, warm, meandering rivers, and still waters of northern Canada. Habitats of the ecologically tolerant *Callibaetis* include clear ponds and lakes, warm desert springs, tide pools, and even sewage treatment ponds.

Riffle species generally orient head first into the current (positively rheotactic) and are able to maintain themselves in strong rapids. Small minnow mayflies are among the best swimmers and most fishlike of the aquatic insects. Larvae of one eastern species of the common genus *Pseudocloeon*, however, are atypically somewhat dorsoventrally flattened, as are all species of *Baetodes*, a genus that occurs primarily in southwestern states and Texas. These more flattened forms are generally slow crawlers and poor swimmers. Some small minnow mayfly larvae are among the most commonly taken species in drift samples, and *Baetis* species are well known for their consistent diel drift periodicity.

Adults sometimes swarm over roads or open fields near the place of emergence. The females of some species enter the water to oviposit eggs. Adults often characteristically wag their abdomen from side to side when at rest. Parthenogenesis is known in several species of *Baetis* and is especially well developed in some northern populations, where only females are known. Some *Cloeon* and *Centroptilum* are also parthenogenetic.

Five genera of Baetidae are important to fly fishermen: *Baetis*, *Callibaetis*, *Centroptilum*, *Cloeon*, and *Pseudocloeon*. At least nine species of *Baetis* (Plate I, Figs. 1–3) have been used as models for tied flies. Imitations of *Baetis tricaudatus* (Plate I, Figs. 2, 3) are used on trout streams throughout North America. This species (which can vary considerably in size and color depending on stream temperatures and time of emergence) goes by such names as the Light Blue Dun and Little Iron Blue Quill for the subimagos and by Light Rusty Spinner, Dark Rusty Spinner, and Rusty Spinner for adults. The small imitations of this species can be fished wet. The Minute Bluewinged Olive (*Baetis flavistriga*), which is also known as the Dark Bluewinged or Graywinged Olive, is another important species on eastern streams. Both of the above species have been known in the fishing literature by a number of scientific names that have recently been synonymized (see Appendix). The name Bluewinged Olive has generally been applied to several of the small subimagos of *Baetis* and *Pseudocloeon*.

Imitations of *Callibaetis*, both larval (Fig. 7.13) and winged, are excellent flies throughout the season for fishing lakes and ponds of the Rockies and western coastal states. A number of hatches can be expected, and larvae are often very abundant. *Callibaetis coloradensis* (the Little Specklewinged Quill, Speckled Spinner, or Specklewinged Dun or Spinner) is perhaps the most popular, although *Callibaetis pacificus* (Plate I, Fig. 4) (the Medium Specklewinged Quill) and *pallidus* (the Pale Specklewinged Sulphur) are also popular. Other species of *Callibaetis* are important in eastern and central regions.

BRUSHLEGGED MAYFLIES
(Oligoneuriidae)

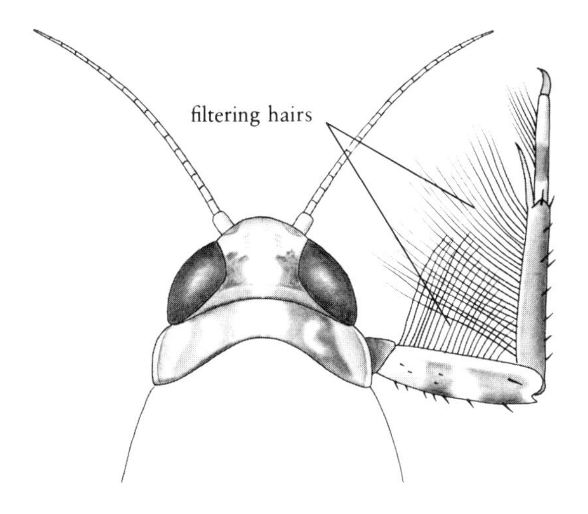

MP fork

Figure 7.17. *Isonychia* hind wing

Figure 7.16. *Isonychia* larva

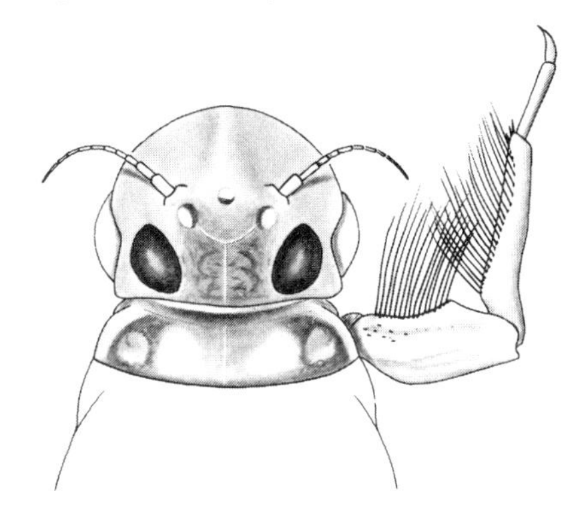
Figure 7.18. *Lachlania* larva

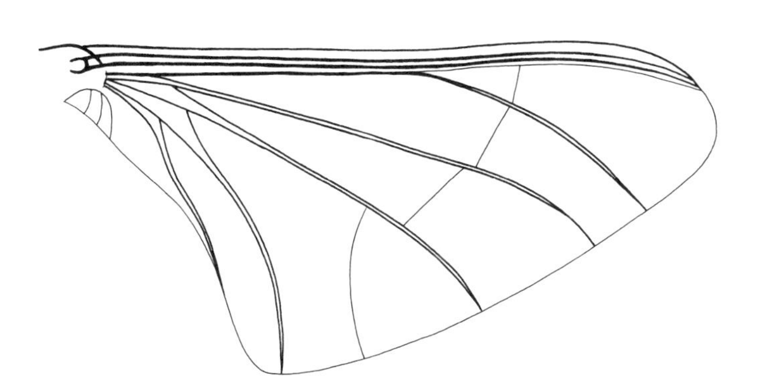
Figure 7.19. *Lachlania* fore wing

BRUSHLEGGED MAYFLIES
(Family Oligoneuriidae)

LARVAL DIAGNOSIS: (Figs. 7.16, 7.18; Plate I, Fig. 6) Mature individuals measure 8–17 mm excluding tails. Gill tufts are present at the base of the maxillary mouthparts. Two rows of conspicuous, long hairs are present along the inner surface of fore legs. Otherwise, the group comprises rather diverse larval forms that are either minnowlike with a vertically oriented head and three tails or flattened with a horizontally oriented head and two tails.

ADULT DIAGNOSIS: (Figs. 7.17, 7.19; Plate I, Fig. 5) There are two groups that differ markedly in appearance. Wing venation is highly reduced or similar to that of the primitive minnow mayflies except that vein MP of the hind wings is forked near the outer margin. Two or three tails are present.

DISCUSSION: The long rows of filtering hairs of the fore legs of the larvae are the most diagnostic feature of the brushlegged mayflies. The group is very interesting because in many characters it is intermediate between the minnowlike mayflies and the flatheaded mayflies.

This diverse family is represented by about 30 species and three genera in North America. Several species of the genus *Isonychia* are common and abundant, primarily in the eastern half of the continent.

The familiar minnowlike larvae of *Isonychia* (subfamily Isonychiinae) are usually found in riffle areas of streams and rivers, often among leaf packs and other detritus. Individuals act much like "fish out of water" when collected. They are good swimmers and are in part carnivorous.

The subfamily Oligoneuriinae includes both minnowlike and flattened, sprawling larvae, which inhabit sandy river bottoms. The adults of this group are very atypical mayflies in that they are very fast and elusive fliers that possess highly modified wings.

Three species of *Isonychia* are important to eastern and central fly fishermen: *Isonychia bicolor* (Plate I, Figs. 5, 6), *harperi,* and *sadleri.* The three species are closely related morphologically and ecologically. Names applied to the subimagos of *bicolor* include Dun Variant, Great Leadwing Coachman, Large Mahogany Dun, Leadwing Coachman, Mahogany Dun, and Slate Drake. Some of these names have also been used interchangeably for the other mentioned species. The name Whitegloved Howdy refers to the adults of all three species. The large larval imitations of these species are excellent for nymph fishing on moderate to swift waters. Dun imitations are also very productive during hatches, which commonly occur in both early and late summer.

SUBFAMILY ISONYCHIINAE: **Larvae** (Fig. 7.16; Plate I, Fig. 6) possess gill tufts at the bases of fore legs. First pair of gills is dorsal. The group is widespread in North America but most common in eastern and central states and eastern and central Canada.

SUBFAMILY OLIGONEURIINAE: **Larvae** (Fig. 7.18) lack gill tufts at the bases of fore legs. First pair of gills is ventral. Species are somewhat rare, and the group is *not* known from eastern Canada, western Canada and Alaska, or northeastern states.

FLATHEADED MAYFLIES
(Family Heptageniidae)

LARVAL DIAGNOSIS: (Figs. 7.20, 7.22–7.32; Plate II, Fig. 8) These are generally flattened and measure 5–20 mm excluding tails when mature. They have sprawling legs and a horizontally oriented head. Gills are present on abdominal segments 1–7 or rarely on 1–6. All or most of the gills have tufts at their base. Two or three tails are present.

ADULT DIAGNOSIS: (Fig. 7.21; Plate II, Figs. 7, 9) Two pairs of cubital intercalary veins are present in fore wings. Two tails are present.

DISCUSSION: The larvae of flatheaded mayflies may superficially resemble some of the pronggills or even spiny crawlers that are somewhat flattened. Nonetheless, there are fundamental differences in the gill structure between these groups. Single platelike gills (usually with basal tufts) on all anterior abdominal segments are typical of the flatheaded mayflies.

This is a large, widespread, and important group of aquatic insects. The vast majority of the more than 100 species belong to the subfamily Heptageniinae in North America.

Heptageniinae larvae are common and occur in a variety of lotic habitats and in the shallow littoral areas of lakes. One species of *Heptagenia*, however, has been found at depths of 25–100 meters in Lake Superior. Species occurring in mountainous streams may be found clinging

FLATHEADED MAYFLIES
(Heptageniidae)

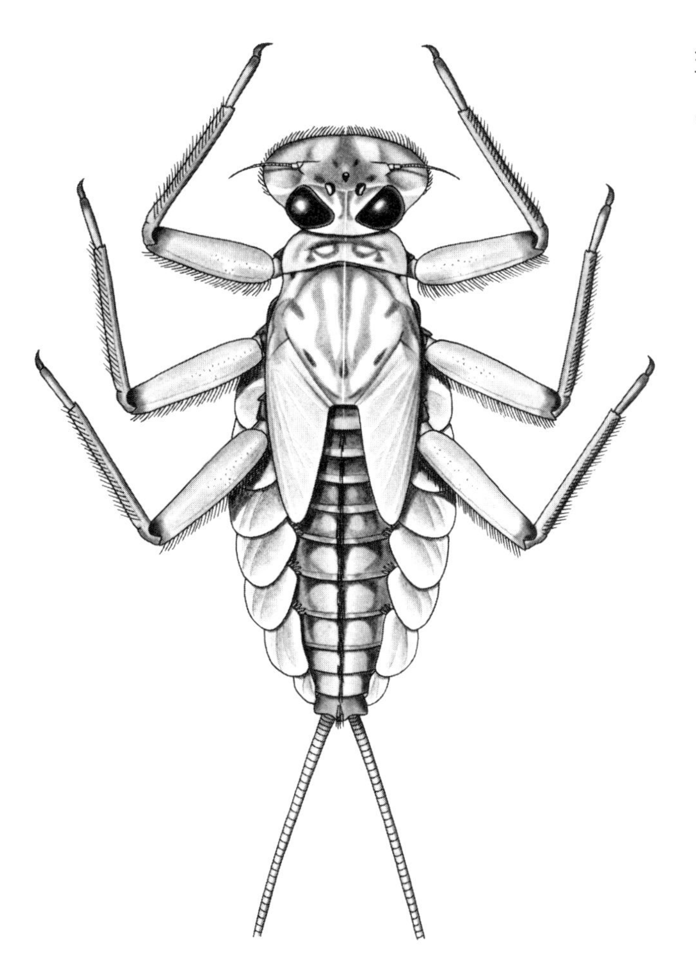

Figure 7.20. *Epeorus* (Heptageniinae) larva

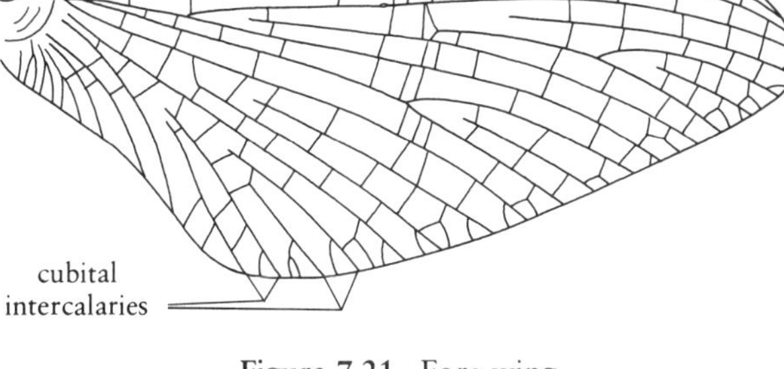

Figure 7.21. Fore wing

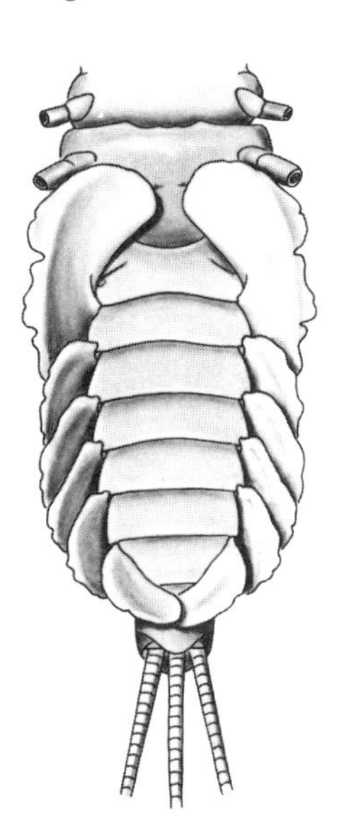

Figure 7.22. *Rhithrogena* (Heptageniinae), larval abdomen (ventral)

to various substrates in rapids or moderately fast waters. Others, including many *Stenonema*, are often found under flat stones or among detritus of moderately flowing rivers and quiet streams. Some occur on gravelly substrates of rivers and streams.

Other subfamilies are much rarer. The Arthropleinae are known from ponds and bogs, often found clinging to vegetation. The Pseudironinae are known from sandy bottoms of streams and rivers. The Anepeorinae are known from sand and gravel substrates of large rivers. Little is yet known of the big-river environment of the Spinadinae.

Heptageniinae larvae are mostly sprawlers, and graze on the algal and detrital material found in association with the substrate. A few have gills adapted to form a ventral attachment or friction disc, such as in *Epeorus* and *Rhithrogena* (Fig. 7.22). Subfamilies other than Heptageniinae are omnivorous or primarily predators. All members of the family, with the exception of *Pseudiron*, are relatively poor swimmers.

Certain species transform to the subimago under water, and the subimago must swim or float to the surface. Hatches of these species thus can induce tremendous feeding activity in trout. Adults of many species are relatively long-lived by mayfly standards. When not flying, adults often walk sideways or backwards if they are disturbed. The color and size of adults of some species vary considerably, depending on the geographic region and/or the time of season that emergence takes place.

This is an important family for fly fishermen, since some 45 or more species in 10 different genera of Heptageniinae have provided patterns for flies. Nymph, emerger, and dun patterns of flatheaded mayflies are the most productive at times when subimagos are hatching under water. Some of the best fishermen's mayflies are in the trout stream genus *Epeorus* (some species have previously been placed under the generic name *Iron*). Particularly noteworthy in the West are *Epeorus albertae* (variously known as the Graywinged Pink Quill, Pink Lady, Salmon Spinner, Slate Cream Dun, and Western Graywinged Yellow Quill) and *Epeorus longimanus* (Plate II, Fig. 7) (variously known as the Medium Blue Quill, Quill Gordon, Red Quill Spinner, Slate Brown Dun, and Western Gordon Quill). A popular eastern species is *Epeorus pleuralis* (variously known as the Quill Gordon, Gordon Quill, Dark Gordon Quill, Iron Dun, Light Blue Quill, and Red Quill Spinner).

A brown red fly at morning grey,
A darker dun in clearer day
George Washington Bethune

Rhithrogena is also a trout stream genus and is important throughout North America: for example, *Rhithrogena hageni* (commonly known as the Western Red Quill) of the West, and *Rhithrogena pellucida* (the Dark Blue Upright) of the East.

In eastern and central regions, species of the genera *Stenonema* and *Stenacron* can be very important; these genera occur both in trout waters and in warmer waters that do not support trout. The familiar name Light Cahill applies primarily to *Stenacron interpunctatum* (Plate II, Fig. 9). Since several other scientific names that were previously applied to this species have recently been synonymized, the reader should consult the Appendix. Important *Stenonema* species include, among others, the northeastern species *Stenonema vicarium* and *ithaca*. Both have been referred to by a number of fishermen's names. Perhaps the most common are Ginger Quill or March Brown for *vicarium* and Light Cahill, Ginger Quill, or Gray Fox for *ithaca*. Again, the Appendix should be consulted for a complete cross index of names.

SUBFAMILY HEPTAGENIINAE: **Larvae** (Figs. 7.20, 7.22–7.24; Plate II, Fig. 8) have gills that originate dorsally or laterally, but sometimes extend ventrally to form a ventral disc. Two or three tails are present. The group is widespread in North America.

SUBFAMILY ARTHROPLEINAE: **Larvae** (Figs. 7.25, 7.26) have conspicuously long palps that extend beyond the margin of the head as seen from above. The abdomen possesses a series of large leaflike gills. Three tails are present. The group is known only from central and eastern Canada and northeastern and central states.

SUBFAMILY PSEUDIRONINAE: **Larvae** (Figs. 7.27, 7.28) have legs that are all long, slender, and posteriorly directed. Gills are narrow and pointed. Three tails are present. These are rare, and the group is *not* known from western Canada and Alaska, eastern Canada, or northeastern states.

SUBFAMILY ANEPEORINAE: **Larvae** (Figs. 7.29, 7.30) have ventral gills, but conspicuous filamentous gill tufts are usually visible from above. Three tails are present. These are rare, and the group is known only from scattered

FLATHEADED MAYFLIES
(Heptageniidae)

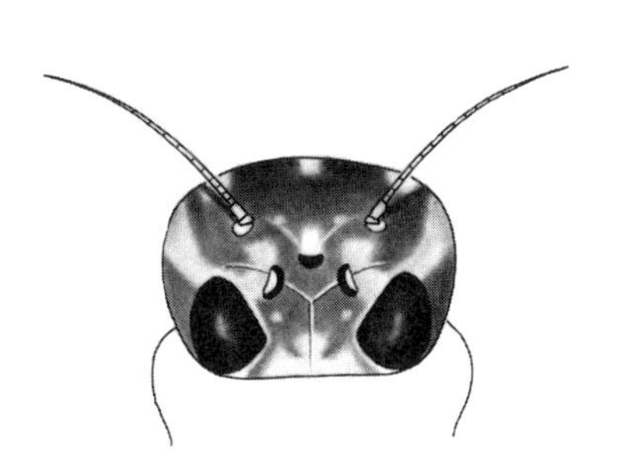

Figure 7.23. Heptageniinae larva, dorsal head

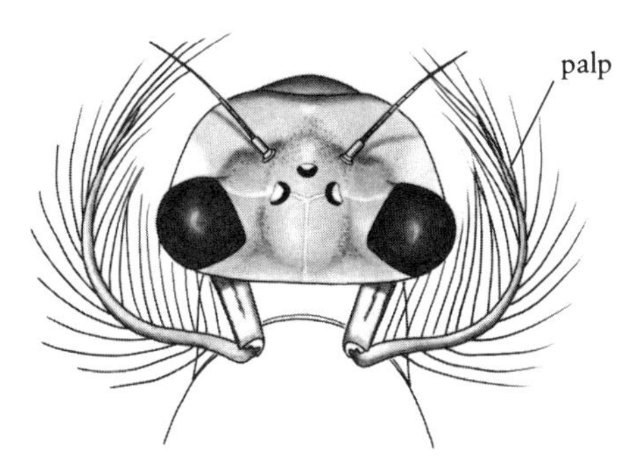

Figure 7.25. Arthropleinae larva, dorsal head

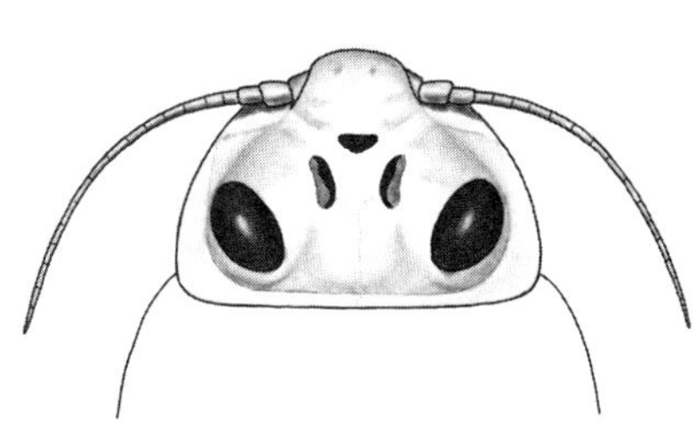

Figure 7.27. Pseudironinae larva, dorsal head

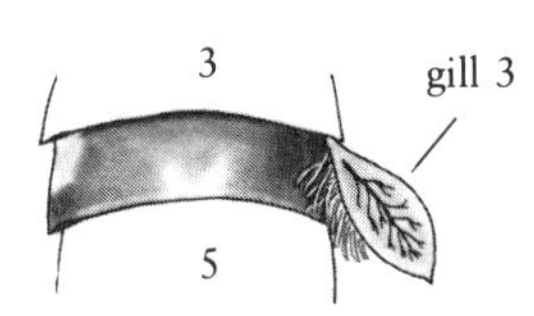

Figure 7.24. Heptageniinae larva, dorsal abdomen (in part)

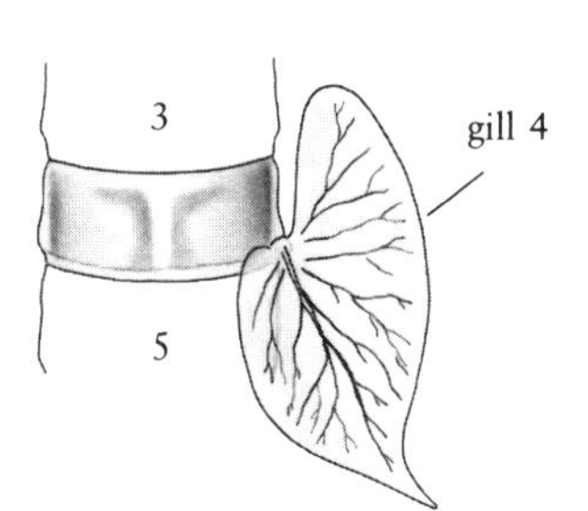

Figure 7.26. Arthropleinae larva, dorsal abdomen (in part)

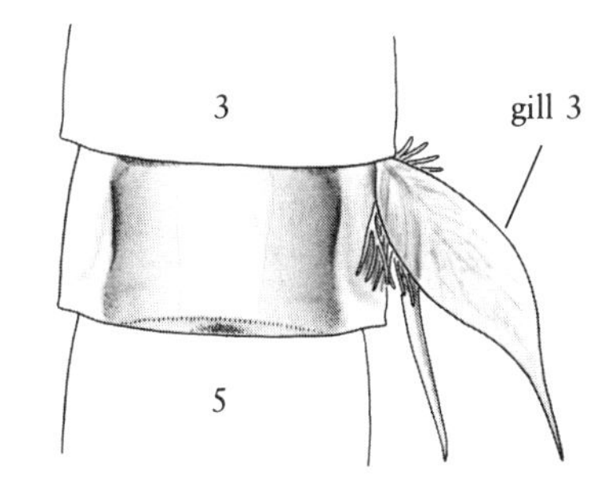

Figure 7.28. Pseudironinae larva, dorsal abdomen (in part)

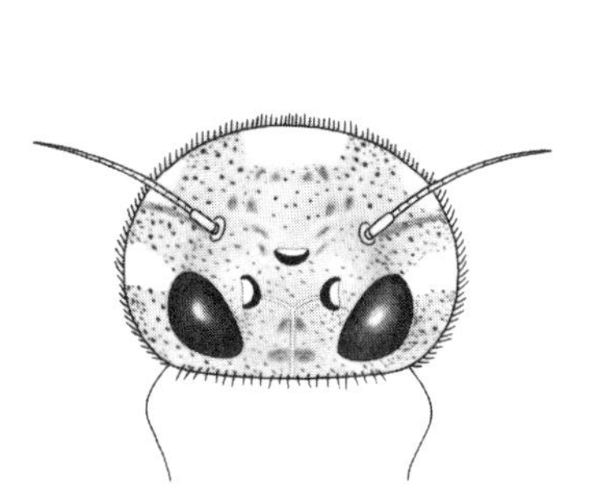

Figure 7.29. Anepeorinae larva, dorsal head

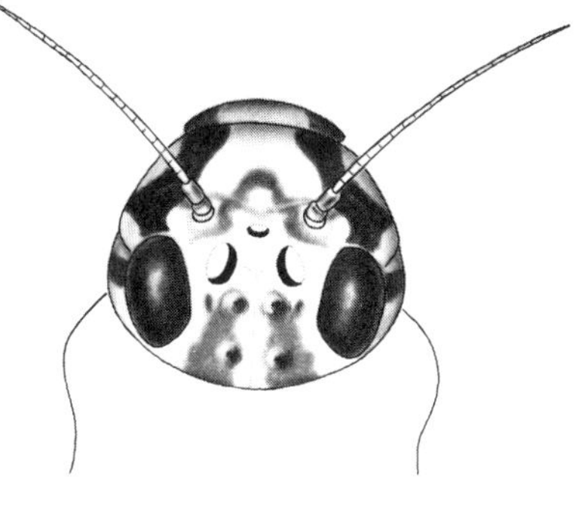

Figure 7.31. Spinadinae larva, dorsal head

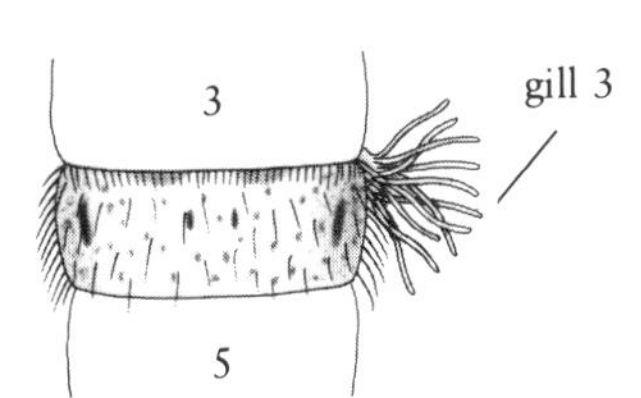

Figure 7.30. Anepeorinae larva, dorsal abdomen (in part)

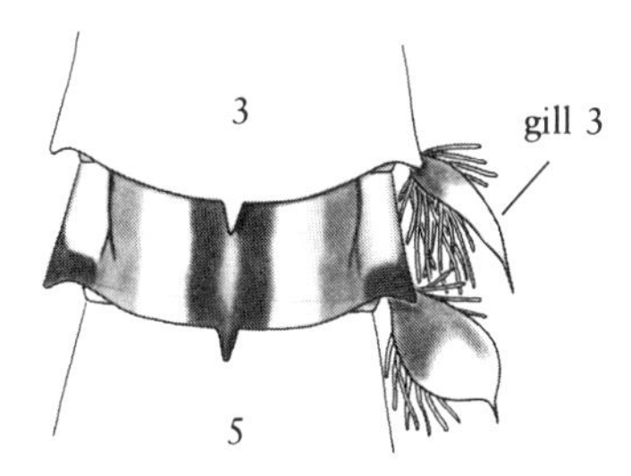

Figure 7.32. Spinadinae larva, dorsal abdomen (in part)

localities in central Canada and northwestern, central, and southeastern states excluding Florida.

SUBFAMILY SPINADINAE: **Larvae** (Figs. 7.31, 7.32) have conspicuous, blunt, dorsal spines on the body. Proceeding from anterior to posterior, gills become increasingly more platelike and lateral. Two tails are present. These are very rare, and the group is known only from central and southeastern states excluding Florida.

SUPERFAMILY LEPTOPHLEBIOIDEA

PRONGGILLS
(Family Leptophlebiidae)

LARVAL DIAGNOSIS: (Figs. 7.33, 7.35–7.40; Plate II, Fig. 10) Body is 4–15 mm in length when mature. These are somewhat cylindrical to flattened forms. Long axis of head ranges from vertical to horizontal. Gills are present on abdominal segments 1–6 or 1–7, are double or forked on segments 2–6, and are variously pointed or bear fingerlike projections. Three tails are present.

ADULT DIAGNOSIS: (Fig. 7.34; Plate II, Fig. 11) One or two pairs of main cubital intercalary veins are present in fore wings. Three tails are present.

DISCUSSION: The larvae of those pronggills that have somewhat flattened bodies may sometimes be mistaken for flatheaded mayflies, but their mandibles are conspicuously visible from above when viewing the head. Some spiny crawlers and pronggills appear superficially alike, but they are in different suborders and have distinctly different gill structures and wing pads. A few western species of pronggills possess tusks reminiscent of those of burrowing mayflies.

Although there are only about 70 species in North America, the pronggills are a large and sometimes dominant group of mayflies in many other areas of the world, where they are extremely diverse in form and ecology.

Larvae are often associated with porous rocks, gravel, woody debris, or rooty banks of streams. Some species occur in still waters, even temporary marshes or ditches during certain times of the year. Interestingly, larvae of some *Paraleptophlebia* feed on stream-bottom detritus during the day and on more exposed algae during the night.

Pronggills are awkward swimmers, but some species of *Leptophlebia* actively crawl upstream to seek out small tributaries of streams before hatching. After hatching from these areas, adult females return to the main stream to oviposit, thus completing a migratory cycle.

Early spring nymph fishing with larval imitations of *Leptophlebia* (Fig. 7.33) can be especially productive along the edges of medium-sized streams and rivers. Dry fly fishing with imitations of the Early Brown Spinner (adult *Leptophlebia cupida*), the Jenny Spinner (adult *Leptophlebia johnsoni*), or Borcher's Drake (*Leptophlebia nebulosa*) also can provide early-season success on medium-sized streams.

Paraleptophlebia (Plate II, Figs. 10, 11) must be considered an important trout stream genus because at least 13 species have served as models for dun and spinner imitations. Two noteworthy species in eastern and central North America are *Paraleptophlebia adoptiva* (variously known as the Blue Dun, Blue Quill, Dark Blue Quill, Dark Brown Spinner, Early Blue Quill,

PRONGGILLS
(Leptophlebiidae)

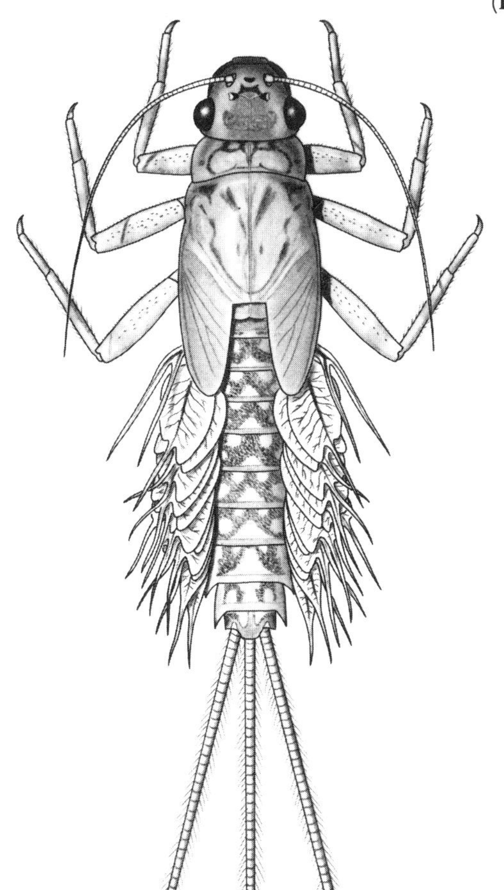

Figure 7.33.
Leptophlebia larva

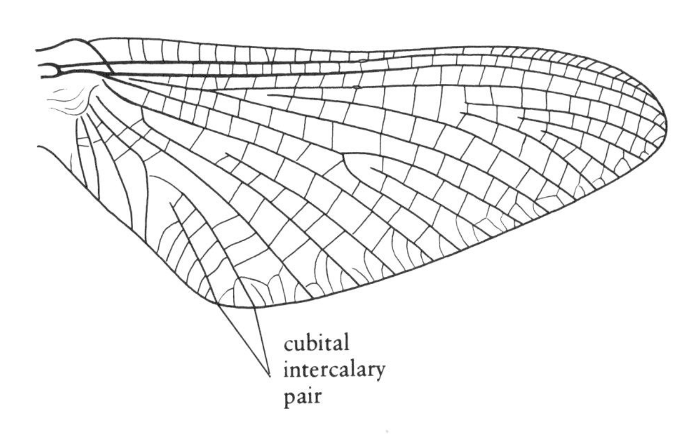

Figure 7.34. Fore wing

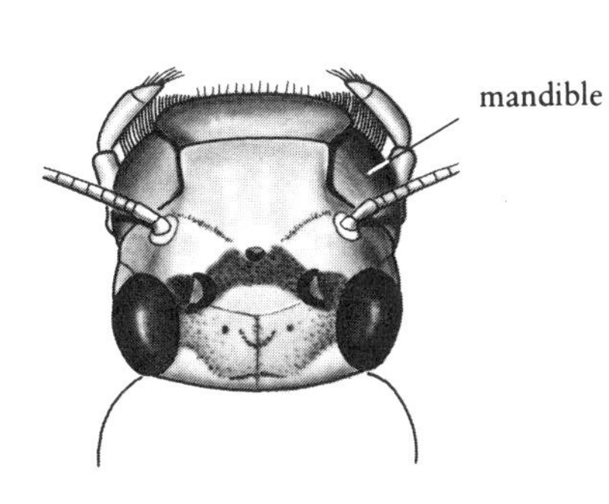

Figure 7.35. *Thraulodes* larva, dorsal head

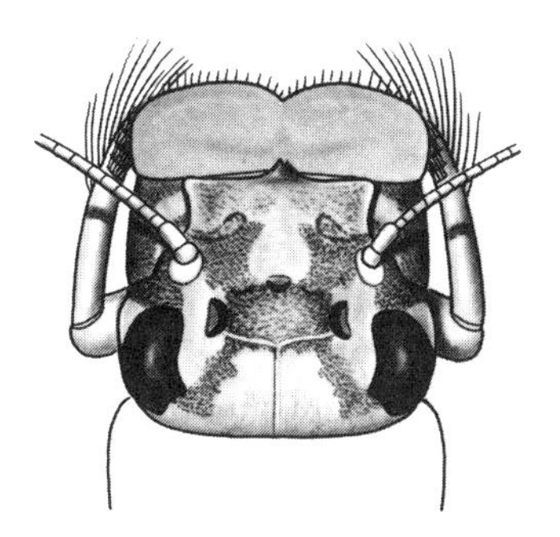

Figure 7.37. *Traverella* larva, dorsal head

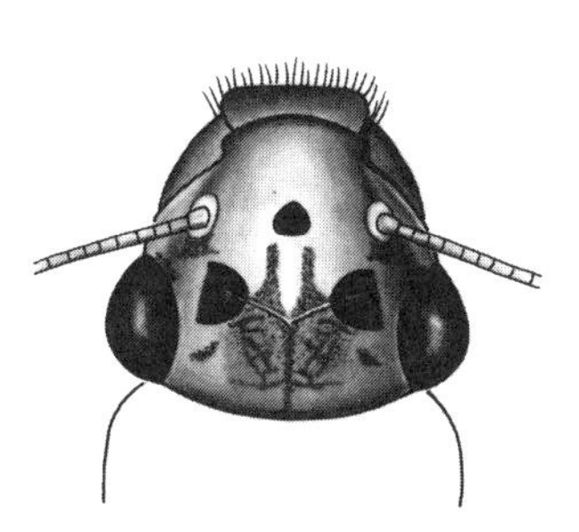

Figure 7.39. *Paraleptophlebia* larva, dorsal head

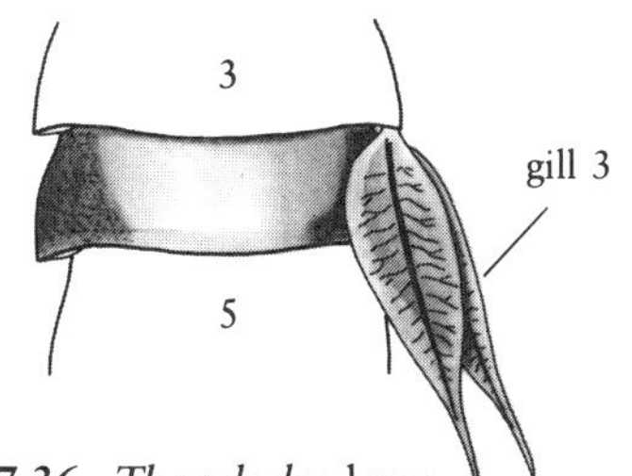

Figure 7.36. *Thraulodes* larva, dorsal abdomen (in part)

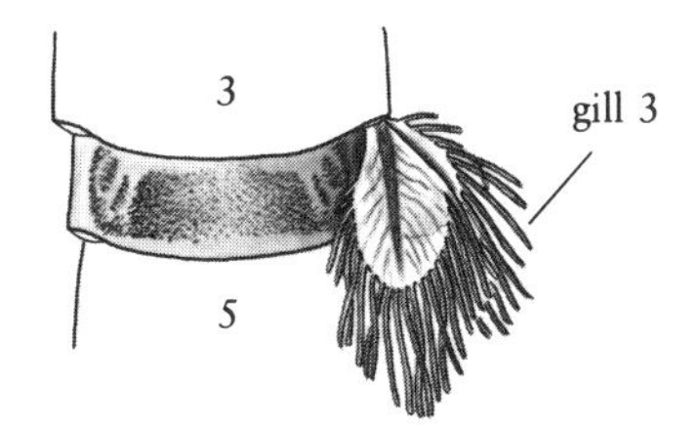

Figure 7.38. *Traverella* larva, dorsal abdomen (in part)

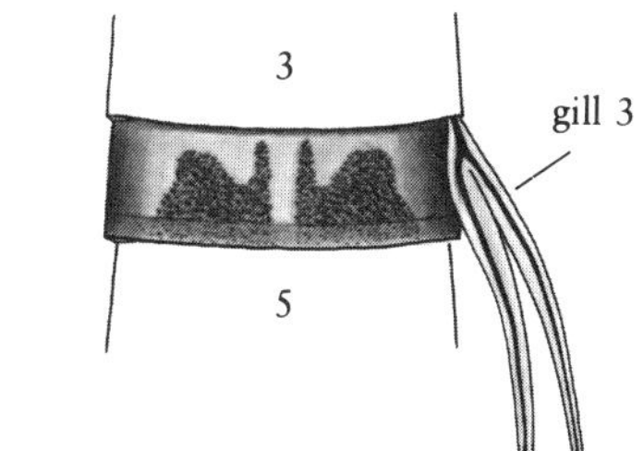

Figure 7.40. *Paraleptophlebia* larva, dorsal abdomen (in part)

TUSKLESS BURROWER
(Behningiidae)

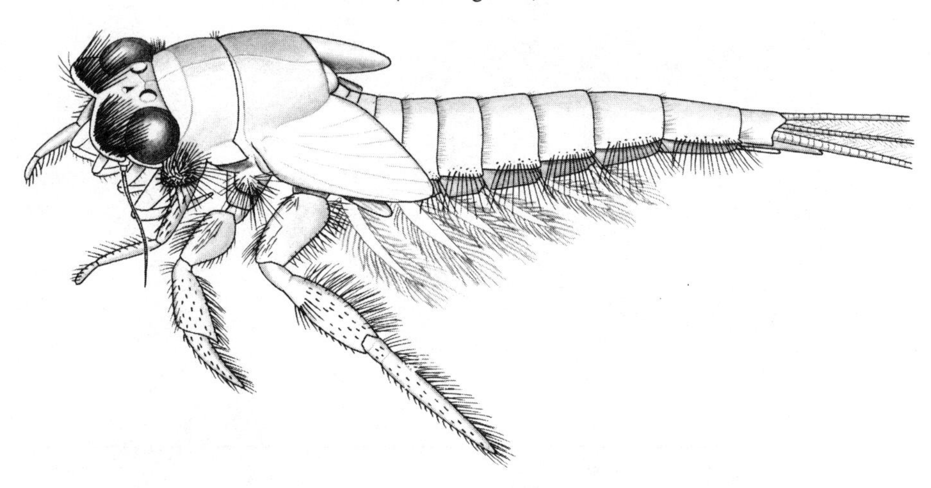

Figure 7.41. *Dolania* larva

Iron Blue Dun, or Slatewinged Mahogany Dun) and *Paraleptophlebia mollis* (variously known as the American Iron Blue Quill, Dark Blue Quill, Dark Brown Spinner, or Jenny Spinner). Several of the western species are also known by the names Dark Blue Quill and Dark Brown Spinner.

BURRROWING AND TUSKED MAYFLIES (Superfamily Ephemeroidea)

Larvae possess gills on abdominal segments 2–7 that are double, elongate, and fringed with slender filaments. All but one North American species possess tusks that protrude forward from the head.

TUSKLESS BURROWERS (Family Behningiidae)

LARVAL DIAGNOSIS: (Fig. 7.41) These are approximately 13 mm excluding tails when mature. Head possesses two conspicuous crowns of bristles, as does the pronotum. Legs are highly modified and have no claws. Gills on abdominal segments 2–7 are ventral. Three tails are present.

ADULT DIAGNOSIS: Veins MP and CuA of the fore wings originate from vein CuP. All legs are feeble. Middle tail is poorly developed.

DISCUSSION: *Dolania americana* is the only North American species of this very unusual group of mayflies. It is known locally only from southeastern states including Florida. Larvae burrow into clean, shifting sands of rivers. The species has an unusual two-year life cycle in that most of the first year is spent in the egg stage.

HACKLEGILLS (Family Potamanthidae)

LARVAL DIAGNOSIS: (Fig. 7.42; Plate III, Fig. 12) These have a flattened, sprawling body and measure up to 8–15 mm excluding tails. Tusks generally curve inward and possess scattered dorsal spines. Legs are outspread and gills laterally oriented. Three tails are present.

ADULT DIAGNOSIS: (Fig. 7.43) Veins MP_2 and CuA of the fore wings are arched at their bases away from vein MP_1. Vein A_1 of the fore wings is

HACKLEGILLS
(Potamanthidae)

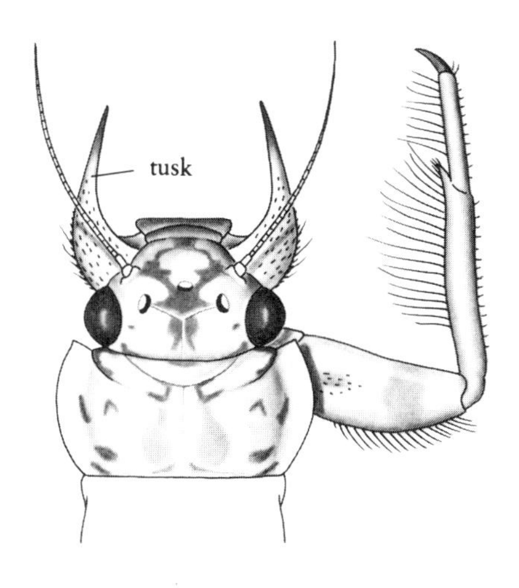

Figure 7.42. *Potamanthus* larva

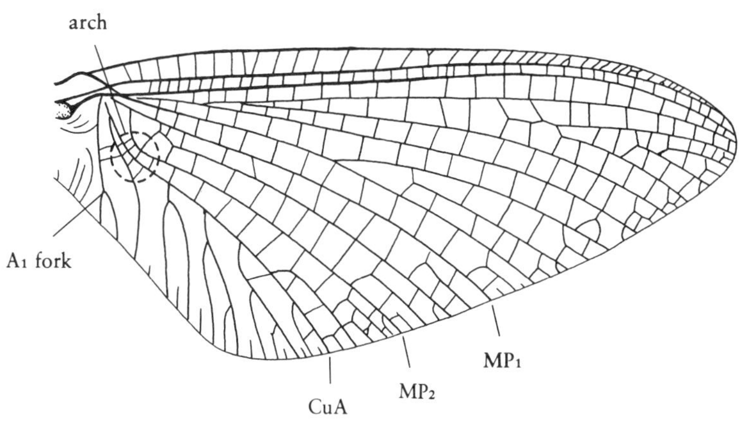

Figure 7.43. Fore wing

forked. Hind wings have a distinct costal projection. Three tails are present.

DISCUSSION: Larval hacklegills are superficially similar to sprawling flatheaded mayflies in form and sometimes habit. Gills, however, are fundamentally different, and the flatheaded mayflies never possess tusks. The single genus *Potamanthus* consists of about eight species and is distributed throughout the eastern half of North America excluding Florida.

Hacklegills occur in rivers and moderately sized streams, where they sprawl on rocks, in mixed substrate, or on woody debris, often in relatively shallow water. They generally do not burrow, but the very young larvae are sometimes found within the substrate.

Species in this family that have been imitated by fishermen's flies include *Potamanthus rufous* (Plate III, Fig. 12), *diaphanus*, and *distinctus*. The last species, which is variously known as the Cream Dun, Cream Variant, Evening Dun, Golden Drake, Golden Spinner, Pale Golden Drake, or Yellow Drake, has proved to be an important mayfly on northeastern trout waters.

PALE BURROWERS
(Family Polymitarcyidae)

LARVAL DIAGNOSIS: (Figs. 7.44–7.46) These are cylindrical burrowing forms. They measure 12–35 mm excluding tails when mature. Tusks are variously spined and curve inward and downward at the tips. Fore legs are modified for digging. Gills are dorsally oriented. Three tails are present.

ADULT DIAGNOSIS: (Fig. 7.47) These have a light, patternless body and somewhat cloudy wings. Veins MP_2 and CuA of the fore wings are arched at their bases away from vein MP_1. At least the middle and hind legs and sometimes the fore legs are poorly developed. Two or three tails are present.

DISCUSSION: Larvae of pale burrowers can be distinguished from those of other, similar burrowers by their hind tibiae, among other traits. The hind tibiae do not end in a distinct process as is the case with the common or spinyheaded burrowers. Each of the head and tusk types found among the pale burrowers is unique.

PALE BURROWERS
(Polymitarcyidae)

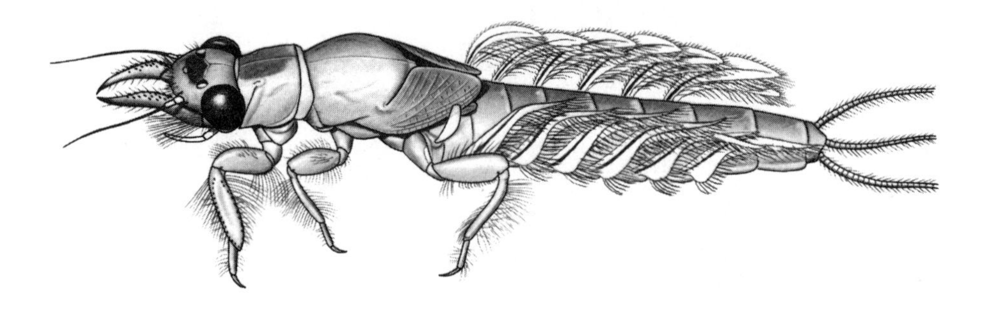

Figure 7.44. *Ephoron* larva

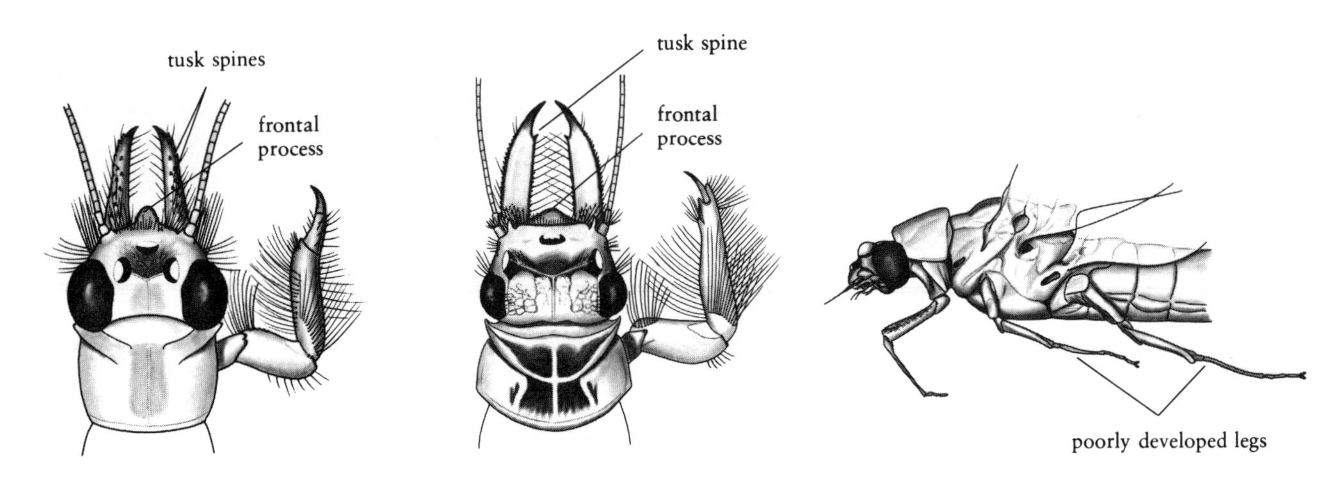

Figure 7.45. Polymitarcyinae larva

Figure 7.46. Campsurinae larva

Figure 7.47. *Ephoron* adult

Three genera and six species of pale burrowers occur in eastern and central regions of North America. Only one of the species, *Ephoron album*, is commonly found in parts of the West.

Larvae occur in silt, silt-gravel mixtures, or clay bottoms and banks of streams, rivers, and lakes. Synchronized mass emergences are typical of pale burrowers, and the winged stages (known by some as "willowflies" or "white mayflies") may be extremely short-lived and have functionless legs. Interestingly, the females of some species never molt to the adult stage but are sexually mature as subimagos.

SUBFAMILY POLYMITARCYINAE: **Larval** head (Fig. 7.45) possesses a narrow, rounded frontal process. Tusks have scattered spines. The single genus *Ephoron* is relatively widespread but is not known from Texas or Florida.

SUBFAMILY CAMPSURINAE: **Larval** head (Fig. 7.46) does not possess a distinct narrow frontal process. Tusks possess spines only along their inner margin. These are rarely encountered as larvae, but adults are occasionally abundant. The group is known from central Canada, Texas, and central and southeastern states including Florida.

He sets, and each ephemeral insect then
Is gathered into death without a dawn

Percy Bysshe Shelley

COMMON BURROWERS
(Family Ephemeridae)

LARVAL DIAGNOSIS: (Figs. 7.48–7.50; Plate III, Fig. 13) These are typical burrowing forms with the body measuring up to 12–32 mm. Tusks lack spines or possess only basal spines. Tusk tips curve upward and outward. Three tails are present.

ADULT DIAGNOSIS: (Fig. 7.51; Plate III, Figs. 14, 15) These have large bodies, and many have somewhat patterned wings. Fore wing venation is similar to that of hacklegills, except that vein A_1 of the fore wings is simple and attached to the hind margin by two or more veinlets. Two or three tails are present; the middle tail, when present, is as long as the body or longer.

DISCUSSION: The tusk type of the larvae of common burrowers will distinguish them from other tusked burrowers. These are the best known and most widespread of the burrowing mayflies in North America, where about 13 species occur.

Larvae generally burrow in silt, silt-marl, or silt-sand substrates in rivers, streams, ponds, and lakes. They are particle feeders, and at least some are thought to ingest sediment nonselectively. They may leave their burrows at night. Individuals of *Hexagenia* are excellent subjects for laboratory experimentation because they are relatively large and can be successfully handled and raised in rearing tanks and aquaria.

The numbers of individuals and the total biomass of common burrowers in some central and eastern North American freshwater habitats can be very large in comparison with those of most other aquatic insects. Spectacular hatches are often common along large rivers and lakes.

Often very colorful, the common burrowers are also the biggest and sometimes the best trout flies on eastern and central waters. *Ephemera guttulata* (Plate III, Fig. 15), better known as the Green Drake in the subimaginal stage or the Coffinfly in the adult stage, induces tremendous trout feeding activity on eastern streams during hatch periods (especially June). *Ephemera simulans* (the Brown Drake or Chocolate Dun) (Fig. 7.48) is a much more widespread species and is more typical of slower waters. *Ephemera varia* (variously known as the Cream Variant, White Dun, Yellow Drake, or Yellow Dun) is another important eastern trout stream species.

Litobrancha recurvata (the Brown Drake, Dark Green Drake, Drakefly, or Great Dark Green Drake) occurs primarily in northeastern states and eastern Canadian trout streams. Several species of *Hexagenia* have provided patterns for fly fishermen. These species, however, are also common in warmer waters that do not support trout. The best use of these imitations is on larger streams, over pools, or on pond and lake waters. The best known species is *Hexagenia limbata* (Plate III, Figs. 13, 14), also known as the Burrowing Mayfly, Fishfly, Giant Michigan Mayfly, Great Olivewinged Drake, Michigan Caddis, Michigan Spinner, and Sandfly. The larva of this particular mayfly is sold as bait in many states, and is a favorite of north-central ice fishermen. Bait fishing with mayflies is sometimes

COMMON BURROWERS
(Ephemeridae)

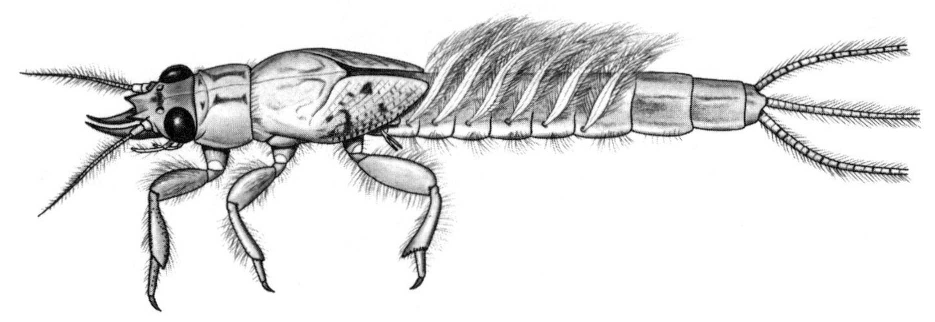

Figure 7.48. *Ephemera* larva

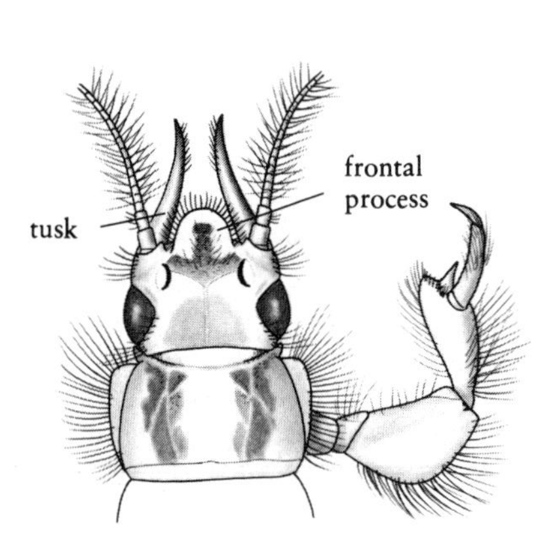

Figure 7.49. *Hexagenia* larva

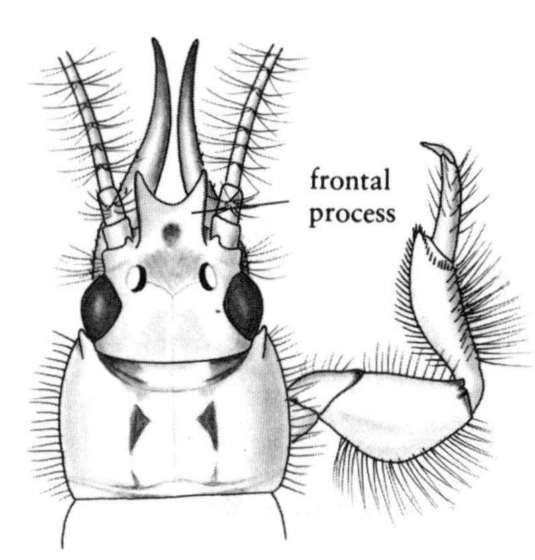

Figure 7.50. *Ephemera* larva

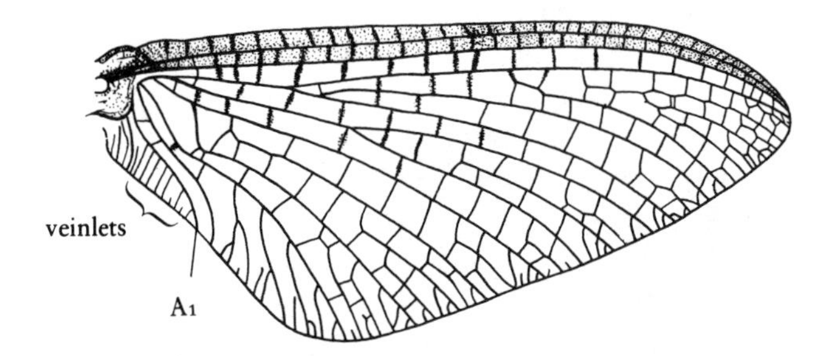

Figure 7.51. Fore wing

known as "dapping the mayfly"; however, to most fly fishermen, "dapping" means repeatedly dropping a dry fly onto the water surface directly from the rod without casting.

SPINYHEADED BURROWERS (Family Palingeniidae)

LARVAL DIAGNOSIS: (Fig. 7.52) These are generally similar to common burrowers. Tusks and fore legs possess a distinct outer ridge with small teeth and spines. Three tails are present.

ADULT DIAGNOSIS: (Fig. 7.53) These are generally similar to common burrowers. Vein A_1 of fore wings has no more than three veinlets. Male pronotum is very short. Middle tail is distinctly shorter than the body (more so among females).

SPINYHEADED BURROWERS
(Palingeniidae)

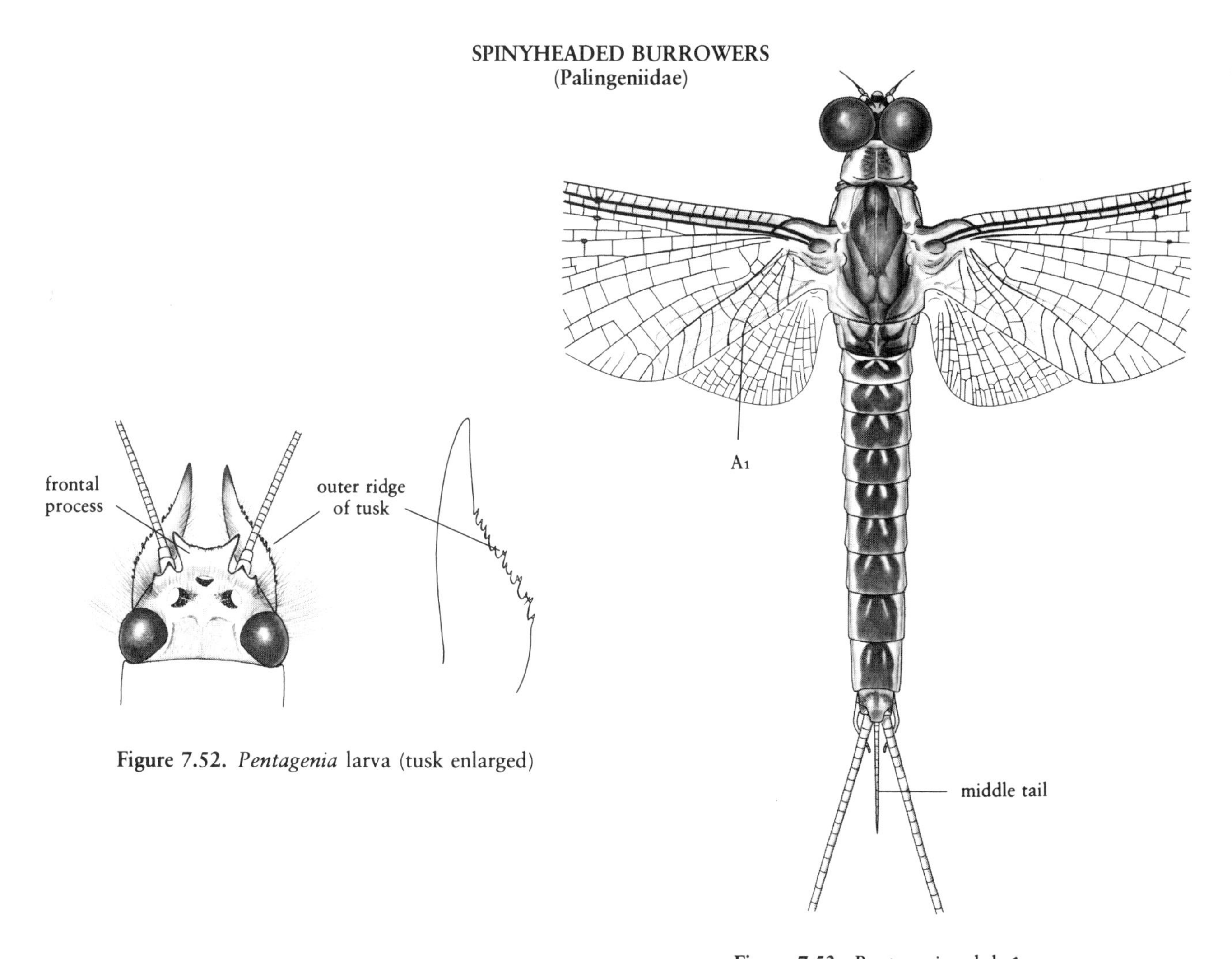

Figure 7.52. *Pentagenia* larva (tusk enlarged)

Figure 7.53. *Pentagenia* adult ♂

DISCUSSION: This is primarily an Old World family and is represented in North America by only two described species of the genus *Pentagenia*. Larvae are rarely encountered. Adults are sometimes common in central Canada, Texas, and central and southeastern states including Florida. Larvae are big-river burrowers, and in Texas, for example, may be the predominant insects in some silt- and clay-bottomed rivers.

Fusedback Mayflies (Suborder Pannota)

Larvae have fore wing pads fused with the thorax for at least half of the pad length. Thorax generally appears robust. Gill series are reduced, lay upon lateral shelves of the abdomen, and/or are covered by enlarged operculate gills or a thoracic shield. These are usually relatively slow and inactive forms, but may become quite active before transforming to the subimago.

Mark well the various seasons of the year,
How the succeeding insect race appear;
In this revolving moon one colour reigns,
Which in the next the fickle trout disdains.

John Gay

CRAWLERS
(Superfamily Ephemerelloidea)

SPINY CRAWLERS
(Family Ephemerellidae)

LARVAL DIAGNOSIS: (Figs. 7.54–7.57; Plate III, Fig. 16, Plate IV, Fig. 18) Body measures 5–15 mm excluding tails when mature and is somewhat cylindrical or flattened. Long axis of head tends to be vertically oriented. Blunt to sharp spines are often present dorsally on head, thorax, and/or abdomen. Abdominal segments usually have lateral spines. Gills are never present on abdominal segment 2. Platelike gills of abdominal segments 3–7 or 4–7 lie upon the abdomen. Three tails are present.

ADULT DIAGNOSIS: (Fig. 7.58; Plate III, Fig. 17, Plate IV, Fig. 19) Series of short intercalary veins present along the outer margin of the fore wings are not attached to other veins.

DISCUSSION: Larvae of spiny crawlers can be told from all other mayfly larvae by the absence of any gills on abdominal segment 2. They are easily told from superficially similar schistonote mayflies by fundamental differences of the thorax. Some larvae resemble little stout crawlers and exhibit similar behavior, but gill operculation is never as extremely developed as it is in the little stout crawlers.

In North America this widespread group consists of about nine genera and over 80 species.

Larvae occur in many lotic habitats and in lakes with considerable wave action. In streams and rivers they are often associated with moss mats, tangled roots, filamentous plant growths, or crevices of woody debris or porous rocks. Some species are also known from rapid zones of mountain streams, and a few are associated with silt or sand bottoms. Larvae are generally awkward swimmers, and many perfer to hide by clinging to debris and vegetation. In some western species that inhabit swift streams, the abdomen is adapted to form an attachment disc. Spiny crawlers, as with other pannote mayflies, sometimes raise the tip of the abdomen in scorpion-like fashion when disturbed or removed from water. Adults of many species swarm at unusual heights, often above tree canopies.

The spiny crawlers constitute one of the most important families of mayflies from the standpoint of fly fishing. No less than 30 species among the genera *Attenella, Dannella, Drunella, Ephemerella, Eurylophella, Serratella,* and *Timpanoga* have been used for patterning imitations. These genera were previously considered subgenera of *Ephemerella.*

Of particular note in the genus *Drunella* are the western species *Drunella grandis* (Plate III, Figs. 16, 17) (variously known as the Dark Morning Olive, Great Leadwinged Olive Drake, Great Red Spinner, Green Drake, and Western Green Drake), *Drunella flavilinea* (the Western Slate Olive Dun, Bluewinged Olive Dun, Dark Olive Spinner, or Dark Slatewinged Olive), and *Drunella coloradensis* (the Dark Olive Dun or Light Slatewinged Olive). Fishermen have long been attuned to hatches of these attractive species and their close relatives from middle to late summer on western rivers.

Perhaps the two most important species of *Ephemerella* in central and eastern regions are *Ephemerella subvaria* (Plate IV, Figs. 18, 19) and *Ephemerella dorothea.* The former is also known as the Beaverkill,

SPINY CRAWLERS
(Ephemerellidae)

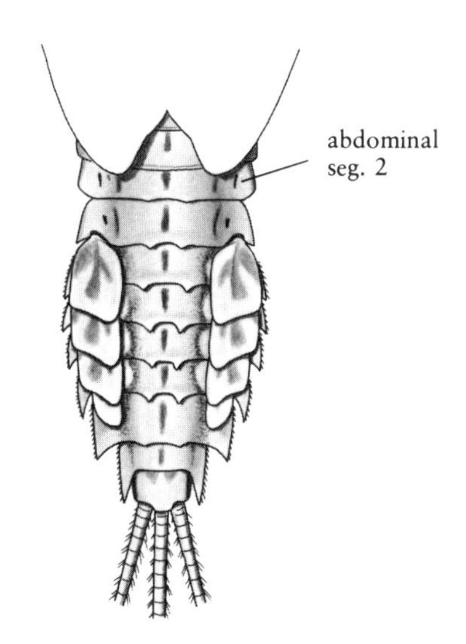

Figure 7.54. *Serratella* larva, dorsal abdomen

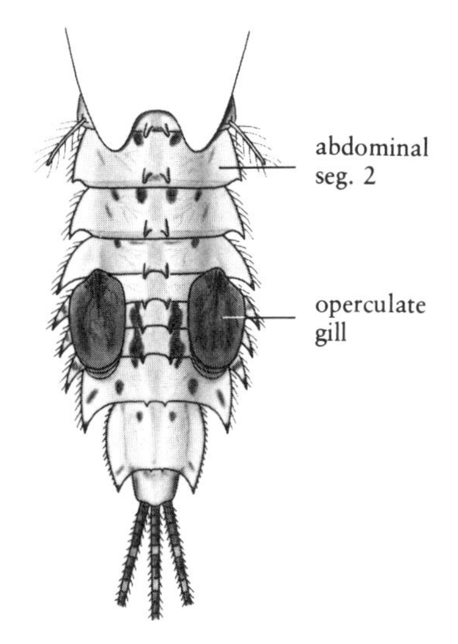

Figure 7.55. *Eurylophella* larva, dorsal abdomen

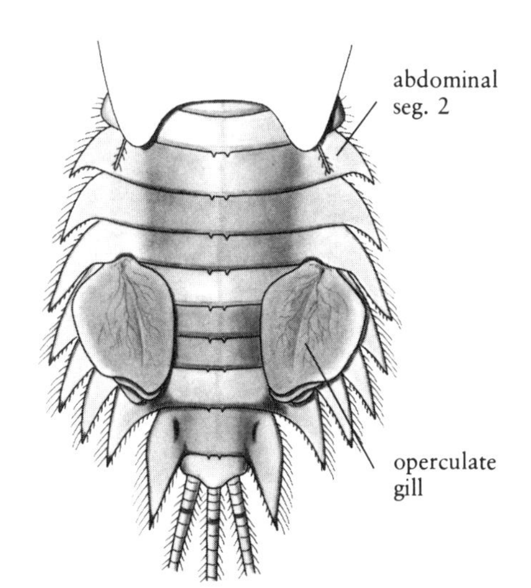

Figure 7.56. *Timpanoga* larva, dorsal abdomen

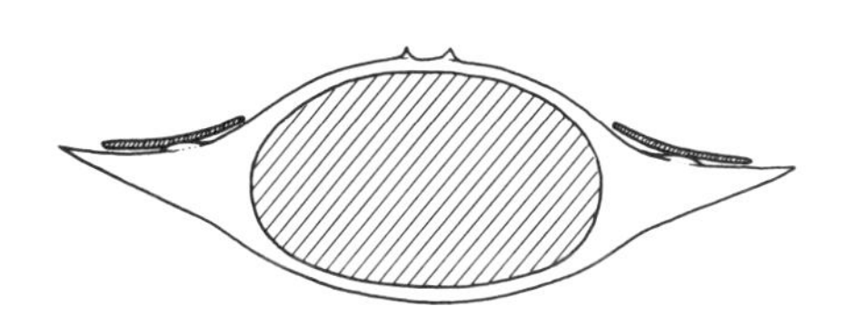

Figure 7.57. Larval abdomen, schematic cross section

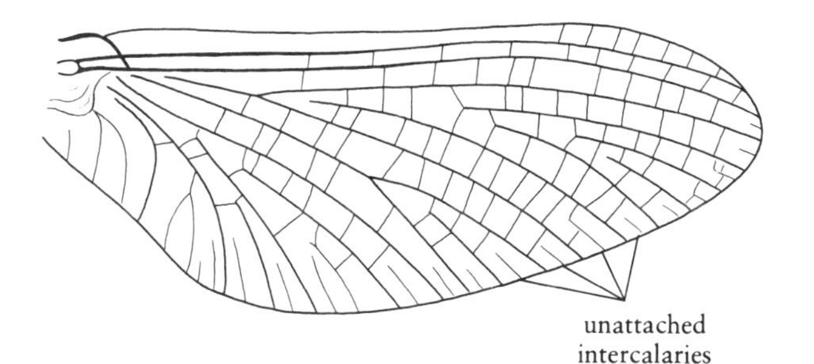

Figure 7.58. Fore wing

Bluewinged Hendrickson, Borcher's, Brown Hen Spinner, Hendrickson, Lady Beaverquill, and Red Quill. *Ephemerella dorothea* is also known as the Brown Hen Spinner, Little Marryat, Pale Evening Dun, Pale Sulphur Dun, Pale Watery Dun, Sulphur, Sulphur Dun, Sulphury Dun, and Small Cream Variant. The importance of both of these species appears to be directly porportional to the large number of names by which they are known, which in turn indicates the widespread and common use of imitations patterned after them.

LITTLE STOUT CRAWLERS
(Family Tricorythidae)

LARVAL DIAGNOSIS: (Fig. 7.59) These measure 3–10 mm excluding tails when mature. Thorax appears stout in relation to abdomen. Hind wing pads are absent or minute. Gills are present on abdominal segments 2–6. Gills on abdominal segment 2 are operculate and oval to triangular. Three tails are present.

ADULT DIAGNOSIS: (Figs. 7.60, 7.61; Plate IV, Fig. 20) These are generally blackish forms. Thorax is stout. Veins MA_1 and MA_2 of the fore wings form a symmetrical fork. Hind wings are absent or in some males are minute with long costal projections. Three tails are present.

LITTLE STOUT CRAWLERS
(Tricorythidae)

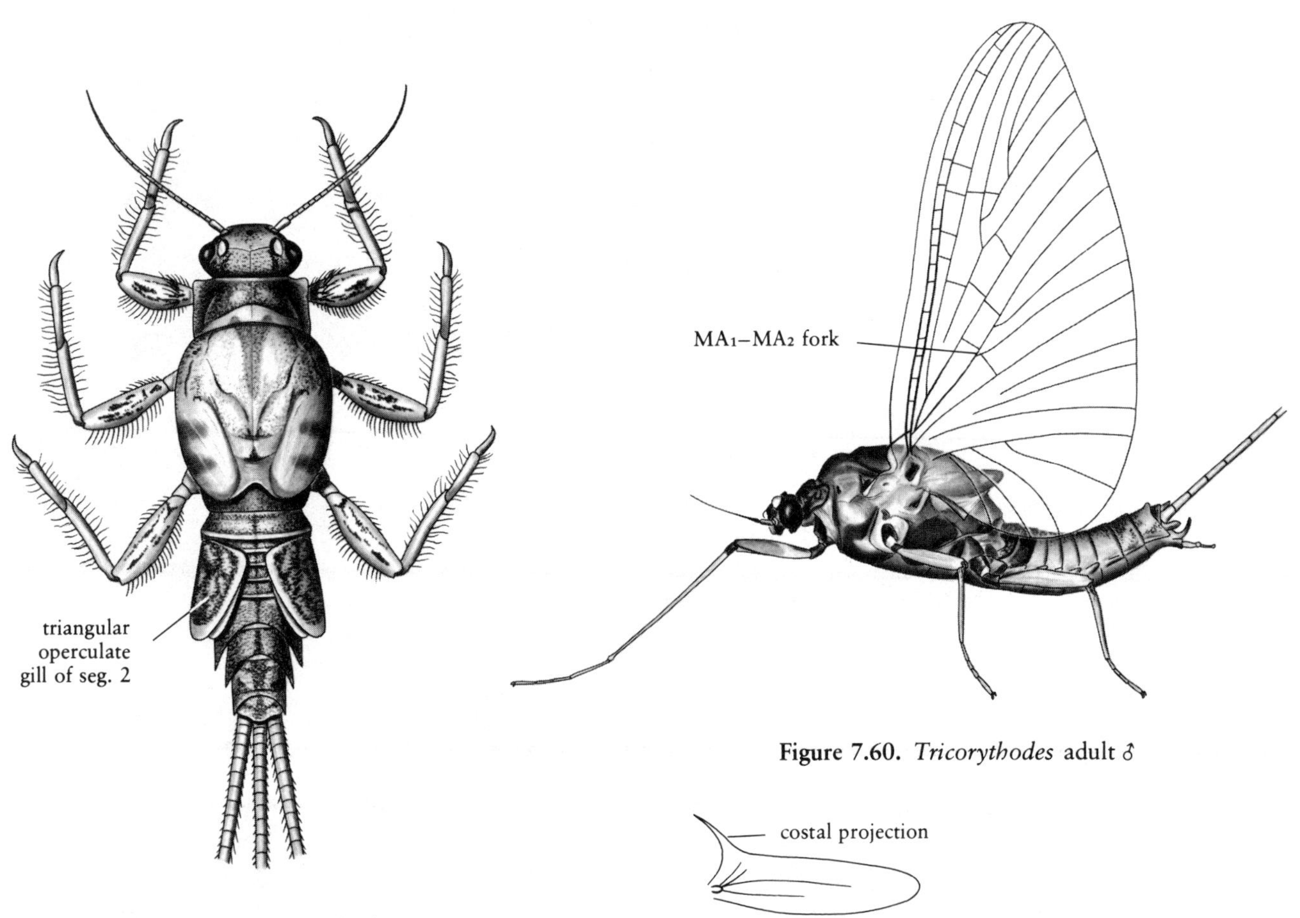

Figure 7.59. *Tricorythodes* larva

Figure 7.60. *Tricorythodes* adult ♂

Figure 7.61. *Leptohyphes* adult ♂, hind wing

DISCUSSION: Larvae of little stout crawlers are somewhat similar to those of spiny crawlers and small squaregills, but the position and shapes of the operculate gills are unique to the little stout crawlers. There are two genera and about 21 species in North America. The family is widespread but somewhat uncommon across Canada and northwestern states.

Larvae can be found associated with vegetation, detritus, or silty and gravelly bottoms of streams and rivers. Little stout crawlers are among the most tolerant of hydropneustic stream insects, sometimes found alone with such groups as water boatmen and scuds. Some *Tricorythodes* live interstitially in silt and are taken most easily by kicking or digging a few centimeters into the substrate. The larvae themselves sometimes become substrate for flocculent matter, including fungal growths.

The three most important species of this group in North America for fly fishing are *Tricorythodes atratus* (Figs. 7.59, 7.60; Plate IV, Fig. 20) and *stygiatus* in central and eastern North America and *Tricorythodes minutus* in western North America. These are all small, dark species that sometimes form clouds above small trout streams because they hatch in such large numbers. The fishermen's names, Dark Brown Spinner, Pale Olive Dun, and Reverse Jenny Spinner, all refer to these species. The name Tiny Whitewinged Black Quill is also used for *stygiatus*. Early- and late-season emergences are common for these species.

SQUAREGILLED MAYFLIES
(Superfamily Caenoidea)

Larvae possess gills on abdominal segment 2 that are square-shaped and operculate.

LARGE SQUAREGILLS
(Family Neoephemeridae)

LARVAL DIAGNOSIS: (see key Fig. 7.5) These measure up to 8–17 mm excluding tails. Operculate gills are fused together and do not overlap. Three tails are present.

ADULT DIAGNOSIS: Basal costal crossveins of the fore wings are either weak or absent. As among burrowing mayflies, veins MP_2 and CuA of the fore wings are arched at their bases away from vein MP_1. Three tails are present.

DISCUSSION: Large squaregills can easily be told from small squaregills by the presence or absence of fused gills and usually also by size. Less common than the small squaregills, the group includes one genus and four or more species, which are restricted to eastern Canada and central and southeastern states including Florida. Larvae cling to vegetation, debris, or the undersides of flat rocks in streams.

SMALL SQUAREGILLS
(Family Caenidae)

LARVAL DIAGNOSIS: (Fig. 7.62; Plate IV, Fig. 22) These are generally small, 2–8 mm excluding tails when mature. Hind wing pads are absent. Operculate gills are not fused but do overlap each other slightly. Three tails are present.

ADULT DIAGNOSIS: (Fig. 7.63; Plate IV, Fig. 21) These are among the smallest adult mayflies. They are generally yellowish and have a relatively robust thorax and small abdomen. Vein MA_2 of the fore wings is attached to vein MA_1 at more-or-less right angles by a crossvein; vein MP_2 is almost as long as vein MP_1. Hind wings are absent.

SMALL SQUAREGILLS
(Caenidae)

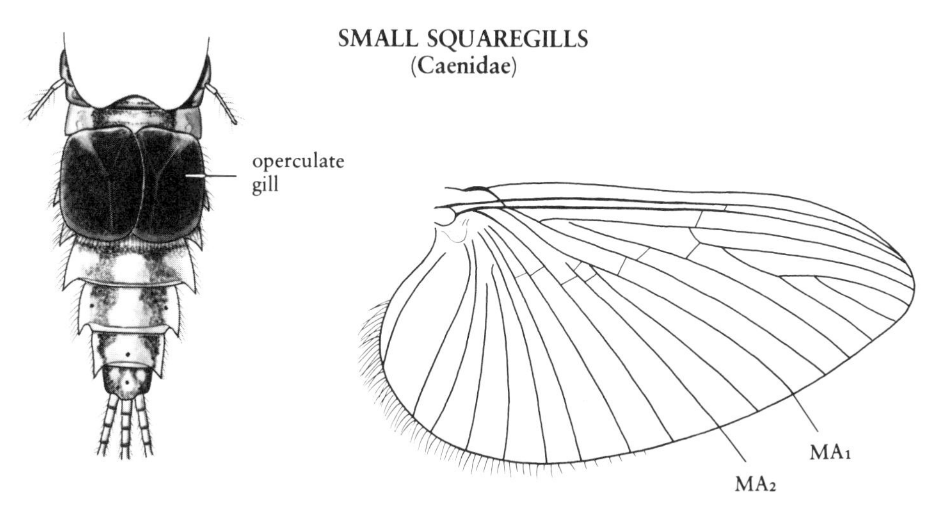

Figure 7.62. *Caenis* larva, dorsal abdomen

Figure 7.63. Fore wing

DISCUSSION: The small squaregills are sometimes confused with the little stout crawlers; however, notice should be taken of the gill shape. The group comprises two widespread genera and about 18 species in North America. *Caenis* is particularly abundant in the eastern half of the continent.

In streams, the larvae of small squaregills are found near silt-bottomed edges, often among vegetation, or within the silt itself. They are common in ponds and small lakes among dense filamentous growths of plants. Larvae are often secretive and difficult to locate (even when in high densities) because of their size and habitat. All active stages of these mayflies may be attracted to artificial lights at night.

Both the tiny *Caenis* and *Brachycercus* have provided useful patterns for fly imitations. *Caenis simulans* (Plate IV, Figs. 21, 22) (the Tiny Whitewinged Sulphur) is very common throughout North America. Subimagos are extremely short-lived, and adults of *Caenis* may live for only a few hours. Trout will selectively feed on *Caenis* during a hatch, and for the fisherman who does not have the proper imitation to match the hatch it could be a disaster. This is perhaps why the small *Caenis* and some of the small midges that have the same effect are sometimes referred to as the "curse."

SHIELDBACK MAYFLIES
(Superfamily Prosopistomatoidea)

ARMORED MAYFLIES
(Family Baetiscidae)

LARVAL DIAGNOSIS: (Fig. 7.64; Plate IV, Fig. 23) Body is very stout and measures 4–14 mm when mature. Thorax is developed into a large carapacelike shield that extends over abdominal segment 5. Three tails are present.

ADULT DIAGNOSIS: (Fig. 7.65) Thorax is stout. Abdomen distinctly tapers posteriorly. Vein A_1 of the fore wings ends in outer margin. Two tails are present.

DISCUSSION: Larvae of armored mayflies are striking and unusual. *Baetisca*, the only genus, is relatively widespread in North America, being absent only in the southwestern states. Between 10 and 20 species are currently recognized.

Larvae are usually only locally abundant in streams, rivers, and occasionally along the edges of lakes. They inhabit sand and gravel, the bases of rooted vegetation, or woody debris. They are more prone to swim than most other pannote mayflies.

More Information about Mayflies

Hatches by Caucci and Nastasi (1975). This is just one of several (see references) good books dealing with the study of mayflies from the fly fisherman's point of view; it includes very good photographs of several mayfly species.

The mayflies of North and Central America by Edmunds et al. (1976). This book provides up-to-date keys to the genera of North American mayflies (both larvae and adults) as well as review discussions of each of the genera.

Environmental requirements and pollution tolerance of Ephemeroptera by Hubbard and Peters (1978). This is an extensive tabular summary and bibliography of the freshwater ecology of North American mayfly species.

ARMORED MAYFLIES
(Baetiscidae)

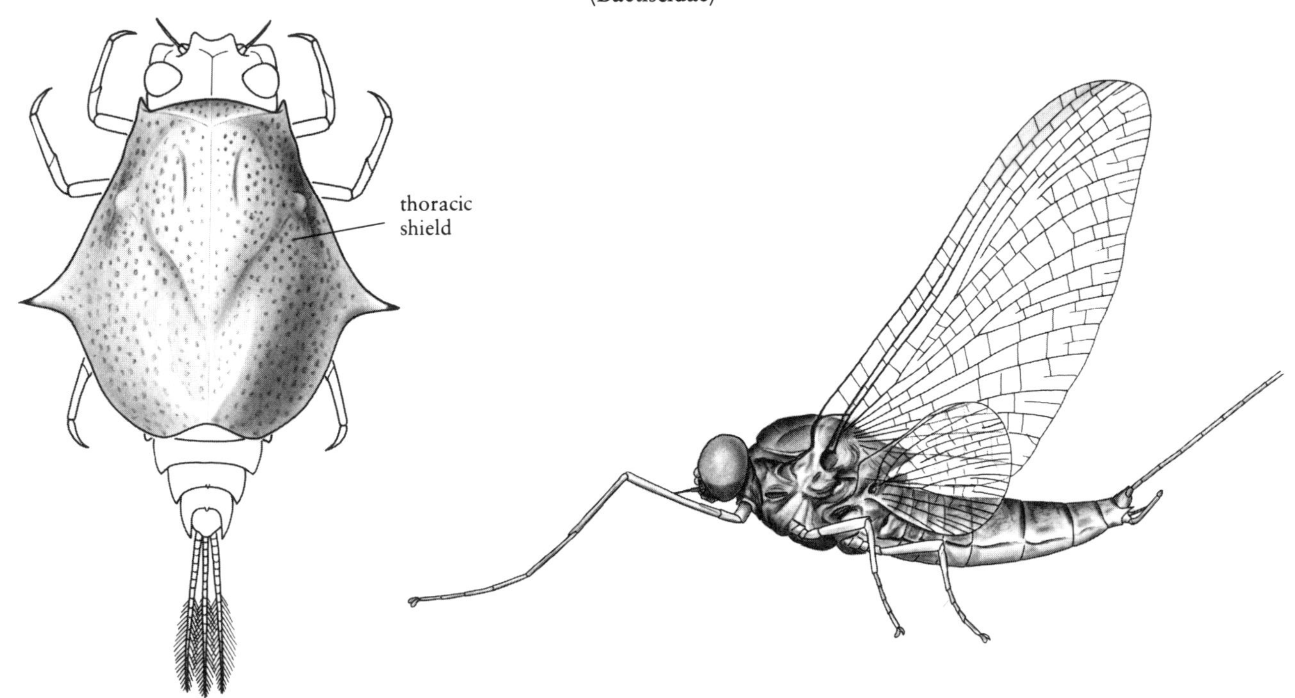

Figure 7.64. *Baetisca* larva

Figure 7.65. *Baetisca* adult ♂

"The burrowing mayflies (Ephemeroptera: Ephemeroidea) of the United States" by McCafferty (1975). This is an up-to-date review of the known biology and distribution of each of the burrowing mayfly species; it contains a key to species in the larval stage and whole-larval illustrations of each genus.

The biology of mayflies by Needham et al. (1935). This recently reprinted book, although out-of-date in several respects, particularly in regard to species taxonomy, remains a valuable introduction to the study of mayflies.

References

Adamus, P. R., and A. R. Gaufin. 1976. A synopsis of Nearctic taxa found in aquatic drift. *Amer. Midl. Natural.* 95:198–204.

Bednarik, A. F., and W. P. McCafferty. 1979. *Biosystematic revision of the genus Stenonema (Ephemeroptera: Heptageniidae).* Canad. Bull. Fish. Aq. Sci. 201. 73 pp.

Berner, L. 1977. *Distributional patterns of southeastern mayflies (Ephemeroptera).* Bull. Fla. St. Mus. Biol. Sci. 22. 55 pp.

Britt, N. W. 1967. *Biology of two species of Lake Erie mayflies, Ephoron album (Say) and Ephemera simulans Walker.* Bull. Ohio Biol. Surv. 1. 70 pp.

Caucci, A., and B. Nastasi. 1975. *Hatches.* Comparahatch Ltd., New York. 320 pp.

Chapman, D. W., and R. L. Demory. 1963. Seasonal changes in the food ingested by aquatic insect larvae and nymphs in two Oregon streams. *Ecology* 44:140–146.

Clifford, H. F., H. Hamilton, and B. A. Killins. 1979. Biology of the mayfly *Leptophlebia cupida* (Say) (Ephemeroptera: Leptophlebiidae). *Canad. J. Zool.* 57:1026–1045.

Coleman, M. J., and H. B. N. Hynes. 1970. The vertical distribution of the invertebrate fauna in the bed of a stream. *Limnol. Oceanogr.* 15:31–40.

Downes, J. A. 1978. *Feeding and mating in the insectivorous Ceratopogonidae (Diptera).* Mem. Entomol. Soc. Canada 104. 62 pp.

Edmunds, G. F., Jr., S. L. Jensen, and L. Berner. 1976. *The mayflies of North and Central America*. Univ. Minn. Press, Minneapolis. 330 pp.

Flick, A. 1966. *Art Flick's new streamside guide to naturals and their imitations.* Crown, New York. 173 pp.

Fremling, C. R. 1970. *Mayfly distribution as a water quality index*. Wat. Poll. Cont. Res. Ser. U.S.E.P.A., Wash., D. C. 39 pp.

Gerlach, R. 1974. *Creative fly tying and fly fishing*. Winchester Press, New York. 231 pp.

Gibbs, K. E. 1977. Evidence for obligatory parthenogenesis and its possible effect on the emergence period of *Cloeon triangulifer* (Ephemeroptera: Baetidae). *Canad. Entomol.* 109:337–340.

Gilpin, B. R., and M. A. Brusven. 1970. Food habits and ecology of mayflies of the St. Maries River in Idaho. *Melanderia* 4:19–40.

Harvey, R. S., R. L. Vannote, and B. W. Sweeney. 1980. Life history, developmental processes, and energetics of the burrowing mayfly *Dolonia americana,* pp. 211–230. In *Advances in Ephemeroptera biology* (J. F. Flannagan and K. E. Marshall, eds.). Plenum, New York.

Hubbard, M. D., and W. L. Peters. 1978. *Environmental requirements and pollution tolerance of Ephemeroptera.* Environ. Monit. Sup. Lab., Off. Res. Dev. U.S.E.P.A., Cincinnati. 460 pp.

Hynes, H. B. N. 1974. Further studies on the distribution of stream animals within the substratum. *Limnol. Oceanogr.* 19:92–99.

Hynes, H. B. N., N. K. Kaushik, M. A. Lock, D. L. Lush, Z. S. J. Stocker, R. R. Wallace, and D. D. Williams. 1974. Benthos and allochthonous organic matter in streams. *J. Fish. Res. Brd. Canada* 31:545–553.

Leonard, J. W., and F. A. Leonard. 1962. *Mayflies of Michigan trout streams.* Cranbrook Inst. Sci., Bloomfield Hills, Mich. 139 pp.

McCafferty, W. P. 1975. The burrowing mayflies (Ephemeroptera: Ephemeroidea) of the United States. *Trans. Amer. Entomol. Soc.* 101:447–504.

McCafferty, W. P. 1979. Swarm-feeding by the damselfly *Hetaerina americana* on mayfly hatches. *Aq. Ins.* 1:149–151.

McCafferty, W. P., and G. F. Edmunds, Jr. 1979. The higher classification of the Ephemeroptera and its evolutionary basis. *Ann. Entomol. Soc. Amer.* 72:5–12.

McCafferty, W. P., and B. L. Huff, Jr. 1978. The life cycle of the mayfly *Stenacron interpunctatum* (Ephemeroptera: Heptageniidae). *Gr. Lakes Entomol.* 11:209–216.

McCafferty, W. P., and D. K. Morihara. 1979. The male of *Baetis macdunnoughi* and notes on parthenogenetic populations within *Baetis. Entomol. News* 90:26–28.

McClane, A. J. (ed.). 1965. *McClane's standard fishing encyclopedia.* Holt, Rinehart and Winston, New York. 1057 pp.

Meck, C. R. 1977. *Meeting & fishing the hatches.* Winchester Press, New York. 194 pp.

Morihara, D. K., and W. P. McCafferty. 1979. The *Baetis* larvae of North America (Ephemeroptera: Baetidae). *Trans. Amer. Entomol. Soc.* 105:139–221.

Needham, J. G., J. R. Traver, and Y-C Hsu. 1935. *The biology of mayflies.* Comstock, Ithaca, N.Y. 759 pp. (Reprint: 1972, E. W. Classey Ltd., London.)

Schwiebert, E. 1973. *Nymphs.* Winchester Press, New York. 339 pp.

Selgeby, J. H. 1974. Immature insects from deep water in western Lake Superior. *J. Fish. Res. Brd. Canada* 31:109–111.

Swisher, D., and C. Richards. 1971. *Selective trout.* Crown, New York. 184 pp.

CHAPTER 8

Damselflies and Dragonflies

(ORDER ODONATA)

DRAGONFLIES AND DAMSELFLIES are unable to fold their four elongate wings back over the abdomen when at rest. This primitive insect condition is shared by the mayflies; and like the mayflies, the Odonata are very distinctive among the insect groups today. They are one of the fundamentally aquatic insect orders, with all known North American larvae developing in aquatic environments. The presence of approximately 450 species in North America belies the fact that on a world-wide basis and especially in the tropics, the order is among the largest of aquatic groups. The terrestrial adults are well known for their beautiful colors, flying ability, and curious habits. For centuries they have been favorite subjects of poets, naturalists, and collectors; and fishermen have no doubt often pondered the fascinating adults.

Fly fishermen have patterned a few tied flies after Odonata larvae, and certain ones, such as the Beaverpelt Nymph and Heather Nymph, can be highly productive on some waters. Larvae are also familiar to many fishermen as bait. They provide an important food source for many game fishes, particularly those in ponds and lakes.

In North America the common names "damselfly" and "dragonfly" have come to denote two distinct suborders, although "dragonfly" has historically been applied to the entire order. A specialized type of reproductive behavior and associated structural modifications are unique to the Odonata. Certain species may be of some economic consequence, since they feed on large numbers of certain insect pests, such as mosquitoes.

Larval Diagnosis

These are elongate, slender, or robust forms, usually of medium to large size (between 10 and 60 mm). Head possesses moderately developed eyes and variously developed, conspicuous antennae. Chewing mouthparts are uniquely modified with a highly developed, prehensile labium. When not in use, this labium masks the bottom of the head and other mouthparts. Thoracic legs are well developed. Two pairs of wing pads are present. Gills are not present along the body. There are no tails; however, the abdomen terminates in either three short, wedge-shaped structures (dragonflies) or three elongate, leaflike caudal lamellae (damselflies).

Adult Diagnosis

These are elongate, often very colorful insects. They have four elongate, membranous, profusely veined wings that are either held outspread horizontally or held together somewhat vertically above the body. Head appears to be quite wide because of the very large eyes. Antennae are short and slender. Thorax is robust and has well-developed legs that originate relatively anteriorly. Abdomen is long and slender and has no tails but does have variously developed anal appendages.

Similar Orders

Only the very early larval instars or young larvae of dragonflies, or more possibly damselflies, might be confused with young larvae of mayflies or stoneflies. For instance, the caudal lamellae of young damselflies are sometimes thin and tail-like as in mayflies, and lateral abdominal gills of some very young mayflies are not yet developed. Some obvious, unique traits of each of these groups (such as the specialized mouthparts of the Odonata) are discernible even in young individuals.

The adults of dragonflies and damselflies are distinctive and do not resemble adults of other aquatic insects, either in flight or in the hand. Elongate, net-veined wings of Megaloptera may superficially resemble those of the Odonata. Terrestrial insects that resemble the Odonata, such as the ant lion adults (Neuroptera), which can sometimes be found in sandy areas near ponds and streams, are distinguished by their much more highly developed antennae.

Life History

Metamorphosis is incomplete, and structural differences between aquatic larvae (sometimes referred to as nymphs or naiads) and terrestrial adults range from slight, as among some damselfly species, to substantial, as among many dragonfly species. Larval development usually involves 10 to 12 instars, but sometimes as many as 15 or as few as 8. Many species have one generation per year with overwintering larvae. Many species also commonly have generation times of two, three, or four years. Only in a few instances do species have more than one generation per year. Some species pass the winter as eggs. The actual duration of larval development and the number of instars can vary within a species, being subject to water temperatures and food availability.

Adults emerge in spring, summer, or fall, with some species having short periods of emergence and others emerging throughout much longer

At sunset on this flat terrace,
Inking my pen on the slanting stone balustrade,
I sit down to write a poem on the wu-tung leaf,
Here is a kingfisher singing on a bamboo clothes rack,
There is a dragonfly clinging to a fishing line,
Now that I know what quiet enjoyment is,
I shall come here again and again.

Tu Fu (eighth century A.D.)

periods. Emergence generally takes place after the last larval instar has crawled from the water to cling to some object, usually in a vertical, upright position.

Adults may live from a few weeks to a few months. For a period of time following emergence, the adult is neither fully pigmented nor sexually mature. These young adults, known as tenerals, are awkward and easy targets for predators. Flight activity, which is predominantly diurnal, is related to feeding and reproductive behavior and, oftentimes, to dispersal or territoriality, depending on the species. Adults have a well-earned reputation of being voracious predators, capturing other insects in flight. They are commonly called "mosquito hawks" for this reason. Males of many species establish territories and partition their flying space for reproduction, feeding, or both. They aggressively defend such territories from a perch or by patrolling. Flying heights and patterns tend to be very specific.

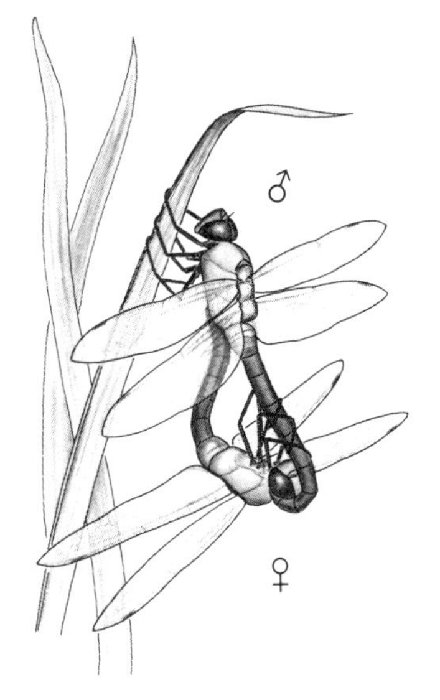

Figure 8.1. Mating pair

Mating is most intricate and in some species involves well-developed courtship behavior. Before mating, the male deposits sperm in unique secondary genitalia located on the underside of his second and third abdominal segments. This is accomplished by bending the end of the abdomen, where the primary genitalia are located, under and forward to the secondary position. The male approaches the female from above and grasps her behind the head with the grasping appendages at the end of his abdomen. Adults are often seen coupled in such a manner and are said to be "in tandem." Mating takes place when the female, as she is held, loops her abdomen down and forward and engages the male's secondary genitalia (accessory genitalia). This particular coupling behavior is unique (Fig. 8.1). Interestingly, in at least some species, the male accessory genitalia may also be used to remove sperm in the female (sperm that has been stored from previous matings with another male).

Damselflies and some dragonflies oviposit eggs in emergent or floating vegetation or debris by puncturing the plant tissue with a well-developed ovipositor. This may be accomplished above or below the waterline and is often done in tandem, the male sometimes anchoring the female during oviposition. In more advanced dragonflies, the female (sometimes flying in tandem) usually repeatedly strikes the end of the abdomen on the water surface in a rhythmic and patterned manner. A few species oviposit directly into the substrate of shallow waters. Some others oviposit overwintering eggs on marginal dry land that will subsequently be flooded in the spring, at which time the eggs hatch.

Occasionally, numbers of adults of a few species partake in mass flights for considerable distances. This is a phenomenon that is not clearly understood. Swarms of young adults are commonly encountered in a few species, and some species regularly disperse or migrate.

TABLE 8.1 COMMON HABITATS AND HABITS OF ODONATA LARVAE

Family	Lotic			Lentic		
	CLIMBERS	SPRAWLERS	INTERSTITIAL[1]	CLIMBERS	SPRAWLERS	INTERSTITIAL[1]
Calopterygidae	X[2]					
Lestidae				X		
Coenagrionidae	X	X		X	X	
Protoneuridae	X[2]					
Petaluridae		X[2]				
Gomphidae		X	X		X	X
Aeshnidae	X	X		X		
Cordulegastridae			X[2]			
Macromiidae		X[3]	X[3]			X
Corduliidae[4]	X	X		X	X	
Libellulidae[4]	X	X		X	X	

[1]Includes burrowers and those that tend to become partially covered with soft substrate.
[2]Primarily streams.
[3]Primarily rivers.
[4]Predominantly pond and marsh inhabitants; only rarely interstitial.

Frogs and birds take a heavy toll on dragonflies and damselflies, which are especially susceptible during emergence periods and are highly attractive as adults. Adults are often found with brightly colored water mites attached to their bodies. Since game fishes sometimes feed heavily on larvae, a few nymph imitations can be valuable additions to a fisherman's tackle box.

Aquatic Habitats

Damselfly and dragonfly larvae inhabit a variety of aquatic environments (Table 8.1) but are most common in ponds, marshes, lake margins, shallow areas of streams, and the slower reaches of rivers and streams. A few species occur in somewhat brackish pools or estuarine habitats. A few dragonfly species are able to exist for extended periods in moist substrate underneath rocks in otherwise dry stream beds or ponds. Many species occur slightly beneath the substrate surface or burrow as deep as several centimeters within soft substrates.

Aquatic Adaptations and Behavior

Larvae are hydropneustic, and although cutaneous respiration may be somewhat general over thin-walled regions of the entire body, the anal area of the larva is especially adapted for both respiration and osmoregulation. They are able to pump water in and out of a rectal chamber at the end of the abdomen, wherein gas exchange—and chloride uptake when needed for internal salt balance—can readily take place.

The caudal lamellae of damselflies increase the available surface area and, when thin-walled, may function as gills (which they are sometimes called). These lamellae may also function as fins or stabilizers in the water. The force and frequency of water pumping in the rectal chamber is generally not as great as in dragonflies. Interestingly, the rectal chamber of certain damselflies harbors a specialized euglenoid flagellate during the winter months. It is likely that these unicellular organisms provide a beneficial source of oxygen for damselfly respiration while gaining a protective site for overwintering and reproduction.

Larvae are, without exception, carnivorous and opportunistically feed on organisms that attract them visually by movement. The likelihood of prey being taken appears to be directly related to the degree to which the prey move about in the environment. Some attack any object within an appropriate size range; some actively stalk their prey, whereas others lie in wait and ambush their prey. Many are cannibalistic.

Larvae are primarily sprawlers on substrate or climbers among vegetation and debris, but some show adaptations for burrowing (Table 8.1). Interestingly, dragonflies may be propelled or aided in their forward locomotion by a forceful expulsion of water from the rectal chamber (jet propulsion).

Thus the dragon-fly enters upon a more nobler life than that it had hitherto led in the water, for in the latter it was obliged to live in misery, creeping and swimming slowly, but now it wings the air.

Jan Swammerdam

Classification and Characters

The damselflies belong to the suborder Zygoptera and in North America consist of four families. The dragonflies belong to the suborder Anisoptera and are here classified into three superfamilies and seven families. These categories are diagnosed by examining eye and wing characters of the adults, and primarily antennal, labial, and terminal abdominal characters of the larvae.

The labium is highly modified in Odonata larvae as an extensible prehensile organ (Fig. 8.23) for catching prey. The shape of the main portion (known as the **median lobe**), which lies below the head when at rest, and the nature of the modified labial palps (known as **lateral lobes**) are of great taxonomic importance. The characters of the median lobe are best examined by holding the specimen dorsal side up and then gently pulling the labium forward to the extended position. In this way one may view the dorsal characters of the median lobe without dissecting it.

As larvae, species and sometimes genera can be difficult to identify. Structural changes occur at each molt and sometimes involve taxonomic characters. Coloration of larvae is highly variable and often related to the habitat and kind of food they have been eating. More mature larvae with elongate wing pads are the most reliably identifiable.

Coloration can be of use in adult identification, but patterns and color are sometimes obscured in pruinose individuals. These are individuals that, upon reaching sexual maturity, exude a whitish fluid over parts of the body and tend to lose some color intensity.

Odonata are one of the few groups of aquatic insects in which nonscientific, or common, names have been liberally applied to species. This is undoubtedly due to the interest that naturalists have long had for them. Most common names are based either on characteristics of the terrestrial adults or on transliterations of the latinized scientific names (e.g., *Gomphus olivaceus* or the Olive Clubtail; *Cordulegaster sayi* or Say's Biddie). Table 8.2 lists some common names of genera and higher groups.

Species names are usually modifiers of the generic names (e.g., *Celithemis* are the spotted skimmers, and *Celithemis ornata* is thus the Ornate Spotted Skimmer; *Macromia* are the river skimmers, and *Macromia magnifica* is thus the Magnificent River Skimmer). There are, however, a few exceptions to the above rule (e.g., although the genus *Lestes* is known as the marsh spreadwings, species of *Lestes* are simply known as spreadwings, such as the Green Spreadwing or the Texas Spreadwing). Moreover, when only one species is known for a genus in North America, the same common name may be applied to both genus and species (e.g., *Hagenius* and *Hagenius brevistylus* are both known as the Black Dragon, and *Chromagrion* and *Chromagrion conditum* are both known as the Variegated Damsel).

TABLE 8.2 SYNOPSIS OF NORTH AMERICAN ODONATA GENERA AND THEIR COMMON NAMES

Scientific Names	Common Names
Zygoptera	Damselflies
Calopterygidae	Broadwinged Damselflies
Calopteryx	Bandwings
Hetaerina	Ruby Spots
Lestidae	Spreadwinged Damselflies
Archilestes	Spreadwings
Lestes	Marsh Spreadwings
Coenagrionidae	Narrowwinged Damselflies
Acanthagrion	Damsels
Amphiagrion	Bog Damsels
Anomalagrion	Speartails
Apanisagrion	Damsels
Argia	Dancers
Argiallagma	Bog Dancers
Chromagrion	Variegated Damsels
Coenagrion	Damsels
Enallagma	Bluets
Hesperagrion	Western Damsels
Ischnura	Forktails
Nehalennia	Green Damsels
Telebasis	Flappers
Protoneuridae	Protoneurid Damselflies
Neoneura	Neoneurans
Protoneura	Protoneurans
Anisoptera	Dragonflies
Petaluridae	Graybacks
Tachopteryx	Eastern Graybacks
Tanypteryx	Western Graybacks
Gomphidae	Clubtails
Aphylla	Clubtails
Dromogomphus	Spinylegged Clubtails
Erpetogomphus	Snake Darners
Gomphus	Common Clubtails
Hagenius	Black Dragons
Lanthus	Clubtails
Octogomphus	Eighttoothed Clubtails
Ophiogomphus	Snake Darners
Phyllogomphoides	Clubtails
Progomphus	Clubtails
Stylogomphus	Clubtails
Aeshnidae	Darners
Aeshna	Blue Darners
Anax	Green Darners
Basiaeschna	Darners
Boyeria	Stream Darners
Coryphaeschna	Giant Darners
Epiaeschna	Hero Darners
Gomphaeschna	Bog Darners
continued	*continued*

Scientific Names	Common Names
Anisoptera (continued)	Dragonflies (continued)
Aeshnidae (continued)	Darners (continued)
Gynacantha	Twospined Darners
Nasiaeschna	Bluenosed Darners
Oplonaeschna	Southwestern Darners
Triacanthagyna	Threespined Darners
Cordulegastridae	Biddies
Cordulegaster	Biddies
Macromiidae	Belted and River Skimmers
Didymops	Belted Skimmers
Macromia	River Skimmers
Corduliidae	Greeneyed Skimmers
Cordulia	Skimmers
Dorocordulia	Skimmers
Epitheca	Skimmers
Helocordulia	Taxonomists' Skimmers
Neurocordulia	Twilight Skimmers
Somatochlora	Bog Skimmers
Williamsonia	Williamson's Skimmers
Libellulidae	Common Skimmers
Belonia	Skimmers
Brachymesia	Cloudywings
Brechmorhoga	Mendacious Skimmers
Cannaphila	Reedloving Skimmers
Celithemis	Spotted Skimmers
Dythemis	Skimmers
Erythemis	Skimmers
Erythrodiplax	Skimmers
Idiataphe	Skimmers
Ladona	Corporals
Lepthemis	Skimmers
Leucorrhinia	Whitefaced Skimmers
Libellula	Skimmers
Macrodiplax	Girdled Skimmers
Macrothemis	Whitebanded Skimmers
Miathyria	Skimmers
Micrathyria	Skimmers
Nannothemis	Dwarf Skimmers
Orthemis	Skimmers
Pachydiplax	Blue Pirates
Paltothemis	Southwestern Skimmers
Pantala	Globe Skimmers
Perithemis	Amberwings
Plathemis	Whitetailed Skimmers
Pseudoleon	Skimmers
Sympetrum	Red Skimmers
Tauriphila	Skimmers
Tramea	Raggedy Skimmers

Adapted primarily from Borror (1963).

Figure 8.2. MATURE ODONATA LARVAE

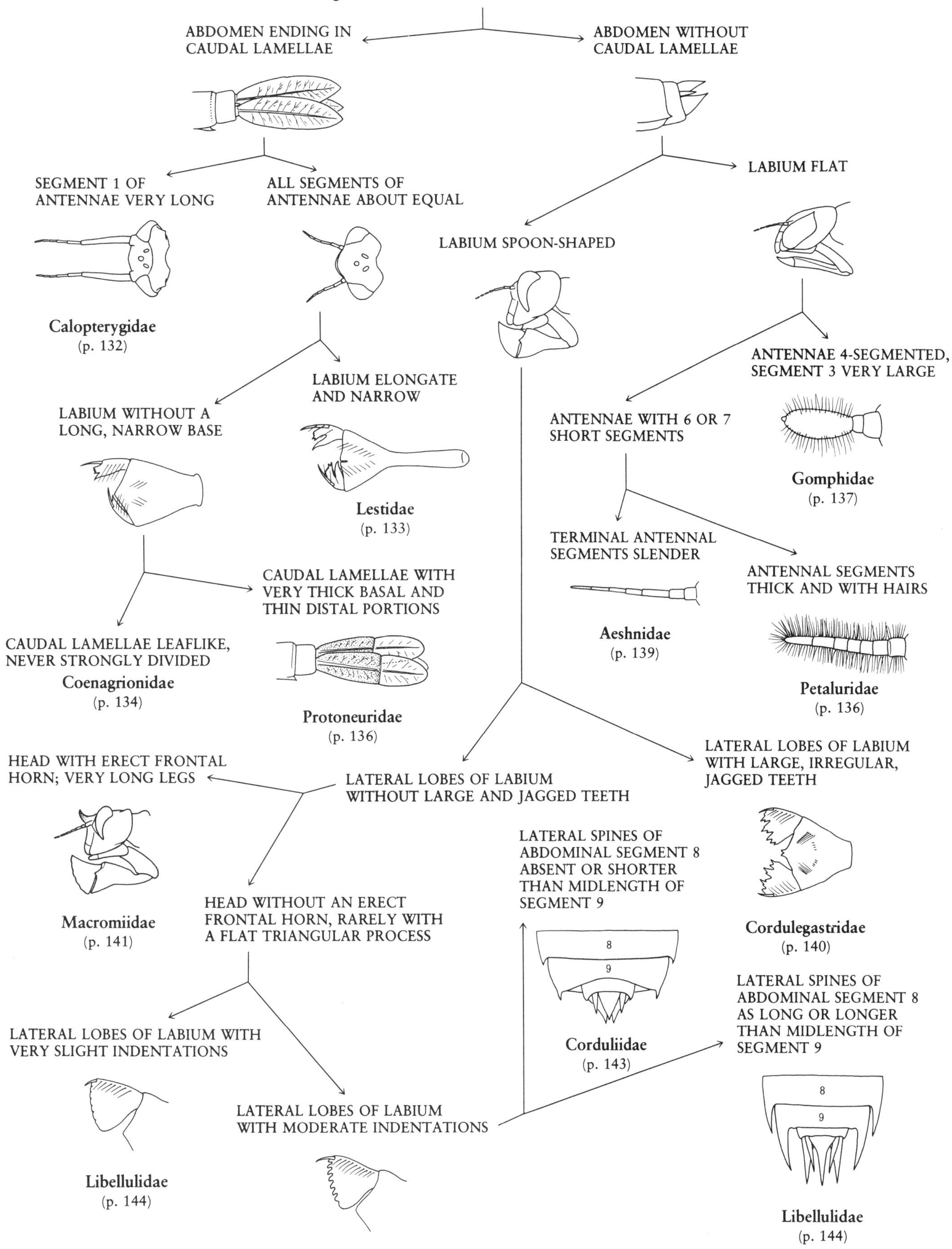

BROADWINGED DAMSELFLIES
(Calopterygidae)

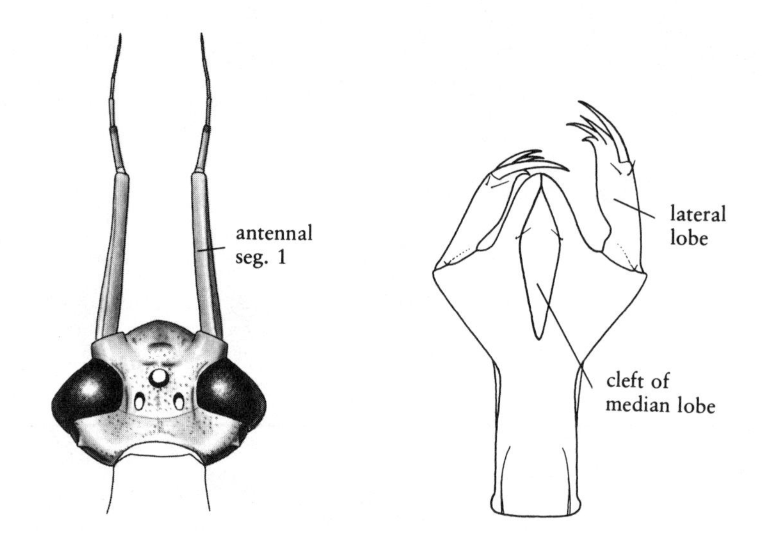

Figure 8.3. Larval head **Figure 8.4.** Larval labium

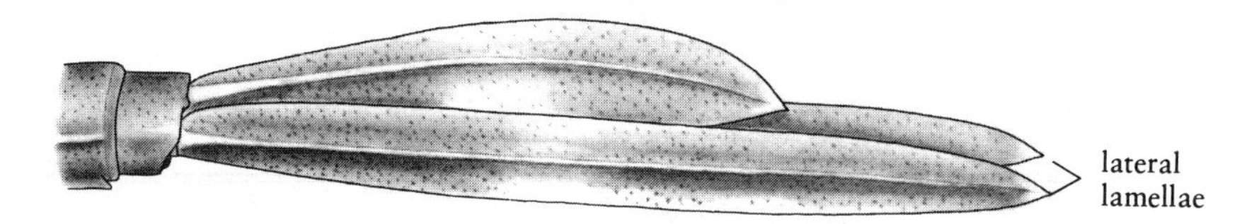

Figure 8.5. Larval caudal lamellae

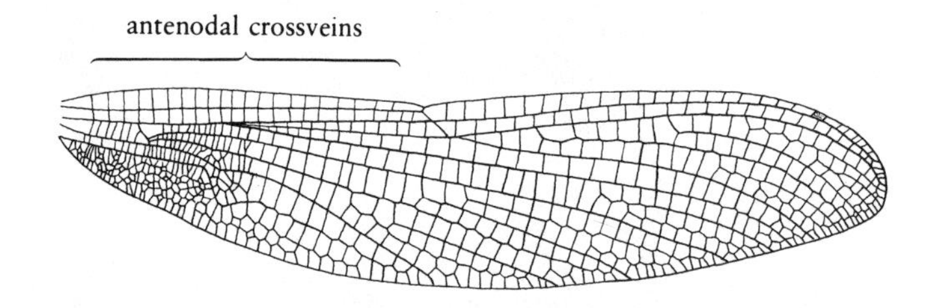

Figure 8.6. *Hetaerina* fore wing

Damselflies
(Suborder Zygoptera)

Larvae are slender, and their usually cylindrical abdomen terminates in three well-developed caudal lamellae (Figs. 8.5, 8.11).

Adults are slender-bodied. Eyes are large and widely separated. Wings, when at rest, are usually held above the body (Plate V, Figs. 25, 27, 29).

BROADWINGED DAMSELFLIES
(Family Calopterygidae)

LARVAL DIAGNOSIS: (Figs. 8.3–8.5; Plate V, Fig. 24) These are long, slender, long-legged forms. When mature they measure 25–50 mm excluding caudal lamellae. Segment 1 of the antennae is conspicuously long—longer than all other antennal segments combined. Median lobe of labium is produced distally and deeply cleft. Caudal lamellae are elongate and slender; the lateral lamellae are more slender and elongate than the middle lamella.

ADULT DIAGNOSIS: (Fig. 8.6; Plate V, Fig. 25) Wings are usually darkly pigmented or grayish and not strongly narrowed at their bases; wings possess several antenodal crossveins.

DISCUSSION: The extremely long first antennal segment (Fig. 8.3) and the generally stiltlike legs of broadwinged damselfly larvae will serve to distinguish them from larvae of other North American damselflies. This small family is represented in North America by about eight species and two common genera, *Hetaerina* and *Calopteryx*. The group is relatively widespread in North America, being absent only from western Canada and Alaska.

Larvae are stream and river insects that crawl stiffly among vegetation, debris, and the root tangles of banks. The metallic-bodied and dark-winged adults are perhaps the most striking of North American damselflies. The ruby spot damselflies (*Hetaerina*) are well known to many naturalists and fishermen, and the American Ruby Spot (*H. americana*) (Plate V, Fig. 25) is an accomplished predator of swarming mayflies and midges.

A fly pattern known as the Great Olive Damselfly Nymph is actually based on the larvae of the American Ruby Spot (Plate V, Fig. 24). This fly can be used for Smallmouth Bass or trout on midwestern and eastern rivers and streams. The Blackwinged Damselfly, *Calopteryx maculatum,* is another common species that may provide a useful pattern for fly fishermen.

SPREADWINGED DAMSELFLIES
(Family Lestidae)

LARVAL DIAGNOSIS: (Figs. 8.7, 8.8; Plate V, Fig. 26) These are slender forms and when mature generally measure 20–29 mm excluding caudal lamellae. Segments of antennae are all about the same length. Labium is conspicuously long and stalklike basally; median lobe is neither produced nor deeply cleft. Abdomen is relatively long and slender. Caudal lamellae are somewhat leaflike, and their lateral tracheal branches are nearly at right angles to their central tracheal trunk.

SPREADWINGED DAMSELFLIES
(Lestidae)

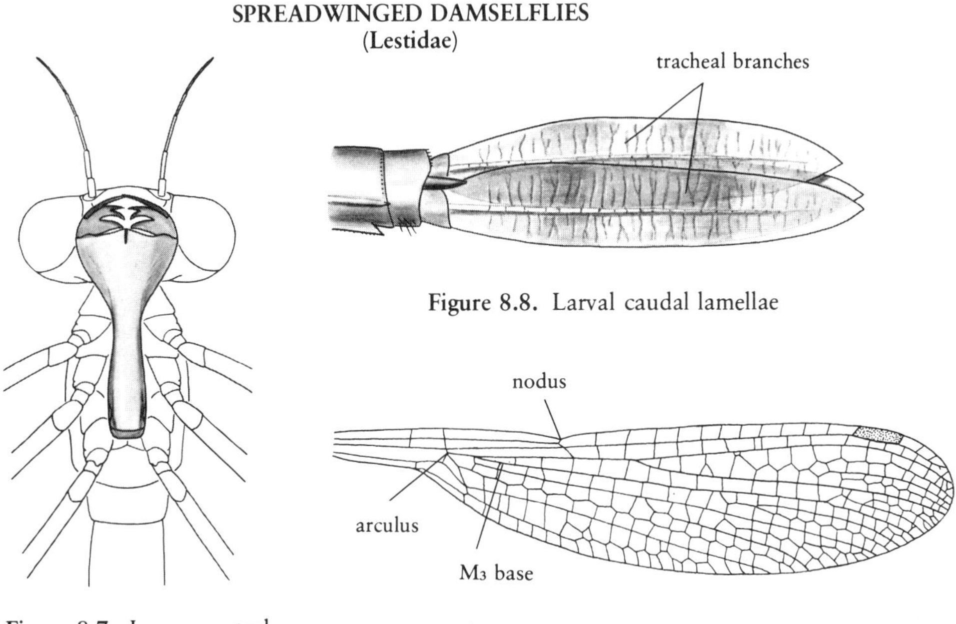

Figure 8.8. Larval caudal lamellae

Figure 8.7. Larva, ventral head and thorax (labium shaded)

Figure 8.9. Fore wing

ADULT DIAGNOSIS: (Fig. 8.9; Plate V, Fig. 27) Wings are narrowly constricted basally. Vein M_3 of the wings originates closer to the arculus than to the nodus.

DISCUSSION: Larvae of spreadwinged damselflies are distinguished from all other damselflies by their highly elongate labium. This family is widespread throughout North America, where it includes two genera and about 19 species.

Larvae of the genus *Lestes* usually climb in the thick aquatic vegetation of marshes, swamps, and small, temporary or permanent, sheltered lakes, ponds, or ditches. Some, however, also inhabit more open waters and are able to move quickly with a sweeping motion of the caudal lamellae. *Archilestes* occurs mostly in streams and may be particularly conspicuous in some desert streams. Adults are unique among North American damselflies because, as intimated by the family's common name, they hold the wings partially outspread when at rest; and interestingly, the females of *Archilestes* oviposit their eggs in trees.

Two patterns used for nymph fishing can be especially useful when fishing for Bluegills in ponds or shallow lakes: These are the Speckled Olive Damselfly Nymph, patterned after the larva of *Lestes disjunctus* (also known as the Disjunct Spreadwing), and the Dark Olive Damselfly Nymph, patterned after the larva of *Lestes inaequalis* (also known as the Unequal Spreadwing).

NARROWWINGED DAMSELFLIES
(Family Coenagrionidae)

LARVAL DIAGNOSIS: (Figs. 8.10–8.13; Plate V, Fig. 28) These are relatively short damselflies and when mature measure 13–25 mm excluding caudal lamellae. Segments of antennae are all about the same length. Labium does not have a long, stalklike base; median lobe is somewhat produced distally but never cleft. Caudal lamellae are variously shaped but often broad, leaflike, and pointed at the tips; lateral tracheal branches of the caudal lamellae, when apparent, are themselves usually profusely branched.

ADULT DIAGNOSIS: (Fig. 8.14; Plate V, Fig. 29) Wings are narrowly constricted basally. Vein M_3 of the wings originates closer to the nodus than to the arculus.

DISCUSSION: The above combination of characteristics will readily distinguish narrowwinged damselflies from the spreadwinged damselflies. A relatively robust body and broad caudal lamellae (Fig. 8.11) are typical of most narrowwinged damselfly larvae. Most North American damselflies belong to this widespread group, which is composed of about 15 genera and over 90 species.

Larvae of a common genus, *Argia*, occur in a variety of habitats, but most *Argia* species dwell primarily in streams among rocks and vegetation in moderate currents. Other narrowwinged damselflies are usually associated with lentic environments. Species of narrowwinged damselflies that are adapted to warm water have expectedly more southern distributions in the United States, but of these, several are known also from northern, isolated, warm-spring habitats.

Adults of *Argia* are somewhat unusual in that they often prefer to rest on open ground or rocks rather than vegetation. Adults of *Enallagma* are very commonly encountered: These are the small, bright blue and black

NARROWWINGED DAMSELFLIES
(Coenagrionidae)

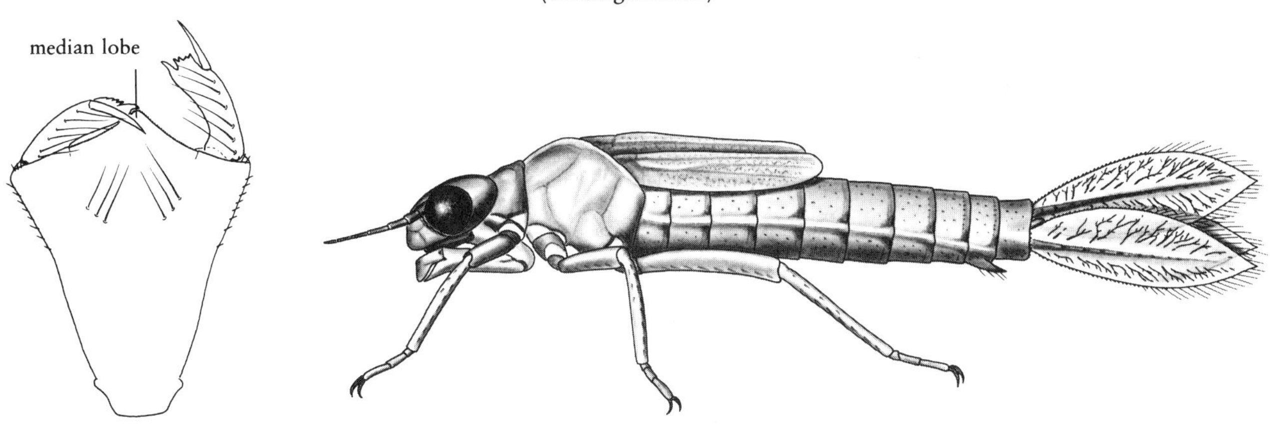

Figure 8.10. Larval labium

Figure 8.11. *Enallagma* larva

Figure 8.12. *Ischnura* larva, caudal lamellae

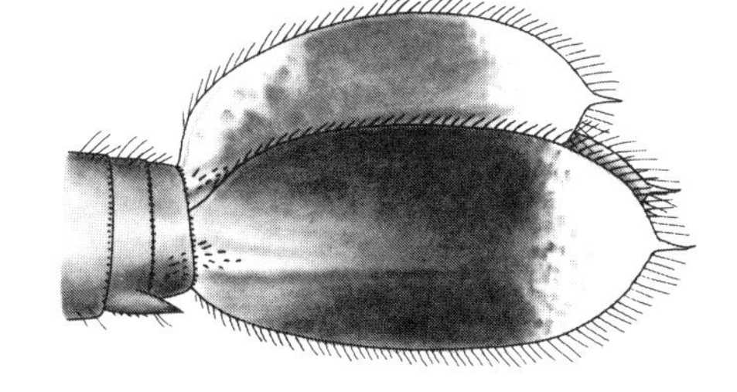

Figure 8.13. *Argia* larva, caudal lamellae

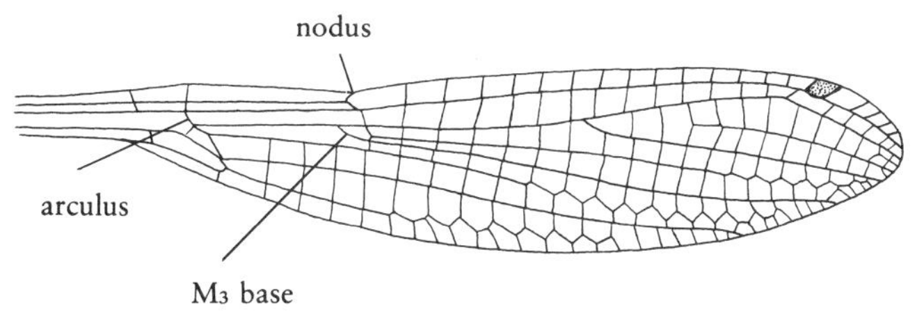

Figure 8.14. Fore wing

damselflies seen among the marginal vegetation of ponds and slow streams. In a few southern species, blue is replaced by other bright colors, such as reds and yellows. Sexes of the same species often differ markedly in color. *Argia* (Plate V, Fig. 29) males may be blue or violet, whereas females are tan; *Enallagma* males may be blue and black, whereas the females have similar patterns but less intense colors; and *Ischnura* males are black and green, whereas most females are black and orange.

Several species of this family have provided patterns for nymph fishing. These flies may be used throughout North America, most profitably in slow or slack waters. Some examples include the Pale Olive Forktailed Nymph, patterned after the larvae of *Ischnura barberi*, or Barber's Forktail; the Bright Olive Damselfly Nymph, patterned after the larvae of *Enallagma civile*, or the Civil Bluet; and the Purple Damselfly Nymph, patterned after *Argia violacea*, or the Violet Dancer. The last-named fly can be fished on or near the bottom of a variety of small streams.

PROTONEURID DAMSELFLY
(Protoneuridae)

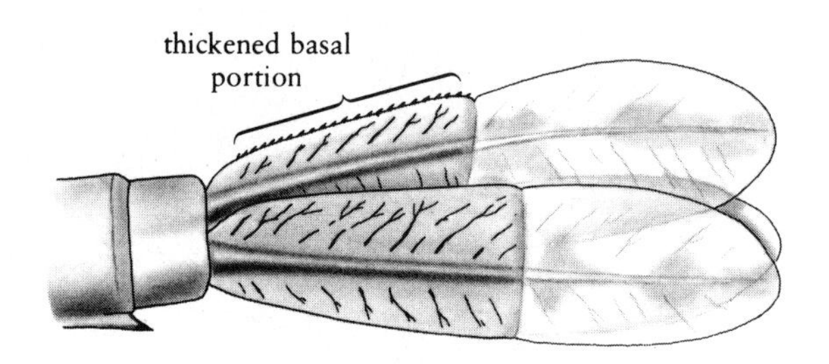

Figure 8.15. Larval caudal lamellae

PROTONEURID DAMSELFLIES
(Family Protoneuridae)

LARVAL DIAGNOSIS: (Fig. 8.15) These are generally similar to narrowwinged damselflies, except that each caudal lamella is divided into a thick basal portion for one-half or slightly more of its length, and a much thinner, more leaflike distal portion.

ADULT DIAGNOSIS: These are generally similar to narrowwinged damselflies except that posterior longitudinal veins are reduced in the wings, vein Cu_2 of the fore wings is very short or absent, and vein Cu_1 is short.

DISCUSSION: This tropical family is represented in North America by two genera and two species from stream habitats in parts of Texas.

Dragonflies
(Suborder Anisoptera)

Larvae are slender to robust. Abdomen broadens from the base and lacks caudal lamellae (Fig. 8.16).

Adults have large, closely set eyes (Figs. 8.18, 8.22). Wings at rest are held outspread laterally from the body (Plate V, Fig. 31; Plate VI, Figs. 32, 34, 35).

SUPERFAMILY AESHNOIDEA

GRAYBACKS
(Family Petaluridae)

LARVAL DIAGNOSIS: (Fig. 8.16) Body is about 38 mm at maturity. Antennae have six or seven short, stout segments. Labium is flat. Middle and posterior segments of abdomen each have patches of black bristles dorsally.

ADULT DIAGNOSIS: These are large, grayish or blackish forms. Eyes are separated dorsally. Most costal crossveins of the fore wings are not in line with subcostal crossveins. Stigma of fore wings measures 8 mm or more in length.

Nevertheless at length, O reedy shallows,
Not as a plodding nose to the slimy stem,
But as a brazen wing with a spangled hem,
Over the jewel-weed and the pink marshmallows,
Free of these and making a song of them,
I shall arise, and a song of the reedy shallows.

Edna St. Vincent Millay

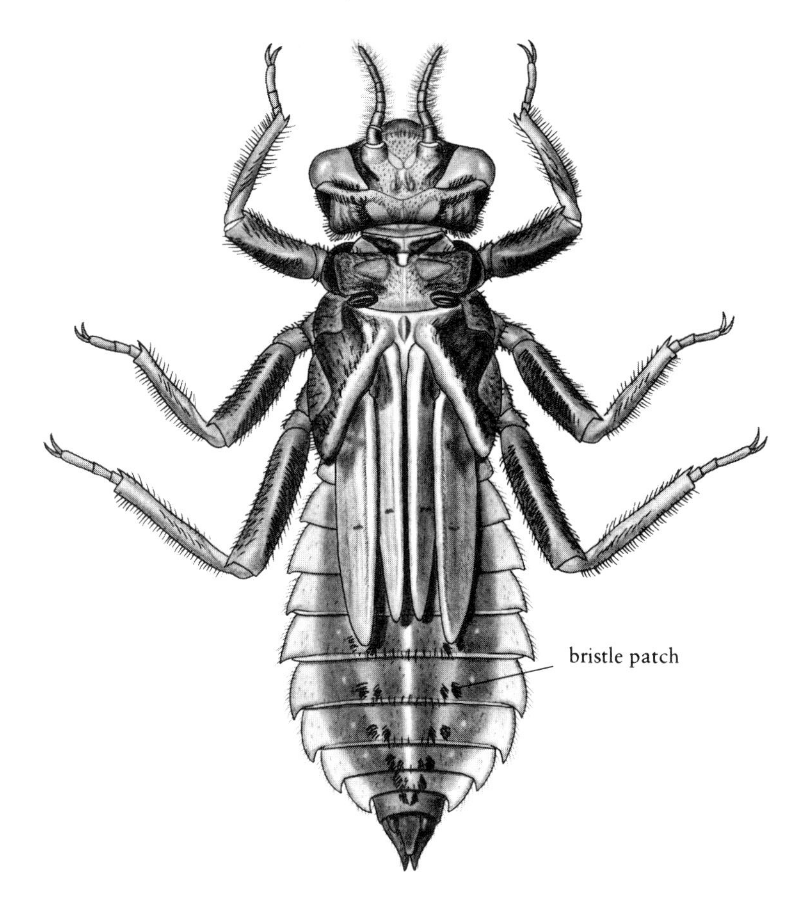

Figure 8.16. *Tachopteryx* larva

DISCUSSION: One western species (the Western Grayback) and one eastern species (the Eastern Grayback) are known only very locally in North America. The family is *not* known from central Canada. Larvae occur in western mountain bogs and in spring bogs and seepage areas of eastern forests.

CLUBTAILS
(Family Gomphidae)

LARVAL DIAGNOSIS: (Figs. 8.17, 8.20; Plate V, Fig. 30) These are variable in form and usually measure 23–40 mm, rarely to 65 mm when mature. Antennae are four-segmented (segment 3 is large and robust, and segment 4 is relatively small and sometimes inconspicuous). Labium is flat. Tarsus of middle legs is two-segmented.

ADULT DIAGNOSIS: (Figs. 8.18, 8.19; Plate V, Fig. 31) These are dark forms that usually have greenish or yellowish markings. Eyes are distinctly separated dorsally. Most costal crossveins of the fore wings are not in line with subcostal crossveins. Stigma of fore wings is less than 8 mm in length. Posterior segments of abdomen are sometimes enlarged.

DISCUSSION: The uniquely robust antennae of clubtail larvae allow them to be readily distinguished from other dragonfly larvae, such as the

CLUBTAILS
(Gomphidae)

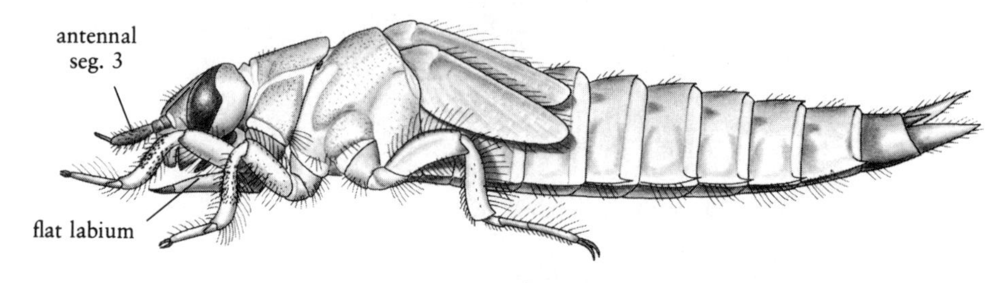

Figure 8.17. *Progomphus* (Gomphinae) larva

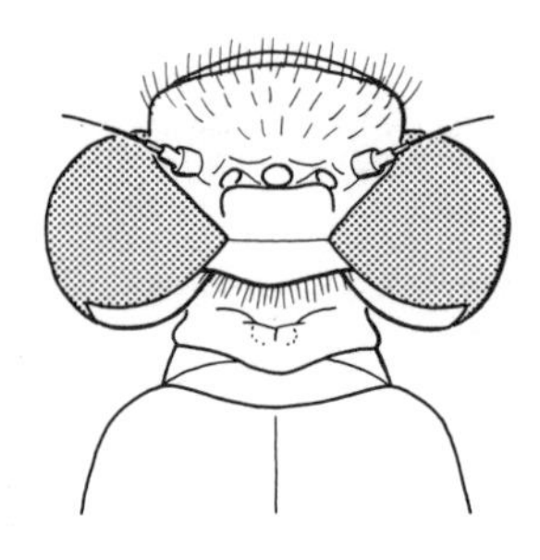

Figure 8.18. Adult head, dorsal

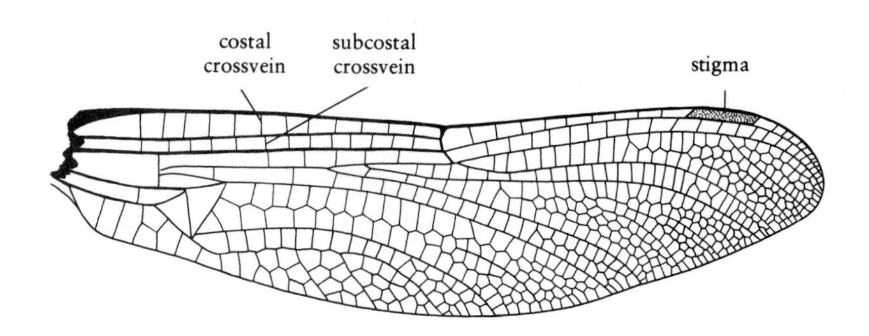

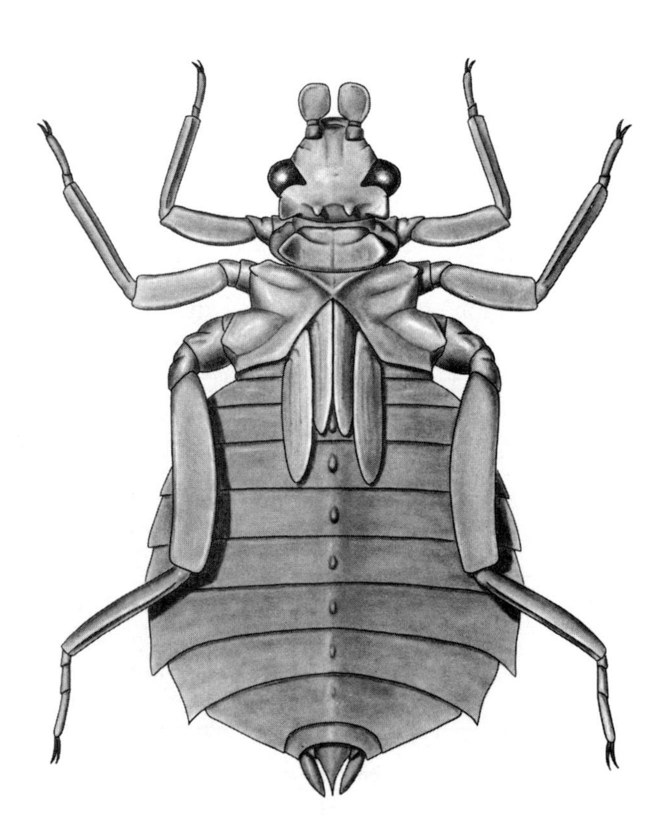

Figure 8.20. *Hagenius* (Hageniinae) larva

darners and graybacks. Somewhat elongate ninth and tenth abdominal segments are also typical of several clubtail species.

This is a moderately large, widespread family represented by about 16 genera and over 80 species in North America.

Larvae of Gomphinae burrow in the sand, silt, or sometimes gravel or debris-strewn substrates of streams, rivers, lakeshores, and ponds. They lie in wait for prey while partially buried. Notable adaptations for this type of microhabitat are the digging spurs on the legs and the elongate posterior abdominal segments that extend the posterior tip of body above the level of the substrate (as in some *Gomphus*; highly developed in *Aphylla*). The highly flattened and broad larvae of the single species of *Hagenius* (Hageniinae) are not burrowers, like the other clubtails, but are adapted more for sprawling. These "Black Dragons" (Fig. 8.20) are most unusual in appearance. *Ophiogomphus* can generally tolerate only very clean water of unaltered streams.

Emergence of clubtails typically takes place on the ground or on rocks near the water's edge. Short, low flights in bright sunlight are the general rule. Some species commonly prey on other dragonflies. Females oviposit somewhat irregularly, directly on the water.

SUBFAMILY HAGENIINAE: **Larval** abdomen (Fig. 8.20) is extremely broadened and flat. Only one species occurs in North America, and it is known from eastern and central areas but not from western states or western Canada.

SUBFAMILY GOMPHINAE: **Larval** abdomen (Fig. 8.17; Plate V, Fig. 30) is elongate to somewhat oval. This group is widespread in North America.

DARNERS
(Family Aeshnidae)

LARVAL DIAGNOSIS: (Fig. 8.21; Plate VI, Fig. 33) These are somewhat elongate, usually patterned forms with a spindle-shaped abdomen that is tapered at the end. When mature they measure 31–50 mm, sometimes more. Antennae have six or seven small, slender segments. Labium is flat.

ADULT DIAGNOSIS: (Fig. 8.22; Plate VI, Fig. 32) These are generally large, dark forms with bluish or greenish markings. Eyes are broadly joined dorsally. Most costal crossveins of the fore wings are not in line with subcostal crossveins.

DARNERS
(Aeshnidae)

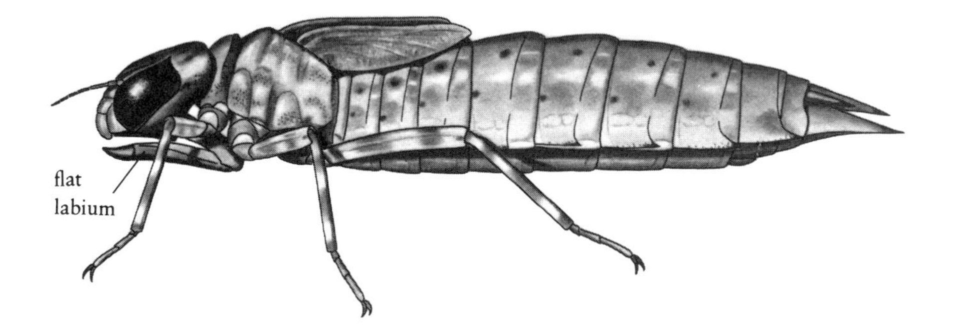

Figure 8.21. *Aeshna* larva

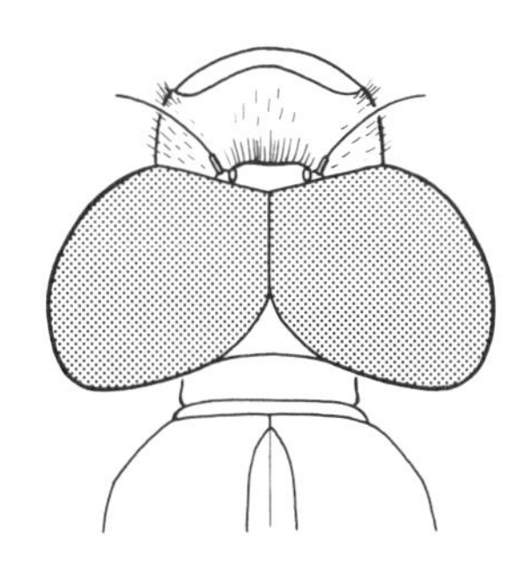

Figure 8.22. Adult head, dorsal

DISCUSSION: Most larvae of darners can readily be distinguished by the long, tapering, and somewhat cylindrical abdomen. Their slender antennae and legs are distinct from those of clubtails, with which they share a characteristically flat labium.

This is a moderately sized group in North America, where about 38 species are known. These species are classified in about 10 small genera, each with few species, and *Aeshna,* which is by far the largest and most widespread genus, with about 20 species.

The larvae of most species actively climb among vegetation, tangled roots, and woody debris in marshes, ponds, and the slower reaches of streams and rivers. They actively stalk their prey—a wide array of small animals, including fishes. They are sometimes highly cannibalistic. A few species are atypically sluggish and are found under rocks or logs along lake shores or in moderate currents of streams.

Adults are large, powerful fliers. They are primarily diurnal, although some species are early evening fliers. *Aeshna* adults may often be seen flying back and forth, commonly pausing to hover in mid air for short periods of time. When at rest, most species perch on vegetation in a vertical position. Eggs are oviposited into vegetation or woody debris. Some *Anax* populations are migratory, with spring emergers flying north to Canada and northern states and late summer and fall emergers returning south.

Large tied flies for nymph fishing have been patterned after a few species of darners. The Green Darner (Plate VI, Fig. 33) is patterned after the larvae of the common *Anax junius*, and the Giant Dragonfly Nymph is patterned after *Aeshna constricta* (the Eastern Paddletail) or other species of *Aeshna*. When used for trout, these flies should be weighted and fished with a slight periodic jerking movement on the bottom of slow or moderately flowing reaches of streams.

SUPERFAMILY CORDULEGASTROIDEA

BIDDIES
(Family Cordulegastridae)

LARVAL DIAGNOSIS: (Figs. 8.23, 8.24) These are hairy, elongate forms that measure 33–43 mm when mature. Antennae are slender and seven-segmented. Labium is spoon-shaped, cupping the bottom of head. Lateral lobes of labium possess large teeth of different sizes along the distal margin.

ADULT DIAGNOSIS: These are very dark and large forms with bright yellow markings. Eyes nearly touch or touch at a single point dorsally. Most costal crossveins of the fore wings are not in line with subcostal crossveins. Brace vein is not present in fore wings.

DISCUSSION: The larvae of biddies are easily distinguished from other dragonfly larvae with spoon-shaped labia by the large jagged teeth of the lateral lobes (Fig. 8.24). This is a small group, represented in North America by seven species of the genus *Cordulegaster.* This genus is widespread but rare in Texas and some central areas, and may be only locally common in other parts of North America.

Unlike the clubtails, the larvae of biddies do not burrow, but they do cover themselves with silt up to their eyes. They lie in wait for prey in this fashion at the bottoms of pools of streams. The stream habitats are located primarily in forested areas. Larvae take three or four years to develop. Adults often fly slowly for considerable distances along the stream course, and females are known to oviposit upstream far from their emergence site.

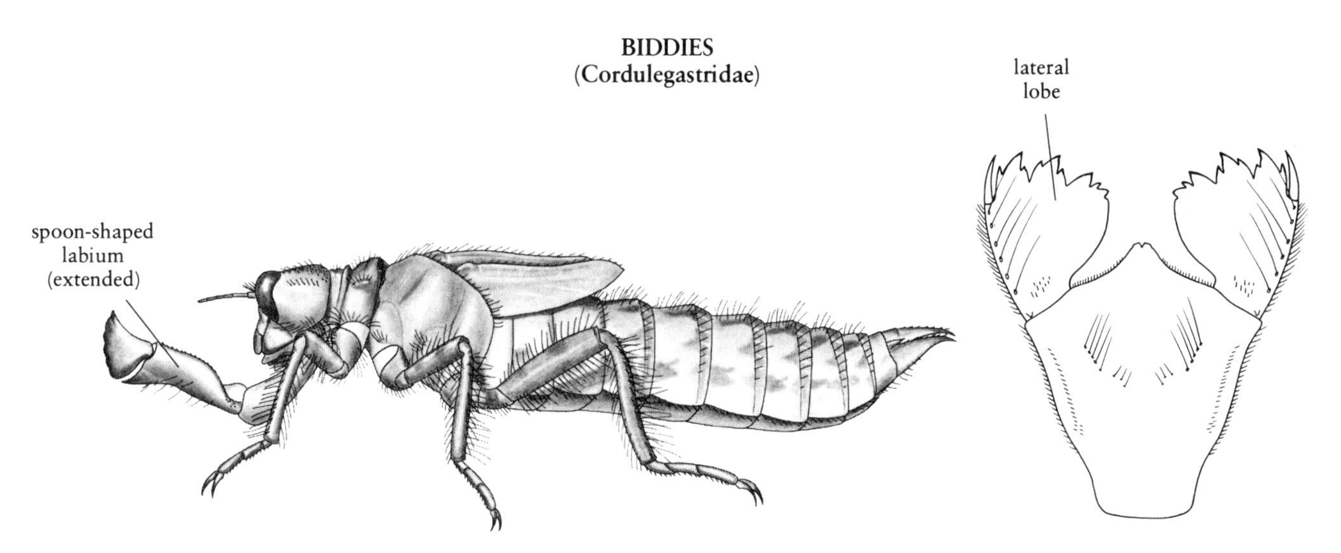

Figure 8.23. *Cordulegaster* larva

Figure 8.24. Larval labium

SKIMMERS
(Superfamily Libelluloidea)

BELTED AND RIVER SKIMMERS
(Family Macromiidae)

LARVAL DIAGNOSIS: (Fig. 8.25) These are robust forms with very long and slender legs and a broad and flat abdomen. They usually measure 25–40 mm when mature. Head possesses a distinctive and well-developed frontal horn arising between antennae. Antennae are slender and seven-segmented. Labium is spoon-shaped and cups the bottom of head.

ADULT DIAGNOSIS: (Fig. 8.26, 8.27) These are large, dark forms with a yellow stripe encircling the thorax and passing between the fore and hind wings. Eyes touch for a short distance dorsally. Spots and bands are not present on wings. Most costal crossveins of the fore wings are in line with subcostal crossveins. Anal loop of hind wings is short and rounded.

DISCUSSION: The combination of long legs, broadly rounding abdomen (as seen from above), and distinctive frontal horn on the head (Fig. 8.25) distinguishes the larvae of this group from all other skimmers. A few species of *Neurocordulia* (a genus of greeneyed skimmers) also possess frontal horns on the head, but do not possess the long legs typical of belted and river skimmers. In macromiids the hind femora typically reach as far back as abdominal segment 8. Other skimmers' hind femora usually do not reach beyond abdominal segment 6.

And the dragonfly in light
Burnished armor shining bright,
Came tilting down the river
In a wild bewildered flight.

James Whitcomb Riley

The Macromiidae are primarily tropical, represented in North America by only about 10 species and two genera, *Macromia* (the river skimmers) and *Didymops* (the belted skimmers). The belted skimmers and most species of river skimmers occur in the eastern half of North America. The family is absent from much of central Canada.

Larvae usually occur on sand and silt bottoms of large streams and rivers or lakes with considerable wave action. As they lie in wait for prey, they may be camouflaged by a thin layer of silt or presumably simply by their color patterns, which act to break up the outline of their bodies against the substrate. Adults are usually high flying and elusive.

BELTED AND RIVER SKIMMERS
(Macromiidae)

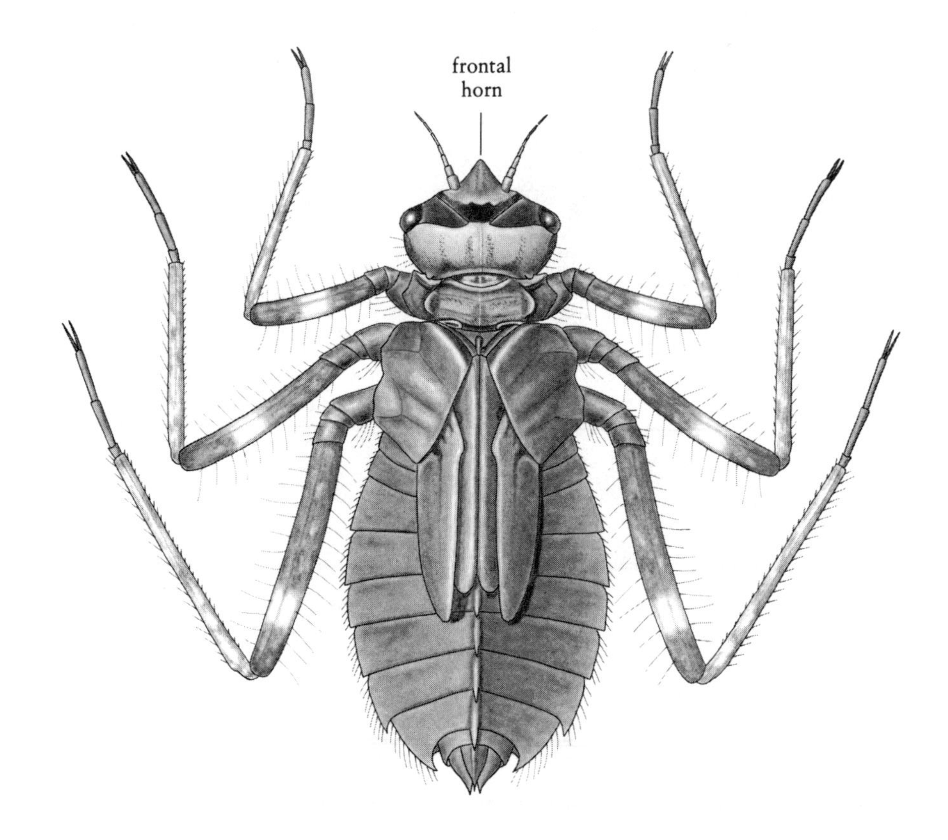

Figure 8.25. *Macromia* larva

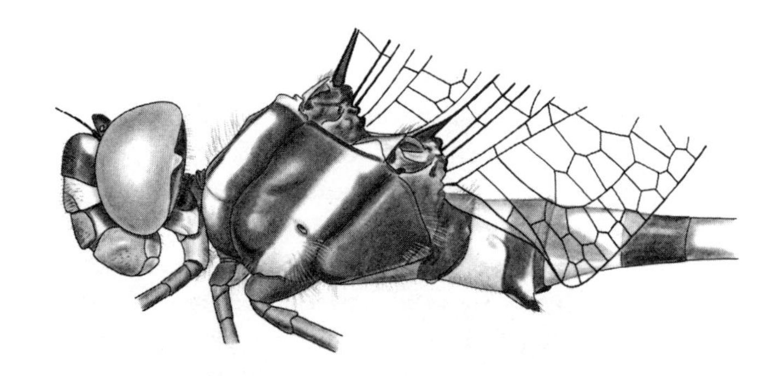

Figure 8.26. *Macromia* adult, lateral

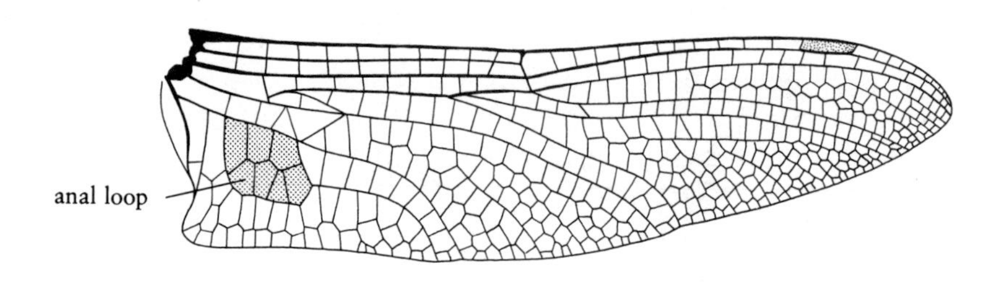

Figure 8.27. Hind wing

GREENEYED SKIMMERS
(Corduliidae)

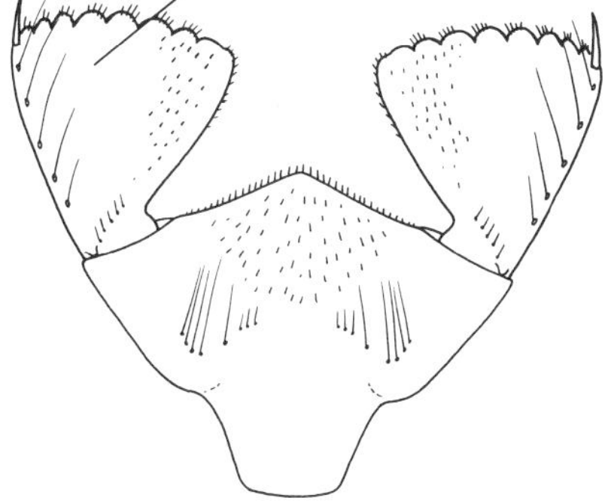

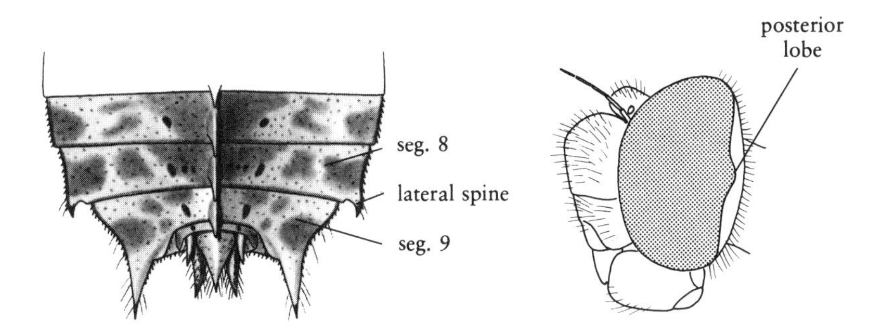

Figure 8.28. *Epitheca* larva, labium

Figure 8.29. *Epitheca* larva, end of abdomen

Figure 8.30. Adult head, lateral

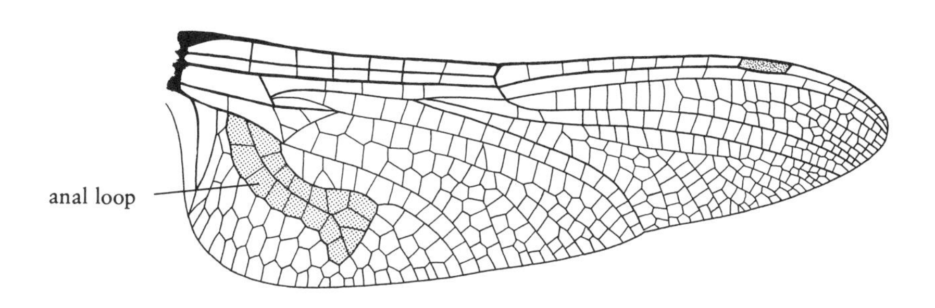

Figure 8.31. Hind wing

GREENEYED SKIMMERS
(Family Corduliidae)

LARVAL DIAGNOSIS: (Figs. 8.28, 8.29) These are relatively small to medium-sized, usually hairy forms with somewhat broadened abdomens. They measure 16–28 mm when mature. Antennae are slender and seven-segmented. Labium is spoon-shaped, cupping the bottom of head. Lateral lobes of labium have a series of distinct, moderately sized indentations. Lateral spines of abdominal segment 8 are either absent or shorter than midlength of abdominal segment 9.

ADULT DIAGNOSIS: (Figs. 8.30, 8.31; Plate VI, Fig. 34) These are usually medium-sized, slender, dark, and somewhat metallic forms. Eyes touch dorsally. Most costal crossveins of fore wings are in line with the subcostal crossveins. Anal loop of hind wings is elongate and not distinctly boot-shaped.

DISCUSSION: Larvae of greeneyed skimmers are most commonly confused with those of common skimmers, and characters must be examined in detail to ensure correct identification. In the greeneyed skimmers (Fig. 8.28), the series of rounded teeth along the edge of the lateral lobes of the labium are usually one-fourth to one-half as long as they are wide. Except for two species, the teeth of common skimmers are not nearly as deeply cut (typical structure shown in Fig. 8.33). Those common skimmers (i.e., genus *Pantala*) that may be confused with greeneyed skimmers on the basis of the lateral lobes can be told by their relatively long lateral spines on abdominal segments 8 and 9 (see Figs. 8.35, 8.36). Some Corduliidae larvae possess a frontal horn on the head, but they are otherwise easily distinguished from those of the Macromiidae.

At least seven genera and about 50 species of greeneyed skimmers are recognized in North America. The group is a bit unusual for Odonata in that as a whole it tends to become better represented at more northern latitudes in North America. Although widespread, the group is most common in the eastern states and Canada.

Larvae often occur in cool ponds, lakes, streams, and a wide variety of bog habitats. They crawl about on bottom substrate and in vegetation. Emergence periods of some *Neurocordulia* correspond closely to emergences of common burrowers (mayfly genera *Ephemera* and *Hexagenia*), which often make up a primary food source of the adults.

A number of behaviorally distinct flight patterns are identifiable in adults of some species of *Epitheca*, among others. These include "patrolling" for territorial purposes, which involves much hovering; "feeding," which involves quick vertical and horizontal flight away from the water to procure food; "copulatory," which involves a linear flight of both sexes above the water for considerable distances; "swarm feeding," which involves aggregating prey insects and predator dragonflies (usually more than one species of dragonfly); and "preoviposition," which involves site selection by females.

COMMON SKIMMERS
(Family Libellulidae)

LARVAL DIAGNOSIS: (Figs. 8.32–8.36; Plate VI, Fig. 36) These are similar to greeneyed skimmers. They measure 8–28 mm when mature. Lateral lobes of labium usually possess only slight indentations; if indentations are deeper and similar to those of greeneyed skimmers, then lateral spines of abdominal segment 8 are as long as or longer than midlength of abdominal segment 9.

ADULT DIAGNOSIS: (Figs. 8.37, 8.38; Plate VI, Fig. 35) These are variably colored forms that often have patterned wings. Eyes almost always lack a lobe on their posterior margin. Anal loop of hind wings is usually boot-shaped and has a distinctly developed, pointed "toe."

DISCUSSION: Larvae of common skimmers are generally similar to those of greeneyed skimmers, and two species of the genus *Pantala* can be confused on the basis of labial characteristics. See the comparative discussion under greeneyed skimmers for details, and compare Figs. 8.28, 8.29, 8.33–8.36.

This is the largest group of dragonflies in North America with over 90 species, including many that are common and widespread.

Larvae are common pond inhabitants. They also occur in swamps, marshes, and the still edgewaters of streams and rivers, and sometimes in springs or ditches (especially in the western states). They are sprawlers along the bottom and climbers among debris and vegetation. Many species (such as those of *Sympetrum*) are adaptable to a wide variety of habitat types and are very tolerant of low dissolved oxygen levels or highly eutrophic environments. A few species are occasionally associated with alkaline pools, estuaries, or salt marshes.

COMMON SKIMMERS
(Libellulidae)

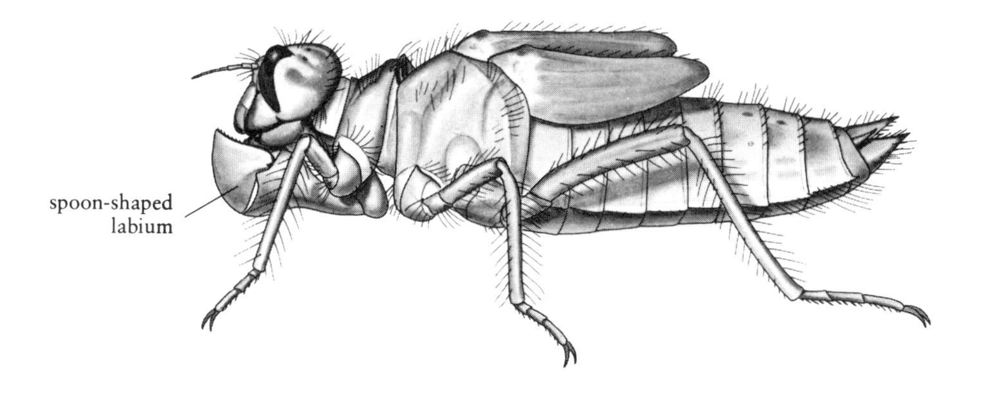

Figure 8.32. *Libellula* larva

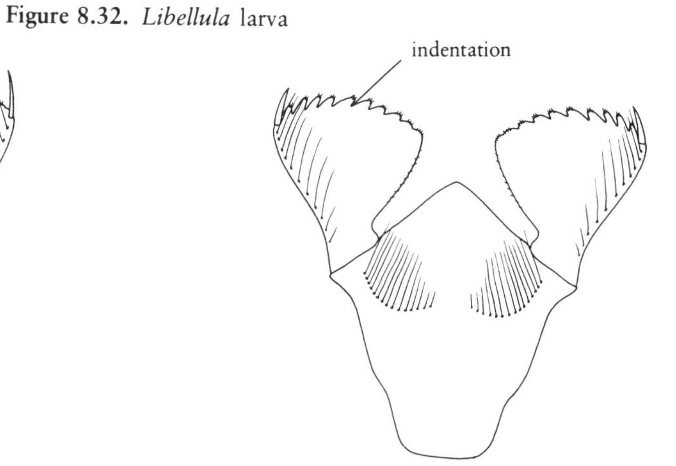

Figure 8.33. Larval labium

Figure 8.35. Larval labium

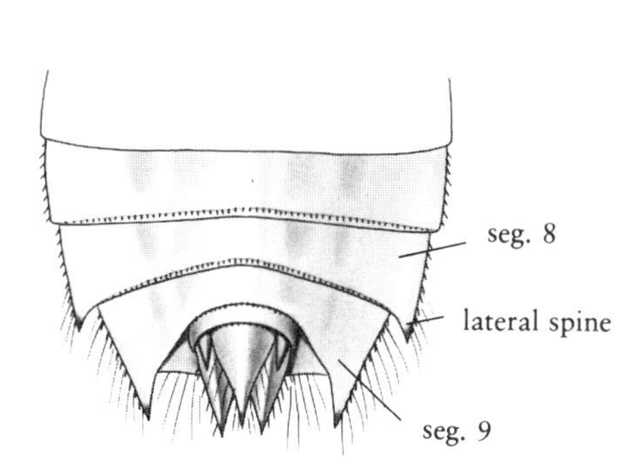

Figure 8.34. *Sympetrum* larva, end of abdomen

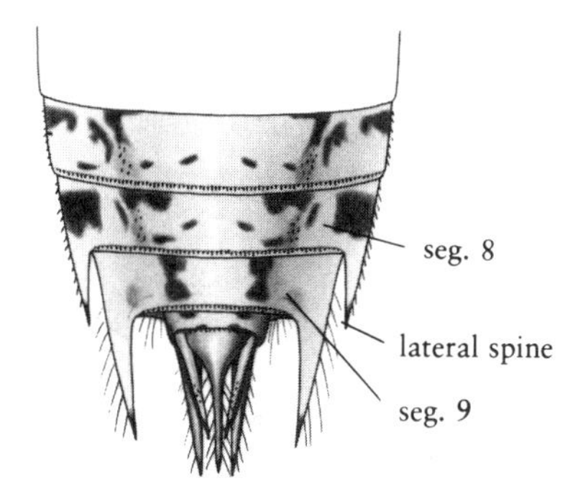

Figure 8.36. *Pantala* larva, end of abdomen

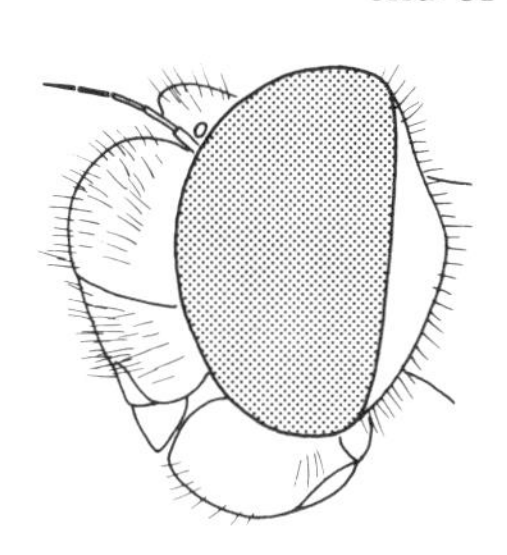

Figure 8.37. Adult head, lateral

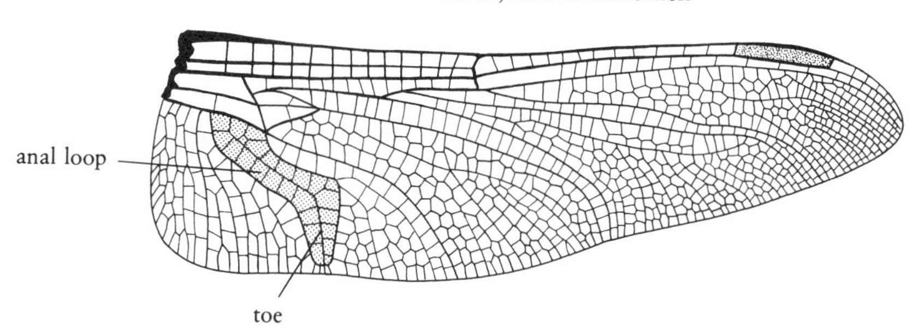

Figure 8.38. Hind wing

The common, rather erratic fliers seen around ponds are usually common skimmers, especially *Libellula* and *Celithemis*. These genera are well known for their beautifully patterned wings. Adults of *Sympetrum* and *Leucorrhinia* are more commonly found flying about marsh habitats and have mostly clear wings. Territorialism is sometimes highly developed among common skimmers. Some species have very long emergence periods, occurring as adults from spring through early fall. Females of most species oviposit by striking the water periodically and rhythmically with the tip of the abdomen. Certain *Sympetrum* females oviposit on marginal land, where the eggs overwinter and do not hatch until flooded in the following spring. A common behavioral trait of the males of common skimmers is to come to rest repeatedly on the same branch or twig between periods of flight.

Larvae of some species, such as *Libellula pulchella* (Plate VI, Figs. 35, 36) (the Ten-Spot Skimmer) and *Sympetrum ribicundulum* (the Common Red Skimmer), can be used for bait fishing or as fly patterns for nymph fishing, especially in ponds and lakes.

More Information about Odonata

A biology of dragonflies by Corbet (1963). This is a most comprehensive review and treatment of the bionomics, ecology, and behavior of the Odonata.

"The damselflies (Zygoptera) of Texas" by Johnson (1972). This work will be of considerable help to those who wish to identify adult damselflies from the southern states.

A manual of the dragonflies of North America by Needham and Westfall (1955). This is perhaps the single best book for taxonomic and distributional information on dragonflies (Anisoptera) of the United States.

The Odonata of Canada and Alaska, Vol. 1 by Walker (1953), Vol. 2 by Walker (1958), and Vol. 3 by Walker and Corbet (1975). This three-volume set is indispensible for the serious student of the Odonata, since it contains a wealth of taxonomic (keys to adults and larvae, descriptions, and distributions), biological, and ecological information on much of the North American fauna. Volume 1 contains a general treatment of the Odonata and specific treatment of the damselflies. Volume 2 contains a specific treatment of Aeshnidae, Petaluridae, Gomphidae, and Cordulegastridae. Volume 3 contains a specific treatment of the Macromiidae, Corduliidae, and Libellulidae.

References

Benke, A. C., and S. S. Benke. 1975. Comparative dynamics and life histories of coexisting dragonfly populations. *Ecology* 56:302–317.

Borror, D. J. 1963. Common names for Odonata. *Proc. N.C.B. Entomol. Soc. Amer.* 18:104–107.

Borror, D. J., D. M. DeLong, and C. A. Triplehorn. 1976. *An introduction to the study of insects* (4th ed.). Holt, Rinehart and Winston, New York. 852 pp.

Byers, C. F. 1930. *A contribution to the knowledge of Florida Odonata.* Univ. Fla. Biol. Sci. Ser. 1. 327 pp.

Corbet, P. S. 1963. *A biology of dragonflies*. Quadrangle Books, Chicago. 247 pp.

Donnelly, T. W. 1970. *The Odonata of Dominica British West Indies*. Smith. Contr. Zool. 37. 20 pp.

Fraser, F. C. 1957. *A reclassification of the order Odonata*. Roy. Zool. Soc. N.S.W. Handbook 12. 133 pp.

Johnson, C. 1972. The damselflies (Zygoptera) of Texas. *Bull. Fla. St. Mus. Biol. Sci.* 16:55–128.

Komnick, H. 1977. Chloride cells and chloride epithelia of aquatic insects. *Inter. Rev. Cyt.* 49:285–329.

Kormondy, E. J. 1959. *The systematics of Tetragoneuria, based on ecological, life history, and morphological evidence (Odonata: Corduliidae)* Misc. Publ. Mus. Zool. Univ. Mich. 107. 79 pp.

McCafferty, W. P. 1979. Swarm-feeding by the damselfly *Hetaerina americana* on mayfly hatches. *Aq. Ins.* 1:149–151.

Needham, J. G., and H. B. Heywood. 1929. *A handbook of the dragonflies of North America.* Charles C. Thomas, Springfield, Ill. 378 pp.

Needham, J. G., and M. T. Westfall, Jr. 1955. *A manual of the dragonflies of North America (Anisoptera).* Univ. Calif. Press, Berkeley. 615 pp.

Provonsha, A. V. 1975. The Zygoptera (Odonata) of Utah with notes on their biology. *Gr. Bas. Natural.* 35:379–390.

Provonsha, A. V., and W. P. McCafferty. 1973. Previously unknown nymphs of western Odonata (Zygoptera: Calopterygidae, Coenagrionidae). *Proc. Entomol. Soc. Wash.* 75:449–454.

Provonsha, A. V., and W. P. McCafferty. 1977. Odonata from Hot Brook, South Dakota with notes on their distribution patterns. *Entomol. News* 88:23–28.

Schwiebert, E. 1973. *Nymphs.* Winchester Press, New York. 339 pp.

Waage, J. K. 1979. Dual function of the damselfly penis: sperm removal and transfer. *Science* 203:916–918.

Walker, E. M. 1953. *The Odonata of Canada and Alaska, Vol. 1: General. The Zygoptera—damselflies.* Univ. Toronto Press, Toronto. 292 pp.

Walker, E. M. 1958. *The Odonata of Canada and Alaska, Vol. 2: The Anisoptera—Four Families.* Univ. Toronto Press, Toronto. 318 pp.

Walker, E. M., and P. S. Corbet. 1975. *The Odonata of Canada and Alaska, Vol. 3: The Anisoptera—Three Families.* Univ. Toronto Press, Toronto. 307 pp.

Westfall, M. J., Jr. 1978. Odonata, pp. 81–98. In *An introduction to the aquatic insects of North America* (R. W. Merritt and K. W. Cummins, eds.). Kendall/Hunt, Dubuque.

Willey, R. L. 1972. The damselfly (Odonata) hindgut as host organ for the euglenoid flagellate *Colocium. Trans. Amer. Micr. Soc.* 91:585–593.

Wright, M., and A. Peterson. 1944. A key to the anisopterous dragonfly nymphs of the United States and Canada (Odonata, suborder Anisoptera). *Ohio J. Sci.* 44:151–166.

CHAPTER 9

Stoneflies

(ORDER PLECOPTERA)

THE STONEFLIES are all freshwater inhabitants as larvae. As a group they are close relatives of the cockroaches and have retained the primitive condition of possessing tails but demonstrate the advanced ability to fold their wings over the back of the body. Their common name undoubtedly is derived from the fact that individuals of many common species are found crawling or hiding among stones in streams or along stream banks. Close to 500 species are represented in North America. Many stoneflies are known as clean-water insects, since they are often restricted to highly oxygenated water. As such, some are excellent biotic indicators of water quality, and many are important food for game fishes. Interestingly, adults of stoneflies can be found throughout the year, some being adapted for winter emergence.

Tied imitations of stoneflies are especially valuable to trout fishermen for use in mountain streams and for some early-season fly fishing. In fact, the first recorded artificial fly—made over four and a half centuries ago—was based on a British stonefly. The use of larval imitations (nymph fishing) of stoneflies has become a most popular sport in recent years. The Salmonfly (a giant stonefly) of western rivers is among the most famous and popular of fishermen's flies in North America. Other stonefly patterns include the Golden Stone, Yellow Sally, Bird's Stonefly, Artesan Green, and Sofa Pillow, to name just a few; and Wooly Worms and the western Bucktail Caddis series are also likely to be accepted as stoneflies by trout. A number of stonefly patterns no doubt actually work more as attractors than as imitations. The larger stoneflies have long proven to be excellent bait for both Smallmouth Bass and trout.

Larval Diagnosis

These are elongate, somewhat flattened to cylindrical forms that usually measure (not including tails) 5–35 mm and sometimes up to 60 mm when mature. Head possesses moderately developed and widely separated eyes, long slender antennae, and chewing mouthparts. Legs are well developed, and each ends in two claws. Developing wing pads are usually present in mature larvae, rarely absent or reduced. Fingerlike or filamentous, simple or branched gills are sometimes present on mouthparts, thorax, leg bases, and/or abdomen. Abdomen terminates in two tails.

Adult Diagnosis

Adults are soft-bodied, elongate insects, and usually have membranous (often lightly pigmented), elongate wings that are held relatively flat over the body and extend beyond the tip of the abdomen. Wings are reduced or absent in some species, but when fully developed, the hind portion of the hind wings is usually enlarged as an anal lobe. Head possesses slender antennae. Tarsi are three-segmented. Abdomen ends in two tails that are usually evident from below but are inconspicuous in some.

Similar Orders

Larvae resemble those of certain flattened mayflies, but those mayflies do not possess elongate antennae and two claws on each leg, as do stoneflies; and mayflies often have three tails.

Among adult insects that may be taken near streams or rivers, the fishflies and dobsonflies may superficially resemble some of the larger stoneflies. Fishflies and dobsonflies, however, lack tails, and the tips of their wings when at rest do not overlap, as is typical of larger stonefly adults. Although typically awkward, somewhat jerky fliers, stoneflies may sometimes be mistaken for mayflies or caddisflies when flying over streams at dusk.

Many fly fishermen collectively refer to the stoneflies along with the caddisflies as the "downwings." These are thus distinguished from the mayflies, which are known as "upwings." Downwings give a much different impression to selective fishes than do the upwings.

Life History

Metamorphosis is incomplete, and although the habitats and habits of the adults and larvae (also commonly referred to as nymphs, naiads, or creepers) are obviously different, the structural differences are not as radical as they are among some other aquatic groups with incomplete metamorphosis, such as mayflies. Larval development usually takes from three months to one year and usually involves from 12 to 22 larval instars. Larger species often require up to three years and more than 22 instars. Many species cope with unfavorable seasonal extremes by having a corresponding period of dormancy as young larvae or eggs.

Emergence of adults takes place at all times of the year, different species emerging at different times. Although there are numerous exceptions in North America, euholognathe stoneflies generally emerge in winter and spring, and systellognathe stoneflies generally emerge in summer or sometimes in fall. Emergence time varies somewhat within a species, depending on factors that govern water temperature, such as altitude and latitude. Larvae typically climb out of the water before transforming to the adult, and this is why stonefly imitations generally are not nearly as widely used by fly fishermen as are those of the mayflies. Mayflies typically hatch off the water, thus inducing a good deal of feeding activity among fishes.

Adult longevity is highly variable among species, generally ranging from a few days to about five weeks. Feeding habits also vary. Short-lived adults usually do not take food. Some species take only liquids, and others are herbivores that feed on such items as pollen, plant buds, or the encrusting growths found on tree bark.

Depending on the species, adult stoneflies may be diurnal, crepuscular, or nocturnal. Some frequent elevated structures and vegetation, many remain on the ground, and others hide under stones or driftwood along stream banks. Although some restrict their activities to the immediate area of the aquatic environment, others fly considerable distances from the aquatic environment before returning for egg laying.

Males often precede females in emergence. Mating takes place on the ground or occasionally on vegetation. Males of some species attract or signal females by beating the abdomen against a hard surface (a phenomenon known as drumming behavior). Drumming may elicit similar behavior in the female and eventually lead to mating. Mated females generally are not responsive to drumming and refuse to mate again. Instead of relying on drumming, males of some species actively search out females for mating.

Most females carry eggs as an exposed mass on their abdomen prior to ovipositing them somewhere in the environment. They may oviposit by running on or flying over the water and dipping the tip of the abdomen onto the surface; by dropping eggs into the water from the air; by depositing them along the bank; or by crawling into the water and depositing them on underwater substrate. It is during this oviposition period that fishing with dry fly stonefly imitations is generally most productive, since trout may often gorge themselves when egg-laying females are plentiful.

The manufacture of the green-drake, grey-drake, and stone-fly, in particular, should be well understood, as it is sometimes difficult to procure, or preserve the natural ones; and moreover, a proficiency in the art of making these will enable any person to make a fly to any pattern, an art highly necessary, for it will often happen that Trout will refuse every fly you have with you; and the only recourse then is, to sit down and make one resembling, as much as possible, those which you may find flying about the spot.

Thomas Salter

Noteworthy in terms of insects in general is the discovery that one species of stonefly spends its entire life history at depths of 70 meters or more in Lake Tahoe, the adults having become completely adapted for aquatic life and reproduction. A cold-season terrestrial existence is more generally developed in stoneflies than in perhaps any other large group of insects. These adults are able to withstand periods of extreme cold air temperatures by insulating themselves within ice or snow caverns or under rocks or leaf detritus. At least one species periodically enters the warmer aquatic environment to escape colder air temperatures. Adults of some stoneflies are known to exude body fluid (hemolymph) from parts of the body as a retardative against would-be predators (a phenomenon known as autohemorrhaging or reflex bleeding).

Aquatic Habitats

Stoneflies are found in benthic habitats of flowing waters and occasionally in northern or cold lakes with considerable wave action. Many are relatively restricted in habitat, and organic enrichment or other forms of pollution that will reduce available dissolved oxygen in the water can prohibit their occurrence. The young larvae of many species remain interstitially at depths of a few centimeters or more within soft or gravelly substrates for a period (e.g., aestivation in a drying intermittent stream) before resuming activities on the surface of the stream bottom.

A few species can exist naturally throughout their larval life in hyporheic habitats at considerable distances from associated river beds, and have even been taken in subterranean water supplies. Larvae of many species change their microhabitat as they grow (e.g., from gravel to leaf detritus or from leaf detritus to stones); movement toward shore prior to emergence is very common.

They clamber over the stones, keeping hold by their strong claws, always seeking the dark side of everything. They have a sidling gait and when the stones on which they live are overturned and exposed to the light they scatter like rats and drop off the edges into the water.

Ann Haven Morgan

Aquatic Adaptations and Behavior

Stonefly larvae are hydropneustic and rely on the cutaneous uptake of oxygen dissolved in water. Filamentous gills found on some species provide additional surface and may be especially effective for oxygen uptake. Some of the common stoneflies ventilate by engaging in a form of "push-ups" when stressed for oxygen, thereby increasing the water flow over their bodies. Chloride cells are used for the uptake of ions in osmoregulation; these cells may be distributed ventrally or laterally on the abdomen or thorax, on the intersegmental membranes, and/or on the gills, depending on the species.

For the most part, larvae are either primarily carnivores, as are many systellognathe stoneflies and some euholognathes, or primarily leaf detritivores, as are many euholognathes. These detritivores apparently utilize the fungi and bacteria associated with leaf detritus as their source of protein. Some species feed primarily on periphyton. Preferred or available food resources may change as larvae develop and seasons change.

Generally, stoneflies are not active swimmers but are adapted more for crawling among stones, gravel, and detritus. Many are able to maintain themselves in rapid zones. A number are known to drift.

Classification and Characters

On the basis of feeding habits and other biological characteristics, stoneflies of North America have historically been divided into two suborders, the Filipalpia and Setipalpia. Recently, it has been shown that this division is not fundamental and that presumed differences do not always hold. Stoneflies are here divided into the Euholognatha (Filipalpia, in part), which includes four North American families, and the Systellognatha (Filipalpia, in part, and Setipalpia), which includes five North American families.

Many species of stoneflies are now classified in genera that have traditionally been relegated only to subgeneric status. Many species names may thus be unfamiliar under the new generic names. These name changes have been updated, and synonyms that apply to those stoneflies most familiar to North American fishermen are pointed out herein and associated with their common names.

Wings and genitalia are important for adult diagnoses. Brachyptery (short wingedness) (Figs. 9.13, 9.14) is encountered in almost all families. Tails are moderately developed unless otherwise stated in the adult diagnoses.

Figure 9.2. MATURE PLECOPTERA LARVAE

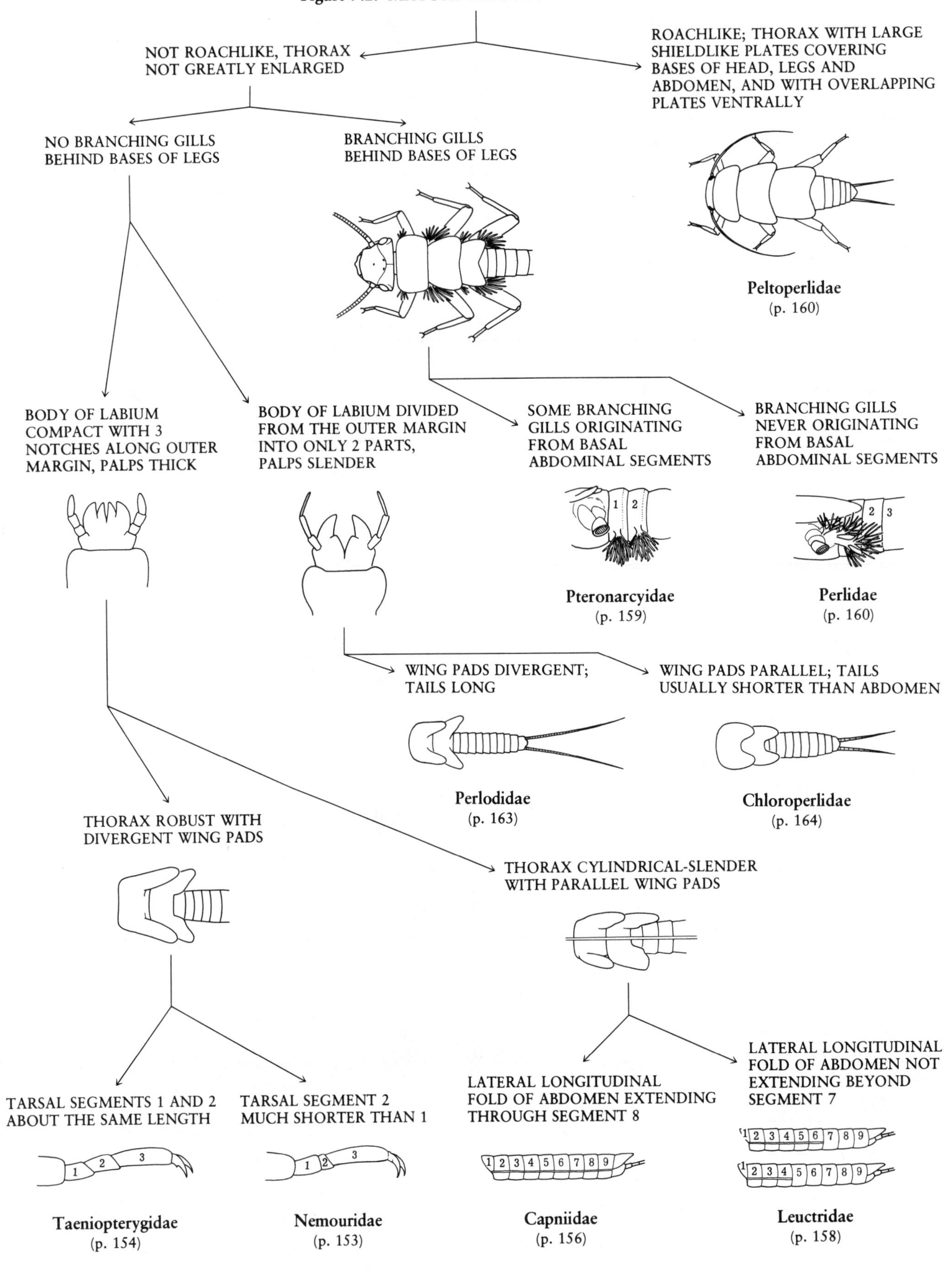

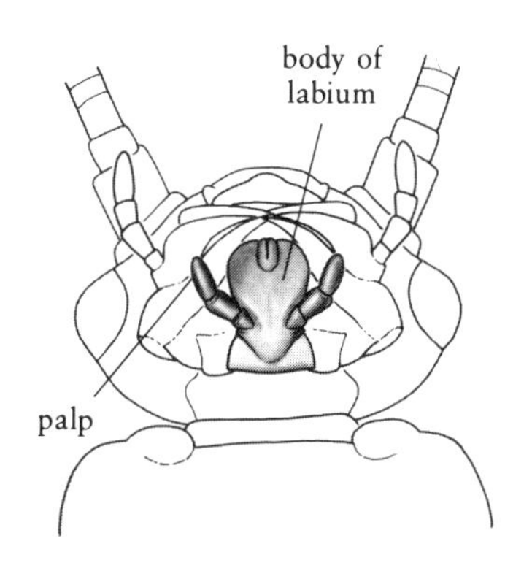

Figure 9.1. Stonefly head, ventral (labium shaded)

Gill location, wing pad shape, mouthpart structure, and hairs and spination are among the important taxonomic characters of larvae. The most reliable identifications of larvae are made using mature individuals (those in which folds of developing adult wings are sometimes detectable, or in which the thin body wall sometimes reveals underlying adult coloration or developing genitalia). Relative divergence of wing pads—an expression of the angle formed by the median axes of the wing pads—is generally indicated by the direction in which the wing pad tips are pointed (either parallel with pad tips directed straight back or divergent with each pad tip directed slightly outward).

The labium (Fig. 9.1), which for simplicity is here divided into the **body of the labium** and the **palps** (the lateral arms), can be viewed without dissection, since it is the posteriormost whole mouthpart and is usually unobstructed. Because of the similarity of larvae and adults, characteristics of mouthparts, tarsal ratios, and gills often hold for both stages.

Winter and Spring Stoneflies
(Euholognatha)

Larvae and **adults** have thick labial palps that are rounded at the tips. Body of labium is compact with three notches located along the outer margin (Figs. 9.2, 9.16, 9.17); there is never only a single notch or a deep medial incision from the outer margin. These are usually small, drab or dark forms. **Larvae** either have conspicuously flanged wing pads (Figs. 9.6, 9.9) or their bodies are entirely slender and cylindrical (Figs. 9.10, 9.18). **Adults** usually emerge in fall, winter, or spring.

NEMOURID BROADBACKS
(Family Nemouridae)

LARVAL DIAGNOSIS: (Figs. 9.4, 9.6; Plate VII, Fig. 37) These are relatively robust forms with many hairs on the body. Mature individuals usually measure 3–8 mm, sometimes longer, excluding tails. Gills are completely absent or restricted to the cervical (neck) area. Hind wing pads are strongly divergent. Hind legs usually can be extended beyond the tip of abdomen. Tarsal segment 2 is shorter than tarsal segment 1.

ADULT DIAGNOSIS: (Figs. 9.3, 9.5) An apparent **X** is usually located in the distal half of the fore wings, formed in part by the apical costal crossvein. Wings lie flat on the body and do not curve around the sides of the body. Tails are inconspicuous and reduced to one full segment.

DISCUSSION: Larvae of nemourids are similar to those of taeniopterygids. The latter, however, have a distinctive tarsus (Fig. 9.9), with segments 1 and 2 short and subequal, and commonly have dorsal color patterning. When cervical gills are present, they will readily distinguish the nemourids.

Over 60 species and 12 genera are known from North America, and the genus *Amphinemura* is widespread.

Small rivers, streams, and springs are the most common habitats of larvae. Some prefer soft substrates, and some commonly are associated with leaf packs and other detritus. Mountain species often occur in cold, rocky-bottomed streams. They are primarily detritivores, devouring leaves and the resident microorganisms of those leaves.

Although sometimes known as spring stoneflies, this common name is in part a misnomer, since adults may emerge throughout the year, depending on the species and local climate.

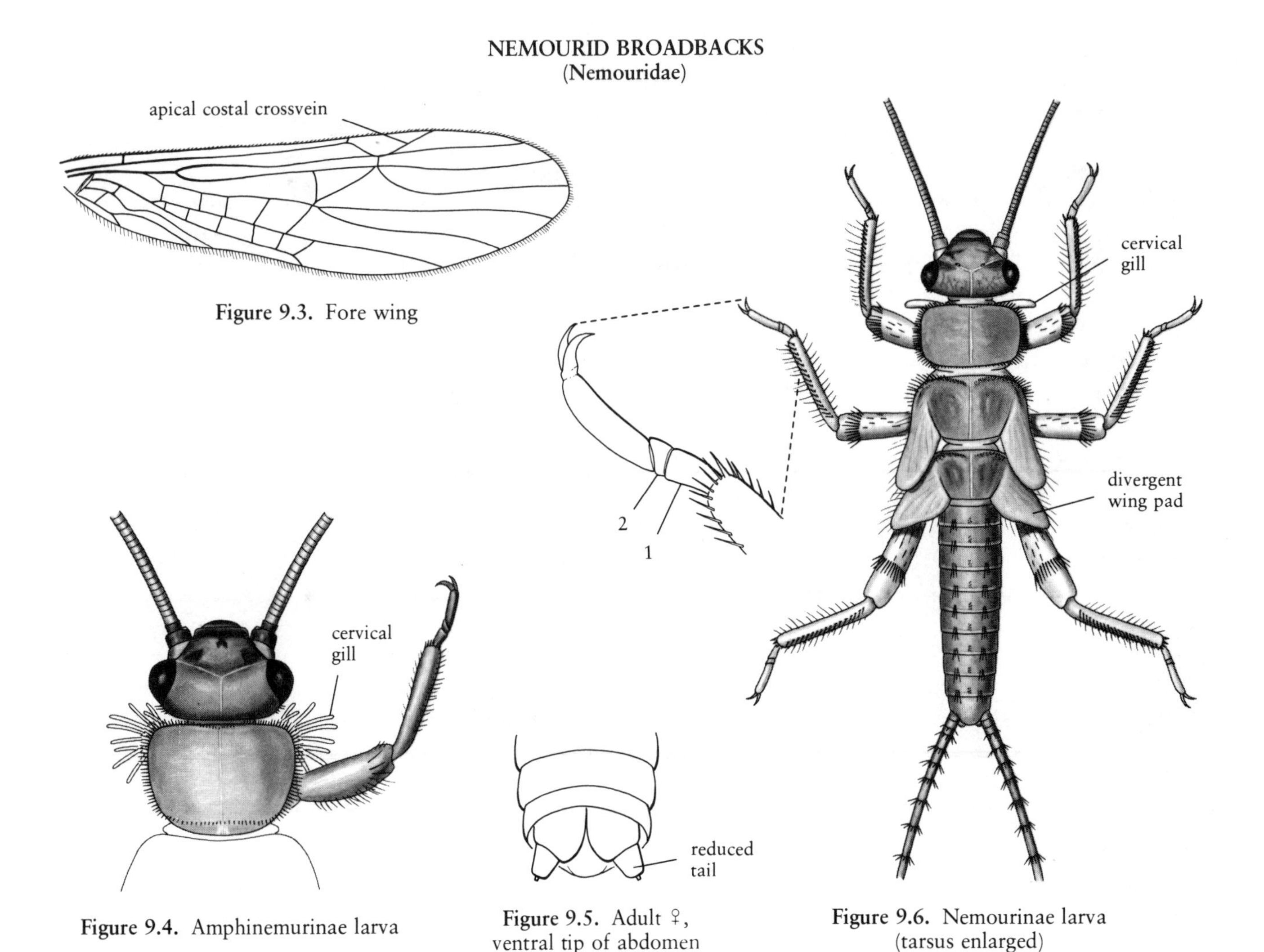

Figure 9.3. Fore wing

Figure 9.4. Amphinemurinae larva

Figure 9.5. Adult ♀, ventral tip of abdomen

Figure 9.6. Nemourinae larva (tarsus enlarged)

The Little Sepia Stonefly *(Zapada cinctipes)* and the Little Western Stonefly *(Malenka californica)* are used for nymph patterns. Some highly productive fishing can be had using such imitations on Pacific and Rocky Mountain streams.

SUBFAMILY AMPHINEMURINAE: **Larvae** (Fig. 9.4; Plate VII, Fig. 37) possess highly branched cervical gills. The group is widespread in North America.

SUBFAMILY NEMOURINAE: **Larvae** (Fig. 9.6) may lack cervical gills, have unbranched cervical gills, or more rarely possess cervical gills with less than six branches. The group is relatively widespread but is not known from Texas or Florida.

TAENIOPTERYGID BROADBACKS
(Family Taeniopterygidae)

LARVAL DIAGNOSIS: (Figs. 9.2, 9.8, 9.9) These are similar to nemourid larvae. Some possess simple fingerlike gills at the base of the legs, but none possess cervical gills. Tarsal segments 1 and 2 are about the same length.

ADULT DIAGNOSIS: (Fig. 9.7) Tarsal segments 1 and 2 are about the same length. There is no apparent X in the fore wings, since the apical costal crossvein is absent or distally located. Tails may be inconspicuous or conspicuous.

TAENIOPTERYGID BROADBACKS
(Taeniopterygidae)

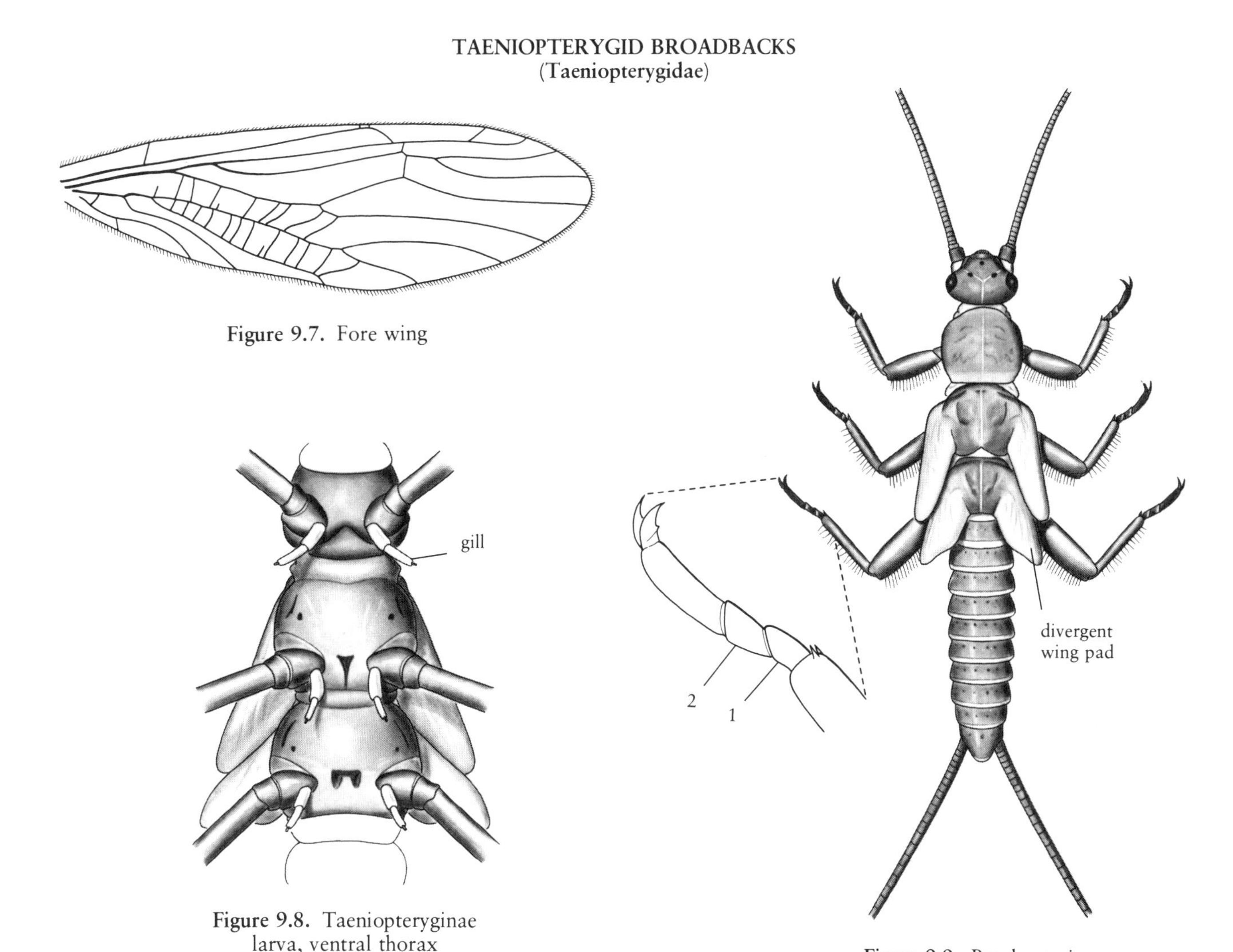

Figure 9.7. Fore wing

Figure 9.8. Taeniopteryginae larva, ventral thorax

Figure 9.9. Brachypterinae larva (tarsus enlarged)

DISCUSSION: Close examination of size ratios of tarsal segments will sometimes be required to distinguish the larvae of this group (Fig. 9.9) from the nemourid broadbacks (Fig. 9.6). Six genera and about 30 species occur in North America.

Larvae often occur in large rivers, and *Taeniopteryx* (one of the most ecologically tolerant groups of North American stoneflies) occurs in habitats ranging from cool springs to warm, sometimes intermittent, sand-bottomed streams, especially in southern states. Larvae are often difficult to find among debris because they are sluggish and sometimes covered with sediment or flocculent matter. Larval specimens have an interesting tendency to curl up after they are placed in fluid preservatives. Feeding habits tend to be more diverse than in other euholognathe families.

Strophopteryx fasciata (well known to many eastern fly fishermen as either the Early Brown Stonefly or the Little Red Stonefly) is an early emerger in eastern trout streams. This is a reddish brown stonefly that measures between 10 and 15 mm including the wings. Imitations of larvae or adults (preferably fished under water) often rank as some of the best early-season flies. The bright red Ready Nymph pattern, although supposedly patterned after *S. fasciata*, probably acts more as an attractor than an imitation. *Taeniopteryx nivalis* is also known as the Early Brown Stonefly.

Adults of one species of *Taenionema* are occasional pests of fruit crops in northwestern states, where they feed on the young buds of trees.

SUBFAMILY BRACHYPTERINAE: **Larvae** (Fig. 9.9) do not possess gills. The group is widespread but is not known from Florida.

SUBFAMILY TAENIOPTERYGINAE: **Larvae** (Fig. 9.8) have simple gills at leg bases. The group is widespread in North America.

SLENDER WINTER STONEFLIES
(Family Capniidae)

LARVAL DIAGNOSIS: (Figs. 9.10, 9.11; Plate VII, 38, 40) These are small, slender and cylindrical, often dark forms. When mature they measure only 3–6 mm, sometimes larger (not including tails). Hind wing pads are nearly parallel. Wing pads are occasionally absent. Anal lobe of hind wing pads of some species extends over half the length of wing pad, thus giving the appearance of a thickened pad. Lateral margins of abdomen as viewed from above appear zigzagged.

ADULT DIAGNOSIS: (Figs. 9.12–9.14; Plate VII, Fig. 39) These are small, dark forms. No long series of crossveins are present in the wings. Wings are rarely absent. Tails are conspicuous.

DISCUSSION: Slender winter stoneflies are easily confused with the rolledwinged stoneflies, and all larval characters should be examined very closely. See the discussion under the rolledwinged stoneflies.

This widespread family includes about 90 species in two genera and another 40 or more in seven other genera.

SLENDER WINTER STONEFLIES
(Capniidae)

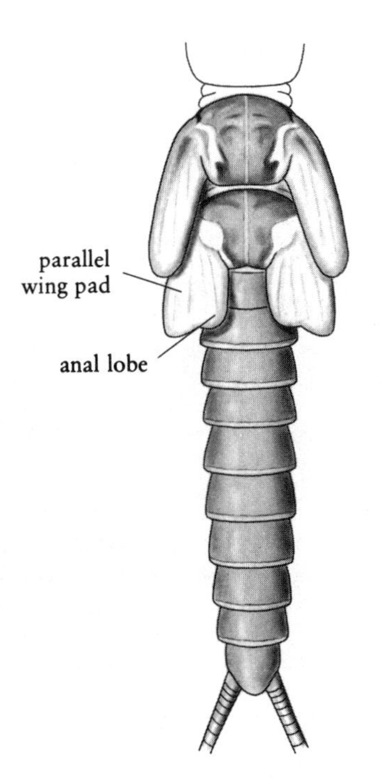

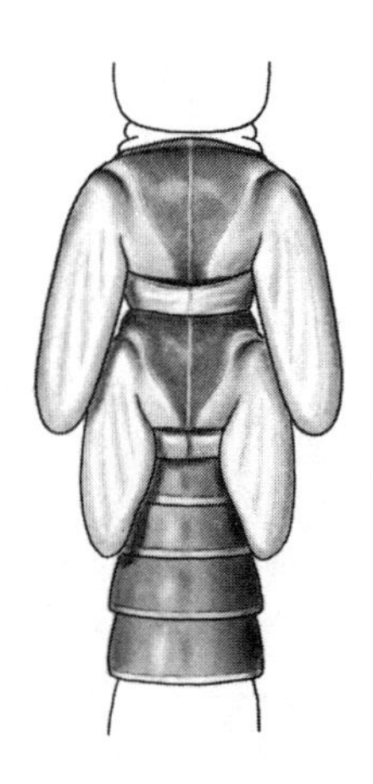

Figure 9.11. *Capnia* larva, thorax and base of abdomen

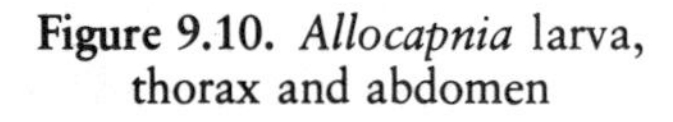

Figure 9.10. *Allocapnia* larva, thorax and abdomen

SLENDER WINTER STONEFLIES
(Capniidae)

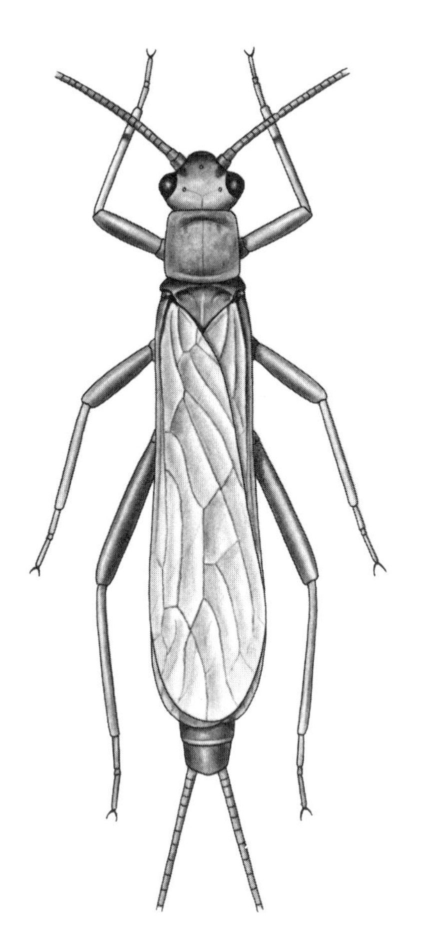

Figure 9.12. Fully winged adult

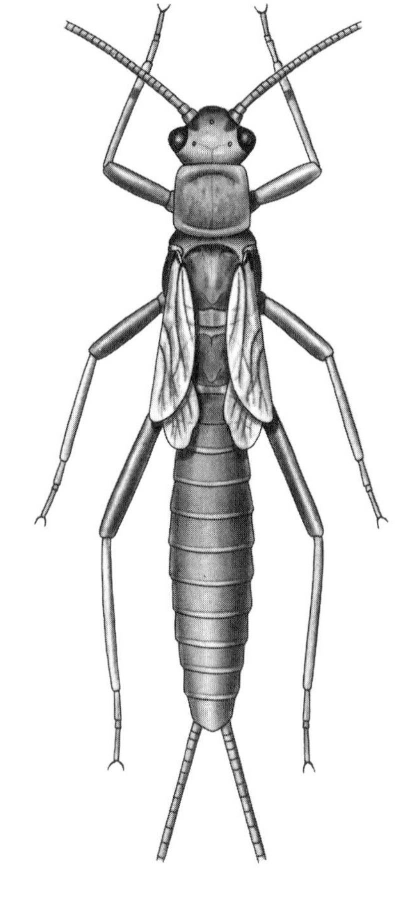

Figure 9.13. Short-winged adult

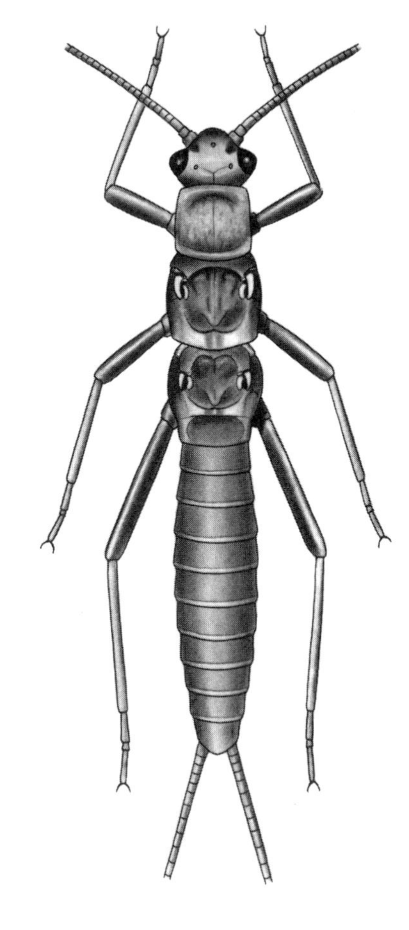

Figure 9.14. Micro-winged adult

Larvae occur in habitats ranging from tiny springs to moderately sized streams, often among detritus or mixed substrate. Species of *Allocapnia* are often closely associated with springs. Some species have small larvae that spend spring and summer resting within the substrate and undergoing little development, whereas other species undergo continual development and remain active. They are essentially detritivores, some having distinct preferences for certain leaf types. Some species have been found in deep hyporheic environments, and one species of the common western and northern genus *Capnia* spends its entire life cycle (including adults) submerged in Lake Tahoe.

In this lonely glen, with its brook draining the slopes, with its creased ice and crystals of all hues, where the spruces and hemlocks stand up on either side, and the rush and sere wild oats in the rivulet itself, our lives are more serene and worthy to contemplate.

Henry David Thoreau

As adults, slender winter stoneflies are the little "flies" that may be seen walking on snow and ice near streams. Adults may rely on algae as a food, and may withstand severe cold within tiny caverns of snow and ice that insulate them. Fishermen have generally referred to this group as the little black stoneflies or early black stoneflies. Some fishermen associate the latter name particularly with the species *Capnia vernalis* (Plate VII, Figs. 38, 39), after which some larval imitations for north central and eastern trout stream fishing have been patterned.

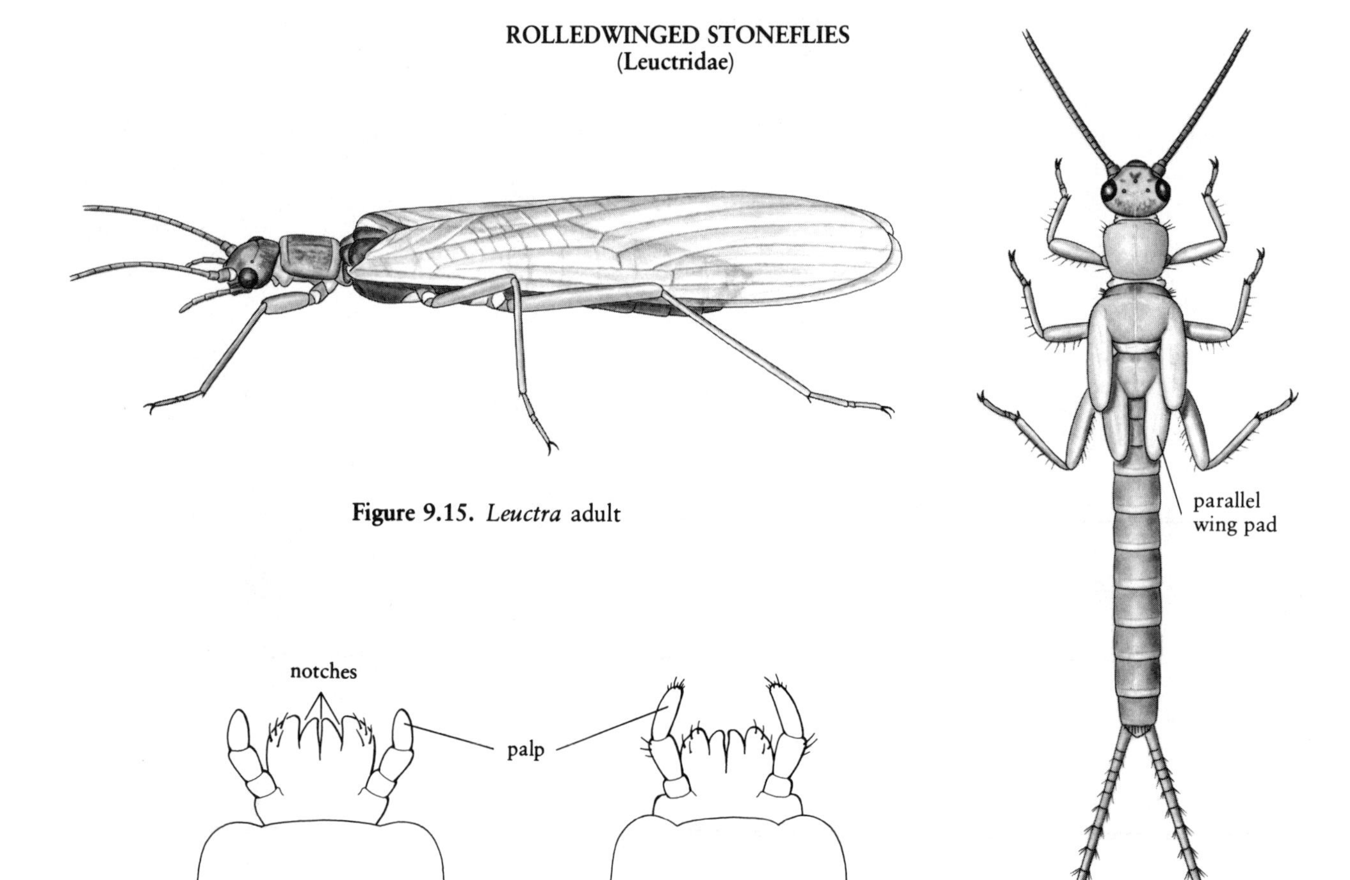

Figure 9.15. *Leuctra* adult

Figure 9.16. *Zealeuctra* labium

Figure 9.17. *Leuctra* labium

Figure 9.18. *Leuctra* larva

ROLLEDWINGED STONEFLIES
(Family Leuctridae)

LARVAL DIAGNOSIS: (Figs. 9.16–9.18) These larvae are generally similar to slender winter stoneflies. Their labial palps sometimes extend beyond the body of the labium. Lateral margins of abdomen, as viewed from above, are relatively smooth and not conspicuously zigzagged.

ADULT DIAGNOSIS: (Fig. 9.15) Wings at rest conspicuously roll over the sides of the body. Tails are reduced and inconspicuous.

DISCUSSION: Close examination of the abdominal characters (compare Figs. 9.10 and 9.18 and see picture key) should distinguish the larvae of rolledwinged stoneflies from the slender winter stoneflies. If this is not clear when comparing these two families, other characters should be used as follows: If the labial palps extend well beyond the body of the labium (Fig. 9.17), they are most likely rolledwinged stoneflies; if not, they could be either family. If hind wing pads possess anal lobes that extend beyond half the pad length (Fig. 9.10), they are slender winter stoneflies; if not, they could be either family.

Six genera and about 40 or more species of Leuctridae occur in North America, and although the family is widespread, species are usually only very locally common.

Larvae of many species occur in swift, rocky-bottomed streams, and some inhabit intermittent streams. Adults have been dubbed "needleflies" by fishermen because the wings are very often held tightly against the abdomen (some larval imitations of leuctrids are known as Needle Nymphs). Any of the rolledwinged stoneflies are sometimes known simply as "willows."

Summer Stoneflies
(Systellognatha)

Larvae and **adults** have labial palps that are slender at least terminally (Figs. 9.24, 9.30, 9.31) and commonly extend beyond the body of the labium. Body of labium either has a single incision from the outer margin that divides (usually deeply) the labium into two parts or is compact and has three short notches along the outer margin. These are variously sized (often medium to large), patterned forms. **Adults** are generally spring, summer, and fall emergers.

GIANT STONEFLIES
(Family Pteronarcyidae)

LARVAL DIAGNOSIS: (Fig. 9.20; Plate VII, Fig. 41) These are large forms and when mature generally measure 15–50 mm excluding tails. Head is somewhat vertically oriented. Body of labium is compact and has three notches along the outer margin. Pronotum is large and is usually well developed or projecting at the corners. Thoracic segments and at least abdominal segments 1 and 2 possess highly branched gills.

ADULT DIAGNOSIS: (Figs. 9.19, 9.21; Plate VII, Fig. 42) Wings are highly crossveined. Gill remnants are evident on abdominal segments 1 and 2.

GIANT STONEFLIES
(Pteronarcyidae)

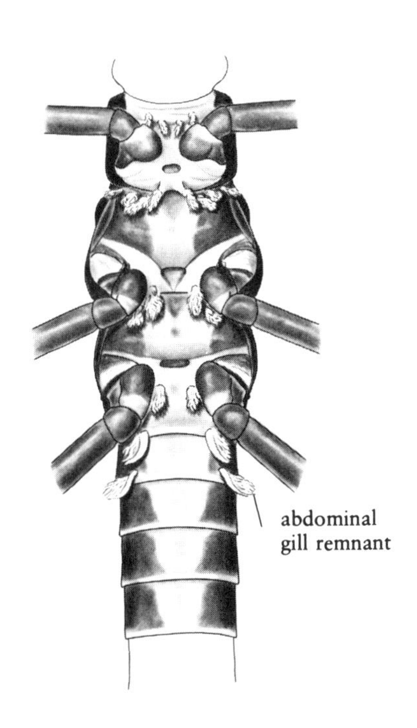

Figure 9.19. *Pteronarcys* adult, ventral thorax and base of abdomen

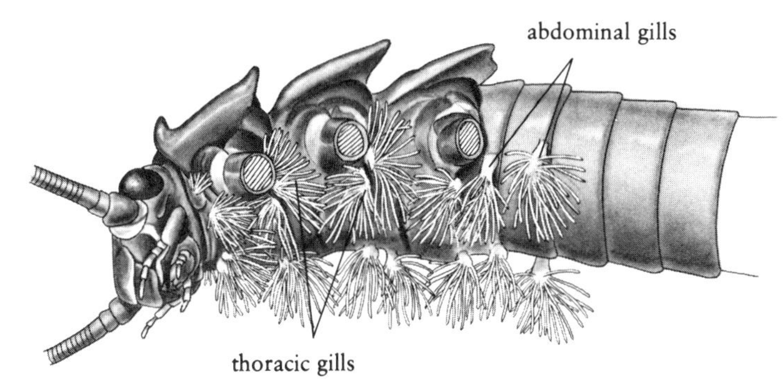

Figure 9.20. *Pteronarcys* larva

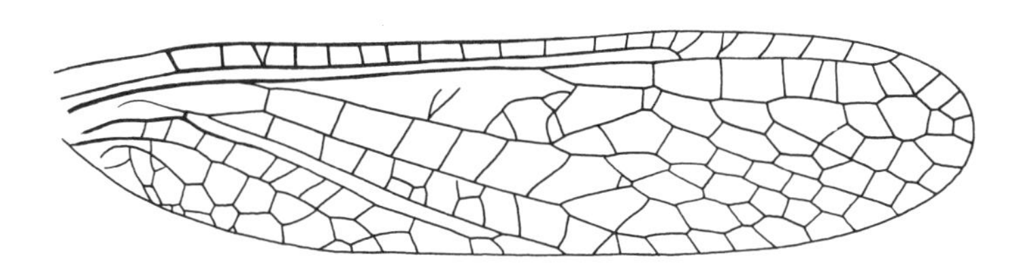

Figure 9.21. Fore wing

DISCUSSION: Larvae should not be confused with those of the larger common stoneflies (Perlidae), since apparent abdominal gills in some of the latter are actually only large, extended thoracic gills (Fig. 9.23). It will also help to keep in mind that the hind legs of giant stoneflies usually cannot be extended beyond the tip of the relatively long abdomen.

This is a small group (about 10 species) of conspicuously large stoneflies that is widespread but absent from Texas.

Larvae are primarily detritivores and herbivores in mixed substrate, detritus, and woody debris of streams and rivers. *Pteronarcys* may occur at considerable depths, and some species are more tolerant of warmer waters than most other systellognathe stoneflies. Some species require three years to complete the life cycle. Adults tend to be nocturnal and probably do not feed.

There are two giant stoneflies that are especially noteworthy for fly fishermen. The first, and perhaps the most important stonefly species as far as the fisherman is concerned, is the Salmonfly (or more specifically, the Western Salmonfly or Lumbering Salmonfly), *Pteronarcys californica* (Plate VII, Figs. 41, 42). The Salmonfly is indispensable on northwestern rivers when the females are ovipositing in large numbers. Larval imitations, such as Bird's Stonefly Nymph, are productively used for nymph fishing on these same rivers throughout the fishing season because of the multiyear life cycle of this species (not all individuals emerge in the same year and thus there may always be some good-sized larvae present in the water). Some bait fishermen refer to the larvae as "hellgrammites," but that name is best reserved for the larvae of dobsonflies (Megaloptera; Chapter 11). *Pteronarcys princeps* of the West is also sometimes referred to as the Salmonfly.

A considerable number of adults of both sexes taken together in Logan Canyon . . . enabled us to learn something of the variability . . . some were emerging from their nymphal skins on the leaf drifts at the edges of the stream and were still pale; some were mature and well colored; and others were old and black. They were found clinging low in the willows and on the weeds by the waterside, and, being indisposed toward flight by day, were picked with the fingers.

James G. Needham and P. W. Claassen

The second noteworthy species is the Giant Black Stonefly, *Pteronarcys dorsata* (Fig. 9.20). This is the largest of the North American stoneflies and is an important insect, especially in the East and Midwest. Unlike the Salmonfly, it generally does not occur in large numbers.

ROACHLIKE STONEFLIES
(Family Peltoperlidae)

LARVAL DIAGNOSIS: (Fig. 9.22; Plate VIII, Fig. 43) Mature larvae usually measure 8–15 mm excluding tails. Body of labium is compact and has three notches along the outer margin. Thorax is large and conspicuous, and its dorsal plates cover some of the vertical head and basal abdominal segments. Thorax also has overlapping ventral plates.

ADULT DIAGNOSIS: Adults are roachlike in appearance. Head is short and possesses two ocelli as seen from above.

DISCUSSION: This distinctive group is represented by about five genera and about 12 species in North America. The group is *not* known from central Canada, central states, Texas, or Florida. Larvae are herbivore-detritivores and occur primarily in springs and streams of mountains or hilly regions, commonly among leaf packs.

COMMON STONEFLIES
(Family Perlidae)

LARVAL DIAGNOSIS: (Figs. 9.23, 9.24, 9.26–9.29; Plate VIII, Fig. 44) Larvae are generally brown with yellow markings and when mature measure 8–35 mm excluding tails. Body of labium has a single incision from the outer margin that divides it into two distally rounded lobes. Branched thoracic gills are present.

ROACHLIKE STONEFLY
(Peltoperlidae)

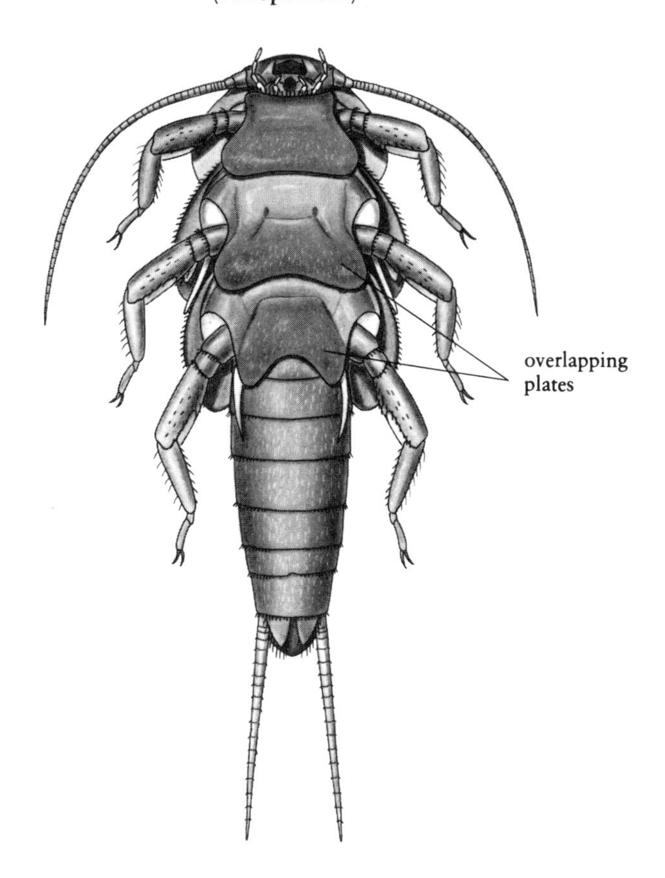

Figure 9.22. *Peltoperla*
larva, ventral

ADULT DIAGNOSIS: (Fig. 9.25; Plate VIII, Fig. 45) Labium is similar to that of the larva (Fig. 9.24). Branched gill remnants are located behind leg bases.

DISCUSSION: The combination of branched thoracic gills and a horizontal head (long axis in a horizontal plane) and the absence of basal abdominal gills distinguishes this group from all other North American stoneflies.

Approximately 40 species and several genera, most of which are either eastern or western in distribution, occur in North America. Among some of the more common genera, *Acroneuria* is most common in the eastern half of the continent, *Perlesta* is very abundant in some southern regions, and *Hesperoperla* is locally common in the West.

Larvae may occur in many lotic habitats, often under stones in riffles and sometimes in sandy substrates. The older larvae can be highly predaceous, utilizing mayflies, midges, and small caddisflies as food sources. Living or decaying plant material sometimes makes up a portion of the diet, especially of young larvae. Some exhibit aggressive territorial behavior by swinging their tails into other larvae and then turning on them. Several species of common stonefly larvae can sometimes be found carrying midges on their bodies. The midges are apparently not parasitic but may simply be along for the ride (a phenomenon known as phoresy).

COMMON STONEFLIES
(Perlidae)

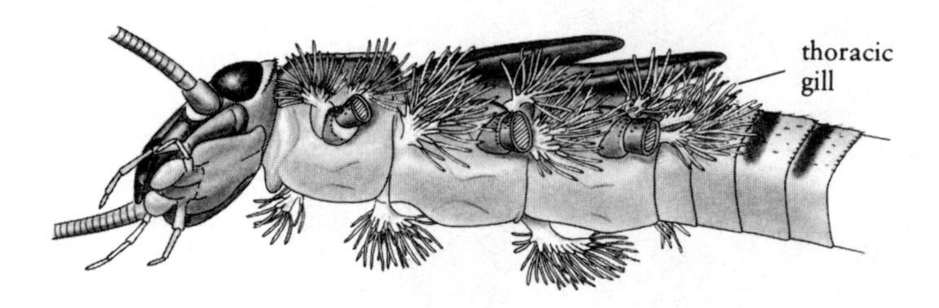

Figure 9.23. *Acroneuria* larva

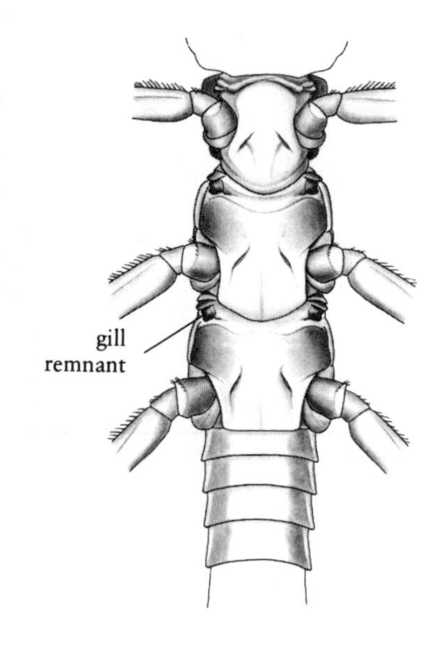

Figure 9.25. Adult, ventral thorax and base of abdomen

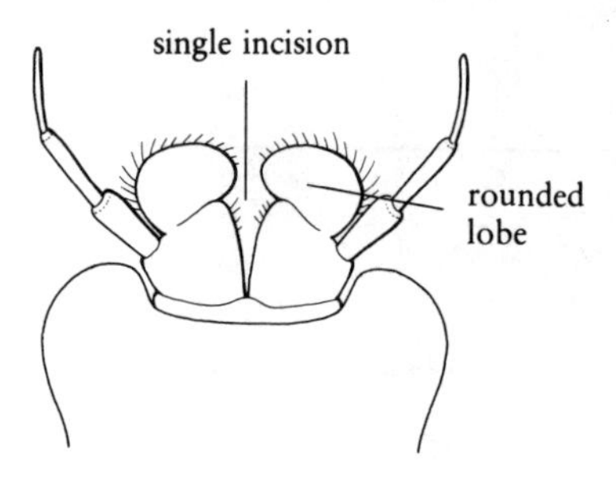

Figure 9.24. Larval labium

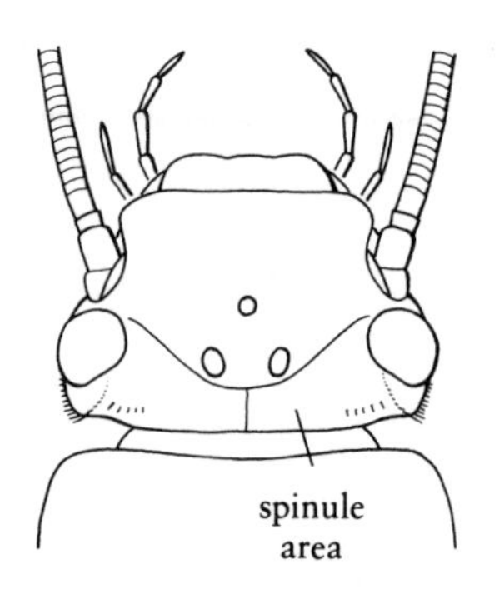

Figure 9.26. Acroneuriinae, larval head

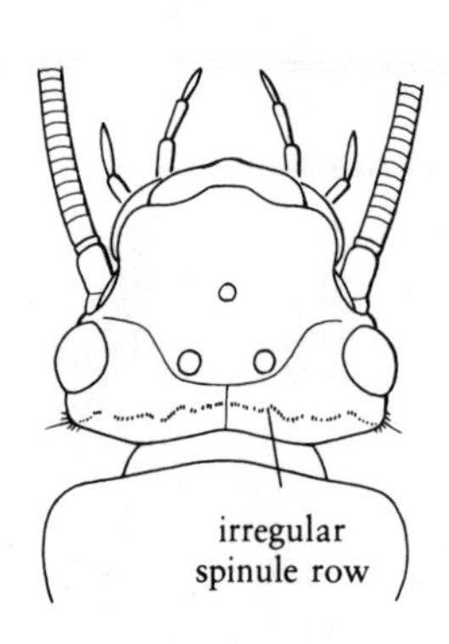

Figure 9.27. Acroneuriinae, larval head

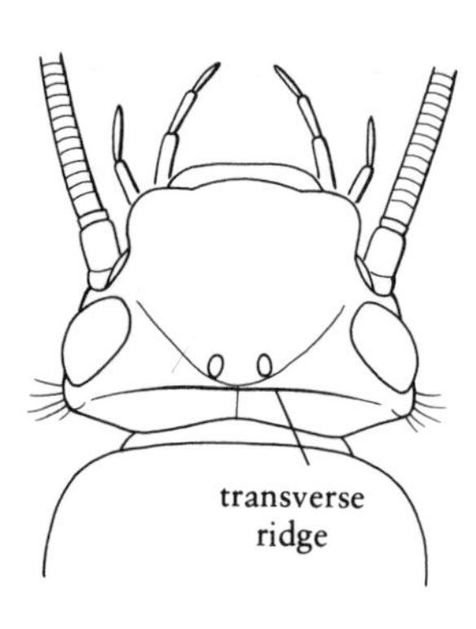

Figure 9.28. Perlinae, larval head

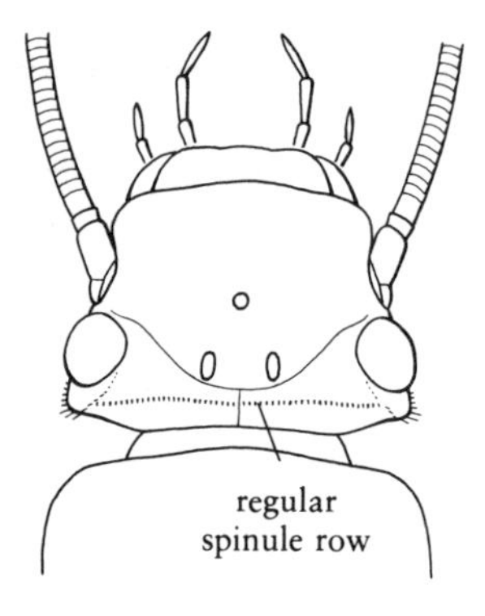

Figure 9.29. Perlinae, larval head

Life cycles sometimes require more than a year. Adults emerge primarily in spring and summer and probably do not take solid foods.

A number of common stonefly species are important for nymph fishing. In the West, these include the Golden Stonefly, or California Salmonfly *(Calineuria californica)*, and the Western Orange Stonefly *(Pictetiella expansa)*. Also in the West, the names Willowfly, Black Willowfly, and Brown Stonefly all refer to *Hesperoperla pacifica*. Some fishermen know this species by its synonyms, *Acroneuria pacifica* and *Acroneuria nigrita*. Larvae of the western perlids are sometimes known as water crickets to bait fishermen.

Other eastern and central species important to fly fishermen include the Great Brown Stonefly (*Acroneuria lycorias*), the Yellowlegged Stonefly (*Attaneuria ruralis*), and the Great Stonefly (*Phasganophora capitata*) (Fig. 9.29). Larvae of *P. capitata* are the common Eastern Stonefly Creepers that have previously been classified as *Perla capitata*. Another popular pattern of the larva of this species is Darbee's Stonefly Nymph. Some fly fishermen believe that night fishing with the larval pattern of this creeper is most productive.

SUBFAMILY ACRONEURIINAE: **Larvae** (Figs. 9.23, 9.24, 9.26, 9.27; Plate VIII, Fig. 44) lack transverse head spinules behind the eyes or possess a transverse row of head spinules behind the eyes that may be incomplete, irregularly spaced, or curving. This group is widespread in North America.

SUBFAMILY PERLINAE: **Larvae** (Figs. 9.28, 9.29) either have an elevated transverse ridge behind the eyes or a transverse row of head spinules behind the eyes that is complete, regularly spaced, and almost straight. The group is *not* known from northwestern states, western Canada, or Alaska.

PERLODID STONEFLIES
(Family Perlodidae)

LARVAL DIAGNOSIS: (Figs. 9.30, 9.31, 9.33–9.35; Plate VIII, Fig. 46) These are often highly patterned forms and when mature usually measure 8–16 mm excluding tails, although size is extremely variable. Body of

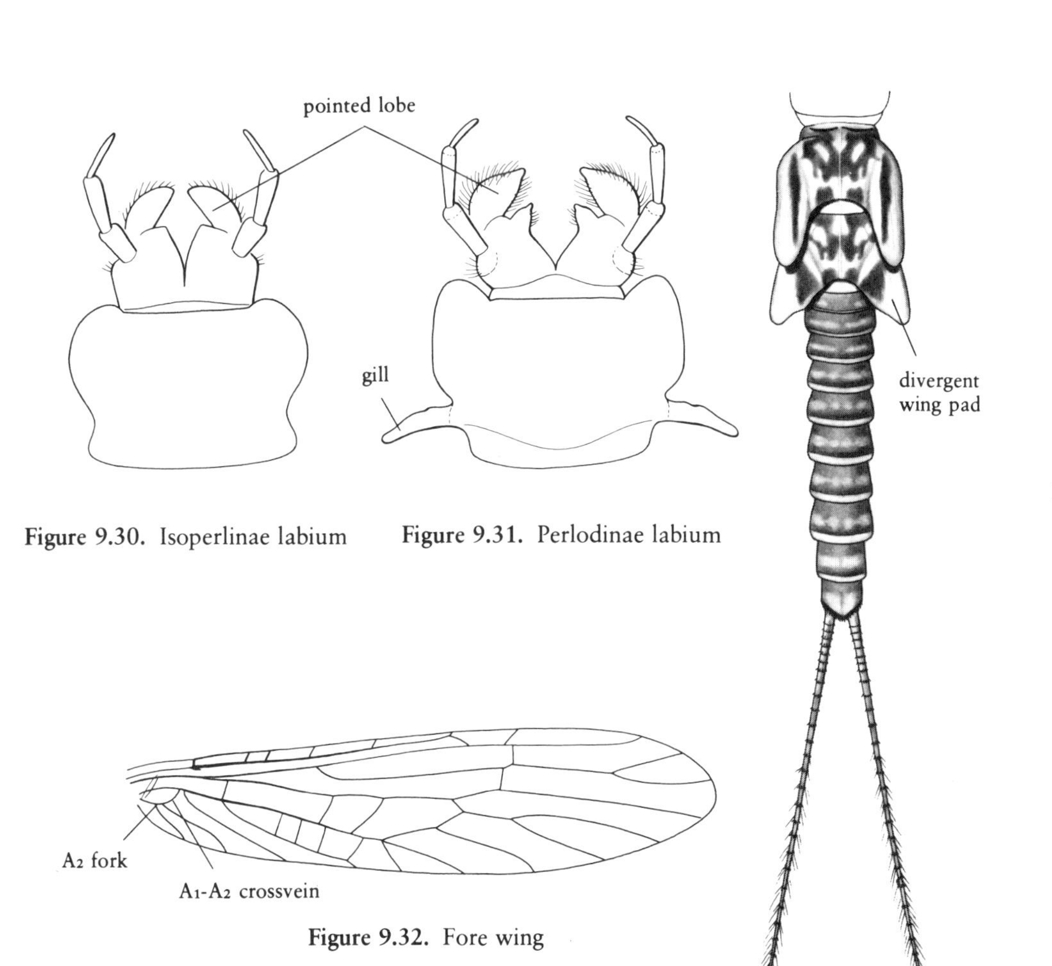

Figure 9.30. Isoperlinae labium

Figure 9.31. Perlodinae labium

Figure 9.32. Fore wing

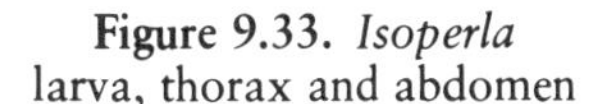

Figure 9.33. *Isoperla* larva, thorax and abdomen

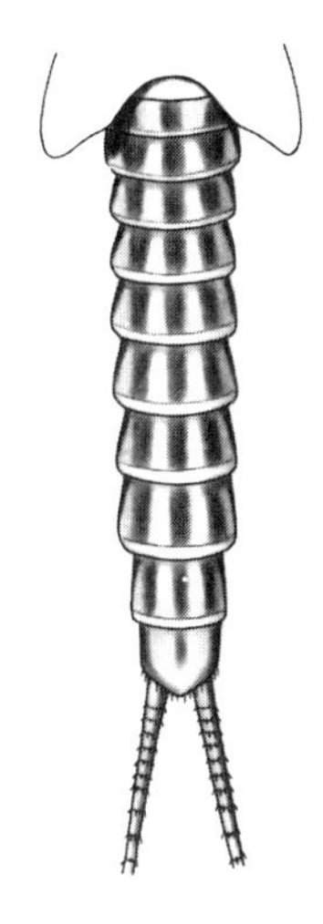

Figure 9.34. *Isoperla* larva, abdomen

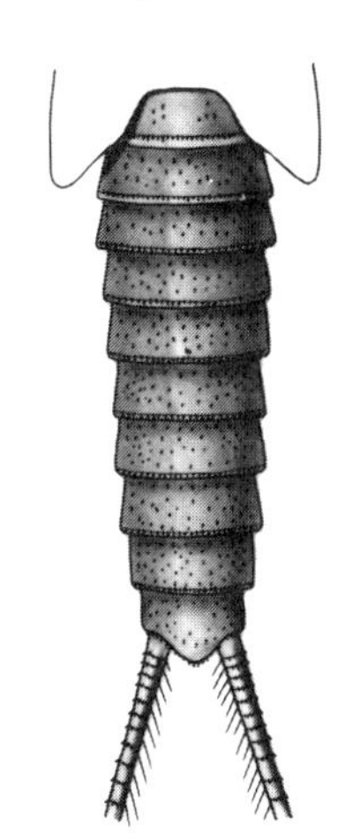

Figure 9.35. *Isogenoides* larva, abdomen

labium is deeply divided, and each half appears somewhat pointed at the outer margin. Branched thoracic gills are absent. Hind wing pads are divergent.

ADULT DIAGNOSIS: (Fig. 9.32; Plate VIII, Fig. 47) These are similar to common stoneflies. They are often greenish or yellowish but sometimes darker. Labium is similar to that of the larva. Gill remnants are absent or unbranched. Vein A_2 of the fore wings forks at a point basal to the A_1–A_2 crossvein.

DISCUSSION: Larvae resemble common stoneflies but lack the branched thoracic gills. One hundred or more species occur in North America, and the genus *Isoperla* is among the largest of North American stonefly genera, containing over 50 species.

Habits and habitats of perlodid larvae are often similar to those of common stoneflies. Most are predators, at least in part. *Arcynopteryx* may occasionally be taken along wave-beaten shores of Lake Superior. Perlodid larvae are relatively well represented in the drift fauna of streams where they occur.

Adults of a few species are atypically known from late winter. Adults may be pollen feeders or nonfeeders.

Three of the most important perlodid species as far as the fly fisherman is concerned are the Light Stonefly, or Light Brown Stonefly (*Isoperla signata*), from eastern trout streams, the Little Yellow Stonefly (*Isoperla bilineata*) (Plate VIII, Figs. 46 and 47) from eastern trout streams, and the Mormon Stonefly (*Isoperla mormona*; Fig. 9.34), also known as the Western Yellow Stonefly. These species are best used for nymph fishing patterns.

SUBFAMILY ISOPERLINAE: **Larvae** (Figs. 9.30, 9.33, 9.34; Plate VIII, Fig. 46) do not have any gills beneath head and thorax. They commonly have longitudinal stripes or a checkered pattern on the abdomen. The group is widespread in North America.

SUBFAMILY PERLODINAE: **Larvae** (Figs. 9.31, 9.35) usually have simple gills beneath the head and/or the thorax. Abdomen usually lacks longitudinal stripes. The group is widespread in North America.

GREEN STONEFLIES
(Family Chloroperlidae)

LARVAL DIAGNOSIS: (Fig. 9.36; Plate VIII, Fig. 48) These are often patternless, somewhat cylindrical forms. They usually measure 5–12 mm (excluding tails) when mature. Body of labium is deeply divided, and each half appears somewhat pointed at the outer margin. Gills are generally lacking. Hind wing pads are almost always nearly parallel. Tails are almost always shorter than the abdomen.

ADULT DIAGNOSIS: (Fig. 9.37) Adults are similar to those of perlodids, but vein A_2 in the fore wings either forks at a point beyond the placement of the A_1–A_2 crossvein or does not fork at all. Anal lobe of the hind wings is sometimes reduced or absent.

DISCUSSION: This group is distinctive as larvae, and in rare cases among the Paraperlinae, whose wing pads are divergent, the head is elongate behind the eyes. About 10 genera and probably over 60 species of green stoneflies occur throughout North America, excluding Texas and Florida.

Larvae are commonly associated with swift waters and gravelly bottoms, becoming more abundant as a group in colder waters or mountainous

GREEN STONEFLIES
(Chloroperlidae)

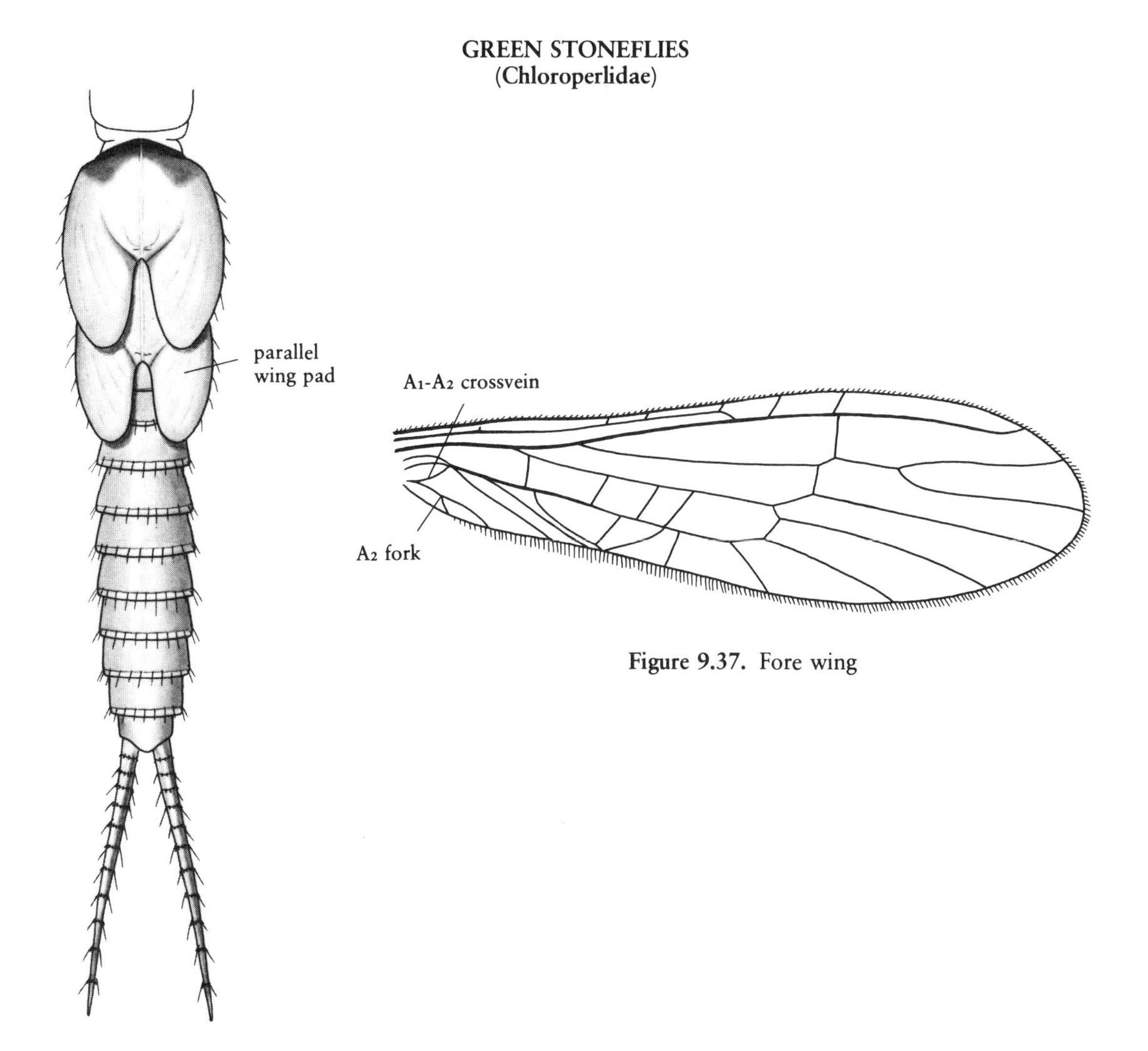

Figure 9.37. Fore wing

Figure 9.36. *Hastaperla* larva, thorax and abdomen

regions. At least one species of *Alloperla* is known to feed on honeydew as an adult. The common name of this family implies that all members of the group are green. It is true that several adults are quite green, but many are yellow or other colors.

The larva of the Little Olive Stonefly (*Sweltsa coloradensis*) (Plate VIII, Fig. 48) provides a productive nymphal pattern for many Rocky Mountain trout streams. Other popular green stonefly species are the Little Green Stonefly (*Alloperla imbecilla*) in the East and the Western Little Green Stonefly (*Sweltsa borealis*). The Yellow Sally patterns are based on a number of Chloroperlinae.

SUBFAMILY CHLOROPERLINAE: **Larval** (Fig. 9.36; Plate VIII, Fig. 48) head is not elongated behind the eyes. Body is not unusually narrow and elongate. The group is widespread in North America but is not known from Texas or Florida.

SUBFAMILY PARAPERLINAE: **Larval** head is sometimes elongated behind the eyes. Body is very narrow and elongate. The group is known only from western states and western Canada, and very rarely from northeastern states and eastern Canada.

More Information about Stoneflies

The stoneflies (Plecoptera) of the Rocky Mountains by Baumann et al. (1977). This is an up-to-date account of most species occurring in western North America, and provides keys to the species, including larvae when possible.

The stoneflies or Plecoptera of Illinois by Frison (1935). This recently reprinted work, although out-of-date in many regards, may be helpful for the central region of North America and includes many full larval illustrations.

The Plecoptera or stoneflies of Connecticut by Hitchcock (1974). This extensive work reviews the biology of stoneflies, gives keys to stonefly species that are most helpful for identification in eastern regions and especially New England and eastern Canada, and provides a checklist of North American species (although an older system of generic names is followed).

Nymphs by Schwiebert (1973). This book reviews those North American stonefly species most often used as patterns for fly fishing, with special emphasis given to nymph fishing.

Environmental requirements and pollution tolerance of Plecoptera by Surdick and Gaufin (1978). This is a tabular summary of the known ecological factors influencing the distribution (in the aquatic environments) of the North American stonefly species. It includes an extensive bibliography.

References

Adamus, P. R., and A. R. Gaufin. 1976. A synopsis of Nearctic taxa found in aquatic drift. *Amer. Midl. Natural.* 95:198–204.

Baumann, R. W. 1976. An annotated review of the systematics of North American stoneflies. *Perla* 2:21–23.

Baumann, R. W., A. R. Gaufin, and R. F. Surdick. 1977. *The stoneflies (Plecoptera) of the Rocky Mountains.* Mem. Amer. Entomol. Soc. 31. 208 pp.

Bednarik, A. F., and W. P. McCafferty. 1977. A checklist of the stoneflies, or Plecoptera, of Indiana. *Gr. Lakes Entomol.* 10:223–226.

Benfield, E. F. 1974. Autohemorrhage in two stoneflies (Plecoptera) and its effectiveness as a defense mechanism. *Ann. Entomol. Soc. Amer.* 67:739–742.

Coleman, M. J., and H. B. N. Hynes. 1970. The vertical distribution of the invertebrate fauna in the bed of a stream. *Limnol. Oceanogr.* 15:31–40.

Cummins, K. W. 1973. Trophic relations of aquatic insects. *Annu. Rev. Entomol.* 18:183–206.

Flick, A. 1966. *Art Flick's new streamside guide to naturals and their imitations.* Crown, New York. 173 pp.

Frison, T. H. 1935. *The stoneflies or Plecoptera of Illinois.* Bull. Ill. Nat. Hist. Surv. 20:281–471. (Reprint: 1975. Entomol. Rep. Specialists, Los Angeles.)

Gaufin, A. R., A. V. Nebeker, and J. Sessions. 1966. *The stoneflies (Plecoptera) of Utah.* Univ. Ut. Biol. Ser. 14. 89 pp.

Gerlach, R. 1974. *Creative fly tying and fly fishing.* Winchester Press, New York. 231 pp.

Harper, P. P., and H. B. N. Hynes. 1971. The Leuctridae of eastern Canada (Insecta; Plecoptera). *Canad. J. Zool.* 49:915–920.

Harper, P. P., and H. B. N. Hynes. 1971. The Capniidae of eastern Canada (Insecta; Plecoptera). *Canad. J. Zool.* 49:921–940.

Harper, P. P., and H. B. N. Hynes. 1971. The nymphs of the Taeniopterygidae of eastern Canada (Insecta; Plecoptera). *Canad. J. Zool.* 49:941–947.

Harper, P. P., and H. B. N. Hynes. 1971. The nymphs of the Nemouridae of eastern Canada (Insecta; Plecoptera). *Canad. J. Zool.* 49:1129–1142.

Harris, J. R. 1952. *An angler's entomology.* Collins, London. 268 pp.

Hitchcock, S. W. 1974. *Guide to the insects of Connecticut: Pt. VII. The Plecoptera or stoneflies of Connecticut.* Bull. St. Geol. Nat. Hist. Surv. Conn. 107. 262 pp.

Hynes, H. B. N. 1974. Further studies on the distribution of stream animals within the substratum. *Limnol. Oceanogr.* 19:92–99.

Hynes, H. B. N. 1976. The biology of Plecoptera. *Annu. Rev. Entomol.* 21:135–153.

Jewett, S. G., Jr. 1959. *The stoneflies (Plecoptera) of the Pacific Northwest.* Ore. St. Monogr. Stud. Entomol. 3. 99 pp.

Jewett, S. G., Jr. 1963. A stonefly aquatic in the adult stage. *Science* 139:484–485.

Komnick, H. 1977. Chloride cells and chloride epithelia of aquatic insects. *Inter. Rev. Cyt.* 49:285–329.

McClane, A. J. (ed.). 1965. *McClane's standard fishing encyclopedia.* Holt, Rinehart and Winston, New York. 1057 pp.

Minshall, G. W., and J. N. Minshall. 1966. Notes on the life history and ecology of *Isoperla clio* (Newman) and *Isogenus decisus* Walker (Plecoptera: Perlodidae). *Amer. Midl. Natural.* 76:340–350.

Richardson, J. W., and A. R. Gaufin. 1971. Food habits of some western stonefly nymphs. *Trans. Amer. Entomol. Soc.* 97:91–121.

Ricker, W. E. 1964. Distribution of Canadian stoneflies. *Gewass. Abwass.* 34/35:50–71.

Ross, H. H., and W. E. Ricker. 1971. *The classification, evolution, and dispersal of the winter stonefly genus Allocapnia.* Ill. Biol. Monogr. 45. 166 pp.

Schwiebert, E. 1973. *Nymphs.* Winchester Press, New York. 339 pp.

Sheldon, A. L. 1969. Size relationship of *Acroneuria californica* and its prey. *Hydrobiologia* 34:85–94.

Stanford, J. A., and A. R. Gaufin. 1974. Hyporheic communities of two mountain rivers. *Science* 185:700–702.

Stark, B. P., and A. R. Gaufin. 1976. *The Nearctic genera of Perlidae (Plecoptera).* Misc. Publ. Entomol. Soc. Amer. 10. 80 pp.

Stark, B. P., and A. R. Gaufin. 1979. The stoneflies (Plecoptera) of Florida. *Trans. Amer. Entomol. Soc.* 104:391–433.

Steffan, A. W. 1967. Larval phoresis of Chironomidae on Perlidae. *Nature* 213:846–847.

Surdick, R. F., and M. R. Cather. 1975. The nymphs of *Utaperla sopladora* Ricker (Plecoptera: Chloroperlidae). *Entomol. News* 86:102–106.

Surdick, R. F., and A. R. Gaufin. 1978. *Environmental requirements and pollution tolerance of Plecoptera.* Environ. Monit. Sup. Lab., Off. Res. Dev. U.S.E.P.A., Cincinnati. 417 pp.

Surdick, R. F., and K. C. Kim. 1976. *Stoneflies (Plecoptera) of Pennsylvania.* Penn. St. Univ. Ag. Exp. Sta. Bull. 808. 73 pp.

Szczytko, S. W., and K. W. Stewart. 1977. The stoneflies (Plecoptera) of Texas. *Trans. Amer. Entomol. Soc.* 103:327–378.

Tozer, W. 1979. Underwater behavioural thermoregulation in the adult stonefly, *Zapada cinctipes. Nature* 281:566–567.

Ziegler, D. D., and K. W. Stewart. 1977. Drumming behavior of eleven Nearctic stonefly (Plecoptera) species. *Ann. Entomol. Soc. Amer.* 70:495–505.

Zwick, P. 1973. *Insecta: Plecoptera, phylogenetisches system und katalog.* Das Tierreich 94, Berlin. 465 pp.

CHAPTER 10

Water Bugs

(ORDER HEMIPTERA)

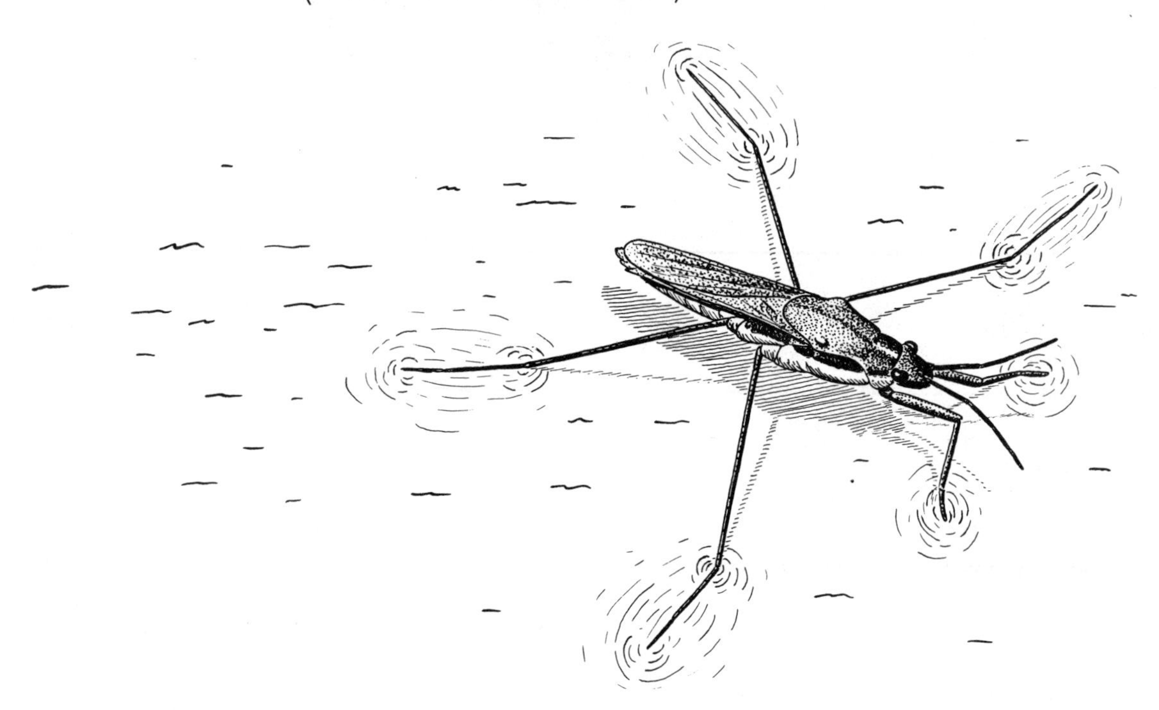

THE RELATIVELY LARGE order Hemiptera contains both terrestrial and aquatic families, with about 300 North American species being adapted to living in or on the water. There are perhaps another 100 that always occur along the edges of water. Some water bugs are well known for their large size and fascinating skating habits. Both immature and mature water bugs are adapted similarly for aquatic existence; both stages have similar life styles; and the two stages are not so radically different from each other in appearance as are the comparable stages of other aquatic insects. Most water bugs are highly predaceous and in many habitats take a heavy toll on other aquatic insects, including such pest species as mosquitoes. Some species are important fish food.

This chapter deals exclusively with that part of the order known as the true bugs (or suborder Heteroptera). The fully developed fore wings of true bugs consist of a thickened basal part and a membranous distal part, hence the name *Hetero*-(different) *ptera* (wing). The closely related group comprising the hoppers (suborder Homoptera), of which there are only a few semiaquatic species (found on emergent aquatic vegetation), is treated in Chapter 18.

Diagnosis

Both adults and larvae are oval to slender elongate forms, many of which are somewhat flattened dorsoventrally. They range in length from 1 to 65 mm. Head is well developed and has well-developed eyes and minute to long antennae. Mouth is modified in the form of a beak arising from the anterior part of the head. Beak is either a conelike structure or a more elongated piercing and sucking structure. Thorax may lack wings, have fully developed wings, or have variously shortened wings or wing pads. Fore wings when at rest are held close over the back, and the two fore wings usually overlap at their tips. Fore wings usually have a thickened or leathery basal region and a membranous distal region. Hind wings when present are completely membranous. Thoracic legs are well developed. Abdomen lacks gills, filaments, and tails, but does end in tubelike or short flaplike structures in some species.

Similar Orders

Some water bugs superficially resemble adult water beetles in body shape and swimming habits. Close examination of the mouthparts and/or the back of the insects will readily distinguish the two groups. Water beetles have chewing mouthparts, and their fore wings, which are always present in adults, are modified into hardened covers over the abdomen; their fore wings do not overlap. Keep in mind that the membranous area of the fore wings of fully winged bugs is sometimes difficult to detect; however, the modified fore wings of bugs do overlap.

Semiaquatic Homoptera also possess a beak, but it is inserted near the rear of the head. They also have fore wings, but these are often held rooflike over the body and are of uniform texture. A few minute, atypical water bugs (pygmy backswimmers) (Fig. 10.5), which cannot be differentiated on the above characters, can be distinguished by their shape and their reduced antennae.

It is interesting to contemplate a tangled bank, clothed with many plants of many kinds, with birds singing on the bushes, with various insects flitting about. . . .

Charles Darwin

Life History

Metamorphosis is incomplete, and adults and larvae (also more commonly known in North America as nymphs or immatures) are quite similar except for size and the relative development of the wings in winged forms. The life cycle generally consists of one generation per year (although several species have more) and five larval instars (a very few have only four larval instars). Growth is usually relatively fast, with each larval instar lasting about a week or two in most species. The adult stage is usually the overwintering stage, although in some species the eggs or rarely the larvae overwinter. Overwintering adults may be active or more or less dormant (hibernating), depending on the species and the local climatic conditions.

Submergent water bugs generally do not leave the water except to disperse by flight or when forced to move (e.g., by drying conditions). Few swarms or mass flights have been reported for water bugs, but individuals are often attracted by lights at night.

Submergent forms generally mate in water, whereas surface and shore bugs mate in their respective habitats. Multiple matings by both the male and female are common in several groups; sometimes a male will attempt to mount another male. Stridulatory sounds in many groups and even specific wave vibrations set up by some water striders can be of behavioral importance for courtship and species recognition.

Oviposition takes place in the spring and early summer among most species. Eggs may be laid on underwater or exposed substrate, in earthen cells or plant tissue, or on a gelatinous pad on the back of the male (some giant water bugs). The open ocean striders lay eggs on flotsam. Eggs normally require from one to a few weeks for incubation.

TABLE 10.1 COMMON HABITATS OF WATER BUGS

Family	Habitats							
	UNDERWATER (SUBMERGED AT LEAST PARTIALLY)			WATER SURFACE		SHORE-DWELLING		ON EMERGENT PARTS OF PLANTS (AT LEAST OCCASIONALLY)
	Freshwater	*Inland Brackish or Salt Water*	*Intertidal*	*Freshwater*	*Marine*	*Inland*	*Coastal Beaches*	
Notonectidae	X							
Pleidae	X							X
Nepidae	X							X
Naucoridae	X	X						
Belostomatidae	X	X						
Corixidae	X	X	X					
Gelastocoridae[1]						X	X	
Ochteridae[1]						X		X
Gerridae				X	X			
Veliidae				X	X	X	X	X
Hydrometridae				X				X
Hebridae				X		X		X
Mesoveliidae				X				X
Macroveliidae[1]						X		
Saldidae[1]						X	X	X
Leptopodidae[1]						X		
Dipsocoridae[1]						X		

[1]See Chapter 17.

Aquatic Habitats

Underwater and surface bugs occur in a wide variety of aquatic environments (Table 10.1). Although most species prefer quiet waters, including the pools of streams and rivers, a few can be found in riffle areas. Habitats also include open ocean areas, hot springs, sewage ponds, and brackish waters. A number of shore species, surface species, and at least two underwater species occur in coastal marine habitats of North America.

Aquatic Adaptations and Behavior

Respiration in water bugs is primarily aeropneustic, although many also utilize hydropneustic respiration to various degrees. Submergent species generally acquire atmospheric oxygen by periodically surfacing or by maintaining some contact between the air-water interface and the posterior end of the body. Air is stored by periodic surfacers as a plastron (thin film of air held by many small hairs or scales) on various parts of the body, especially the ventral abdomen, and also in an underwing chamber in many. Under highly oxygenated conditions, periodic surfacers are able to remain completely submerged for prolonged periods before having to replace their air store, because the plastron is a relatively efficient physical gill. Some creeping water bugs remain submerged indefinitely, utilizing physical gills, and some larvae of creeping water bugs are evidently able to secure oxygen from the water by direct cutaneous uptake. In those instances, the bugs are hydropneustic. Surface dwellers, which generally respire as terrestrial insects, are usually able to utilize plastrons at times when they are subjected to splash or on those rare occasions when they dive. The plastron is particularly important in keeping these bugs from drowning. Sea skaters are able to trap very extensive plastrons, which help to keep them dry in the open-ocean environment.

Chloride cells, which are used for osmoregulation by submergent forms, are distributed over the nonplastron areas of the body, so that they are in direct contact with water.

Water bugs, except for some of the water boatmen, are strictly animal feeders and suck the body fluids from their prey. Most are active predators, but many also feed, or scavenge, on other organisms that have fallen into the water, including a wide variety of terrestrial insects. The mouths (or beaks) of water boatmen are adapted more for rasping than sucking, and these bugs feed on small solid foods (including plants and detritus, depending on the species) rather than, or in addition to, juices. Many of the water bugs will inflict a painful bite if handled recklessly.

Submergent species are generally swimmers; the hind legs, and middle legs in some, are variously adapted for swimming. When not swimming, they may rest on vegetation or other substrates, hang suspended in the water, or become partially or completely buried in soft bottoms. As is befitting of their predaceous nature, the fore legs of many are modified as grasping appendages. Among some groups (e.g., backswimmers), there is apparently a relationship between the length of these appendages and prey preferences. For example, species with long legs are more efficient for capturing small, mobile prey, and those with shorter, more powerful legs tend to take slower or stranded prey that are much larger. Surface bugs are adapted for skating or, in some of the less active species, for walking or running over the water surface.

Other common behavioral traits found among certain water bugs include grooming and cleaning of mouthparts and legs, especially after feeding; rubbing specialized areas of the fore legs over the head to produce sounds; and producing unpleasant odors from well-developed scent glands, most probably as a defense mechanism. A few, such as marsh treaders and giant water bugs, commonly display catalepsy (feigning death) by becoming motionless when disturbed.

Classification and Characters

The water bugs of North America are represented by six families whose members are essentially underwater dwellers, five families whose members spend some or most of their lives on the surface of water, and another six families whose members are shore-dwelling or riparian. The underwater bug families form a more-or-less natural grouping, as do the surface bug families. The shore bugs, however, are a mixture of families with different natural affinities (Dipsocoroidea are allied with the terrestrial bugs, the Ochteroidea with the underwater bugs, and the Saldoidea and the Gerroidea with the surface bugs). For convenience, the families are treated here in two sections: (1) the Underwater Bugs and (2) the Surface and Shore Bugs.

The adults of many water bugs exhibit polymorphism in their wing development; that is, adults may have fully developed wings, reduced wings, or no wings at all. Therefore, it is sometimes difficult to distinguish stages on the basis of wing development. The diagnoses, however, are applicable to both larvae and adults except where specifically noted. The

Figure 10.1. UNDERWATER AND SURFACE HEMIPTERA

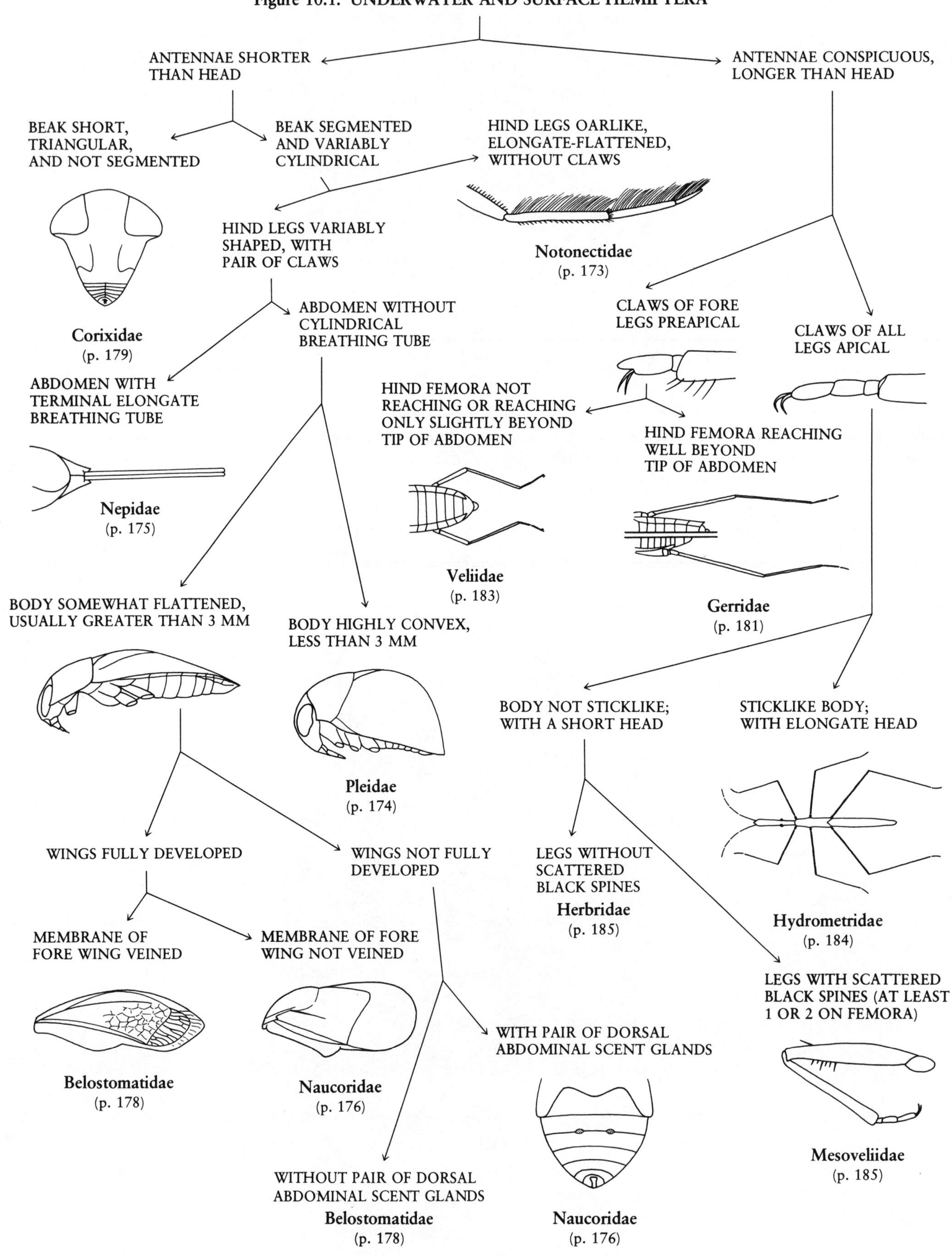

term **hemielytron** (plural, **hemielytra**) is used for the fore wings of bugs. Hemielytra, when fully developed, typically consist of a leathery basal portion and a membranous distal portion (Fig. 10.13). The membranous portion may or may not contain wing veins.

Raptorial fore legs are designed for capturing and holding prey and consist of two sections that are opposable to each other in a somewhat pincerlike fashion (Fig. 10.6). **Swimming** legs are mostly oarlike and have swimming hairs (Figs. 10.2, 10.14). Various degrees of swimming adaptations of the legs are found among the underwater families. **Preapical** claws (Figs. 10.16–10.18) originate just before the tip of the tarsus (end of leg); in checking this character, the fore legs should be used.

Underwater Bugs

Antennae are shorter than the head and are usually concealed. Some members are well adapted for swimming. Underwater bugs usually remain submerged or partially submerged, but some are occasionally found on aquatic plants at the water surface.

SUPERFAMILY NOTONECTOIDEA

BACKSWIMMERS
(Family Notonectidae)

DIAGNOSIS: (Figs. 10.2–10.4; Plate IX, Fig. 49) These are elongate, somewhat slender and deep-bodied forms. Body sometimes has colorful patterns. Adults are 5–16 mm. Segmented beak is stout and reaches to the

BACKSWIMMERS
(Notonectidae)

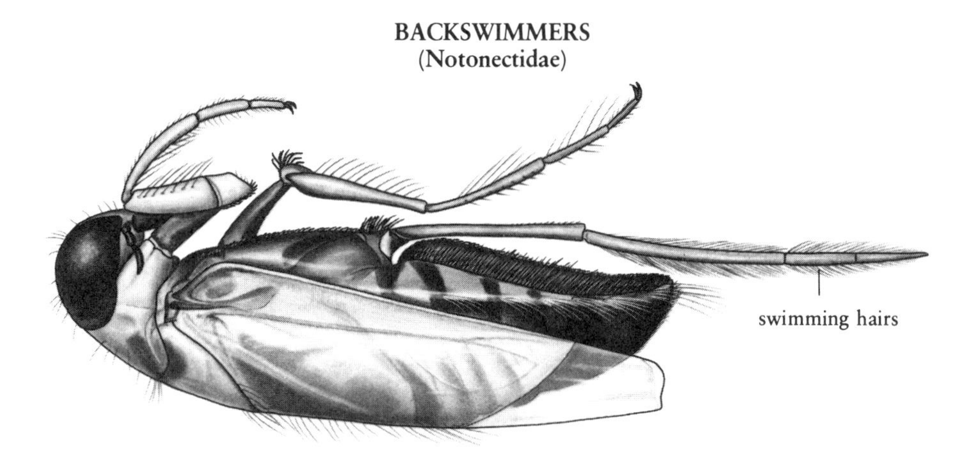

Figure 10.2. *Buenoa* adult

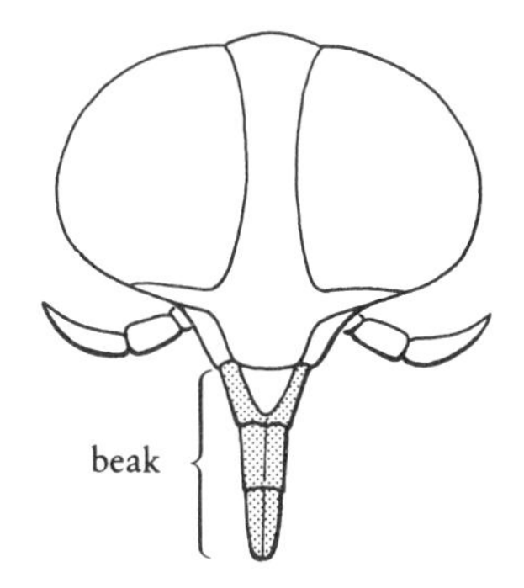

Figure 10.3. Anisopinae head

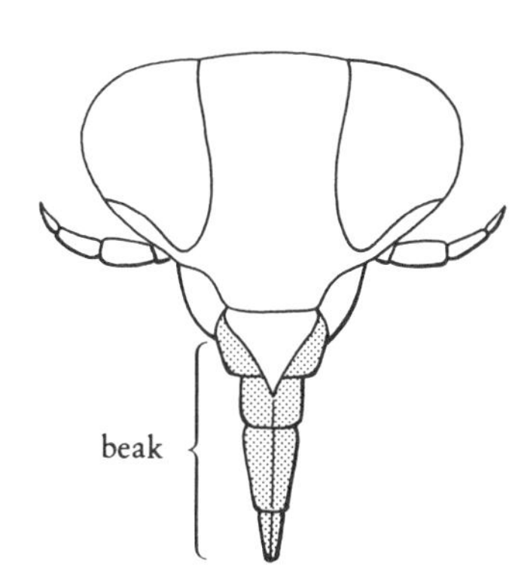

Figure 10.4. Notonectinae head

base of fore legs. Fore legs each have two apical claws. Hind legs are oarlike and possess swimming hairs. Abdomen lacks prominent terminal structures.

DISCUSSION: These rather sleek swimming bugs can be distinguished from water boatmen, which also have well-developed swimming hind legs, by their segmented beak (Figs. 10.3, 10.4), which is more typical of bugs in general, and by their fore legs, which are not the highly modified type possessed by water boatmen (Fig. 10.14). This widespread family comprises two subfamilies, two genera, and about 30 species in North America.

Backswimmers, as the name implies, swim upside down. They usually live in standing water or slower reaches of streams and rivers. Species of *Buenoa* (Anisopinae) tend to be very smooth swimmers that are most efficient in capturing small, fast-moving prey. *Notonecta* (Notonectinae) species tend to be less adept swimmers (some jerking movement), often utilizing larger but slower or immobile prey. Food generally consists of other insects, crustaceans, snails, and small fish. Most overwinter as adults, hibernating at higher latitudes beneath ice and often within the substrate.

A few artificial wet flies have been patterned after the backswimmers for fishing shallow trout ponds or lakes. Such flies as the Graywinged Backswimmer, Pale Moon Backswimmer, and Grousewinged Backswimmer should be fished with more-or-less continuous surfacing and diving motions.

SUBFAMILY ANISOPINAE: (Figs. 10.2, 10.3) Antennae and beak are three-segmented. Adults are 5–8 mm. This group is widespread in North America, but few species are known from Canada.

SUBFAMILY NOTONECTINAE: (Fig. 10.4; Plate IX, Fig. 49) Antennae and beak are four-segmented. Adults are 10–16 mm. This group is widespread in North America.

PYGMY BACKSWIMMERS
(Family Pleidae)

DIAGNOSIS: (Fig. 10.5) These are minute, highly convex forms that are less than 3 mm. Fore wings are *not* modified into hemielytra; wing veins are usually absent. Legs are short.

DISCUSSION: These atypical-appearing bugs include five species among three genera in North America. Of these species, most occur in the lower

PYGMY BACKSWIMMER
(Pleidae)

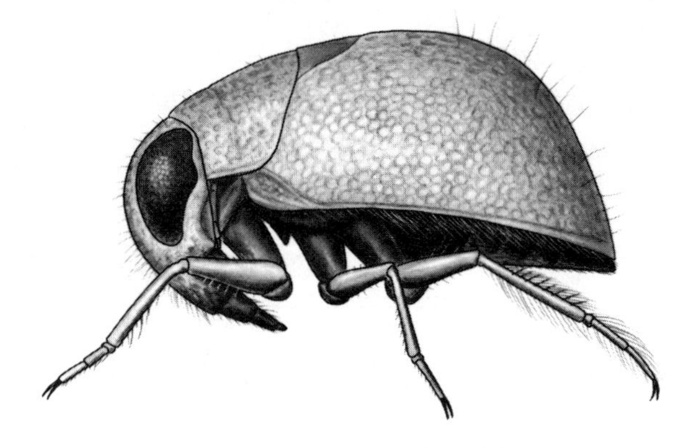

Figure 10.5. *Neoplea* adult

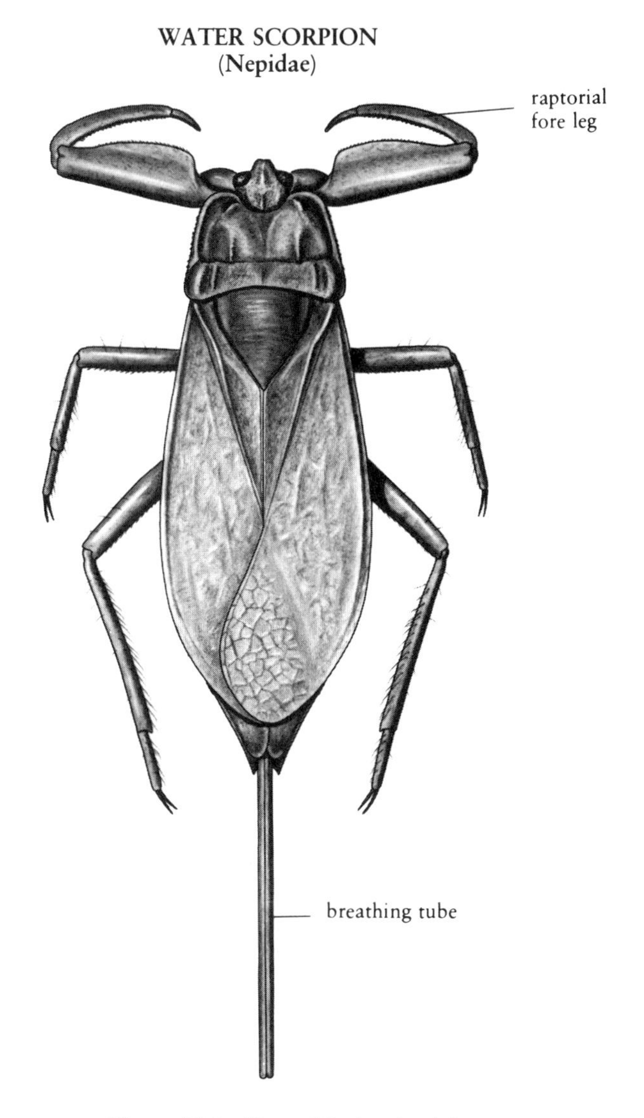

Figure 10.6. *Nepa* (Nepinae) adult

southeastern states including Florida. One species of *Neoplea* is relatively widespread, although rare in the West. Pleids are not known from western Canada or Pacific coastal areas.

Pygmy backswimmers are usually found crawling among aquatic vegetation in lentic habitats; they swim only occasionally.

WATER SCORPIONS
(Family Nepidae)

DIAGNOSIS: (Fig. 10.6; Plate IX, Fig. 54) These are generally large, either narrow and elongate or broad and oval forms. Adults are 15–45 mm (not including breathing tube). Fore legs are raptorial. Abdomen terminates in a well-developed, elongate breathing tube (sometimes over 20 mm) that in adults is composed of two rodlike structures.

DISCUSSION: The terminal breathing tube easily distinguishes the cylindrical, almost sticklike water scorpions (genus *Ranatra*) (Plate IX, Fig. 54) from the somewhat similar but smaller and more delicate marsh treaders (Plate IX, Fig. 52). The terminal breathing tube also distinguishes the oval water scorpions (genera *Nepa*, Fig. 10.6, and *Curicta*) from the somewhat similar giant water bugs (Plate IX, Fig. 53).

The family is widespread and consists of about 13 species in North America.

Water scorpions occur in a variety of streams, ponds, and swamps, usually among debris or vegetation in shallow areas. They are poor swimmers but sometimes clumsily paddle through the water. They breath by maintaining contact with the air-water interface via their elongate breathing tubes. Special static sense organs located on the abdomen may aid water scorpions in their underwater orientation.

SUBFAMILY NEPINAE: (Fig. 10.6) Body is oval. The group is *not* known from northwestern states or western Canada; it is rare or only locally common in other areas.

SUBFAMILY RANATRINAE: (Plate IX, Fig. 54) Body is slender and elongate. The group is widespread in North America.

CREEPING WATER BUGS
(Family Naucoridae)

DIAGNOSIS: (Figs. 10.7–10.10; Plate IX, Fig. 50) These are more-or-less flattened and oval bugs. Adults usually measure 6–15 mm. Anterior margin of the pronotum closely fits the hind margin of the head and usually extends along the sides of the head. Fore legs are raptorial and have a single claw. Hemielytra, when fully developed, lack veins in the membranous area.

DISCUSSION: Creeping water bugs are similar in general shape and fore leg structure to the giant water bugs. When their hemielytra are fully developed, the creeping water bugs can be distinguished by the lack of veins in the membranous area of each hemielytron. In both adults and larvae of creeping water bugs, the anterior margin of the pronotum is somewhat concave to receive the hind margin and usually the sides of the head. The head of giant water bugs is generally more distinctively offset from the pronotum (Fig. 10.11).

About 20 species and five genera occur in North America and are classified in four subfamilies.

Creeping water bugs, despite their common name, often swim as well as crawl about. They occur in a variety of aquatic habitats, usually in quiet waters and rarely in riffles. A few inhabit hot springs and desert pools. A common western genus, *Ambrysus*, usually lives in clear streams or ditches. The only eastern genus, *Pelocoris* (Naucorinae), is most common in ponds.

Very close and diligent looking at living creatures, even through the best microscope, will leave room for new and contradictory discoveries.

George Eliot

Studies with the scanning electron microscope have revealed that the plastrons of some adult creeping water bugs are held by as many as four million hairs per square millimeter. The plastron remains effective as a physical gill for prolonged periods in well-oxygenated water, and individuals at rest sometimes ventilate by creating a current over the ventral plastron with a rowing motion of their hind legs.

SUBFAMILY NAUCORINAE: (Fig. 10.7; Plate IX, Fig. 50) Margin of pronotum between the eyes is straight to slightly curved. Inner margins of eyes converge anteriorly. This group is known from eastern Canada, central and eastern states including Florida, and rarely southwestern states.

SUBFAMILY LIMNOCORINAE: (Fig. 10.8) These are similar to Naucorinae, but inner margins of the eyes are divergent anteriorly. The group is known only from Texas and Nevada.

SUBFAMILY AMBRYSINAE: (Fig. 10.9) Margin of pronotum between the eyes is deeply concave. Abdomen has a covering of short hairs ventrally that

CREEPING WATER BUGS
(Naucoridae)

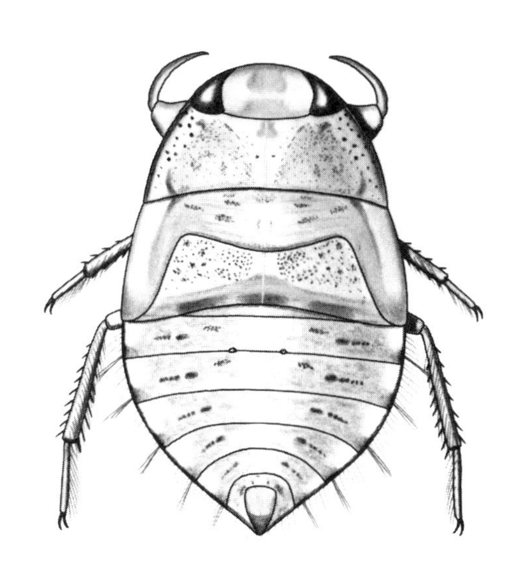

Figure 10.7. Naucorinae immature

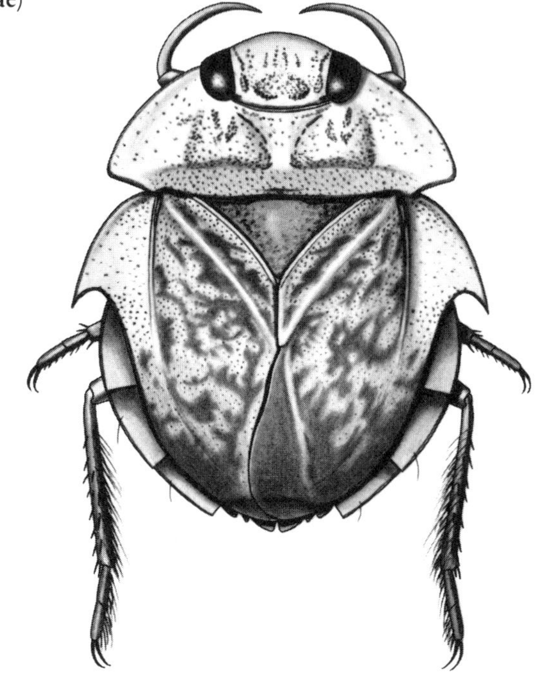

Figure 10.8. Limnocorinae adult

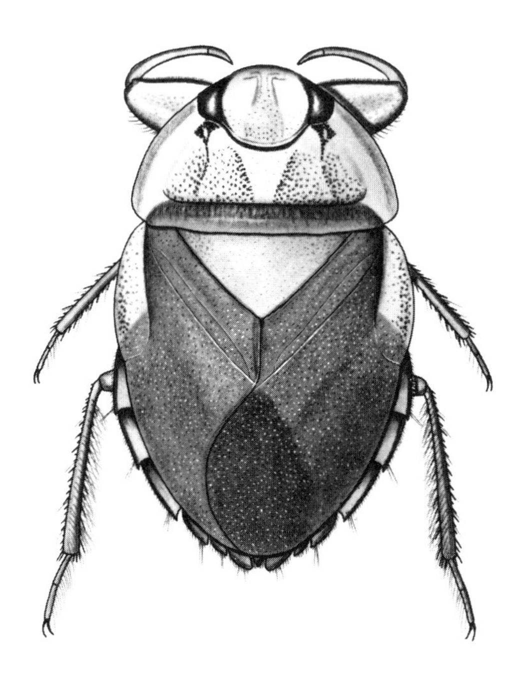

Figure 10.9. Ambrysinae adult

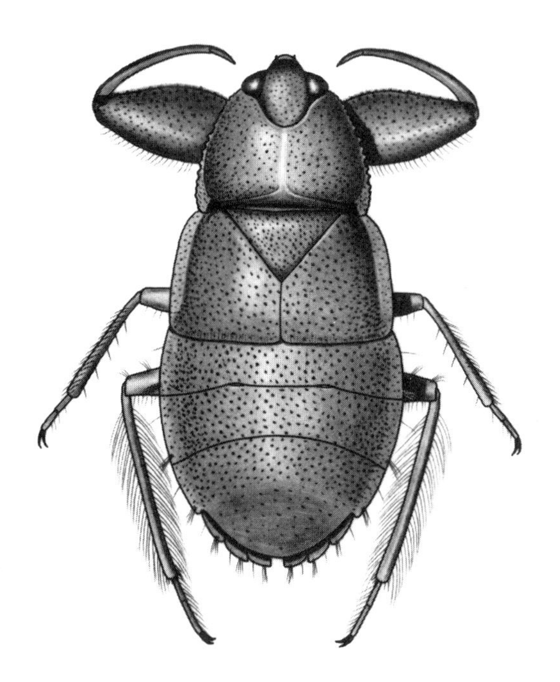

Figure 10.10. Cryphocricinae adult

are especially noticeable on terminal segments. This group is known only from western states.

SUBFAMILY CRYPHOCRICINAE: (Fig. 10.10) These have eyes and pronotum similar to Ambrysinae, but are distinctive otherwise, and abdomen is not covered by short hairs ventrally. The group is known only from swift streams of Texas.

GIANT WATER BUGS
(Belostomatidae)

egg pad

raptorial
fore leg

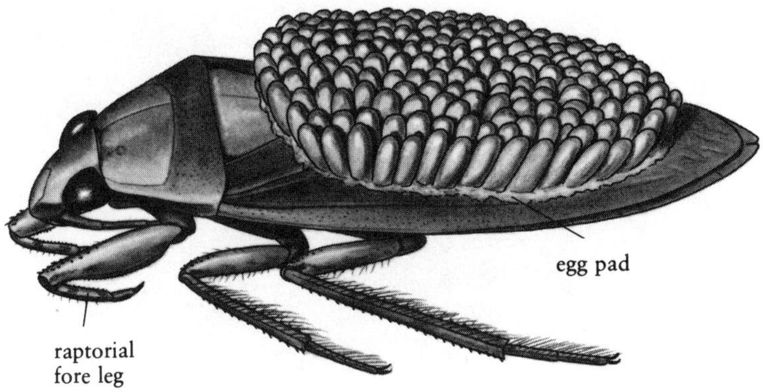

Figure 10.12. *Belostoma* adult ♂

Figure 10.11. *Belostoma* immature

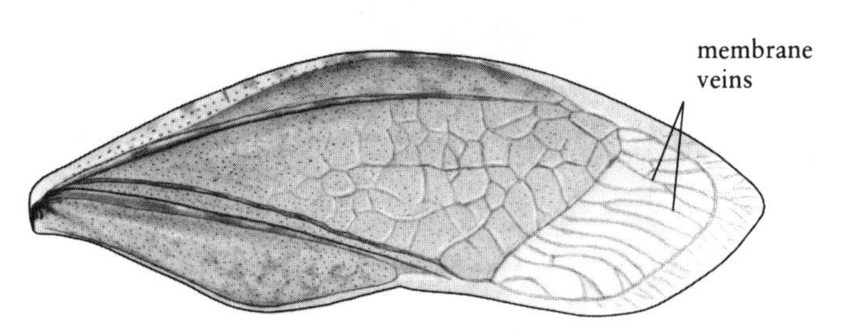

Figure 10.13. Hemielytron (fore wing)

GIANT WATER BUGS
(Family Belostomatidae)

DIAGNOSIS: (Figs. 10.11–10.13; Plate IX, Fig. 53) These are medium- to large-sized bugs, somewhat flattened and oval. Adults are usually 20–65 mm. Fore legs are raptorial, with one or two claws each. Hemielytra possess veins in the membrane. Abdomen never ends in an elongate tube or in rod-like structures, but a terminal pair of short platelike structures is conspicuous in some.

DISCUSSION: Giant water bugs are somewhat similar to some water scorpions and creeping water bugs. All may have similarly adapted legs and be somewhat oval. See previous discussions under those families.

The Belostomatidae comprises two subfamilies, four genera, and approximately 20 species in North America, where the family is widespread.

These spectacular bugs are also known colloquially by several names: (1) fish killers, because they are voracious and will attack small fishes and other small animals such as ducklings; (2) electric light bugs, because dispersing adults are often attracted to lights at night, often far from water; and (3) toe biters, for obvious reasons. Species of *Lethocerus* and *Benacus* truly are giants among aquatic insects. Belostomatids mostly inhabit ponds, pools of streams, or slow ditches, and are often concealed among debris and vegetation. They often spend considerable time as if hanging from the surface of the water by the tip of their abdomen. If disturbed they often feign death, but they will inflict a painful bite if handled carelessly.

Up to 100 eggs or more are glued in sticky egg pads (Fig. 10.12) on the backs of males in the genera *Belostoma* and *Abedus*. Specific brooding behavior in these males contributes to the successful hatching of the eggs. For example, males maintain an intermittent flow of fresh water over the eggs by stroking them with the hind legs. Interestingly, if the egg pad becomes loose or detached, the male may readily cannibalize the eggs. Although this cannibalism would seemingly be a disadvantage to the propagation of the species, or more particularly the contribution of the water bug's own genes to the population, it is not, since dislodged eggs have little if any chance of hatching anyway.

SUBFAMILY LETHOCERINAE: **Adults** (Plate IX, Fig. 53) are over 40 mm. **Larvae** have two claws on the fore leg. The group is widespread in North America.

SUBFAMILY BELOSTOMATINAE: **Adults** (Figs. 10.12, 10.13) are 36 mm or less. **Larvae** (Fig. 10.11) have a single claw on the fore leg. The group is widespread in North America.

The Giant Water Bug: The facile master of the ponds and rivers. . . . Developing in the quiet pools, secreting itself beneath stones or rubbish, it watches the approach of a mud-minnow, frog or other small-sized tenant of the water, when it darts with sudden rapidity upon its unprepared victim, grasps the creature with its strong clasping fore legs, plunges its deadly beak deep into the flesh and proceeds with the utmost coolness to leisurely suck its blood.

P. R. Uhler

SUPERFAMILY CORIXOIDEA

WATER BOATMEN
(Family Corixidae)

DIAGNOSIS: (Fig. 10.14; Plate IX, Fig. 51) These are somewhat flattened, parallel-sided, swimming forms. Adults are 3–11 mm. Beak is short, blunt, triangular, and not distinctly segmented. Fore legs are short; tarsus is modified into a scoop-shaped structure. Hind legs are oarlike and possess swimming hairs.

DISCUSSION: Although the short, modified beak and the short, modified fore legs are found only among water boatmen, these bugs are sometimes confused with backswimmers because of similarities of general shape

WATER BOATMAN
(Corixidae)

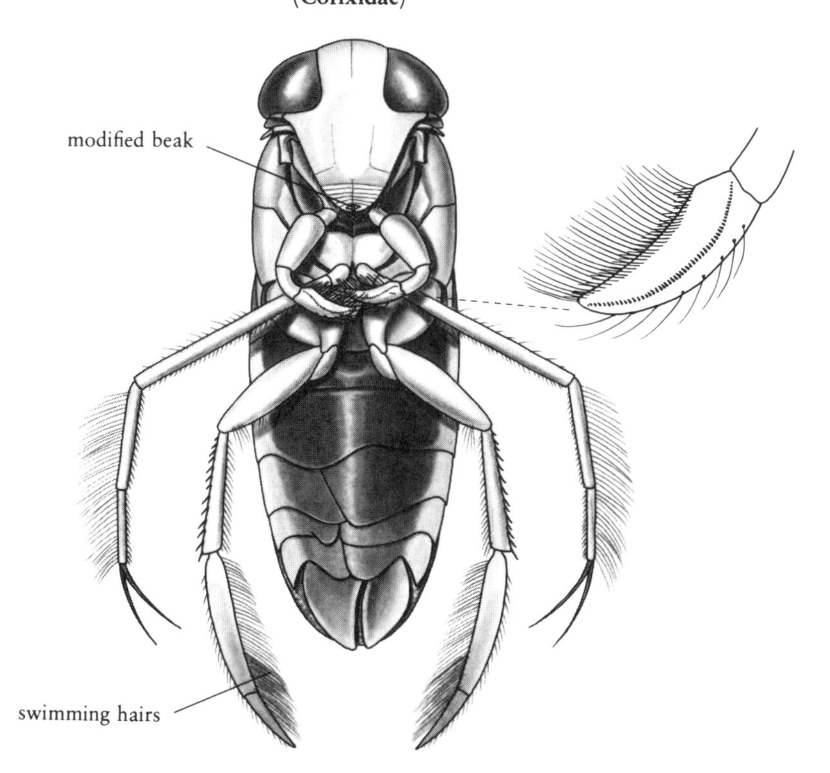

Figure 10.14. *Sigara* adult, ventral (fore tarsus enlarged)

and hind legs. Unlike other water bugs, males of water boatmen are somewhat asymmetrical in respect to certain structures, including their distal abdominal sternites (Fig. 10.14). These sternites are divided into unequally sized sclerites, the larger of which may appear on either the left or right side, depending on the species.

Water boatmen constitute the largest group of water bugs, with well over 100 species occurring in North America.

These free-swimming bugs occur in a number of habitats, including running and quiet waters, brackish pools, and intertidal ocean waters. They are often among the most common aquatic insects of ponds and shallow lakes, and some species are highly tolerant of pollution. Air is stored primarily as a plastron ventrally on the body. The plastron's efficiency as a physical gill is increased by a rowing motion of the hind legs (a type of ventilation). Water boatmen are thus less reliant on atmospheric oxygen than some other underwater bugs, although surfacing for air replenishment may be very frequent in oxygen-poor environments.

Water boatmen differ considerably in their feeding habits from other water bugs. Their beaks are distinctly modified, and they are not restricted to consuming juices but can also grind up small bits of food. They may be classified as herbivore-detritivores, omnivores, predators, or scavengers, depending on the species. Members of the widespread genus *Sigara* are notable as herbivores.

There are generally one or two generations per year, with the larvae or adults overwintering. A species of *Trichocorixa* from saline lakes in central Canada, however, is known to overwinter only in the egg stage. Some species of water boatmen lay their eggs on crayfishes.

Sound production is common. Specific songs are produced (mostly by males) by stridulating the base of the fore femora against the sharp lateral edge of the head. Most produce only one song, which is different for each species and sex, and use it to attract mates, to form aggregations, or to territorially space the individual males within a population.

Water boatmen are an important food item of many fishes, including the Yellow Perch, Largemouth Bass, Bluegill, and White and Black Crappie. They sometimes form a large part of the diet of these fishes, especially in late summer and fall.

SUBFAMILY CORIXINAE: (Fig. 10.14; Plate IX, Fig. 51) Beak possesses distinct transverse grooves. The group is widespread in North America.

SUBFAMILY CYMATIINAE: Beak lacks transverse grooves. The group is known from the northern central states, central and western Canada, and Alaska.

Surface and Shore Bugs

Antennae are longer than the head except among a few of the shore dwellers. These are usually found on top of the water, along the water's edge, or on aquatic or marginal vegetation.

SUPERFAMILY OCHTEROIDEA

TOAD BUGS
(Family Gelastocoridae)

This family is composed of shore-dwelling species and is treated in Chapter 17.

VELVET SHORE BUGS
(Family Ochteridae)

This small family contains only shore-dwelling species and is treated in Chapter 17.

SUPERFAMILY GERROIDEA

WATER STRIDERS
(Family Gerridae)

DIAGNOSIS: (Figs. 10.15–10.17; Plate IX, Fig. 56) These are generally slender to robust, long-legged forms. Adults usually measure 3–20 mm. Adults may be winged, short-winged, or wingless. Preapical claws (see fore legs) arise just before the end of the legs. Femur of hind legs is usually very long and always surpasses the end of the abdomen.

DISCUSSION: Small water striders are relatively similar in form and habit to some of the shortlegged striders. They share the preapical claw condition and are water-surface inhabitants. The hind and middle femora of water striders are usually relatively much longer than those of the Veliidae. Although males of a few species of gerrids have short hind femora,

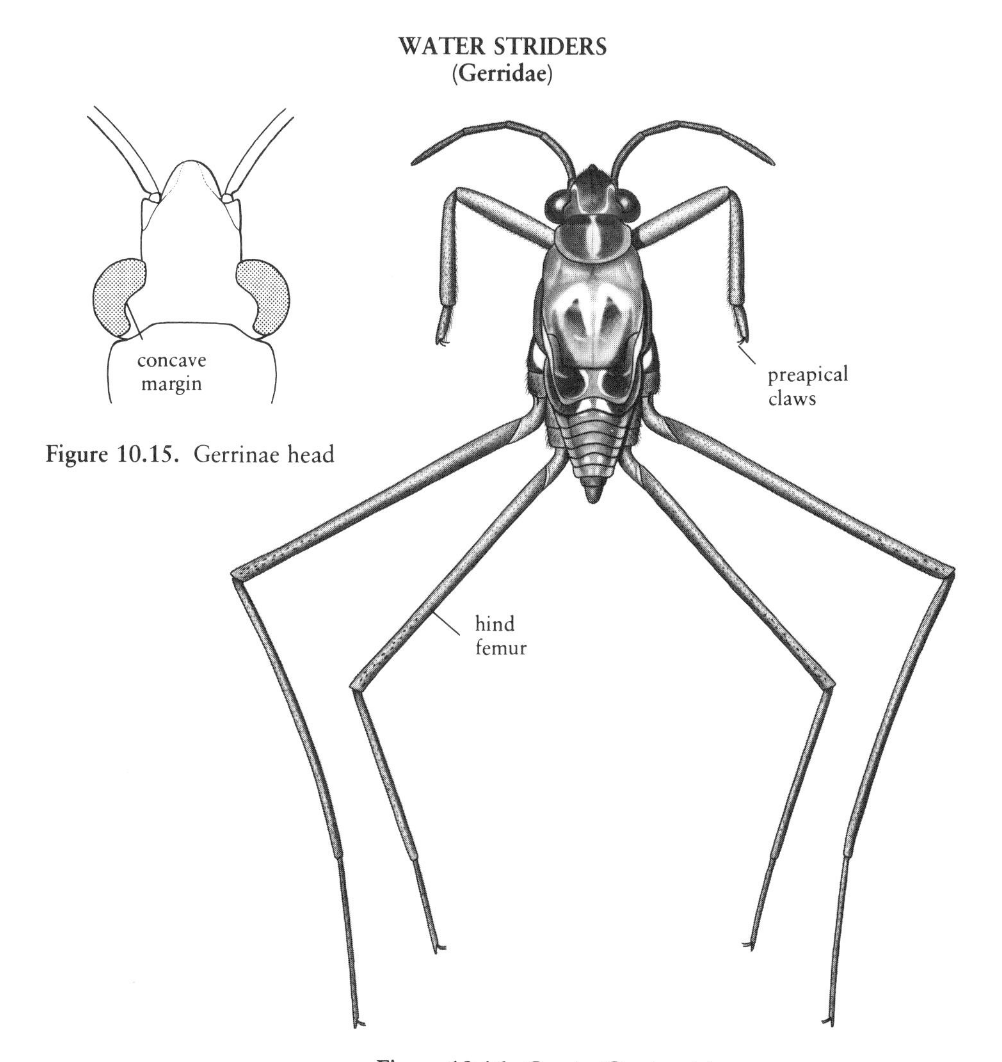

Figure 10.15. Gerrinae head

Figure 10.16. *Gerris* (Gerrinae) immature

these clearly extend beyond the tip of the short abdomen. Because adult water striders often do not possess fully developed wings, they may be confused with larvae; however, adults have two-segmented tarsi, and larvae have one-segmented tarsi.

Between 40 and 50 species of this widespread group occur in North America.

Water striders, which are also commonly known as water skaters or wherrymen, are familiar insects. In parts of Canada, they are sometimes called Jesus bugs because they "walk on water." Their adeptness at skating is sure to catch one's attention; in shallow waters, large round shadows are cast for each slender leg. These shadows result from the depressed but unbroken area of surface film upon which they skate.

They may be encountered in all kinds of aquatic habitats. *Halobates* (sometimes known as the sea skaters) is strictly a marine genus that is known from coastal areas as well as open ocean. *Metrobates* is restricted to riffle areas. Other genera are either primarily lentic or both lentic and lotic.

Of all the bugs I know I can think of none so amorous as our common large water-strider, Gerris remigis *Say. From the earliest days of spring, when the Frost King releases the waters from his bondage, till the cloudy days of autumn, when the leaves fall and the winds grow bleak at his return, these beasties are common and familiar sights to the lover of the quiet flowing waters running to the distant seas. In these haunts in some still little bay or moveless backwater, under a bridge, or in the shadow of a tree, or in the cool recesses of an overhanging bank, you may see* remigis *gathered in numbers, rowing silently about, now and again skipping to escape the maw of some greedy fish, or pouncing on some unfortunate insect fallen into the water and struggling to escape from the clutches of that deadly element. Here they rear large families and spend at ease the sultry dog-days. When winter comes again the old generation have passed away and their young descendants, now full grown, seek shelter against frost and snow under nearby logs or stones on the banks, or crevices in them; there to sleep until in the round of days Old Sol routs the chills of winter and spring once more ushers in the leaves and flowers, and vivifies all the reproductive powers of nature in which* Gerris *is not the least factor.*

J. R. de la Torre-Bueno

WATER STRIDER
(Gerridae)

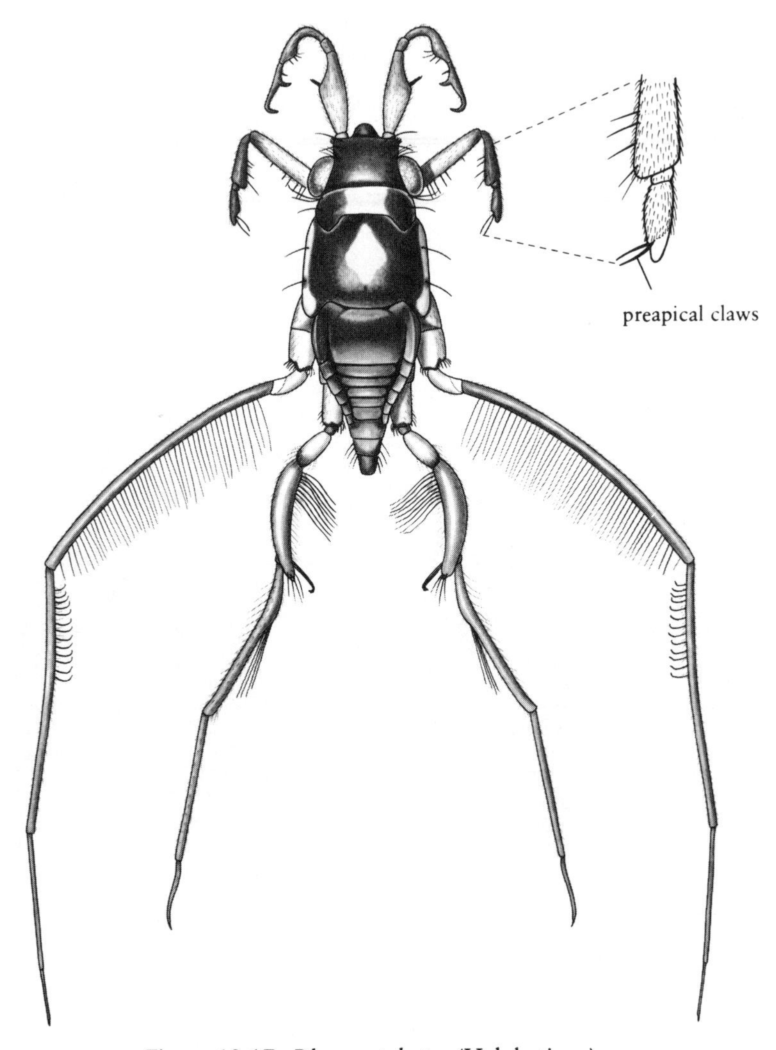

Figure 10.17. *Rheumatobates* (Halobatinae) adult ♂ (fore tarsus enlarged)

Species of the relatively large genus, *Gerris,* are common in the adult stage throughout summer. Those of other genera tend to occur as adults from late summer to late fall.

Water striders are surface dwellers that have a terrestrial type of respiration, but those known to dive occasionally or to be subject to splash and rain benefit by a plastron during such periods. If not for respiratory purposes, the unwettable hairs of these insects at least offer them protection and provide a bubble of air to buoy them to the surface quickly.

Marine water striders are always wingless, but freshwater species are polymorphic, with wings variously developed, sometimes even within the same population. A variety of aquatic and terrestrial insects provide the major food source of water striders, but cannibalism is also common, especially under crowded conditions. The exact manner and frequency with which water striders ripple the surface may possibly have some communicative significance within populations.

SUBFAMILY GERRINAE: (Figs. 10.15, 10.16; Plate IX, Fig. 56) Body is relatively slender. Inner margin of eyes is concave dorsally. The group is widespread in North America.

SUBFAMILY HALOBATINAE: (Fig. 10.17) Body is relatively stout. Inner margin of eyes is straight or evenly convex. The group is widespread in North America.

SHORTLEGGED STRIDERS
(Family Veliidae)

DIAGNOSIS: (Figs. 10.18–10.21; Plate IX, Fig. 55) These are short, stout forms that measure 1–12 mm. Adults may be winged, short-winged, or wingless. Claws are preapical (see fore legs). Femur of hind legs never extends much, if at all, beyond the end of abdomen.

DISCUSSION: These can be distinguished from water striders by their shorter hind femora. In those gerrids that have small, stout bodies (Figs. 10.16, 10.17), the hind femora are usually twice or more the length of the abdomen. There are about 35 species of shortlegged striders in North America, and these are classified in five genera and three subfamilies.

The shortlegged striders are sometimes known as the broadshouldered water striders and other common names. However, as can be seen below, differences in habitat preference within the family have precluded the use of any one common name suggestive of a special habitat for the family as a whole. Open-water preferences are found among the Rhagoveliinae: *Rhagovelia* (the so-called riffle bugs) live on riffles or surface water below riffle areas of streams and rivers, or rarely on the surface water of lakes; *Trochopus* species live in warm-water bays and estuaries. Edgewater preferences are found among the Microveliinae: *Microvelia* (the so-called pond bugs) are usually found along the margins of pools, ponds, and lakes, and only infrequently in open water; and *Husseyella* inhabit salt marshes and marginal marine environments. *Paravelia* (Veliinae) are usually found on emergent vegetation and occasionally on vegetation some distance from water. As is common among the gerrids, the shortlegged striders often detect would-be prey by wave vibrations.

SUBFAMILY RHAGOVELIINAE: (Fig. 10.19; Plate IX, Fig. 55) Tarsus of middle legs is deeply cleft and possesses long plumose hairs. The group is *not* known from western Canada or Alaska and is rare in northwestern states.

SHORTLEGGED STRIDERS
(Veliidae)

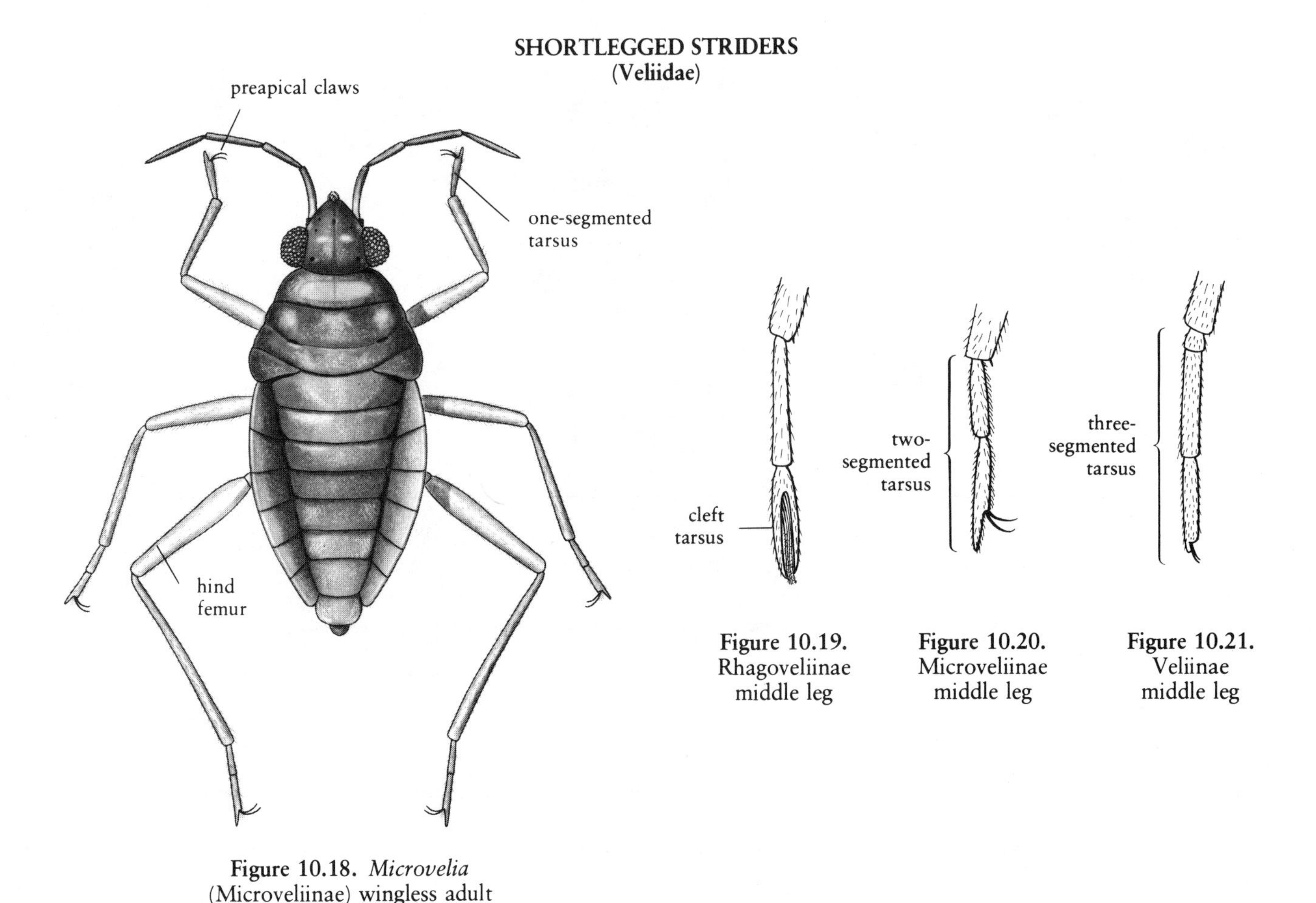

Figure 10.18. *Microvelia* (Microveliinae) wingless adult

Figure 10.19. Rhagoveliinae middle leg

Figure 10.20. Microveliinae middle leg

Figure 10.21. Veliinae middle leg

SUBFAMILY MICROVELIINAE: (Figs. 10.18, 10.20) Tarsus of middle legs is not cleft. **Adults** have fore legs with a one-segmented tarsus and middle and hind legs with a two-segmented tarsus. The group is widespread in North America.

SUBFAMILY VELIINAE: (Fig. 10.21) Tarsus of middle legs is not cleft. **Adults** have three-segmented tarsi on all legs. The group is known only from southeastern states including Florida, southwestern states, and Texas.

MARSH TREADER
(Hydrometridae)

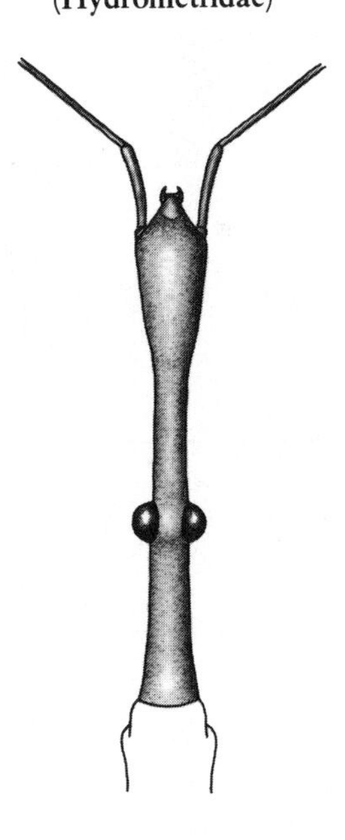

Figure 10.22. *Hydrometra* head

MARSH TREADERS
(Family Hydrometridae)

DIAGNOSIS: (Fig. 10.22; Plate IX, Fig. 52) These are very slender forms that measure 8–11 mm as adults. Head is highly elongate. Some adults are short-winged. Claws are apical. Legs are very slender.

DISCUSSION: Marsh treaders are very delicate bugs. Other water bugs having a somewhat sticklike appearance are the much larger and robust, elongate water scorpions. Only about nine species of the widespread genus *Hydrometra* are found in North America.

Marsh treaders (also known as water measurers) hide among emergent vegetation along the edges of quiet waters and swamps. Although their claws are apical, they are able to walk sluggishly on the water without breaking the surface tension. They lie in wait or stalk their prey in slow motion, with their antennae helping to sense food sources. They also commonly scavenge, and two or more individuals sometimes share food, such as a dead insect.

VELVET WATER BUGS
(Hebridae)

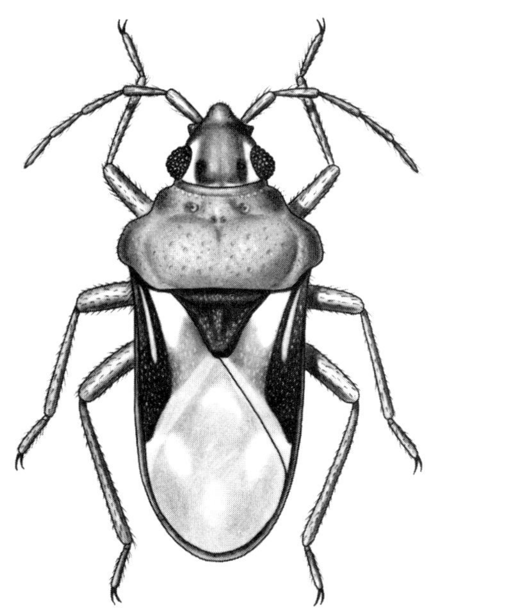

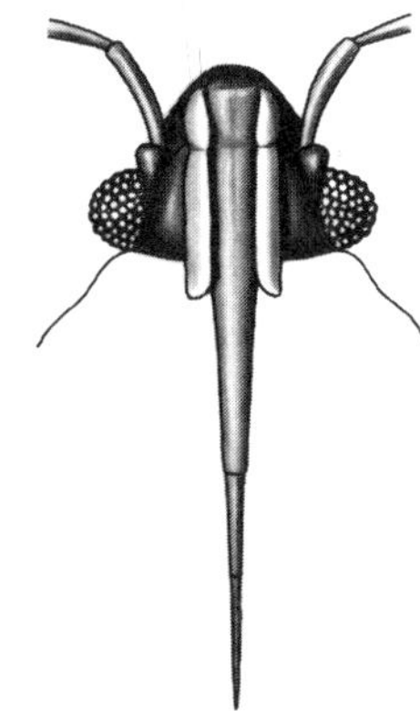

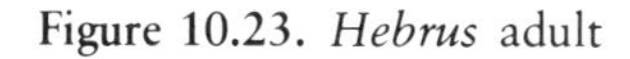

Figure 10.23. *Hebrus* adult

Figure 10.24. *Hebrus* head, ventral

VELVET WATER BUGS
(Family Hebridae)

DIAGNOSIS: (Figs. 10.23, 10.24) These are minute, stout forms that measure 1–2 mm. Body is covered with short, dense hairs. Head possesses a ventral groove for receiving the beak. Pronotum is as broad or broader than the rest of body. All claws are apical.

DISCUSSION: These rather secretive water bugs lack the wing venation of the shore bugs (Saldidae) and the preapical claws of the small pond bugs (Veliidae). The widespread family is represented by two genera and about 15 species in North America. Individuals usually occur in floating algae, weeds, or debris near the shores or on the shores of lentic habitats. They occasionally move about on the water surface.

WATER TREADERS
(Family Mesoveliidae)

DIAGNOSIS: (Fig. 10.25) These are small and moderately slender forms that measure 2–4 mm as adults. They may be winged or wingless. Legs are slender and have conspicuous, scattered black spines. Claws are apical.

DISCUSSION: Water treaders are usually greenish or light brown, whereas pond bugs (*Microvelia*: Veliidae) are usually darkly colored. The black spines along the legs and their apical claws also distinguish the water treaders from pond bugs. This small, widespread family consists of only three species of the genus *Mesovelia* in North America.

Water treaders occur on the surface of ponds, lakes, and marshes, and are often associated with floating or emergent vegetation. They are commonly taken in surface water samples of ponds in many regions. They feed primarily by scavenging.

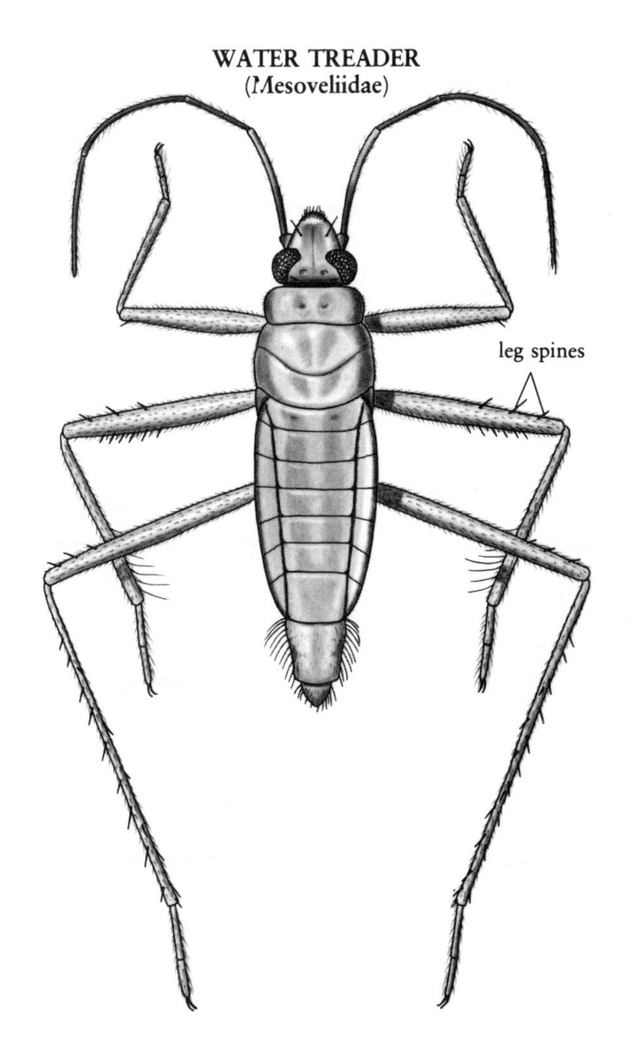

Figure 10.25. *Mesovelia* wingless adult

MACROVELIID SHORE BUGS
(Family Macroveliidae)

This family is treated in Chapter 17.

SUPERFAMILY SALDOIDEA

SHORE BUGS
(Family Saldidae)

This family is composed of species that are commonly found in habitats adjacent to bodies of water, either on the shore or on marginal vegetation. They are treated with the shore-dwelling insects; see Chapter 17.

SPINY SHORE BUGS
(Family Leptopodidae)

This small family is treated in Chapter 17.

SUPERFAMILY DIPSOCOROIDEA

JUMPING GROUND BUGS
(Family Dipsocoridae)

This small group, which contains species sometimes found adjacent to streams, is treated in Chapter 17.

More Information about Water Bugs

The aquatic and semi-aquatic Hemiptera of Virginia by Bobb (1974). This comprehensive treatment of the Virginia fauna contains keys, descriptions, and bionomics of species, and is useful for identification in much of eastern North America.

Aquatic and semiaquatic Heteroptera of Alberta, Saskatchewan, and Manitoba (Hemiptera) by Brooks and Kelton (1967). Those interested in waters of Canada and north-central United States will be greatly aided by this work, which contains keys and discussions of species and higher groups.

The biology and ecology of aquatic and semi-aquatic Hemiptera by Hungerford (1920). This is a large, comprehensive treatment of North American water bug biology, and although many years old, it is still of much use.

The semiaquatic and aquatic Hemiptera of California edited by Menke (1979). This very excellent, well-illustrated work, including contributions by four specialists, should prove indispensible for the study of water bugs in western North America; it is also a fine introduction to the group as a whole.

References

Andersen, N. M. 1979. Phylogenetic inference as applied to the study of evolutionary diversification of semiaquatic bugs (Hemiptera: Gerromorpha). *Syst. Zool.* 8:554–578.

Andersen, N. M., and J. T. Polhemus. 1976. Water-striders (Hemiptera: Gerridae, Veliidae, etc.), pp. 187–224. In *Marine insects* (L. Cheng, ed.). North-Holland, Amsterdam.

Applegate, R. L., and R. W. Kieckhefer. 1977. Ecology of Corixidae (water boatman) in Lake Poinsett, South Dakota. *Amer. Midl. Natural.* 97:198–208.

Bobb, M. L. 1974. *The insects of Virginia: No. 7. The aquatic and semi-aquatic Hemiptera of Virginia.* Va. Poly. Inst. St. Univ. Res. Div. Bull. 87. 195 pp.

Brooks, A. R., and L. A. Kelton. 1967. *Aquatic and semiaquatic Heteroptera of Alberta, Saskatchewan, and Manitoba (Hemiptera).* Mem. Entomol. Soc. Canad. 51. 92 pp.

Chapman, H. C. 1958. Notes on the identity, habitat and distribution of some semi-aquatic Hemiptera of Florida. *Fla. Entomol.* 41:117–124.

Cheng, L. 1973. Marine and freshwater skaters: Differences in surface fine structures. *Nature* 242:132–133.

DeCoursey, R. M. 1971. Keys to the families and subfamilies of the nymphs of North American Hemiptera-Heteroptera. *Proc. Entomol. Soc. Wash.* 73: 413–428.

Drake, C. J., and H. C. Chapman. 1953. Preliminary report on the Pleidae (Hemiptera) of the Americas. *Proc. Biol. Soc. Wash.* 66:53–60.

Drake, C. J., and H. C. Chapman. 1963. A new genus and species of water-strider from California (Hemiptera: Macroveliidae). *Proc. Biol. Soc. Wash.* 76: 227–234.

Gittelman, S. H. 1977. Leg segment proportions, predatory strategy and growth in backswimmers (Hemiptera: Pleidae, Notonectidae). *J. Kans. Entomol. Soc.* 50:161–171.

Hinton, H. E. 1976. Plastron respiration in bugs and beetles. *J. Ins. Physiol.* 22:1529–1550.

Hungerford, H. B. 1920. *The biology and ecology of aquatic and semi-aquatic Hemiptera.* Univ. Kans. Sci. Bull. 21. 341 pp.

Kittle, P. D. 1977. The biology of water striders (Hemiptera: Gerridae) in northwest Arkansas. *Amer. Midl. Natural.* 97:400–410.

Komnick, H. 1977. Chloride cells and chloride epithelia of aquatic insects. *Inter. Rev. Cyt.* 49:285–329.

Lauck, D. R., and A. S. Menke. 1961. The higher classification of the Belostomatidae (Hemiptera). *Ann. Entomol. Soc. Amer.* 54:644–657.

Maier, C. T. 1977. The behavior of *Hydrometra championiana* (Hemiptera: Hydrometridae) and resource partitioning with *Tenagogonus quadrilineatus* (Hemiptera: Gerridae). *J. Kans. Entomol. Soc.* 50:263–271.

Menke, A. S. (ed.). 1979. *The semiaquatic and aquatic Hemiptera of California (Heteroptera: Hemiptera)*. Bull. Calif. Ins. Surv. 21. 166 pp.

Polhemus, J. T. 1976. A reconsideration of the status of the genus *Paravelia* Breddin, with other notes and a check list of species (Veliidae: Heteroptera). *J. Kans. Entomol. Soc.* 49:509–513.

Polhemus, J. T. 1978. Aquatic and semiaquatic Hemiptera, pp. 119–131. In *An introduction to the aquatic insects of North America* (R. W. Merritt and K. W. Cummins, eds.). Kendall/Hunt, Dubuque.

Schwiebert, E. 1973. *Nymphs.* Winchester Press, New York. 339 pp.

Smith, C. L., and J. T. Polhemus. 1978. The Veliidae (Heteroptera) of America north of Mexico—Key and check list. *Proc. Entomol. Soc. Wash.* 80:56–68.

Smith, R. L. 1976. Brooding behavior of a male water bug *Belostoma flumineum* (Hemiptera: Belostomatidae). *J. Kans. Entomol. Soc.* 49:333–343.

Smith, R. L. 1976. Male brooding behavior of the water bug *Abedus herbeti* (Hemiptera: Belostomatidae). *Ann. Entomol. Soc. Amer.* 69:740–747.

Smith, R. L., and J. B. Smith. 1976. Inheritance of a naturally occurring mutation in a giant water bug. *J. Hered.* 67:182–185.

Tones, P. I. 1977. The life cycle of *Trichocorixa verticalis interiores* Sailer (Hemiptera, Corixidae) with special reference to diapause. *Freshwat. Biol.* 7:31–36.

Usinger, R. L. 1941. Key to the subfamilies of Naucoridae with a generic synopsis of the new subfamily Ambrysinae (Hemiptera). *Ann. Entomol. Soc. Amer.* 34:5–16.

Usinger, R. L. 1956. Aquatic Hemiptera, pp. 182–228. In *Aquatic insects of California* (R. L. Usinger, ed.). Univ. Calif. Press, Berkeley.

Wilcox, R. S. 1972. Communication by surface waves. Mating behavior of a water strider (Gerridae). *J. Comp. Physiol.* 80:255–266.

Wilcox, R. S. 1979. Sex discrimination in *Gerris remigis*: Role of a surface wave signal. *Science* 206:1325–1326.

CHAPTER **11**

Fishflies, Dobsonflies, and Alderflies

(ORDER MEGALOPTERA)

THE LARVAE OF Megaloptera are primarily aquatic, whereas all other life stages are terrestrial. Although small in numbers of species (about 50 in North America), the larvae of this order, especially those of the large dobsonflies (hellgrammites), are well known to fishermen and students of aquatic life because of their size and highly active, rather ferocious nature. Adults, although also striking in appearance, are usually secretive and short-lived and thus are not as commonly seen. As is evidenced by an early occurrence in the fossil record and retention of a primitive type of wing venation, the order is perhaps the oldest among those insects having an advanced type of metamorphosis (with a transitional pupal stage). Because the larvae are major predators on other aquatic insects, they form an important link in the aquatic food chain.

Larval Diagnosis

Larvae are elongate and slightly flattened forms that measure 10–90 mm at maturity. Head possesses well-developed chewing mouthparts, filamentous antennae, and poorly developed eyes. Thoracic legs are present. Wing pads are absent. Abdominal segments 1–7 or 1–8 each have a pair of lateral filaments. Abdomen terminates either in a single unbranched filament or in a pair of anal prolegs (each proleg has a pair of terminal hooks and a dorsal filament).

Pupal Diagnosis

These are nonaquatic forms that are over 12 mm in length. Appendages are distinct and are not fused to the body. Mandibles are well developed and project forward. Developing fore wings are not thickened more than the developing hind wings. See Figure 11.9.

Adult Diagnosis

These are medium-sized to large insects, with two pairs of elongate, net-veined wings (sometimes darkly pigmented) that are held rooflike over the abdomen when at rest. Hind wings are folded (pleated) lengthwise when not in use. Abdomen lacks tail-like structures.

Similar Orders

Those water beetle larvae that possess lateral filaments along the abdomen are the aquatic insects most likely to be confused with dobsonfly, alderfly, or fishfly larvae. The end of the abdomen should be examined closely until the groups have become familiar. Beetle larvae with lateral abdominal filaments never have a single completely undivided terminal filament, nor do they have a pair of anal prolegs that each possesses a pair of hooks. Caddisfly larvae with distinctive anal prolegs have only one terminal hook per proleg and never have large segmented lateral abdominal filaments like the Megaloptera.

The adults of Megaloptera resemble some of the larger stoneflies in flight. Generally they can be distinguished from stoneflies by their lack of tails and the more vertical orientation of the wings when at rest. In the field, the dark-winged alderflies superficially resemble some caddisflies. Close examination, however, will clearly distinguish them.

Life History

Metamorphosis is complete, but larvae and adults are not nearly as morphologically distinct from each other as are most insects having complete metamorphosis. Larvae generally go through 10 or 11 instars or fewer and require almost a year to more than three years to complete growth.

Larvae that are ready to pupate (known as prepupae) migrate to the shore and carefully choose a site a few centimeters to 10 meters from the water to prepare a cell in which they undergo transformation to the pupal stage. The cell is usually formed 1 to 10 cm deep in the soil but may be formed under mossy or decaying vegetation or even in decayed driftwood or tree stumps. Among a few species that live in intermittent streams, pupation is synchronized to seasonal drying periods and takes place in dried stream beds under rocks.

The prepupal form of the insect lasts from a day to two weeks and is followed by a brief pupal stage. Pupae are generally quiescent, but some become quite active if necessary in order to move or defend themselves. Prior to adult emergence, the pupa (actually the pharate adult within the pupal skin) crawls from its unlined cell. The adult, upon emergence from the pupal skin, requires about 20 to 60 minutes for drying and final maturation.

Adults of most species live for only a few days. Emergence generally takes place in the spring and summer; however, some fishflies emerge year round in southern states, and a very few fishflies are known to emerge from

A pond's a mirrored world, where strong on weak,
Cunning on simple prey.

W. S. Blatchley

March to November in northern regions. Adults are weak, almost cumbersome fliers. Alderflies are typically active during warm midday times. They are excellent runners but sluggish fliers. Dobsonflies and fishflies tend to be active primarily at twilight or later. In nature, adults probably do not take solid food (although some have well-developed mandibles); in captivity, they will drink honey-water and some other sweet juices.

Mating takes place on the ground or on vegetation. Among the dobsonflies, mating is occasionally preceded by fighting between males if more than one is in the vicinity of the female. The male dobsonfly (in at least one species) places his enlarged mandibles over the wings of the female for a brief period before mating with her. Mating behavior is not presently known for other species of Megaloptera.

Oviposition always takes place out of the water, often on bridge abutments, overhanging vegetation, or exposed rocks. Laid egg masses consist of a large number of eggs that take from a few days to two weeks to hatch. Such eggs are often parasitized by wasps. Egg hatching takes place at night, and the new larvae drop into the water. The larvae of those fishflies that pupate and emerge in the beds of dry streams or ponds may live as active terrestrial insects or burrow into the substrate until the bed is again inundated with water.

Aquatic Habitats

Larvae of most species occur as benthos in streams and rivers and may be associated with a variety of substrate and current types. Some also inhabit ponds, forest pools, and littoral lake environments. A few regularly occur within soft substrate, and some may overwinter in the substrate.

Aquatic Adaptations and Behavior

Larvae are capable of hydropneustic and aeropneustic respiration to some degree. For most, the utilization of dissolved oxygen is the sole method of respiration while in the water. Lateral filaments in all species and the tufted gills of the dobsonflies may greatly enhance oxygen uptake. In some fishflies spiracles are located at the ends of a pair of elongated breathing tubes that arise from near the end of the abdomen (Fig. 11.3). These allow the larvae to make contact with the air-water interface and utilize atmospheric oxygen while otherwise remaining submerged. Larvae in general also retain a full complement of spiracles along the body. These are probably always closed when submerged but are normally used by prepupae when they leave the water to pupate and by young larvae that have not yet entered the water or have not been inundated in intermittent environments.

Means of osmoregulation are poorly understood in Megaloptera, but alderflies, at least, are capable of ion uptake in the midgut and may actively swallow water for such purposes.

Larvae are, without exception, carnivorous and primarily prey on other insects, although they are also cannibalistic. Some alderfly larvae are known to ingest prey without chewing. Feeding activities may be more in-

The bright little wink under water!
Mysterious wink under water!
Delightful to ply
The subaqueous fly
And watch for the wink under water!

G. E. M. Skues

tensive at night. Netspinning caddisflies and black flies often account for a major proportion of the food of dobsonflies, and some species may be somewhat selective of their prey, although this is also related to the seasonal abundance of prey species; others are indiscriminate predators. These relatively voracious larvae are strong sprawlers or clingers, and they can also swim either forwards or backwards with undulating motions. They will readily bite would-be collectors or fishermen if given the opportunity.

Megaloptera larvae are not taken as commonly in the drift as are many other lotic insects, possibly because they simply do not drift frequently or are strong enough to avoid or escape the sample nets.

Classification and Characters

The order Megaloptera consists of two families, the Sialidae and Corydalidae, both of which are well represented in most areas of North America.

The primary distinguishing features of the larvae are their abdominal appendages, which make recognition easy. As adults, individuals of each family can be distinguished by their size, among other traits.

ALDERFLIES
(Sialidae)

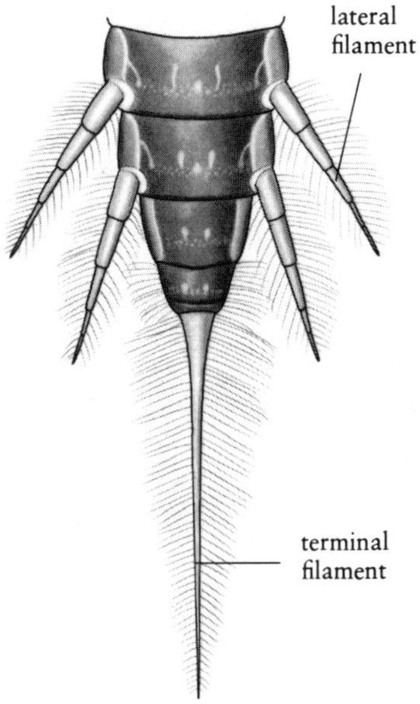

lateral filament

terminal filament

Figure 11.1. *Sialis* larva, end of abdomen

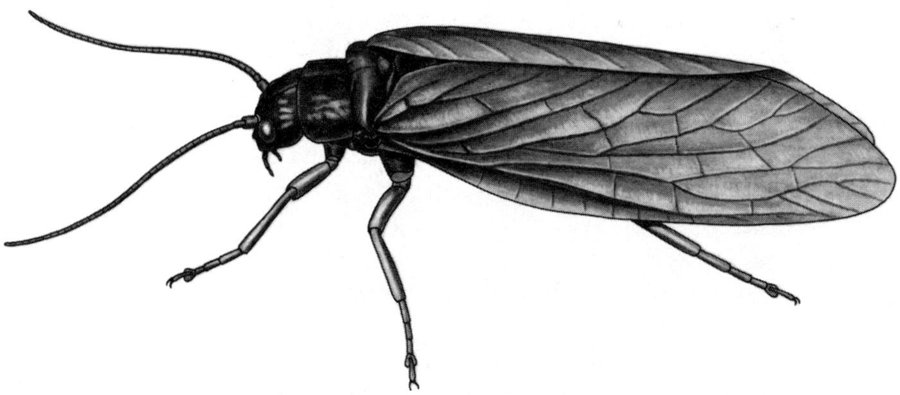

Figure 11.2. *Sialis* adult

ALDERFLIES
(Family Sialidae)

LARVAL DIAGNOSIS: (Fig. 11.1; Plate XI, Fig. 74) These larvae, when full grown, are usually 10–25 mm in length, rarely as much as 30 mm including the terminal filament. Abdomen possesses seven pairs of four- to five-segmented lateral filaments and a single unbranched terminal filament.

ADULT DIAGNOSIS: (Fig. 11.2) These measure less than 20 mm in length. They sometimes have blackish or dark brown wings.

DISCUSSION: Only the genus *Sialis* (with about 20 or more species) occurs in North America. Although alderflies are found in all regions of North America, their relative abundance varies considerably from place to place. For example, they are very rarely taken in Florida and are only very locally common in much of the West.

Larvae inhabit both lotic and lentic environments. Many species seem to prefer quiet waters or pools where detritus and debris have accumulated on the soft bottoms. Some are often found on the bottom sides of rocks in rivers and streams. They can also be found slightly within the substrate.

The adults are very awkward fliers but good runners, and they generally remain very close to the freshwater habitat of the larvae.

FISHFLIES AND DOBSONFLIES
(Family Corydalidae)

LARVAL DIAGNOSIS: (Figs. 11.3–11.5; Plate XI, Fig. 72) These measure 25–90 mm at maturity. Abdomen possesses eight pairs of two-segmented lateral filaments and ends in one pair of anal prolegs, each of which has two terminal hooks.

FISHFLIES AND DOBSONFLIES
(Corydalidae)

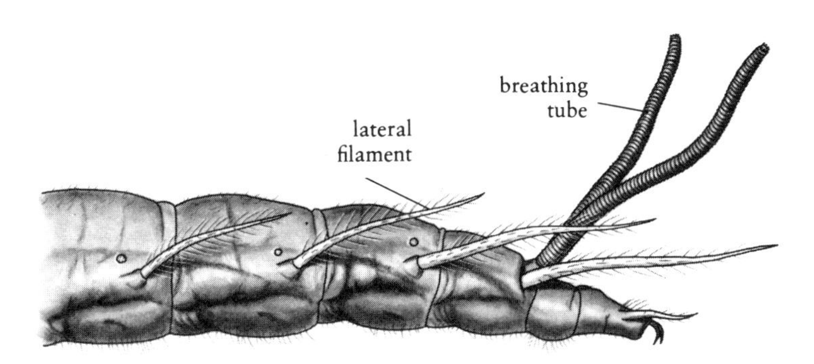

Figure 11.3. *Chauliodes* larva, end of abdomen (lateral)

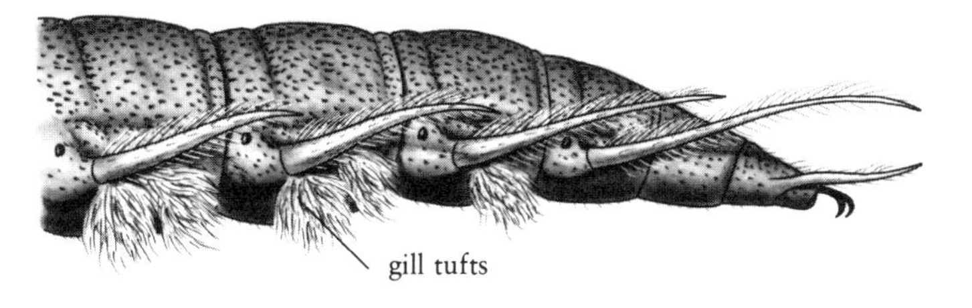

Figure 11.4. *Corydalus* larva, end of abdomen (lateral)

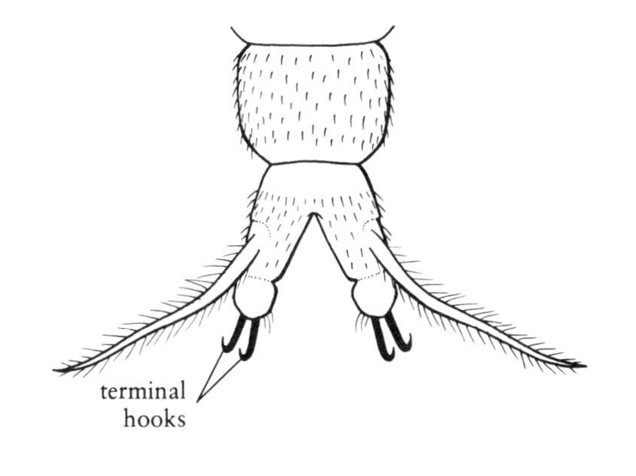

Figure 11.5. End of larval abdomen (dorsal)

FISHFLIES AND DOBSONFLIES
(Corydalidae)

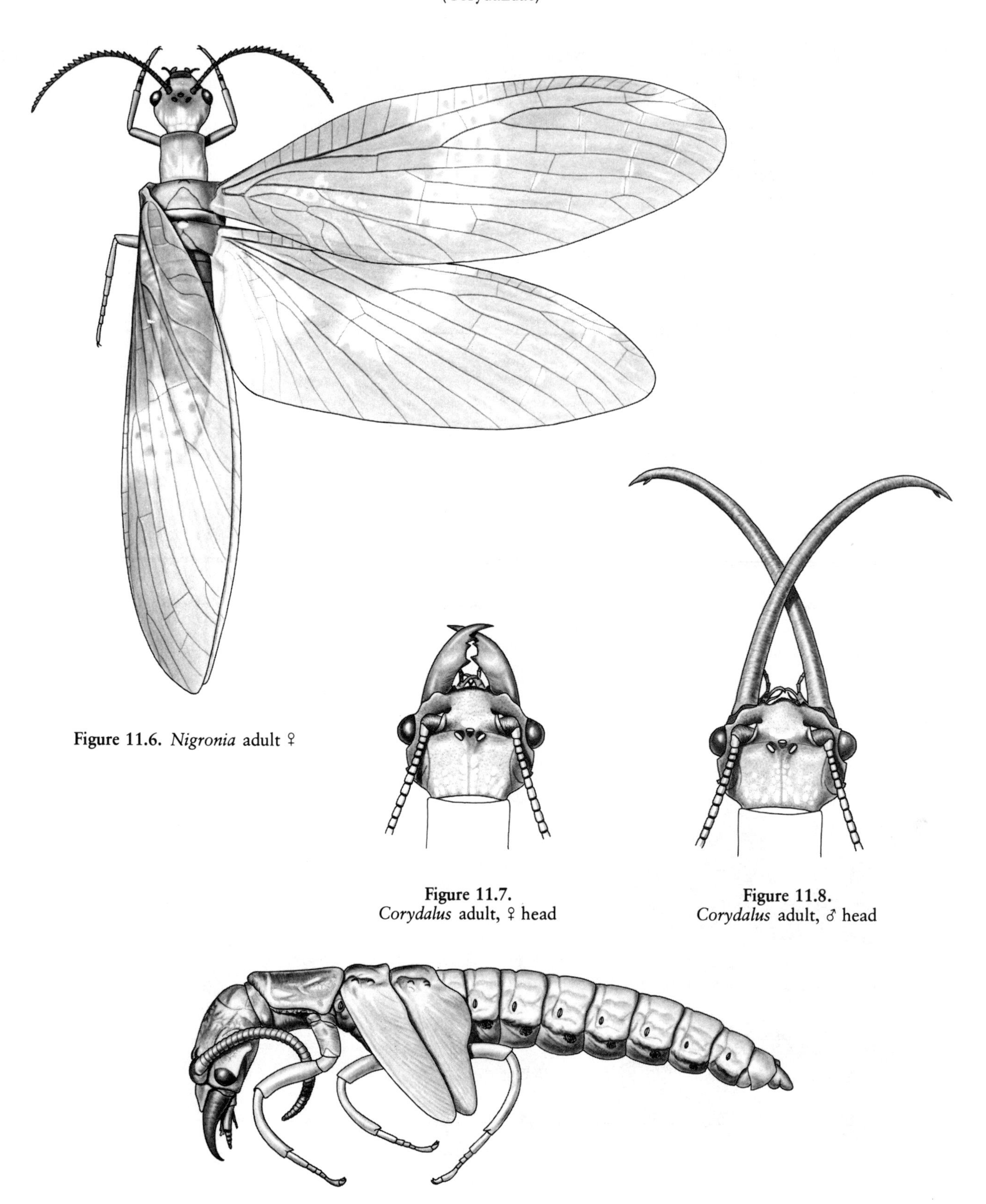

Figure 11.6. *Nigronia* adult ♀

Figure 11.7. *Corydalus* adult, ♀ head

Figure 11.8. *Corydalus* adult, ♂ head

Figure 11.9. *Corydalus* pupa

ADULT DIAGNOSIS: (Figs. 11.6–11.8; Plate XI, Fig. 73) These are large and usually measure well over 35 mm. They often have mottled wings.

DISCUSSION: Fishflies and dobsonflies as a group are distinctive both as larvae and adults. Two subfamilies are widespread in North America but are conspicuously absent from the Great Basin region of the western United States.

The Chauliodinae, or fishflies, comprise five genera and about 18 species, whereas the Corydalinae, or dobsonflies, comprise only one genus, *Corydalus*, and two species.

Dobsonfly larvae (also commonly known as hellgrammites and less commonly as bass bait or crawlers) and most fishfly larvae usually occur in well-oxygenated flowing water. Some larvae of the fishfly genus *Nigronia* may be taken under the bark or decaying parts of driftwood. Some fishfly larvae (e.g., in the eastern genus *Chauliodes*; Plate XI, Fig. 72) more typically occur in quiet pools of streams, in ponds, intermittent forest pools under leaf detritus, and in shallow lakes. *Chauliodes* is able to obtain atmospheric oxygen via a pair of elongate breathing tubes that contact the air-water interface. At the other extreme, hellgrammites possess well-developed tufted gills along the abdomen.

Larvae are prized bait for fishermen who require something with a great deal of action. Male adults of dobsonflies have greatly enlarged sicklelike mandibles that are used in courtship behavior, for defending themselves, and for aggressive behavior.

SUBFAMILY CHAULIODINAE (fishflies): **Larvae** (Fig. 11.3; Plate XI, Fig. 72) lack abdominal gill tufts. The group is widespread but patchy in distribution in western North America.

SUBFAMILY CORYDALINAE (dobsonflies): **Larvae** (Figs. 11.4, 11.5; chapter opening) possess abdominal gill tufts at the base of lateral filaments 1–7. The group is widespread but patchy in distribution in western North America; it is absent from much of the intermountain area of the United States and Canada.

More Information about Fishflies, Dobsonflies, and Alderflies

"The egg masses, eggs, and first instar larvae of eastern North American Corydalidae" by Baker and Neunzig (1968). This most interesting article comparatively treats the very early stages of five species of Corydalidae.

"Megaloptera" (Chapter 8 in *Aquatic insects of California*) by Chandler (1956). This work is useful for keying western adults to species and North American larvae to genus.

"Nearctic alder flies of the genus *Sialis* (Megaloptera, Sialidae)" by Ross (1937). This is a good introduction to the alderflies and provides comprehensive information as well as keys for adults.

References

Azam, K. M., and N. H. Anderson. 1969. Life history and habits of *Sialis rotunda* and *S. californica* in western Oregon. *Ann. Entomol. Soc. Amer.* 62:549–558.

Baker, J. R., and H. H. Neunzig. 1968. The egg masses, eggs, and first-instar larvae of eastern North American Corydalidae. *Ann. Entomol. Soc. Amer.* 61:1181–1187.

Chandler, H. P. 1956. Megaloptera, pp. 229–233. In *Aquatic insects of California* (R. L. Usinger, ed.). Univ. Calif. Press, Berkeley.

Cuyler, R. D. 1958. The larvae of *Chauliodes* Latreille (Megaloptera: Corydalidae). *Ann Entomol. Soc. Amer.* 51:582–586.

Flint, O. S., Jr. 1965. The genus *Neohermes* (Megaloptera: Corydalidae). *Psyche* 72:255–263.

Kevan, D. K. McE. 1979. Megaloptera, pp. 351–352. In *Canada and its insect fauna* (H. V. Danks, ed.). Mem. Entomol. Soc. Canada 108.

Komnick, H. 1977. Chloride cells and chloride epithelia of aquatic insects. *Inter. Rev. Cyt.* 49:285–329.

Neunzig, H. H. 1966. Larvae of the genus *Nigronia* Banks. *Proc. Entomol. Soc. Wash.* 68:11–16.

Parfin, S. I. 1952. The Megaloptera and Neuroptera of Minnesota. *Amer. Midl. Natural.* 47:421–434.

Ross, H. H. 1937. Studies of Nearctic aquatic insects. I. Nearctic alder flies of the genus *Sialis* (Megaloptera, Sialidae). *Bull. Ill. Nat. Hist. Surv.* 21:57–78.

Stewart, K. W., G. P. Friday, and R. E. Rhame. 1973. Food habits of hellgrammite larvae, *Corydalus cornutus* (Megaloptera: Corydalidae), in the Brazos River, Texas. *Ann. Entomol. Soc. Amer.* 66:959–963.

Tarter, D. C., W. D. Watkins, and M. L. Little. 1975. Life history of the fishfly, *Nigronia fasciatus* (Megaloptera: Corydalidae). *Psyche* 82:81–88.

Tarter, D. C., W. D. Watkins, M. L. Little, and D. L. Ashley. 1977. Seasonal emergence patterns of fishflies east of the Rocky Mountains (Megaloptera: Corydalidae). *Entomol. News* 88:69–76.

CHAPTER 12

Spongillaflies

(ORDER NEUROPTERA, FAMILY SISYRIDAE)

INSECTS BELONGING TO the order Neuroptera are all terrestrial except for the small family Sisyridae, the spongillaflies. The order Neuroptera is closely related to the entirely aquatic order Megaloptera, and the two groups are sometimes classified together in the order Neuroptera. Two genera (*Sisyra* and *Climacia*) and six species of spongillaflies are currently recognized in North America. The aquatic larvae are associated with freshwater sponges, on which they feed. Other life stages of spongillaflies are terrestrial. The group is generally widespread in North America but rare in the southwestern states and western intermountain areas, being limited by the relative abundance of their sponge hosts.

Larval Diagnosis

These are small, soft-bodied, bristled forms that usually measure 3–8 mm in length (Fig. 12.1; Plate XI, Fig. 75). Eyes are moderately developed. Antennae are long. Mouthparts are highly modified into a needlelike sucking apparatus. Thoracic legs are present, and each possesses a single claw. Wing pads are absent. All but the very young larvae have ventral pairs of segmented filaments on the abdomen. No tail-like structures are present.

I see men fishing beside quiet streams, and those who do not are pursued by collectors, and plastered with liens.

Kenneth Fearing

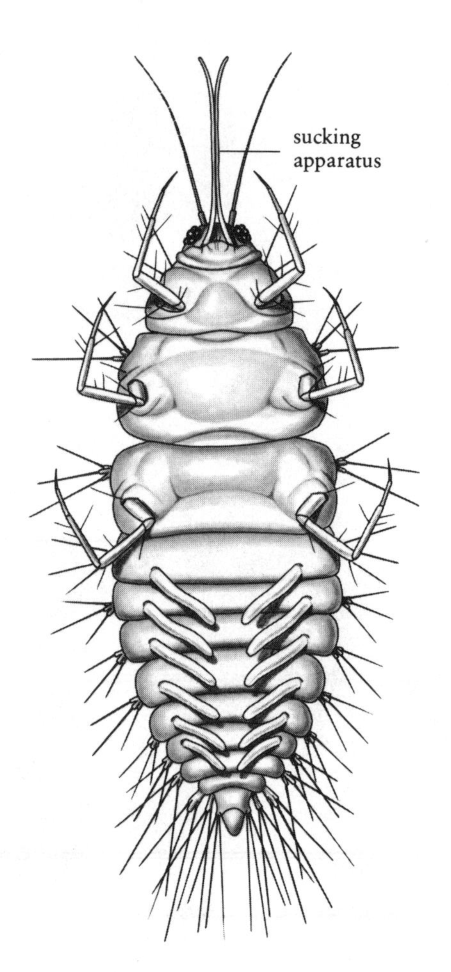

Figure 12.1. *Climacia* larva (ventral)

Pupal Diagnosis

Pupae are terrestrial, quiescent forms that occur within silken cocoons (Fig. 12.2). They measure less than 10 mm. Appendages are not fused to the body. Mandibles are stout but not curved or crossing.

Adult Diagnosis

These are small insects that possess two pairs of brown or mottled, net-veined wings held rooflike over the body when at rest (Fig. 12.3). Hind wings are never pleated lengthwise. No tails are present.

Similar Orders

Spongillafly larvae should not be confused with other aquatic larvae, because their suctorial mouthparts and peculiar ventral filaments are highly diagnostic. At a distance, however, they may resemble some beetle larvae.

Adults are typical of Neuroptera in general and, except for their smaller size, are similar to fishflies and alderflies, which frequent similar habitats. Hind wings of the spongillaflies are never folded accordionlike in the hind region like those of the Megaloptera.

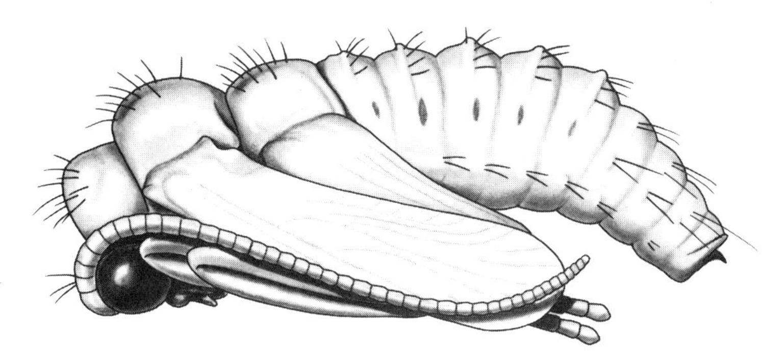

Figure 12.2. *Climacia* pupa

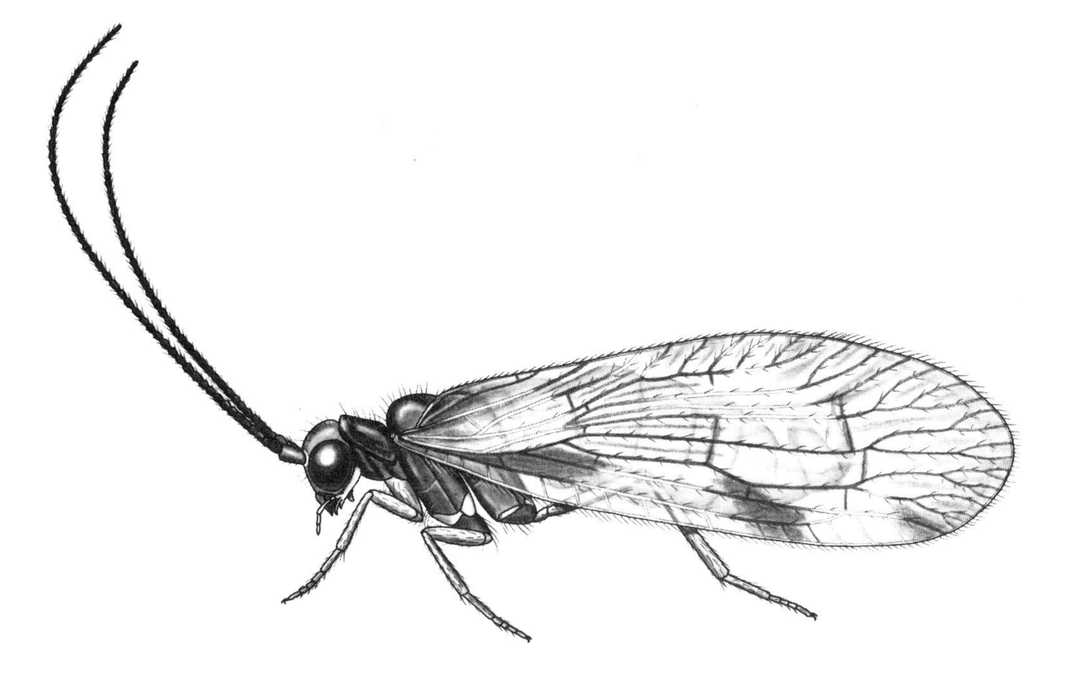

Figure 12.3. *Climacia* adult

Life History

Metamorphosis is complete, and there are one or more generations per year. The larval stage consists of three instars. Spongillaflies overwinter as prepupae (larvae within the pupal cocoons) or as aquatic larvae if host sponges remain available (e.g., in warmer climates).

Pupation occurs up to several meters from the water in various crevices, on plant parts, or even in shed skins of mayfly subimagos or dragonfly larvae. A double-walled silken cocoon is spun by the larva. This activity generally takes place at night and can require several hours. The outer wall of the cocoon is often a distinct, loose netlike structure but sometimes is difficult to separate from the inner, tightly woven wall. The end of the larval abdomen is modified into a spinneret that spins the silk from the rectal area, which is closed off from the remainder of the alimentary system and is modified for silk production. The pupal stage lasts from five days to two weeks. Just before emergence, stout mandibles are used to free the individual from the cocoon. Adult emergence (a period when the insects would be highly susceptible to predation) takes place during the night.

Soon after adults emerge, they excrete a fecal pellet that incorporates all of the waste products accumulated up to this time (the alimentary tract of the larvae having been closed off). Adults are short-lived and emerge from spring to fall in northern latitudes and more generally in warmer climates. Adults are active at night and are probably primarily nectar feeders. They are generally attracted to lights.

Mating is preceded by a male approaching a female from the side and evidently eliciting some courtship response with his antennae. The male then deposits a small case of sperm (spermatophore) on the tip of the female abdomen.

The female oviposits at night, laying 1–20 eggs at a time. The eggs are placed in dry crevices overhanging the water and are covered with a silken web. After one or two weeks, young larvae hatch from the eggs and fall onto the water. After becoming submerged (sometimes requiring considerable exertion), they immediately seek out their sponge hosts.

Aquatic Habitats

Spongillaflies are generally found in all of the freshwater environments that their sponge hosts inhabit. These include ponds, lakes, streams, and rivers. Most of the sponges prefer moderate to little current and are attached to underwater plant parts, rocks (commonly on the undersides), or debris. Larvae live on or within the sponge host. At least nine genera of freshwater sponges are known to be parasitized by spongillaflies in North America, and there is seemingly little or no host specificity.

Aquatic Adaptations and Behavior

Aquatic respiration of spongillaflies is hydropneustic, and uptake of dissolved oxygen takes place cutaneously over the general body surface. The ventral abdominal filaments, which become developed only in the second and third larval instars, may aid respiration by adding surface area and, when vibrated periodically, by ventilating the body with oxygenated water.

This ought to convince us of our ignorance of the mutual relationships of all organic beings; a conviction as necessary as it is difficult to acquire.

Charles Darwin

Needlelike mouthparts are used to suck the juices and small cells from sponge tissue. Larvae generally remain with their hosts throughout their aquatic existence or until the host dies, at which time they seek out a new host. Although they crawl about on their hosts, larvae are able to swim by repeatedly flexing their bodies into a C-shape and then straightening out again.

More Information about Spongillaflies

"The life history of *Climacia areolaris* (Hagen), a neuropterous 'parasite' of fresh water sponges" by Brown (1952). This paper remains the most comprehensive biological study of any North American spongillafly.

"The spongilla-flies, with special reference to those of the Western Hemisphere (Sisyridae, Neuroptera)" by Parfin and Gurney (1956). This is a rather complete taxonomic treatment of the family and gives a good overview; however, keys for larvae and discussions of biology are now somewhat out-of-date.

"Studies on southern Sisyridae (spongilla-flies) with a key to the third-instar larvae and additional sponge-host records" by Poirrier and Arceneaux (1972). This short paper is valuable for its key to larvae.

References

Brown, H. P. 1952. The life history of *Climacia areolaris* (Hagen), a neuropterous 'parasite' of fresh water sponges. *Amer. Midl. Natural.* 47:130–160.

Grigarick, A. A. 1975. The occurrence of a second genus of spongilla-fly (*Sisyra vicaria* [Walker]) at Clear Lake, Lake County, California. *Pan-Pac. Entomol.* 51:296–297.

Old, M. C. 1933. Observations on the Sisyridae (Neuroptera). *Pap. Mich. Acad. Sci. Arts Lett.* 17:681–684.

Parfin, S. I., and A. B. Gurney. 1956. The spongilla-flies, with special reference to those of the Western Hemisphere (Sisyridae, Neuroptera). *Proc. U.S. Nat. Mus.* 105:421–529.

Poirrier, M. A. 1969. Some fresh-water sponge hosts of Louisiana and Texas spongilla-flies, with new locality records. *Amer. Midl. Natural.* 81:573–575.

Poirrier, M. A., and Y. M. Arceneaux. 1972. Studies on southern Sisyridae (Spongilla-flies) with a key to the third-instar larvae and additional sponge-host records. *Amer. Midl. Natural.* 88:455–458.

Steffan, A. W. 1967. Ectosymbiosis in aquatic insects, pp. 207–289. In *Symbiosis*, Vol. II (S. M. Henry, ed.). Academic Press, New York.

White, D. S. 1976. *Climacia areolaris* (Neuroptera: Sisyridae) in Lake Texoma, Texas and Oklahoma. *Entomol. News* 87:287–291.

CHAPTER 13

Water Beetles

(ORDER COLEOPTERA)

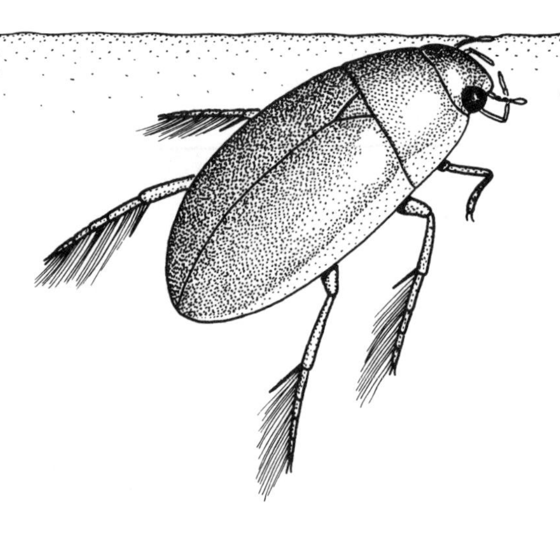

THE BEETLES CONSTITUTE the largest and perhaps most highly advanced group of insects. Their diverse adaptations are reflected by some 30,000 known species in North America. Of these, over 1,000 are aquatic or semiaquatic. Partial or complete aquatic adaptation has occurred independently several times in the history of beetle evolution. As a result, some families are primarily aquatic, and other families contain only certain species that are aquatic or semiaquatic. In some families, both adults and larvae are aquatic, and in others only one or the other of these stages is aquatic.

Beetles in general have received considerable attention from researchers; however, the larvae of water beetles remain poorly known. The many predatory species in this group are undoubtedly important in freshwater ecosystems, especially ponds. Some stream species are good indicators of water quality. A few plant-feeding water beetles are injurious to rice, and some are useful in controlling aquatic weeds. Behaviorally, the water beetles are a most fascinating group of organisms.

Adult Diagnosis

These are distinctive, generally hard-bodied, oval to elongate insects that measure 1–40 mm. Head possesses chewing mouthparts, well-developed eyes, and variably shaped antennae. Fore wings are modified into hardened covers (elytra) that usually meet and cover the abdomen, part of the thorax, and the hind wings when they are present. Tarsi are three- to five-segmented.

Pupal Diagnosis

This stage is generally not aquatic. Pupae are mummylike forms with appendages that are distinct and not fused to the body. Developing fore wings are thickened. Antennae usually possess 11 or fewer segments.

Larval Diagnosis

These are variously shaped and measure 2–60 mm at maturity. Head is usually distinct and possesses chewing mouthparts (sometimes partially modified for sucking), poorly developed eyes, and variously developed antennae. Thoracic legs are usually present but absent in a few species. Wing pads are absent. Abdomen is eight- to ten-segmented and, depending on the species, may possess lateral platelike expansions, lateral and/or terminal filaments, tail-like structures, gills, or four terminal hooks; it never possesses a pair of terminal prolegs, and it never possesses lateral filaments and a single undivided terminal filament in combination.

The naturalist-angler is a common species. . . . Fishing takes him bird's-nesting, insect-watching, flower-gathering, into places where otherwise he would be a trespasser.

J. C. Mottram

Similar Orders

Adult water beetles are sometimes confused with aquatic bugs (Hemiptera) by those not thoroughly familiar with the two groups. At a glance, beetles can be seen to have the elytra (hardened fore wings) meeting along the midline of the body, whereas the fore wings of true bugs overlap and are not hardened distally. Mouthparts in the two groups are fundamentally different. Those of true bugs are specialized to form a single piercing and sucking tube or conelike structure.

The various larvae of water beetles are most easily confused with the larvae of such groups as dobsonflies, fishflies, alderflies, caddisflies, or possibly even spongillaflies, true flies, or moths. Those water beetles that have lateral abdominal filaments never have a single, undivided terminal filament as in alderflies, and they never have a pair of anal prolegs each with two hooks as in dobsonflies. Certain beetle larvae, however, have four terminal hooks or a double or split terminal filament. Caddisfly larvae are distinguished from those of beetles primarily by their highly reduced antennae and pair of anal prolegs each with a single hook. Spongillafly larvae can be distinguished by their highly modified sucking mouthparts. Legless aquatic beetle larvae (weevils) are distinguished from fly larvae that have distinctive heads by their crescent shape, the beetles being thickest in the middle and tapering toward the ends. Aquatic caterpillars have a series of ventral abdominal prolegs with small hooklets that are generally distinct.

Life History

Metamorphosis is complete, and morphological differences between the adults and larvae are extreme. Most water beetles have three larval instars and one generation per year, the most common exception being six instars and a two-year developmental period (e.g., among riffle beetles). Generally, water beetles overwinter as adults, but some overwinter as larvae or, less commonly, as eggs or pupae, depending on the species or group. Overwintering aquatic larvae are usually quiescent, and a few enter moist soil and remain dormant.

Pupation of aquatic species is terrestrial with rare exceptions. Most pupate in moist earthen cells adjacent to their aquatic environment. Some

TABLE 13.1 COMMON AQUATIC HABITATS AND HABITS OF BEETLES

Within Freshwater, Not Having to Surface		
CUTANEOUS AND/OR GILL RESPIRATION	UTILIZING PERMANENT PHYSICAL GILL	UTILIZING OXYGEN FROM UNDERWATER PLANT PARTS
Haliplidae larvae	Some Dryopidae adults	Few Dytiscidae adults
Some Dytiscidae larvae	Many Elmidae adults	Some Noteridae? larvae
Gyrinidae larvae	Few Ptilodactylidae? larvae	Several Chrysomelidae larvae
Hydroscaphidae larvae		
Some Hydrophilidae larvae		
Psephenidae larvae		
Some Ptilodactylidae larvae		
Some Limnichidae larvae		
Elmidae larvae		
Within Freshwater, Having to Contact Surface for Air		
SWIMMERS	CRAWLERS	SWIMMERS/CRAWLERS
Most Dytiscidae adults	Amphizoidae adults and ? larvae	Haliplidae adults
Noteridae adults	Some Haliplidae? larvae	Most Dytiscidae larvae
Most Hydrophilidae adults	Hydroscaphidae adults	Most Hydrophilidae larvae
	Few Hydrophilidae adults	Several Curculionidae adults
	Helodidae larvae	
At the Surface of Freshwater		
SWIMMERS	OTHERS	
Gyrinidae adults	Some Staphylinidae[1] adults	
	Miscellaneous terrestrial beetles that have fallen onto the water	

others attach silken cocoons to emergent vegetation. The pupae of only a few species are completely or partially submerged, and those usually develop within an air-filled cocoon (e.g., among the burrowing water beetles). Pupation is sometimes initiated by lowered water levels or drying conditions.

Aquatic adults of many species disperse by flight, either immediately following emergence from the pupal stage or at various times after emergence. In many species (e.g., some riffle beetles), the hind wings atrophy or become nonfunctional after an initial flight period and after the individual has re-entered the water. Some adults leave the water to seek overwintering sites or to aestivate in soil during dry periods. Overwintering adults of most species in colder regions undergo dormancy.

Mating takes place in the water or adjacent to it, but not in flight. Some species of water scavenger beetles use calling signals to find mates under water. Other types of courtship behavior, such as stroking, take place in some species. Individuals generally mate more than once.

Oviposition often takes place in water. The females of a few species of water beetles that are otherwise not aquatic enter the water to oviposit. Eggs are deposited in aquatic plant tissue, on algae or floating vegetation, on aquatic substrates, or in adjacent moist soil. Eggs of some species are attached to vegetation by a silken web or within a silken case; the eggs of a few species of water scavenger beetles remain within a case attached to the underside of the female. Eggs of most species hatch within a few weeks.

TABLE 13.1 CONTINUED

Crawling Above or Occasionally Below the Water Line on Substrate Emerging from Freshwater	
EMERGENT VEGETATION ASSOCIATES	EMERGENT SUBSTRATE: STONES, WOODY DEBRIS, WET MOSSY, & ALGAL MATS
Some Chrysomelidae adults and rarely larvae	Amphizoidae
Several Curculionidae adults	Some Hydrophilidae
	Hydraenidae larvae
	Rarely Helodidae adults
	Some Psephenidae adults
	Few Ptilodactylidae adults
	Few Limnichidae adults
	Few Dryopidae adults
	Few Elmidae adults
Within Tree Holes or Flower Cups	
Rarely Hydrophilidae larvae	Some Helodidae larvae
Wet Areas Along Margins of Freshwater	
Several Carabidae[1]	Some Helodidae adults
Sphaeridae[1]	Most Psephenidae adults
Some Hydrophilidae	Some Ptilodactylidae adults
Most Hydraenidae	Heteroceridae[1]
Georyssidae[1]	Some Limnichidae adults
Several Staphylinidae[1]	Few Elmidae adults
Within Nonoceanic Brackish Waters	
Rarely Haliplidae	Several Hydrophilidae
Several Dytiscidae	Some Hydraenidae adults
Rarely Noteridae	
Coastal Marine, Intertidal Zones	
Few Carabidae[1]	Some Melyridae[1]
Some Hydrophilidae	Few Rhizophagidae[1]
Some Hydraenidae	Few Tenebrionidae[1]
Many Staphylinidae[1]	Few Salpingidae[1]
Few Heteroceridae[1]	Few Curculionidae
Few Limnichidae	

[1]See Chapter 17.

Aquatic Habitats

Water beetles occur as larvae, adults, or both in a wide variety of aquatic and semiaquatic environments. The habitats and habits summarized in Table 13.1 can be consulted to determine the relative aquatic adaptiveness among the families and life stages of North American beetles. The quality of water is not as restrictive as it is to some aquatic insects because many water beetles use atmospheric rather than dissolved oxygen for respiration. They are known from brackish waters, hot springs, intertidal zones, aquifers, and tree holes, for example, as well as virtually all common freshwater habitats. Depending on the species, they can be found in or on the substrate, in or on the aquatic plants, or swimming at or beneath the surface of the water. Some species enter and leave the water at will, and many species frequent environments that are marginal between terrestrial and aquatic habitats. There is a tendency among a few species of aquatic beetles to become less aquatic with maturity.

Aquatic Adaptations and Behavior

Most aquatic adults rely on atmospheric oxygen to various degrees, and most species must regularly surface to replenish their air supply. Most swim, but some crawl, and a few float to the surface. In general, adults of the suborder Adephaga obtain air by breaking the surface film with the tip of the abdomen. The air is stored directly beneath the elytra in an underwing chamber where it is readily available to the abdominal spiracles.

In the suborder Polyphaga, the surface film is commonly broken with the unwettable antennae, and air proceeds via a funnel-like connection to the underwing chamber or other areas of the body. Many polyphagans have a covering of minute, unwettable hairs or scales over the ventral surface of the body, and these hairs are able to hold a thin film of air (plastron). The plastron acts as a physical gill in that dissolved oxygen is diffused into it from the water. Plastron respiration of many bottom-dwelling adults such as riffle beetles is highly advanced; in well-oxygenated water, individuals are able to remain submerged for extended or indefinite periods of time by utilizing this physical gill. These adult beetles are therefore essentially hydropneustic.

In the deep places, the Water-beetle dives, carrying with him his reserves of breath: an air bubble at the tip of the wing cases and, under the chest, a film of gas that gleams like a silver breastplate.

J. Henri Fabre

Swimming adults move their highly modified hind legs and often their middle legs (rarely only the middle legs, as exemplified by some water weevils) in an oarlike fashion, either alternating left and right legs (e.g., water scavenger beetles) or using both in unison (e.g., predaceous diving beetles). Certain species are aided in surfacing by a bubble of gas. Bottom dwellers usually have long grasping claws, such as those of longtoed water beetles and riffle beetles. Feeding habits of adults differ considerably from group to group. As is typical of many beetles, some aquatic adults emit defensive secretions or gases when disturbed. Secretions released into water by some predaceous diving beetles can inactivate or even kill certain fishes.

Aquatic larvae may be hydropneustic or aeropneustic, depending on the species. A few species (e.g., some crawling water beetles) develop functional spiracles only in later instars. Among hydropneustic forms, oxygen may be taken up directly through membranous areas of the body, the gills, or both, or via a plastron. Aquatic leaf beetles are examples of aeropneustic larvae that are endophytic breathers. They tap the oxygen available in aquatic plants with their specialized terminal spiracles. Other aeropneustic beetle larvae generally must make periodic or continual contact with surface air by way of functional spiracles at the end of the abdomen. Certain of these species (e.g., marsh beetle larvae) have internal trachea that are greatly enlarged for temporary air storage while submerged.

The anal papillae found on the larvae of some species may possibly play a primary role in osmoregulation rather than respiration. At least some of the predaceous diving beetles, however, absorb ions internally in the gut from ingested water.

Some larvae are adapted for clinging to substrates with well-developed claws, others with terminal abdominal hooks or attachment areas (e.g., skiff beetles), and still others with a disc-shaped body (e.g., water pennies) (Figs. 13.41, 13.42). At least some predaceous diving beetle larvae that surface regularly orient to the surface by a positive response to the source of light (phototaxis). In the absence of natural light as a stimulus, these larvae are unable to orient properly.

Although beetle larvae exhibit a wide variety of feeding habits, a large number of species are highly predaceous and have various adaptations for this mode of life. Some species, upon attacking their prey, inject or apply a fluid with their mouthparts. This fluid has a poisonous and/or digestive function. Mouthparts are used also for chewing, holding the prey, and ingesting predigested portions of the prey.

Classification and Characters

Three suborders of Coleoptera contain aquatic species in North America. Within the suborder Adephaga, five families are primarily aquatic, and an additional family, which is primarily terrestrial, contains several species that are typically found adjacent to water. The suborder Myxophaga contains one small aquatic family and one small semiaquatic family. Within the suborder Polyphaga, 17 families, which include a considerable mixture of aquatic, semiaquatic, and terrestrial species, are treated. Ten of the polyphagan families include freshwater species, and seven of these families are considered primarily aquatic. Diagnoses and comparisons of groups are based primarily on aquatic species and therefore may not always hold for terrestrial species. Coleoptera families that do not contain any fully aquatic freshwater species as adults or larvae, but otherwise contain species that occur in marginal freshwater or coastal marine habitats, are discussed in Chapter 17.

Individuals of many terrestrial species of beetles fall or are blown into water. Therefore, when specimens do not fit any of the family discussions given in this book, they are most likely terrestrial forms. If beetles will not key out by using the picture keys (Figs. 13.1, 13.2), they are probably not fully aquatic, although they may occur in marginal wet habitats.

The family identification of adult beetles is based to a great extent on the form of the antennae, the number and shape of the antennal segments, and the number and shape of tarsal segments. A **clubbed** antenna, as the term is used here, is any antenna with enlarged terminal segments. Thus the club may be bulbous, elongated, or laterally expanded into various shapes (e.g., Figs. 13.29, 13.34, 13.50). An "apparently four-segmented tarsus" is one that is actually five-segmented, but the fourth segment is hidden somewhat between the lobes of the third segment. Segment 1 always refers to the basal segment.

Swimming adult beetles can usually be distinguished by the form of the body and legs. The body is streamlined; when viewed laterally, it appears nearly uniformly convex from end to end, and when viewed dorsally, the pronotum and **elytra** (modified fore wings) usually form a more-or-less continuous, similar curvature of the body (e.g., Figs. 13.17, 13.20, 13.22, 13.35). The hind legs, and usually the middle legs, are oarlike and held close to the body and/or they possess long **swimming hairs.** Water beetles that crawl rather than swim are usually more typical of terrestrial forms. The elytra and pronotum may be variously shaped, but together usually do not form a continuous, similar curvature either dorsally or laterally (e.g., Figs. 13.3, 13.51, 13.53–13.55). Swimming hairs are absent on the legs, although some species have small, fine hairs.

The most difficult adult characters to interpret are often those dealing with ventral structures such as coxal shapes or sternites and their processes. The illustrations should be used liberally when dealing with these characters. Because the bodies of beetles are hard, shiny, and usually dark, care should be taken in examining minute characters. Segment lines, for example, are best seen in dry specimens or specimens completely submerged

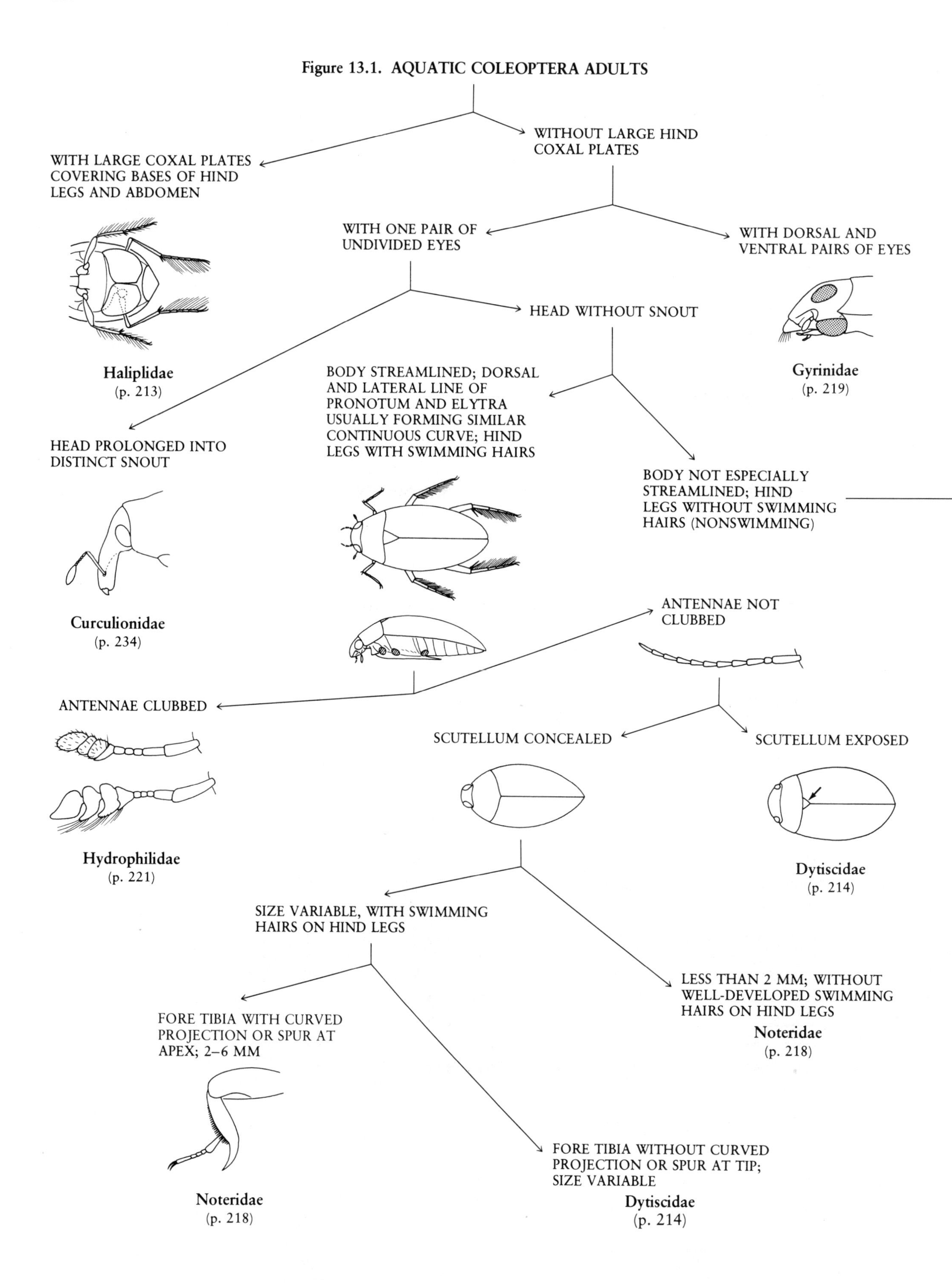
Figure 13.1. AQUATIC COLEOPTERA ADULTS
WITH LARGE COXAL PLATES COVERING BASES OF HIND LEGS AND ABDOMEN
Haliplidae
(p. 213)
WITHOUT LARGE HIND COXAL PLATES
WITH ONE PAIR OF UNDIVIDED EYES
WITH DORSAL AND VENTRAL PAIRS OF EYES
Gyrinidae
(p. 219)
HEAD PROLONGED INTO DISTINCT SNOUT
Curculionidae
(p. 234)
HEAD WITHOUT SNOUT
BODY STREAMLINED; DORSAL AND LATERAL LINE OF PRONOTUM AND ELYTRA USUALLY FORMING SIMILAR CONTINUOUS CURVE; HIND LEGS WITH SWIMMING HAIRS
BODY NOT ESPECIALLY STREAMLINED; HIND LEGS WITHOUT SWIMMING HAIRS (NONSWIMMING)
ANTENNAE CLUBBED
Hydrophilidae
(p. 221)
ANTENNAE NOT CLUBBED
SCUTELLUM CONCEALED
SCUTELLUM EXPOSED
Dytiscidae
(p. 214)
SIZE VARIABLE, WITH SWIMMING HAIRS ON HIND LEGS
LESS THAN 2 MM; WITHOUT WELL-DEVELOPED SWIMMING HAIRS ON HIND LEGS
Noteridae
(p. 218)
FORE TIBIA WITH CURVED PROJECTION OR SPUR AT APEX; 2–6 MM
Noteridae
(p. 218)
FORE TIBIA WITHOUT CURVED PROJECTION OR SPUR AT TIP; SIZE VARIABLE
Dytiscidae
(p. 214)

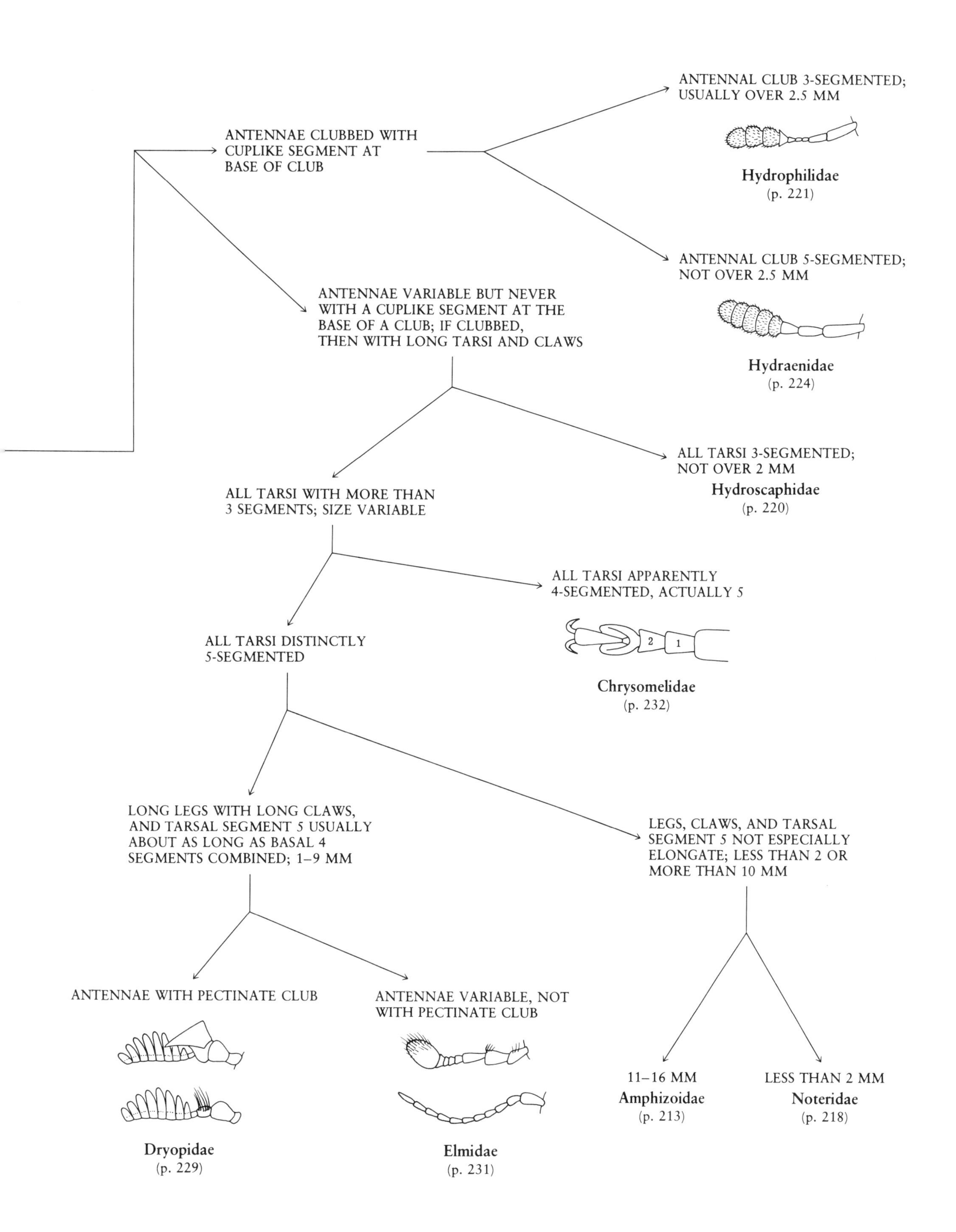

ANTENNAE CLUBBED WITH CUPLIKE SEGMENT AT BASE OF CLUB
ANTENNAL CLUB 3-SEGMENTED; USUALLY OVER 2.5 MM
Hydrophilidae
(p. 221)
ANTENNAL CLUB 5-SEGMENTED; NOT OVER 2.5 MM
Hydraenidae
(p. 224)
ANTENNAE VARIABLE BUT NEVER WITH A CUPLIKE SEGMENT AT THE BASE OF A CLUB; IF CLUBBED, THEN WITH LONG TARSI AND CLAWS
ALL TARSI 3-SEGMENTED; NOT OVER 2 MM
Hydroscaphidae
(p. 220)
ALL TARSI WITH MORE THAN 3 SEGMENTS; SIZE VARIABLE
ALL TARSI APPARENTLY 4-SEGMENTED, ACTUALLY 5
2
1
Chrysomelidae
(p. 232)
ALL TARSI DISTINCTLY 5-SEGMENTED
LONG LEGS WITH LONG CLAWS, AND TARSAL SEGMENT 5 USUALLY ABOUT AS LONG AS BASAL 4 SEGMENTS COMBINED; 1–9 MM
LEGS, CLAWS, AND TARSAL SEGMENT 5 NOT ESPECIALLY ELONGATE; LESS THAN 2 OR MORE THAN 10 MM
ANTENNAE WITH PECTINATE CLUB
Dryopidae
(p. 229)
ANTENNAE VARIABLE, NOT WITH PECTINATE CLUB
Elmidae
(p. 231)
11–16 MM
Amphizoidae
(p. 213)
LESS THAN 2 MM
Noteridae
(p. 218)

Figure 13.2. AQUATIC COLEOPTERA LARVAE

BODY DISCLIKE WITH DORSAL PLATES COVERING HEAD AND LEGS

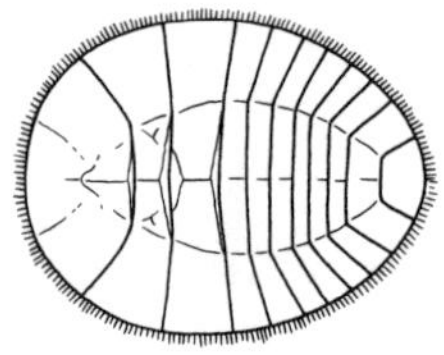

Psephenidae
(p. 226)

BODY MORE-OR-LESS ELONGATE, NOT DISCLIKE, NOT WITH CONCEALED HEAD AND LEGS WHEN PRESENT

THORACIC LEGS PRESENT

NO LEGS
Curculionidae
(p. 234)
See also: Shore-Dwelling Hydrophilidae, Chap. 17

LEGS 6-SEGMENTED (INCLUDING CLAW AS SEGMENT)

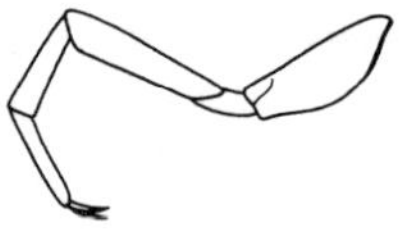

LEGS 5-SEGMENTED (INCLUDING CLAW)

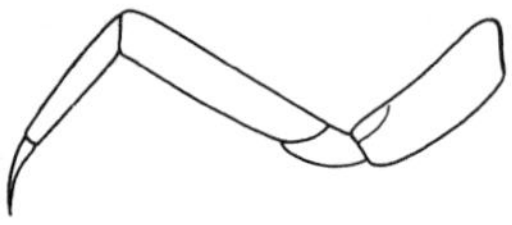

ABDOMEN 8-SEGMENTED

ABDOMEN 9- TO 10-SEGMENTED

BODY WITH LATERALLY EXPANDED PLATES

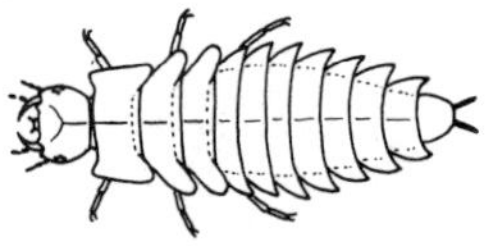

Amphizoidae
(p. 213)

BODY WITHOUT LATERALLY EXPANDED PLATES

LEG CLAWS DOUBLE

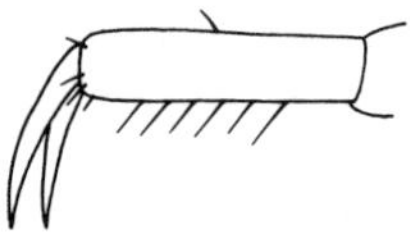

Gyrinidae
(p. 219)

LEG CLAWS SINGLE

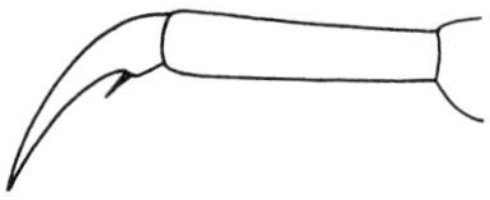

Haliplidae
(p. 213)

BODY USUALLY TAPERED AT ENDS, WITH SLENDER LEGS; OFTEN WITH VARIOUS ABDOMINAL FILAMENTS AND SICKLE-SHAPED MANDIBLES
Dytiscidae
(p. 214)

BODY ELONGATE, PARALLEL SIDED WITH SHORT THICK LEGS

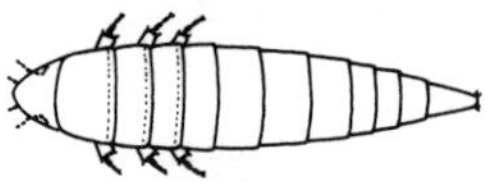

Noteridae
(p. 218)

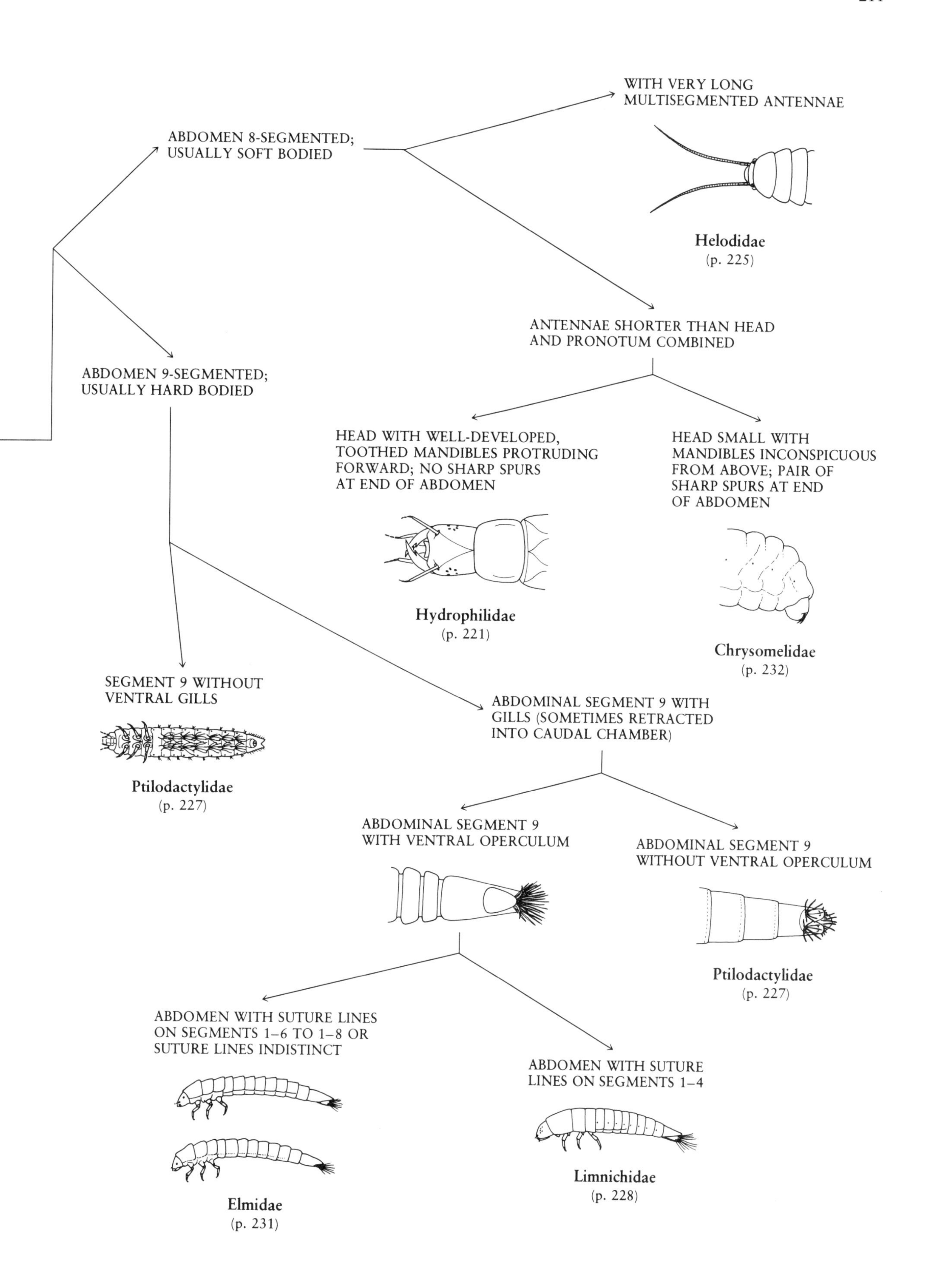
ABDOMEN 8-SEGMENTED; USUALLY SOFT BODIED
WITH VERY LONG MULTISEGMENTED ANTENNAE
Helodidae
(p. 225)
ANTENNAE SHORTER THAN HEAD AND PRONOTUM COMBINED
HEAD WITH WELL-DEVELOPED, TOOTHED MANDIBLES PROTRUDING FORWARD; NO SHARP SPURS AT END OF ABDOMEN
Hydrophilidae
(p. 221)
HEAD SMALL WITH MANDIBLES INCONSPICUOUS FROM ABOVE; PAIR OF SHARP SPURS AT END OF ABDOMEN
Chrysomelidae
(p. 232)
ABDOMEN 9-SEGMENTED; USUALLY HARD BODIED
SEGMENT 9 WITHOUT VENTRAL GILLS
Ptilodactylidae
(p. 227)
ABDOMINAL SEGMENT 9 WITH GILLS (SOMETIMES RETRACTED INTO CAUDAL CHAMBER)
ABDOMINAL SEGMENT 9 WITH VENTRAL OPERCULUM
ABDOMINAL SEGMENT 9 WITHOUT VENTRAL OPERCULUM
Ptilodactylidae
(p. 227)
ABDOMEN WITH SUTURE LINES ON SEGMENTS 1–6 TO 1–8 OR SUTURE LINES INDISTINCT
Elmidae
(p. 231)
ABDOMEN WITH SUTURE LINES ON SEGMENTS 1–4
Limnichidae
(p. 228)

in fluid so as to decrease any glare of light reflecting from the body. Swimming hairs are commonly matted against the legs in dried specimens and thus may be difficult to detect.

Larvae are characterized according to family in very general terms whenever possible. Except for the branching gills, most filamentous or fleshy processes are simply termed "filaments." The number of abdominal segments is sometimes important, and segments should be counted with care, beginning basally. For convenience, the claw, whether single or paired, is to be counted as a segment when counting leg segments of beetle larvae. Adephagan larvae, therefore, have six segments (see Fig. 13.6) as follows: coxa, trochanter, femur, tibia, tarsus, and claw or claws. Aquatic polyphagan and myxophagan larvae have five segments (when legs are present) as follows: coxa, trochanter, femur, tibia, and the fused tarsus and claw. The trochanter is usually a small jointlike segment. If it is divided by a suture line, as it is in some species, it should still be counted only as a single segment.

Suborder Adephaga

Adults have the posterior margin of first abdominal sternite divided by the hind coxae (Fig. 13.8); or the base of abdomen and base of hind legs are covered by enlarged hind coxal plates (Fig. 13.5); or the eyes are divided into dorsal and ventral pairs (Fig. 13.24).

Larvae have six-segmented legs (Fig. 13.6) (counting the claw as a segment), but fore legs are modified from this in some. Claws may be double or single on each leg.

TROUTSTREAM BEETLES
(Amphizoidae)

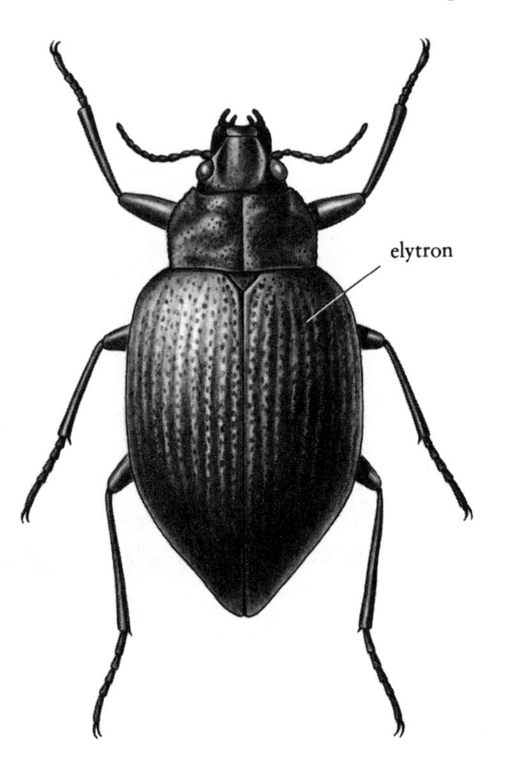

Figure 13.3. *Amphizoa* adult

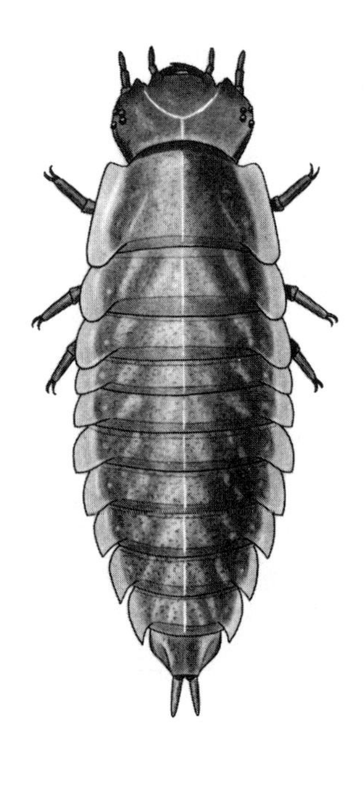
Figure 13.4. *Amphizoa* larva

GROUND BEETLES
(Family Carabidae)

This family contains some shore-dwelling insects; see Chapter 17.

TROUTSTREAM BEETLES
(Family Amphizoidae)

ADULT DIAGNOSIS: (Fig. 13.3) These are somewhat elongate forms that measure 11–16 mm. Body and legs are not adapted for swimming.

LARVAL DIAGNOSIS: (Fig. 13.4) These are somewhat flattened and elongate forms. Body segments are expanded slightly into lateral, platelike projections. Abdomen is eight-segmented.

DISCUSSION: The rare troutstream beetles are more typical of ground beetles than they are of other water beetles. The adults differ basically from the ground beetles (Carabidae) in that the hind coxae contact the lateral region of the body and divide the abdomen from the thorax ventrally. Larvae differ from those of ground beetles in having a reduced number of abdominal segments.

There is one genus with about four species in the western states and western Canada.

Adults and larvae occur under stones and debris of slackwater areas of streams and springs, and they also commonly crawl about on wet surfaces of emergent objects or margins of these habitats. Both active stages are predaceous, and the larvae are not adapted for aquatic respiration.

CRAWLING WATER BEETLES
(Family Haliplidae)

ADULT DIAGNOSIS: (Figs. 13.5; Plate X, Fig. 57) These are generally oval, usually spotted species that measure 2–6 mm. Elytra usually possess rows of tiny holes. Hind legs possess swimming hairs. Hind coxae are enlarged into conspicuous plates that cover bases of legs and much of abdomen.

LARVAL DIAGNOSIS: (Figs. 13.6, 13.7; Plate X, Fig. 58) These are elongate. Tarsal claws are single. Abdomen is 9- or 10-segmented, and terminal segment is often elongate. Body may possess long dorsal and lateral filaments.

DISCUSSION: Adults of crawling water beetles may be mistaken for small predaceous diving beetles when viewed dorsally. The enlarged coxal plates of haliplids, however, are unique; and the common spotted pattern as seen dorsally is useful much of the time. The six-segmented legs and single tarsal claws always distinguish the larvae of this group.

Of four recognized genera, *Haliplus* and *Peltodytes* are widespread and common. Over 60 species occur in North America.

Adults and larvae commonly inhabit pools, ponds, or shallow regions of lakes among aquatic vegetation. A few species live in rocky-bottomed streams, especially in the West. Adults swim and crawl, and most regularly surface to replenish their air supply. In addition to the underwing chamber, the coxal plates serve to hold reservoirs of air.

Larvae are strictly hydropneustic when young and later develop functional spiracles. The larvae of some species may overwinter in damp soil. Both adults and larvae are usually omnivorous feeders. Those that eat algae vary in their feeding habits, some piercing cells and sucking juices, others biting and chewing algae.

CRAWLING WATER BEETLES
(Haliplidae)

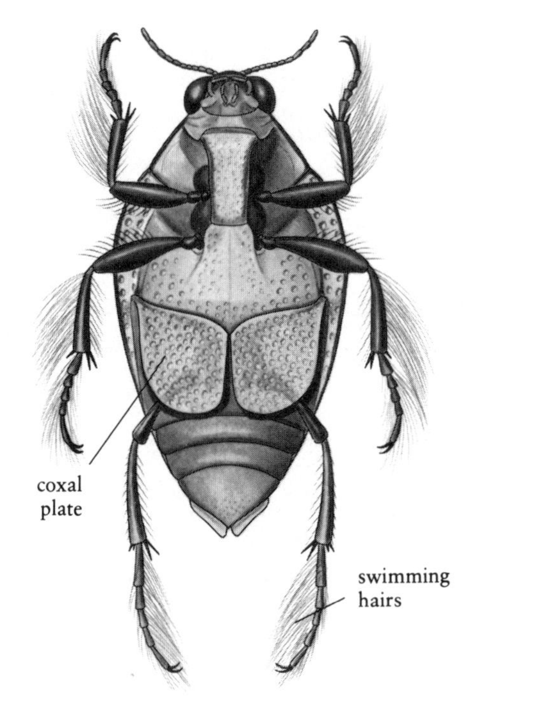

Figure 13.5.
Haliplus adult (ventral)

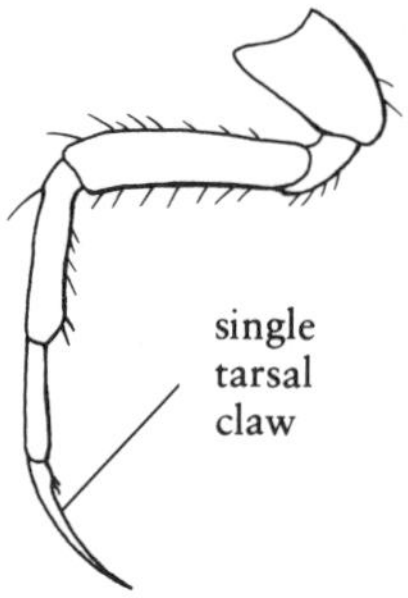

single
tarsal
claw

Figure 13.6.
Larva, middle leg

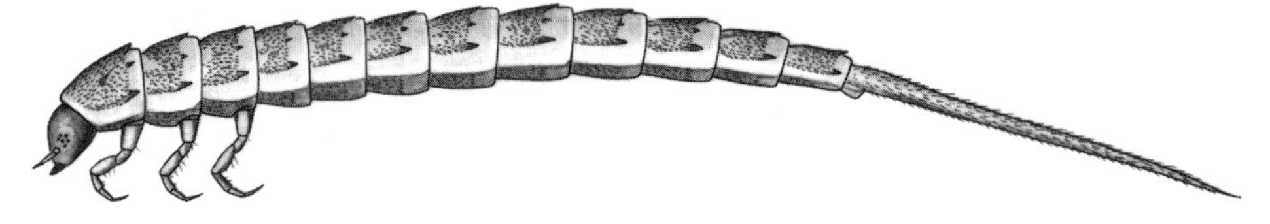

Figure 13.7. *Haliplus* larva

PREDACEOUS DIVING BEETLES
(Family Dytiscidae)

ADULT DIAGNOSIS: (Figs. 13.8, 13.11–13.19; Plate X, Figs. 59, 60) These adults have oval, streamlined bodies. Their length may range from 1 to 40 mm, but they usually measure 3–25 mm. Antennae are slender. Scutellum is exposed or concealed. Fore tibiae do not have a large, curved spur at the apex. Hind legs, and usually middle legs, possess long swimming hairs and, in some, pads of hairs.

LARVAL DIAGNOSIS: (Figs. 13.9, 13.10; Plate X, Fig. 61) These elongate larvae measure 5–70 mm at maturity. Legs are slender. Abdomen is eight-segmented and is usually strongly tapered at the end. If lateral filaments are present on the abdomen, then there are no terminal hooks at the end of the abdomen.

DISCUSSION: Adults of this common group of water beetles are distinguished from adults of water scavenger beetles (also very common) by many characters. Along with subordinal characters that separate the Adephaga and Polyphaga, the water scavenger beetles are typified by their usually shorter, clubbed antennae and by their usually deeper, more convex

PREDACEOUS DIVING BEETLES
(Dytiscidae)

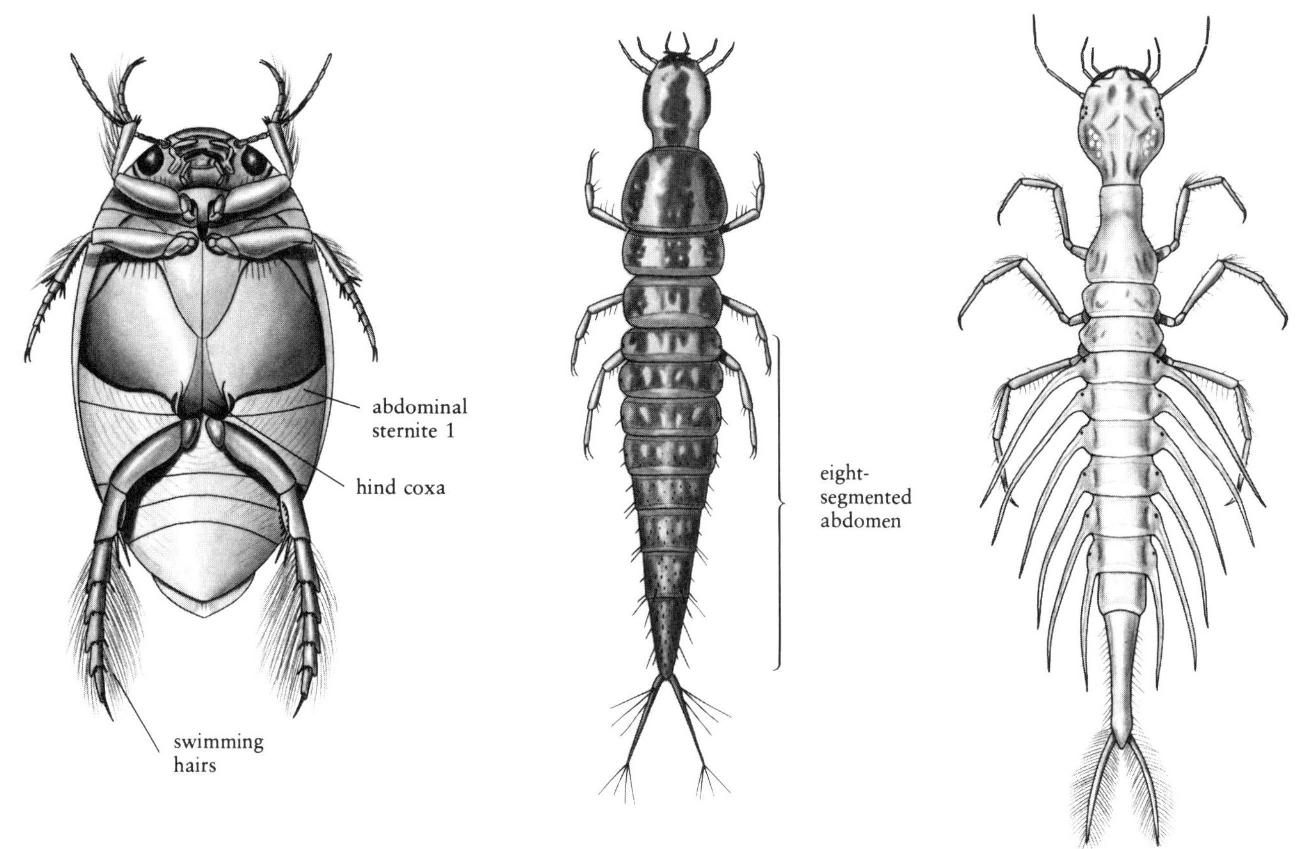

Figure 13.8. *Laccophilus* (Laccophilinae) adult (ventral)

Figure 13.9. *Agabus* larva

Figure 13.10. *Coptotomus* larva

shape as viewed laterally. Also, they surface head first for air rather than with the tip of the abdomen, as is the case with predaceous diving beetles. Small adults of predaceous diving beetles are sometimes confused with those of burrowing water beetles or crawling water beetles. See the discussions under the last two groups.

Larvae of predaceous diving beetles usually have a slightly to strongly elongated prothorax and sickle-shaped, toothless mandibles that aid in their field identification. The larvae of water scavenger beetles, on the other hand, usually have robust mandibles with one or more teeth. Those few species with lateral abdominal filaments possess a pair of terminal filaments rather than hooks, such as those of whirligig beetle larvae.

Well over 400 North American species and approximately 35 genera make this the largest of the water beetle families for the region.

Predaceous diving beetles live in almost every kind of aquatic habitat in North America, including brackish waters and tiny seepages. These beetles are most prevalent in quiet freshwaters and far less common in swift currents of streams and rivers. Nevertheless, some genera, such as *Hydroporus* and *Oreodytes,* are common in running waters. A blind species of *Haideoporus* is completely adapted to living in artesian wells in Texas.

Adults must regularly surface to renew air supplies. They are adept swimmers and move (row) both hind legs in unison. Some are able to secure

The oval lurks, hanging,
As if by some invisible thread,
At the watery rim, waiting,
For below, unknowing oft do tread.

This aqueous tiger, with skill—
Anticipates battle, inevitably to win;
Then, to enjoin nature's cruel will,
It strikes, hawk-like upon the moving fin!

McCafferty

air from bubbles formed on the underwater parts of plants. Most larvae must surface to obtain air, but those of a few species are fully submergent and hydropneustic.

Both active stages are highly predaceous and sometimes cannibalistic. Larvae are commonly called "water tigers" because of their voraciousness. Depending on the size of the species, a variety of aquatic animals may be attacked, including small fishes. The larvae possess mandibles with specialized channels through which they are able to draw the internal fluids of their prey.

Many species are excellent fliers and dispersers. Species of *Laccophilus* are often the first insects to inhabit newly formed pools and other lentic environments. Some *Agabus* species fly shortly after emergence but then become flightless. *Dytiscus* are often found at lights at night (as are many predaceous diving beetles) and are striking because of their large size.

SUBFAMILY LACCOPHILINAE: **Adults** (Figs. 13.8, 13.11) have a scutellum that is concealed or barely visible. Fore and middle tarsi are distinctly five-segmented. Viewed laterally, the ventral thorax appears more or less straight and not greatly angled immediately behind head. This group is widespread in North America.

SUBFAMILY HYDROPORINAE: **Adults** (Figs. 13.12–13.14) have a scutellum that is usually concealed. Fore and middle tarsi are usually apparently four-segmented. If all tarsi are distinctly five-segmented, then when the adult is viewed laterally, the ventral area immediately behind the head forms a distinct angle with the remaining ventral surface of the thorax. These never possess the combination of exposed scutellum and tarsi that are all five-segmented. The group is widespread in North America.

SUBFAMILY COLYMBETINAE: **Adults** (Fig. 13.15; Plate X, Fig. 59) have eyes that are notched above antennae. Scutellum is exposed. All tarsi are five-segmented. The group is widespread in North America.

SUBFAMILY DYTISCINAE: **Adults** (Fig. 13.16; Plate X, Fig. 60) are relatively large. Scutellum is exposed. All tarsi are five-segmented. Outer and inner apical spurs of the hind tibiae are slender. Hind tarsi lack rows of short golden hairs on posterior margin of the four basal tarsal segments. The group is widespread except for Florida.

SUBFAMILY HYDATICINAE: **Adults** (Figs. 13.17, 13.18) are similar to those of Dytiscinae except they are smaller (less than 15 mm) and hind tarsi possess a fringe of short, flat, golden hairs on the posterior margin of the four basal tarsal segments. The group is widespread in North America.

SUBFAMILY CYBISTRINAE: **Adults** (Fig. 13.19) are similar to those of Dytiscinae except the outer apical spur of the hind tibiae is much broader than the inner spur. This group is widespread except for northwestern states and western Canada and Alaska.

PREDACEOUS DIVING BEETLES
(Dytiscidae)

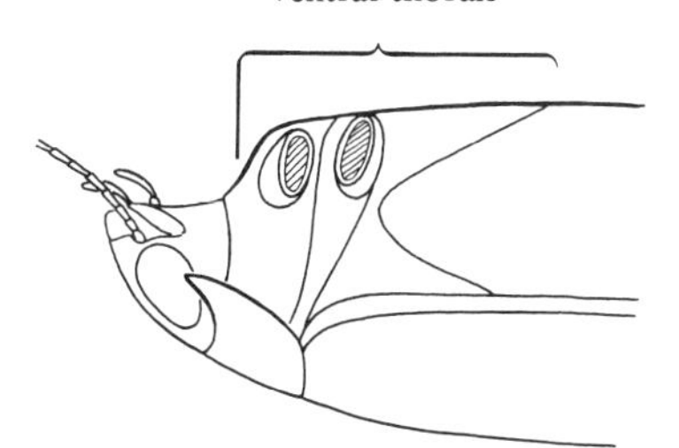

Figure 13.11.
Laccophilinae adult

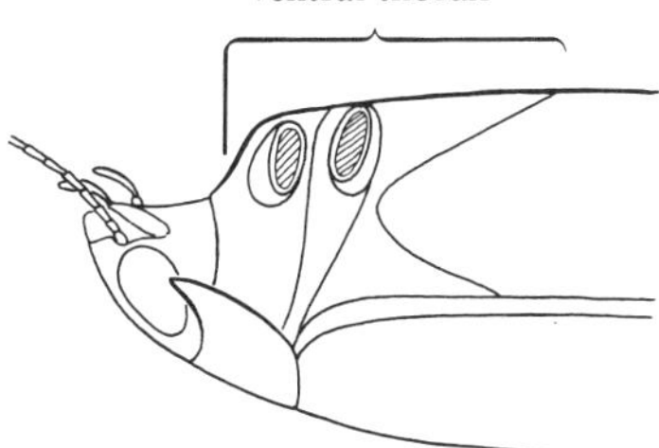

Figure 13.12.
Hydroporinae adult

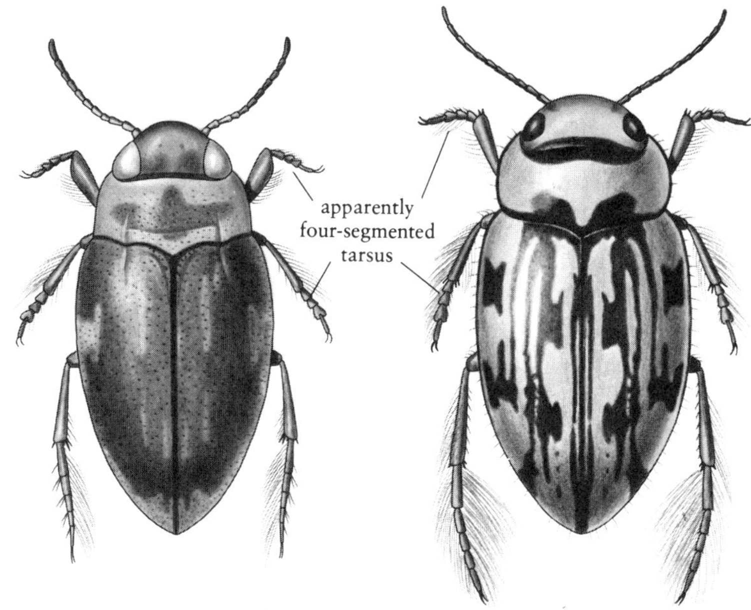

Figure 13.13. *Uvarus*
(Hydroporinae) adult

Figure 13.14. *Deronectes*
(Hydroporinae) adult

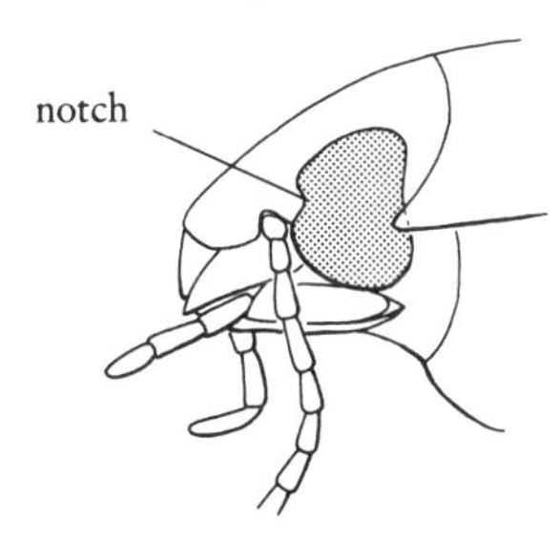

Figure 13.15.
Colymbetinae adult head

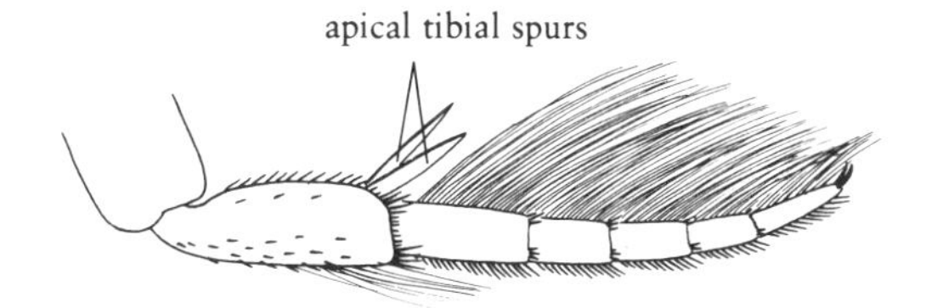

Figure 13.16. Dytiscinae adult,
right hind leg (ventral)

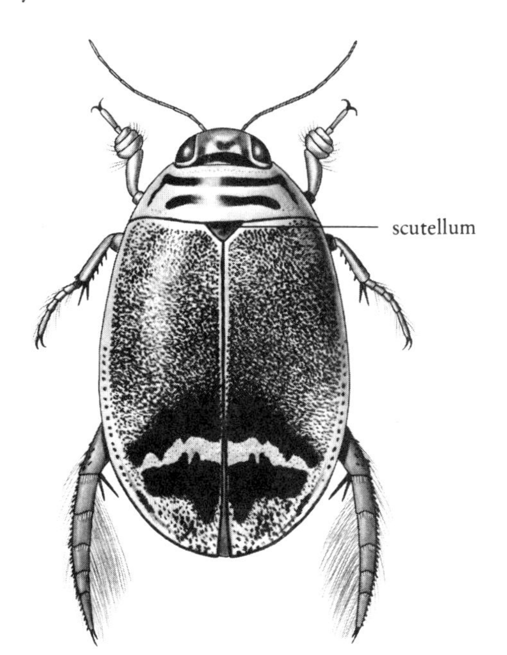

Figure 13.17. *Acilius*
(Hydaticinae) adult ♂

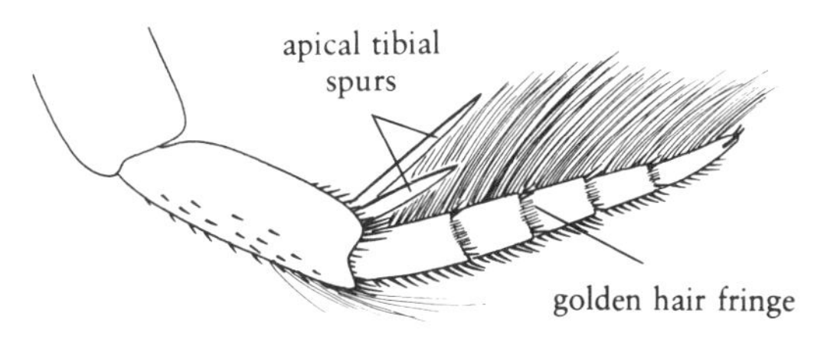

Figure 13.18. Hydaticinae adult,
right hind leg (ventral)

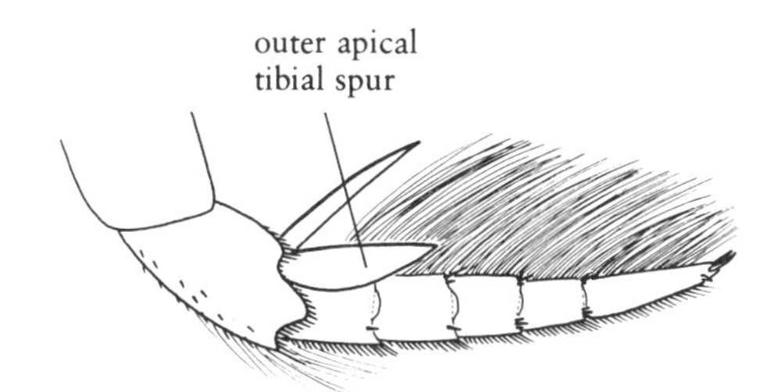

Figure 13.19. Cybistrinae adult,
right hind leg (ventral)

BURROWING WATER BEETLES
(Noteridae)

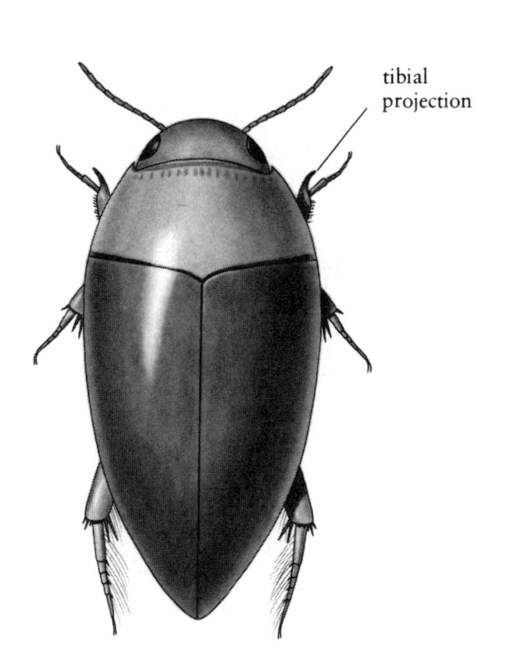

Figure 13.20. *Hydrocanthus* adult

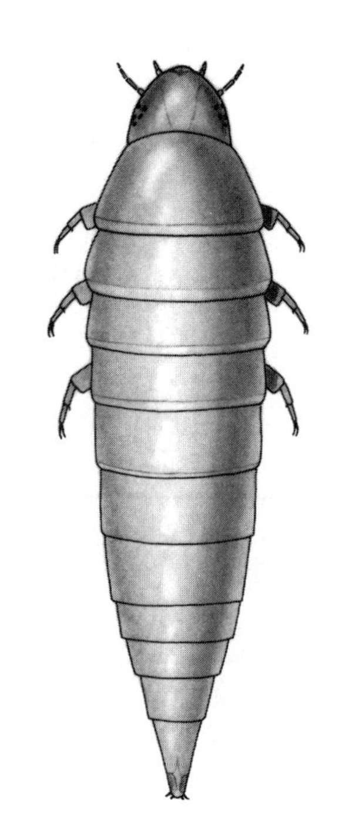

Figure 13.21. *Hydrocanthus* larva

BURROWING WATER BEETLES
(Family Noteridae)

ADULT DIAGNOSIS: (Fig. 13.20) These are basically similar to predaceous diving beetles. Their length is 1.2–5.5 mm. Scutellum is concealed. Fore tibiae usually have a large, curved projection at apex. Claws of hind legs are double.

LARVAL DIAGNOSIS: (Fig. 13.21) Body is elongate and parallel-sided for much of its length. Legs are short and stout, being adapted for burrowing.

DISCUSSION: Burrowing water beetle adults closely resemble those predaceous diving beetles that have a concealed scutellum. The latter either have a single claw on the hind legs or have apparently four-segmented fore and middle tarsi. These characteristics are not found among burrowing water beetles.

Embracing about five genera and 18 species in North America, this group is known from central and eastern states, California, and extreme southern Ontario. It is uncommon in California and becomes scarce in the northern parts of its range.

These beetles occur in weedy ponds and lakes where the predaceous adults swim or crawl about. Larvae are poorly known in North America, but on the basis of known habitats of European species, they are presumed to be adapted for burrowing into mud around the roots of aquatic plants. Some utilize plants for their source of oxygen. Pupation may take place in an air-filled cocoon in the pond substrate.

WHIRLIGIG BEETLES
(Gyrinidae)

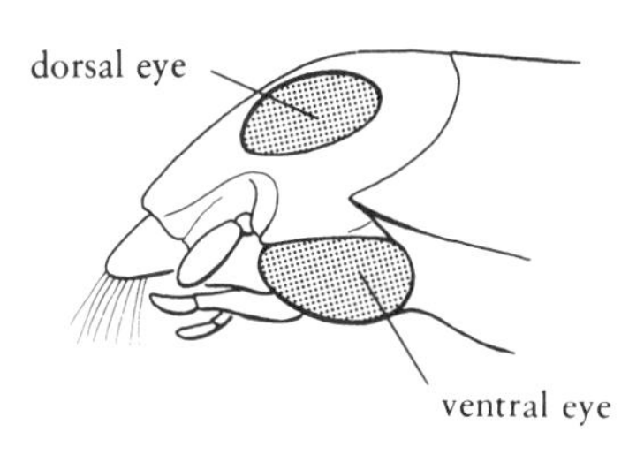

Figure 13.24. Adult, lateral head

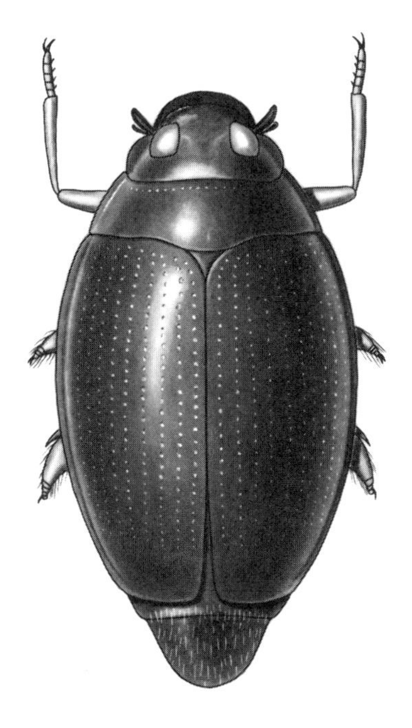

Figure 13.22. *Gyrinus* adult

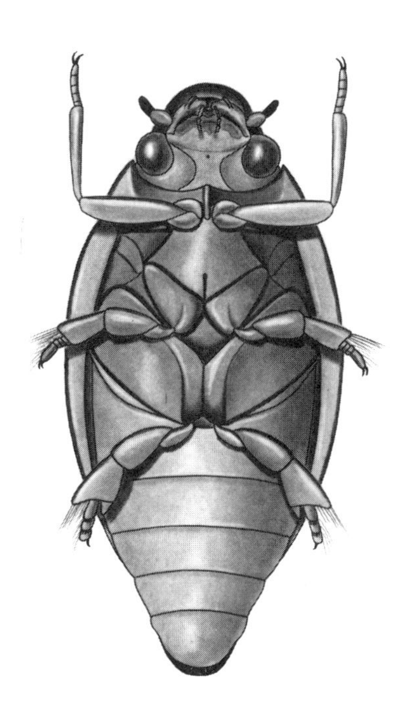

Figure 13.23. *Gyrinus* adult

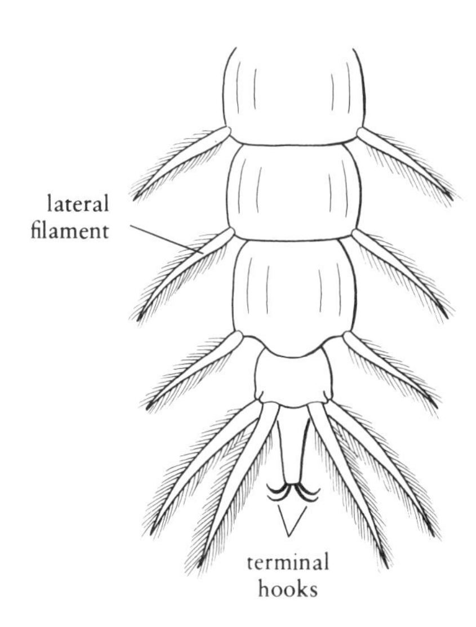

Figure 13.25. End of larval abdomen

WHIRLIGIG BEETLES
(Family Gyrinidae)

ADULT DIAGNOSIS: (Figs. 13.22–13.24; Plate X, Fig. 63) The body is oval, flattened, and 3–15 mm in length or slightly less. Eyes are divided into dorsal and ventral pairs. Fore legs are relatively long, and middle and hind legs are short and usually flattened.

LARVAL DIAGNOSIS: (Fig. 13.25; Plate X, Fig. 62) Larvae are elongate and measure up to 30 mm. Abdomen possesses lateral filaments and terminates in four small hooks.

DISCUSSION: Adults of whirligig beetles are quite distinctive among water beetles, both in form and habit. The four terminal hooks of the larvae distinguish this group from other beetle larvae that also have lateral abdominal filaments. Over 50 species among four North American genera can be classified into two subfamilies.

"Gigs are whirly,
Cues are curly
And the dew is pearly early."
James Thurber

These beetles occur in a variety of lentic and lotic habitats, especially ponds and streams. The adults are usually found on the surface of quiet waters (or swift waters rarely) and are specially adapted for this habitat. A ventral pair of eyes serves for vision in water, and a dorsal pair simultaneously serves for aerial vision. This is a unique adaptation for optimal vision while at the surface of the water.

Individuals often congregate in so-called schools that sometimes consist of more than one species. When disturbed, whirligig beetles swim erratically or dive while emitting defensive secretions. Species of *Dineutus*, when handled, secrete a milky substance, the odor of which resembles that of ripe apples, and as a consequence they have been referred to as "apple bugs." Species of *Gyrinus* are quite vile smelling. Defensive secretions apparently render the adults distasteful or otherwise repel would-be predators, such as fishes. Whirligigs are exceedingly fast and effective surface swimmers (up to extremes of one meter per second); their speed is due partly to the manner in which they generate waves while swimming. Waves may also function as echolocators; that is, the reflection of waves off objects may transmit certain information to the beetles about size and distance of the objects.

Adults feed primarily by preying on small organisms or by scavenging on small bits of floating material. In colder regions some adults overwinter submerged. Larvae are hydropneustic and predaceous. Pupation occurs in cocoons on emergent vegetation or along the shore.

SUBFAMILY SPANGLEROGYRINAE: **Adults** have their dorsal and ventral eyes contacting each other along the margin of head and separated by a narrow ridge. Middle and hind legs are not flattened. These small beetles are currently known only from southern Alabama.

SUBFAMILY GYRININAE: **Adults** (Figs. 13.22–13.24; Plate X, Fig. 63) have their dorsal and ventral eyes distinctly separated from each other. The group is widespread in North America and contains three genera that had been regarded in separate subfamilies until the recent discovery of the primitive group Spanglerogyrinae.

Suborder Myxophaga

Adults are minute, and all tarsi are three-segmented.

Larvae possess five-segmented legs including claw. Claws are single.

MINUTE BOG BEETLES
(Family Sphaeridae)

A treatment of this semiaquatic family can be found in Chapter 17.

SKIFF BEETLES
(Family Hydroscaphidae)

SKIFF BEETLE
(Hydroscaphidae)

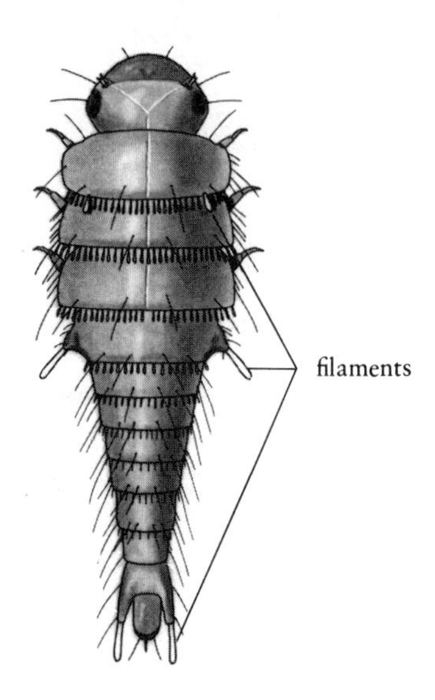

Figure 13.26.
Hydroscapha larva

ADULT DIAGNOSIS: These are minute forms, approximately 1.5 mm in length. Elytra are truncate and expose three abdominal segments.

LARVAL DIAGNOSIS: (Fig. 13.26) These are minute and possess thin-walled balloonlike filaments (probably spiracular gills) on prothorax and abdominal segments 1 and 8. Abdomen is evenly tapered from anterior to posterior and terminates in a small attachment disc ventrally.

DISCUSSION: This unique family is represented in North America by one species of the genus *Hydroscapha*. It is known only from southwestern states, where it is rarely taken. Both adults and larvae inhabit algal growths in streams and are tolerant of a wide range of water temperatures.

Suborder Polyphaga

Adults have the posterior margin of the first abdominal sternite undivided by the hind coxae (Fig. 13.32). Eyes never consist of dorsal and ventral pairs.

Larvae are legless or have five-segmented legs including claw. Claws are single.

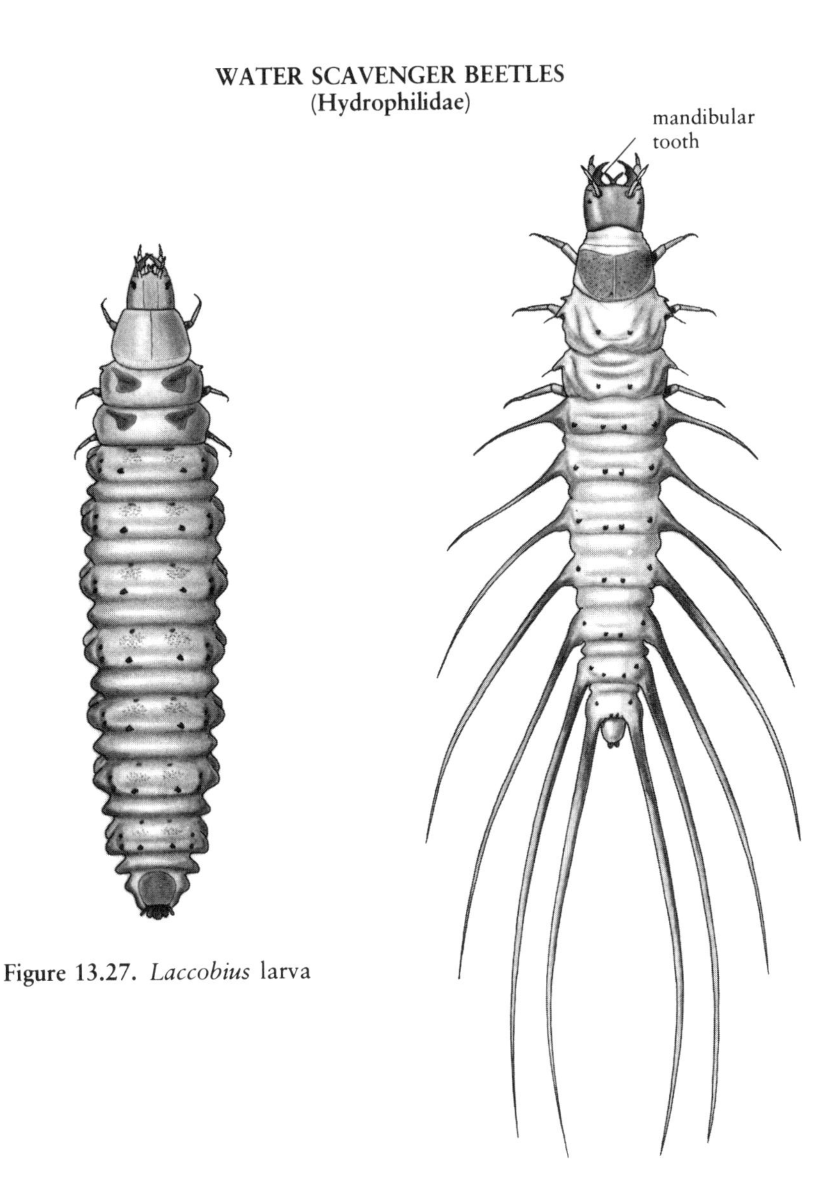

Figure 13.27. *Laccobius* larva

Figure 13.28. *Berosus* larva

WATER SCAVENGER BEETLES
(Family Hydrophilidae)

ADULT DIAGNOSIS: (Figs. 13.29–13.35; Plate X, Figs. 64, 66) These are oval-elongate, usually highly convex forms that range from 1 to 40 mm. Antennae possess a three-segmented club and usually a cuplike segment at base of the club; antennae are sometimes hidden beneath head. Thorax of some has a ventral spine that extends between bases of legs. Hind legs of many have swimming hairs. Abdomen has five well-developed sternites; sometimes a sixth sternite is either membranous or tucked into the fifth.

LARVAL DIAGNOSIS: (Figs. 13.27, 13.28; Plate X, Fig. 65) These measure 4–60 mm at maturity. Mandibles are usually toothed and clearly visible from above. Abdomen is eight-segmented (nine in some semiaquatic species), often has a wrinkled appearance, and may or may not possess filaments. Terminal filaments are commonly short and retracted into abdominal segment 8.

DISCUSSION: The distinctively clubbed antennae of adult water scavenger beetles distinguish them from most other commonly encountered aquatic beetles, including the predaceous diving beetles. Caution should be taken not to mistake their well-developed maxillary palps for the antennae. Behavioral characteristics of adult water scavenger beetles that are useful in field identification include their habits of surfacing head first and swimming by moving the hind legs alternately instead of together. The number of leg segments and the toothed mandibles distinguish the larvae from those predaceous diving beetles that also have well-developed, projecting mandibles.

Among North American water beetles, this family is second in size only to the Dytiscidae and contains about 200 aquatic species classified in seven subfamilies.

Adults and larvae of most water scavenger beetles live in ponds, the shallow regions of lakes, and the pools and quieter waters of streams and rivers. One terrestrial subfamily with legless larvae is associated with dung and rotting vegetation; however, a few species in this otherwise terrestrial subfamily are known from semiaquatic coastal and/or inland habitats. The adults of all other subfamilies (discussed below) are aquatic for at least part of their lives.

Adults are primarily aeropneustic and must surface periodically; but when the water is sufficiently oxygenated, submergence time is increased because the ventral plastron acts as a physical gill. This plastron sometimes gives the beetles a silvery appearance in the water. Larvae are either aquatic or semiaquatic. Aquatic larvae of most species are aeropneustic, although those of the genus *Berosus* are hydropneustic.

Feeding habits are diverse in adults, with many species being omnivore-detritivores and a few strictly predaceous. Larvae are usually predaceous, although some species feed on algae.

Pupation is generally terrestrial, although cocoons of some species of *Enochrus* are made with algal fragments and occur in floating algal mats. Some of these beetles (e.g., *Tropisternus*) use chirping calls to locate mates. Oviposition is highly varied, with the females of some species carrying the eggs on their bodies.

Each of the six aquatic subfamilies can be identified by a unique combination of characteristics, as indicated in the following diagnoses.

SUBFAMILY HELOPHORINAE: **Adults** (Fig. 13.29) have a pronotum with five distinct longitudinal grooves; pronotum covers much of head. This group is generally widespread except for some southeastern states including Florida.

SUBFAMILY EPIMETOPINAE: **Adults** (Fig. 13.30) have the pronotum narrowly projecting over much of the head. This group is known only from southwestern states including Texas.

SUBFAMILY HYDROCHINAE: **Adults** (Fig. 13.31) have a pronotum that is much narrower than base of elytra. Scutellum is very small. This group may superficially resemble the Helophorinae, but lacks longitudinal grooves of the pronotum. The group is widespread but known only very locally in many parts of its range; it is common in Texas, and southeastern states including Florida.

SUBFAMILY HYDROPHILINAE: **Adults** (Fig. 13.32; Plate X, Fig. 66) have a distinct ventral spine (keel) extending between hind coxae. The group is widespread in North America.

WATER SCAVENGER BEETLES
(Hydrophilidae)

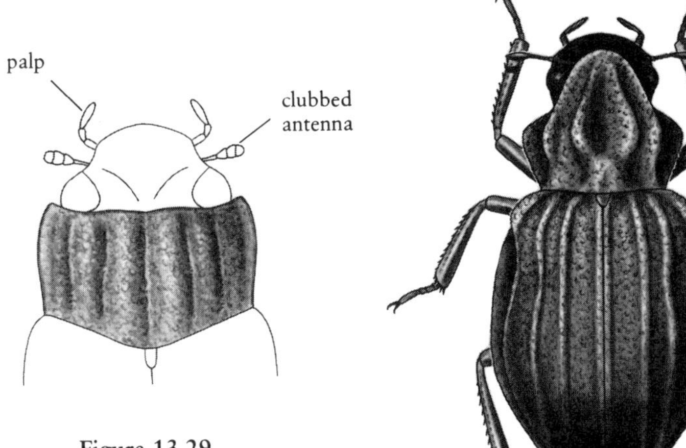

Figure 13.29.
Helophorinae adult

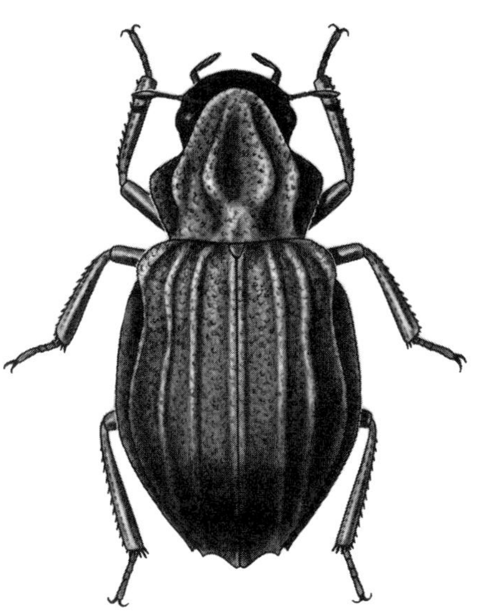

Figure 13.30.
Epimetopinae adult

Figure 13.31.
Hydrochinae adult

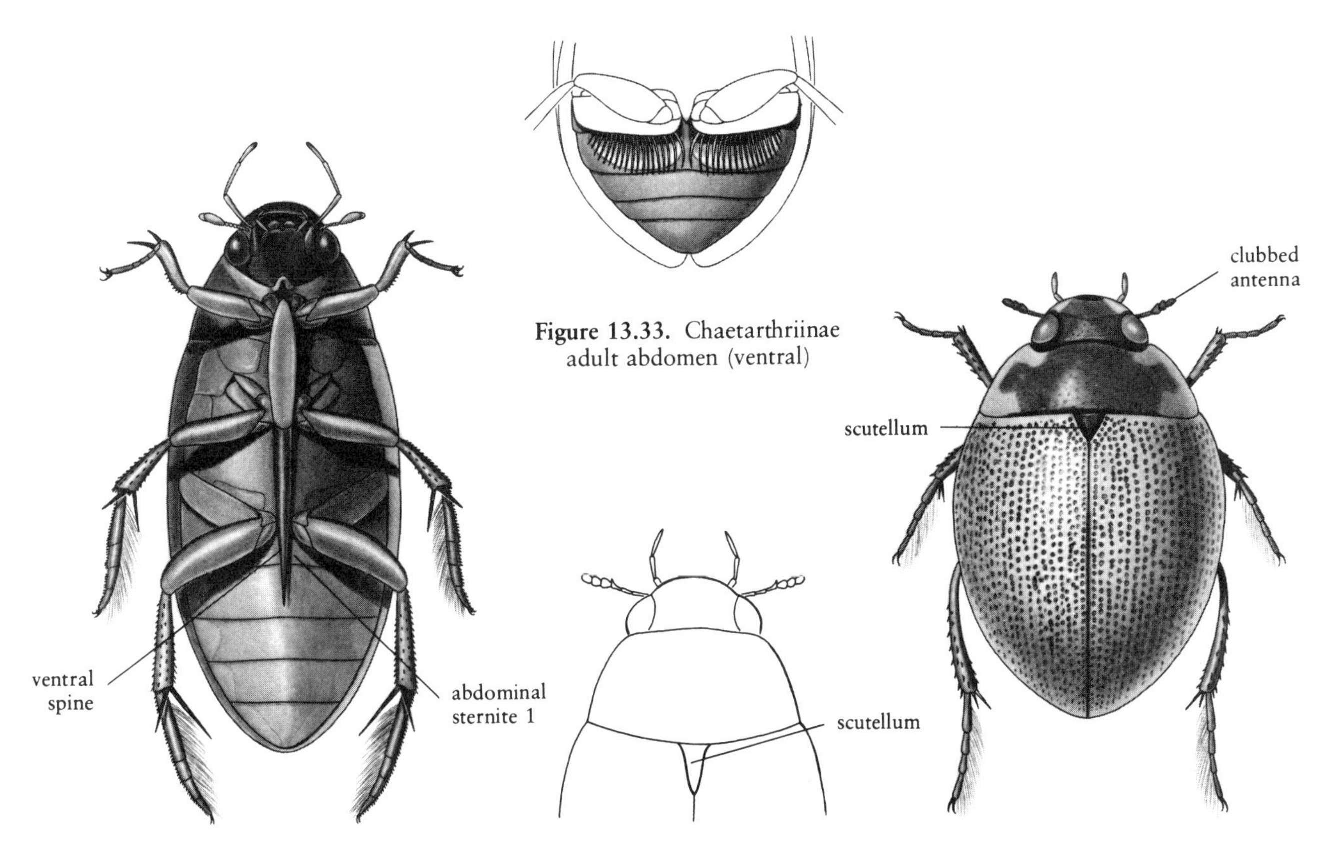

Figure 13.33. Chaetarthriinae adult abdomen (ventral)

Figure 13.32.
Hydrophilinae adult (ventral)

Figure 13.34.
Berosinae adult

Figure 13.35.
Hydrobiinae adult

SUBFAMILY CHAETARTHRIINAE: **Adults** (Fig. 13.33) possess depressions on each side of abdominal sternites 1 and 2; these depressions are each covered by a close-set fringe of long golden hairs. This group is widespread except for eastern Canada.

SUBFAMILY BEROSINAE: **Adults** (Fig. 13.34; Plate X, Fig. 64) possess a head that is strongly bent downward. Scutellum is much longer than its basal width. The group is widespread in North America.

SUBFAMILY HYDROBIINAE: **Adults** (Fig. 13.35) possess a head that is not strongly bent downward. Scutellum is shorter or subequal in length to its basal width. This group is widespread in North America.

MINUTE MOSS BEETLES
(Family Hydraenidae)

ADULT DIAGNOSIS: (Figs. 13.36, 13.37) These are somewhat similar to Hydrophilidae. They are minute, measuring 1.2–2.5 mm. Antennae are clubbed, and each club is five-segmented. Legs lack swimming hairs. Abdomen has six or seven segments ventrally.

LARVAL DIAGNOSIS: (Fig. 13.38) Abdomen has 10 complete segments and two short, posteriorly directed filaments originating on segment 9.

DISCUSSION: Larvae may be confused with those of rove beetles, but the basal molar region of the mandibles is more highly developed in minute moss beetles. At least 30 species occur in North America.

Minute moss beetles primarily crawl along the margins of streams, often in tangled roots and debris, or in or around freshwater and brackish pools, intertidal areas, or warm springs. They are often associated with filamentous algae and leaf detritus. Larvae are not adapted for aquatic respiration. As a group, minute moss beetles are probably more semiaquatic than aquatic.

MINUTE MOSS BEETLES
(Hydraenidae)

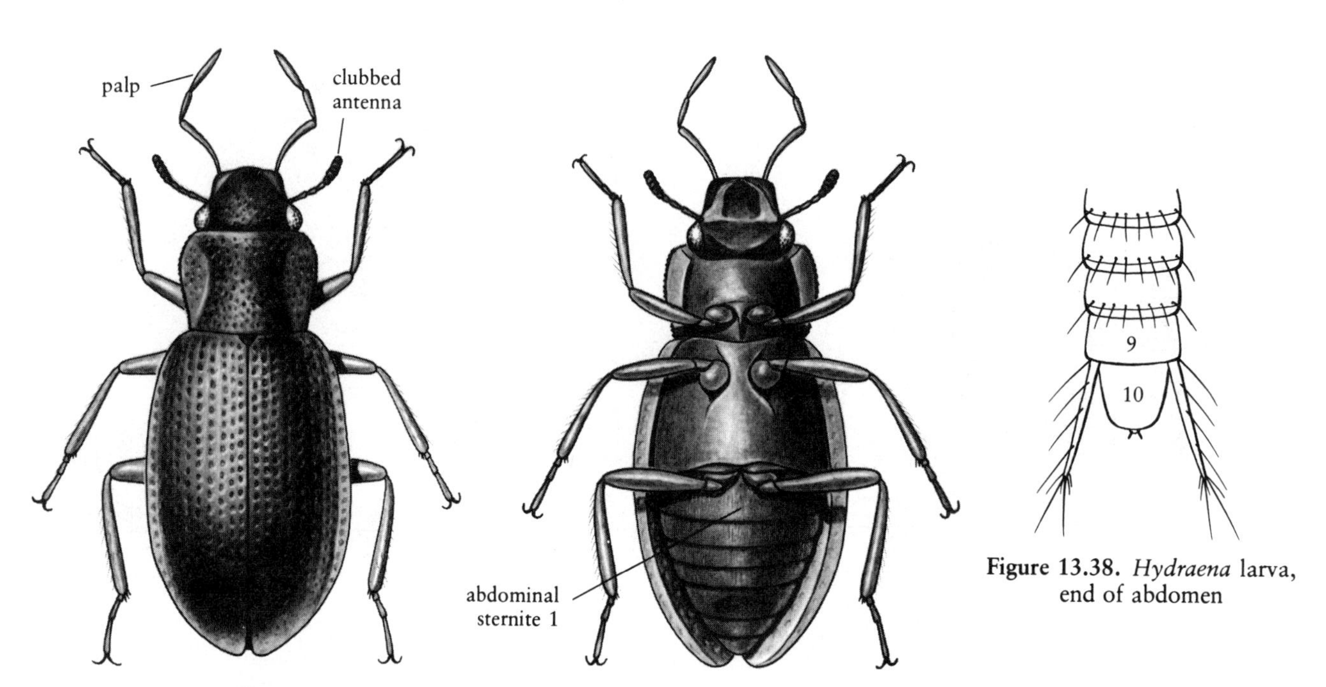

Figure 13.36. *Hydraena* adult

Figure 13.37. *Hydraena* adult (ventral)

Figure 13.38. *Hydraena* larva, end of abdomen

SUBFAMILY LIMNEBIINAE: **Adults** have hind tarsi in which segment 2 is distinctly longer than segment 3. This group is generally widespread but usually only very locally common.

SUBFAMILY HYDRAENINAE: **Adults** (Figs. 13.36, 13.37) have hind tarsi in which segment 2 is subequal to or shorter than segment 3. This group is widespread in North America.

MINUTE MUDLOVING BEETLES
(Family Georyssidae)

This small semiaquatic family is treated in Chapter 17.

ROVE BEETLES
(Family Staphylinidae)

This family, which contains a number of shore-dwelling species (both marine and inland), is treated in Chapter 17.

MARSH BEETLES
(Family Helodidae)

ADULT DIAGNOSIS: (Fig. 13.39) These are small oval forms, about 2–4 mm in length. Posterior margin of pronotum is smooth. Fore coxae are conical and projecting. Hind femora are usually expanded. Segment 5 of tarsi is not longer than the basal four segments.

LARVAL DIAGNOSIS: (Fig. 13.40) These are somewhat flattened and elongate forms. Antennae have many segments and are usually longer than head and thorax combined.

MARSH BEETLES
(Helodidae)

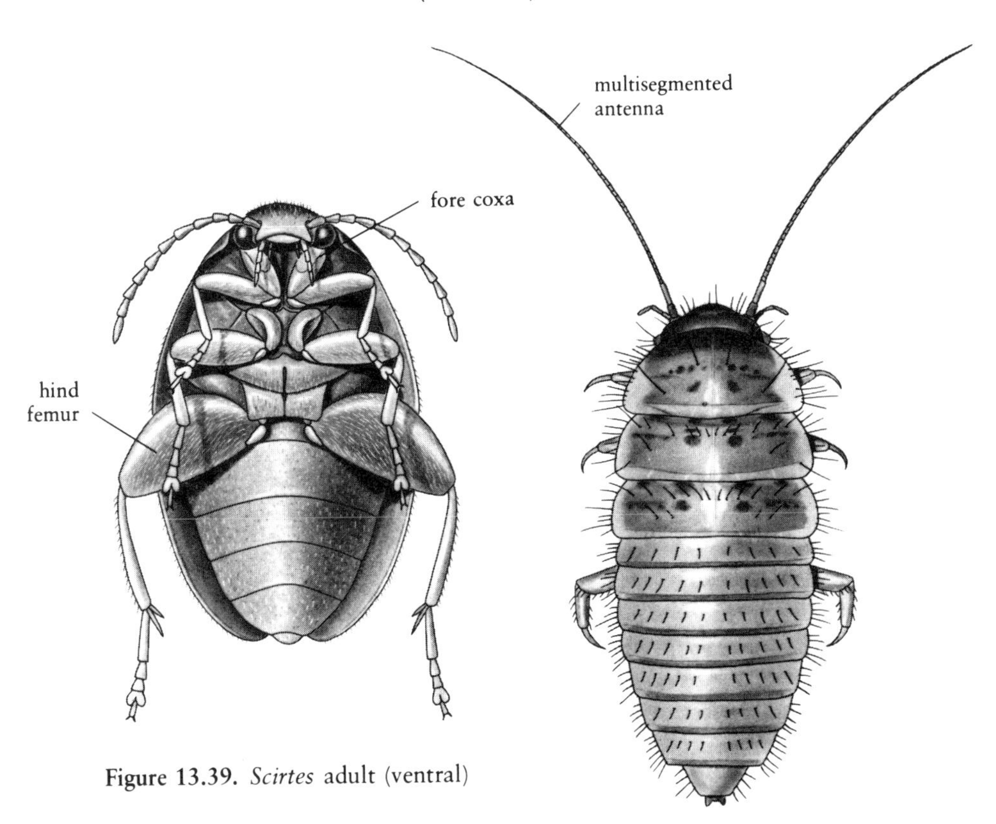

Figure 13.39. *Scirtes* adult (ventral)

Figure 13.40. *Scirtes* larva

And the poor beetle that we tread upon
In corporal sufferance finds a pang as great
As when a giant dies

William Shakespeare

DISCUSSION: The uniquely long and multisegmented antennae of the larvae of marsh beetles readily distinguish them from other beetle larvae. The family is widespread in North America. The approximately 35 species and seven genera that occur in North America are usually only very locally common.

Adults of most species are terrestrial, but some regularly occur in the vicinity of water. Adults of *Elodes* have rarely been seen entering water for brief periods (perhaps to oviposit). The larvae are aquatic and aeropneustic, and some species have large tracheal reservoirs for storing air. They inhabit a variety of lentic and lotic habitats, and are frequently found in tree holes or various small containers that hold water. Larvae are detritivores.

WATER PENNIES
(Family Psephenidae)

ADULT DIAGNOSIS: (Fig. 13.43) These are somewhat flattened, variously shaped, soft-bodied forms that measure 4–6 mm. Mandibles are concealed from above when closed. Posterior margin of pronotum is jagged or beaded in some and smooth in others; if smooth as in most families, then tarsal segment 5 is as long as the basal four segments combined, and fore coxae are conical and projecting.

LARVAL DIAGNOSIS: (Figs. 13.41, 13.42; Plate XI, Fig. 70) These are conspicuously flattened and disclike forms. They are almost as broad as they are long. Dorsal platelike expansions conceal head and legs from above.

WATER PENNIES
(Psephenidae)

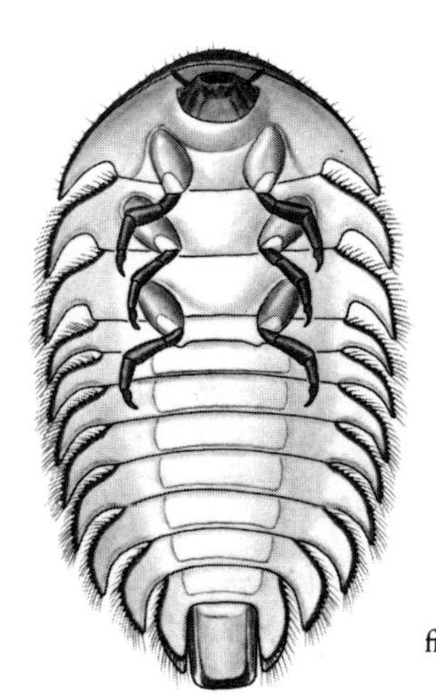

Figure 13.41. Eubriinae larva (ventral)

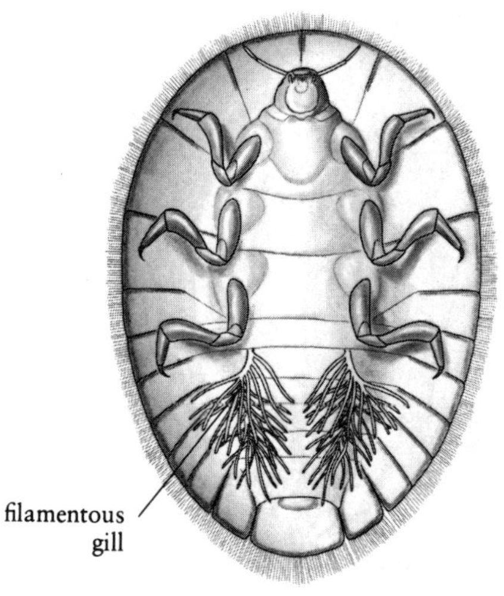

Figure 13.42. Psepheninae larva (ventral)

Figure 13.43. *Psephenus* adult

DISCUSSION: Water penny larvae are some of the most distinctive of aquatic insects. Adults, which are sometimes found along stream margins, must be carefully identified using a combination of characters. About nine species, classified in three subfamilies, occur in North America.

Adults are not aquatic, but may be seen along streams or on stones just above the waterline; they may enter the water to oviposit. Larvae are hydropneustic and live attached to stones in streams and rivers or in lakes with considerable wave action. They are usually found in moderate to swift currents but are occasionally found in well-oxygenated pools. Larvae are highly adapted for adhering to stones and for feeding on the periphyton and encrusted materials associated with this substrate. Interestingly, foraging trout apparently have some propensity for plucking these strongly adhering larvae from the substrate. Pupation usually occurs in protected areas slightly above the waterline but rarely in water.

SUBFAMILY EUBRIINAE: **Larvae** (Fig. 13.41) have expanded abdominal segments that are somewhat separated laterally. This group is known from western and central states excluding Texas, eastern states including Florida, and eastern Canada.

SUBFAMILY EUBRIANACINAE: **Larvae** have expanded abdominal segments that are held tightly together. Abdomen possesses four pairs of ventral gills. This group is known only from western states.

SUBFAMILY PSEPHENINAE: **Larvae** (Fig. 13.42; Plate XI, Fig. 70) are similar to those of Eubrianacinae except abdomen possesses five pairs of ventral gills. This group is relatively widespread but is not known from Florida or western Canada and Alaska.

PTILODACTYLID BEETLES
(Family Ptilodactylidae)

ADULT DIAGNOSIS: These are somewhat elongate, relatively soft-bodied forms, measuring 4–6 mm. Mandibles project beyond margin of head and are usually conspicuous. Fore coxae are conical and projecting.

PTILODACTYLID BEETLES
(Ptilodactylidae)

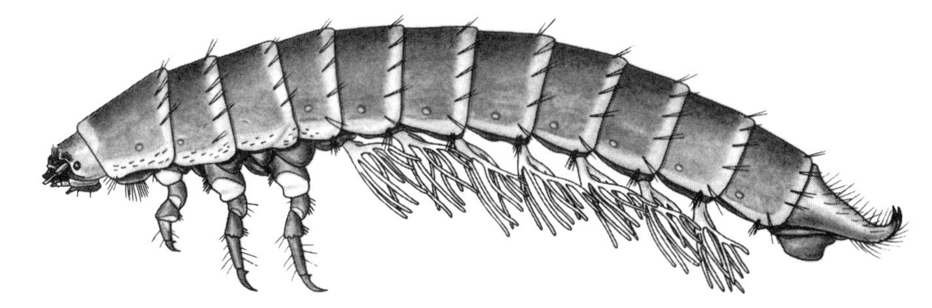

Figure 13.44. *Stenocolus* larva

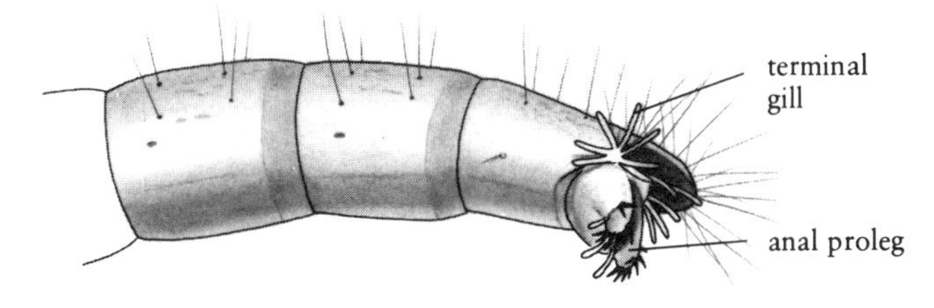

Figure 13.45. *Anchytarsus* larva, end of abdomen

LARVAL DIAGNOSIS: (Figs. 13.44, 13.45) These are elongate and relatively hard-bodied. Abdomen is nine-segmented. Abdominal segment 9 does not have a ventral lidlike operculum at base of caudal gill chamber. Either abdominal segments 1–7 each have a pair of tufted, ventral gills or only segment 9 has ventral gills.

DISCUSSION: Ptilodactylid larvae are all easily distinguishable from riffle beetle larvae. Those with tufted gills on abdominal segments 1–7 are distinct, and those with gills only on abdominal segment 9 are also distinct because of a unique pair of spiny anal prolegs found at the end of the body.

This is a small and rare family consisting of three genera and three species in North America. They are known only very locally from southwestern states and rarely from eastern Canada and eastern states excluding Florida. Larvae occur in streams where they may burrow in soft substrate. Adults are not aquatic.

VARIEGATED MUDLOVING BEETLES
(Family Heteroceridae)

For a treatment of this family see Chapter 17.

MINUTE MARSHLOVING BEETLES
(Family Limnichidae)

ADULT DIAGNOSIS: These are minute, oval, and convex, and measure 1–2 mm. Antennae are 10-segmented. Claws are elongate and prominent. Tarsal segment 5 is not as long as the basal four tarsal segments.

LARVAL DIAGNOSIS: (Figs. 13.46, 13.47) These are minute, elongate, and robust. Thorax is membranous ventrally or lacks discernible sternites. Abdomen possesses lateral suture lines on segments 1–4. Abdominal segment 9 has a ventral operculum that possesses a pair of internally attached hooks. Filamentous gills originate in caudal gill chamber of abdominal segment 9. End of abdomen appears rounded as seen from above.

DISCUSSION: The larvae of minute marshloving beetles are poorly known. Those of the genus *Lutrochus* are quite similar to those of riffle beetles. Four genera in two subfamilies occur in North America, but only three species (in *Lutrochus*) are fully aquatic as larvae. These larvae have been found in calcareous incrustations on stones in streams.

Adults are generally terrestrial or semiaquatic, occurring along stream margins or on rocks and debris projecting from the water, on low overhanging branches, and on streamside grasses. Adults of a few species of *Throscinus* occur on tidal mud flats.

MINUTE MARSHLOVING BEETLES
(Limnichidac)

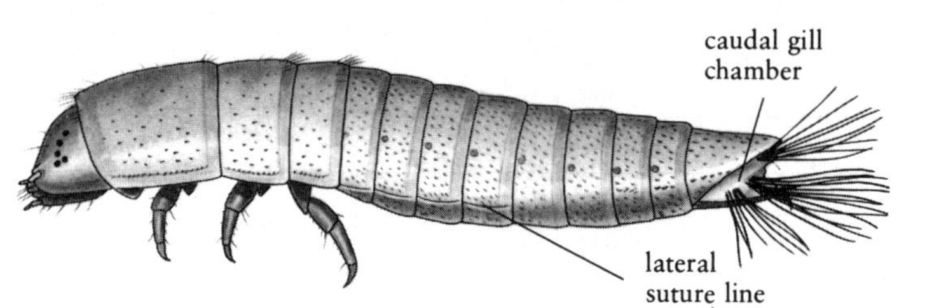

Figure 13.46. *Lutrochus* larva

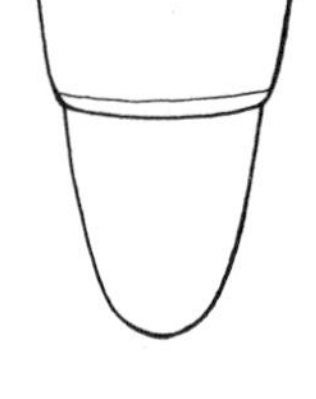

Figure 13.47. End of larval abdomen (dorsal)

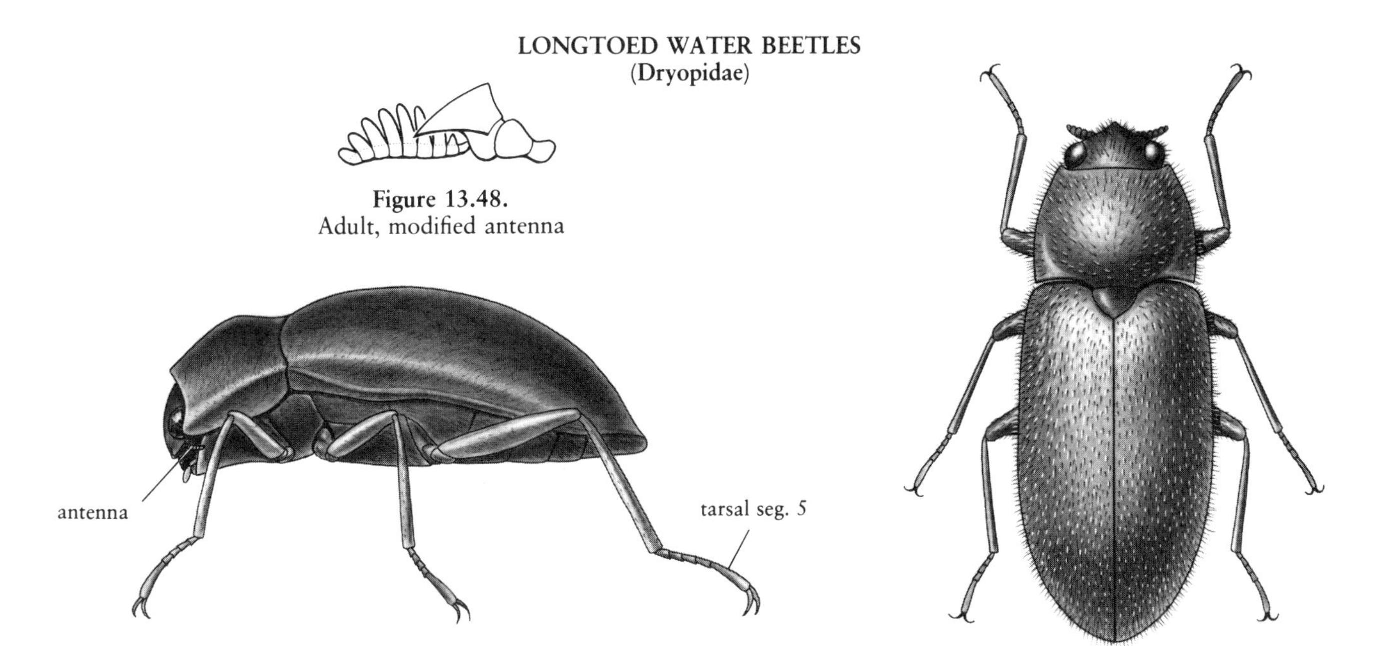

Figure 13.48. Adult, modified antenna

Figure 13.49. *Helichus* adult

Figure 13.50. *Pelonomus* adult

SUBFAMILY LIMNICHINAE: (Figs. 13.46, 13.47) These are usually inland aquatic, semiaquatic, or terrestrial forms. Aquatic larvae are known from central states, eastern states including Florida, western states including Texas, and southern Canada. One semiaquatic species is known from coastal intertidal areas of Texas.

SUBFAMILY CEPHALOBYRRHINAE: These include intertidal species known from Texas and California.

LONGTOED WATER BEETLES
(Family Dryopidae)

ADULT DIAGNOSIS: (Figs. 13.48–13.50; Plate XI, Fig. 67) These are somewhat elongate forms measuring 4–10 mm. Antennae are short, thick, and 11-segmented; segments 4–11 are laterally expanded. Tarsal claws are long and prominent. Tarsal segment 5 is at least as long as the basal four tarsal segments.

LARVAL DIAGNOSIS: These are cylindrical, elongate, and relatively hard-bodied. Abdominal segment 9 has a ventral operculum that possesses internally attached hooks, but filamentous gills are not present in caudal chamber.

DISCUSSION: The adults of longtoed water beetles can usually be distinguished from those of riffle beetles by the antennal differences. Larvae are similar to those of riffle beetles, but longtoed water beetle larvae are more-or-less circular in cross section and lack gills.

The family is relatively widespread except for western and central Canada and Alaska. The group consists of three genera—two with few species each and the genus *Helichus* with about 10 species in North America.

Larvae are terrestrial. Adults of several species are found in streams and rivers, where they crawl among rocks and debris. They are relatively rare in ponds or swamps, and some species are entirely terrestrial. Adults may utilize plastron respiration for prolonged periods of time. Both larvae and adults are herbivorous.

RIFFLE BEETLES
(Elmidae)

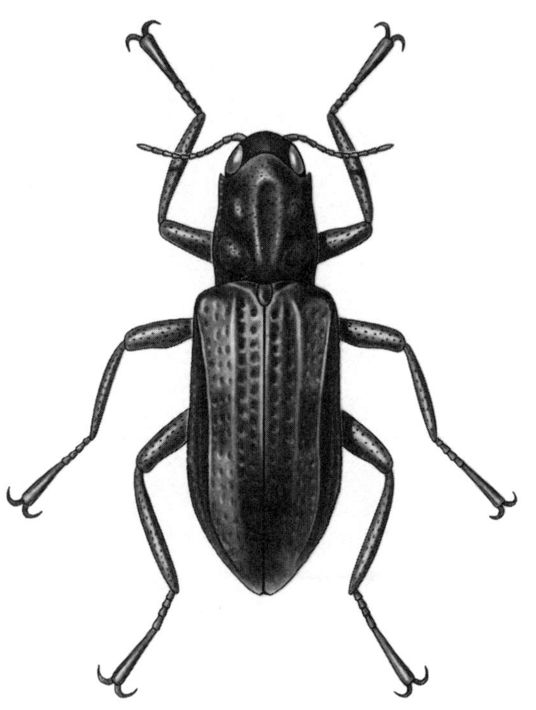

Figure 13.51. *Stenelmis* adult

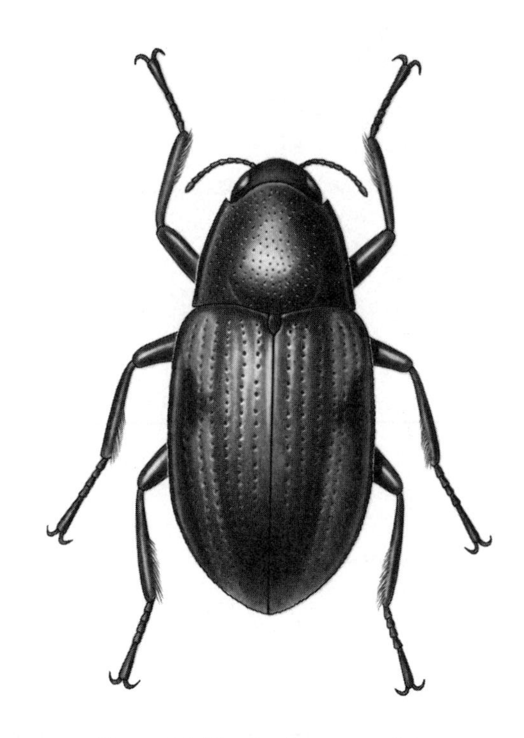

Figure 13.52. *Optioservus* adult

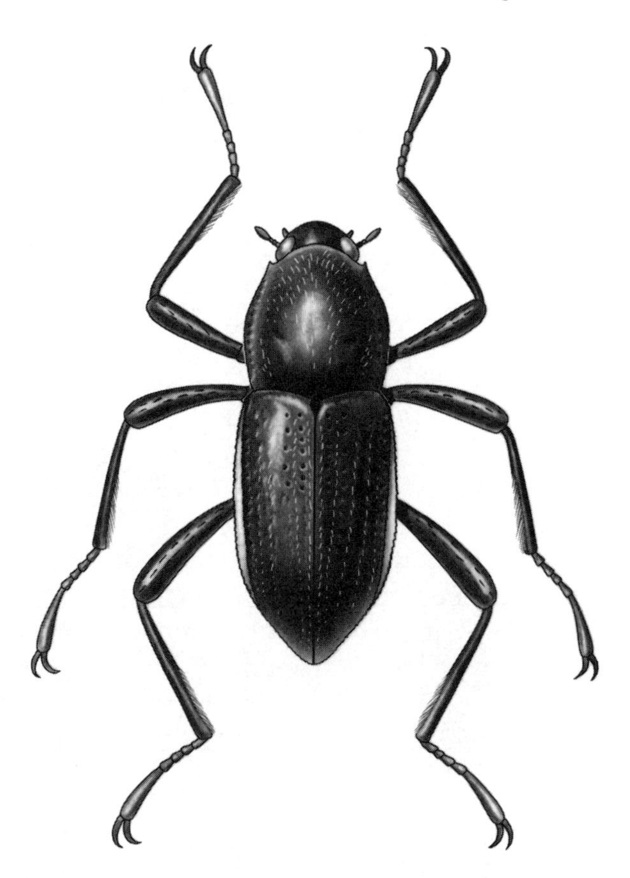

Figure 13.53. *Macronychus* adult

RIFFLE BEETLES
(Elmidae)

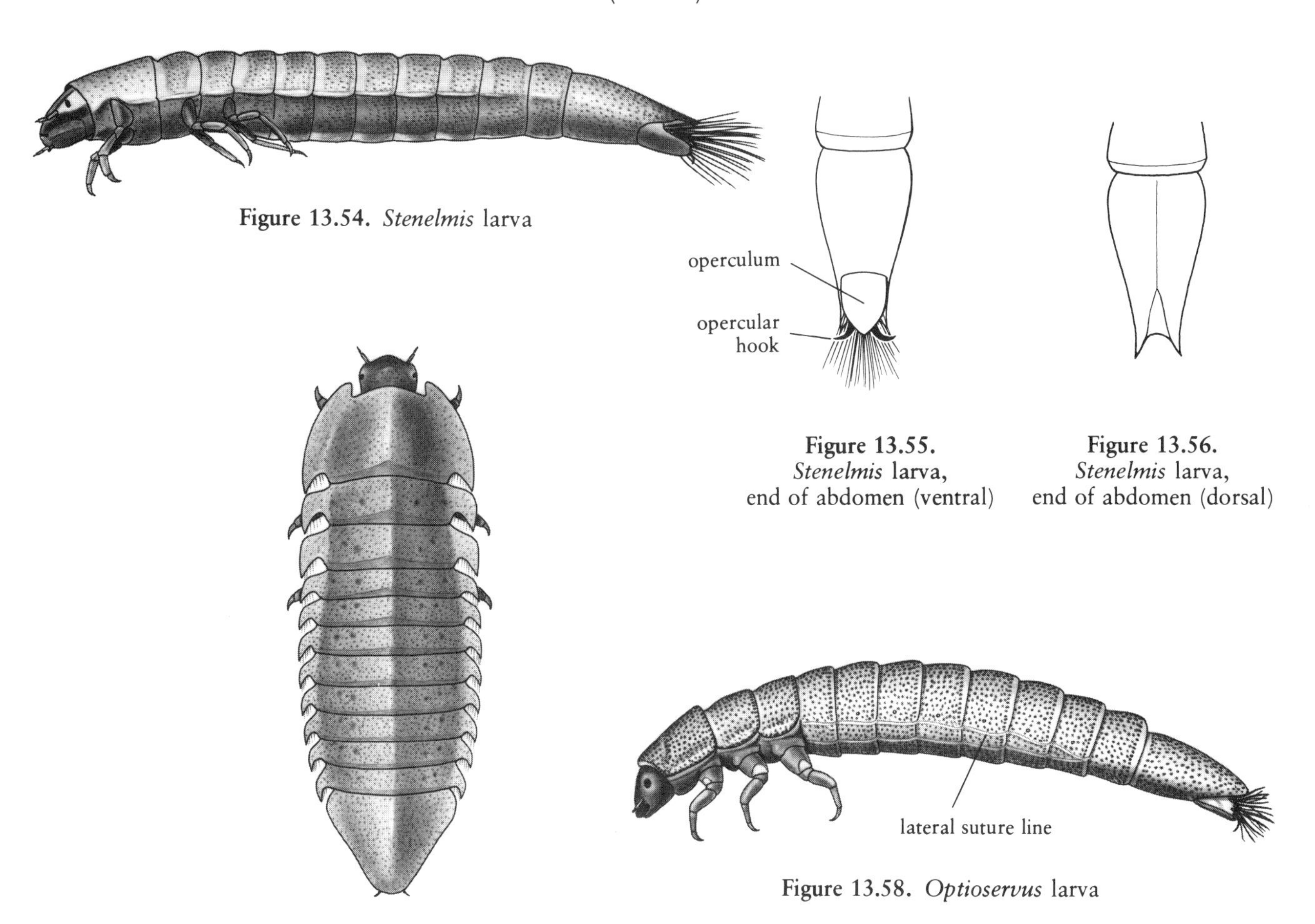

Figure 13.54. *Stenelmis* larva

Figure 13.55. *Stenelmis* larva, end of abdomen (ventral)

Figure 13.56. *Stenelmis* larva, end of abdomen (dorsal)

Figure 13.57. *Phanocerus* larva

Figure 13.58. *Optioservus* larva

RIFFLE BEETLES (Family Elmidae)

ADULT DIAGNOSIS: (Figs. 13.51–13.53; Plate XI, Fig. 68) These are small, somewhat elongate, long-legged forms that measure 1–8 mm. Antennae are usually slender but are sometimes clubbed (but never as in Dryopidae). Tarsal claws are long. Tarsal segment 5 is at least as long as the basal four tarsal segments combined.

LARVAL DIAGNOSIS: (Figs. 13.54–13.58; Plate XI, Fig. 69) These are elongate and relatively hard-bodied. Body is usually somewhat hemispherical in cross section, being flattened or slightly concave ventrally in many; a few are cylindrical or flattened. Lateral suture lines are present on abdominal segments 1–6 to 1–8 or are indistinct. Abdominal segment 9 has a ventral operculum that possesses hooks. Caudal chamber has filamentous gills.

DISCUSSION: These small, hard-bodied, crawling aquatic beetles are easily distinguished from most other aquatic beetles by both their form and habits. Riffle beetle adults do not have a highly modified type of clubbed antennae like those of the longtoed water beetles; and most riffle beetle larvae have the apex of their abdomen slightly forked as seen from above rather than rounded as in the aquatic larvae of minute marshloving beetles.

There are over 80 species in about 24 genera in North America. The entire family is aquatic, although the adults of the tribe Larini are usually found out of the water. An initial post-emergence flight period is common in many species.

Habitats include gravelly and rocky bottoms of riffles and rapids of streams and rivers. Some species also occur in lakes with considerable wave action. Others inhabit sandy bottoms of slower reaches of streams, and a few inhabit ponds, commonly amongst aquatic vegetation. Riffle beetles are often found in the crevices or under the bark of decaying woody debris. The relatively diverse genus *Stenelmis* is known from a variety of these habitats, whereas several species of *Microcylloepus* are restricted to warm springs in the West. *Optioservus* is a typical genus of riffles in the East. Many riffle beetle species are important indicators of water quality.

Adults generally utilize highly developed plastrons, and some need not resurface in well-oxygenated habitats. Plastron-bearing adults often fly after emergence but evidently either lose this ability or become obligatory aquatics once they enter the water. Larvae use gills in respiration.

Adults and larvae are crawlers and are able to cling steadfastly to the substrate. Most species feed on periphyton; others are detritivores. Riffle beetles are atypical of most water beetles in that they generally have six larval instars and a highly variable developmental period.

SOFTWINGED FLOWER BEETLES
(Family Melyridae)

This family includes a few species that are found in coastal marine habitats; it is treated in Chapter 17.

ROOTEATING BEETLES
(Family Rhizophagidae)

At least one species of this family is found in coastal areas; see Chapter 17.

DARKLING BEETLES
(Family Tenebrionidae)

At least one species of this family is known from coastal habitats; see Chapter 17.

NARROWWINGED BARK BEETLES
(Family Salpingidae)

A few members of this family are known from coastal habitats; see Chapter 17.

AQUATIC LEAF BEETLES
(Family Chrysomelidae)

ADULT DIAGNOSIS: (Plate XI, Fig. 71) These are somewhat elongate and measure 5–10 mm. All tarsi are apparently four-segmented but actually five-segmented. Tarsal segment 4 is hidden between the lobes of segment 3. There are no apparent adaptations for swimming.

LARVAL DIAGNOSIS: (Figs. 13.59, 13.60) These are robust and somewhat grublike. The body curls slightly at the posterior end. Head and legs are small. Mouthparts are directed ventrally. Abdomen is eight- or nine-segmented; if eight-segmented, as in most, then it possesses a pair of sharp spurs terminally.

AQUATIC LEAF BEETLES
(Chrysomelidae)

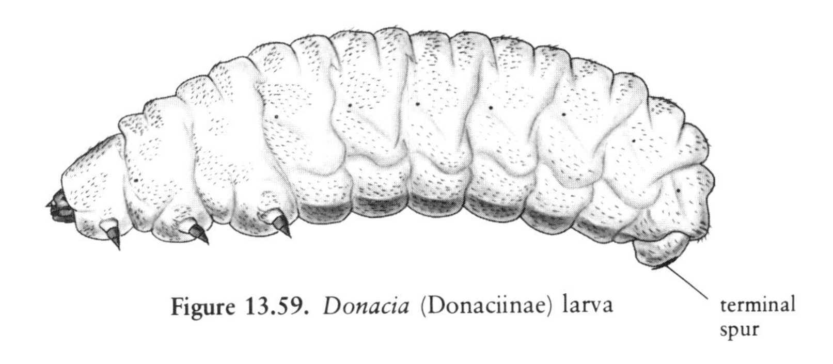

Figure 13.59. *Donacia* (Donaciinae) larva

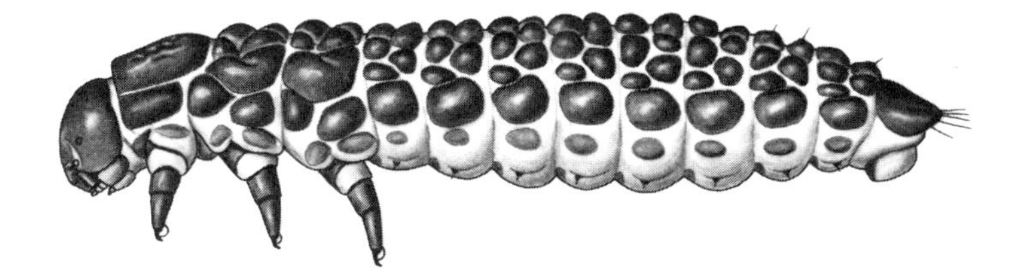

Figure 13.60. *Pyrrhalta* (Galerucinae) larva

DISCUSSION: The aquatic leaf beetles are distinctive both as larvae and adults and should not be confused with other beetles found in water or on emergent vegetation. Within this large, primarily terrestrial family, perhaps over 40 either aquatic or semiaquatic species (mostly of the widespread genus *Donacia*) occur in North America.

Larvae of the subfamily Donaciinae are aquatic and live in association with the underwater parts of floating or emergent vegetation, on which they feed. Some inhabit the substrate, where they feed on roots. They are primarily aeropneustic and obtain air from their host plant by inserting the sharp terminal spurs, which are modified spiracles, into the plant tissue. Early instar larvae are limited to feeding on thin-walled parts of the plant because of their smaller spurs. Last instar larvae have functional thoracic and abdominal spiracles that are used in the cocoon prior to pupation. Cutaneous respiration is possible but probably insignificant.

The pupae, which live within silken cocoons, are generally found in the same place as the larvae and are also aeropneustic.

The adults of the Donaciinae are usually found on emergent parts of the host plant but occasionally crawl beneath the surface of the water. Eggs are laid on the bottom of floating leaves or slightly below the waterline on stems. The adults and larvae of one species of the subfamily Galerucinae occur on the upper surfaces of floating leaves.

SUBFAMILY DONACIINAE: **Larvae** (Fig. 13.59) possess terminal breathing spurs on abdominal segment 8. The group is widespread in North America.

SUBFAMILY GALERUCINAE: **Larvae** (Fig. 13.60) lack terminal breathing spurs. Abdomen has tiny ventral prolegs. The group is widespread in North America.

WATER WEEVIL
(Curculionidae)

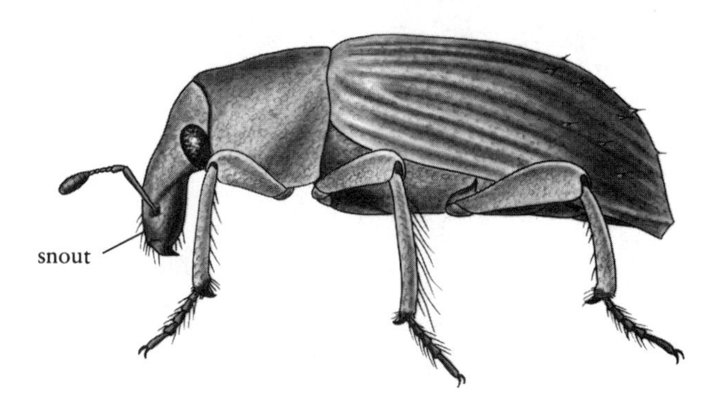

Figure 13.61. *Lissorhoptrus* adult

WATER WEEVILS
(Family Curculionidae)

ADULT DIAGNOSIS: (Fig. 13.61) These measure 1–12 mm but most commonly 2–4 mm. Head is elongated into a distinct snout. Antennae arise on the snout, are elbowed, and possess a terminal club. Swimming hairs, when present, are located on middle legs only.

LARVAL DIAGNOSIS: These are legless, fleshy forms, usually with a curved and crescent-shaped body. Mouthparts are directed ventrally.

DISCUSSION: Both adults and larvae of water weevils are very distinctive. Although this large and very important family is primarily terrestrial, perhaps 100 or more North American species in about 33 genera and seven subfamilies are aquatic, semiaquatic, or both. About 30 species in five genera are fully aquatic.

Water weevils occur in most habitats where there are aquatic plants, and are most common in ponds, marshes, and along the edges of lakes and streams. The adults are generally found crawling on emergent vegetation above or below the waterline, and some occasionally swim. Primarily the middle legs, rather than the hind legs (as is the case for other swimming beetles), are used to propel the body. Although most water weevil adults are semiaquatic and live on plant parts above the water, a number are able to remain in water for hours or even days. These aquatic species utilize plastron respiration while under water, but if the oxygen supply in the water becomes limiting, as it often does in many lentic habitats, they can easily leave the water and resume terrestrial respiration.

Larvae are not aquatic but live within plant stems, often below the waterline. Larvae seldom occur within the water. Pupation and oviposition take place in or on the host plant.

The rice water weevil (genus *Lissorhoptrus*) is an important pest of rice. Some water weevils have been introduced into North America to help control aquatic weeds, and some species have been accidentally introduced.

More Information About Water Beetles

The beetles of the United States by Arnett (1963). This book contains a family by family treatment of the beetles and includes general discussions at the family level, keys to the genera (adults), and distributional information on each genus.

Larves et nymphes des Coleopteres aquatiques du globe by Bertrand (1972). This book, written in French, is the most comprehensive treatment of water beetle larvae and pupae to date; it contains excellent family keys and many drawings.

Aquatic dryopoid beetles (Coleoptera) of the United States by Brown (1972). For those who require information on the bottom-dwelling beetles of our streams and rivers, this work, with its keys to species (both adults and larvae), is most useful.

"Aquatic Coleoptera" (Chapter 13 in *Aquatic Insects of California*) by Leech and Chandler (1956). This work contains considerable information on water beetles and includes keys to the larvae of many of the North American genera.

The water beetles of Florida by Young (1954). This book contains a comprehensive treatment of Florida species of water beetles and in many respects is useful for work with beetles throughout much of North America.

References

Anderson, R. D. 1962. The Dytiscidae (Coleoptera) of Utah: Keys, original citation, types, and Utah distribution. *Gr. Basin Natural.* 22:54–75.

Arnett, R. H. 1963. *The beetles of the United States.* Catholic Univ. Amer. Press, Wash., D.C. 1112 pp.

Bertrand, H. P. I. 1972. *Larves et nymphes des Coleopteres aquatiques du globe.* Centre National de Recherche Scientifique, Paris. 804 pp.

Brown, H. P. 1972. *Aquatic dryopoid beetles (Coleoptera) of the United States.* Wat. Poll. Cont. Res. Ser. U.S.E.P.A., Wash., D.C. 82 pp.

Campbell, J. M. 1979. Coleoptera, pp. 357–363. In *Canada and its insect fauna* (H. V. Danks, ed.). Mem. Entomol. Soc. Canada 108.

Doyen, J. T. 1976. Marine beetles (Coleoptera excluding Staphylinidae), pp. 497–519. In *Marine insects* (L. Cheng, ed.). North-Holland, Amsterdam.

Evans, G. 1975. *The life of beetles.* Allen and Unwin Ltd., London. 232 pp.

Folkerts, G. W. 1979. *Spanglerogyrus albiventris,* a primitive new genus and species of Gyrinidae from Alabama. *Coleop. Bull.* 33:1–8.

Hilsenhoff, W. L. 1975. *Aquatic insects of Wisconsin.* Wisc. Dept. Nat. Res. Tech. Bull. 89. 52 pp.

Hinton, H. E. 1976. Plastron respiration in bugs and beetles. *J. Ins. Physiol.* 22:1529–1550.

Houlihan, D. F. 1969. Respiratory physiology of the larva of *Donacia simplex,* a root-piercing beetle. *J. Ins. Physiol.* 15:1517–1536.

Komnick, H. 1977. Chloride cells and chloride epithelia of aquatic insects. *Inter. Rev. Cyt.* 49:285–329.

Leech, H. B., and H. P. Chandler. 1956. Aquatic Coleoptera, pp. 293–371. In *Aquatic insects of California* (R. L. Usinger, ed.). Univ. Calif. Press, Berkeley.

Miller, J. R., and R. O. Mumma. 1973. Defensive agents of the American water beetles *Agabus seriatus* and *Graphoderus liberus. J. Ins. Physiol.* 19:917–925.

Murvosh, C. M., and H. P. Brown. 1977. Underwater pupation by a psephenid beetle. *Ann. Entomol. Soc. Amer.* 70:975–976.

Nilakhe, S. S. 1977. Reproductive status of overwintering rice water weevils. *Ann. Entomol. Soc. Amer.* 70:599–601.

Perkins, P. O. 1976. Psammophilous aquatic beetles in southern California: A study of microhabitat preferences with notes on responses to stream alteration (Coleoptera: Hydraenidae and Hydrophilidae). *Coleop. Bull.* 30:309–324.

Ryker, L. C. 1975. Calling chirps in *Tropisternus natator* (D'orchymont) and *T. lateralis nimbatus* (Say) (Coleoptera: Hydrophilidae). *Entomol. News* 86: 179–186.

Schöne, H. 1951. Die lichtorientierung der larven von *Acilius sulcatus* L. und *Dytiscus marginalis* L. *Z. Vergleich. Physiol.* 33:63–98.

Seeger, W. 1971. Morphologie, bionomie und ethologie von halipliden, unter besonderer berücksichtigung funktionsmorphologischer gesichtspunkte (Haliplidae: Coleoptera). *Arch. Hydrobiol.* 68:400–435.
Shepard, W. D. 1979. Co-occurrence of a marine and a freshwater species of Limnichidae (Coleoptera). *Entomol. News* 90:88.
Spangler, P. J. 1980. Chelonariid larvae, aquatic or not? (Coleoptera: Chelonariidae). *Coleop. Bull.* 34:105–114.
Tucker, V. A. 1969. Wave making by whirligig beetles (Gyrinidae). *Science* 166:897–899.
Young, F. N. 1954. *The water beetles of Florida.* Univ. Fla. Press, Gainesville. 238 pp.
Young, F. N., and G. Longley. 1976. A new subterranean aquatic beetle from Texas (Coleoptera: Dytiscidae-Hydroporinae). *Ann. Entomol. Soc. Amer.* 69:787–792.
Zimmerman, J. R. 1970. *A taxonomic revision of the aquatic beetle genus Laccophilus (Dytiscidae) of North America.* Mem. Entomol. Soc. Amer. 26. 275 pp.

CHAPTER 14

Caddisflies

(ORDER TRICHOPTERA)

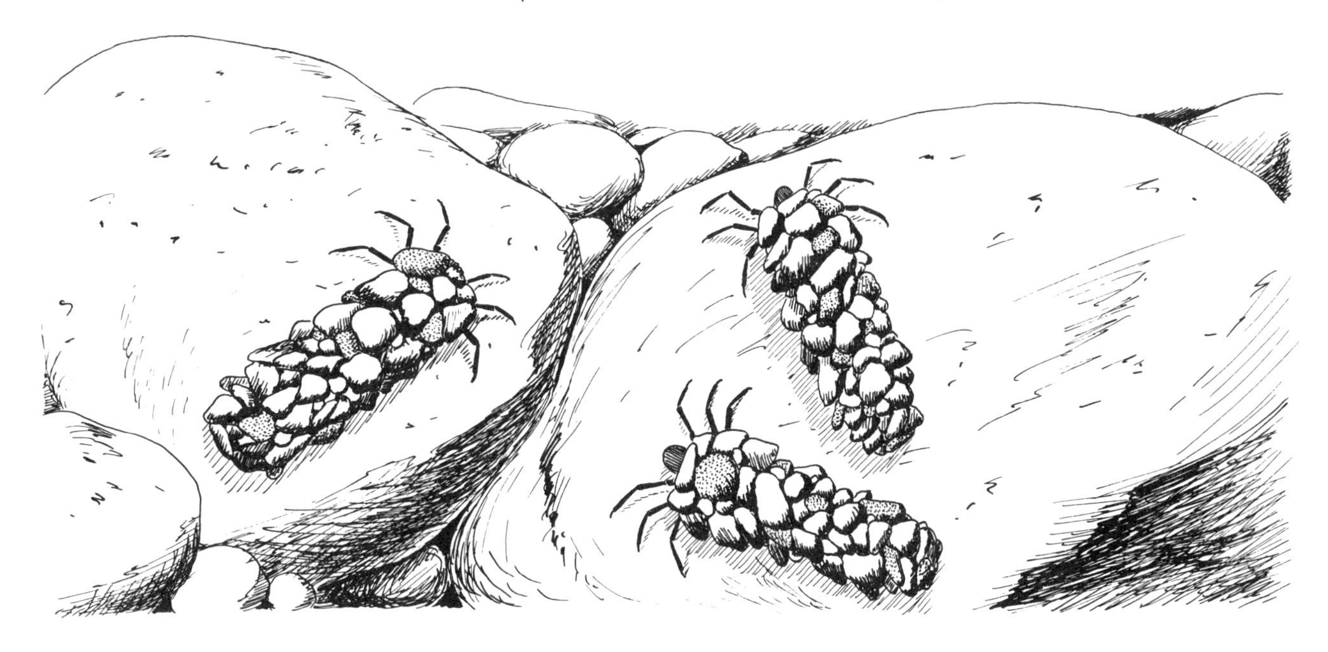

CADDISFLIES CONSTITUTE a highly advanced order of insects, closely related to the Lepidoptera (moths and butterflies), but adapted for aquatic life in the immature stages. Although adult caddisflies are essentially mothlike, the scientific name (from *trichos* meaning "hair," and *pteron* meaning "wing") attests to their characteristic wings, which usually possess hairs rather than the scales typical of moth wings. And although larval caddisflies are essentially caterpillarlike, the common name "caddis" was probably derived from an allusion to their general habit of constructing either nets or cases. The most studied and fascinating aspects of caddisfly biology have always involved the forms and behavior related to these intriguing cases or retreats in which most larvae live.

Caddisflies are a relatively large group of aquatic insects in North America, being represented by over 1,200 species. Adults of some species occur in large numbers, and many are strongly attracted to lights. These insects have been known by a number of colloquial names other than "caddisflies." The names "sedges" and "caddisworms" for the adults and larvae, respectively, are probably the most commonly encountered. Other terms such as "shadflies" and "sandflies" for adults and "periwinkles" for larvae are less frequently used, at least in North America.

Larvae are often a large and important component of benthic communities. They contribute significantly to the food chain of freshwater ecosystems and thus, directly or indirectly, provide food for fishes. In addition, caddisflies are one of the aquatic insect groups that provide fly fishermen with models for tied fly patterns. Although not nearly as important for fly fishing as the mayflies, a few caddisfly patterns, such as the Grannom, White Miller, Adams, and Grizzly King, are productively used on North American waters. Generally speaking, these flies, unlike some others, are most productive when fished wet, since they are much more apt to be taken by trout if they are perceived as pupae surfacing for emergence. The popular fishermen's fly known as the Michigan Caddis is not a caddisfly but actually a burrowing mayfly.

Larval Diagnosis

These are elongate, more-or-less caterpillarlike forms that measure 2–40 mm or more when mature. Head is distinctive, and chewing mouthparts are located at the end of the head capsule. Eyes are small and simple. Antennae are reduced and often inconspicuous. Thoracic legs are well developed. Wing pads are absent. End of abdomen possesses a pair of variously developed (often basally fused) anal prolegs. Each anal proleg terminates in an anal claw that is usually hook-shaped. No tails are present. Filamentous gills are sometimes present.

Pupal Diagnosis

Pupae are enclosed within a sealed cocoon except immediately prior to adult emergence. Appendages are not fused to body, but wings are held very tightly against body. Antennae generally are very long. Mandibles are usually stout and often cross each other.

Adult Diagnosis

These are mothlike insects with two pairs of hairy (more rarely scaled) wings that are held rooflike over the body when at rest. Head possesses relatively long, slender antennae and well-developed mouth palps. Wings are rarely reduced or absent in females. Tails are absent.

Similar Orders

As larvae, caddisflies might initially be confused with dobsonfly or alderfly larvae, some aquatic beetle larvae, or even some aquatic caterpillar larvae. Caddisflies never possess any elongate processes at the end of the abdomen, and the claws of both their thoracic legs and anal prolegs are single rather than double. Moreover, their prothorax is not greatly enlarged, as among dobsonflies and some predaceous beetle larvae. The aquatic pupae of caddisflies are commonly taken, but should be easily distinguished because they have relatively robust mandibles. Many are associated with a tubecase.

Adult caddisflies are distinguished from similar-looking moths, which may occasionally be taken near water, by their mouth, which never consists of a coiled siphoning tube, and generally by their very long antennae and elongate appearance when at rest with their wings folded. Mating swarms of caddisflies sometimes resemble those of mayflies or midges.

Life History

Metamorphosis is complete, including very different larval and adult stages and a transitional pupa. The life cycle typically includes five larval instars and one generation per year. Some species, however, have multiple

I wound myself in a white cocoon of singing,
All day long in the brook's uneven bed,
Measuring out my soul in a mucous thread;
Dimly now to the brook's green bottom clinging
Men behold me, a worm spun out and dead,
Walled in an iron house of silky singing.

Edna St. Vincent Millay

broods per year, which often overlap, and a few require more than a year for development. The larvae of some species undergo periods of non-development in damp substrate during the summer, when intermittent streams and ponds dry up or when leaf detritus is at a minimum, and then resume development as a response to shortening daylength in late summer or fall. This larval dormancy may also act to ensure synchronized adult emergence.

Pupation is almost always aquatic, and transformation of the last larval instar to the pupal stage takes place in a sealed cocoon (pupal case) that is usually some modification of the larval retreat. Pupal cases are generally fixed to some object. Development of adult structures during the pupal stage requires approximately three weeks in most species. The individual then cuts itself free, crawls from the water or swims to the surface, and the adult emerges from the pupal skin. It is at this time, when the caddisfly is relatively exposed, that it is most vulnerable to fish predation.

Adults generally fly very quickly from the water. This knowledge is of importance to fly fishermen because it means that dry or surface fly fishing with caddisfly imitations is usually not very productive. The sight of a caddisfly adult resting on the water surface is apparently one to which most trout would be unaccustomed. It is for this reason and because of the vulnerability of the pupae that caddisfly imitations should be fished beneath the surface with a rising motion. The downwing and tailless appearance of adult imitations seemingly works well for pupal imitations.

Adults may be either strong or weak fliers, and are usually agile runners. They generally live between one and two months, sometimes much more or less, and they are herbivorous liquid feeders. Although most species are nocturnal or crepuscular, many are diurnal.

Mating generally takes place on the ground or on vegetation. It is often preceded by swarming activity, and certain species rely on swarm markers (particularly shaped or positioned objects that orient their swarming activity). The adults of a few species that inhabit intermittent waters undergo extended periods of dormancy prior to mating and oviposition, ensuring that oviposition corresponds with wetter seasonal periods.

Oviposition behavior is varied, but females generally lay strands or masses of eggs in the water by dipping the abdomen or by crawling or diving into the water. The term "traveling sedge," which is common to many fishermen, actually refers to those ovipositing females that skim the water surface or fly up and down the stream, periodically descending to the water surface and briefly touching the surface with the end of their abdomen.

Eggs of a few species are deposited near the water's edge or above the water in vegetation. Dew, fog, or rain liquefies the gelatinous material within which the eggs are laid, allowing eggs or newly hatched larvae to

At eve when twilight shades prevail,
Try the hackle white and snail;

George Washington Bethune

drop or migrate into the water. Eggs of some species remain dormant during dry periods in intermittent streams.

Among the family Limnephilidae, one species that occurs in the northwestern states is secondarily adapted to a completely terrestrial life history. Another species from the northeastern states can be found in damp leaf litter near the intermittent pool environments of the young larvae.

Aquatic Habitats

Caddisflies are common bottom fauna in most freshwater lotic and lentic environments and are found in an amazing array of microhabitats. They occur in association with all substrate types, and some can live in seepage areas or, rarely, on wet terrestrial surfaces adjacent to water. Some commonly occur in association with the underwater parts of aquatic vegetation.

Aquatic Adaptations and Behavior

Caddisflies are hydropneustic, and a cutaneous type of respiration takes place over the soft surfaces of the body. Filamentous gills, when present, may be particularly important for aquatic respiration. Case-making species are able to increase the amount of dissolved oxygen passing over the body by undulating the abdomen within the case and increasing the stream of water over their bodies. Such ventilation may allow some species to inhabit oxygen-poor microhabitats, including deep lake environments.

Osmoregulation via the uptake of chloride ions from the environment takes place through specialized soft structures (papillae) at the end of the body or at special areas of abdominal segments 2–7 that are known as chloride epithelia.

Very generally, the netspinning caddisflies obtain a variety of suspended food materials by using fixed, silken nets; freeliving caddisflies are often highly predaceous; and caddisflies with portable cases utilize a variety of food resources, depending on the species. Interestingly, some northern case makers (Limnephilidae) are known to feed opportunistically on dead fishes that come to rest in the slower reaches of streams.

Silk is emitted from near the tip of the labium of the larvae. The construction of cases, retreats, nets, cocoons, and so forth is all dependent on the production and spinning of silk by caddisflies. The utilization of this silk is ultimately related to important adaptations involving feeding behavior, camouflage, shelter, respiration, microhabitat partitioning, and others.

The form and function of the cases of the case-making caddisflies (sometimes referred to as "rockrollers" or "stickbait" by fishermen) are extremely interesting. Portable cases are architecturally well adapted for specific habitats. For example, in swift currents one is more apt to find cases that are either streamlined, relatively resistant to crushing, equipped with ballast stones, or equipped with trailing twigs that act as rudders. Although trout will eat larva-filled cases, the cases and retreats in general are seemingly well adapted for camouflaging the caddisflies and protecting

Stretched over the brooks, in the midst of the frost-bound meadows, we may observe the submarine cottages of the caddis-worms, the larvae of the Plicipennes; their small cylindrical cases built around themselves, composed of flags, sticks, grass, and withered leaves, shells, and pebbles, in form and color like the wrecks which strew the bottom,—now drifting along over the pebbly bottom, now whirling in tiny eddies and dashing down steep falls, or sweeping rapidly along with the current, or else swaying to and fro at the end of some grass-blade or root. Anon they will leave their sunken habitations, and, crawling up the stems of plants, or to the surface, like gnats, as perfect insects henceforth, flutter over the surface of the water, or sacrifice their short lives in the flame of our candles at evening.

Henry David Thoreau

TABLE 14.1 COMMON LARVAL RETREATS OF CADDISFLY FAMILIES

		Type of Retreat	Philopotamidae	Psychomyiidae	Polycentropodidae	Hydropsychidae	Rhyacophilidae	Glossosomatidae	Hydroptilidae	Phryganeidae	Brachycentridae	Limnephilidae	Lepidostomatidae	Beraeidae	Sericostomatidae	Odontoceridae	Molannidae	Helicopsychidae	Calamoceratidae	Leptoceridae
With a portable case	Made entirely or mostly of plant material.	Excavated twig.										X							X	
		Four-sided "log cabin" type.									X									
		Four-sided, of blocklike pieces.											X							
		Flattened, of large pieces.										X							X	
		Cylindrical, wound with thin strips.									X									
		Somewhat irregular piles of twigs or bark.										X	X							X
		Cylindrical, with spiraling or ringed layers of uniform pieces.								X			X							X
		Other, often, with twigs extending posteriorly.								X		X								
	Incorporating small snail shells.											X								
	Made entirely or mainly of rock particles.	Helical snailcase.																X		
		Flanged hoodcase.															X			X
		Tortoiseshell saddlecase.						X												
		With twigs or needles extending anteriorly.																		X
		Small, flattened.							X											
		Stout, cylindrical, not gradually tapered or curved; commonly with many sizes of rock particles.							X			X								
		Cylindrical, gradually tapering and/or curved.								X	X	X	X	X	X	X				X
	Mostly silken and cylindrical.								X		X									X
	Silken, flattened, or bivalved, small, sometimes covered with particles or incorporating plant material.								X											
Without a portable case	Within a fixed retreat.	Within a silken net.	X		X															
		Within particle-covered tubes.		X																
		Within interstitial tubes.			X															
		Variable retreats with silken catchnets.			X	X														
	Freeliving.						X		X*											

*During only part of their lives.

them from would-be predators. It is in part for this reason that bottom fishing with larval patterns of caddisflies may not be as productive as bottom fishing with mayfly or stonefly larval imitations.

Most caddisflies are crawlers or somewhat sedentary, but some are active swimmers or burrowers. Larvae are relatively well known from the drift, and this fact leads to an exception to the above generalization about nymph fishing with caddisflies. Nymph fishing can be productive at night, especially shortly after complete darkness in order to take advantage of peak drift time. An examination of trout stomach contents occasionally reveals that larvae are being taken in abundance.

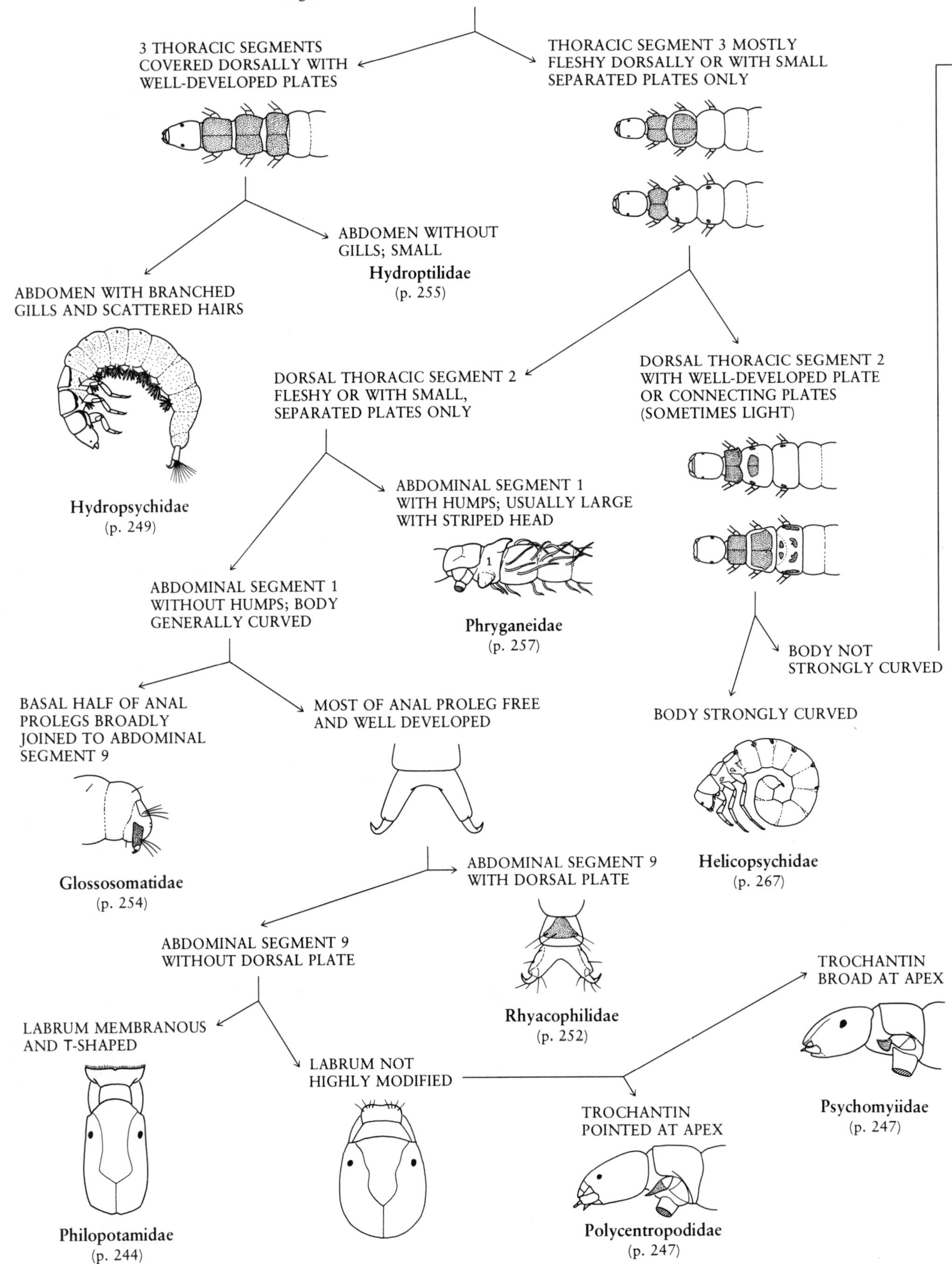
Figure 14.1. MATURE TRICHOPTERA LARVAE
3 THORACIC SEGMENTS COVERED DORSALLY WITH WELL-DEVELOPED PLATES
THORACIC SEGMENT 3 MOSTLY FLESHY DORSALLY OR WITH SMALL SEPARATED PLATES ONLY
ABDOMEN WITHOUT GILLS; SMALL
Hydroptilidae
(p. 255)
ABDOMEN WITH BRANCHED GILLS AND SCATTERED HAIRS
Hydropsychidae
(p. 249)
DORSAL THORACIC SEGMENT 2 FLESHY OR WITH SMALL, SEPARATED PLATES ONLY
DORSAL THORACIC SEGMENT 2 WITH WELL-DEVELOPED PLATE OR CONNECTING PLATES (SOMETIMES LIGHT)
ABDOMINAL SEGMENT 1 WITH HUMPS; USUALLY LARGE WITH STRIPED HEAD
1
Phryganeidae
(p. 257)
ABDOMINAL SEGMENT 1 WITHOUT HUMPS; BODY GENERALLY CURVED
BODY NOT STRONGLY CURVED
BODY STRONGLY CURVED
Helicopsychidae
(p. 267)
BASAL HALF OF ANAL PROLEGS BROADLY JOINED TO ABDOMINAL SEGMENT 9
Glossosomatidae
(p. 254)
MOST OF ANAL PROLEG FREE AND WELL DEVELOPED
ABDOMINAL SEGMENT 9 WITH DORSAL PLATE
Rhyacophilidae
(p. 252)
ABDOMINAL SEGMENT 9 WITHOUT DORSAL PLATE
LABRUM MEMBRANOUS AND T-SHAPED
Philopotamidae
(p. 244)
LABRUM NOT HIGHLY MODIFIED
TROCHANTIN BROAD AT APEX
Psychomyiidae
(p. 247)
TROCHANTIN POINTED AT APEX
Polycentropodidae
(p. 247)

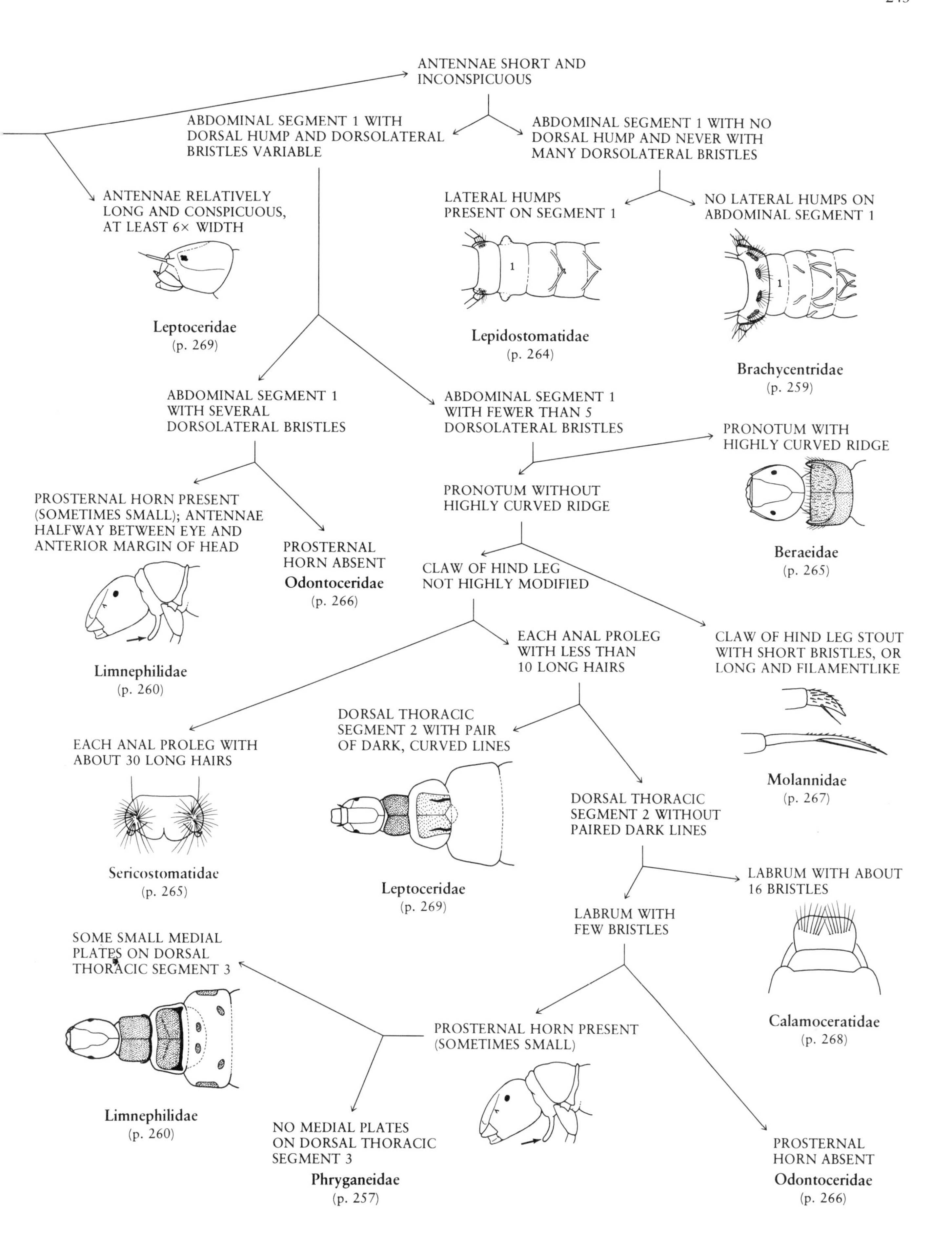

ANTENNAE SHORT AND INCONSPICUOUS
ABDOMINAL SEGMENT 1 WITH DORSAL HUMP AND DORSOLATERAL BRISTLES VARIABLE
ABDOMINAL SEGMENT 1 WITH NO DORSAL HUMP AND NEVER WITH MANY DORSOLATERAL BRISTLES
ANTENNAE RELATIVELY LONG AND CONSPICUOUS, AT LEAST 6× WIDTH
Leptoceridae
(p. 269)
LATERAL HUMPS PRESENT ON SEGMENT 1
1
Lepidostomatidae
(p. 264)
NO LATERAL HUMPS ON ABDOMINAL SEGMENT 1
1
Brachycentridae
(p. 259)
ABDOMINAL SEGMENT 1 WITH SEVERAL DORSOLATERAL BRISTLES
ABDOMINAL SEGMENT 1 WITH FEWER THAN 5 DORSOLATERAL BRISTLES
PRONOTUM WITH HIGHLY CURVED RIDGE
Beraeidae
(p. 265)
PRONOTUM WITHOUT HIGHLY CURVED RIDGE
PROSTERNAL HORN PRESENT (SOMETIMES SMALL); ANTENNAE HALFWAY BETWEEN EYE AND ANTERIOR MARGIN OF HEAD
Limnephilidae
(p. 260)
PROSTERNAL HORN ABSENT
Odontoceridae
(p. 266)
CLAW OF HIND LEG NOT HIGHLY MODIFIED
CLAW OF HIND LEG STOUT WITH SHORT BRISTLES, OR LONG AND FILAMENTLIKE
Molannidae
(p. 267)
EACH ANAL PROLEG WITH LESS THAN 10 LONG HAIRS
EACH ANAL PROLEG WITH ABOUT 30 LONG HAIRS
Sericostomatidae
(p. 265)
DORSAL THORACIC SEGMENT 2 WITH PAIR OF DARK, CURVED LINES
Leptoceridae
(p. 269)
DORSAL THORACIC SEGMENT 2 WITHOUT PAIRED DARK LINES
LABRUM WITH ABOUT 16 BRISTLES
Calamoceratidae
(p. 268)
LABRUM WITH FEW BRISTLES
SOME SMALL MEDIAL PLATES ON DORSAL THORACIC SEGMENT 3
Limnephilidae
(p. 260)
PROSTERNAL HORN PRESENT (SOMETIMES SMALL)
NO MEDIAL PLATES ON DORSAL THORACIC SEGMENT 3
Phryganeidae
(p. 257)
PROSTERNAL HORN ABSENT
Odontoceridae
(p. 266)

Classification and Characters

Five natural groups of caddisflies are placed in three superfamilies. In North America, these consist of (1) the netspinning or fixed-retreat forms, including four families of the superfamily Hydropsychoidea; (2) the freeliving forms, saddlecase makers, and pursecase makers, each consisting of one family in the superfamily Rhyacophiloidea; and (3) the tubecase makers, including 11 families of the superfamily Limnephiloidea.

Characteristics used in identifying adults to family include the location of warts on the dorsal surface of the thorax, the location of spurs and spines on the legs, and the segmentation of the maxillary palps. **Warts** are bumps that usually possess more hair than surrounding areas. The **scutum** is the relatively large area of the thorax posterior to the head and small pronotum; it is shortest in the middle and in some species is longitudinally divided by a suture or furrow; and it continues posteriorly at the sides where it borders the smaller, medial **scutellum. Spines,** as the term is used here, are small, sharp, dark, bristlelike structures on the legs. **Spurs** are much larger and may appear flatter. Two pairs of palps are usually evident on the head. The maxillary palps are generally the longest and can be assumed to have five segments unless otherwise stated.

Pairs of minute, hardened patches occur dorsally on some of the abdominal segments of the pupae. The distribution of these so-called **hook plates** and the number of tiny hooks present on each are important in identifying pupae to the correct family.

Among larvae, dorsal surfaces (nota) of the three thoracic segments and abdominal segment 9 may have areas that are hardened (or sclerotized), and these sclerotized areas are referred to as **plates.** Although plates are usually conspicuously pigmented, they may be light and difficult to detect. **Setal areas** of the thoracic nota (abbreviated *sa*1, *sa*2, *sa*3) are labeled on drawings when it is necessary to locate them for identification purposes. The **trochantin** is a small sclerotized plate at the side of the prothorax just above and slightly to the front of the base of the fore legs. The labrum, the "upper lip," articulates with the head.

When case-making larvae are collected, especially by kicking or disrupting the substrate, they are sometimes knocked free of their cases. Some also actively evacuate their cases when disturbed. The absence of an associated case in collected material does not automatically indicate that a larva is freeliving or netspinning. Although very useful, Table 14.1 is best used only to verify family identifications after a careful examination of larval morphology. Case types are most consistent at the generic level.

Netspinning Caddisflies
(Superfamily Hydropsychoidea)

Larvae either have well-developed, sclerotized dorsal plates on all three thoracic segments (Fig. 14.14) or have a well-developed plate only on the pronotum (Figs. 14.3, 14.8, 14.10). Anal prolegs are free and well developed and do not form an apparent abdominal segment 10. These are netspinning forms associated with a fixed retreat.

FINGERNET CADDISFLIES
(Family Philopotamidae)

LARVAL DIAGNOSIS: (Figs. 14.3–14.5) These are usually 10–12 mm when mature, but some range up to 16.5 mm. Head is elongate. Labrum is

FINGERNET CADDISFLY
(Philopotamidae)

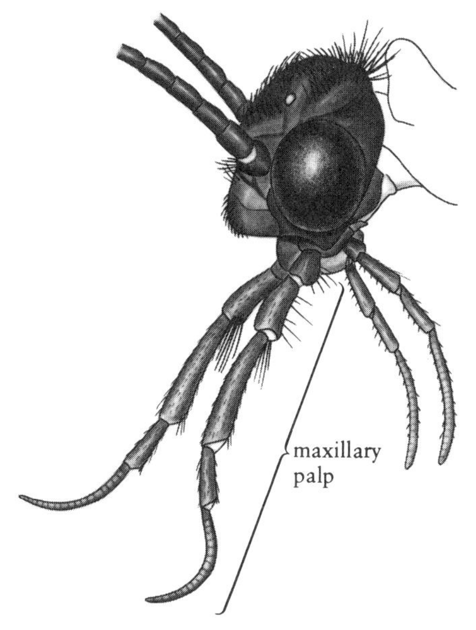

Figure 14.2. Adult head

membranous and laterally expanded along its distal margin. Pronotum alone possesses a dorsal plate. Abdomen tends to have a strong ventral curl and does not possess gills.

PUPAL DIAGNOSIS: (Figs. 14.6, 14.7) Abdomen lacks lateral fringe of hairs and abdominal gills; one pair of hook plates is located on abdominal segment 4. Mandibles have inner teeth.

ADULT DIAGNOSIS: (Fig. 14.2) These are generally small and dark forms. Ocelli are present. Some females are wingless. Maxillary palps are five-segmented, and segment 5 is twice or more the length of segment 4.

DISCUSSION: The larvae of fingernet caddisflies are most easily confused with the generally similar nettube, trumpetnet, and tubemaking caddisflies. The usually more elongate head is typically more deflexed in fingernet caddisflies, and their distally expanded, membranous labrum is unique. Sometimes, however, this mouthpart is retracted into the head and inconspicuous.

Three genera and about 40 species are known in North America.

Larvae occur only in riffle areas of streams. Their open-ended, elongate nets are attached at the upstream end and are distended by the current. These fingernets have the smallest mesh size known among netspinning caddisflies. The food of the larvae consists of both plant and animal matter. *Dolophilodes* emerge at all times of the year in the eastern half of the continent; adult females are sometimes wingless.

Chimarra larvae (Fig. 14.3) sometimes appear yellow or orange, and tied fly imitations patterned after them include the Yellow Caddis Worm and Orange Caddis Worm. These are best used for night nymph fishing. *Chimarra atterrima* is the Little Black Caddis. This small caddisfly is abundant on small streams throughout the eastern states, and when adults are emerging, the imitation can be valuable. The Pale Microcaddis Pupa is patterned after *Wormaldia gabriella,* a small western fingernet caddisfly.

FINGERNET CADDISFLIES
(Philopotamidae)

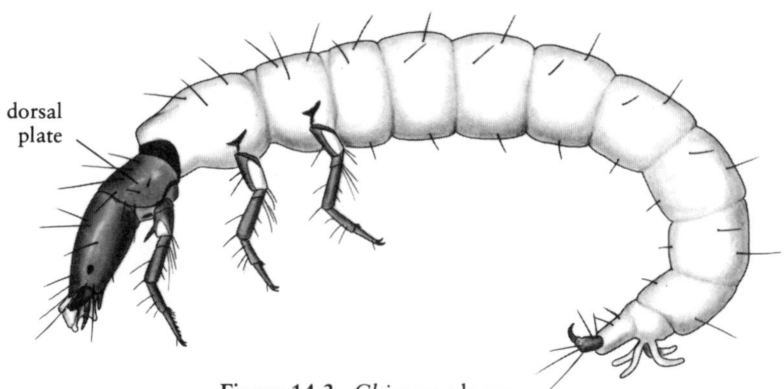

Figure 14.3. *Chimarra* larva

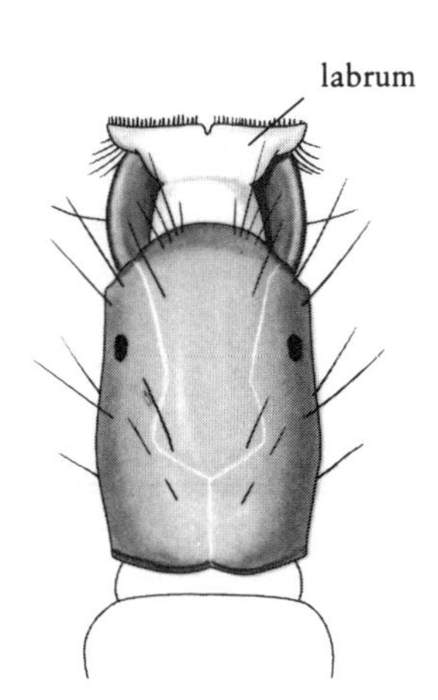

Figure 14.4. Philopotaminae larva, head

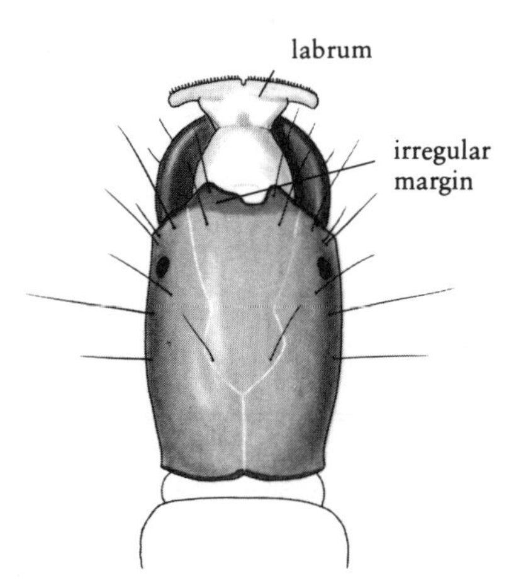

Figure 14.5. Chimarrinae larva, head

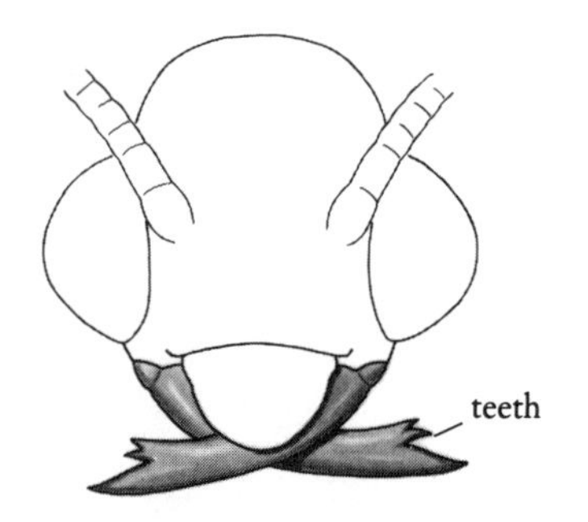

Figure 14.6. Pupal head (mandibles shaded)

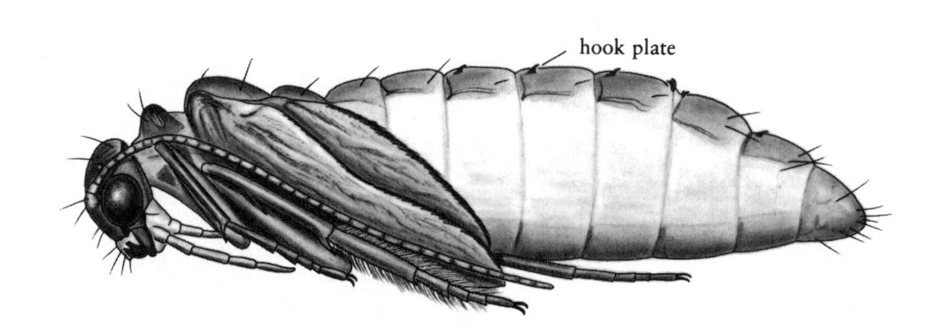

Figure 14.7. Pupa

SUBFAMILY PHILOPOTAMINAE: **Larvae** (Fig. 14.4) have the anterior margin of the head symmetrical or only slightly asymmetrical as seen from above. This group is generally widespread in North America.

SUBFAMILY CHIMARRINAE: **Larvae** (Figs. 14.3, 14.5) have the anterior margin of the head capsule irregularly sculptured with a distinct notch as seen from above. This is a widespread group in North America, and although it can be very common in western states, it is not yet known from western Canada.

NETTUBE CADDISFLY
(Psychomyiidae)

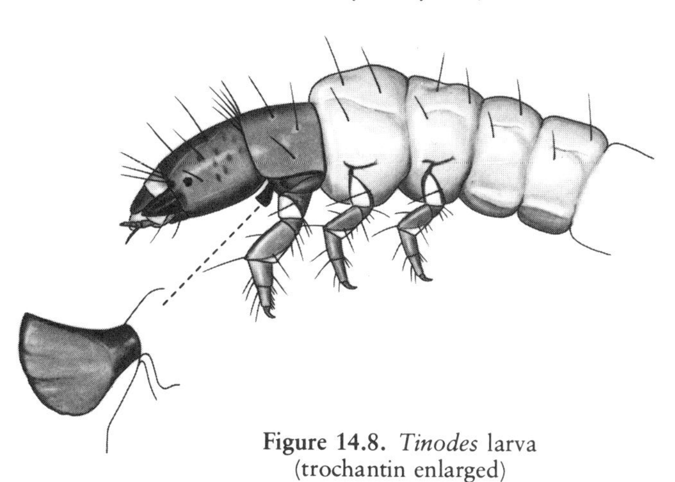

Figure 14.8. *Tinodes* larva
(trochantin enlarged)

NETTUBE CADDISFLIES
(Family Psychomyiidae)

LARVAL DIAGNOSIS: (Fig. 14.8) These are generally similar to the fingernet caddisflies, but labrum is not modified and prothorax possesses a unique, broad and flange-shaped trochantin. Pronotum is not bordered posteriorly by a black line.

PUPAL DIAGNOSIS: Mandibles lack inner teeth. Abdomen lacks a lateral fringe of hairs; two pairs of hook plates are located only on abdominal segment 5 (each of the posterior pair of hook plates possesses six or more hooks); hook plates are present on abdominal segments 2–7 or segments 2–8.

ADULT DIAGNOSIS: Ocelli are absent. Maxillary palps are five-segmented; segment 5 is striated with a series of crosslines and is usually twice or more the length of segment 4; and segments 2 and 3 are usually subequal in length. Scutum possesses warts. Tibia of fore legs lacks a preapical spur.

DISCUSSION: Five genera and about 15 species are known from North America, where larvae occur in streams and construct tubelike, sometimes branching retreats on hard substrate surfaces or in crevices. The silken roofs of the retreats are usually camouflaged with bits of detritus or rock fragments.

SUBFAMILY PSYCHOMYIINAE: **Larvae** (Fig. 14.8) have distinct tibiae and tarsi that are not fused together. The group is widespread in North America.

SUBFAMILY PADUNIELLINAE: Only the adults of this group are known (exclusively from Arkansas).

SUBFAMILY XIPHOCENTRONINAE: **Larvae** possess tibiae and tarsi that are fused together. This group is rare and known only from Texas.

TRUMPETNET AND TUBEMAKING CADDISFLIES
(Family Polycentropodidae)

LARVAL DIAGNOSIS: (Figs. 14.10–14.13; Plate XII, Fig. 77) Larvae are generally similar to fingernet and nettube caddisflies but are often larger,

TRUMPETNET AND TUBEMAKING CADDISFLIES
(Polycentropodidae)

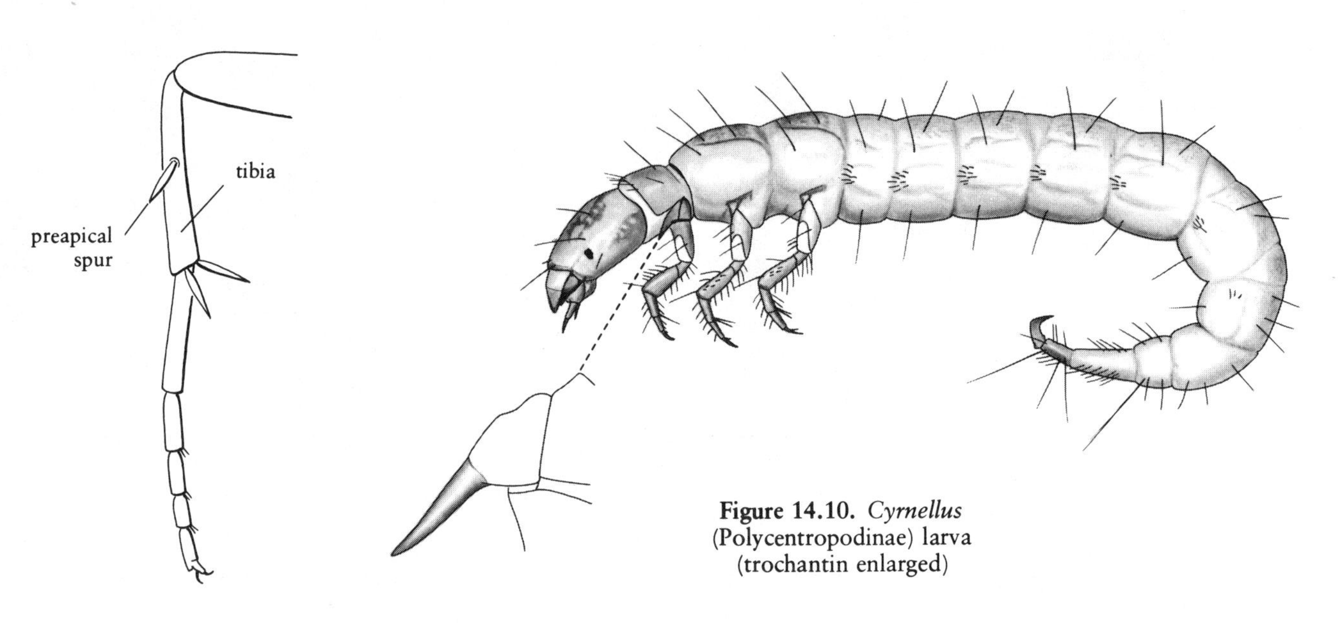

Figure 14.9. *Polycentropus* adult, fore leg

Figure 14.10. *Cyrnellus* (Polycentropodinae) larva (trochantin enlarged)

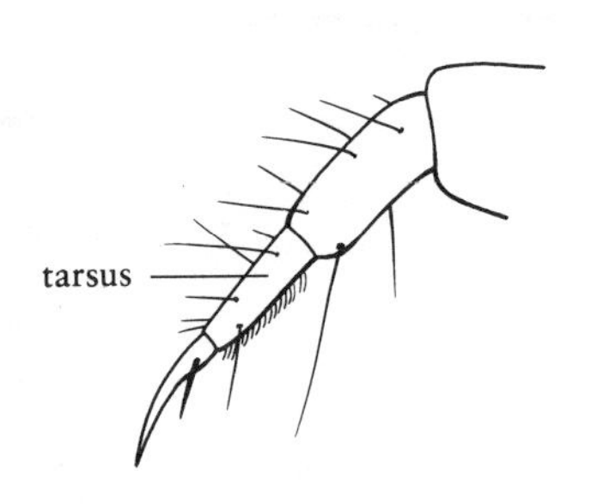

Figure 14.11. Polycentropodinae larva, leg

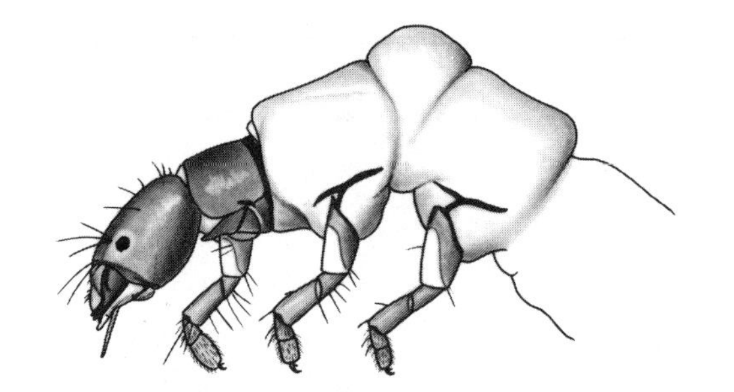

Figure 14.12. *Phylocentropus* (Dipseudopsinae) larva

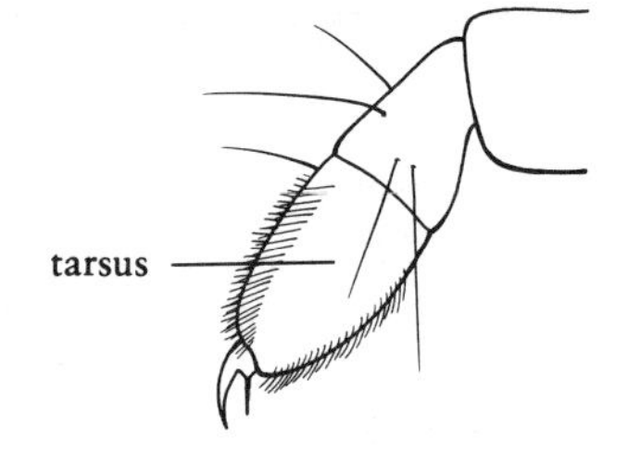

Figure 14.13. Dipseudopsinae larva, leg

up to 25 mm. Labrum is not modified. Trochantin is pointed. Abdomen usually has a lateral fringe of short hairs but never possesses gills.

PUPAL DIAGNOSIS: These are similar to nettube caddisfly pupae but lack hook plates on abdominal segment 2.

ADULT DIAGNOSIS: (Fig. 14.9) Ocelli are absent. Segments 4 and 5 of maxillary palps are similar to those of other netspinners. Scutum possesses warts. Tibia of fore legs usually has a preapical spur. Those few without a preapical spur are distinguished from nettube caddisflies by the relative lengths of segments 2 and 3 of maxillary palps; segment 2 is shorter than segment 3.

DISCUSSION: Larvae can be distinguished from nettube caddisflies by the shape of the trochantin (compare Figs. 14.8 and 14.10). The genus *Polycentropus,* with its approximately 40 species, along with about six smaller genera are known from North America.

Larvae occur in a variety of lentic environments and lotic environments that usually have a moderate or slight current. A few species are lake-bottom dwellers. Larval retreats consist of trumpetnets, which have a large opening and then taper to the end, where a larva resides; long branching tubes, mostly within soft substrate; net-covered depressions; or short, cylindrical nets open equally at both ends. Larvae that construct the last type of retreat are predaceous and can detect their prey by vibrations of the silken tube caused by contact of prey organisms.

SUBFAMILY POLYCENTROPODINAE: **Larvae** (Figs. 14.10, 14.11; Plate XII, Fig. 77) have tarsi that are not broad and flat. The group is widespread in North America.

SUBFAMILY DIPSEUDOPSINAE: **Larvae** (Figs. 14.12, 14.13) have broad and flat tarsi. This group is restricted to eastern and central Canada and central and eastern states excluding Florida.

COMMON NETSPINNERS
(Family Hydropsychidae)

LARVAL DIAGNOSIS: (Figs. 14.14, 14.20–14.25; Plate XII, Fig. 78) These have a generally strongly curved body that usually measures 10–16 mm when mature, sometimes up to 30 mm. All three thoracic segments have well-developed dorsal plates. Branched gills are present on the venter of the second and third thoracic segments and at least abdominal segments 1–6. Abdomen is usually covered with small hairs. Anal prolegs usually have a tuft of long hairs.

PUPAL DIAGNOSIS: (Fig. 14.17; Plate XII, Fig. 79) These are similar to fingernet caddisfly pupae, but abdominal gills are present. Two pairs of hook plates are present on abdominal segments 3 and 4.

ADULT DIAGNOSIS: (Figs. 14.16, 14.18, 14.19; Plate XII, Figs. 80, 81) These are generally similar to other netspinning caddisflies, having a typically long and striated segment 5 of the maxillary palps. They are sometimes larger than many other netspinner adults, and some have brownish, mottled wings. Ocelli, warts of scutum, and preapical spurs of fore legs are all absent.

DISCUSSION: The larvae of this group are unique and readily identifiable in the field. When collected, they typically retreat backwards with great agility by using their abdomen and well-developed anal prolegs. This large, important group includes the most commonly and consistently taken caddisflies in North America, where about 150 species in about 13 genera and four subfamilies are presently recognized.

Common netspinners are often abundant in streams and rivers and sometimes along shorelines of lakes with considerable wave action. They construct a catchnet (Fig. 14.15) in front of or adjacent to various fixed retreats. Their retreats are located on rocks (often algae covered) and woody debris or within crevices or depressions of the same objects. They are usually camouflaged by pieces of detritus or rock.

Larvae are omnivore-detritivores that, during at least part of their lives, obtain suspended food (organic fraction of the seston) caught on the strands of the catchnet. The proportion of animal versus plant material in the diet of common netspinner larvae usually changes with age and season. The ecology of these caddisflies as it relates to suspension feeding is very interesting. Differences in catchnet area, strand size, and mesh size among different species may be related to their functional adaptation to preferred current speeds and amount of suspended resources, to some preference for

COMMON NETSPINNERS
(Hydropsychidae)

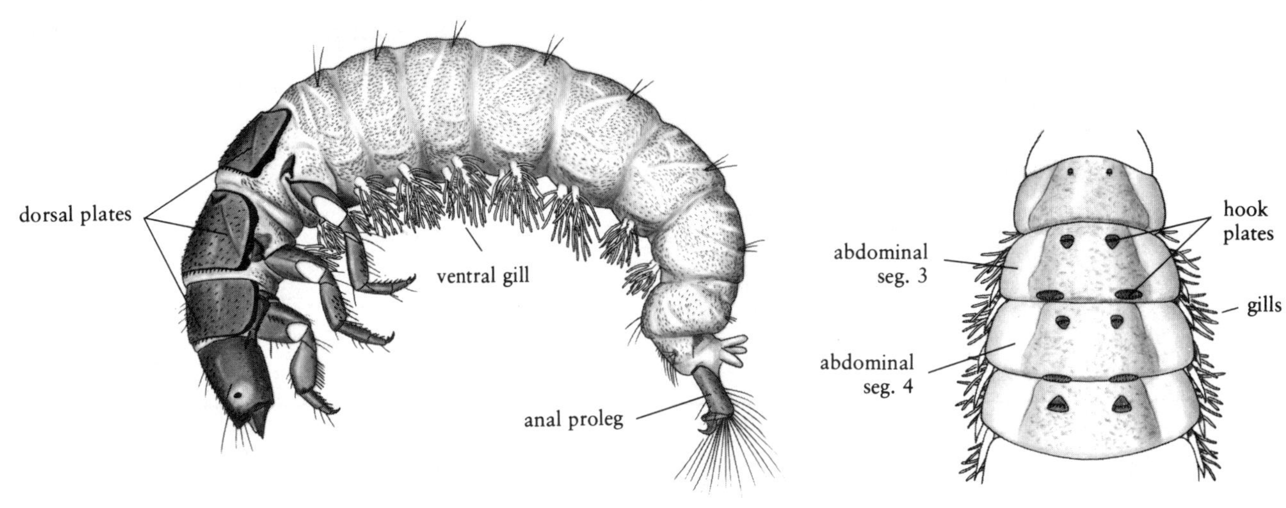

Figure 14.14. *Hydropsyche* larva

Figure 14.17. Pupal abdomen (dorsal)

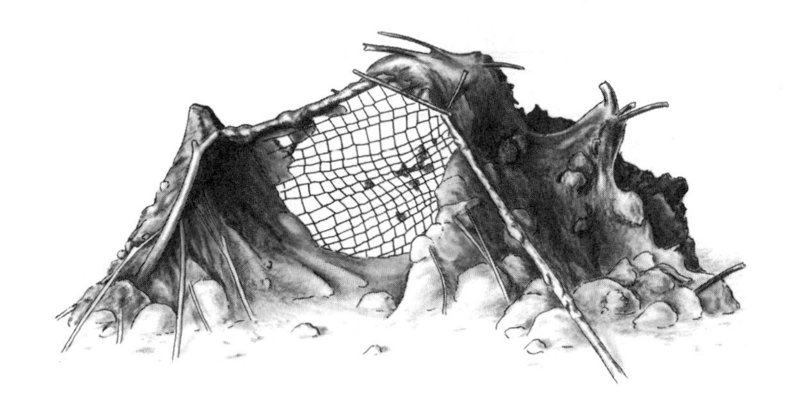

Figure 14.15. *Arctopsyche* catchnet

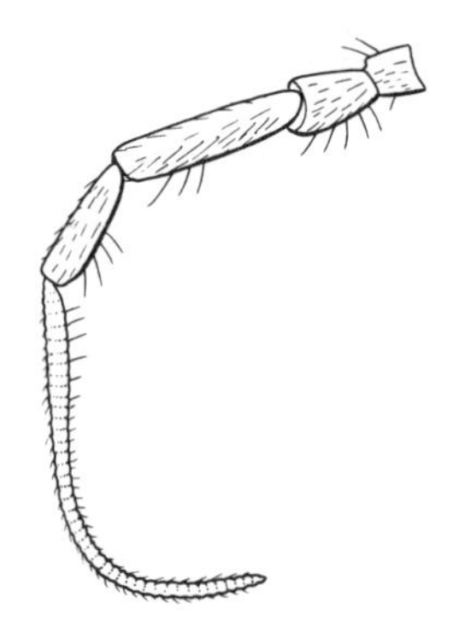

Figure 14.18. Adult maxillary palp

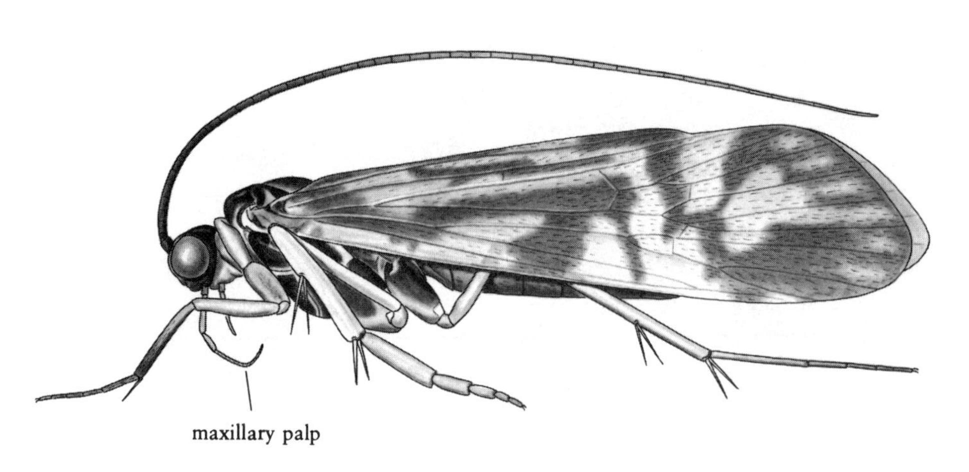

Figure 14.16. *Macronema* adult

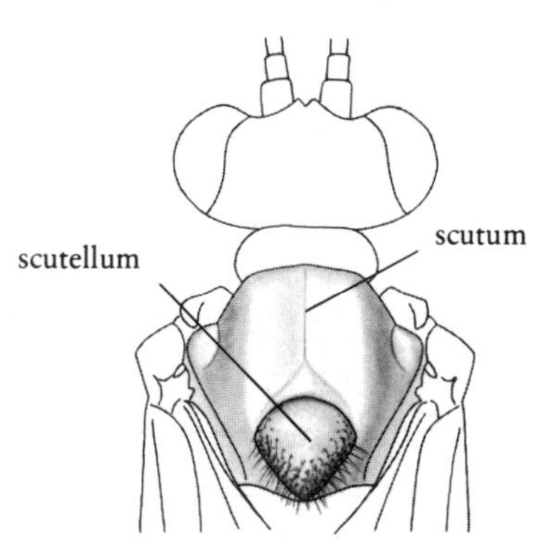

Figure 14.19. *Hydropsyche* adult,
head and thorax

COMMON NETSPINNERS
(Hydropsychidae)

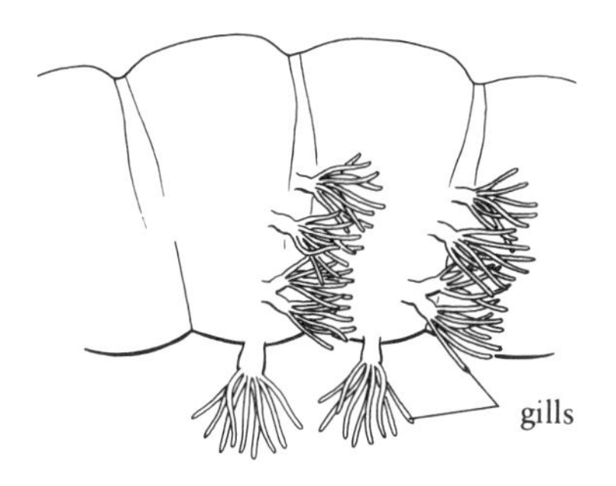

Figure 14.20. Arctopsychinae larva, abdomen in part (lateral)

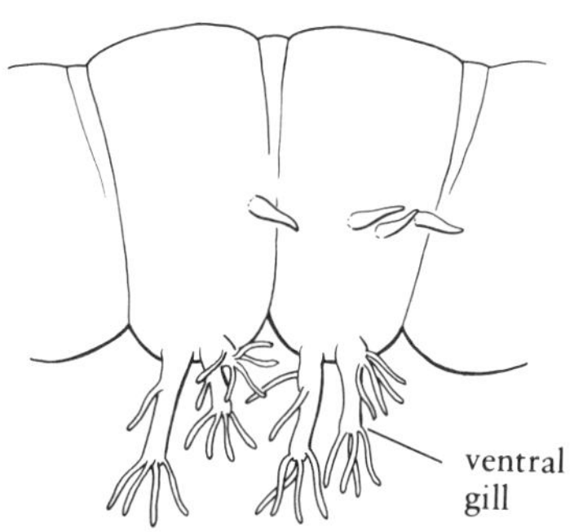

Figure 14.21. Diplectroninae larva, abdomen in part (lateral)

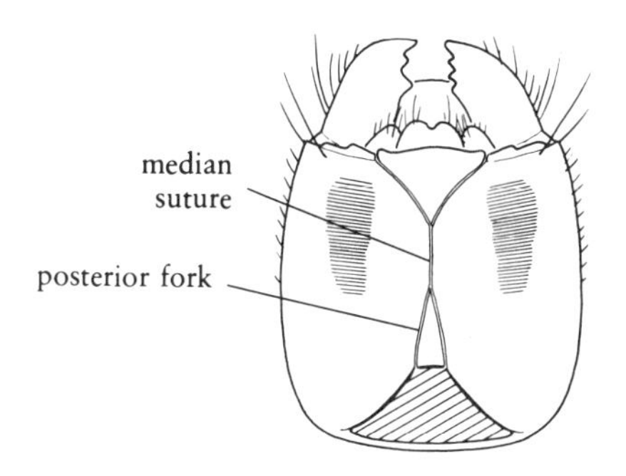

Figure 14.24. Diplectroninae larva, head (ventral)

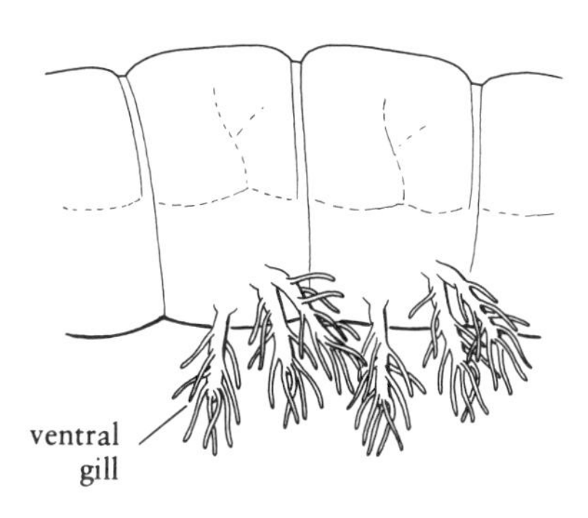

Figure 14.22. Hydropsychinae larva, abdomen in part (lateral)

Figure 14.23. Macronematinae larva, abdomen in part (lateral)

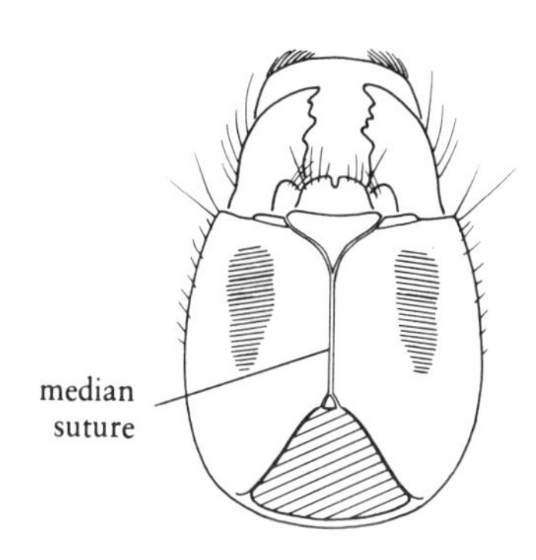

Figure 14.25. Hydropsychinae larva, head (ventral)

food particle size, or to both. Early instars of some species do not construct catchnets at all, and some other species do not construct catchnets during the winter in colder areas.

When larvae become dislodged from their retreats, they may search for new retreats by crawling upstream or by drifting downstream. Oftentimes they will attempt to enter the retreat of another netspinning larva (either the same or different species). A fight usually ensues between the intruder and the defender of the retreat. After an initial encounter of this sort, and if aggression continues, the individual in the retreat will begin to stridulate. Stridulation is carried out by clinging tightly to the net with the fore legs and then moving the head back and forth against these legs. Special areas on the ventrolateral margins of the head (files) are rubbed against roughened areas of the femora (scrapers). The sound or more probably just the vibration caused by such stridulation usually has the effect of warding off the intruder, and thus greatly increases the success of the defender.

Common netspinners are the most abundant caddisflies (especially *Hydropsyche, Symphitopsyche,* and *Cheumatopsyche*) in many parts of North America. The rivers of the Midwest yield tremendous numbers of adults (e.g., of *Potamyia;* Plate XII, Fig. 81) that are often attracted to lights and sometimes become a nuisance because of their numbers.

Although many Hydropsychidae species are typical of warmer waters, which do not support trout, several also occur in trout waters, and three tied fly patterns based on common netspinners are worthy of mention. These are the Spotted Sedge, based on *Symphitopsyche slossanae* (previously classified in the genus *Hydropsyche*) (Plate XII, Fig. 80) of the northeastern states and central states and Canada; the Little Western Sedge, based on *Cheumatopsyche gracilis* of western North America; and the Small Spotted Sedge, based on *Hydropsyche alternans*. The last named fly can be especially good on eastern Canadian trout streams.

SUBFAMILY ARCTOPSYCHINAE: **Larvae** (Fig. 14.20) have ventral and lateral gills that are branched only at the apex of their central stalk. The group is generally widespread except for Texas and Florida and is sometimes rare or absent in some central states.

SUBFAMILY DIPLECTRONINAE: **Larvae** (Figs. 14.21, 14.24) have the ventral median suture of the head forked posteriorly; posterior forks are at least half as long as the undivided part of the suture. Ventral gills are branched at the apex and very sparse along their central stalk; lateral gills are simple or absent. The group is widespread in North America.

SUBFAMILY HYDROPSYCHINAE: **Larvae** (Figs. 14.14, 14.22, 14.25; Plate XII, Fig. 78) have an unforked ventral median suture on the head or one that is forked posteriorly, but the posterior forks are less than half the length of the undivided part of the suture. Ventral gills have numerous scattered lateral branches as well as branches at the apex; lateral gills are simple or absent. The group is widespread in North America.

SUBFAMILY MACRONEMATINAE: **Larvae** (Fig. 14.23) have ventral and lateral gills that are uniformly branched and featherlike. The group is *not* known from western Canada, Alaska, or northwestern states; it is rare in southwestern states.

Superfamily Rhyacophiloidea

Larvae have large, well-developed dorsal plates on all thoracic segments (Figs. 14.33, 14.36, 14.38) or only on the pronotum (Figs. 14.30–14.32). Anal prolegs are either free (Fig. 14.26) or variously fused (Figs. 14.29, 14.33, 14.36). Some of these forms are totally freeliving, some have small tortoiseshell portable cases, and others have tiny, usually flattened and delicate portable cases.

FREELIVING CADDISFLIES
(Family Rhyacophilidae)

LARVAL DIAGNOSIS: (Fig. 14.26; Plate XII, Fig. 82) These usually measure 11–18 mm when mature, rarely up to 32 mm. Pronotum alone possesses a plate. Abdomen usually has deep constrictions between segments as

To furnish the small animal, provide
All the Gay hues that wait on female pride;
Let nature guide thee—sometimes golden wire
The shining bellies of the fly require;
The peacock's plumes thy tackle must not fail,
Nor the dear purchase of the sable's tail.
Each gaudy bird some slender tribute brings
And lends the glowing insect proper wings.
Silks of all colors must their aid impart,
And every fur promotes the fisher's art.

John Gay

FREELIVING CADDISFLIES
(Rhyacophilidae)

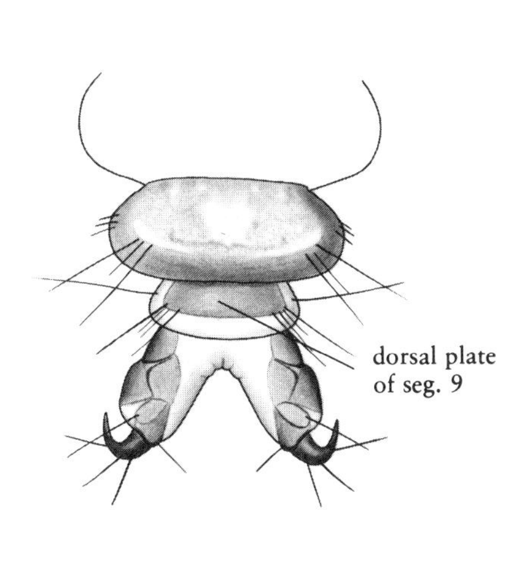

Figure 14.26. *Rhyacophila* larva, end of abdomen (dorsal)

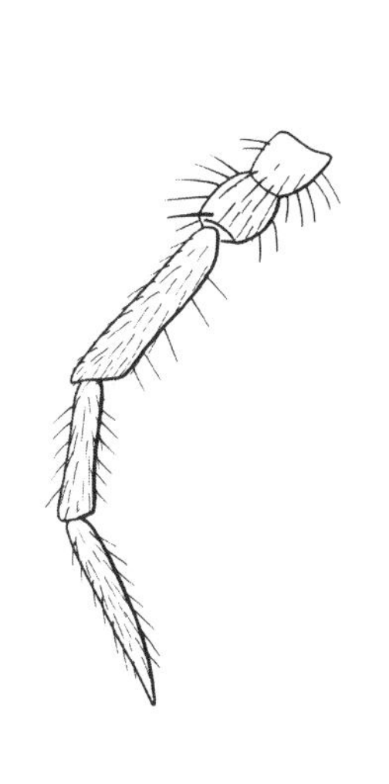

Figure 14.27. Adult maxillary palp

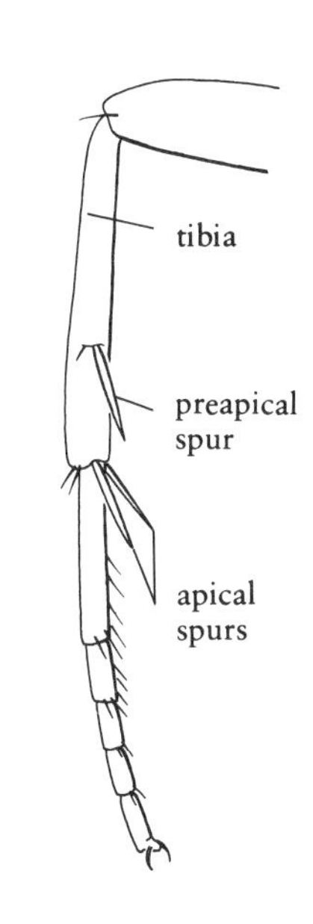

Figure 14.28. Adult fore leg

seen from above. Gills are absent or present in small clusters of simple filaments. Abdominal segment 9 possesses a dorsal plate. Anal prolegs are well developed and free for most of their length.

PUPAL DIAGNOSIS: Mandibles have inner teeth. Abdomen lacks lateral fringe of hairs and abdominal gills. One pair of hook plates is present on abdominal segment 3, two pairs are present on segment 4, but hook plates are absent on segments 8 and 9. No terminal bristles are present on abdomen.

ADULT DIAGNOSIS: (Figs. 4.27, 4.28) Ocelli are present. Segment 5 of maxillary palps is not much longer than segment 4, and segments 1 and 2 are short and subequal in length. Tibia of fore legs has apical and preapical spurs.

DISCUSSION: The plate of abdominal segment 9 in combination with the free and well-developed anal prolegs will distinguish this group. The medium to large size range and generally robust nature of freeliving caddisflies may also be helpful. Their abdomens may curl when they are preserved in fluid, as is also normal for the netspinning caddisflies.

Most of the 100 or more species in North America belong to the widespread genus *Rhyacophila.*

Larvae are known primarily from cool, clean streams. Each species tends to have a small geographic range. Most larvae are active predators, and some species have distinct prey preferences. Adults are generally secretive and do not fly far from the place of emergence.

Freeliving caddisflies are generally referred to as "green caddisflies" by fly fishermen because of the predominant color of the larvae. Larval or adult imitations fished submerged are of great value on certain streams. Being freeliving, the larvae are not as well protected from fish predation as are most caddisfly larvae. Fishes may actively seek them out, even though they may not be as numerous as some other caddisflies. Fly patterns include the Green Caddis (*Rhyacophila lobifera*) of central North America

(Plate XII, Fig. 82), the Western Olive Sedge (*Rhyacophila grandis*) of western North America, and the Little Olive Sedge (*Rhyacophila basalis*) of Pacific Coast streams.

SUBFAMILY RHYACOPHILINAE: **Larvae** (Fig. 14.26; Plate XII, Fig. 82) have generalized fore legs that are never pincerlike. The group is widespread in North America.

SUBFAMILY HYDROBIOSINAE: **Larvae** have fore legs that are modified into a chelate (pincerlike) form. The group is restricted to southwestern states.

SADDLECASE MAKERS
(Family Glossosomatidae)

LARVAL DIAGNOSIS: (Figs. 14.29–14.32; Plate XII, Fig. 83) When mature these usually measure 3–6.5 mm, occasionally up to 9 mm or more. Pronotum alone possesses a well-developed plate. Abdomen lacks gills. Abdominal segment 9 has a dorsal plate. Anal prolegs are fused for approximately half of their length.

PUPAL DIAGNOSIS: These are similar to freeliving caddisfly pupae, but hook plates may be present on abdominal segments 8 and/or 9. Terminal bristles may be present.

SADDLECASE MAKERS
(Glossosomatidae)

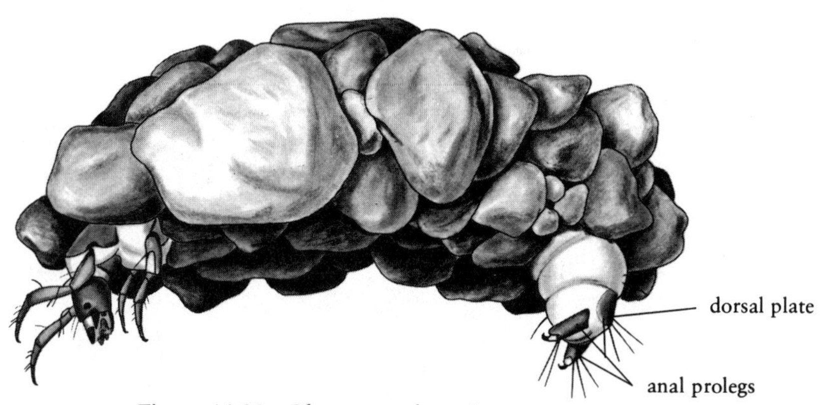

Figure 14.29. *Glossosoma* larva in case

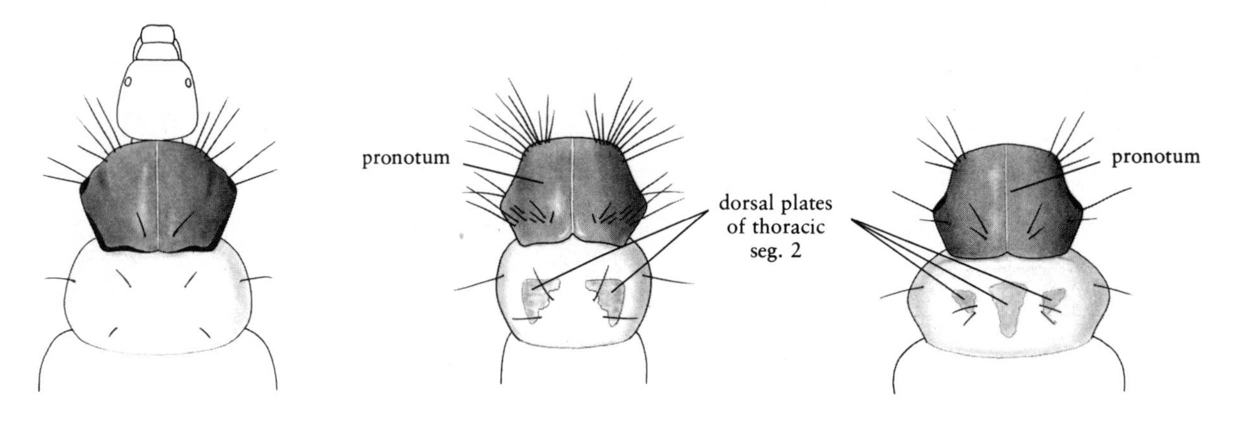

Figure 14.30. Glossosomatinae larva

Figure 14.31. Agapetinae larva

Figure 14.32. Protoptilinae larva

ADULT DIAGNOSIS: These are similar to freeliving caddisfly adults but are generally smaller, and the tibia of the fore legs lacks either apical or preapical spurs or both.

DISCUSSION: Saddlecase makers are intermediate in characteristics between freeliving caddisflies and other case makers, and can be distinguished from both by the partially free anal prolegs, singular notal plate, and their unique portable saddlecase. About 80 species in five genera occur in North America.

The saddlecase resembles a tortoiseshell made of heterogeneous rock material, oval at the top and flat at the bottom (Fig. 14.29). It is open ventrally at both ends, from which both ends of the larva protrude. The anal prolegs are thus free to help anchor the body and case. Saddlecase makers abandon their cases readily, especially under stress conditions, and rebuild them later. Unlike the tubecase makers, they remake their cases after each larval molt. Larvae occur primarily in cool streams with considerable current, where they feed on periphyton and fine detritus from the substrate.

SUBFAMILY GLOSSOSOMATINAE: **Larvae** (Figs. 14.29, 14.30; Plate XII, Fig. 83) have no plates on the second thoracic segment. This group is locally common in the eastern half of the continent and rare in the West.

SUBFAMILY AGAPETINAE: **Larvae** (Fig. 14.31) have two small plates on the second thoracic segment. The group is generally found in mountainous or hilly regions and is rare in the central area of the continent.

SUBFAMILY PROTOPTILINAE: **Larvae** (Fig. 14.32) have three small dorsal plates on the second thoracic segment. This group is widespread in North America.

MICRO CADDISFLIES
(Family Hydroptilidae)

LARVAL DIAGNOSIS: (Figs. 14.33–14.39; Plate XII, Fig. 84) These are very small forms, usually 1–4 mm and never more than 6.5 mm. Sometimes at least middle segments of the abdomen are gradually or abruptly expanded. All thoracic segments have well-developed plates. Gills are generally absent. Anal prolegs are either free or fused for most of their length.

PUPAL DIAGNOSIS: Abdomen lacks a lateral fringe of hairs. Mandibles lack inner teeth. Body is less than 5 mm.

ADULT DIAGNOSIS: (Fig. 14.40; Plate XII, Fig. 85) These are very small and hairy forms, never measuring over 6 mm. Antennae are shorter than fore wings. Wings are narrow and have long fringes of hairs.

DISCUSSION: The small size, the plates of the thorax, and the sometimes enlarged abdomen make the larvae of micro caddisflies difficult to confuse with those of other groups. The only other mature caddisfly larvae that are ever as small include some saddlecase makers and nettube caddisflies.

This is a relatively diverse group with about 180 species presently recognized in North America. Individuals are often overlooked because of their small size.

Larvae are freeliving for part of their lives, reserving the case-making activity until the final larval instar. It is only during this later part of development that the abdomen becomes conspicuously enlarged. Cases vary somewhat but are most commonly purselike, such as those of the genera *Hydroptila* and *Ochrotrichia*. These cases are laterally flattened, bivalved

MICRO CADDISFLIES
(Hydroptilidae)

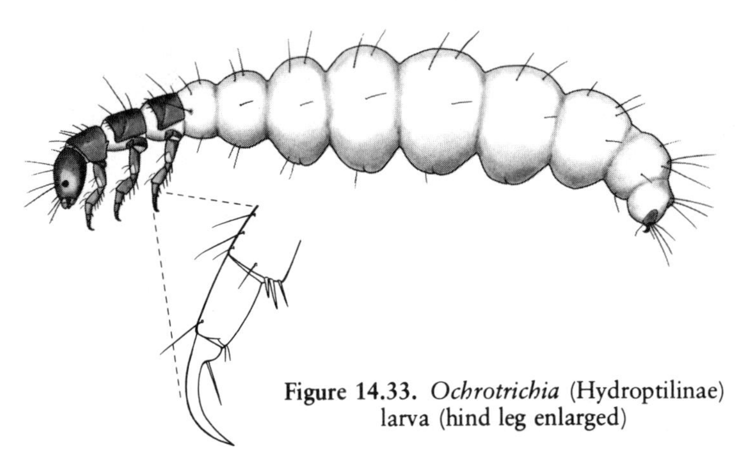

Figure 14.33. *Ochrotrichia* (Hydroptilinae) larva (hind leg enlarged)

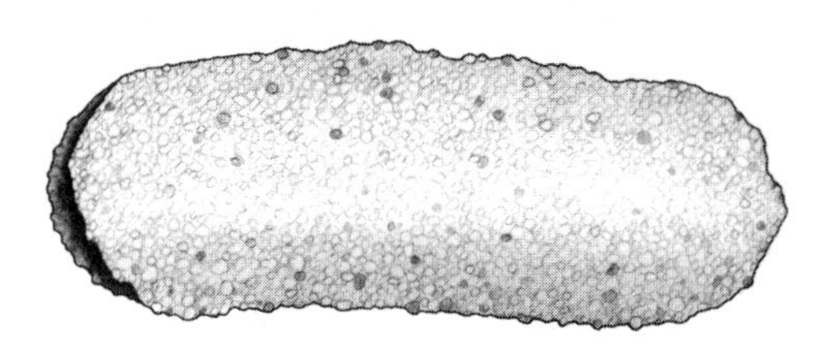

Figure 14.34. *Ochrotrichia* larval case

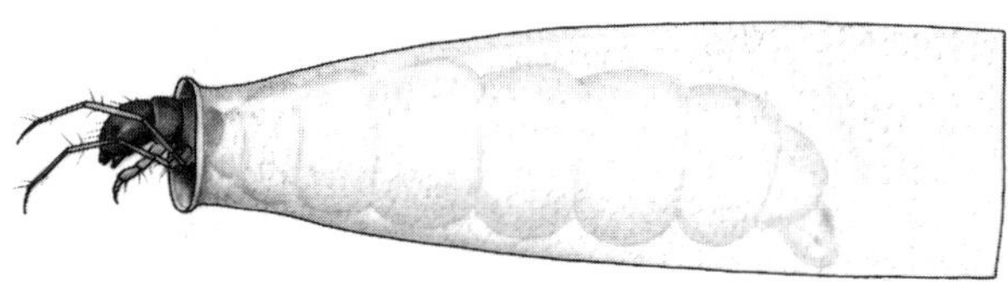

Figure 14.35. *Oxyethira* (Hydroptilinae) larva in case

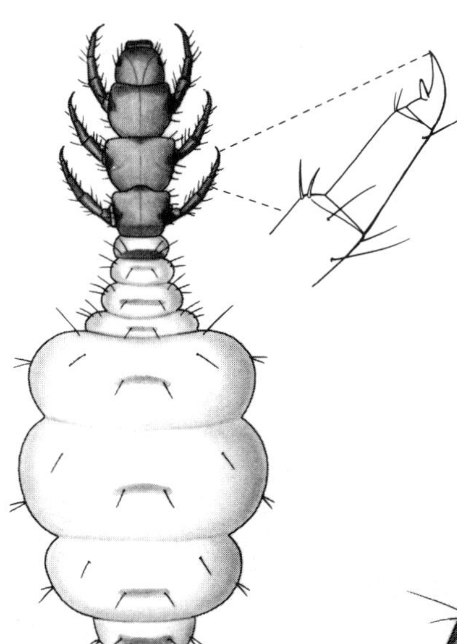

Figure 14.36. *Leucotrichia* (Leucotrichiinae) larva (hind leg enlarged)

Figure 14.37. *Leucotrichia* larval case

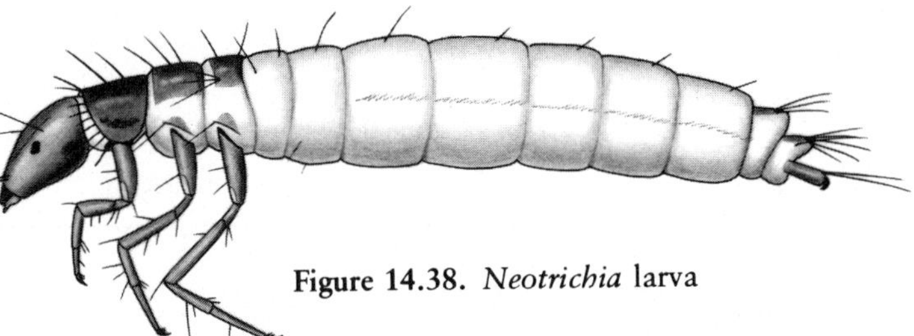

Figure 14.38. *Neotrichia* larva

Figure 14.39. *Neotrichia* larval case

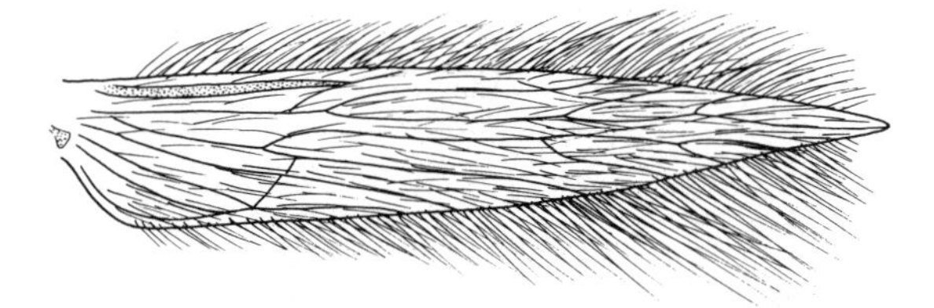

Figure 14.40. Adult fore wing

structures made of silk, although they are often encrusted with other materials. Cases of others, such as *Oxyethira,* can resemble flattened milk bottles. Still others, such as *Neotrichia,* construct cylindrical cases of rock particles that are more typical of those of tubecase makers.

Larvae feed mainly on filamentous and/or unicellular algae. They occur in a variety of lentic and lotic habitats, often where little current exists and sometimes in highly eutrophic environments. Many species overwinter as final instar larvae. Adults are often abundant at lights, and some species fly very near the water surface.

The trout swallow the animals case and all, doubtless being unable to get them apart. The case persists after the animal within has been disintegrated, but the sand grains gradually fall off. . . .

James G. Needham et al.

Fifth instar larvae may be identified as follows:

SUBFAMILY PTILOCOLEPINAE: **Larvae** have a body that is dorsoventrally flattened and with lateral fleshy tubercles on abdominal segments 1–8. The group is known only very locally and is restricted to eastern and western Canada and the western and southeastern states excluding Florida.

SUBFAMILY HYDROPTILINAE: **Larvae** (Figs. 14.33–14.35; Plate XII, Fig. 84) have a body that is laterally flattened. Tarsus and claw of hind legs are about the same length. This is a relatively common and widespread group in North America.

SUBFAMILY ORTHOTRICHIINAE: **Larvae** have a body that is flattened either dorsoventrally or laterally. Middle and hind legs are longer than fore legs. Tarsus of hind legs is about twice the length of the claw. The group is widespread except for Florida.

SUBFAMILY LEUCOTRICHIINAE: **Larvae** (Figs. 14.36, 14.37) have a body that is dorsoventrally flattened. All legs are subequal in length. Tarsi are about twice the length of the claws. The group is locally known from most of North America.

UNPLACED GENERA (e.g., *Neotrichia*): **Larvae** (Figs. 14.38, 14.39) have a body that is uniformly cylindrical. Anal prolegs are distinct.

Tubecase Makers
(Superfamily Limnephiloidea)

Larvae have a well-developed plate on the first thoracic segment; may or may not have a well-developed plate on the second thoracic segment; and have inconspicuous plates or small plates that never connect on the third thoracic segment (e.g., Figs. 14.41, 14.47). Bases of anal prolegs are fused to form an apparent abdominal segment 10 (e.g., Fig. 14.47). Spacing humps (e.g., Fig. 14.53) are often present on abdominal segment 1. Larvae construct portable cases, usually incorporating a variety of substrate or detrital material.

GIANT CASE MAKERS
(Family Phryganeidae)

LARVAL DIAGNOSIS: (Fig. 14.41; Plate XII, Fig. 86) These are large, elongate forms that measure 20–40 mm when mature. Head is elongate and usually possesses conspicuous, longitudinal stripes dorsally. Second segment of thorax lacks a completely well-developed plate; a ventral median spur is present on the first thoracic segment. Abdominal segment 1 has ventral gills, other segments may or may not have gills.

PUPAL DIAGNOSIS: (Fig. 14.43) Abdomen has a lateral fringe of hairs and short anal processes; abdominal segment 1 has dorsal spiny lobes.

ADULT DIAGNOSIS: (Figs. 14.44–14.46) These are generally large and often have patterned wings. Ocelli are present. Male maxillary palps are

GIANT CASE MAKER
(Phryganeidae)

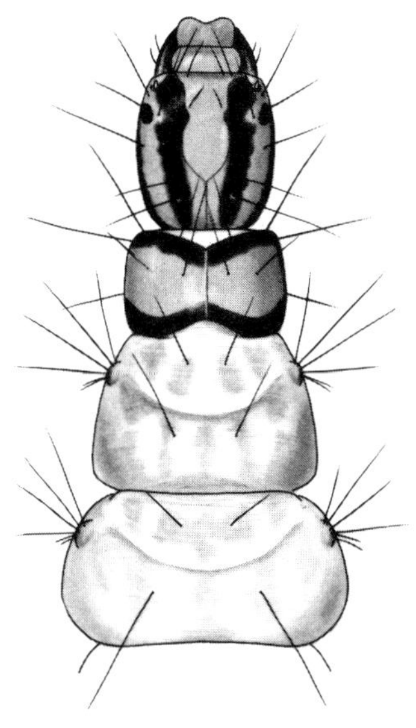

Figure 14.41.
Agrypnia (Phryganeinae) larva, head and thorax

GIANT CASE MAKERS
(Phryganeidae)

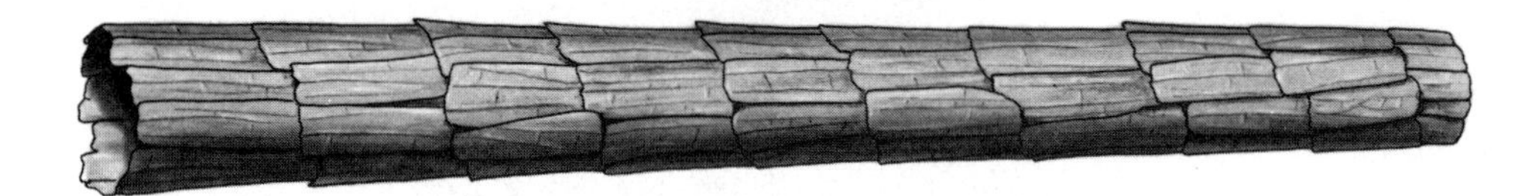

Figure 14.42. *Agrypnia* larval case

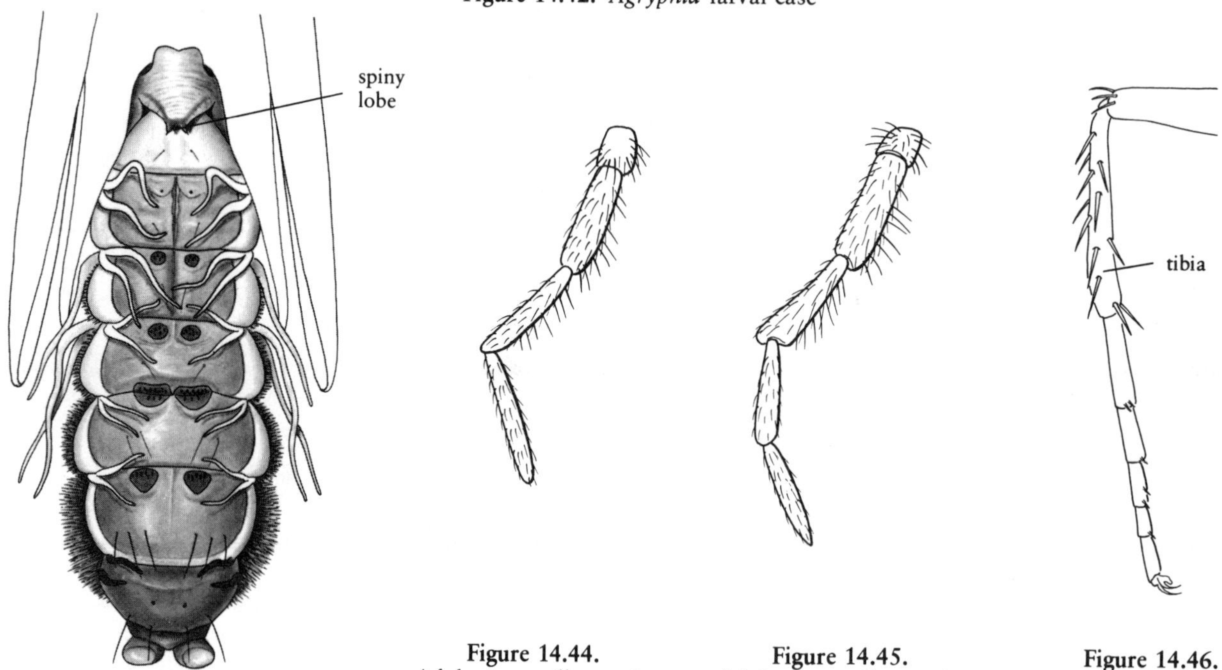

Figure 14.43. Pupal abdomen (dorsal)

Figure 14.44. Adult ♂ maxillary palp

Figure 14.45. Adult ♀ maxillary palp

Figure 14.46. *Agrypnia* adult, fore leg

four-segmented; female maxillary palps are five-segmented, segment 2 being much longer than segment 1, and segment 5 being not much longer than segment 4. Tibia of fore legs has two or more spurs.

DISCUSSION: Other larvae that are generally similar to the giant case makers have well-developed plates on thoracic segment 2 or relatively long antennae that distinguish them. About 30 species are known in North America.

Giant case makers commonly occur in marshes, backwaters, slow-flowing ditches, and ponds; however, they also occur in streams, rivers, and occasionally the deep reaches of lakes. Most species are omnivores and are generally more agile than most other tubecase makers. Some *Banksiola* feed primarily on filamentous algae as early larval instars, and in the final larval instar, become entirely predaceous. Cases are usually constructed of bits of wood and leaves, either in uniform series of end-to-end rings or spiraling the entire length of the case (Fig. 14.42).

SUBFAMILY YPHRIINAE: **Larvae** have a somewhat large medial plate on the second thoracic segment. Rock fragments in addition to plant detritus are used in case construction. The group is known only from northwestern states.

SUBFAMILY PHRYGANEINAE: **Larvae** (Fig. 14.41; Plate XII, Fig. 86) lack a medial plate on the second thoracic segment. The group is widespread in North America.

HUMPLESS CASE MAKERS
(Family Brachycentridae)

LARVAL DIAGNOSIS: (Fig. 14.47; Plate XIII, Fig. 87) These are usually 6–12 mm when mature. Abdominal segment 1 lacks both dorsal and lateral humps. Gills are simple or absent.

PUPAL DIAGNOSIS: These are somewhat similar to the pupae of giant case makers, but the anal processes are long and slender.

ADULT DIAGNOSIS: (Figs. 14.51, 14.52; Plate XIII, Fig. 89) Ocelli are absent. Male maxillary palps are three-segmented, and female palps are five-segmented. Scutum possesses a pair of small, moderately separated

HUMPLESS CASE MAKERS
(Brachycentridae)

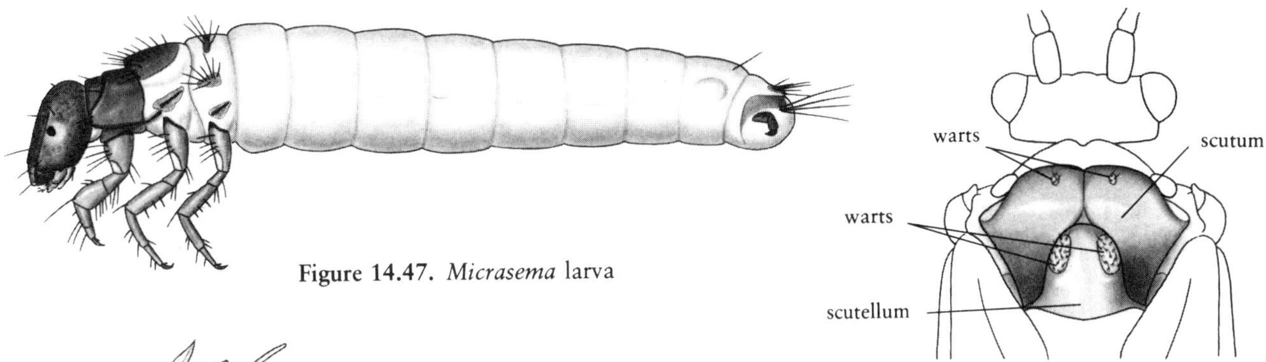

Figure 14.47. *Micrasema* larva

Figure 14.51. *Brachycentrus* adult, head and thorax

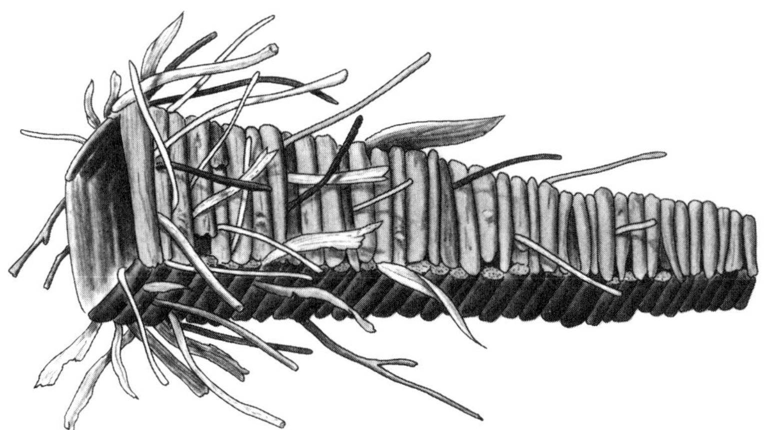

Figure 14.48. *Adicrophleps* larval case

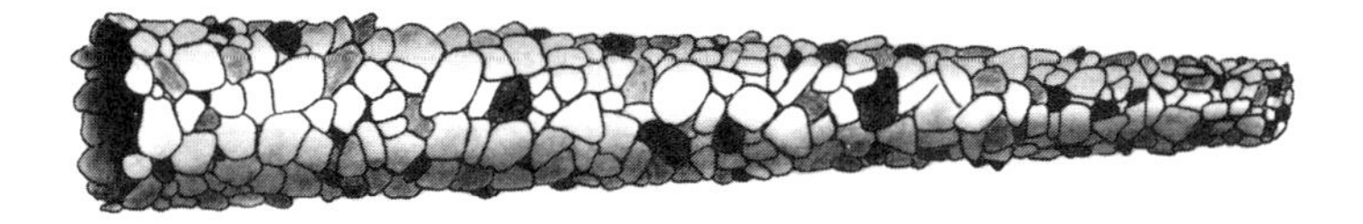

Figure 14.49. *Oligoplectrum* larval case

Figure 14.50. *Micrasema* larval case

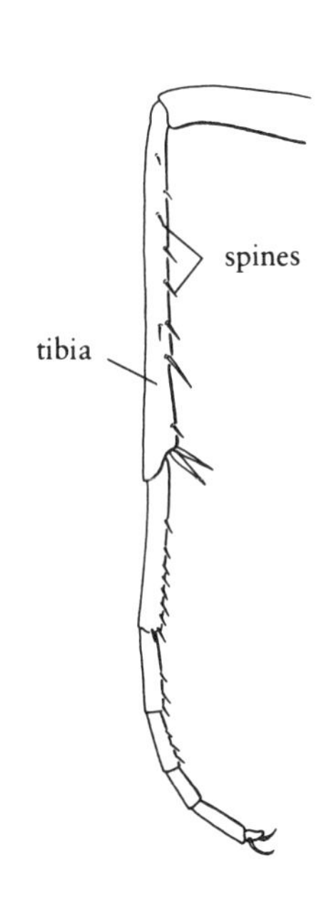

Figure 14.52. Adult middle leg

warts, and scutellum has a larger pair of warts. Tibia of middle legs has an irregular row of spines.

DISCUSSION: Larvae of brachycentrids are the only tubecase makers that lack both the dorsal and lateral humps of the first abdominal segment. This widespread group consists of six genera, each with few species in North America. Larvae occur in a variety of flowing-water habitats. Cases are four-sided, "log-cabin" or chimneylike types (Fig. 14.48; Plate XIII, Fig. 88) or cylindrical (Figs. 14.49, 14.50).

Some of the most successful fishermen's caddisflies are patterned after the brachycentrids. Flies such as the Grannom, Dark Gray or Dark Brown Caddis, and Shadfly are generally patterned after adults of *Brachycentrus*. The American Sedge is an imitation of *Brachycentrus americanus* (Plate XIII, Fig. 89) and is fished on northern trout streams from coast to coast. Another widely used fly is the American Grannom, which is based on *Brachycentrus fuliginosus*.

NORTHERN CASE MAKERS
(Family Limnephilidae)

LARVAL DIAGNOSIS: (Figs. 14.53, 14.66–14.71; Plate XIII, Fig. 90) Larvae are wide-ranging in size, generally 6–30 mm when mature. A median ventral spur is present on the first thoracic segment but is sometimes inconspicuous. Abdominal segment 1 almost always has an array of scattered bristles. Gills are simple, branched, or absent.

PUPAL DIAGNOSIS: (Fig. 14.54) The lateral abdominal fringe of hairs is present but does not extend anteriorly farther than abdominal segment 5. The anterior hook plates on each segment possess two to four hooks each; the posterior hook plates on segment 5 have more than three hooks each.

ADULT DIAGNOSIS: (Figs. 14.55–14.57; Plate XIII, Fig. 92) Ocelli may be present or absent. Male maxillary palps are three-segmented, and female palps are five-segmented and similar to those of female giant case makers. Tibia of fore legs has less than 2 spurs; or if more, then scutellum possesses a single, oval wart.

DISCUSSION: Northern case makers are generally similar to many other tubecase makers, and the combination of characteristics given above should be used. Many kinds of cases are made by the limnephilids (Figs. 14.58–14.65).

Northern case makers constitute one of the largest groups of aquatic insects in North America, being represented by over 300 species and 50 genera. The group occurs throughout Canada and the northern half and mountainous regions of the United States. Fewer species occur in the southern states or southern Midwest and are often most common from hilly areas.

Although larvae of northern case makers are found in a wide variety of habitats, individual species are often very restricted in habitat preference and may not be taken in any abundance. Common habitats and habits of the larvae are summarized here for each subfamily. The Dicosmoecinae usually inhabit cool streams and rivers, rarely intermittent streams and ponds. They construct variously shaped cases made from plant and rock materials (Fig. 14.58). They feed primarily by scraping rocks. Some may scavenge on dead fishes when available. The Pseudostenophylacinae are restricted to small streams, construct rock particle cases, and eat plant detritus. The Limnephilinae occur in a large variety of habitats. Many of those

NORTHERN CASE MAKERS
(Limnephilidae)

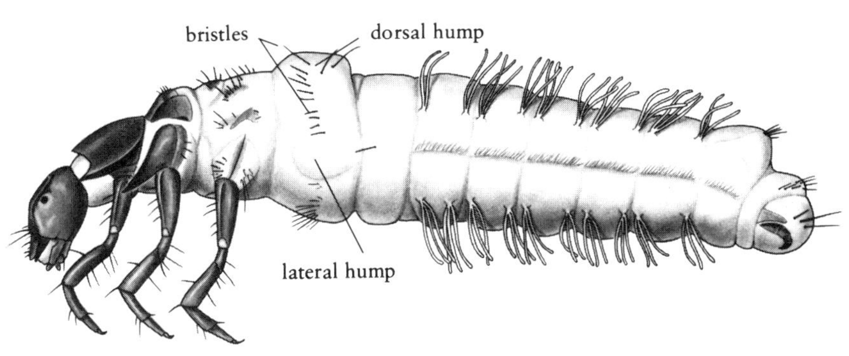

Figure 14.53. *Goera* larva

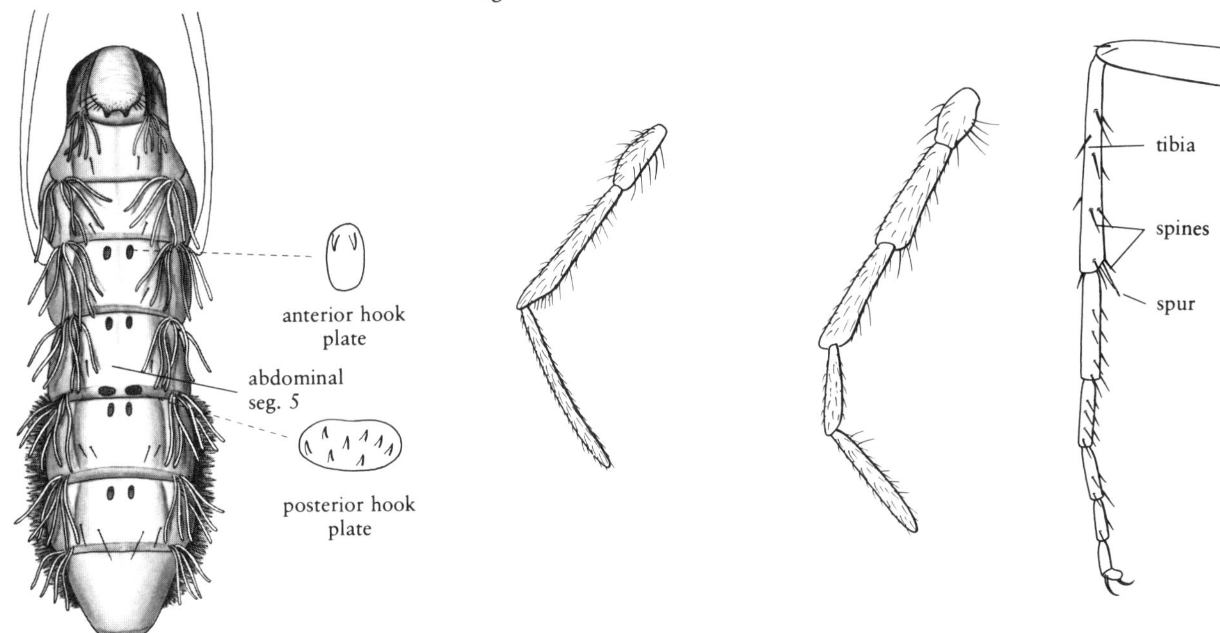

Figure 14.54. Pupa, dorsal abdomen (hook plates enlarged)

Figure 14.55. Adult ♂ maxillary palp

Figure 14.56. Adult ♀ maxillary palp

Figure 14.57. Adult fore leg

species found in somewhat warm or standing water have relatively well-developed gills. Cases are most commonly made of plant material (Figs. 14.59–14.61). They generally feed on large detritus, but feeding habits are quite diverse. The Apataniinae live in cool, mountainous streams, or cold, northern lakes. They construct cases from rock particles (Fig. 14.62) and feed by scraping. The Neophylacinae occur in flowing waters. They, too, construct cases of rock particles (Figs. 14.63, 14.64) and feed by scraping. The Goerinae occur from small seepages to small rivers. Their cases are made from rock particles, and they feed either by scraping rocks or by devouring plant detritus.

A few fly patterns are based on northern case makers. These include such imitations as the Autumn Phantom (*Dicosmoecus atripes*), the Orange Sedge (*Limnephilus coloradensis*) (Plate XIII, Fig. 92), and the Giant Red Sedge (*Pycnopsyche scabripennis,* not in *Limnephilus* or

NORTHERN CASE MAKERS
(Limnephilidae)

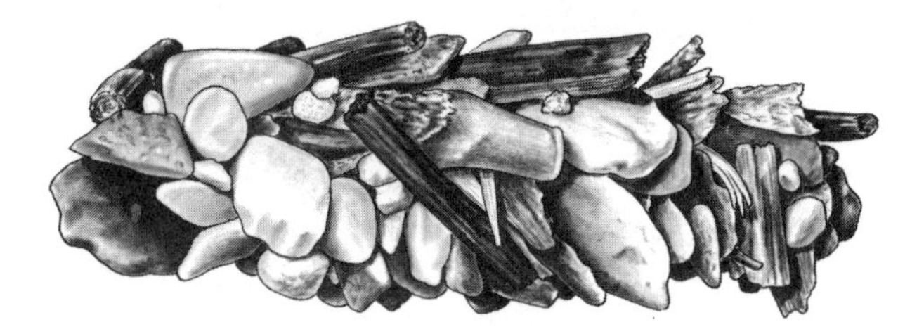

Figure 14.58. *Dicosmoecus* larval case

Figure 14.59. *Limnephilus* larval case

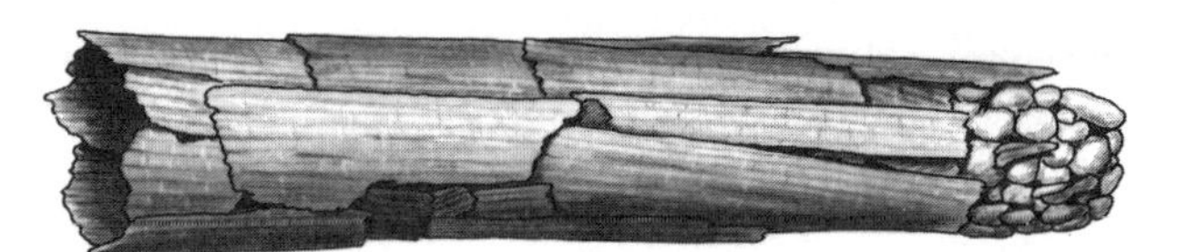

Figure 14.60. *Limnephilus* larval case

Figure 14.61. *Pycnopsyche* larval case

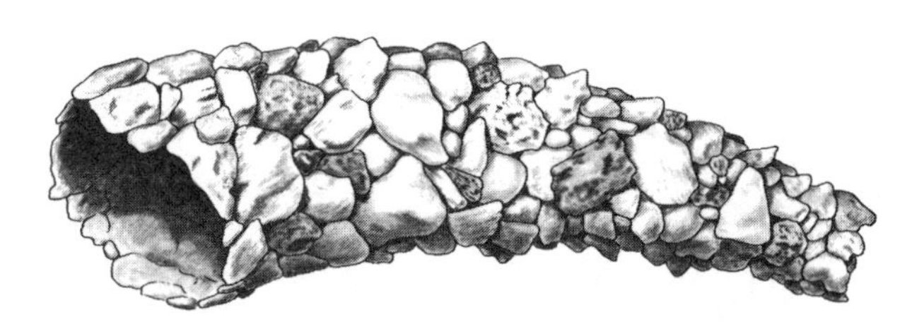

Figure 14.62. *Apatania* larval case

Figure 14.63. *Neophylax* larval case

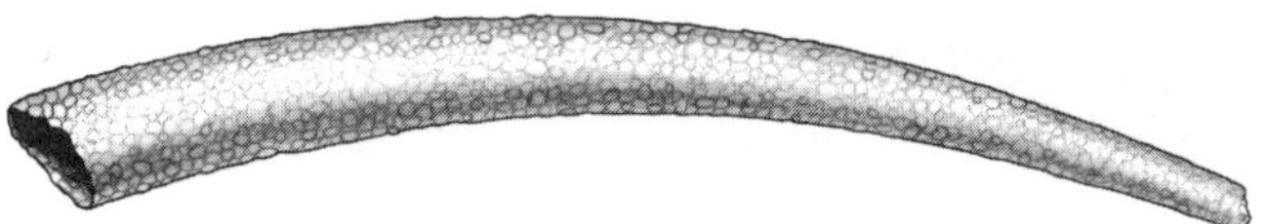

Figure 14.64. *Farula* larval case

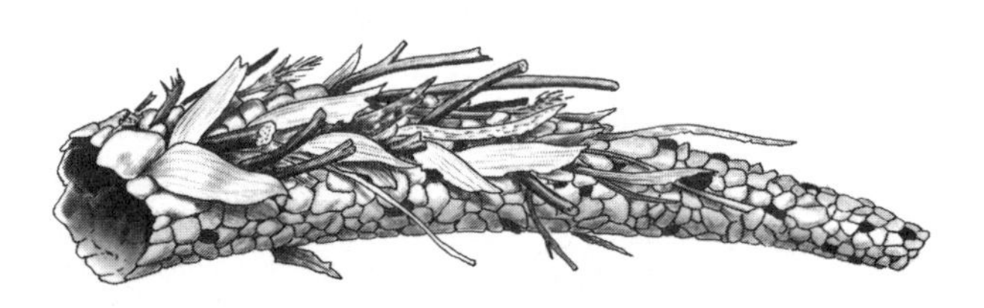

Figure 14.65. *Manophylax* larval case

NORTHERN CASE MAKERS (Limnephilidae)

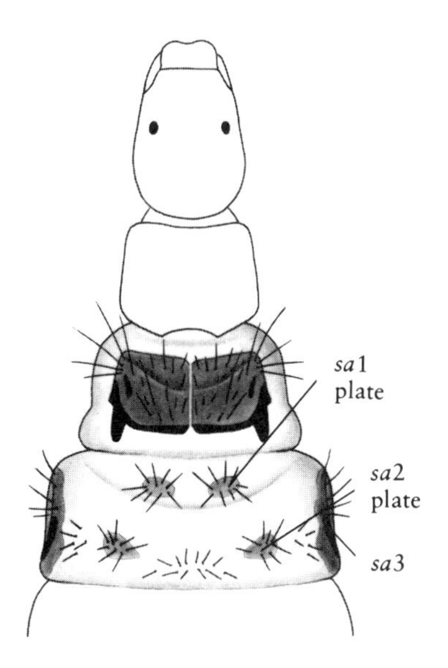

Figure 14.66. *Dicosmoecus* (Dicosmoecinae) larva, head and thorax (dorsal)

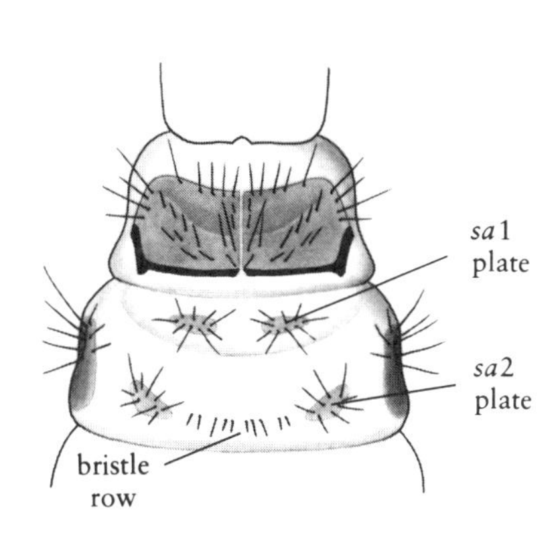

Figure 14.67. *Pseudostenophylax* (Pseudostenophylacinae) larva, thorax (dorsal)

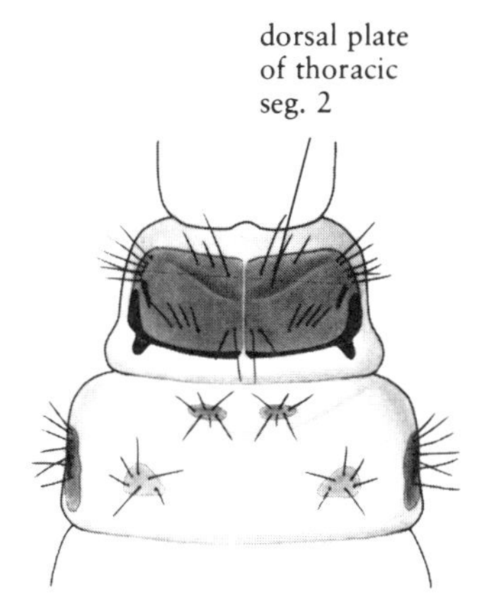

Figure 14.68. *Limnephilus* (Limnephilinae) larva, thorax (dorsal)

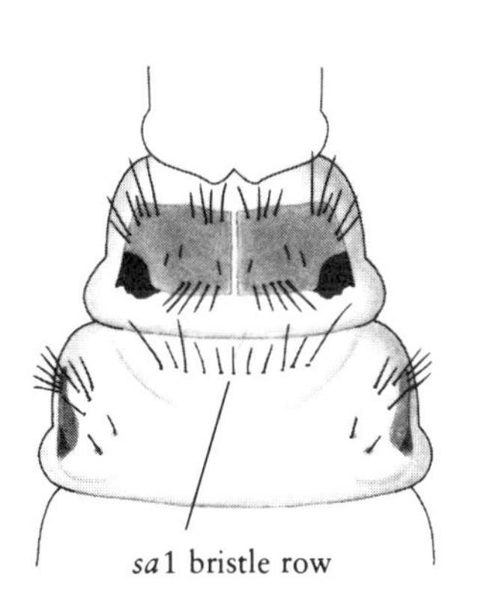

Figure 14.69. *Apatania* (Apataniinae) larva, thorax (dorsal)

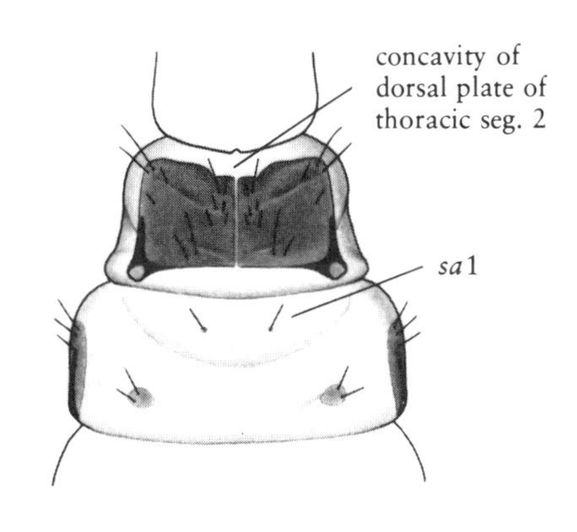

Figure 14.70. *Neophylax* (Neophylacinae) larva, thorax (dorsal)

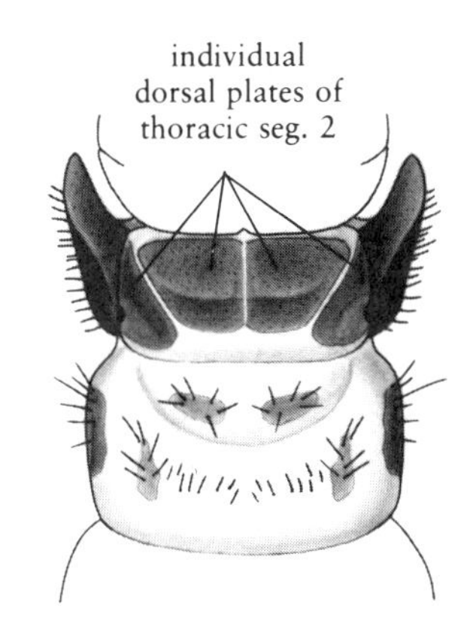

Figure 14.71. *Goera* (Goerinae) larva, thorax (dorsal)

Stenophylax as cited in some fishing literature). The Autumn Phantom and Orange Sedge should be used on cold headwater streams in the West, whereas the Giant Red Sedge is designed for eastern streams.

SUBFAMILY DICOSMOECINAE: **Larvae** (Fig. 14.66) have bristles generally scattered over a large plate of the second thoracic segment. Notal areas *sa*1 and *sa*2 of the third thoracic segment possess small plates with bristles. The *sa*1 plates are occasionally fused. Bristles may or may not be present between *sa*2 plates. Abdominal gills are usually branched. The group is widespread in North America.

SUBFAMILY PSEUDOSTENOPHYLACINAE: **Larvae** (Fig. 14.67) are similar to Dicosmoecinae, but unlike that subfamily have unbranched abdominal gills and a row of several bristles between *sa*2 plates of the third thoracic segment. The group is widespread in North America.

SUBFAMILY LIMNEPHILINAE: **Larvae** (Fig. 14.68; Plate XIII, Fig. 90) are similar to Dicosmoecinae, but bristles of the second thoracic plate are reduced in number and more restricted to *sa* areas. This group is widespread in North America.

SUBFAMILY APATANIINAE: **Larvae** (Fig. 14.69) are very distinct in that the *sa*1 area of the third thoracic segment lacks plates but has a transverse row of bristles. The group is widespread in North America.

SUBFAMILY NEOPHYLACINAE: **Larvae** (Fig. 14.70) have the large plate of the second thoracic segment concave medially along the anterior margin. Notal area of the third thoracic segment has no plates and few bristles in the *sa*1 area. The group is widespread in North America.

SUBFAMILY GOERINAE: **Larvae** (Fig. 14.71) have a second thoracic segment that is distinctive and possesses four or more plates. Third thoracic segment is variable. The group is widespread in North America.

LEPIDOSTOMATID CASE MAKERS
(Family Lepidostomatidae)

LARVAL DIAGNOSIS: (Fig. 14.72) When mature these are 7–11 mm. They are generally similar to northern case makers, but lack the dorsal hump of abdominal segment 1, and bristles are not well developed on abdominal segment 1.

PUPAL DIAGNOSIS: These are similar to pupae of giant case makers, but lack the dorsal spiny lobe of abdominal segment 1.

ADULT DIAGNOSIS: (Fig. 14.74) These are similar to humpless case makers, but maxillary palps sometimes appear one-segmented, and the tibia of middle legs lacks the row of spines.

DISCUSSION: The presence of only the lateral humps on the first abdominal segment distinguishes the larvae of this group of tubecase makers. Two genera, including *Lepidostoma* with over 60 species, are currently recognized in North America. The group is most common in eastern and western mountainous regions.

Larvae are found in small, cool streams and occasionally along lake shores or in large rivers. They are detritivores. Cases are diverse, but many

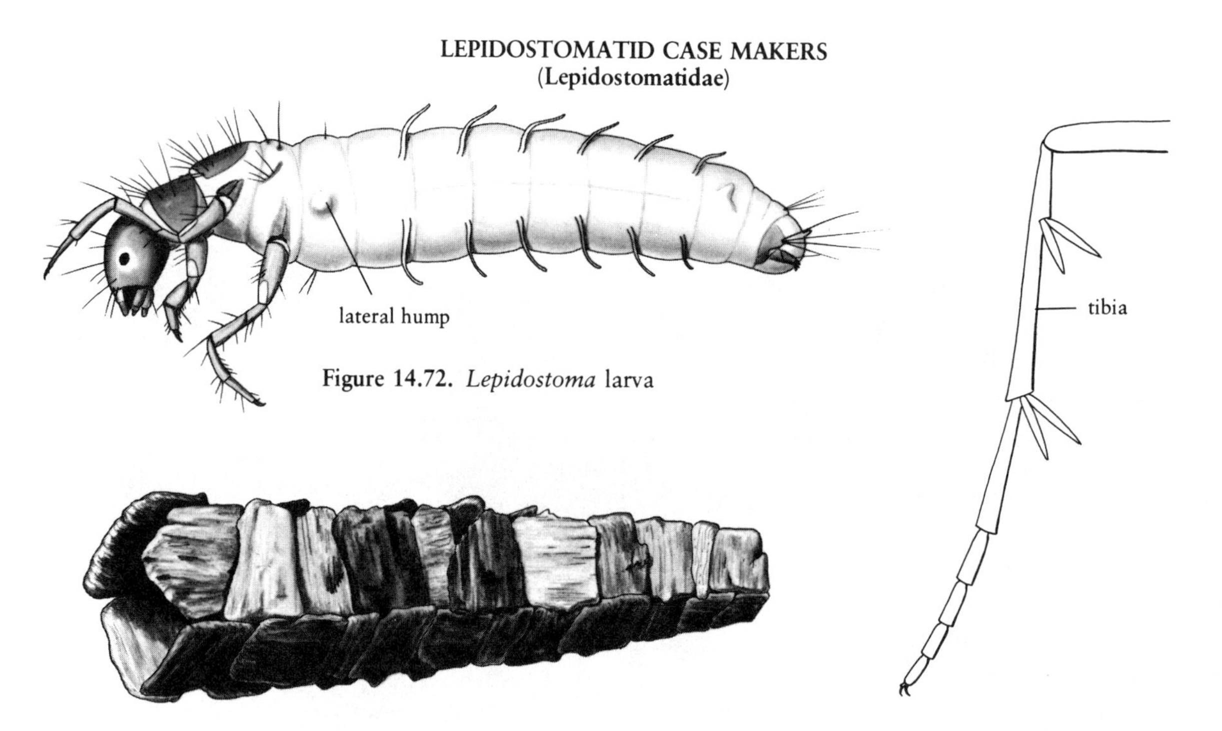

Figure 14.72. *Lepidostoma* larva

Figure 14.73. *Lepidostoma* larval case

Figure 14.74. Adult middle leg

species construct four-sided cases made of quadrate pieces of plant material (Fig. 14.73). Certain species of humpless case makers use much thinner pieces to construct somewhat similar cases.

Interestingly, the adults of some species have leaflike legs; some have atypically wide wings; some have the wing tips folded back to form pockets of black scales; and some possess highly modified, spoon-shaped maxillary palps.

BERAEID CASE MAKERS
(Family Beraeidae)

LARVAL DIAGNOSIS: (See key Fig. 14.1) These are about 6 mm when mature. Pronotum has a medial transverse ridge that curves forward laterally. Lateral plate of anal prolegs is pointed and possesses a long apical bristle.

PUPAL DIAGNOSIS: Mandibles lack inner teeth. Abdomen lacks lateral fringe of hairs. Two pairs of hook plates occur on abdominal segment 5 only; hook plates are absent posterior to abdominal segment 6. Anal processes are divergent and forked.

ADULT DIAGNOSIS: Scutum lacks warts. Tarsal segments of middle and hind legs have a crown of black apical spines; a few preapical spines are present on the first tarsal segment only.

DISCUSSION: Three species are known only very locally from seepage habitats of eastern Canada and eastern states excluding Florida.

BUSHTAILED CASE MAKERS
(Family Sericostomatidae)

LARVAL DIAGNOSIS: (Fig. 14.75) These measure up to 19 mm. First thoracic segment lacks a median ventral spur. Anal prolegs have a distinctive cluster of 30 or more long hairs dorsally.

PUPAL DIAGNOSIS: These are similar to pupae of northern case makers, but posterior hook plates on segment 5 have two or three hooks each.

ADULT DIAGNOSIS: (Fig. 14.76) Scutum has a deep medial furrow and a pair of warts nearly touching the furrow.

DISCUSSION: Three genera and about 12 species are known very locally from western, central, and southeastern states including Florida, and from parts of eastern and central Canada. Larvae usually occur in small springs and sometimes in other flowing waters or along lake shores. They usually burrow in sandy substrates. Cases are usually constructed of very small, uniform pieces of rock.

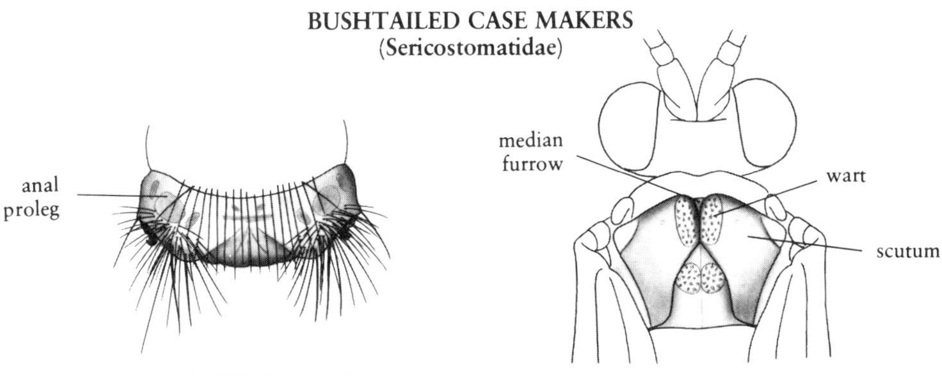

Figure 14.75. *Fattigia* larva, end of abdomen (dorsal)

Figure 14.76. *Fattigia* adult, head and thorax

STRONGCASE MAKERS
(Odontoceridae)

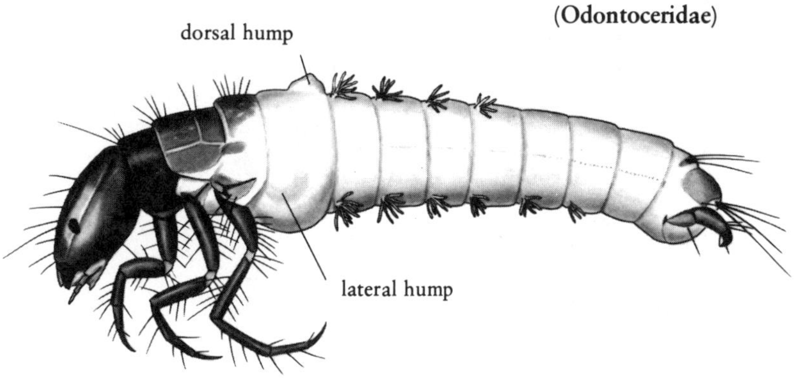

dorsal hump

lateral hump

Figure 14.77. *Marilia* larva

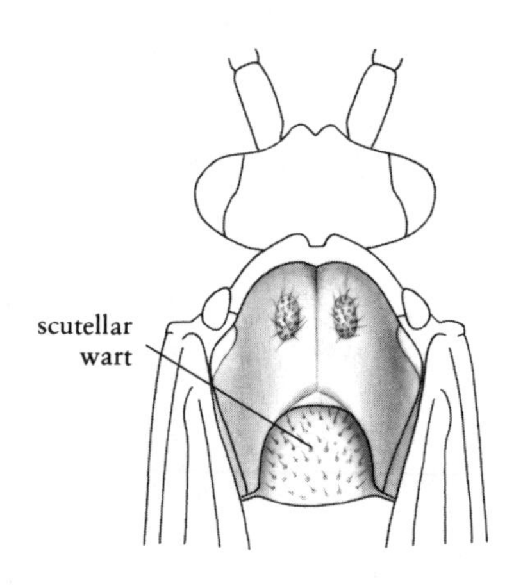

Figure 14.79. Adult head and thorax

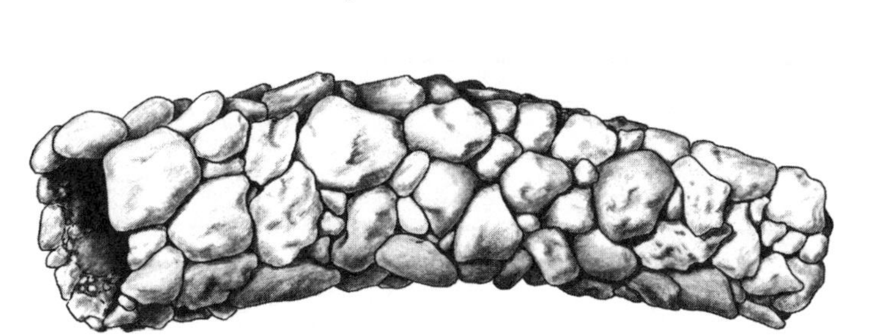

Figure 14.78. *Psilotreta* larval case

STRONGCASE MAKERS
(Family Odontoceridae)

LARVAL DIAGNOSIS: (Fig. 14.77) These are generally similar to the bushtailed case makers, but anal prolegs have no more than five long hairs (usually some small spines are also present). Gills consist of clusters of many small filaments.

PUPAL DIAGNOSIS: (Plate XIII, Fig. 93) These are similar to northern case makers and bushtailed case makers, but most of the anterior pairs of hook plates on each segment have only one hook each; posterior hook plates on abdominal segment 5 have two or three hooks each.

ADULT DIAGNOSIS: (Fig. 14.79) Scutellum is domelike, and a single wart occupies most of its area and its entire length.

DISCUSSION: This group consists of about 15 species in North America. Larvae are generally omnivores; they burrow in soft substrate or hide in mossy areas of small spring-fed streams or shallow rivers. Cases are made of sand and other rock particles, and are atypical in that they are not lined with silk. The cases are very strongly cemented together.

Psilotreta adults sometimes emerge in masses. It is during these hatches on northern waters that the Dark Blue Sedge, an imitation based on *Psilotreta frontalis* (pupa depicted in Plate XIII, Fig. 93), can be fished with a good probability of success.

SUBFAMILY ODONTOCERINAE: **Larvae** (Fig. 14.77) have the tibia of fore legs about the same length as the tarsus. The group is known only very locally throughout much of North America; it is not known from western Canada, Alaska, or Florida.

SUBFAMILY PSEUDOGOERINAE: **Larvae** have the tibia of fore legs much longer than the tarsus. The group is known only from southeastern states excluding Florida.

HOODCASE MAKERS
(Molannidae)

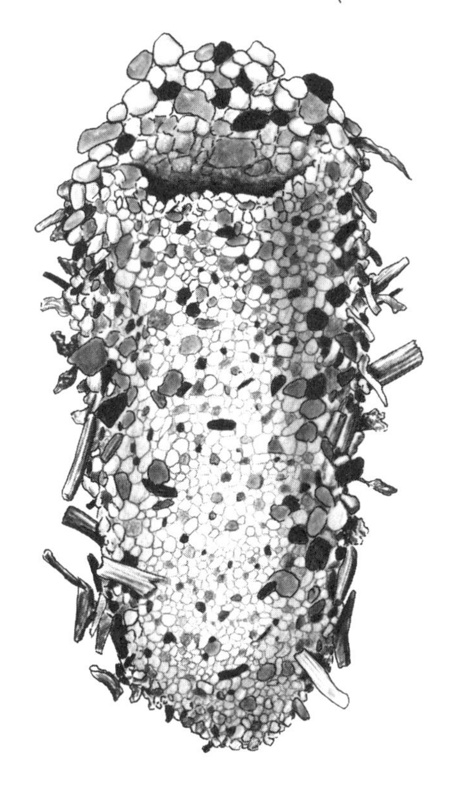

Figure 14.80.
Molanna larval case (ventral)

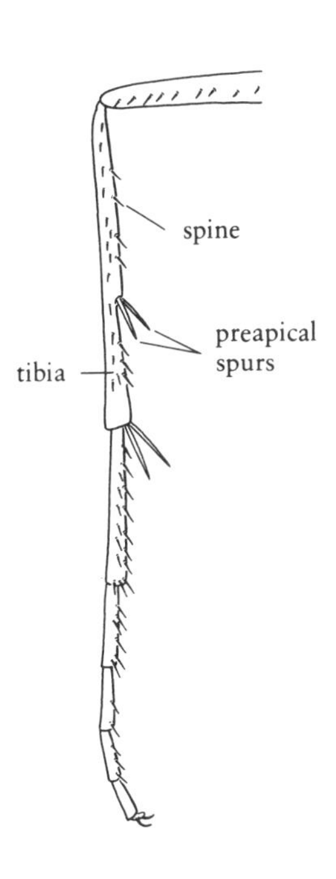

Figure 14.81.
Adult middle leg

HOODCASE MAKERS
(Family Molannidae)

LARVAL DIAGNOSIS: These measure up to 10–20 mm when mature. Plate of second thoracic segment is lightly sclerotized. Claw of hind legs is either short and stout and covered with tiny stout bristles or long and filamentlike.

PUPAL DIAGNOSIS: Abdomen has a lateral fringe of hairs extending anteriorly to abdominal segment 4. Anal processes are long and slender. Dorsal hairs on the abdomen do not occur in clusters.

ADULT DIAGNOSIS: (Fig. 14.81) Tibia of middle legs has a pair of preapical spurs and a row of 6 to 10 spines on the back side.

DISCUSSION: The few North American species in this widespread family are most easily distinguished by their somewhat flattened and flanged hoodcase (Fig. 14.80), which entirely covers the larva from above, and blends with the sand or mud substrate of the slow-flowing or standing water in which larvae live. Similar cases of some longhorned case makers have only a very tiny and smooth hole at the end and are generally much smaller (not over 13 mm when the larva is fully developed).

SNAILCASE MAKERS
(Family Helicopsychidae)

LARVAL DIAGNOSIS: (Plate XIII, Fig. 94) These have a slightly spiraling, medium-sized body. Claw of anal prolegs is distinctively comblike.

PUPAL DIAGNOSIS: These are similar to beraeid case makers, but anal processes are neither divergent nor apically forked.

SNAILCASE MAKERS
(Helicopsychidae)

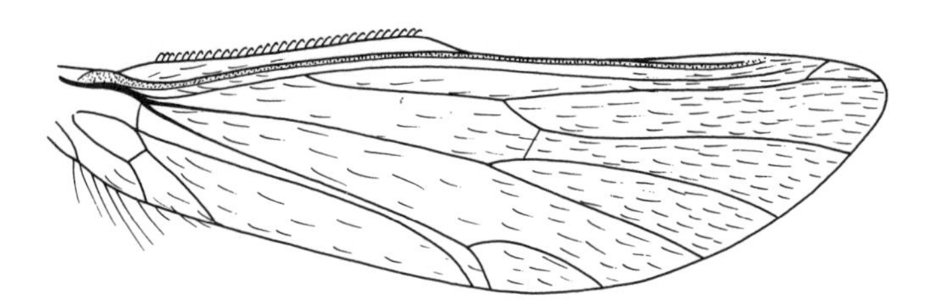

Figure 14.82. Adult hind wing

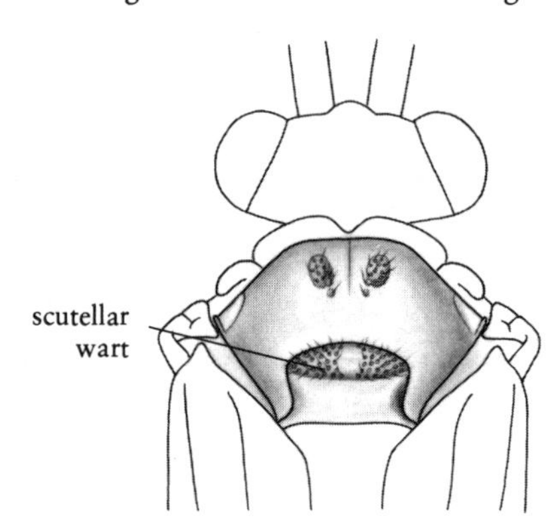

Figure 14.83. Adult head and thorax

ADULT DIAGNOSIS: (Figs. 14.82, 14.83) Scutellum has a single, narrow, transverse wart. Anterior margin of hind wings has a row of modified hairs in the basal half and a slight concavity in the distal half.

DISCUSSION: Only the widespread genus *Helicopsyche* (about four species) occurs in North America, and its case-making habit is unique among the caddisfly fauna. The larval cases are made of rock particles and resemble snail shells. The larvae are omnivorous and occur in a variety of habitats, including some relatively warm waters that other caddisflies cannot generally tolerate.

COMBLIPPED CASE MAKERS
(Family Calamoceratidae)

LARVAL DIAGNOSIS: (Figs. 14.84, 14.85) These measure up to 19–25 mm. Labrum has a row of about 16 long bristles. Anterior corners of the pronotum project forward. Lateral humps of abdominal segment 1 are more ventral than in other tubecase makers.

PUPAL DIAGNOSIS: These are similar to hoodcase makers, but the dorsal abdominal hairs occur in clusters.

ADULT DIAGNOSIS: (Fig. 14.86) Scutum is elongate, and each side of it has a simple longitudinal row of small warts.

DISCUSSION: The highly bristled labrum of the larvae is unique to this small group. None of the approximately five species are known from central Canada, central states, or Florida. Larvae utilize large leaf pieces or hollowed twigs (Fig. 14.84) to construct cases in slow-flowing areas of small streams.

The Medium Brown Sedge is a tied fly based on *Heteroplectron americanum* (previously *Ganonema americana*).

LONGHORNED CASE MAKERS
(Family Leptoceridae)

LARVAL DIAGNOSIS: (Fig. 14.87; Plate XIII, Fig. 95) These are usually 7–15 mm when mature. Head almost always has distinctive antennae that are at least six times longer than they are wide; head very rarely has short antennae. Second thoracic segment often has a light plate. Claw of hind legs is never filamentlike, although in many species it is very long and projects forward. Lateral humps of abdominal segment 1 are often covered with a cluster of small bristles.

PUPAL DIAGNOSIS: Antennae are much longer than the body and are coiled around the anal processes. Anal processes are long and slender with very short terminal bristles. Abdomen has a continuous lateral fringe of hairs.

ADULT DIAGNOSIS: (Fig. 14.92; Plate XIII, Fig. 96) These are very slender forms with long, slender antennae. Scutum has warts scattered in wide rows on either side.

DISCUSSION: The very long antennae are unique to this group of caddisflies. This is a large, common, and widespread family with seven genera and over 100 species in North America.

Most larvae are omnivore-detritivores, but *Oecetis* species are apparently highly predaceous. Some *Ceraclea* burrow into sponge colonies, where they feed on the sponges. Habitats include ponds, lake shores, and

COMBLIPPED CASE MAKERS
(Calamoceratidae)

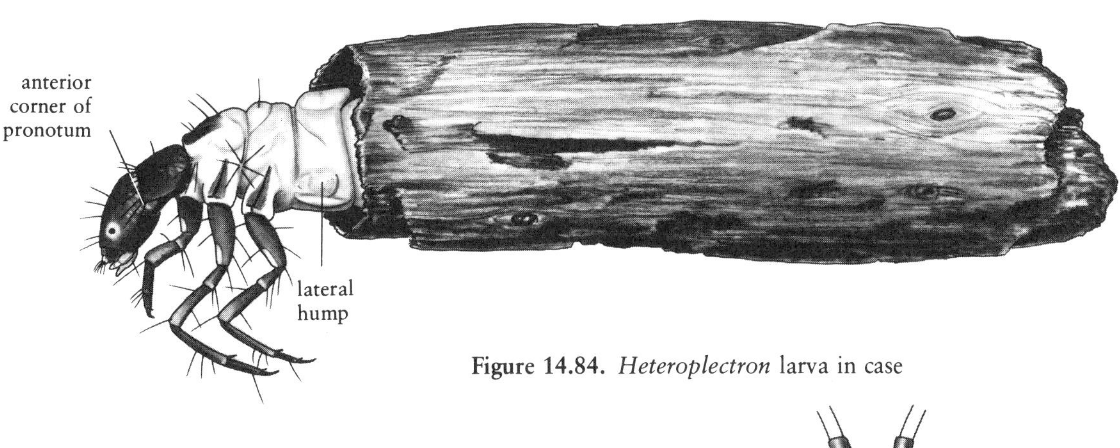

Figure 14.84. *Heteroplectron* larva in case

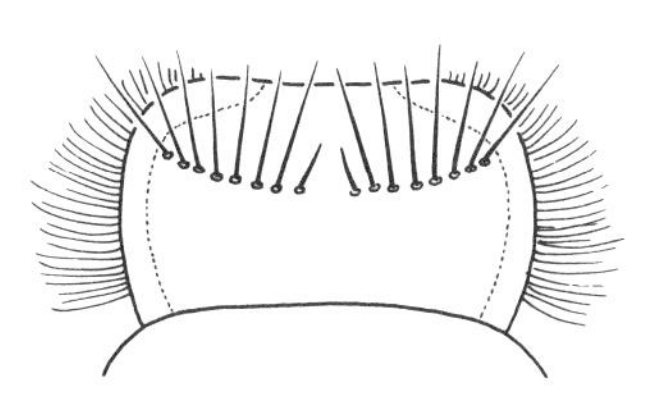

Figure 14.85. Larval labrum (dorsal)

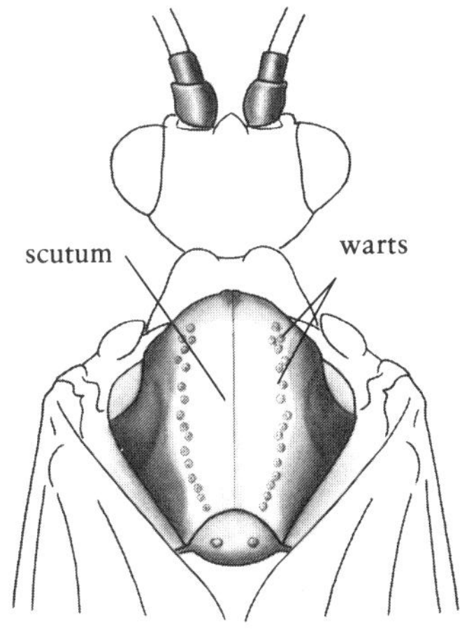

Figure 14.86. Adult head and thorax

LONGHORNED CASE MAKERS
(Leptoceridae)

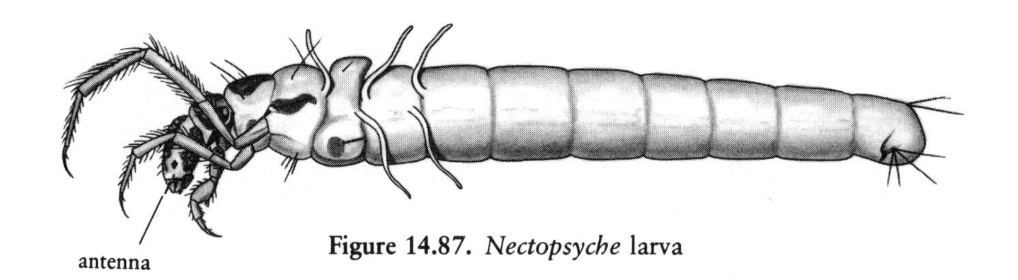

Figure 14.87. *Nectopsyche* larva

Figure 14.88. *Nectopsyche* larval case

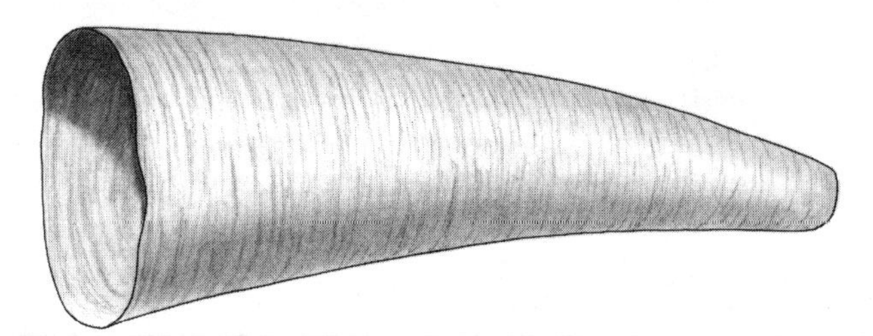

Figure 14.89. *Ceraclea* larval case

Figure 14.90. *Triaenodes* larval case

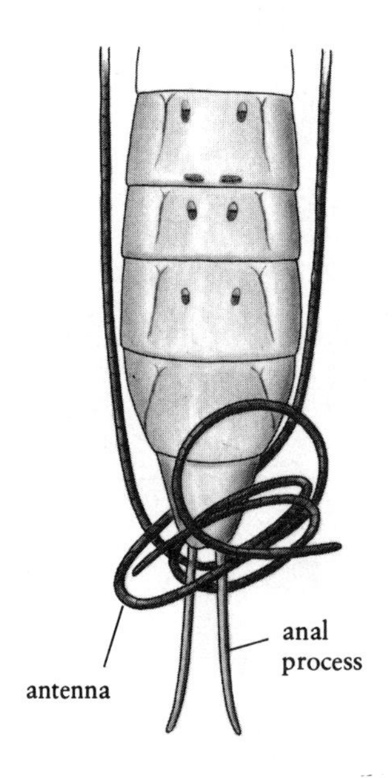

Figure 14.91.
Pupa, end of abdomen

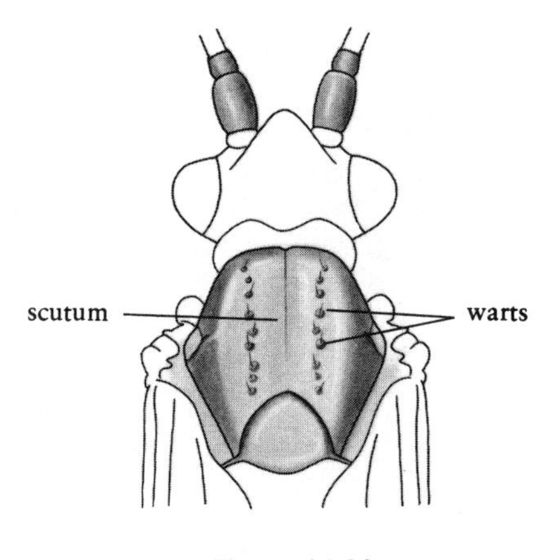

Figure 14.92.
Adult head and thorax

generally slow-flowing areas of streams and rivers. Many species may be found swimming about with their cases, often among aquatic vegetation. The long hind legs are used to help propel them. Cases (Figs. 14.88–14.90; Plate XIII, Fig. 95) may be flanged hoodcases; long cylindrical tubes made entirely of silk; long, rock-particle cases with twigs or conifer needles attached; or spiraling cases similar to those of the giant case makers.

The adults of many leptocerid species are day-fliers with very specific kinds of swarming behavior. Swarming patterns are of two general types. Certain species (e.g., those of *Ceraclea*) swarm near the water surface with a horizontal zig-zag flight pattern and orient according to the position of the shoreline. Other species (e.g., those of *Mystacides* and *Oecetis*) swarm high above the water with a vertical up-and-down pattern similar to that of many mayfly species. These species tend to orient their swarms according to the height of the adjacent vegetation. Very large swarms of some species are common along big rivers of the Midwest and lakes in the eastern half of the continent.

A couple of fishermen's flies patterned after longhorned case-maker species are noteworthy. The Black Dancer, an imitation of *Mystacides alafimbriata,* can be used on slower waters of the West. The White Miller is one of the most famous and successful of all fly imitations that have been based on caddisflies. It is usable on most North American trout streams as well as lakes. The imitation may be patterned specifically after *Nectopsyche albida* or the genus *Nectopsyche* (Plate XIII, Fig. 96) in general. Many fishermen know this genus by its junior synonym *Leptocella.*

More Information About Caddisflies

Environmental requirements and pollution tolerance of Trichoptera by Harris (1978). This is a tabular summary of the known habitat ecology of North American caddisfly larvae.

"Ecological diversity in Trichoptera" by Mackay and Wiggins (1979). This is a useful and up-to-date review of the ecological literature on caddisflies, particularly from the viewpoint of evolutionary ecology.

The caddis flies, or Trichoptera, of Illinois by Ross (1944). This recently reprinted book is an absolute necessity for the student of caddisflies in North America, particularly if concerned with adults. Although the book deals primarily with midwestern fauna, it also provides an introduction and North American checklist.

Larvae of the North American caddisfly genera by Wiggins (1977). This book has made the study of caddisfly larvae approachable and represents a giant stride in our knowledge of these forms. Each known North American genus is keyed and beautifully illustrated.

References

Adamus, P. R., and A. R. Gaufin. 1976. A synopsis of Nearctic taxa found in aquatic drift. *Amer. Midl. Natural.* 95:198–204.

Alstad, D. N. 1980. Comparative biology of the common Utah Hydropsychidae (Trichoptera). *Amer. Midl. Natural.* 103:167–174.

Anderson, N. H. 1967. Life cycle of a terrestrial caddisfly, *Philocasca demita* (Trichoptera: Limnephilidae). *Ann. Entomol. Soc. Amer.* 60:320–323.

Anderson, N. H. 1976. *The distribution and biology of the Oregon Trichoptera.* Ore. St. Univ. Ag. Exp. Sta. Tech. Bull. 134. 152 pp.

Blickle, R. L. 1979. *Hydroptilidae (Trichoptera) of America north of Mexico.* New Hampshire Ag. Exp. Sta. Bull. 509. 97 pp.

Brusven, M. A., and A. C. Scoggan. 1969. Sarcophagous habits of Trichoptera larvae on dead fish. *Entomol. News* 80:103–105.

Dick, L. 1966. *The art and science of fly fishing.* Citadel Press, Secaucus, N.J. 169 pp.

Etnier, D. A. 1965. An annotated list of the Trichoptera of Minnesota with a description of a new species. *Entomol. News* 76:141–152.

Flint, O. S., Jr. 1958. The larva and terrestrial pupa of *Ironoquia parvula* (Trichoptera, Limnephilidae). *J. N. Y. Entomol. Soc.* 66:59–62.

Flint, O. S., Jr. 1960. *Taxonomy and biology of Nearctic limnephilid larvae (Trichoptera), with special reference to species in eastern United States.* Entomologica Am. 40. 117 pp.

Gallepp, G. W. 1977. Responses of caddisfly larvae (*Brachycentrus* spp.) to temperature, food availability and current velocity. *Amer. Midl. Natural.* 98: 59–84.

Harris, T. L. 1978. *Environmental requirements and pollution tolerance of Trichoptera.* Environ. Monit. Sup. Lab., Off. Res. Dev. U.S.E.P.A., Cincinnati. 309 pp.

Jansson, A., and T. Vuoristo. 1979. Significance of stridulation in larval Hydropsychidae (Trichoptera). *Behaviour* 71:167–186.

Lehmkuhl, D. M. 1970. A North American trichopteran larva which feeds on freshwater sponges (Trichoptera: Leptoceridae, Porifera: Spongillidae). *Amer. Midl. Natural.* 85:514–515.

Mackay, R. J. 1969. Aquatic insect communities of a small stream on Mont St. Hilaire, Quebec. *J. Fish. Res. Brd. Canada* 26:1157–1183.

Mackay, R. J., and J. Kalff. 1973. Ecology of two related species of caddisfly larvae in the organic substrate of a woodland stream. *Ecology* 54:499–511.

Mackay, R. J., and G. B. Wiggins. 1979. Ecological diversity in Trichoptera. *Annu. Rev. Entomol.* 24:185–208.

Malas, D., and J. B. Wallace. 1977. Strategies for coexistence in three species of net-spinning caddisflies (Trichoptera) in second-order southern Appalachian streams. *Canad. J. Zool.* 55:1829–1840.

McClane, A. J. (ed.). 1965. *McClane's standard fishing encyclopedia.* Holt, Rinehart and Winston, New York. 1057 pp.

Meck, C. R. 1977. *Meeting & fishing the hatches.* Winchester Press, New York. 194 pp.

Nimmo, A. P. 1971. *The adult Rhyacophilidae and Limnephilidae (Trichoptera) of Alberta and eastern British Columbia and their post-glacial origin.* Quaest. Entomol. 7. 234 pp.

Nimmo, A. P. 1974. The adult Trichoptera (Insecta) of Alberta and eastern British Columbia, and their post-glacial origins. II. The families Glossosomatidae and Philopotamidae. *Quaest. Entomol.* 10:315–349.

Nimmo, A. P. 1977. The adult Trichoptera (Insecta) of Alberta and eastern British Columbia, and their post-glacial origins. I. The families Rhyacophilidae and Limnephilidae. Supplement 1. *Quaest. Entomol.* 13:25–67.

Ross, H. H. 1944. *The caddis flies, or Trichoptera, of Illinois.* Bull. Ill. Nat. Hist. Surv. 23. 326 pp. (Reprint: 1972. Entomol. Rep. Specialists, Los Angeles.)

Ross, H. H. 1956. *Evolution and classification of the mountain caddisflies.* Univ. Ill. Press, Urbana. 213 pp.

Schwiebert, E. 1973. *Nymphs.* Winchester Press, New York. 339 pp.

Solem, J. O. 1978. Swarming and habitat segregation in the family Leptoceridae (Trichoptera). *Norw. J. Entomol.* 25:145–148.

Wallace, J. B. 1975. The larval retreat and food of *Arctopsyche;* with phylogenetic notes on feeding adaptations in hydropsychid larvae. *Ann. Entomol. Soc. Amer.* 68:167–173.

Wiggins, G. B. 1977. *Larvae of the North American caddisfly genera.* Univ. Toronto Press, Toronto. 401 pp.

Wiggins, G. B. 1978. Trichoptera, pp. 147–185. In *An Introduction to the aquatic insects of North America* (R. W. Merritt and K. W. Cummins, eds.). Kendall/Hunt, Dubuque.

Wiggins, G. B. 1979. Trichoptera, pp. 482–484. In *Canada and its insect fauna* (H. V. Danks, ed.). Mem. Entomol. Soc. Canada 108.

Winterbourn, M. J. 1971. An ecological study of *Banksiola crotchi* Banks (Trichoptera, Phryganeidae) in Marion Lake, British Columbia. *Canad. J. Zool.* 49:637–645.

CHAPTER 15

Aquatic Caterpillars

(ORDER LEPIDOPTERA)

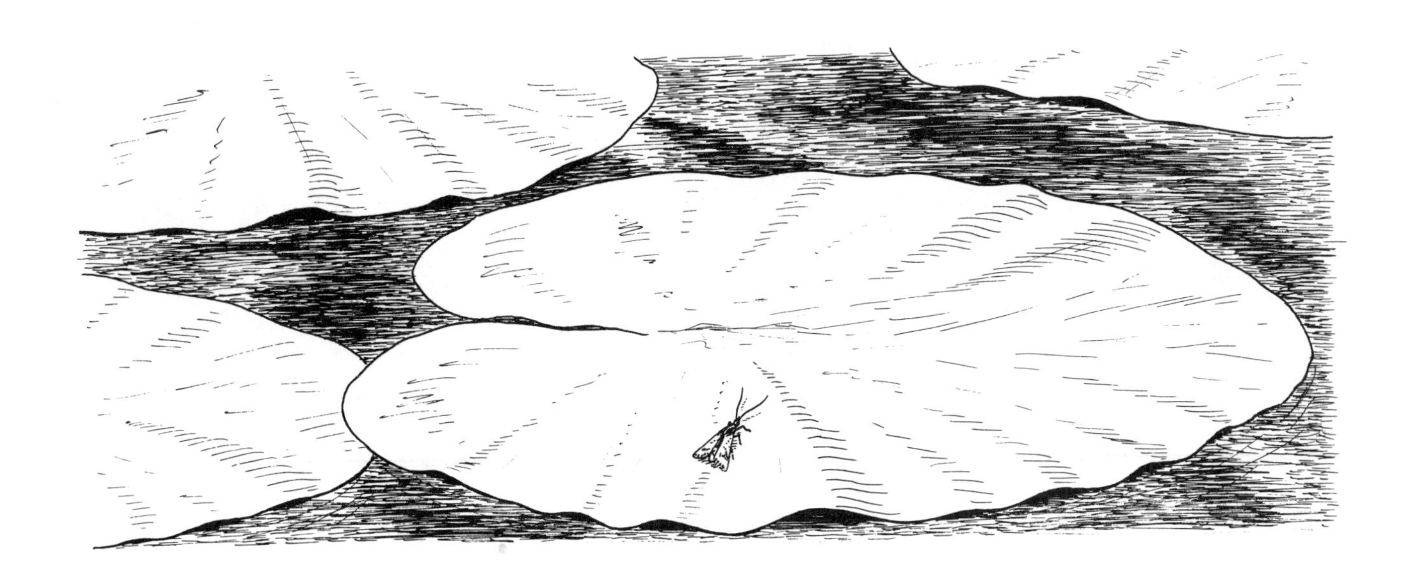

THE ORDER LEPIDOPTERA comprises the butterflies and moths, the larvae of which are known as caterpillars. Although primarily a terrestrial group, it is closely related to the caddisflies, a primary aquatic group that probably originated from mothlike insects. Over 10,000 species of Lepidoptera occur in North America. Of these only about 50 species of moths are known to be aquatic (adapted for spending most or all of their lives as immatures in freshwater). Certain female adults of one exceptional species remain aquatic, otherwise all Lepidoptera adults are terrestrial, even though a few occur in water briefly at emergence or oviposition times.

Caterpillars are herbivores and commonly have close associations with their host plants. Some aquatic species feed internally or externally on aquatic vascular plants, and some graze on the microflora associated with substrates. Perhaps as many as another 100 species in North America are regarded as semiaquatic (primarily miners and borers in emergent plants). Some of these are treated in more detail in Chapter 18. Caterpillars are important members of many littoral ecosystems, and some are injurious to such plants as rice and water lilies. Some have the potential of being biological control agents for several aquatic weeds.

Larval Diagnosis

Larvae are elongate, more-or-less cylindrical forms that measure between 3 and 75 mm when mature. Head is distinctive, with at most simple eye spots. Mouthparts are adapted for chewing. Thorax has three pairs of legs that are usually well developed and five-segmented. Wing pads are not present. Abdomen possesses pairs of short ventral prolegs on segments 3–6 and 10, or 5, 6, and 10, or rarely 2–7. Each proleg usually has a crochet (series of minute curved hooks terminally). Filamentous gills may or may not be present.

Pupal Diagnosis

These (Fig. 15.7) are quiescent forms that usually develop within cocoons. Appendages are fused to the body. Two pairs of developing wings are evident. Gills and respiratory horns are absent.

Adult Diagnosis

Body and two pairs of wings (rarely reduced) are covered with scales. Mouthparts include a tubelike siphon (reduced in a few species). Antennae are well developed but are usually distinctly shorter than body.

Similar Orders

Aquatic and semiaquatic caterpillars can be distinguished from the larvae of water beetles by the presence of abdominal prolegs and from aquatic fly larvae by the three pairs of thoracic legs. Other aquatic insects are not apt to be confused with caterpillars. The fact that all developing appendages are apparent but fused to the body of the pupae of aquatic moths (Fig. 15.7) render these pupae distinct from all but some of the aquatic fly pupae. Those fly pupae, however, have only one pair of developing wings and commonly possess thoracic respiratory horns of various types.

Moths are distinctive insects as adults, and aquatic species generally remain in the vicinity of the aquatic environment. Some terrestrial species of moths, however, may also be taken near water. Distinguishing field characteristics of the somewhat similar caddisfly adults include the generally much longer antennae and the rooflike position of both pairs of wings when not flying.

Life History

Metamorphosis is complete. Larval growth includes five to seven instars in some species, but others have highly variable, indefinite numbers of molts. Some species have one generation per year, and others have two. Generally, the larva is the overwintering stage, and many semiaquatic forms become dormant during the cold season (some species swim to shore to hibernate).

Silk, which is spun from the lower lip of caterpillars, is often used to build protective retreats (hibernacula), and pupation usually takes place in silken cocoons or silk-lined retreats. Pupae are usually attached securely to the silken case by hooklike processes (known as cremasters) at the end of the abdomen. Species with submerged larvae usually also have submerged

Above the wet and tangled swamp
White vapor's gathered thick and damp,
And through their cloudy curtaining
Flapped many a brown and dusky wing—

John Greenleaf Whittier

pupae, which are attached to rocky substrates or plant parts. Others pupate on exposed emergent plant parts or within burrows in stems or petioles. Some species plug these burrows with silk. The pupal stage generally requires only a month or less for transition to adult.

Adults of aquatic species float, swim, or crawl to the water surface upon emergence; some are aided by swimming hairs on the middle and hind legs. Emergence takes place in the spring, summer, and early fall, generally at night. Thirty minutes or more may be necessary for wing drying and spreading.

Adults live from about 24 hours to two months, depending on the species and sex. Females of some species live approximately twice as long as males. Adult activity takes place near the emergence site and is generally nocturnal (some are attracted to lights). Most moths feed on plant liquids, but some, including very short-lived forms, have highly reduced mouthparts and evidently do not feed.

Males are probably attracted to females by a pheromone, a common phenomenon among terrestrial moths. Some semiaquatic species mate soon after emergence of the female, even while her wings are still drying. Males of at least one species search out a female among vegetation and, upon locating one, repeatedly and gently tap her abdomen with their palps previous to mating. Mating often takes place on the host plant and may last several hours. Some species mate several times. Interestingly, the short-winged, fully aquatic female adults of the unusual species *Acentropus niveus* lift their abdomen out of the water in order to mate with the winged, terrestrial males. Winged females of this species can fly for considerable distances after mating.

Females of benthic stream species enter the water to deposit eggs on the rocky substrate. Females of other aquatic species usually lay rows or rings of eggs on the undersides of floating leaves by bending the abdomen over the edge of the leaf or inserting it through holes cut by aquatic leaf beetles or through other slits and holes. Females of many semiaquatic species lay egg masses on exposed parts of host plants (sometimes in the larval retreats) and cover them with hairs, scales, and froth. Eggs of most species hatch in one or two weeks.

The best flies for moonlight fishing are the white and brown, and cream-colored moths. The white are made: Body, *white ostrich herl, and a white cock's hackle over it; the* wings *from the feather of the white owl. The brown:* Body, *dark bear's hair and a brown cock's hackle over it;* wings *from the wing-feather of the brown owl. Cream-colored moth:* Body, *fine cream-colored fur, with pale yellow hackles;* wings, *feather of the yellow owl of the deepest cream-color.*

George Washington Bethune

Aquatic Habitats

Moth larvae and pupae live on the rocky substrates of streams and rivers and on or in various parts of aquatic vascular plants that grow in ponds, lakes, or streams or adjacent to them. Some larvae occasionally swim at or just below the water surface, or rarely deeper. Larvae and pupae of semiaquatic species occur on emergent plant parts, mining within leaves or boring into petioles, stems, or rootstocks. Plant burrows may be dry or filled with water, depending on the species.

Aquatic Adaptations and Behavior

Most aquatic caterpillars are hydropneustic, and many possess filamentous gills over much of the body. These gills become more numerous with age in some species, sometimes reaching 400 or more. Some hydropneustic aquatic caterpillars lack gills, and spiracles of some become functional as they grow older. Larvae of at least one species have minute unwettable hairs that can hold a thin film of air over much of the body. Most semiaquatic species make use of a general type of terrestrial respira-

But see! a wandering Night-moth enters,
Allured by taper gleaming bright;
Awhile keeps hovering round, then ventures
On Goethe's mystic page to light.

Thomas Carlyle

tion, but those that live submerged in water within a plant burrow (e.g., some noctuid moths) periodically expose a pair of specialized spiracles at the end of the abdomen in order to capture air.

Rock-dwelling species feed on algae, including diatoms. Other species feed on the emergent plants with which they are associated. Some larvae use leaf pieces to cover themselves or to form tubular or flat larval cases (Fig. 15.12) on the underwater parts of plants. Some larvae spin silken tentlike structures under which they feed. Larvae that burrow into plants may do so either to feed or to seek protective retreats while not feeding.

Completely submerged caterpillars are benthic and sometimes relatively sedentary. Semiaquatic species are poor to proficient swimmers, depending on the species; they generally use this type of locomotion to locate new host plants. Young larvae of at least one species remain anchored to the host plant by silken threads as they move about on emergent leaves.

Classification and Characters

No moth family is strictly aquatic, but the subfamily Nymphulinae of the family Pyralidae is primarily aquatic. The family Pyralidae (including both aquatic and semiaquatic members) is treated in detail in this chapter. Some other families have a very few species that are associated with emergent vegetation to various degrees and are considered semiaquatic. These families in North America include the Cosmopterygidae, Nepticulidae, and Noctuidae and are treated in more detail in Chapter 18.

Important characters for the identification of larvae include the placement and form of the abdominal prolegs. Caterpillar prolegs usually possess a series of minute hooks known as a **crochet**. The shape of the crochet and the relative lengths of the tiny hooks must sometimes be examined very closely (Figs. 15.8–15.10). The tubelike mouthpart of the adults, sometimes known as a **siphon**, is usually long and coiled when fully developed (Fig. 15.2).

AQUATIC PYRALID MOTHS (Family Pyralidae)

LARVAL DIAGNOSIS: (Figs. 15.5, 15.6, 15.8–15.12; Plate XI, Fig. 76) These are 3–35 mm at maturity. Gills are present or absent on thorax and abdomen. Prolegs are present on abdominal segments 3–6 and 10. Crochet is in the form of a circle (incomplete in some species) or two curved rows and is composed of hooks of at least two distinct sizes.

ADULT DIAGNOSIS: (Figs. 15.1–15.4) Most of these small and otherwise drab moths possess finely patterned hind wings. Fore wings are usually less than 15 mm long, occasionally highly reduced. Hind wings are sometimes held rooflike over the abdomen. Mouth siphon is well developed and coiled. Middle and hind legs of some possess swimming hairs.

DISCUSSION: In North America, any caterpillars with gills and any living on underwater parts of vascular plants (some in cases and some also

AQUATIC PYRALID MOTHS
(Pyralidae)

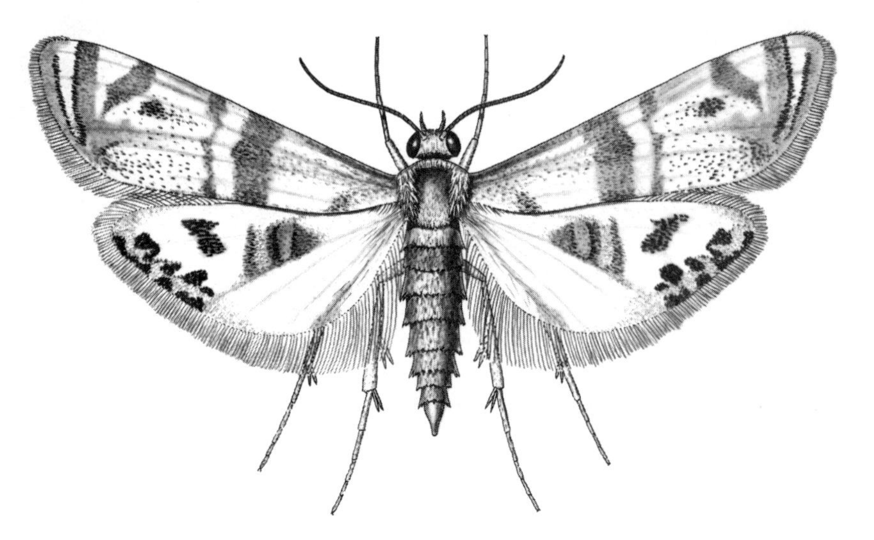

Figure 15.1. *Parargyractis* adult

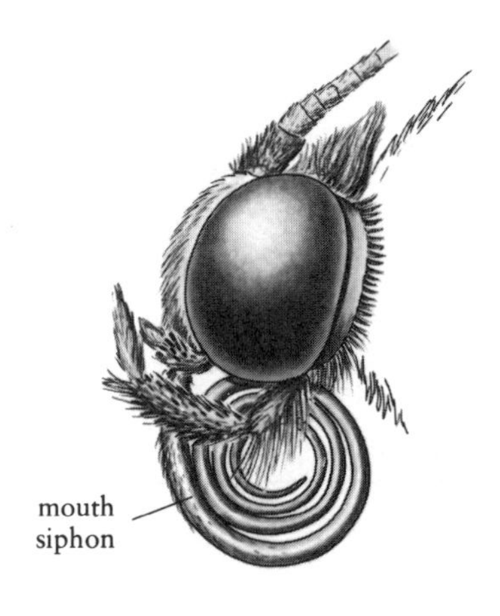

Figure 15.2. *Parargyractis* adult, head (lateral)

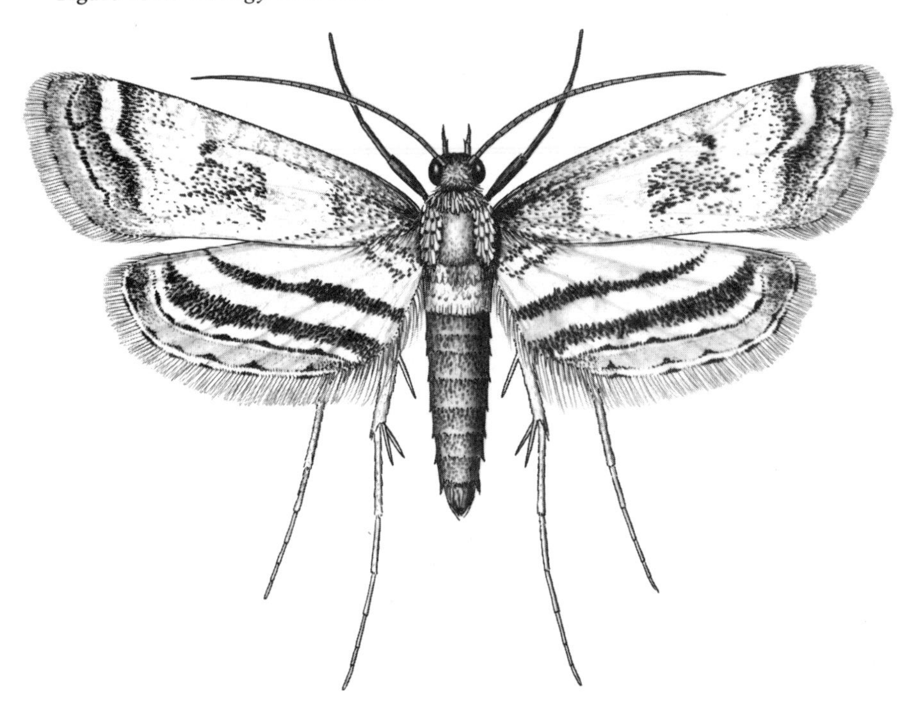

Figure 15.3. *Parapoynx* adult

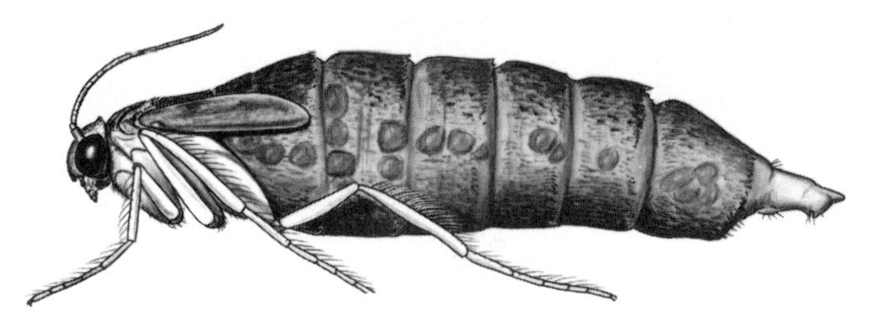

Figure 15.4. *Acentropus* adult, short-winged ♀

having gills) are pyralids. Semiaquatic pyralid caterpillars are distinguished from semiaquatic noctuid caterpillars by their relatively smaller size when mature, and from other semiaquatic caterpillars in general by their crochets, each of which is made up of different-sized hooks. The Pyralidae is the only group of moths to have naturally submerged pupae (most species of Nymphulinae), although pupae of some other moths float in water-filled plant burrows.

Over 20 genera (most with few species) in four subfamilies of aquatic or semiaquatic pyralid moths occur in North America.

Most benthic stream and river species belong to the genus *Parargyractis* or its close relatives (subfamily Nymphulinae). These caterpillars feed on microflora and reside within a silk-covered (sometimes encrusted), tentlike retreat. They are often only locally abundant and may have very specific environmental requirements. Most other species of Nymphulinae live and feed on aquatic vascular plants in ponds, lakes, bogs, and slower reaches of streams and rivers. Nymphuline caterpillars of such genera as *Parapoynx* and *Synclita* construct cases made from pieces of their host plants (e.g., pond weeds, water lilies, or duck weed).

A few species of the Schoenobiinae are semiaquatic stem borers, primarily in sedges and bulrushes; however, in eastern and central North America, *Acentropus niveus,* which was probably recently introduced into this continent from Europe, is aquatic and utilizes plastron respiration. Interestingly, in this species, adult females with reduced wings are also aquatic; fully winged females are terrestrial. A few species of the subfamilies Crambinae (e.g., the Rice Stalk Borer, *Chilo plejadellus*) and Pyraustinae are miners and borers in various emergent vegetation. Mature caterpillars of *Ostrinia* (Pyraustinae), such as the Lotus Borer, *Ostrinia penitalis,* are adept swimmers, moving easily from one host plant to the next.

SUBFAMILY NYMPHULINAE: **Larvae** (Figs. 15.5, 15.8, 15.11, 15.12; Plate XI, Fig. 76) are aquatic and may or may not possess gills. Some larvae with gills are bottom dwellers in flowing water. Other larvae (without gills as well as some with gills) construct underwater cases and are associated with vascular host plants, including various water lilies, pondweed, watermilfoil, water shield, eel grass, duckweed, and sedges. The group is widespread in North America.

SUBFAMILY SCHOENOBIINAE: **Larvae** (Fig. 15.9) are aquatic or semiaquatic, possess functional spiracles, and lack gills. Crochets are shaped in an incomplete ellipse. The later instar larvae of submergent forms (*Acentropus*) live within retreats constructed of leaf parts. These are associated with such plants as hornwort, waterwort, waterweed, *Hydrilla,* watermilfoil, and pondweed. Semiaquatic forms are miners or borers, mostly in sedges, bulrush, and spikerush. The group is widespread in North America.

SUBFAMILY CRAMBINAE: **Larvae** are semiaquatic, have functional spiracles, and lack gills. Crochets are shapd in a circle. They are miners and borers in such plants as rice, rush, bulrush, and water hyacinth. The group is widespread in North America.

SUBFAMILY PYRAUSTINAE: **Larvae** (Figs. 15.6, 15.10) are semiaquatic, have functional spiracles, and lack gills. Crochet forms a circle, or nearly so. They are leaf-surface feeders and borers in various parts of host plants of the water lily family.

AQUATIC PYRALID MOTHS
(Pyralidae)

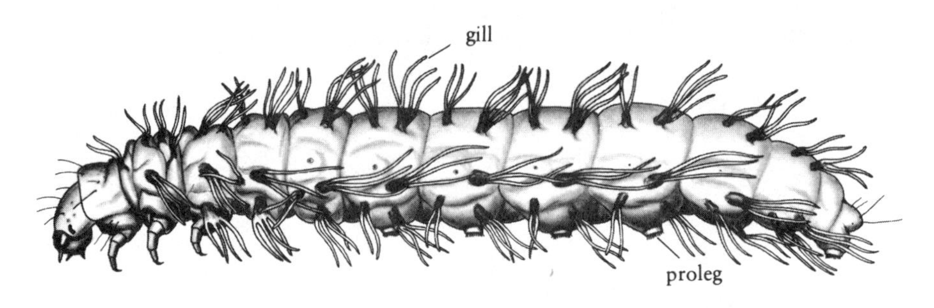

Figure 15.5. *Parapoynx* (Nymphulinae) larva

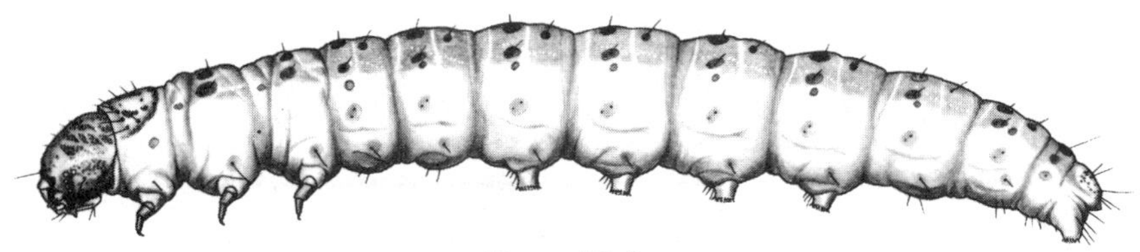

Figure 15.6.
Ostrinia (Pyraustinae) larva

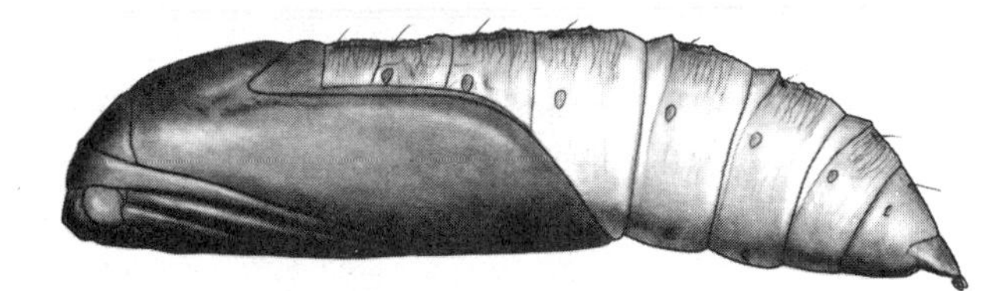

Figure 15.7. *Ostrinia* pupa

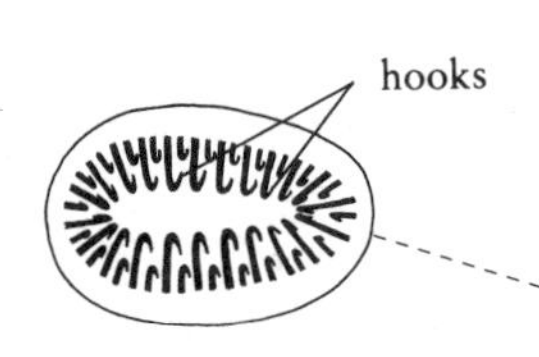

Figure 15.8.
Nymphulinae larva, crochet

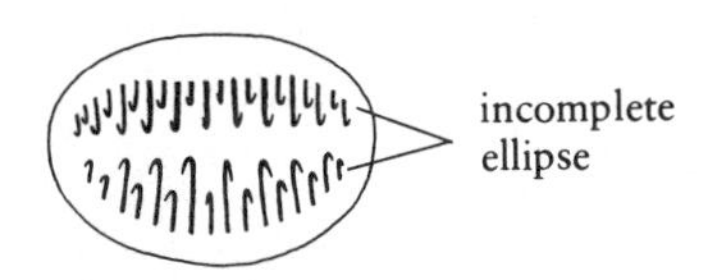

Figure 15.9.
Schoenobiinae larva, crochet

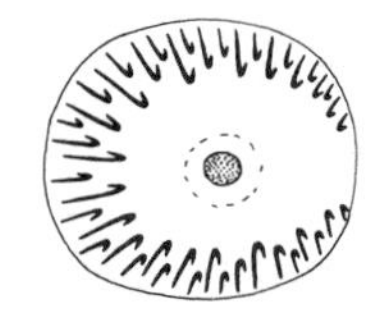

Figure 15.10.
Pyraustinae larva, crochet

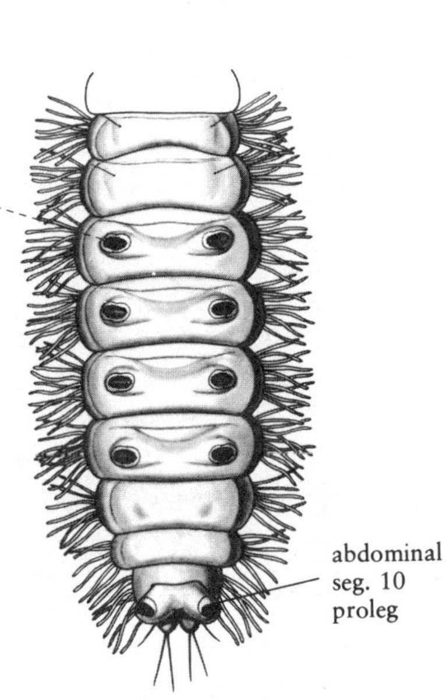

Figure 15.11. *Parargyractis* larva, abdomen (ventral)

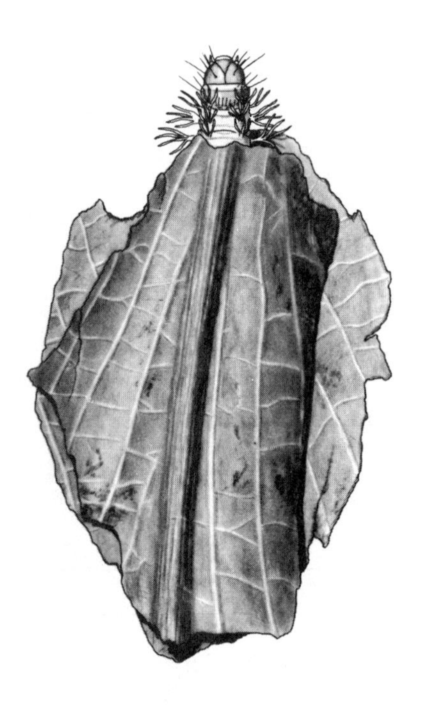

Figure 15.12. *Parapoynx* larva in case

COSMOPTERYGID MOTHS
(Family Cosmopterygidae)

A very few species of this terrestrial family are miners in emergent vegetation; see Chapter 18.

SERPENTINE MOTHS
(Family Nepticulidae)

A very few species of this terrestrial family are miners in emergent vegetation; see Chapter 18.

NOCTUID MOTHS
(Family Noctuidae)

This large and otherwise terrestrial family contains a few species associated with pond vegetation. These are further detailed in Chapter 18.

More Information About Aquatic Caterpillars

Pyraloidea. Pyralidae (in part), Fascicle 13.1 of *The moths of America north of Mexico* by Munroe (1972–1973) and edited by R. B. Dominick. This is the most comprehensive treatment of the nymphuline moths in North America; the work includes keys and descriptions to adults and reviews known biologies of species. Unfortunately, no single comprehensive key to species in the larval stage is presently available.

References

Batra, S. W. T. 1977. Bionomics of the aquatic moth, *Acentropus niveus* (Olivier), a potential biological control agent for Eurasian watermilfoil and Hydrilla. *J. N.Y. Entomol. Soc.* 85:143–152.

Berg, C. O. 1949. Limnological relations of insects to plants of the genus *Potamogeton. Trans. Amer. Micr. Soc.* 68:279–291.

Berg, C. O. 1950. Biology of certain aquatic caterpillars (Pyralidae: *Nymphula* spp.) which feed on Potamogeton. *Trans. Amer. Micr. Soc.* 69:254–266.

Frohne, W. C. 1938. Contribution to knowledge of the limnological role of the higher plants. *Trans. Amer. Micr. Soc.* 57:256–268.

Frohne, W. C. 1938. Biology of *Chilo ferbesellius* Fernald, an hygrophilous crambine moth. *Trans. Amer. Micr. Soc.* 58:304–326.

Frohne, W. C. 1939. Observations on the biology of three semiaquatic lacustrine moths. *Trans. Amer. Micr. Soc.* 58:327–348.

Lange, W. H. 1956. A generic revision of the aquatic moths of North America: (Lepidoptera: Pyralidae, Nymphulinae). *Wasmann J. Biol.* 14:59–144.

Lange, W. H. 1978. Aquatic and semiaquatic Lepidoptera, pp. 187–201. In *An introduction to the aquatic insects of North America* (R. W. Merritt and K. W. Cummins, eds.). Kendall/Hunt, Dubuque.

Levine, E., and L. Chandler. 1976. Biology of *Bellura gortynoides* (Lepidoptera: Noctuidae), a yellow water lily borer, in Indiana. *Ann. Entomol. Soc. Amer.* 69:405–414.

McCafferty, W. P., and M. C. Minno. 1979. The aquatic and semiaquatic Lepidoptera of Indiana and adjacent areas. *Gr. Lakes Entomol.* 12:179–187.

McGaha, Y. J. 1952. The limnological relations of insects to certain aquatic flowering plants. *Trans. Amer. Micr. Soc.* 71:355–381.

McGaha, Y. J. 1954. Contribution to the biology of some Lepidoptera which feed on certain aquatic flowering plants. *Trans. Amer. Micr. Soc.* 73:167–177.

Munroe, E. G. 1972–1973. Pyraloidea. Pyralidae (in part), pp. 72–134, Fasc. 13.1. In *The moths of America north of Mexico* (R. B. Dominick, ed.) E. W. Classey Ltd., London.

Munroe, E. G. 1976. Pyraloidea. Pyralidae (in part), pp. 1–78, Fasc. 13.2. In *The moths of America north of Mexico* (R. B. Dominick, ed.). E. W. Classey Ltd., London.

Treat, A. E. 1955. Flightless females of *Acentropus niveus* reared from Massachusetts progenitors. *Lep. News* 9:69–73.

Welch, P. S. 1914. Habits of the larva of *Bellura melanopyga* Grote (Lepidoptera). *Biol. Bull.* 27:97–114.

Welch, P. S. 1916. Contribution to the biology of certain aquatic Lepidoptera. *Ann. Entomol. Soc. Amer.* 9:159–187.

Welch, P. S. 1919. The aquatic adaptations of *Pyrausta penitalis* Grt. (Lepidoptera). *Ann. Entomol. Soc. Amer.* 12:213–226.

Welch, P. S. 1924. Observations on the early larval activities of *Nymphula maculalis* Clemens (Lepidoptera). *Ann. Entomol. Soc. Amer.* 17:395–402.

CHAPTER 16

Midges, Mosquitoes, Aquatic Gnats and Flies

(ORDER DIPTERA)

THE INSECT ORDER DIPTERA, the true flies, derives its scientific name from the fact that adults possess only two (*Di-*) wings (*ptera*) rather than four, as do most other winged insects. This is one of the largest, most highly evolved, and most biologically diverse of the insect orders. It includes a number of families that are primarily aquatic or aquatic at least in part. It is the immature forms that are adapted for aquatic life. Adults are terrestrial, although some frequent the water surface or margin. There are an estimated 3,500 or more species of aquatic and semiaquatic Diptera in North America, with the extraordinary family Chironomidae (the midges) having more aquatic species than any one of the previously discussed aquatic orders. Aquatic Diptera constitute a large proportion of the entire insect fauna of the arctic tundra and other cold regions of North America.

Owing to the size, adaptability, and feeding habits of many of its members, this group is ominous in terms of its influences on the health and well being of humans and other animals. Adults of most mosquitoes, black flies, biting midges, deer flies, horse flies, and some other aquatic species feed on the blood of people, livestock, and wild animals. They can be generally pestiferous because of their biting habits or sheer numbers, and they can pose a direct health hazard by taking quantities of blood, inflicting pain, and sometimes causing allergic reactions. The presence of adult flies in some regions (especially northern) inhibit human habitation during certain

When if an insect fall, (his certain guide)
He gently takes him from the whirling tide;
Examines well his form, with curious eyes,
His gaudy vest, his wings, his horns and size:
Then round his hook the chosen fur he winds,
And on the back a speckled feather binds;
So just the colours shine through ev'ry part,
That nature seems to live again in art.

John Gay

times of the year. More importantly, some species, especially among mosquitoes, are potential transmitters of serious diseases, such as encephalitis, malaria, and yellow fever, to name only a few. Vast amounts of human effort continue to be devoted to the control of these aquatic insects. The larvae of some flies cause serious damage to rice and other aquatic plants, and some are major predators of other aquatic invertebrates such as snails.

Midges and other aquatic flies are of general importance to freshwater ecosystems, often being a major fish food and an important part of bottom-dwelling communities. Game fishes feed on many kinds of Diptera that live in and on the water. Fishermen have long noted that when trout are rising for no readily apparent reason, particularly in still waters, midge pupae often prove to be the instigators of such activity. Although the most successful imitations of Diptera for fly fishing have been patterned after the midges, a few imitations of black flies, crane flies, and even mosquitoes and biting midges have also been tied.

Larval Diagnosis

Larvae are usually elongate, maggotlike forms that measure 1–100 mm. Head may or may not be well developed and conspicuous. Eyes are poorly developed. Antennae are variously developed. Mouthparts are often highly modified. Thorax never possesses three pairs of segmented legs, although in some one or two anterior prolegs are present. Wing pads are absent. Abdomen is 8- to 10-segmented (these segments are occasionally fused into fewer apparent segments or appear to be subdivided); abdomen may possess variously developed prolegs (lobelike legs or welts) on various segments; terminal segment may have prolegs or other fleshy or filamentous processes.

Pupal Diagnosis

Pupae may be active or quiescent, and many are aquatic. Some are enclosed within a hardened puparium (capsulelike case) that conceals developing wings and legs. Others are either free-living or live within a structured cocoon. Thorax has one pair of developing wings. Appendages are either fused to the body or held rigidly apart from it. The puparium or pupa often possesses a pair of dorsal thoracic respiratory horns or gills. Free-living pupae commonly also possess a terminal breathing apparatus.

Adult Diagnosis

These diverse forms generally have a single pair of membranous wings. Hind wings are modified into small peglike structures known as halteres. Fore wings are rarely highly reduced or absent; hairs are sometimes variously developed on the wings. Eyes are large. Antennae are variably shaped. Mouthparts are often reduced or modified for piercing and sucking or cutting and lapping. Thorax is well developed and somewhat humplike. Legs of some are very long. Abdomen does not possess tails.

Similar Orders

Aquatic fly larvae are easily distinguished from all other aquatic insects by the absence of three pairs of thoracic legs. Although beetle larvae that lack thoracic legs are rarely found free in water, it should be remembered that they have well-developed external mandibles and lack terminal breathing tubes or gills. Among pupae apt to be found in aquatic environments (primarily flies and caddisflies), the flies are distinguished by their puparium (when present) or by the presence of only one pair of developing wings. Moreover, the developing wings are completely fused to the body of certain fly pupae.

The adults are sometimes confused with several other orders of aquatic insects, especially from a distance, but close examination reveals only one pair of wings. Those species that swarm are sometimes confused with other groups, such as mayflies, particularly since the long, trailing legs of some superficially resemble the tails of mayflies in flight.

Of all the many forms of life which exist upon the surface of this old earth of ours, and which are our daily companions for good or ill during our few years' stay thereon, none are more numerous or less known than insects. Not only are they abundant as individuals, but the number of species is many fold greater than that of all other animals taken together. Both on the land and water they occur by millions, yet the life history of even the house-fly is known to but few.

W. S. Blatchley

Life History

Metamorphosis is complete, and the form and habits of immatures and adults are very different. Larval instars number three or four, although some species have more (black flies usually have six or seven). Larval development takes as little as one week in some mosquitoes, biting midges, or black flies, and as much as a year or more in some crane flies and others. Extreme multiple-year life cycles (perhaps as many as seven years) are known in some arctic midges. Larval growth of many species slows or ceases as a reaction to seasonal increases or decreases in temperature. It is interesting that among the midges, a few species can survive extreme dehydration (a phenomenon known as cryptobiosis), freezing, or heat during unfavorable growing conditions. Many Diptera species have both summer and overwintering generations.

Several modes of pupation occur. Some species with aquatic pupae (e.g., many midges, black flies, and some crane flies) construct a cocoon from substrate or detrital particles. Occasionally, pupal cases of other species are used. For example, dance fly larvae may be found in abandoned black fly cases, and marsh flies may be found in snail shells. Some other inactive, aquatic pupae are more-or-less exposed. A number of species are active and free-living in water as pupae (e.g., mosquitoes, phantom midges, and some midges). Many species leave the water to pupate in marginal areas, usually in thick-matted vegetation or in mud or sand. Examples of this behavior are found among some crane flies, deer flies, horse flies, and the dixid midges. Pupae of some species can be intermittently exposed or submerged without detrimental effects. Many of the higher Diptera (suborder Brachycera) form a puparium in which to pupate. The capsulelike puparium is formed from the hardened skin of the last larval instar and often retains features of the larva.

The pupal stage lasts for only a short time, with some exceptions. When the adult has become fully formed within the pupal skin, it is called a "pharate adult." Just prior to emergence, the pupa or pharate adult rises to the surface of the water, if it is not already there. Buoyancy is often provided by pockets of gas within the pupal skin or the puparium.

In some newly emerged adults it takes several minutes or more before structures become hardened enough for flight. These adults usually rest on the discarded pupal skin or puparium. Species that inhabit rapid or rough waters (e.g., netwinged midges, some black flies, and some midges) are necessarily adapted for flying immediately upon emergence. Emergence time is often highly synchronized, especially in arctic species and littoral marine species, to ensure reproduction during the very limited favorable times available in such environments. Males of numerous species emerge before females. Mass emergences are common among some species and are particularly noteworthy among some moth flies and midges.

Most midges do not feed as adults. Some adult flies are predaceous and feed on a wide variety of small organisms, mostly other insects. Some feed on algae and detritus at the water surface. Some are piercing and sucking feeders that utilize the blood of vertebrates or body fluids of other insects. Among the so-called biting flies (e.g., most mosquitoes, black flies, biting midges), the females are blood feeders, whereas the males are nectar feeders or nonfeeders.

Adults live from a few hours to several months, depending on the species. Among many groups, such as the crane flies, females are generally longer-lived than males. Some adults of shore flies, aquatic longlegged flies, a few midges, and certain others regularly skate on the water. Some species of mosquitoes and a few other biting groups make long-range flights. Adults of several species are inadvertent transporters of a number of algal and protozoan species from one aquatic environment to another.

Mating occurs in flight, on the ground, on vegetation, or even occasionally on the water. The mating positions of males and females vary from group to group, and a twisting or torsion of the male abdomen is common. In many species mating is preceded by various kinds of swarming or some other courtship behavior. Swarms of males are very common among midges and some other groups. These species often use swarm markers for orientation. Among some flies (e.g., some of the biting midges), mate recognition leading to the capture of females by males takes place as a response to the wing beat tone made by the female when flying into the swarm. Among many other Diptera, mate recognition is apparently visual. Females of some species mate with only one male. Females of some biting flies require a blood (protein) meal before they can lay eggs.

Oviposition behavior is extremely variable. Females oviposit directly on the water, on submerged substrate, on shoreline substrate, or on aquatic vegetation, depending on the species. Some other sites for laying eggs include overhanging vegetation, as demonstrated by the watersnipe flies and aquatic soldier flies, and exposed intertidal substrate during low tide, as demonstrated by some marine midges.

Aquatic Habitats

It is fair to say that the immature stages of aquatic flies can be expected in virtually every type of aquatic habitat, and as a group the flies are the most widely adapted of all aquatic orders (Table 16.1). Flies occur in torrential mountain streams, deep lakes, littoral marine habitats, brine lakes,

TABLE 16.1 COMMON AQUATIC AND SEMIAQUATIC HABITATS OF DIPTERA

Streams and Rivers

IN MODERATE TO SWIFT CURRENT	IN POOLS, SLOW WATERS, OR STILL SHALLOW MARGINS
Some Tipulidae	Some Tipulidae
Some Psychodidae	Tanyderidae
Nymphomyiidae	Some Psychodidae
Blephariceridae	Rarely Ptychopteridae
Deuterophlebiidae	Rarely Deuterophlebiidae
Thaumaleidae	Some Dixidae
Some Ceratopogonidae	Some Culicidae
Many Chironomidae	Some Ceratopogonidae
Simuliidae	Many Chironomidae
Some Tabanidae	Some Stratiomyidae
Athericidae	Several Tabanidae
Many Empididae	Some Empididae
Few Dolichopodidae	Some Dolichopodidae
Rarely Ephydridae	Rarely Syrphidae
Few Muscidae-Anthomyiidae	Some Sciomyzidae
	Several Ephydridae
	Few Muscidae-Anthomyiidae

Freshwater Ponds, Lakes, Pools, Marshes, and Bogs

MARGINAL AREAS OR SHALLOW BODIES OF WATER	DEEP, OPEN WATERS OR BOTTOMS OF LAKES
Few Tipulidae	Few Chaoboridae
Several Psychodidae	Rarely Culicidae
Ptychopteridae	Rarely Ceratopogonidae
Some Dixidae	Several Chironomidae
Several Chaoboridae	Rarely Ephydridae
Many Culicidae	
Some Ceratopogonidae	
Some Chironomidae	
Some Stratiomyidae	
Some Tabanidae	
Few Empididae	
Some Dolichopodidae	
Rarely Phoridae	
Syrphidae	
Several Sciomyzidae	
Many Ephydridae	

Within Underwater Parts of Emergent Vegetation

Few Tipulidae	Few Dolichopodidae
Few Chironomidae	Several Ephydridae
Rarely Cecidomyiidae[1]	Some Scatophagidae[1]

On or Within Freshwater Invertebrates

Few Chironomidae	Many Sciomyzidae
Few Simuliidae	Few Tachinidae
Few Bombyliidae	

Wet Marginal Areas Along Bodies of Freshwater or on Saturated Surfaces

Many Tipulidae	Some Dolichopodidae
Rarely Tanyderidae	Rarely Phoridae
Some Psychodidae	Rarely Sphaeroceridae[2]
Rarely Ptychopteridae	Rarely Tethinidae[2]
Some Ceratopogonidae	Rarely Milichiidae
Few Chironomidae	Many Sciomyzidae
Many Stratiomyidae	Many Ephydridae
Many Tabanidae	Some Muscidae-Anthomyiidae
Some Empididae	

Wet Marine Beaches or Intertidal Waters

Few Culicidae	Few Sphaeroceridae[2]
Some Ceratopogonidae	Tethinidae[2]
Some Chironomidae	Canaceidae[2]
Few Dolichopodidae	Several Ephydridae
Some Dryomyzidae[2]	Few Muscidae-Anthomyiidae
Coelopidae[2]	

Salt Marshes or Estuaries

Few Tipulidae	Some Tabanidae
Few Psychodidae	Few Dolichopodidae
Some Culicidae	Some Sciomyzidae
Few Ceratopogonidae	Several Ephydridae
Few Chironomidae	

Inland Salt Water or Brackish Pools, Ponds, or Lakes

Rarely Tipulidae	Few Chironomidae
Few Psychodidae	Few Tabanidae
Rarely Chaoboridae	Few Dolichopodidae
Few Culicidae	Few Sciomyzidae
Few Ceratopogonidae	Many Ephydridae

In Tree Holes or Plant Cups Holding Water

Some Culicidae	Few Dolichopodidae
Few Ceratopogonidae	Few Phoridae
Few Chironomidae	Several Syrphidae
Few Tabanidae	Few Sarcophagidae[3]

[1]See Chapter 18.
[2]See Chapter 17.
[3]See Chapter 19.

petroleum pools, sewage waters, desert alkaline pools, arctic bogs, salt marshes, hot springs, and other environments. They are found on or within the substrate, on or in aquatic plants, swimming or floating in open water, or associated with the air-water interface. Small larvae frequently appear in municipal water systems, and others occasionally appear in the drains of kitchens and restrooms.

Aquatic Adaptations and Behavior

Respiration may be hydropneustic or aeropneustic and sometimes differs among species of the same family. Many aeropneustic forms also appear to have some capacity for obtaining oxygen hydropneustically. Hydropneustic respiration generally occurs over the soft body surface, although the posterior end of the body or thoracic processes of some appear to be more specifically adapted for this function. Some lotic forms, such as netwinged midges, possess filamentous gills (Fig. 16.22). Some very small forms living in moist semiaquatic habitats and lacking functional spiracles are evidently able to obtain atmospheric oxygen cutaneously. Some midges contain hemoglobin in their bodies. Since hemoglobin is able to hold oxygen, these midges are often found in habitats with little available dissolved oxygen, and some can survive for short periods in habitats completely lacking dissolved oxygen (e.g., the bottoms of some lakes). Pupae of several families have specialized spiracular gills that allow dissolved oxygen to enter the spiracles via a plastron provided by the gill. These structures are usually highly branched and usually located on the thorax (Fig. 16.61). Spiracular gills allow the pupae to respire hydropneustically in water and aeropneustically when exposed to air. Hemolymph in the spiracular gills of some midges (Fig. 16.55), rather than a plastron, mediates the transfer of oxygen from the water.

Aeropneustic species usually obtain oxygen by maintaining some contact with surface air. In some this is accomplished by means of breathing tubes or siphons at the posterior end of the body (e.g., Figs. 16.18, 16.30, 16.81, 16.92). These breathing tubes are commonly surrounded by unwettable hairs or processes at the tip, and are either fixed or highly extensible into elongate telescoping tubes when in use. Many pupae have a pair of thoracic respiratory horns that make contact with the surface air (e.g., Figs. 16.5, 16.36, 16.79). Among a few species, these horns are adapted for tapping air supplies in the underwater parts of plants. Some species that are partially or only intermittently submerged or that occur in moist marginal areas often show no essential modifications for aquatic respiration and have a more-or-less complete complement of spiracles. Some species surface at intervals for breathing, and a few of these, such as some aquatic soldier flies, also obtain a bubble of air and retain it with hairs that surround terminal spiracular openings (Fig. 16.63). Some aeropneustic species, such as phantom crane flies, can survive prolonged periods of complete submergence, although development may not be completed without the resumption of aeropneustic respiration.

Osmoregulation presents a very serious problem for many species, particularly for those living in environments subject to shifts in the salt content of the water (intermittent ponds, salt marshes, estuaries). Specialized thin-walled papillar structures (anal papillae) for the uptake of chloride ions are present in the posterior region of the body of many Nematocera. Several members of the Brachycera have special anal areas for chloride uptake (chloride epithelia), but some possess papillae. One saltwater mosquito species has the unique capability of both ion absorption in freshwater and ion excretion in salt water via its anal papillae.

Feeding behavior and food preferences are highly variable among aquatic Diptera. Food sources include everything from fine detritus and microorganisms to whole plant parts, decaying wood, and other insects and

Thro' the clear stream the fishes rise,
And nimbly catch the incautious flies.

Charles Bowlker
(attributed to Dr. Jenner)

invertebrates. The food preferences and feeding habits of some larvae (e.g., certain midges) change with the age of the larvae and the season of the year.

Many aquatic flies are bottom dwellers. Some of these burrow and construct tubes within the substrate. Some actively sprawl about, and many are equipped with prolegs or fleshy lobes along the abdomen that aid a peristaltic or wormlike movement. Some construct cases, being somewhat similar to case-making caddisflies in habit. Some are able to cling steadfastly to substrates by means of such adaptations as abdominal prolegs with hooks (e.g., Fig. 16.72), special attachment discs along the bottom of the abdomen (Figs. 16.15, 16.22), and attachment structures at the end of the abdomen (Fig. 16.57). Some midges and black flies, among others, may utilize the body surface of mayflies and other insects as a substrate. Bottom-dwelling aeropneustic species, such as some rattailed maggots, phantom crane flies, and others, are generally restricted to very shallow waters.

Many larvae and active pupae are swimmers in open water or floaters near the surface. A few species, such as the deep-water phantom midges of lakes, live at the bottom or in profundal areas during periods of high light intensity but migrate to the open-water limnetic area during periods of reduced light intensity. Upward movement in water by these species is aided by special tracheal bladders (Fig. 16.28) that act as hydrostatic organs. The aquatic flies of streams and rivers, especially the midges, often make up a substantial portion of the drift.

Classification and Characters

Flies are classified into two major groups: the suborder Nematocera (longhorned flies), having a large representation of primarily aquatic families, and the suborder Brachycera (the higher Diptera), having a smaller proportion of aquatic forms and many semiaquatic forms. The Brachycera can be further divided into two groups, one that includes the Tabanomorpha and Asilomorpha and another that includes the more advanced Cyclorrhapha.

Members of 35 different families of flies in North America are associated with aquatic or semiaquatic environments. Of these, 23 are treated in detail in this chapter, since they contain at least some fully aquatic species; five are treated in Chapter 17, since they contain only shore-dwelling species; two are treated in Chapter 18, since, although essentially terrestrial, they do contain species that are closely associated with aquatic vegetation; and one is treated in Chapter 19, since, although essentially terrestrial, it contains some species that develop in water-filled leaves of pitcher plants. The remaining four families, which are not treated in any detail in this book, are essentially terrestrial. Two of the four contain species that are endoparasites of some other aquatic insects, and the other two occur very sporadically in the wet, organically enriched environment of sewage treatment facilities. Table 16.1 summarizes the aquatic associations of each of the families.

Figure 16.1. AQUATIC DIPTERA IMMATURES

BODY WITH DEVELOPING WINGS; 3 PAIRS OF THORACIC LEGS APPARENT (MAY BE FUSED TO BODY)

BODY WITHOUT 3 PAIRS OF THORACIC LEGS AND WITHOUT DEVELOPING WINGS (ALL AQUATIC FLY LARVAE AND PUPARIA)

DEVELOPING ANTENNAE ELONGATE AND LYING OVER EYES

DEVELOPING ANTENNAE NOT LYING OVER EYES, NOT REACHING BEYOND WING BASES

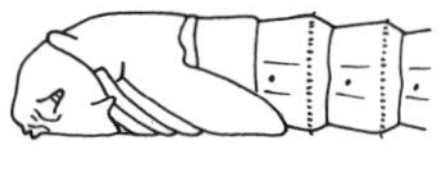

Nematocera Pupae
(See Family Discussions)

Brachycera Pupae

BODY DORSOVENTRALLY FLATTENED WITH 6 DEEP LATERAL CONSTRICTIONS; 6 REGIONS WITH VENTRAL ATTACHMENT DISCS

Blephariceridae
(p. 300)

BODY NOT DIVIDED INTO 6 OR 7 DEEPLY CONSTRICTED REGIONS AND WITHOUT VENTRAL ATTACHMENT DISCS

HEAD FULLY FORMED, HEADLIKE, AND DISTINCT FROM THORAX

PROTHORAX WITH PROLEG(S)

HEAD INCONSPICUOUS, INCOMPLETELY FORMED, OFTEN REPRESENTED BY A MERE TIP OF ANTERIORLY TAPERED BODY, AND/OR RETRACTED INTO THORAX

PROTHORAX WITHOUT PROLEGS

BODY MORE-OR-LESS CYLINDRICAL, NOT LEATHERY

DISTAL THIRD OF ABDOMEN SWOLLEN

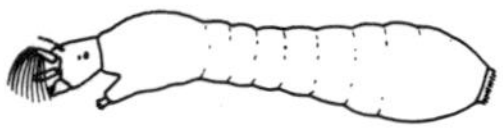

DISTAL THIRD OF ABDOMEN NOT SWOLLEN

Simuliidae
(p. 314)

BODY SOMEWHAT DORSOVENTRALLY FLATTENED AND LEATHERY

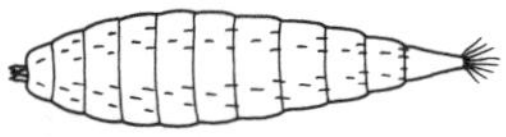

Stratiomyidae
(p. 316)

FLESHY PROCESSES OR BRISTLES DORSALLY ALONG BODY

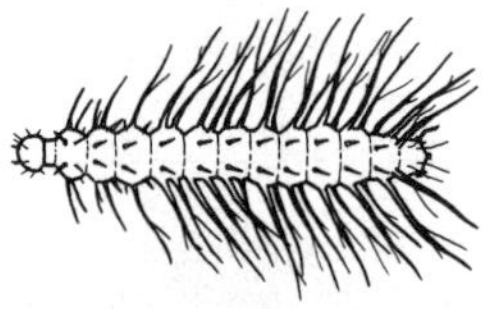

Ceratopogonidae
(p. 307)

THORACIC SEGMENTS DISTINCT, NOT BROADER THAN ABDOMEN

BODY WITHOUT FLESHY PROCESSES, SOME HAIRS AT MOST

THORACIC SEGMENTS FUSED, USUALLY SWOLLEN BUT NOT ALWAYS

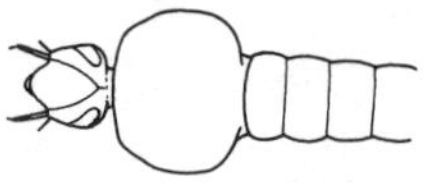

PROLEGS PAIRED (IF ONLY SLIGHTLY AT TIP)

PROLEGS COMPLETELY UNDIVIDED

ANTENNAE WITH SHORT HAIRS ONLY

ANTENNAE WITH TERMINAL BRISTLES

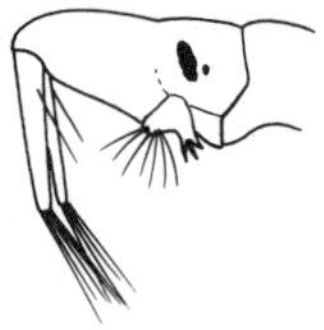

Chironomidae
(p. 309)

Thaumaleidae
(p. 307)

Culicidae
(p. 305)

Chaoboridae
(p. 303)

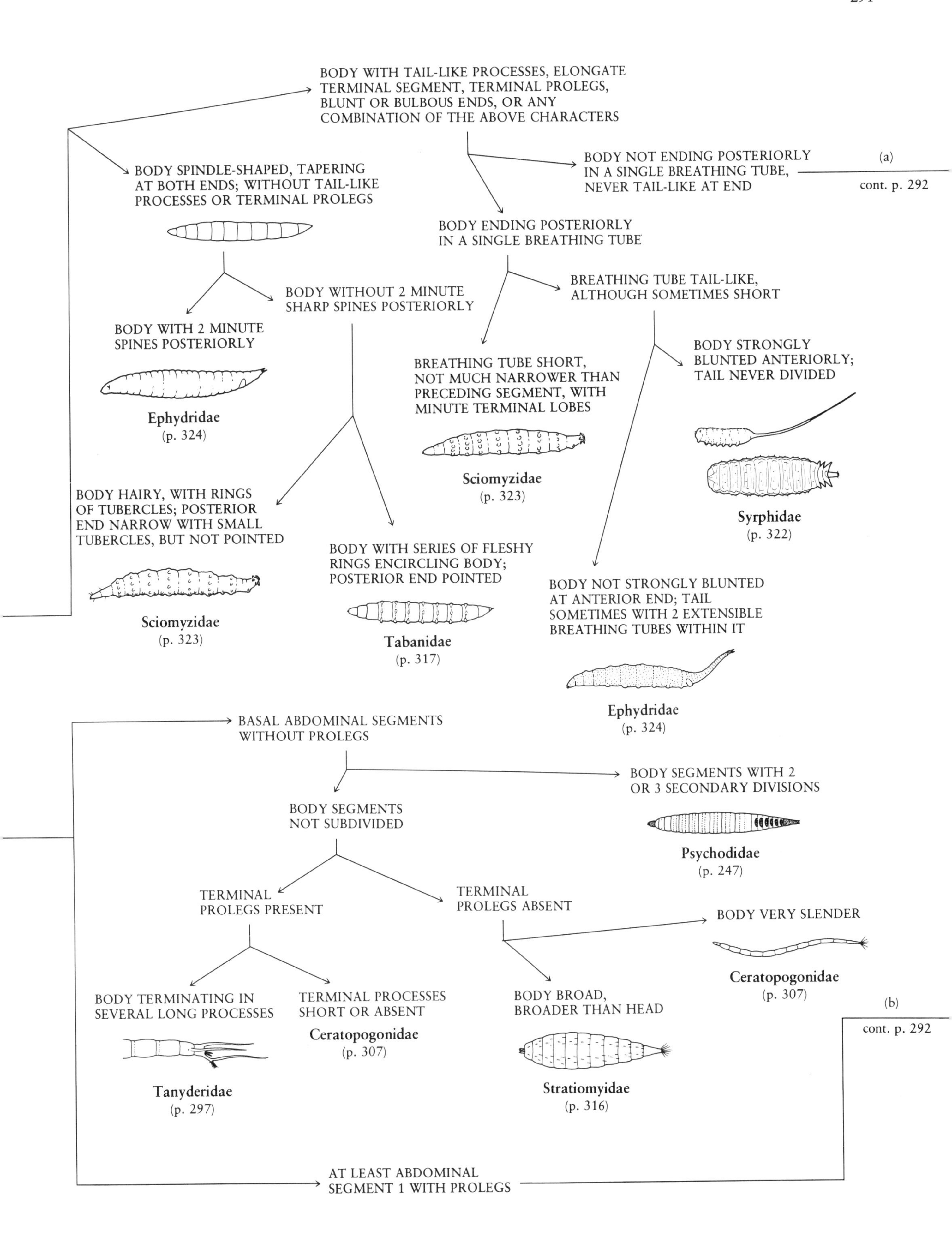
BODY WITH TAIL-LIKE PROCESSES, ELONGATE TERMINAL SEGMENT, TERMINAL PROLEGS, BLUNT OR BULBOUS ENDS, OR ANY COMBINATION OF THE ABOVE CHARACTERS
BODY SPINDLE-SHAPED, TAPERING AT BOTH ENDS; WITHOUT TAIL-LIKE PROCESSES OR TERMINAL PROLEGS
BODY NOT ENDING POSTERIORLY IN A SINGLE BREATHING TUBE, NEVER TAIL-LIKE AT END
(a)
cont. p. 292
BODY ENDING POSTERIORLY IN A SINGLE BREATHING TUBE
BODY WITHOUT 2 MINUTE SHARP SPINES POSTERIORLY
BREATHING TUBE TAIL-LIKE, ALTHOUGH SOMETIMES SHORT
BODY WITH 2 MINUTE SPINES POSTERIORLY
Ephydridae
(p. 324)
BREATHING TUBE SHORT, NOT MUCH NARROWER THAN PRECEDING SEGMENT, WITH MINUTE TERMINAL LOBES
Sciomyzidae
(p. 323)
BODY STRONGLY BLUNTED ANTERIORLY; TAIL NEVER DIVIDED
Syrphidae
(p. 322)
BODY HAIRY, WITH RINGS OF TUBERCLES; POSTERIOR END NARROW WITH SMALL TUBERCLES, BUT NOT POINTED
Sciomyzidae
(p. 323)
BODY WITH SERIES OF FLESHY RINGS ENCIRCLING BODY; POSTERIOR END POINTED
Tabanidae
(p. 317)
BODY NOT STRONGLY BLUNTED AT ANTERIOR END; TAIL SOMETIMES WITH 2 EXTENSIBLE BREATHING TUBES WITHIN IT
Ephydridae
(p. 324)
BASAL ABDOMINAL SEGMENTS WITHOUT PROLEGS
BODY SEGMENTS WITH 2 OR 3 SECONDARY DIVISIONS
Psychodidae
(p. 247)
BODY SEGMENTS NOT SUBDIVIDED
TERMINAL PROLEGS PRESENT
TERMINAL PROLEGS ABSENT
BODY VERY SLENDER
Ceratopogonidae
(p. 307)
BODY TERMINATING IN SEVERAL LONG PROCESSES
Tanyderidae
(p. 297)
TERMINAL PROCESSES SHORT OR ABSENT
Ceratopogonidae
(p. 307)
BODY BROAD, BROADER THAN HEAD
Stratiomyidae
(p. 316)
(b)
cont. p. 292
AT LEAST ABDOMINAL SEGMENT 1 WITH PROLEGS

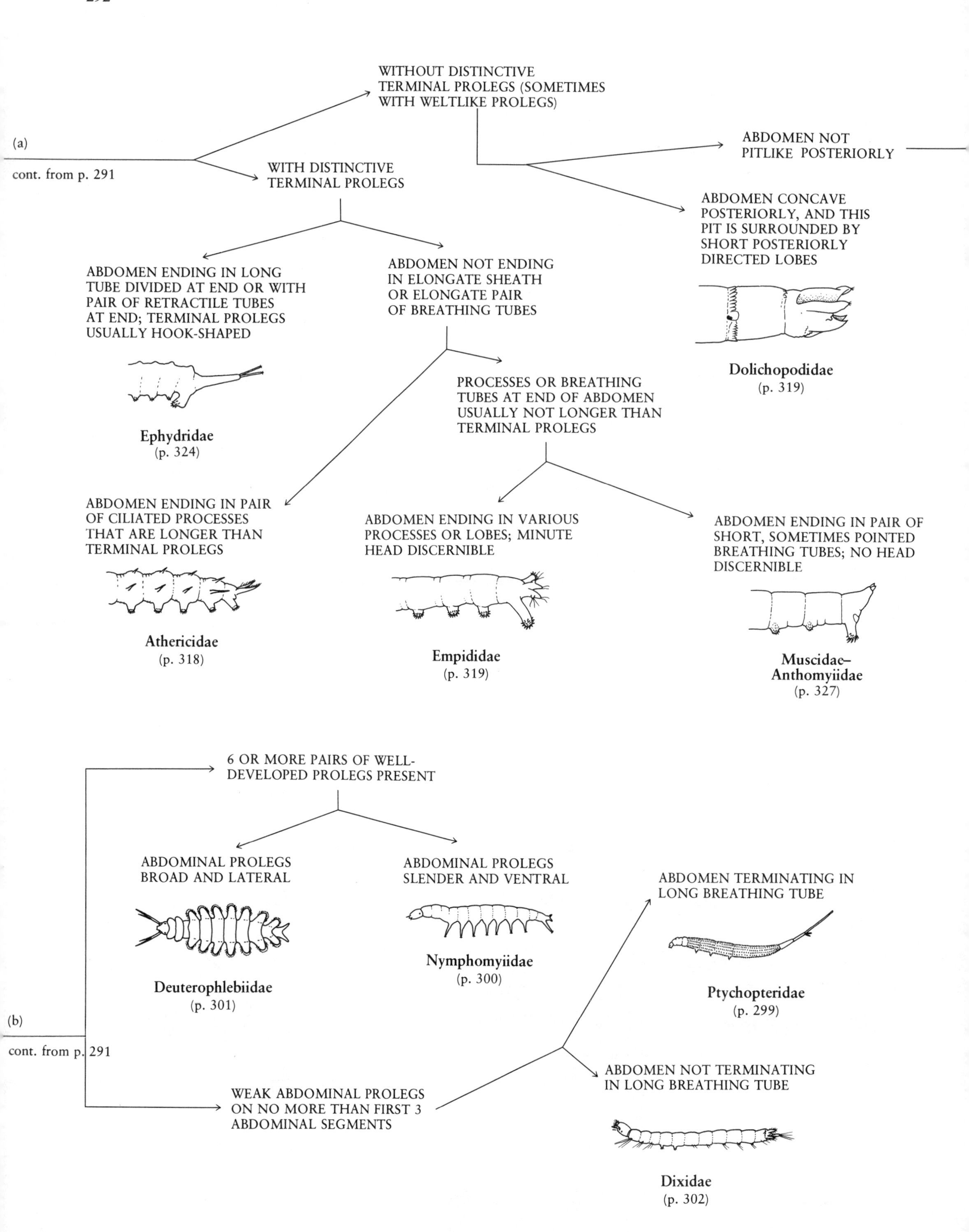
(a)
cont. from p. 291
WITHOUT DISTINCTIVE TERMINAL PROLEGS (SOMETIMES WITH WELTLIKE PROLEGS)
ABDOMEN NOT PITLIKE POSTERIORLY
ABDOMEN CONCAVE POSTERIORLY, AND THIS PIT IS SURROUNDED BY SHORT POSTERIORLY DIRECTED LOBES
Dolichopodidae
(p. 319)
WITH DISTINCTIVE TERMINAL PROLEGS
ABDOMEN ENDING IN LONG TUBE DIVIDED AT END OR WITH PAIR OF RETRACTILE TUBES AT END; TERMINAL PROLEGS USUALLY HOOK-SHAPED
Ephydridae
(p. 324)
ABDOMEN NOT ENDING IN ELONGATE SHEATH OR ELONGATE PAIR OF BREATHING TUBES
PROCESSES OR BREATHING TUBES AT END OF ABDOMEN USUALLY NOT LONGER THAN TERMINAL PROLEGS
ABDOMEN ENDING IN PAIR OF CILIATED PROCESSES THAT ARE LONGER THAN TERMINAL PROLEGS
Athericidae
(p. 318)
ABDOMEN ENDING IN VARIOUS PROCESSES OR LOBES; MINUTE HEAD DISCERNIBLE
Empididae
(p. 319)
ABDOMEN ENDING IN PAIR OF SHORT, SOMETIMES POINTED BREATHING TUBES; NO HEAD DISCERNIBLE
Muscidae–Anthomyiidae
(p. 327)
(b)
cont. from p. 291
6 OR MORE PAIRS OF WELL-DEVELOPED PROLEGS PRESENT
ABDOMINAL PROLEGS BROAD AND LATERAL
Deuterophlebiidae
(p. 301)
ABDOMINAL PROLEGS SLENDER AND VENTRAL
Nymphomyiidae
(p. 300)
WEAK ABDOMINAL PROLEGS ON NO MORE THAN FIRST 3 ABDOMINAL SEGMENTS
ABDOMEN TERMINATING IN LONG BREATHING TUBE
Ptychopteridae
(p. 299)
ABDOMEN NOT TERMINATING IN LONG BREATHING TUBE
Dixidae
(p. 302)

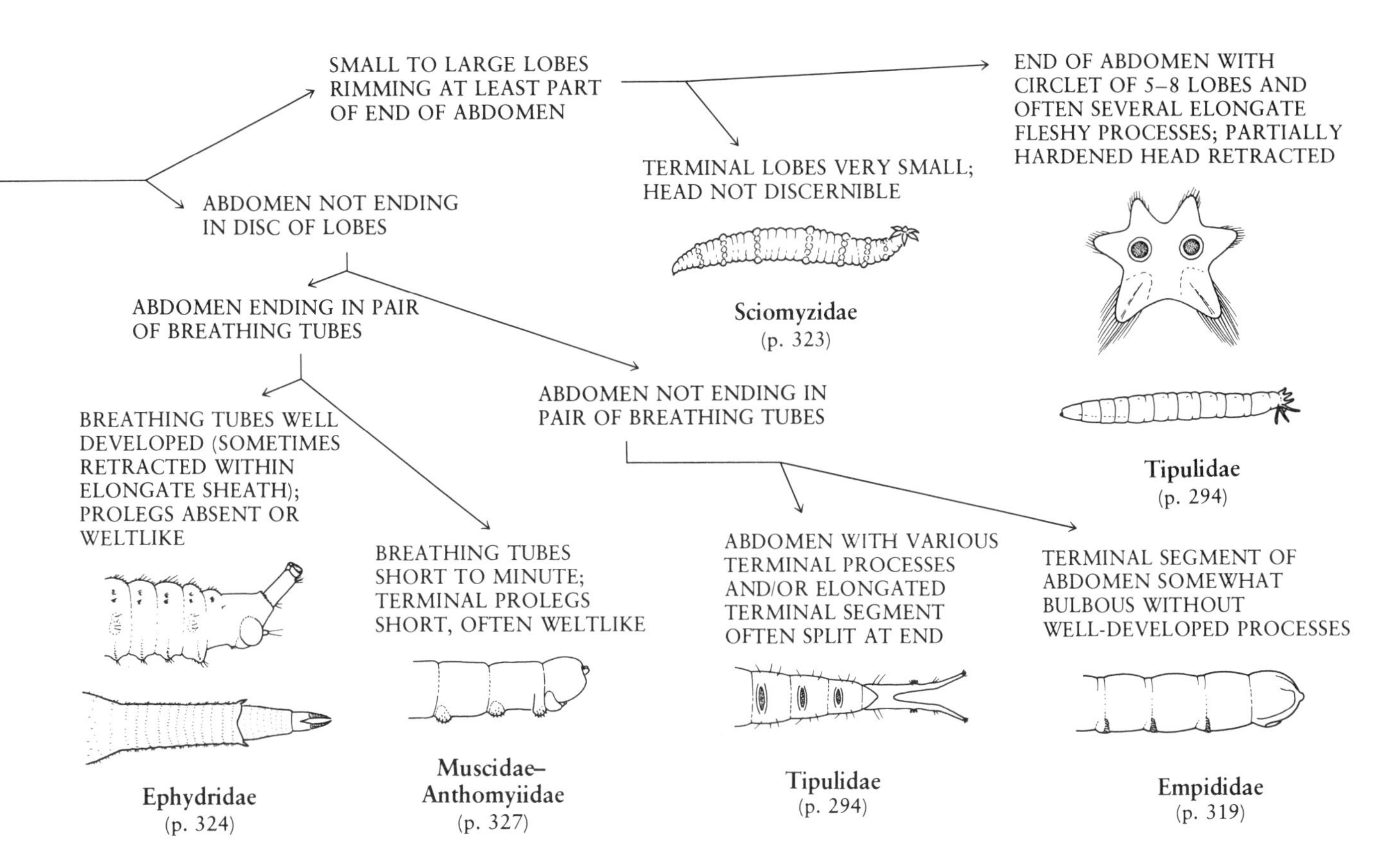

Wing venation (see illustrations) is of primary importance when identifying adults to family, although many can be identified in the field by the general appearance of body and wings. The head and its appendages are also used to some extent. Although adults may fit one of the family diagnoses given, keep in mind that many adults, especially among the higher Diptera, do not necessarily have aquatic immature stages, since many of the families treated here include terrestrial as well as aquatic species.

Larval identification is based mostly on the presence or absence of body appendages and their various modifications. These include such structures as terminal breathing tubes or siphons, mouth brushes, terminal brushes, and prolegs. The term **proleg** refers to the commonly present leglike structures of the posterior end of the abdomen and the anterior end of the thorax as well as other fleshy lobes or leglike structures along the abdomen. When the head capsule is partially or fully retracted into the thorax or when the head is reduced, it may be difficult for the beginner to distinguish the anterior and posterior ends without close examination. If the head is not apparent, the tapered or smoothly rounded end without appendages is the anterior, whereas the posterior is blunt, expanded, or extended and often possesses hairs or tail-like, leglike, lobelike, spinelike, or papillar structures, although these are sometimes minute.

Family characteristics used for identifying larvae are usually recognizable on puparia and often suffice for identifying the pupa of many Brachycera. Pupal characteristics are given in the discussions under each family treatment.

In the higher Diptera, family identification is sometimes difficult and occasionally impossible when limited to larvae and pupae. Do not rely solely on the picture key for identification; always check figures and discussions in the text.

Longhorned Flies
(Suborder Nematocera)

Larvae possess a head that is usually well developed and conspicuous. If the head is incomplete or retracted into the thorax (see crane flies, Figs. 16.2, 16.3, 16.8), then mandibles move in opposition to each other in a horizontal plane, and the body often tapers anteriorly and possesses various lobelike processes posteriorly.

Pupae are not encased in a puparium, and developing adult structures are apparent. Developing antennae usually extend over eyes and reach the bases of developing wings. Thoracic respiratory horns or spiracular gills are usually well developed.

Adults usually have slender, many-segmented, sometimes hairy antennae. Body is often small and gnatlike or midgelike. The medium- and large-sized adults usually have very long legs. Small flies may or may not have long legs.

CRANE FLIES
(Family Tipulidae)

LARVAL DIAGNOSIS: (Figs. 16.2–16.4, 16.7–16.13; Plate XIV, Fig. 97) Crane fly larvae are often peglike (pointed or somewhat rounded anteriorly and more-or-less blunt and expanded posteriorly) and measure 10–25 mm when mature, sometimes less or as much as 100 mm. Head capsule is incomplete posteriorly and is usually retracted partially or fully into thorax. Abdomen is usually cylindrical but may be somewhat flattened; it commonly possesses small lobes or creeping welts and, less commonly, leaflike or long spinelike processes. End of abdomen in most species is a spiracular disc consisting of variously developed lobes and processes surrounding posterior spiracles. In a few species, the abdomen terminates in a pair of elongate processes.

Those who have seen tipula, or long legs, and the larger kind of gnat, have most probably mistaken the one for the other, they have often accused the tipula, a harmless insect, of depredations made by the gnat, and the innocent have suffered for the guilty. . . . The gnat is sanguinary and predaceous, ever seeking out for a place in which to bury its trunk, and pumping up the blood from the animal in large quantities.

Goldsmith

ADULT DIAGNOSIS: (Fig. 16.6; Plate XIV, Fig. 98) These are generally medium-sized to large mosquitolike flies with extremely long legs. Thorax has a V-shaped suture dorsally. Wings have two anal veins reaching margin, but wings of some are reduced or absent.

DISCUSSION: Crane fly larvae (also commonly known as "leather jackets") are distinguished from other aquatic nematocerans by the incomplete or retracted head, and from other superficially similar aquatic larvae of higher Diptera by the opposable mandibles. The general peglike form of the body is typical of most species.

Those few pupae that occur in water usually have short thoracic respiratory horns (Fig. 16.5). These horns are modified into spiracular gills in a few species. The straight legs that extend beyond the developing wings are typical of tipulid pupae.

A number of tipulid genera contain both aquatic and semiaquatic species as well as terrestrial species. Several hundred species in over 30 genera occur in aquatic or marginally aquatic habitats in North America. Several of the genera are widespread, and the large number of species makes this an important group.

Many species are semiaquatic as larvae, preferring marginal and moist areas. Most aquatic species inhabit streams, where such genera as *Hexatoma* and *Tipula* are often common. Stream species are generally bottom dwellers, and many occur amongst algal growths and woody debris. Species of *Helius* and *Phalacrocera,* for example, are found in ponds and

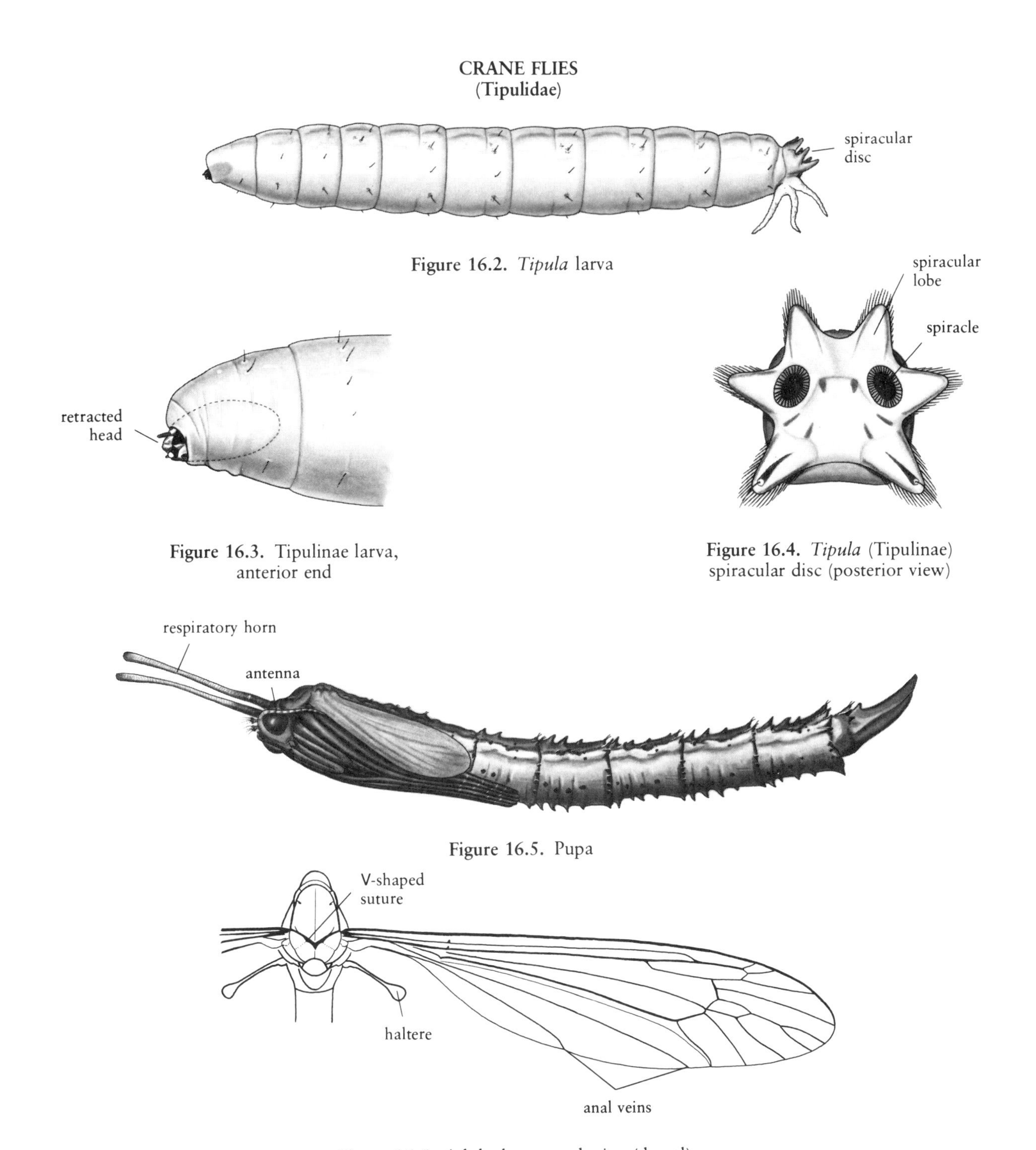

Figure 16.2. *Tipula* larva

Figure 16.3. Tipulinae larva, anterior end

Figure 16.4. *Tipula* (Tipulinae) spiracular disc (posterior view)

Figure 16.5. Pupa

Figure 16.6. Adult thorax and wing (dorsal)

marshes. Larvae of a few species live within the underwater parts of emergent vegetation, and some even occur at the surface of lentic waters.

Most aquatic larvae are aeropneustic, utilizing their posterior spiracles for obtaining air, although those occurring in well-oxygenated water may obtain dissolved oxygen in the water through the body wall for extended periods of time. Some larvae, such as those of the genus *Antocha,* are hydropneustic and lack open spiracles. However, the pupae of *Antocha* and *Lipsothrix* possess spiracular gills, an apparent adaptation for living in stream environments subject to periodic drying or inundation by water.

CRANE FLIES
(Tipulidae)

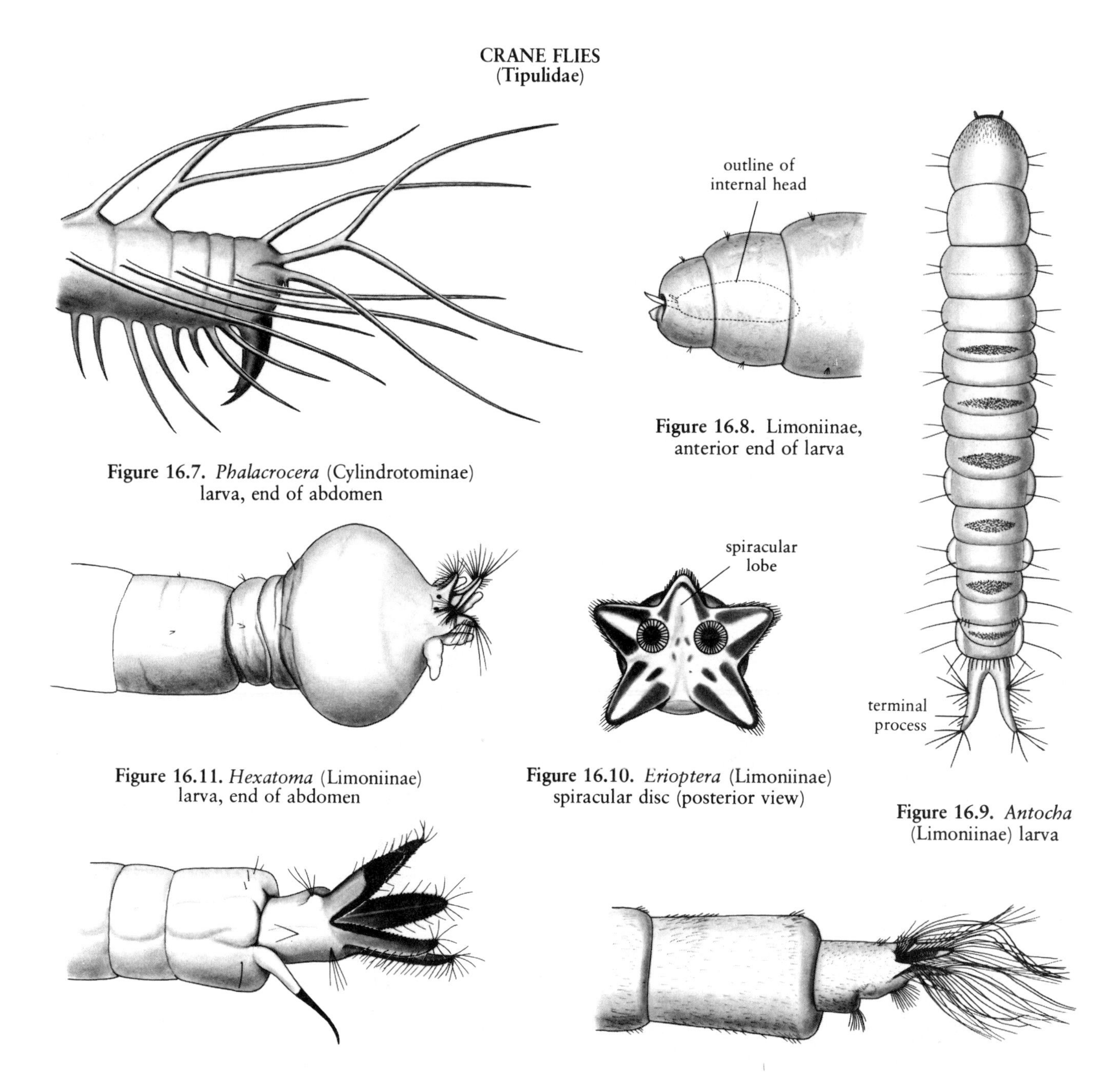

Figure 16.7. *Phalacrocera* (Cylindrotominae) larva, end of abdomen

Figure 16.8. Limoniinae, anterior end of larva

Figure 16.9. *Antocha* (Limoniinae) larva

Figure 16.10. *Erioptera* (Limoniinae) spiracular disc (posterior view)

Figure 16.11. *Hexatoma* (Limoniinae) larva, end of abdomen

Figure 16.12. *Ormosia* (Limoniinae) larva, end of abdomen

Figure 16.13. *Pseudolimnophila* (Limoniinae) larva, end of abdomen

Most species have a one-year life cycle, although two generations a year are also common. Pupation usually takes place in marginal areas. Food habits range from active predation to periphyton feeding. Woody detritus is ingested by a few species, and these evidently have resident gut bacteria that function in the digestion of cellulose. Many are omnivorous, especially when young. Interestingly, a posterior segment of the abdomen of a few species is swollen and bulbous, a characteristic that becomes conspicuous in specimens preserved in fluids.

The large, soft-bodied larvae of crane flies are juicy morsels for trout and other game fishes. They are sold in some Midwestern bait shops for bass fishing and are called "spikes." A number of tied flies are patterned after crane fly larvae and used for nymph fishing by fly fishermen. These wormlike patterns are best fished in spring and early summer, since many of the trout stream tipulids are early emergers. The Orange Crane Fly, a species of *Tipula* that occurs in central and eastern North America, is a crane fly for which both adult and larval patterns have been devised. The Yellow Spider is a common adult crane fly pattern.

SUBFAMILY TIPULINAE: **Larvae** (Figs. 16.2–16.4; Plate XIV, Fig. 97) have a spiracular disc surrounded by six or rarely eight short lobes, and usually two of these lobes are dorsal. The group is widespread in North America.

SUBFAMILY CYLINDROTOMINAE: **Larvae** (Fig. 16.7) have fleshy, tapering protuberances or spinelike structures along the body dorsally and laterally. The group is *not* known from southwestern and south-central states or Florida.

SUBFAMILY LIMONIINAE: **Larvae** (Figs. 16.8–16.13) usually have a spiracular disc surrounded by five or rarely seven variously shaped lobes, and usually only one of these lobes is dorsal. In some, spiracles are absent and lateral terminal lobes are reduced or absent. The group is widespread in North America.

PRIMITIVE CRANE FLIES
(Family Tanyderidae)

LARVAL DIAGNOSIS: (See key Fig. 16.1) These slender and elongate larvae are 12–18 mm when mature. Body lacks processes except terminally, where it possesses six long spinelike filaments, a pair of narrow terminal prolegs, and four thin-walled, shorter filaments.

ADULT DIAGNOSIS: Adults are similar to crane flies, but R vein is five-branched.

DISCUSSION: This is a rare family consisting of two genera and about four species in North America. It is *not* known from central Canada, central states, or Texas. The distinct larvae usually burrow into silt, sand, and gravel in shallow areas of streams and rivers. Pupae are similar to those of crane flies.

MOTH FLIES
(Family Psychodidae)

LARVAL DIAGNOSIS: (Figs. 16.14, 16.15; Plate XV, Fig. 102) Larvae are small, cylindrical or somewhat flattened forms and are usually less than 5 mm long. Body lacks prolegs, but body segments are usually subdivided into annuli; at least some of these annuli possess small dorsal plates. Thorax is not distinctly thicker than rest of body. Flattened forms possess a row of attachment discs on the ventral side.

ADULT DIAGNOSIS: (Fig. 16.17; Plate XV, Fig. 104) Adults are small, mothlike or gnatlike forms. Wings are short, broad, pointed, held rooflike over the body, and usually hairy.

DISCUSSION: Moth fly larvae are distinguished from those biting midge larvae that also entirely lack prolegs by their secondary body segmentation (annuli), a characteristic not present in any other aquatic nematoceran larvae. Although some moth flies possess ventral attachment discs, they are very distinctive from the netwinged midges, which also have a series of ventral attachment discs but whose bodies have several deep lateral constrictions.

MOTH FLIES
(Psychodidae)

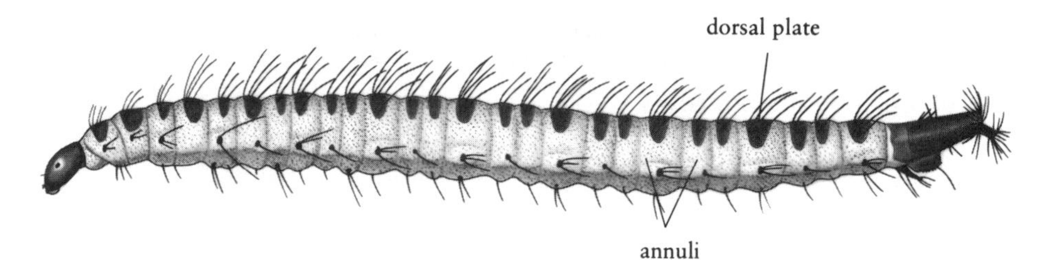

Figure 16.14. *Telmatoscopus* larva

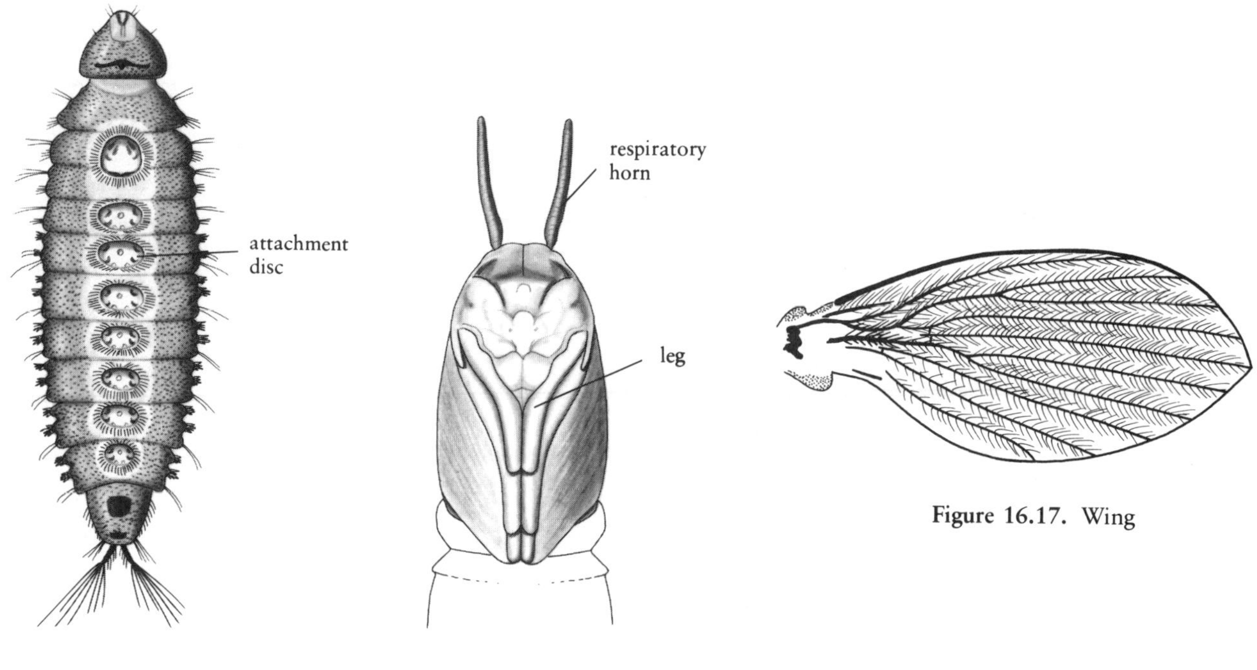

Figure 16.15. *Maruina* larva (ventral)

Figure 16.16. Pupa, head and thorax (ventral)

Figure 16.17. Wing

The aquatic pupae do not have cocoons. Those that attach themselves to the substrate in rivers and streams are flatter than other fly pupae that live in the same habitat. Unattached forms (Fig. 16.16; Plate XV, Fig. 103) possess curved developing legs and a thorax that is not strongly arched.

The widespread subfamily Psychodinae contains four genera with over 50 aquatic or semiaquatic species in North America.

Species of *Maruina* inhabit stream bottoms in western North America. Other species are usually found along the margins of streams and ponds, in floating moss or algae, or rarely in open areas of ponds. Some species of *Psychoda* are commonly found in settling ponds or trickling filters of sewage treatment facilities or in other highly eutrophic environments, and some can tolerate hot water, detergents and harsh chemicals, and even the urinals in men's restrooms.

Both larvae and adults of most species feed on microorganisms and fine detritus. *Phlebotomus* species (sand flies), which develop in moist sand of beaches and similar environments, are serious biting pests in southern states.

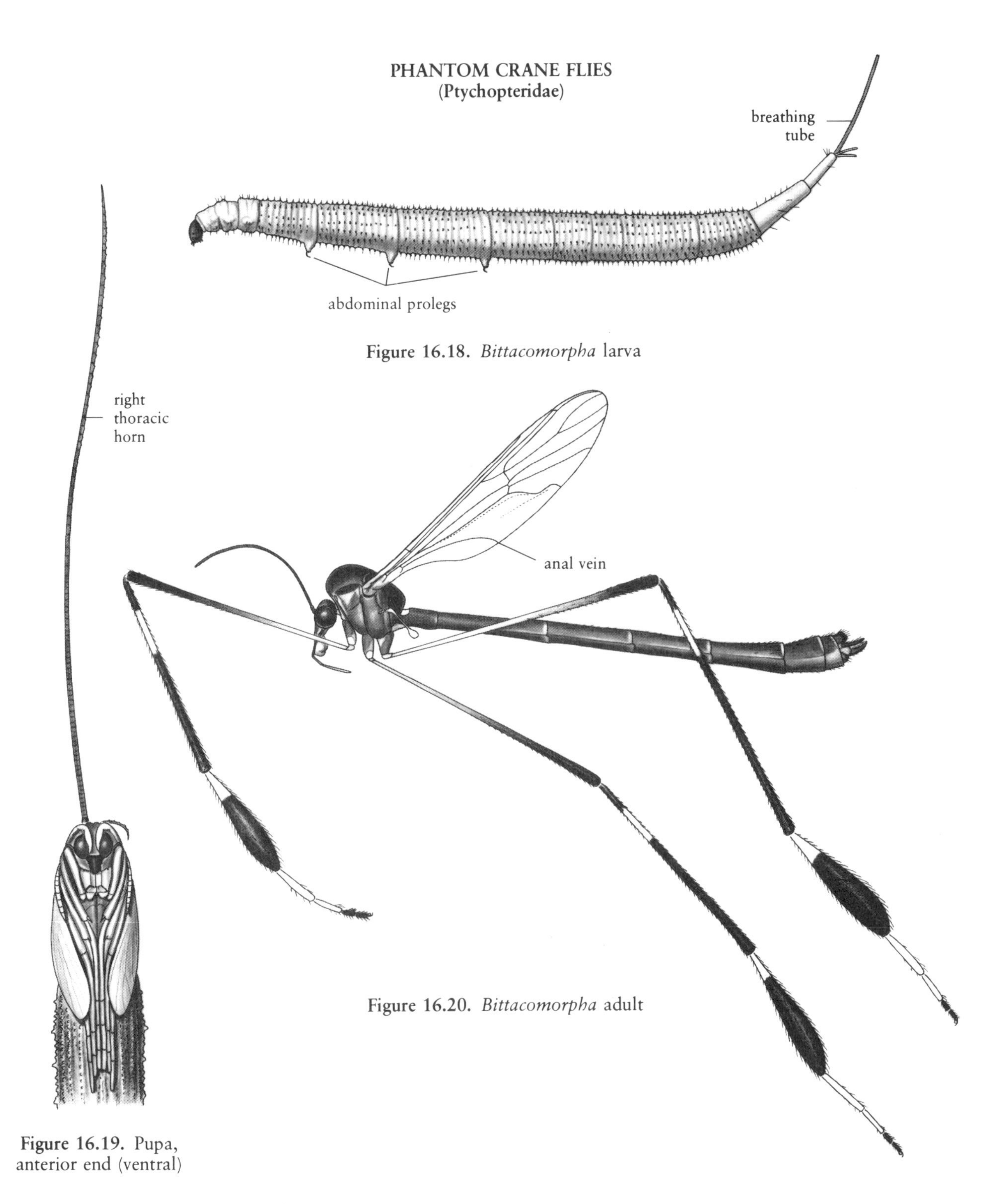

Figure 16.18. *Bittacomorpha* larva

Figure 16.20. *Bittacomorpha* adult

Figure 16.19. Pupa, anterior end (ventral)

PHANTOM CRANE FLIES
(Family Ptychopteridae)

LARVAL DIAGNOSIS: (Fig. 16.18) Body is extensible and measures about 15–60 mm when larvae are mature and fully extended. Thoracic and terminal prolegs are absent. Each of the first three abdominal segments possesses a ventral pair of prolegs. Terminal abdominal segment possesses a long, extensible breathing tube.

ADULT DIAGNOSIS: (Fig. 16.20) These generally resemble crane flies. Thorax has a V-shaped suture dorsally. Wings have only one anal vein that reaches the margin. Those with long legs often have the legs conspicuously banded with black and white.

DISCUSSION: The larvae of this group are unique. The pupae (Fig. 16.19) have a well-developed right thoracic horn and a degenerate left horn. This small but widespread group consists of about 15 species, three genera, and two subfamilies in North America.

Larvae occur in rich detrital areas of shallow lentic environments, usually at depths of less than two or three centimeters. They maintain contact with the surface via the breathing tube. They are also able to breathe hydropneustically, and some burrow into deep bottom sediment for prolonged periods when the water freezes over. The long outstretched legs of the adults of *Bittacomorpha* allow individuals to drift in the wind.

SUBFAMILY PTYCHOPTERINAE: **Larvae** possess weakly developed prolegs. Mandibles have three outer teeth. The group is widespread in North America.

SUBFAMILY BITTACOMORPHINAE: **Larvae** (Fig. 16.18) have well-developed prolegs. Mandibles have only one large outer tooth. The group is widespread in North America.

WATERNYMPH FLIES
(Family Nymphomyiidae)

LARVAL DIAGNOSIS: (See key Fig. 16.1) These are small forms with nine apparent abdominal segments. Abdominal segments 1–7 each have a ventral pair of two-segmented prolegs. Terminal segment possesses a pair of prolegs.

ADULT DIAGNOSIS: Adults are very small, unusual forms. Antennae are clubbed; a long third segment sometimes appears subsegmented; second segment is cuplike. Wings are straplike, much longer than the body, and have long marginal hairs. Wings are sometimes absent or broken off at base.

DISCUSSION: The one aquatic North American species of this rare and recently discovered family occurs in eastern Canada. Larvae are found in fast-flowing streams, clinging to mossy vegetation. Adults shed their wings and may enter the water after mating.

NETWINGED MIDGES
(Family Blephariceridae)

LARVAL DIAGNOSIS: (Figs. 16.21, 16.22) These are dorsoventrally flattened forms, usually 4–12 mm when mature. Body is divided into seven distinct regions (the last less distinct) demarked by six lateral constrictions. The head and thorax are fused with abdominal segment 1 to form the first body region. Each of the first six regions possesses a ventral attachment disc.

ADULT DIAGNOSIS: (Fig. 16.24) Adults are long-legged forms that resemble small, broad-winged crane flies. Angle of wing is distinctly produced at the inner hind margin. Wings often have a secondary netlike venation between major veins.

DISCUSSION: Netwinged midge larvae are distinctive and can be easily distinguished from mountain midges, which also possess lateral constrictions, by the presence of ventral attachment discs in the former group. Pupae (Fig. 16.23), which occur on stream substrates, are convex dorsally and flattened ventrally. The thorax of the pupa possesses a pair of spiracular gills usually consisting of two (four in some) thin plates on each side.

NETWINGED MIDGES
(Blephariceridae)

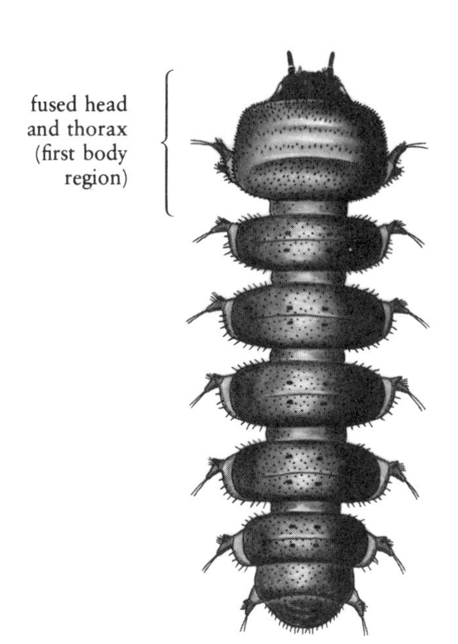

Figure 16.21. *Bibiocephala* larva

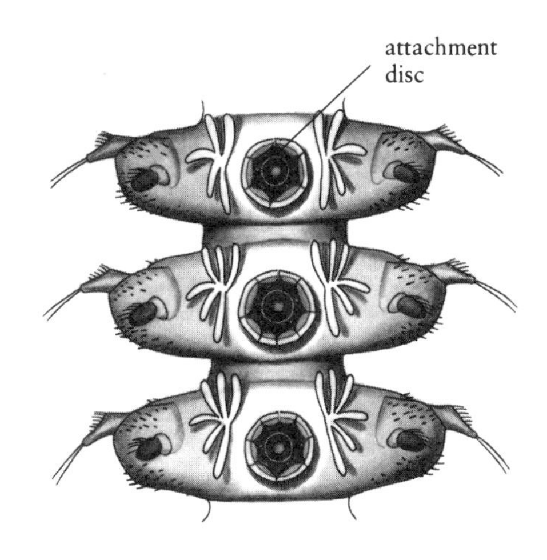

Figure 16.22. *Bibiocephala* larva, abdomen in part (ventral)

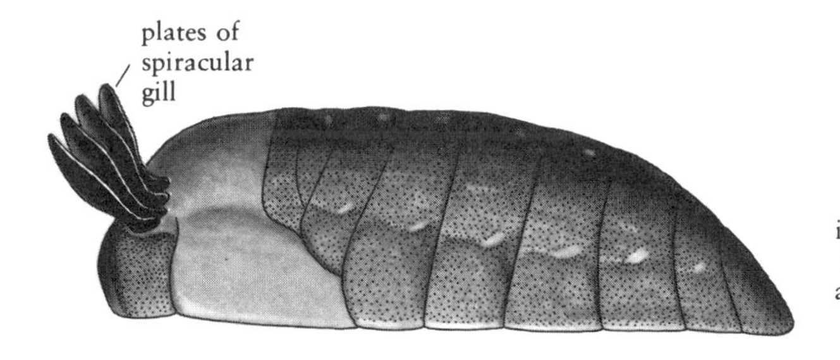

Figure 16.23. *Bibiocephala* pupa (lateral)

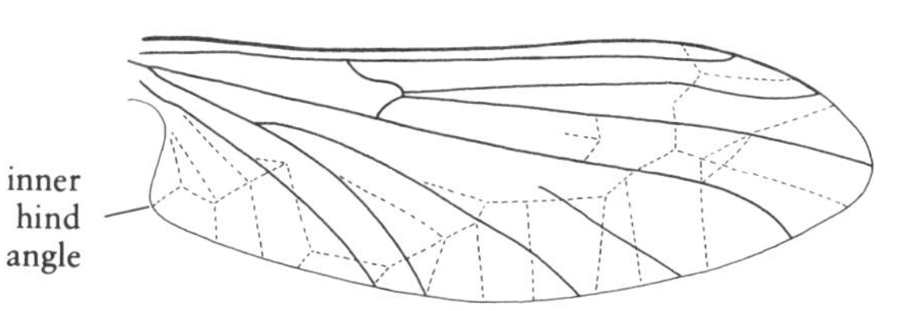

Figure 16.24. Wing

The specialized structure of the spiracular gills of these pupae distinguish them from pupae of mountain midges.

Over 20 species (in five genera) are generally restricted to mountainous stream environments of either eastern or western North America. *Blepharicera* is the only eastern genus. Netwinged midges are *not* known from Florida or Texas or most of the central states. They are very locally known from northern central states and central Canada, where they occur in cold, swift streams.

Larvae feed primarily on diatoms scraped from the rocky substrate to which they cling. They are often an important trout food in mountain streams. Adults emerge under the water in shallow areas. The females prey on other insects, including midges, whereas the males feed on nectar or do not feed at all. Females enter the water to cement their eggs to the substrate.

MOUNTAIN MIDGES
(Family Deuterophlebiidae)

LARVAL DIAGNOSIS: (See key Fig. 16.1) These are small, flattened forms, usually 3–6 mm when mature. Head and all three segments of thorax are distinct. Antennae are branched. Abdomen has seven pairs of lateral prolegs that give an appearance of a strongly constricted body when viewed from above.

ADULT DIAGNOSIS: These are gnatlike forms. Antennae are sometimes extremely long. Wings are large, fanlike, and broadest in the basal region.

DISCUSSION: This small (five species) and rare group is represented by a single genus known only from western Canada and western states. The hydropneustic larvae and pupae may be found clinging to rocky substrates in swift currents or to mossy vegetation in slackwater areas of streams. Hooklets at the ends of the prolegs aid in anchoring the larvae.

Adults usually emerge at dawn and live for only one or two hours.

DIXID MIDGES
(Family Dixidae)

LARVAL DIAGNOSIS: (Figs. 16.25, 16.26) These are elongate, slender forms, usually 3–7 mm at maturity. Head possesses mouth brushes and simple antennae. Thorax has three distinct segments, is not greatly en-

DIXID MIDGES
(Dixidae)

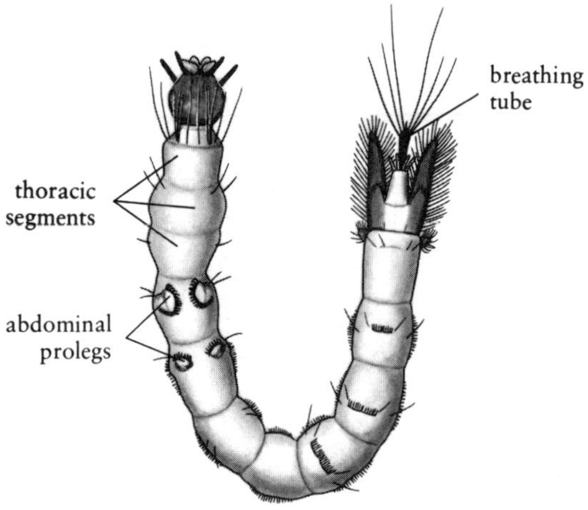

thoracic segments
abdominal prolegs
breathing tube

Figure 16.25. *Dixa* larva (ventral)

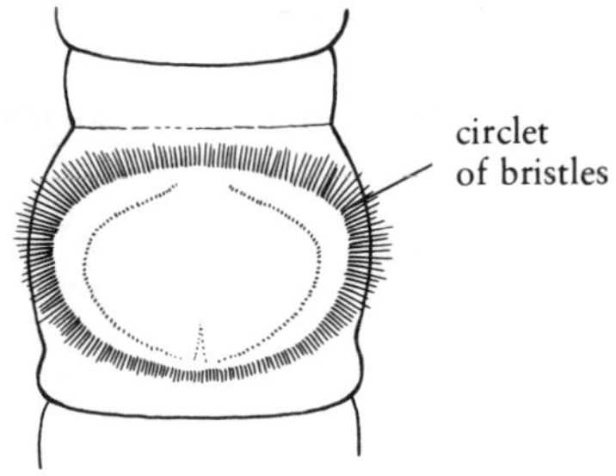

circlet of bristles

Figure 16.26. Dixinae larva, abdominal segment (dorsal)

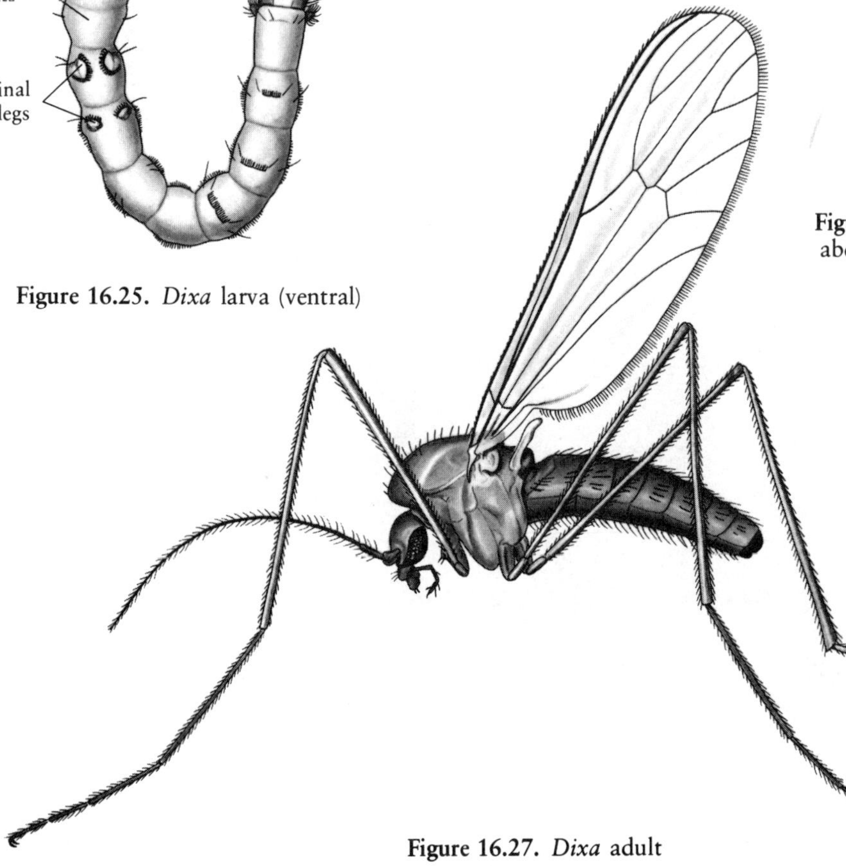

Figure 16.27. *Dixa* adult

larged, and is broadly joined to the abdomen. Abdomen possesses a small pair of ventral prolegs on segment 1 and usually another pair on segment 2, and terminates in a median breathing tube and lateral paddlelike structures.

ADULT DIAGNOSIS: (Fig. 16.27) Adults are mosquitolike. Mouthparts are not developed into a sucking proboscis. Wings are long and narrow, lack scales, and are held flat over the abdomen at rest.

DISCUSSION: The larvae of dixid midges are similar to those of mosquitoes and phantom midges. The thoracic segments of the last two groups, however, are distinctly fused and, in most species, thickened. The pupae are also somewhat similar to those of mosquitoes and phantom midges, but they are often inactive and attached above the waterline, although some are found floating on their sides at the surface.

Three genera occur in North America. The genera *Dixa* and *Dixella* together contain over 40 species.

Larvae usually occur in the slower reaches of small streams and in the shallows of lakes, ponds, and marshes. They bow the body into a U-shape and then straighten out when moving. They remain near the surface of the water and sometimes crawl about on partially exposed substrates.

The adults are short-lived and do not bite.

SUBFAMILY DIXINAE: **Larvae** (Figs. 16.25, 16.26) have two pairs of abdominal prolegs. Abdominal segments possess circlets of bristles dorsally. The group is widespread in North America except for Texas and Florida.

SUBFAMILY PARADIXINAE: **Larvae** have two pairs of abdominal prolegs. Abdominal segments do not possess circlets of bristles dorsally. The group is widespread in North America.

SUBFAMILY MERINGODIXINAE: **Larvae** have only one pair of abdominal prolegs. The group is known only from southwestern states.

PHANTOM MIDGES
(Family Chaoboridae)

LARVAL DIAGNOSIS: (Figs. 16.28–16.30) Larvae are similar to those of mosquitoes. Some, however, are somewhat transparent and do not have a thickened thorax. Antennae are provided with several long spinelike hairs. Mouth brushes are not present.

ADULT DIAGNOSIS: These are similar to mosquitoes, but mouthparts do not form a long proboscis, and wing scales, when present, are mostly located along the margins.

DISCUSSION: The specialized antennae distinguish the larvae of phantom midges from those of mosquitoes. The pupae, which are active and aquatic, are similar to mosquitoes but differ either in having the thoracic respiratory horns closed and somewhat elliptical (Fig. 16.31) or, if open, in having the spiracular openings at about half way up the length of the horns (Fig. 16.32).

Four genera and about 15 species are represented in North America.

Phantom midge larvae inhabit lakes and ponds, small intermittent pools, and snow melt. All are predaceous, feeding on crustaceans and other insects such as mosquitoes. Rotifers may be heavily preyed on, especially by the earlier instars. Larger items such as copepods become more important as prey items for the larger larvae. Aeropneustic species usually float quietly at the surface, maintaining contact with air and then quickly grabbing would-be prey with their raptorial antennae.

PHANTOM MIDGES
(Chaoboridae)

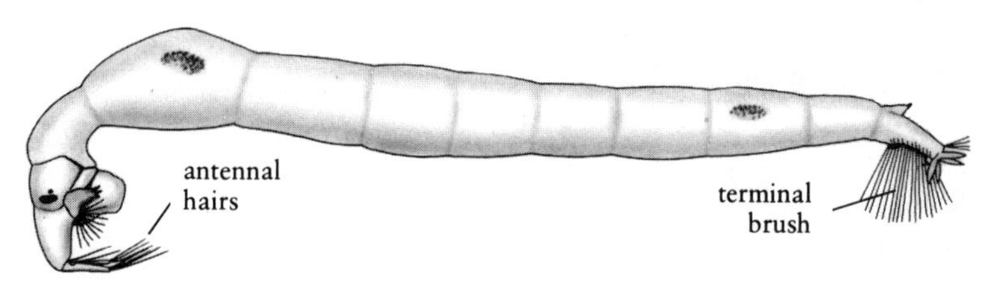

Figure 16.28. *Chaoborus* (Chaoborinae) larva

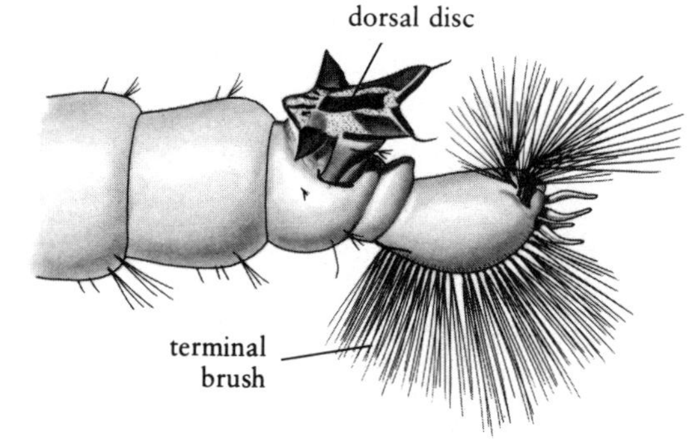

Figure 16.29. Eucorethrinae larva, end of abdomen

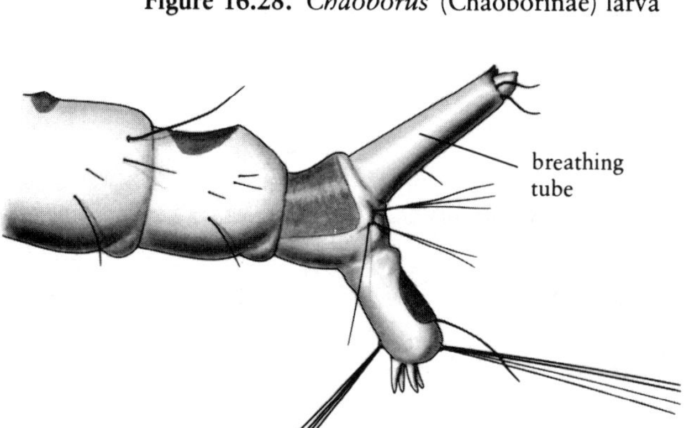

Figure 16.30. Corethrellinae larva, end of abdomen

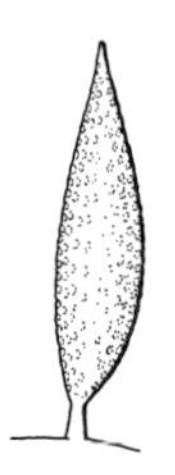

Figure 16.31. Pupal respiratory horn

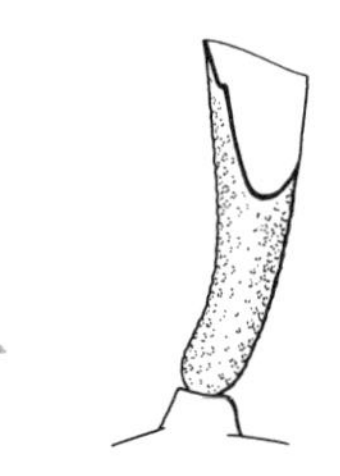

Figure 16.32. Pupal respiratory horn

Some hydropneustic species (e.g., in *Chaoborus*) inhabit deep-water regions of lakes, where they may remain on the bottom or burrow during daylight hours and move to open-water areas at night or when light is reduced by ice or snow cover. Specialized air bladders (serving as hydrostatic organs) aid in the vertical movement of these species. Those occurring in reservoirs have been known to enter drinking water supplies and clog filters.

Adults are short-lived, nonbiting, and sometimes emerge in masses from lakes.

SUBFAMILY CHAOBORINAE: **Larvae** (Fig. 16.28) are generally 6 mm or less and possess a well-developed terminal brush of hairs. Terminal breathing tube is present in a few species. The group is widespread in North America.

SUBFAMILY EUCORETHRINAE: **Larvae** (Fig. 16.29) are about 15 mm when mature. They do not have a terminal breathing tube but do have a dorsal, lobed disc near the end of the abdomen. The group is known from most of North America but is not known from Texas, Florida, or other southeastern states.

SUBFAMILY CORETHRELLINAE: **Larvae** (Fig. 16.30) are usually less than 6 mm. Terminal brush of hairs is not well developed. The group is widespread in North America.

MOSQUITOES
(Family Culicidae)

LARVAL DIAGNOSIS: (Figs. 16.35, 16.38–16.40; Plate XIV, Fig. 99) These small or medium-sized larvae do not have prolegs. Head has mouth brushes and simple antennae. Thoracic segments are fused and are much thicker than the abdomen. Dorsal terminal breathing tube (siphon) is present in most species.

ADULT DIAGNOSIS: (Figs. 16.33, 16.34; Plate XIV, Fig. 101) Mouthparts are formed into an elongate proboscis. Wings have scales along the veins and are held flat over body at rest.

DISCUSSION: Both the larvae (also known as "wrigglers" or "wiggletails") and pupae ("tumblers") of mosquitoes are very distinctive. The fused head and thoracic region of the pupa (Figs. 16.36, 16.37; Plate XIV, Fig. 100) is greatly enlarged.

About 150 species occur in North America and are grouped in 12 genera, including the genus *Aedes* with its approximately 60 species.

Larvae and pupae occur in a wide variety of primarily lentic habitats, including lakes, ponds, marshes, swamps, bogs, tree holes, leaves of pitcher plants, edgewaters or backwaters of streams and rivers, and in fact any depressions or containers where water accumulates. They are often found in brackish waters and salt marshes.

All species are aeropneustic, and most either remain at the surface or surface regularly for air. The anopheline larvae lie in a horizontal position, whereas the culicine forms hang head down from the surface. *Mansonia* larvae and pupae, however, have a modified siphon (Fig. 16.40) with which they tap the underwater stems and roots of plants in order to obtain oxy-

MOSQUITOES
(Culicidae)

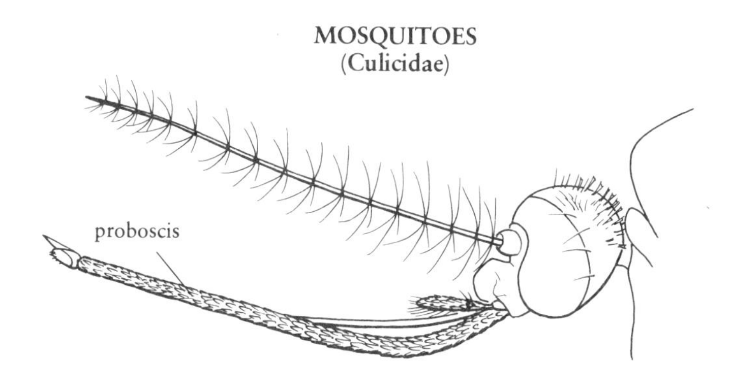

Figure 16.33. Adult head (lateral)

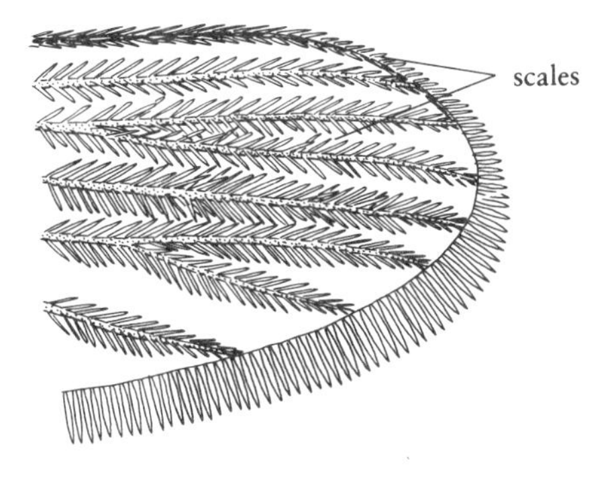

Figure 16.34. Wing tip

MOSQUITOES
(Culicidae)

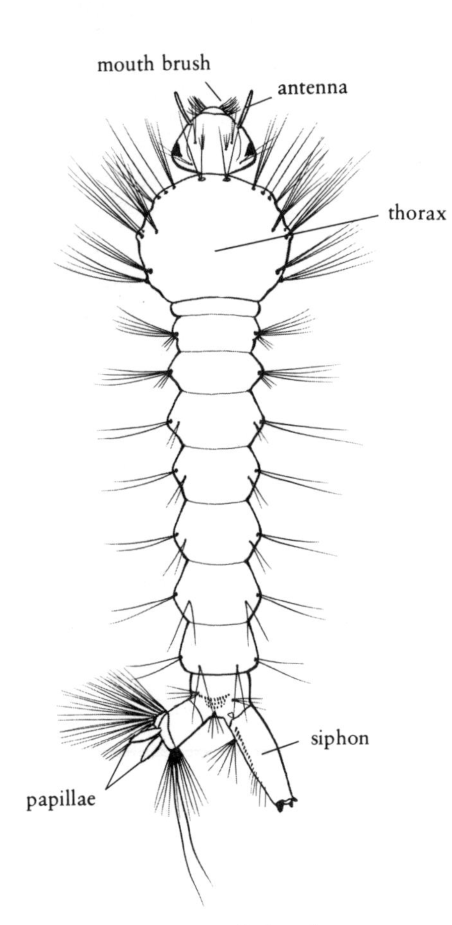

Figure 16.35. Culicinae larva

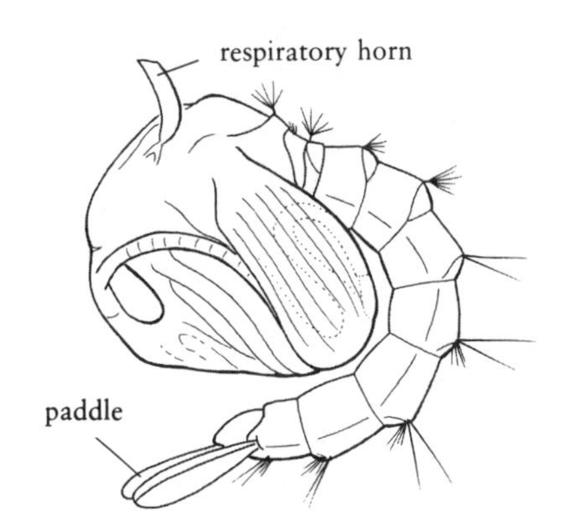

Figure 16.36. Pupa

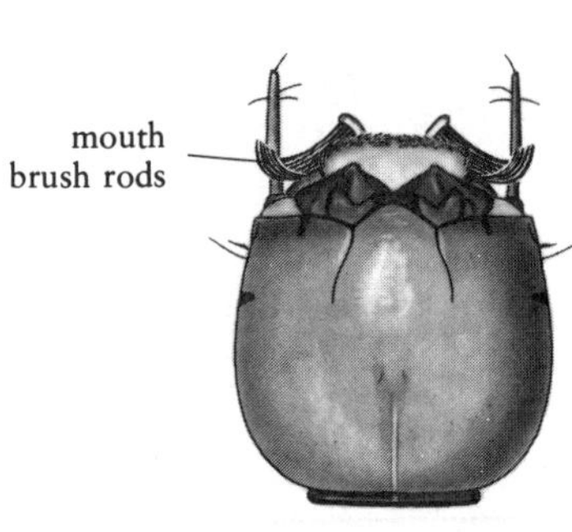

Figure 16.38. Toxorhynchitinae larva, head

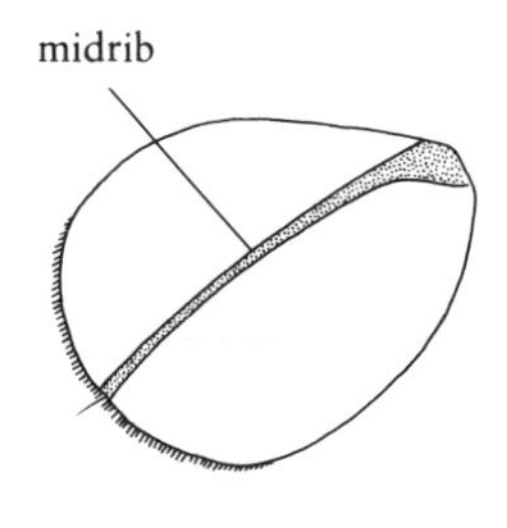

Figure 16.37. Pupal paddle

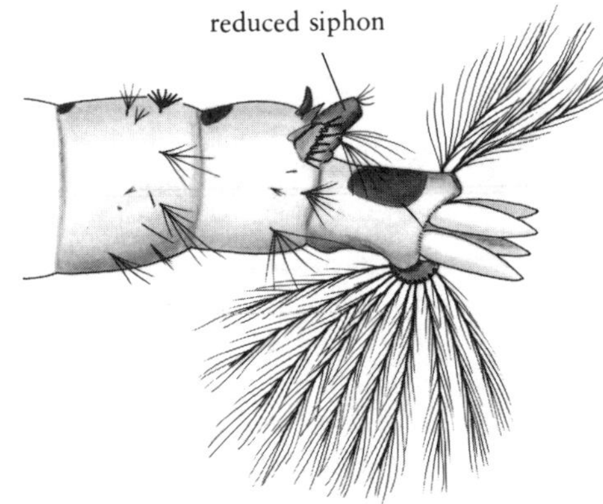

Figure 16.39. Anophelinae larva, end of abdomen

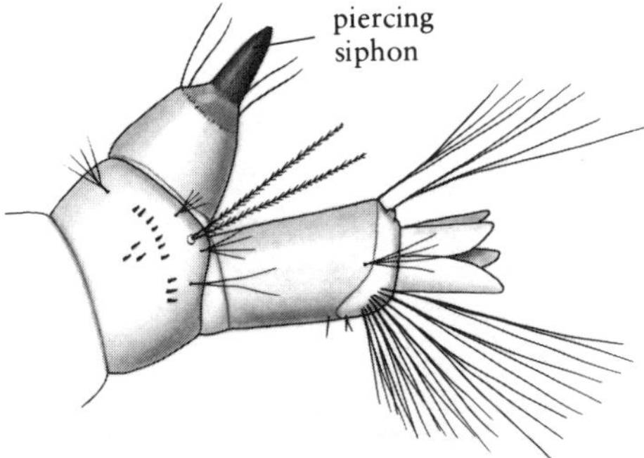

Figure 16.40. *Mansonia* (Culicinae) larva, end of abdomen

gen. Most larvae feed on floating or suspended microorganisms and detritus obtained by sweeping with the mouth brushes. Some *Psorophora* and *Toxorhynchites* are active predators that often feed on other mosquitoes. Most pupae remain near the surface, utilizing their thoracic respiratory horns. When disturbed, they readily swim or dive with a jerking, tumbling motion.

Eggs are oviposited in a variety of ways. Some species that commonly inhabit marshes and other habitats subject to water depth fluctuation lay eggs singly in moist areas where they remain dormant until submerged in water. Many species lay eggs singly or as rafts on the water surface. Many species, especially the *Psorophora* and some *Aedes* of temporary pools, complete development to the adult stage in as short a time as one week. There are often several generations per year. Adult males do not take blood. Females of most species are blood feeders, and many attack humans. Most feed at night or twilight, but a few (e.g., some western *Culex*) readily attack during daylight hours. Some species swarm, and some fly great distances.

Mosquitoes are of enormous medical and economic importance, not only from the standpoint of disease transmission but also as pests of humans and other animals. Species of *Culex* and *Aedes* are vectors of various strains of encephalitis throughout parts of North America. *Anopheles* remains a potential vector of malaria in warmer regions, although this disease has been kept under control in recent times.

SUBFAMILY ANOPHELINAE: **Larvae** (Fig. 16.39) lack a well-developed siphon. The group is widespread in North America.

SUBFAMILY TOXORHYNCHITINAE: **Larvae** (Fig. 16.38) have a siphon. Mouth brushes consist of about 10 stout rods. This group is known only from central and eastern states including Florida.

SUBFAMILY CULICINAE: **Larvae** (Figs. 16.35, 16.40; Plate XIV, Fig. 99) have a well-developed siphon. Mouth brushes usually have 30 or more hairs; if brushes consist of stout rods, then there is a row of hairs in basal half of siphon. The group is widespread in North America.

SOLITARY MIDGES
(Family Thaumaleidae)

LARVAL DIAGNOSIS: (See key Fig. 16.1) The slender, cylindrical larvae are about 12 mm when mature. Prothorax has a single (unpaired) ventral proleg and a pair of minute dorsal respiratory tubercles. Abdomen lacks prolegs along its length but terminates in a single proleg; segment 8 possesses a pair of spiracular openings dorsolaterally near its hind margin.

ADULT DIAGNOSIS: Adults are gnatlike. Head is small and has short antennae. Thorax is robust. Wings lack scales and have seven longitudinal veins reaching the margin.

DISCUSSION: This rare group, represented by about five species in North America, is *not* known from central and southwestern states or Texas. Larvae occur in streams on rocky substrates at or near the surface.

BITING MIDGES
(Family Ceratopogonidae)

LARVAL DIAGNOSIS: (Figs. 16.41–16.43; Plate XV, Fig. 112) Biting midge larvae are very slender, usually cylindrical, and measure 2–15 mm at maturity. Prothoracic and terminal prolegs may be present or absent; when both are present, the body possesses well-developed bristles or spines. Functional spiracles are not present.

ADULT DIAGNOSIS: (Fig. 16.45; Plate XV, Fig. 113) These are small or minute gnatlike forms. Mouthparts are modified for piercing. Antennae are generally longer than head. Legs are of moderate length, hind legs being longest. Wing venation is reduced, scales are absent, and M vein is usually forked.

DISCUSSION: Biting midges are also known as "punkies" or "no-see-ums." Those larvae without prolegs are distinguished from similar moth fly larvae by the total lack of dorsal plates on their abdomens. Those possess-

BITING MIDGES
(Ceratopogonidae)

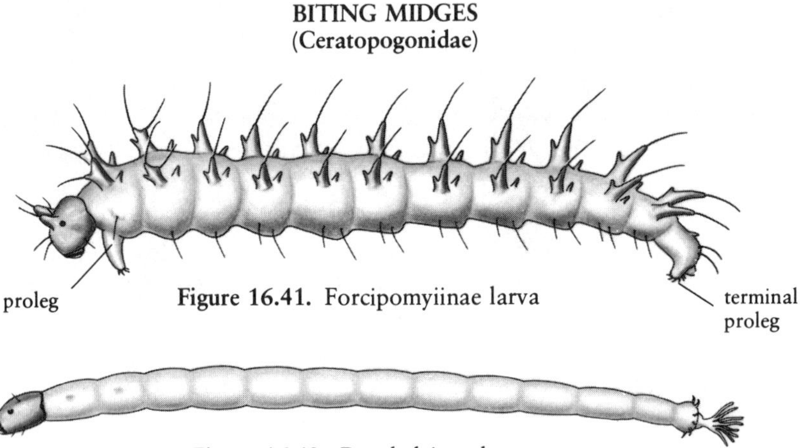

Figure 16.41. Forcipomyiinae larva

Figure 16.42. Dasyheleinae larva

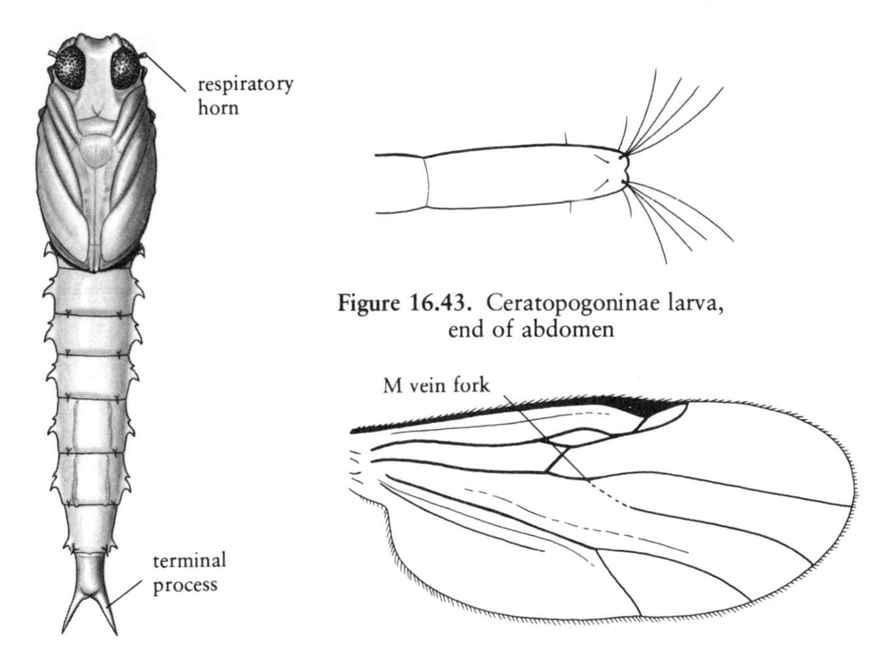

Figure 16.43. Ceratopogoninae larva, end of abdomen

Figure 16.44. Pupa (ventral)

Figure 16.45. Wing

ing prothoracic prolegs are usually only semiaquatic and are distinguished from midges by their large spines or bristles.

Those biting midge pupae (Fig. 16.44) that are aquatic are not found in cocoons. They maintain contact with air at the water surface by means of simple thoracic respiratory horns. The abdomen usually ends in a pair of lateral, pointed processes.

This large family consists of about 34 genera (including *Culicoides,* with over 100 species) in North America. Some aquatic or semiaquatic species are found in most genera.

Semiaquatic larvae inhabit the moist sand of ocean beaches; mud; wet vegetation on rocks along lake, pond, stream, and river margins; swamps; salt marshes; tree holes; and pitcher plants. Some species of Ceratopogoninae occur as benthos among algal growths or occasionally in the open water of lakes and streams. Many aquatic species are carnivorous, sometimes feeding on other aquatic insect eggs (e.g., species of *Bezzia* feed on shore fly eggs). Others feed on periphyton or small detritus, and many are omnivorous.

The feeding habits of adults are varied, but females of most species are blood feeders. *Culicoides, Lasiohelea,* and *Leptoconops* are serious biting pests of humans, especially in mountain and beach resort areas. The name "sand fly" is sometimes applied, as it is in Psychodidae. Some *Culicoides* are transmitters of a number of diseases, including onchoceriasis of livestock, bluetongue of sheep, diseases of birds, and horse sickness. Some species of Forcipomyiinae suck fluids from the wing veins of larger insects, such as dragonflies.

Many of the adult Ceratopogoninae are predaceous on other small insects, but nectar feeding is also common in this subfamily. Insectivorous females (e.g., of *Palpomyia*) feed on small, swarming aquatic insects, including midges, phantom midges, mosquitoes, mayflies, other biting midges, and even their own species. They fly into the prey swarms, sometimes hovering or hunting within the swarm, capturing an individual (usually a male), and devouring its bodily fluids. Females of some species normally feed on the male of their own species while mating.

SUBFAMILY LEPTOCONOPINAE: **Larvae** possess a soft head and a smooth body that lacks prolegs and bristles. These are sand and soil dwellers and are not fully aquatic. The group is widespread but rare in northwestern states and western Canada.

SUBFAMILY FORCIPOMYIINAE: **Larvae** (Fig. 16.41) possess prothoracic prolegs and well-developed bristles or spines along the body. They are never fully aquatic. The group is widespread in North America.

SUBFAMILY DASYHELEINAE: **Larvae** (Fig. 16.42) lack prothoracic prolegs but have a small terminal proleg possessing hooklets. The group is widespread in North America but uncommon in central Canada.

SUBFAMILY CERATOPOGONINAE: **Larvae** (Fig. 16.43; Plate XV, Fig. 112) have a well-developed, somewhat elongate head. Body lacks prolegs and often has hairs, especially at the posterior end. The group is widespread in North America.

MIDGES
(Family Chironomidae)

LARVAL DIAGNOSIS: (Figs. 16.47–16.54; Plate XV, Figs. 105–107) Larvae are slender, commonly cylindrical and slightly curved forms that usually measure 2–20 mm but are occasionally larger. Body has a pair of prothoracic prolegs and a pair of terminal prolegs. Terminal segment usually has a short dorsal pair of tubercles or projections, each with a variable tuft of hairs (dorsal preanal brushes).

ADULT DIAGNOSIS: (Fig. 16.46; Plate XV, Fig. 108) These are minute to medium-sized forms. Antennae are generally longer than the head and often possess long hairs. Mouthparts are generally reduced and not modified for piercing. Legs are long, fore legs usually being longest. Wings, when developed, are slender and lack scales.

DISCUSSION: Larvae of this very large, common, and geographically widespread family are distinctive.

Pupae (Fig. 16.55) of most species live within cylindrical or conical cocoons. Others are free-swimming, and some resemble mosquito larvae. The thoracic respiratory horns are variously modified, being either simple,

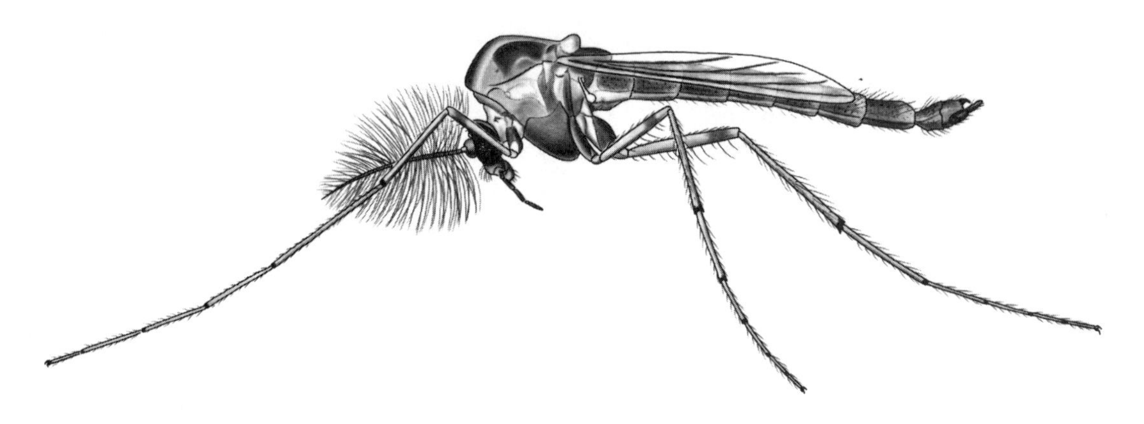

Figure 16.46. *Glyptotendipes* adult ♂

highly branched, or absent. When terminal swimming paddles are present, they do not possess a midrib like those of mosquitoes (Fig. 16.37). Forms with simple, open respiratory horns differ from biting midges in having a terminal bilobed swimming paddle rather than pointed posterolateral processes on the terminal abdominal segment.

Although difficult to estimate, over 100 genera and perhaps as many as 2,000 species occur in North America. New species continue to be discovered at a rapid rate.

An entire chapter could easily be devoted to this important family. Many of the diverse ecological and biological adaptations discussed generally for aquatic flies in the beginning of this chapter are applicable to the midges. The group is possibly the most widely adapted of all aquatic insects. The importance of the larvae of this group as a fish food and as indicators of environmental quality cannot be overemphasized.

Aquatic habitats of the immatures range from littoral marine waters to mountain torrents, from mangrove swamps to arctic bogs, and from clear, deep lakes to heavily polluted waters. They can be expected in almost all inland waters. Most species are bottom-dwelling, and many live within tubes or loosely constructed silk-lined cases in the substrate. A few build distinctive cases (Fig. 16.54). These benthic forms can occur in extremely high densities; their tube cases sometimes cover large areas of the bottom, virtually becoming substrate themselves for other organisms, such as encrusting diatoms.

The larvae of many midge species build winter cocoons in which they reside in a dormant state during the coldest period of the year, sometimes only after ice has formed. These cocoons are closed, and are generally thicker than the tube cases that they live in during warmer periods. The larvae are tightly folded within the winter cocoons, and the manner in which they are folded apparently differs with each species.

Some species that live in lentic situations are generally free-swimming, and even the first instar of some of the otherwise benthic species are free-swimming. Many of the free-swimming and some of the bottom-dwelling species are predaceous at least in part. Many species are herbivore-detritivores and feed off the substrate. Some eat large detritus, such as decomposing leaves with resident microbiota. Some species utilize microorganisms and small detritus in the seston by suspension feeding, and a few of these spin elaborate catchnets. When suspended food becomes limited, some midges change from suspension feeding to grazing.

The substrate ooze of some enriched shallow ponds or lakes consists, to a large degree, of the fecal pellets of midge larvae (especially *Chironomus*). These pellets remain coherent and durable when midges are feeding on blue-green algae. Interestingly, such pelletal oozes form substrates that are similar to those that gave rise to certain oil shale formations (minute fecal pellets and mineralized parts of midge larvae are common in some shale deposits).

The entomology of these lakes I can merely touch upon, mentioning only the most important and abundant insect larvae . . . wriggling through the water or buried in the mud the larvae of Chironomus*—the shallow water species white, and those from the deeper ooze of the central parts of the lakes blood-red and larger.*

Stephen A. Forbes

Some midge larvae are miners in colonial algae or underwater parts of emergent vegetation. The symbiotic relationship of some larvae of the midge genus *Cricotopus* and the colonial alga *Nostoc* is a rare situation in which both algae and insect derive benefit from a direct relationship, the midges obtaining food and shelter and the algae obtaining substrate placement and propagation of reproductive filaments. A few midges are parasitic on other aquatic animals, such as mayflies, caddisflies, clams, or sponges. Symbiotic relationships with other aquatic invertebrates are often intricate, some associated with nutritional needs of the midge, others with substrate or dispersal needs.

A few species of midges are adapted to a terrestrial or semiaquatic habitat. Larvae and almost all pupae of aquatic species are hydropneustic. Most rely on cutaneous respiration; however, many pupae possess thoracic spiracular gills. Some species have very specific environmental requirements, whereas others are relatively tolerant of various pollutants. Members of the Chironominae contain hemoglobin that stores oxygen within their bodies and allows some of them to exist, at least temporarily, in environments with little or no available oxygen. These so-called bloodworms are often bright red in color. Some species of *Chironomus* are thus able to exist in lake bottoms and often in highly polluted or organically enriched water.

Chironomid larvae are generally very important in the efficient operation of sewage oxidation ponds. During the day, algae contribute greatly to the available oxygen supply necessary for the aerobic degradation processes of such facilities. However, the accumulation of dead algal matter is counterproductive to the facilities' operation. Midges thrive in the dense algal growths and detritus and are perhaps the single most important factor in removing them by their feeding.

Stream and river species of midges often show a distinct periodic drifting pattern. Cast skins of midge larvae and especially pupae are commonly taken in drift samples or seen floating on standing waters. Sampling floating pupal skins that have been shed upon emergence is an effective method of determining the chironomid fauna of certain environments. This technique, as well as more standard methods, is often used by specialists.

Adults of midges are generally short-lived and nonfeeding, although there are exceptions. They are often attracted to lights and sometimes emerge in large numbers. All stages of midges can be important in the diet of many fish species. Interestingly, in all these respects, the midges parallel the mayflies.

To the fly fisherman, midges constitute the most important group of aquatic Diptera. Fly patterns referred to as midges come in an array of colors and usually range in size from hook numbers to 20 to 28. It is difficult to determine exactly which species of midges are being imitated, and patterns in general are not as species-specific (except in size) as are those of mayflies and stoneflies, for example. Larval, pupal, and adult patterns are popular, and taking a big fish on such small flies is an exciting challenge. Such flies are particularly productive on lakes and pools during large midge hatches. Some popular patterns include the Black Midge, Blue Dun Midge, Chironomid Killer, Cream Midge, Grey Midge Pupa, and Green Midge Pupa. The Snow Fly is an unusually large midge pattern used in the West, often during colder months.

The following discussion of subfamilies is not intended to be foolproof, and exceptions to generalities can almost always be expected. Furthermore, not all midge specialists are in agreement as to the subfamily classification. Nevertheless it will serve as a useful guide for those interested in an introduction to the major subgroups of midges. Some morphological characteristics cannot be interpreted correctly unless parts are slide mounted and examined under high magnification. Larval color characteristics disappear quickly in fluid preservatives!

SUBFAMILY TANYPODINAE: **Larvae** (Figs. 16.47, 16.48) may be pale brownish, reddish, or rarely greenish in nature. Head is somewhat elongate, and antennae are sometimes withdrawn into the head. Lingua is fork-shaped. Prolegs are stiltlike and have few terminal hooks. Bases of dorsal preanal brushes are at least three times as long as they are wide. These are mostly free-swimming, predaceous forms, occurring primarily in lentic habitats. The group is widespread in North America.

SUBFAMILY PODONOMINAE: **Larvae** (Fig. 16.49) may be whitish, yellowish, or brownish in nature. A fork-shaped lingua is not present. Bases of dorsal preanal brushes are 5 to 10 times as long as they are wide. These are known primarily from streams in western Canada and Alaska, eastern Canada, and western and northeastern states.

SUBFAMILY TELMATOGETONINAE: **Larvae** are usually greenish or brown in nature. These are marine forms with four-segmented antennae. Anal papillae are reduced or absent. The group is known only from coastal regions in North America.

SUBFAMILY DIAMESINAE: **Larvae** (Figs. 16.50, 16.51; Plate XV, Fig. 106) are whitish or yellowish in nature, and often over 9 mm when mature. Eyespots may be joined or separate; if separate, then ventral eyespot is anterior to dorsal eyespot; if joined, then the eyespots are circular in some lentic species. Bases of dorsal preanal brushes are short or absent. These are primarily known from streams. The group is relatively widespread but unknown or uncommon in Texas and southeastern states including Florida.

SUBFAMILY ORTHOCLADIINAE: **Larvae** (Fig. 16.52; Plate XV, Fig. 105) are usually greenish, but may also be brown, violet, or banded in nature, and they are often less than 9 mm when mature. Eyespots are joined or separate; if separate, then ventral eyespot is anterior to dorsal eyespot; if joined, then they are only rarely circular. Third antennal segment never has annulations. Bases of dorsal preanal brushes are short or absent. This is a widespread group, and larvae occur in a variety of aquatic habitats, a few in marine habitats.

MIDGES
(Chironomidae)

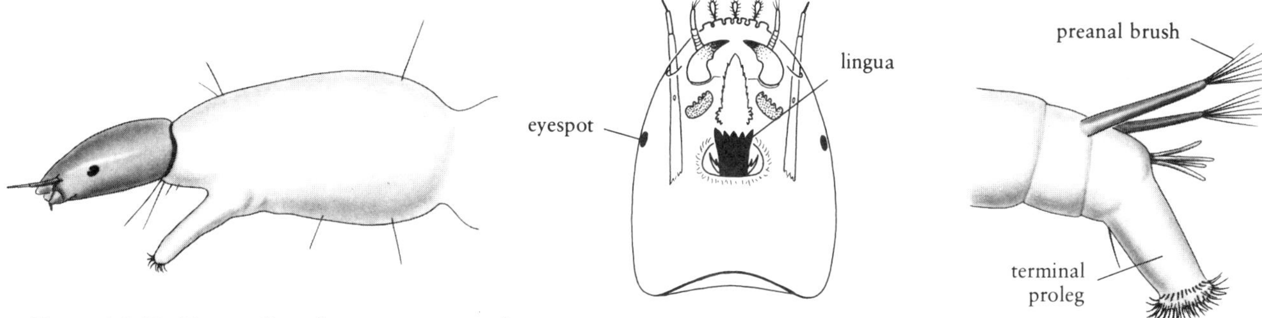

Figure 16.47. Tanypodinae larva, anterior end

Figure 16.48. Tanypodinae larva, head (ventral schematic)

Figure 16.49. Podonominae larva, end of abdomen

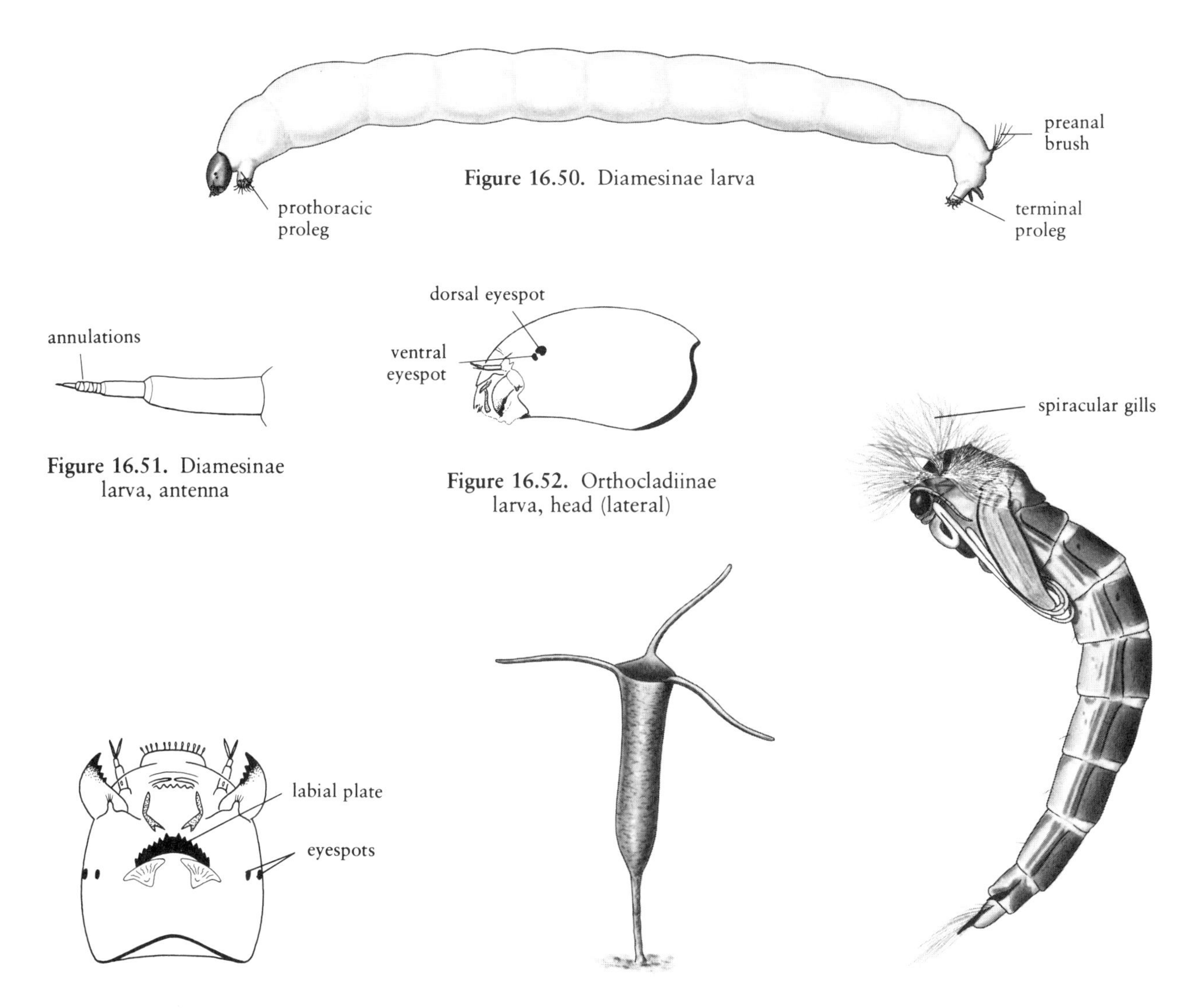

Figure 16.50. Diamesinae larva

Figure 16.51. Diamesinae larva, antenna

Figure 16.52. Orthocladiinae larva, head (lateral)

Figure 16.53. Chironominae larva, head (ventral schematic)

Figure 16.54. *Tanytarsus* larval and pupal case

Figure 16.55. *Chironomus* pupa

SUBFAMILY CHIRONOMINAE: **Larvae** (Fig. 16.53; Plate XV, Fig. 107) are usually reddish in nature and vary in size. Eyespots are usually separate and in line, but dorsal eyespot is anterior to ventral eyespot in some. Bases of dorsal preanal brushes are short or absent. This is a widespread group whose larvae are known from a variety of habitats, including the semiaquatic and marine habitats of some species. They are primarily bottom dwellers in lentic or slow-flowing waters.

BLACK FLIES
(Family Simuliidae)

LARVAL DIAGNOSIS: (Figs. 16.56–16.59; Plate XV, Fig. 109) The larvae are cylindrical and usually measure 3–8 mm or rarely up to 12 mm or more. Head often possesses distinct fanlike mouth brushes. Prothorax has a ventral median proleg; otherwise prolegs are absent. Abdomen is generally swollen posteriorly and terminates in an attachment disc.

ADULT DIAGNOSIS: (Fig. 16.60; Plate XV, Fig. 111) These small, robust forms are often black but occasionally lighter in coloration. Antennae are short and atypical of the longhorned flies. Mouthparts are short and adapted for cutting. Thorax is strongly arched. Legs are short and stout. Wings are broad.

DISCUSSION: The swollen end of the abdomen is perhaps the best field character for identifying black fly larvae.

Pupae (Fig. 16.61; Plate XV, Fig. 110) possess a pair of highly branched spiracular gills on the thorax. They occur in cocoons of various shapes from which the branched spiracular gills protrude. Cocoons of many species are open-ended (facing downstream), slipper-shaped cases and are attached to the surface of rocks, vegetation, or sometimes even flatheaded mayflies. Vacated cases may be commonly encountered in streams.

Over 150 species of black flies occur in North America, with the majority of species placed in the genera *Simulium* and *Prosimulium.*

The hatch of smuts may be very local. Under the shelter of trees or waterside herbage, when they are swarming, i.e., buzzing, round a particular spot, selecting mates or egg-laying, no more than a single fish, or perhaps half a dozen, may be found taking them; much time may be wasted over such local smutting fish by attempting to take them with duns, spinners, or other flies, as a result of failing to recognize that the fish are taking smut.

J. C. Mottram

Larvae are generally attached by the end of their abdomens to rocks, woody debris, or vegetation in the currents of streams and rivers. They hang with the head downstream and filter food particles suspended in the flowing water with their mouth brushes. Thousands of individual larvae sometimes occur in close association on the substrate. Movement is usually accomplished by trailing downstream on anchored silken threads. These threads are as much as a meter in length. Sometimes, loosely constructed webs or cables composed of as many as 50 threads are formed. Downstream movement will occur as a response to adverse conditions, such as crowding or lowering of the water.

Unlike most other nematoceran flies, black flies usually have six or rarely seven larval instars, with one to several generations per year. Adults typically live three weeks or less. Unlike most mosquitoes, the females are biting pests during daylight hours. Black flies, which are sometimes referred to as "buffalo gnats" in North America, often pose a serious problem in the vicinity of rivers or areas with many streams. Their bites can cause serious blood loss or allergic reactions, and even death in vertebrate animals. Certain regions are unfit for habitation when adult black flies are prevalent, and repellents and screens are often ineffective against them. Some are transmitters of blood parasites of waterfowl and turkeys in North America.

The so-called Black Gnat fly patterns of fishermen are most likely accepted as black flies by game fishes. The name "smut" has histori-

BLACK FLIES
(Simuliidae)

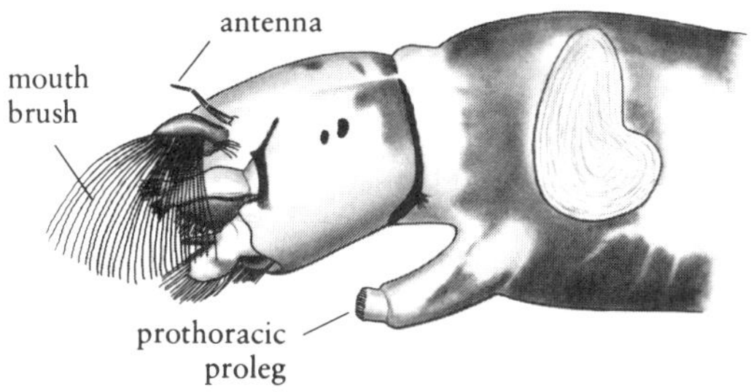

antenna
mouth brush
prothoracic proleg

Figure 16.56. Simuliinae larva, anterior end

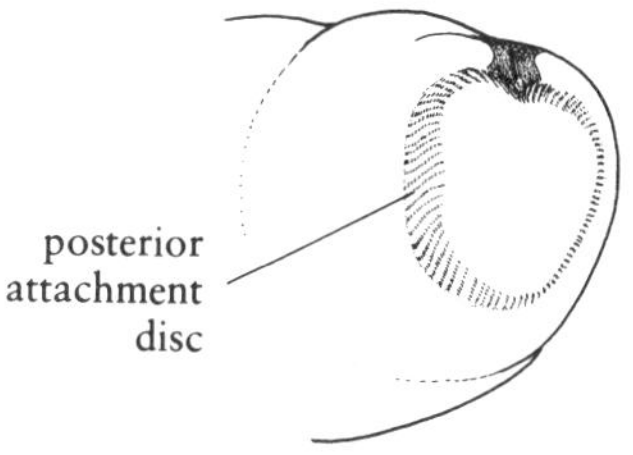

posterior attachment disc

Figure 16.57. Larva, end of abdomen (posterolateral)

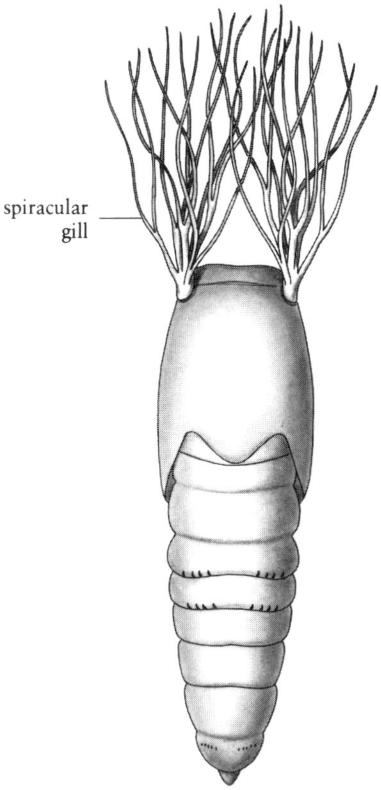

spiracular gill

Figure 16.58. Prosimuliinae larva, antenna

Figure 16.59. Simuliinae larva, antenna

antenna

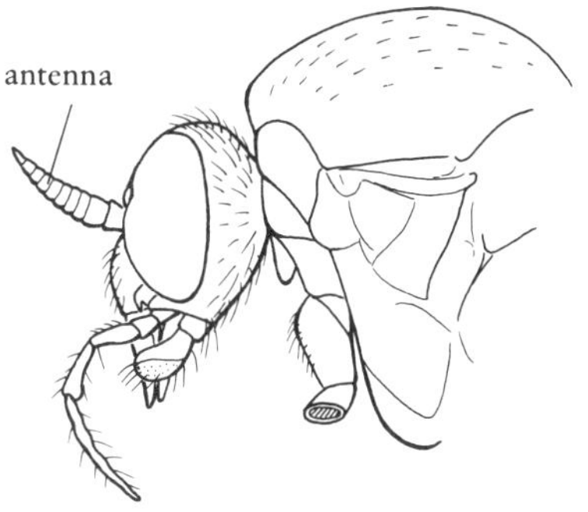

Figure 16.60. Adult, anterior end (lateral)

Figure 16.61. *Simulium* pupa (dorsal)

cally been applied to the black flies by fly fishermen, and in North America, larval and pupal imitations of the Riffle Smut (*Simulium* spp.) (Plate XV, Figs. 109–111) are used for nymph fishing in many streams.

SUBFAMILY PROSIMULIINAE: **Larvae** may lack fanlike mouth brushes; if present, as they are in most, then antennae (Fig. 16.58) have two terminal segments that are dark and contrast strongly with the much paler basal segments. The group occurs throughout much of North America but is not known from Florida or Texas and is uncommon in southeastern states and southern central states.

SUBFAMILY SIMULIINAE: **Larvae** (Figs. 16.56, 16.57, 16.59; Plate XV, Fig. 109) have fanlike mouth brushes. Terminal segments of antennae are not contrastingly darker than basal segments. Each lobe of the eversible preanal papillae is often branched. The group is widespread in North America.

GALL GNATS
(Family Cecidomyiidae)

This family contains some species that live in the plant tissue of aquatic vegetation; see Chapter 18.

Suborder Brachycera—Tabanomorpha and Asilomorpha

Larvae have a body that is more-or-less tapered and pointed anteriorly. Head is usually incomplete and indistinct. If the head capsule is distinct, as seen in dorsal view (Fig. 16.62), then body lacks prolegs. Mandibles are usually sickle-shaped and move parallel to each other in a vertical plane.

Pupae of all but the soldier flies lack a puparium. Puparium of soldier flies is not capsulelike and is very little changed from the larva. Developing antennae do not reach bases of developing wings and do not lie over the eyes.

Adults are usually stout-bodied and medium-sized. Antennae have less than five, usually three segments, although segment 3 may have numerous annulations and a terminal bristle.

AQUATIC SOLDIER FLIES
(Family Stratiomyidae)

LARVAL DIAGNOSIS: (Figs. 16.62, 16.63; Plate XVI, Fig. 114) These larvae are somewhat flattened and broadened forms and when mature measure 3–50 mm, but usually 7–30 mm. Head is distinctive and set off from thorax. Body surface is somewhat hardened and thickened with deposits of calcium carbonate. Prolegs are lacking, but bristles or filaments are often present. Abdomen is terminally fringed with hairs.

ADULT DIAGNOSIS: (Fig. 16.64) Antennae usually appear bent at about midlength. There is a small, distinctive discal cell in the middle of wing.

DISCUSSION: Larvae and puparia are distinctive. The pupae are actually quite small and are contained within the anterior end of a puparium that is similar in shape to the larval form. This family contains both terrestrial and aquatic species.

AQUATIC SOLDIER FLIES
(Stratiomyidae)

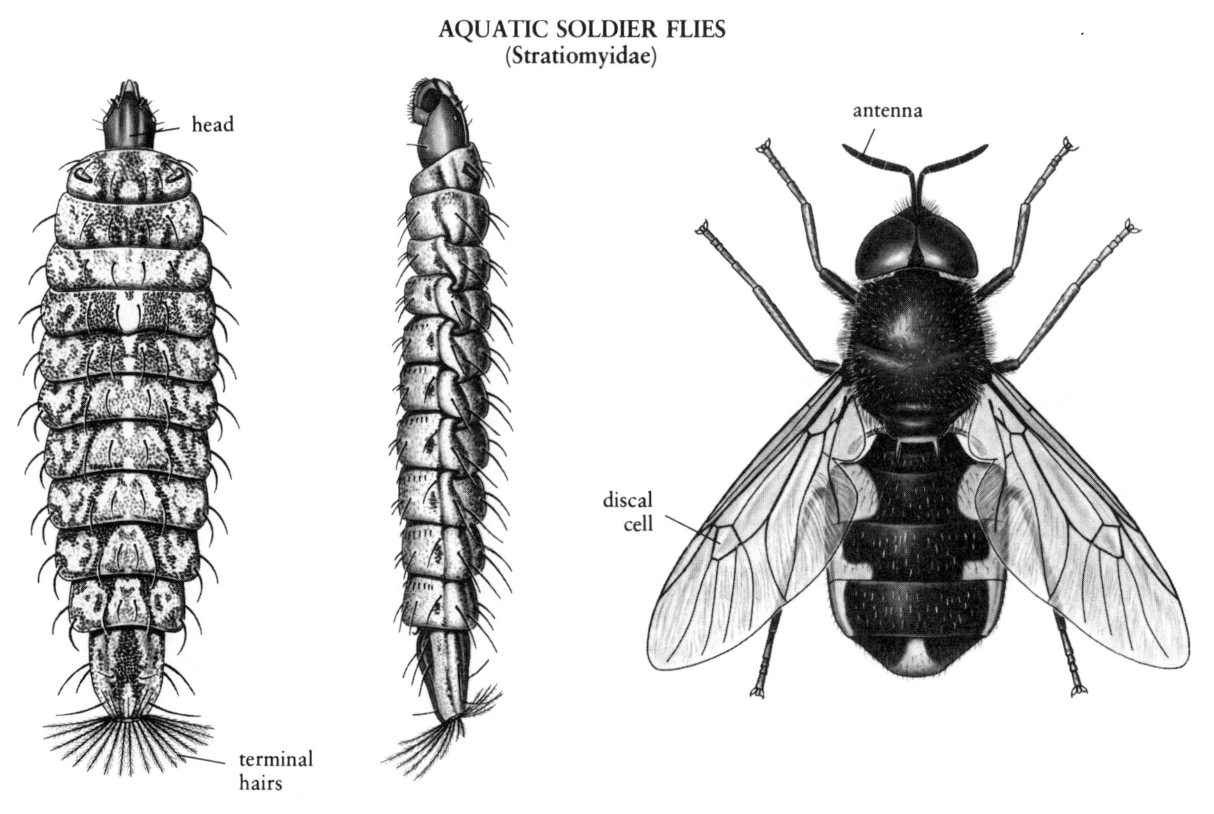

Figure 16.62. *Euparyphus* larva (dorsal)

Figure 16.63. *Euparyphus* larva (lateral)

Figure 16.64. *Stratiomys* adult

Although the family is large, only about eight of the genera and 100 or more species are known to be aquatic or semiaquatic in North America, where the family is widespread.

Many aquatic or semiaquatic larvae occur in ponds, especially in thick vegetation along the margins, where they feed primarily on algae and detritus. Some occur in marginal or shallow areas of streams and large lakes. Larvae are also known from brackish water, hot springs, and bogs. Larvae of *Stratiomys* are sometimes submerged several feet, and a few occur as benthos in riffles. Those larvae that are aeropneustic must either remain in contact with air or surface intermittently. The puparium is buoyant. Eggs are usually deposited on plants overhanging the water.

HORSE AND DEER FLIES
(Family Tabanidae)

LARVAL DIAGNOSIS: (Figs. 16.65–16.67; Plate XVI, Fig. 115) Larvae are generally elongate, cylindrical, spindle-shaped forms that measure 11–55 mm when mature. Body tapers at both ends and possesses a series of fleshy, encircling rings. Abdomen lacks terminal prolegs, but terminates in a small respiratory siphon.

ADULT DIAGNOSIS: (Figs. 16.68, 16.69; Plate XVI, Fig. 116) These are medium-sized to large, stout-bodied flies, some of which possess patterned

HORSE AND DEER FLIES
(Tabanidae)

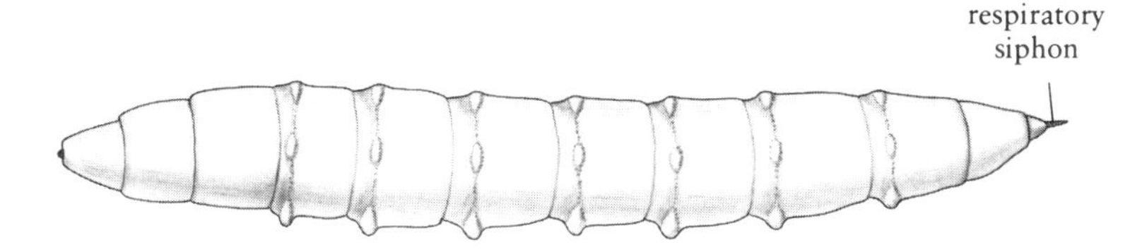

Figure 16.65. *Chrysops* larva

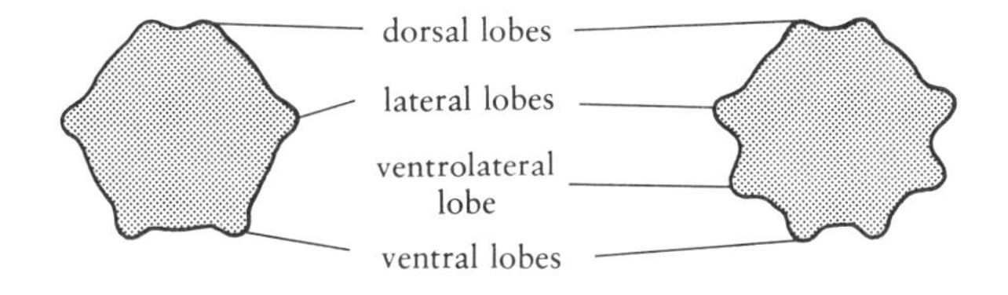

Figure 16.66. *Chrysops* larva, cross section of abdomen

Figure 16.67. *Tabanus* larva, cross section of abdomen

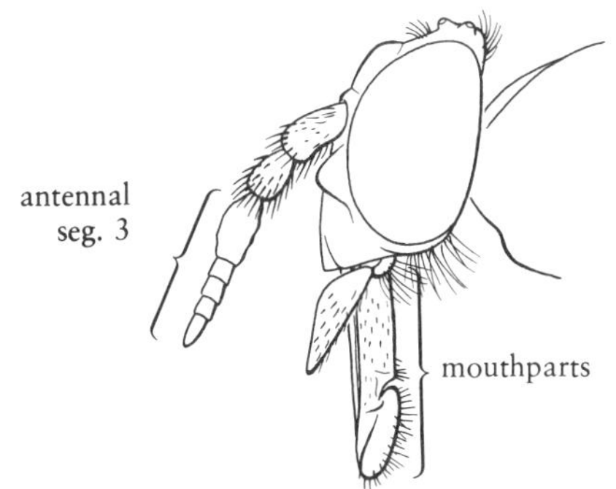

Figure 16.68. *Chrysops* adult, head (lateral)

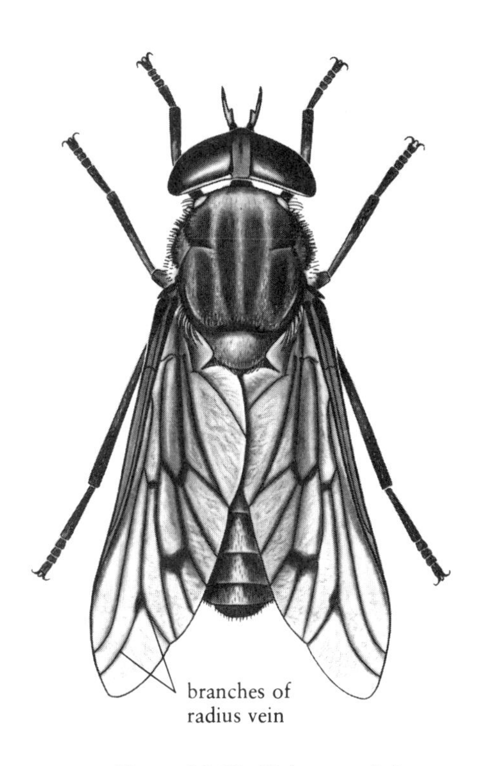

Figure 16.69. *Tabanus* adult

wings and colorful eyes. Head as seen from above is semicircular (straight behind and otherwise rounded). Antennal segment 3 is elongated. Mouthparts are modified for cutting and sucking. Wing tips are enclosed by two short branches of the radius vein.

DISCUSSION: Pupae, which are rarely found in water, possess conspicuous dorsal pits on the thorax, and in most the majority of abdominal segments have transverse fringes of bristles at least dorsally.

About 22 genera occur in North America, where the genera *Chrysops* and *Tabanus* are widespread and together contain nearly 200 species. With few exceptions, the larvae are semiaquatic or aquatic.

Many species, especially of *Chrysops,* occur along the margins of streams and ponds. Many others, especially of *Tabanus,* are more typical of marshes and swamps. Some *Tabanus* and rarely *Chrysops,* inhabit stream bottoms. Several species live in salt marshes or brackish pools. Some species are able to live in either terrestrial or semiaquatic environments. A species of *Haematopota* occurs in moist silt along streams as well as sometimes in rotting vegetation. The larvae of most species are predators; however, *Chrysops* feeds on fine detritus.

Adults are strong fliers, and the females are often persistent and painful biters, especially in wetlands and woodlands. This is particularly true of the deer flies (*Chrysops*), which are generally smaller and more patterned (Plate XVI, Fig. 116) than the larger, drabber or darker horse flies (*Tabanus* and others). The bloodsucking of these flies can seriously interfere with livestock production.

Many larvae of Tabanidae remain unknown, and diagnoses of the subfamilies would be somewhat premature. The most commonly known larvae of *Chrysops* (Chrysopinae) and *Tabanus* (Tabaninae) may be distinguished as follows: *Chrysops* larvae are less than 20 mm, and the fleshy rings on their abdomen consist of three pairs of small lobes (dorsal, lateral, ventral) (Fig. 16.66). *Tabanus* and most other aquatic Tabanidae are usually more than 20 mm when mature, and the rings of the abdomen consist of four pairs of small lobes (dorsal, lateral, ventrolateral, ventral) (Fig. 16.67).

WATERSNIPE FLY
(Athericidae)

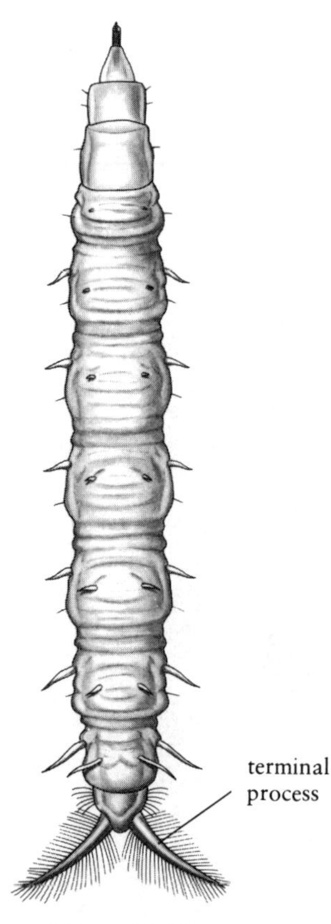

Figure 16.70.
Atherix larva (dorsal)

WATERSNIPE FLIES
(Family Athericidae)

LARVAL DIAGNOSIS: (Fig. 16.70; Plate XVI, Fig. 117) These are 12–18 mm when mature. Abdomen has a series of well-developed pairs of ventral prolegs and short dorsal and lateral filaments on each segment; terminal segment has a pair of divergent processes that are longer than the terminal prolegs.

ADULT DIAGNOSIS: These medium-sized flies have a tapering abdomen and moderately long legs. Antennal segment 3 is rounded and has a long terminal bristle. Each tarsus has three padlike structures distally.

DISCUSSION: Larvae of watersnipe flies can be distinguished from some similar aquatic dance flies, which also possess well-developed ventral prolegs, by their more elongate terminal processes and the presence of hairs on those processes. Pupae are terrestrial.

The genus *Atherix* contains three species in North America, where it is generally widespread, but it is not known from Florida or Texas and is uncommon in some central states. The one North American species of the genus *Suragina* is known from extreme southwestern Texas.

The predaceous, hydropneustic larvae usually inhabit well-oxygenated streams and rivers. Adults of *Atherix* do not bite, but female adults of *Suragina* do. Egg masses are laid on vegetation overhanging the stream; the female dies there after oviposition.

BEE FLIES
(Family Bombyliidae)

The larvae of some species of this family are internal parasites of some aquatic or semiaquatic insects and therefore are only indirectly associated with an aquatic environment.

AQUATIC DANCE FLIES
(Family Empididae)

LARVAL DIAGNOSIS: (Figs. 16.71–16.74; Plate XVI, Fig. 118) These are usually 2–7 mm when mature. They are generally similar to watersnipe flies, but terminal processes are variable in number and shape and are rarely longer than terminal prolegs. If prolegs are weltlike and terminal prolegs not developed, then terminal abdominal segment is somewhat bulbous and slightly pointed.

ADULT DIAGNOSIS: (Fig. 16.75) These flies are usually small and have a tapering abdomen and long legs. Antennae possess a terminal bristle. R-M crossvein is located in the apical three-fourths of the wing. End of the abdomen is sometimes enlarged but never recurved ventrally.

DISCUSSION: If the prolegs and terminal processes are reduced in the larvae, then close examination reveals a minute sclerotized head with antennae (often retracted) that allows dance flies to be distinguished from cyclorrhaphan larvae. Some pupae have terminal processes similar to those of the larvae; others (Figs. 16.76, 16.77) possess hooklike terminal processes, and their abdominal segments have fringes of bristles similar to those of horse and deer flies; all lack dorsal pits between the thoracic spiracles, and some species, at least among *Hemerodromia* and *Chelifera*, possess long spiracular gills along the body (Fig. 16.77).

Most species of this family are terrestrial as larvae; however, the known larvae of the subfamilies Hemerodromiinae and Clinocerinae are aquatic or semiaquatic. These two widespread subfamilies are represented by about 18 genera and over 200 species in North America.

Larvae and pupae of most aquatic species live on rocky bottoms of swift streams. Some live in wet marginal areas of ponds and streams. Pupae of *Roederiodes* sometimes occupy the abandoned slippercases of black fly pupae. Larvae are generally predaceous. The adults often occur in swarms over streams, where they characteristically fly up and down or in a circular pattern.

AQUATIC LONGLEGGED FLIES
(Family Dolichopodidae)

LARVAL DIAGNOSIS: (Fig. 16.78) Larvae are usually 3–10 mm at maturity. Abdomen may or may not have ventral prolegs; when prolegs are present, they are usually in the form of small welts. Abdomen is concave at the end, forming a spiracular pit surrounded by short posteriorly projecting lobes.

ADULT DIAGNOSIS: (Fig. 16.80) These small or medium-sized flies have a tapering abdomen and long legs, and many are metallic green, blue, or

AQUATIC DANCE FLIES
(Empididae)

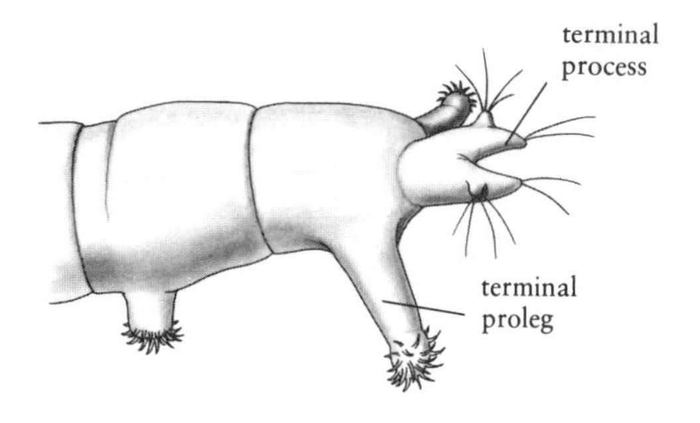

Figure 16.71 *Hemerodromia* larva, end of abdomen (dorsolateral)

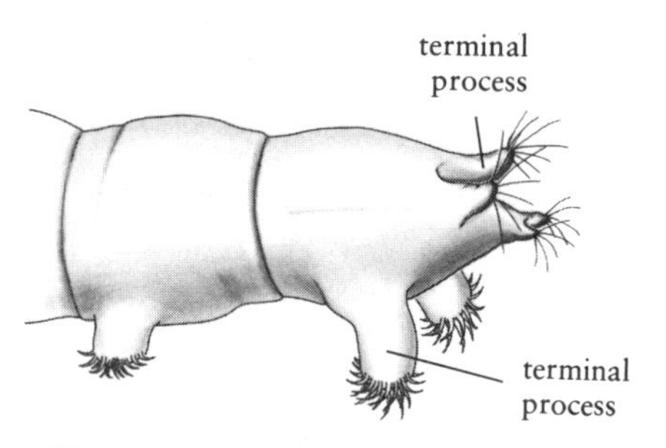

Figure 16.72. *Chelifera* larva, end of abdomen (dorsolateral)

Figure 16.73. *Chelifera* larva, end of abdomen (lateral)

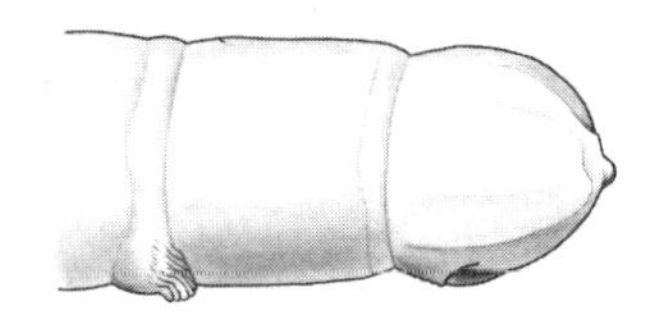

Figure 16.74. *Rhamphomyia* larva, end of abdomen (lateral)

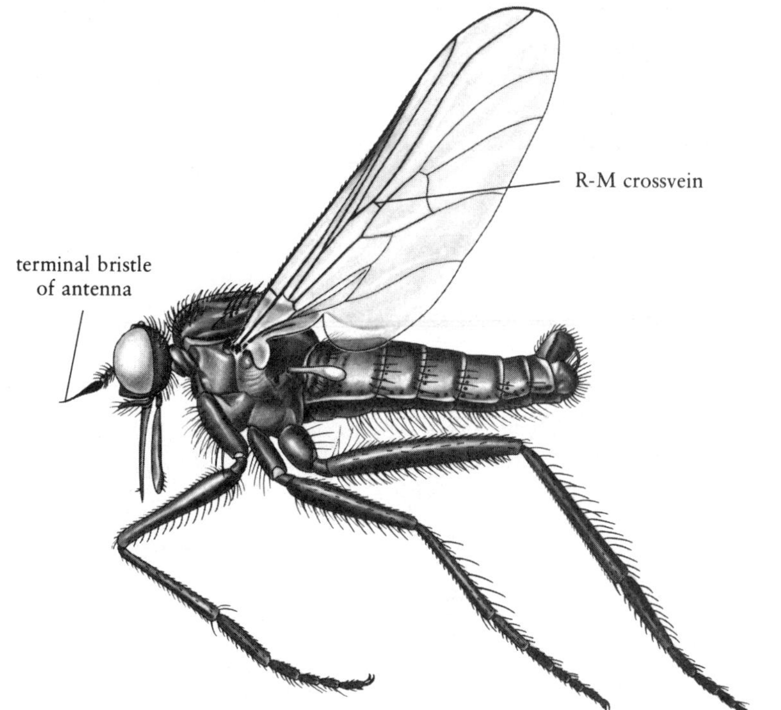

Figure 16.75. *Rhamphomyia* adult

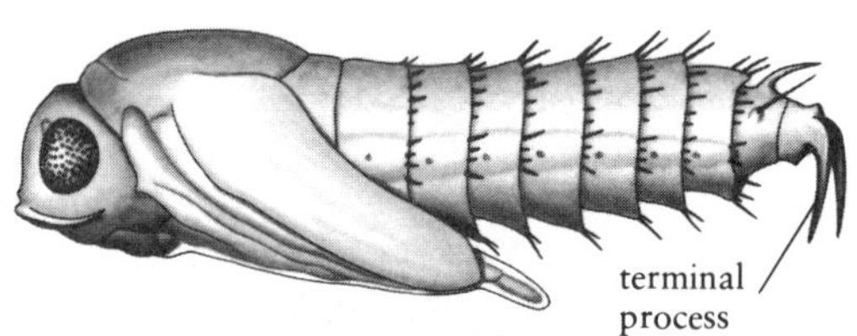

Figure 16.76. *Clinocera* pupa

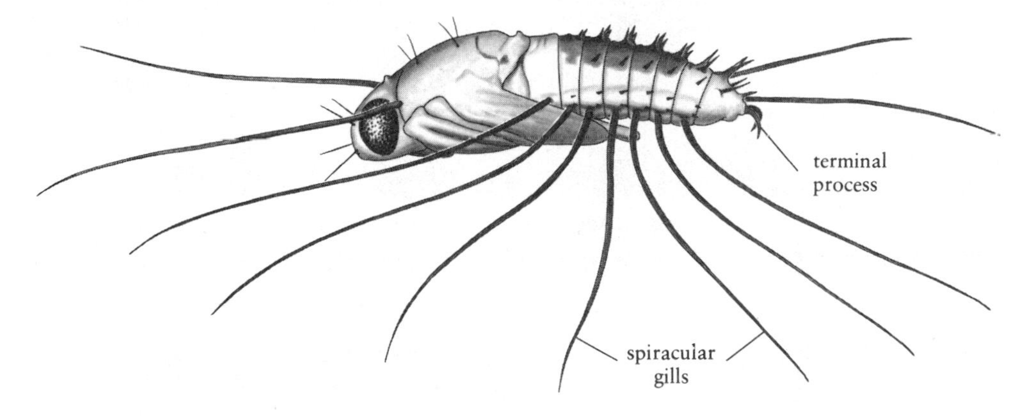

Figure 16.77. *Hemerodromia* pupa

AQUATIC LONGLEGGED FLIES
(Dolichopodidae)

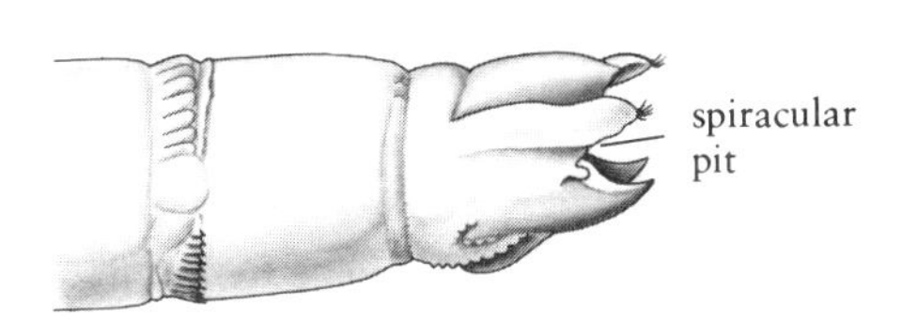

Figure 16.78. Larva, end of abdomen (dorsolateral)

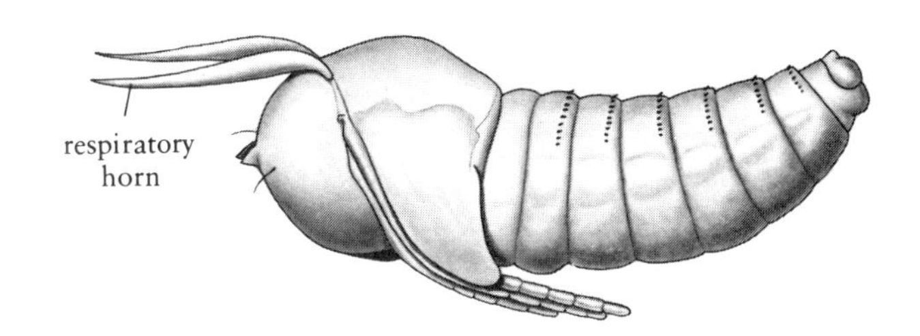

Figure 16.79. Pupa

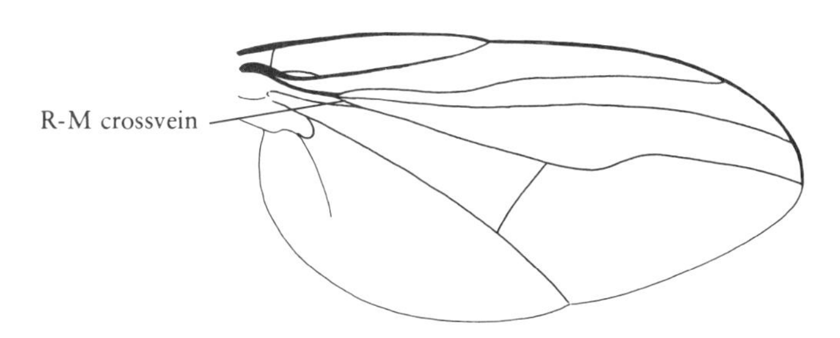

Figure 16.80. Wing

copper colored. Segment 3 of antennae terminates in a long bristle. R-M crossvein is located in the basal fourth of wing. Abdomen is sometimes recurved ventrally.

DISCUSSION: Some aquatic longlegged fly larvae generally resemble larvae of Empididae or Tabanidae. The spiracular pit of the former is, however, distinctive.

Pupae (Fig. 16.79) possess long thoracic respiratory horns, whereas those of horse and deer flies and dance flies do not. They usually occupy cocoons made from plant or substrate material or pieces of shed larval skin.

Only about eight, or approximately one-fifth, of the North American genera of longlegged flies are presently known to have some aquatic or semiaquatic species. Knowledge of aquatic adaptation is very preliminary. The group is widespread geographically.

Larval habitats include ponds and streams and their muddy margins, wet sandy shores and rocky intertidal habitats, tree holes, and cavities within the stems of sedges. The larvae are generally aeropneustic and predaceous. Adults are also predaceous and often skate on the water. Mating dances are reminiscent of those of the dance flies.

Suborder Brachycera—Cyclorrhapha

Larvae either taper or are blunt anteriorly, and are wrinkled or maggotlike. Head is completely reduced. Mouthparts, if evident, are hooklike.

Pupae are encased in a capsulelike puparium formed from the last larval skin and lacking a distinct head.

Adults are typically flylike, stout-bodied, and long-legged. Antennae are three-segmented, and segment 3 has a bristle originating from near its base.

HUMPBACKED FLIES
(Family Phoridae)

Immatures of this family have been taken rarely in sewage treatment lagoons and on moist substrates of trickling filter treatment facilities. Nevertheless, no species appears to be clearly adapted for aquatic environments.

RATTAILED MAGGOTS
(Family Syrphidae)

LARVAL DIAGNOSIS: (Figs. 16.81–16.83; Plate XVI, Fig. 120) Rattailed maggots are soft-bodied, semitransparent forms that usually measure 4–14 mm but may exceed 70 mm in fully extended larger species. Body is blunt anteriorly and wrinkled in most, except when fully extended; it is often clothed with very fine minute hairs, sometimes possesses ventral prolegs, and terminates in a single breathing tube that in most is extensible into an elongate tail-like structure. In a few species, the breathing tube is very short.

ADULT DIAGNOSIS: (Fig. 16.85; Plate XVI, Fig. 119) These flies are commonly brightly colored and patterned. Wings have a unique vein, known as the spurious vein, along with several longitudinal veins that usually form long closed cells.

DISCUSSION: Other aquatic cyclorrhaphan larvae that have the terminal spiracles situated in a single terminal retractile process are either tapered anteriorly or have a terminal proleg that is distinctly larger than the other

RATTAILED MAGGOTS
(Syrphidae)

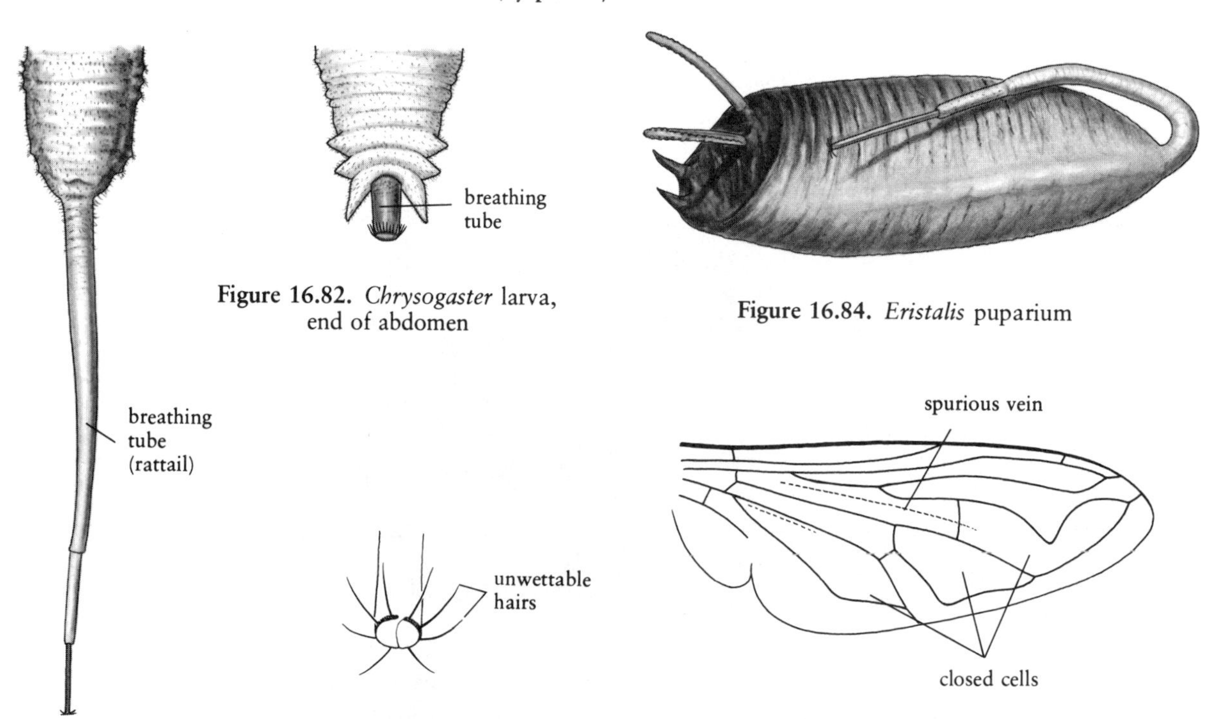

Figure 16.81. *Eristalis* larva, end of abdomen

Figure 16.82. *Chrysogaster* larva, end of abdomen

Figure 16.83. End of larval breathing tube

Figure 16.84. *Eristalis* puparium

Figure 16.85. Wing

abdominal prolegs, and are thus distinguishable from rattailed maggots. In some species, the puparium (Fig. 16.84) possesses well-developed thoracic respiratory horns.

About 11 widespread genera of this large, primarily terrestrial family are presently known to have at least some species that are aquatic or semiaquatic.

The tails of these aeropneustic larvae are used to maintain contact with the air, and when extended in those species with rattails, they allow the larvae to feed and rest at depths of several centimeters. Some species live in the water that accumulates in tree holes. A few species, primarily of *Eristalis*, *Helophorus*, *Chrysogaster*, and *Sphegina*, inhabit the shallow marginal waters of ponds, pools, and marshes. They are commonly associated with organically enriched water; for example, larvae and pupae of *Eristalis* are abundant in sewage water. The mouthparts of some are modified for straining fine detritus and microorganisms. Others eat fine organic ooze without straining it.

Rattailed maggots are sometimes called "mousies" and are popular with bait fishermen in certain regions, where they are available in bait shops.

SEABEACH AND BARNACLE FLIES
(Family Dryomyzidae)

This family contains some shore-dwelling species, and the larvae of one genus are predators of intertidal barnacles; see Chapter 17.

SEAWEED FLIES
(Family Coelopidae)

This family contains shore-dwelling species and is treated in Chapter 17.

MARSH FLIES
(Family Sciomyzidae)

LARVAL DIAGNOSIS: (Figs. 16.86, 16.87) Larvae are cylindrical and usually 4–14 mm. Body tapers at both ends; anterior end is slightly curved ventrally; posterior end may be slightly curved dorsally or ventrally. Scattered, round tubercles usually encircle each body segment. Abdomen may or may not possess a short, tapered breathing tube; however, some short lobes or processes usually surround the terminal spiracular disc.

ADULT DIAGNOSIS: (Fig. 16.88; Plate XVI, Fig. 124) These are small or medium-sized flies that are usually yellowish or brownish and may have patterned wings. Antennae usually project forward. The head is somewhat concave in the area below the antennae. Vein Sc is complete and ends in anterior margin of wing. Femora possess bristles.

DISCUSSION: This widespread group contains possibly as many as 160 aquatic or semiaquatic species in North America.

Larvae most commonly occur in freshwater and saltwater marshes, and along the margins of ponds, lakes, and streams, especially among vegetation and floating debris. Fully aquatic species are aeropneustic and maintain contact with the surface much of the time. They are predators on snails, snail eggs, slugs, and fingernail clams. All but the predators of slugs and land snails are associated with water. The aquatic predators float below the surface film and maintain buoyancy by frequently surfacing and swallowing

MARSH FLIES
(Sciomyzidae)

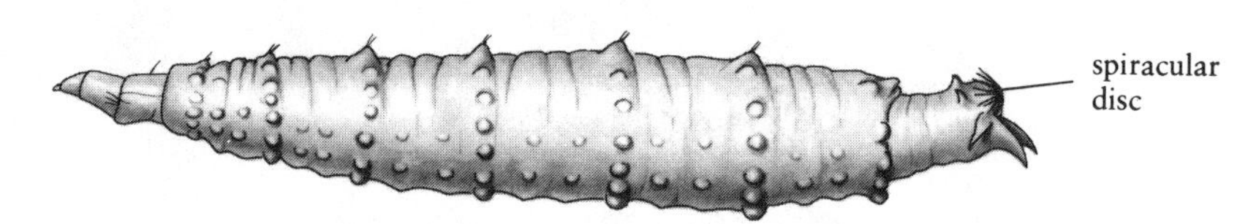

Figure 16.86. *Sepedon* larva

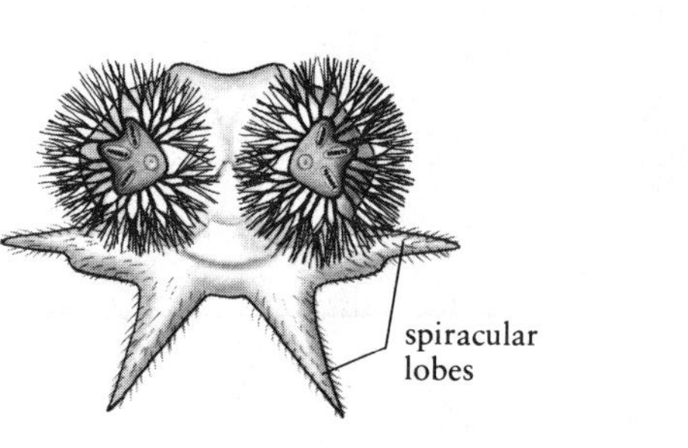

Figure 16.87. *Sepedon* larva, spiracular disc (posterior view)

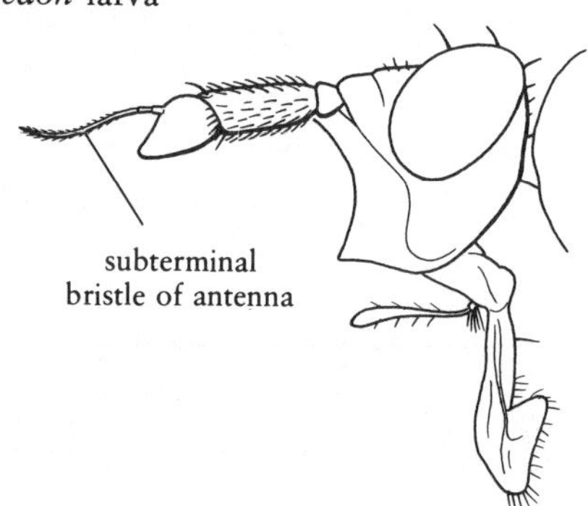

Figure 16.88. Adult head (lateral)

a bubble of air. Prey may be killed immediately or death may take a few days (the larvae are essentially parasitic during this time). They sometimes pupate within the shells of snails, and oviposition sometimes takes place on the snail shell. The group could very well be called the snail feeders or snail killers.

SMALL DUNG FLIES
(Family Sphaeroceridae)

Adults of some species of this family are common along muddy shores, especially seashores; see Chapter 17.

TETHINIDS
(Family Tethinidae)

Most members of this small family occur in seashore environments; see Chapter 17.

MILICHIIDS
(Family Milichiidae)

A species of this family has recently been found to develop on the wet filters of a trickling-filter sewage treatment facility in Indiana.

BEACH FLIES
(Family Canaceidae)

This family contains seashore-dwelling species and is treated in Chapter 17.

SHORE FLIES
(Family Ephydridae)

LARVAL DIAGNOSIS: (Figs. 16.89–16.92; Plate XVI, Fig. 121) Larvae are small, usually 1.2–12 mm or more if extended. Generally, they lack fleshy or tuberculate rings. Body may be wrinkled and extensible and may or may not possess ventral abdominal prolegs. Depending on the genus, the abdomen may terminate in (1) two sharp, minute spines (Fig. 16.89, 16.90), (2) a pair of short to long breathing tubes (Fig. 16.91), (3) a short, single breathing tube without lobes at apex of tube, or (4) a long sheath into which two slender breathing tubes can be retracted (Fig. 16.92). The sheathed structures are commonly found in combination with a more-or-less hook-shaped terminal proleg.

ADULT DIAGNOSIS: (Figs. 16.96, 16.97; Plate XVI, Fig. 122) These are small or minute, mostly blackish or dull, gnatlike forms. Head usually protrudes forward and is concave or angled below the antennae. Anterior border of wing is broken twice along its length. Vein Sc of wing is short and unattached at its end. Wings lack an anal cell.

DISCUSSION: Care should be taken in distinguishing larvae of the shore flies, dance flies, and aquatic muscids. Those shore flies with developed paired breathing tubes are usually quite distinctive. The terminal structures, however, may be reduced in young larvae.

Puparia (Figs. 16.93–16.95) more-or-less resemble the larvae, are commonly boat-shaped, and may have well-developed horns at the anterior end.

The Ephydridae in North America comprise about 65 genera and 400 species. Most species are associated with aquatic or semiaquatic environments.

In the genus *Ephydra,* which includes species sometimes known as brine flies, the immatures commonly occur in salt water or alkaline pools, springs, and lakes, and they are often a major aquatic component of highly saline environments, such as the Great Salt Lake. A few occur in hot springs, and some occur in freshwater ponds and streams. Larvae are usually found in algal mats or silt bottoms of shallow water. Well-developed, hooklike prolegs are generally most common in the algal-mat dwellers. Puparia often attach to debris by their hooklike terminal prolegs or float on the surface, sometimes in large masses.

Other species, especially of the genus *Hydrellia,* are miners in a variety of aquatic plants and are sometimes harmful to rice crops. Species of *Helaeomyia* are remarkable in that they live in pools of crude petroleum and waste oil. Immatures of *Notiphila* live within the substrate of streams, lakes, and ponds, where they pierce the roots of vegetation to obtain oxygen. *Scatella stagnalis* is perhaps the most common and ubiquitous species of shore fly in North American wetlands.

Many species inhabit semiaquatic marginal areas or are terrestrial. Large numbers of different species may coexist in the same mud-shore habitat. The group is also well represented in coastal marine habitats. The larvae of the widespread coastal species, *Dimecoenia spinosa,* live submerged in salt marshes and pierce plant roots to obtain oxygen.

Ephydrid larvae utilize a variety of food, but algae (including bluegreens), and particularly diatoms, are often important in the diet. One case-building species feeds almost exclusively on diatoms.

Adults are often common along shores. They can be seen skating on the water surface, walking along the ground, or swarming, sometimes in thick clouds. Adults of some species of *Ephydra* are commonly encountered on algal mats, where they lay eggs. Shore fly adults feed on such items as floating algae and dead floating insects, including springtails, mayflies, or other shore flies, depending on the species. Some actively prey on surface springtails.

SHORE FLIES
(Ephydridae)

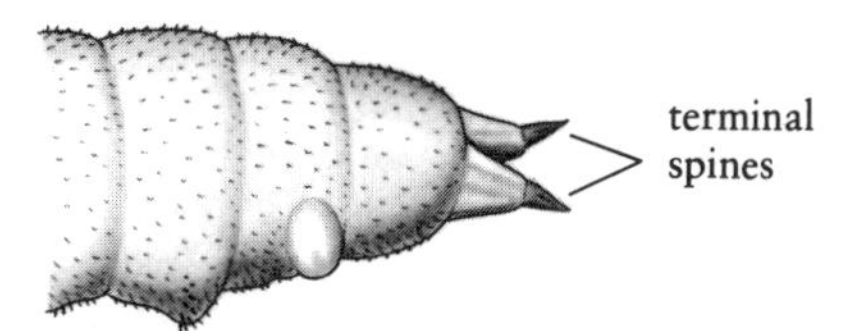

Figure 16.89. *Hydrellia* (Notiphilinae) larva, end of abdomen (dorsolateral)

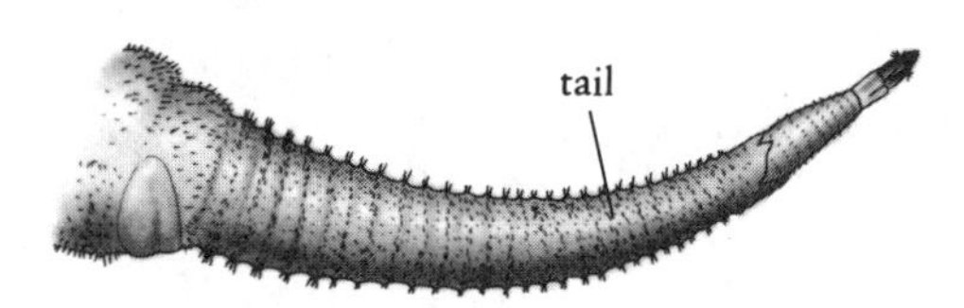

Figure 16.90. *Notiphila* (Notiphilinae) larva, end of abdomen (lateral)

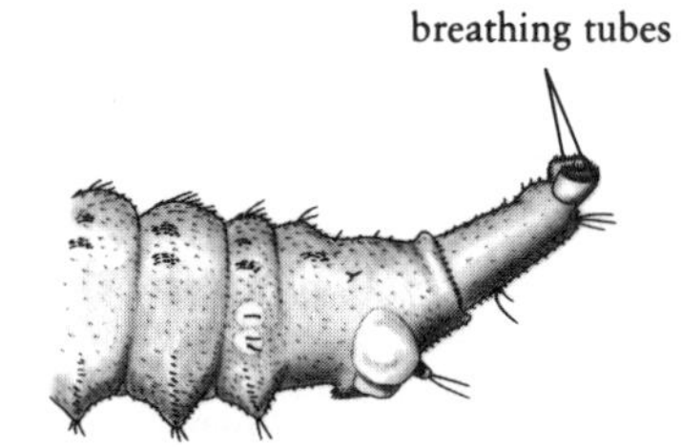

Figure 16.91. *Brachydeutera* (Parydrinae) larva, end of abdomen (lateral)

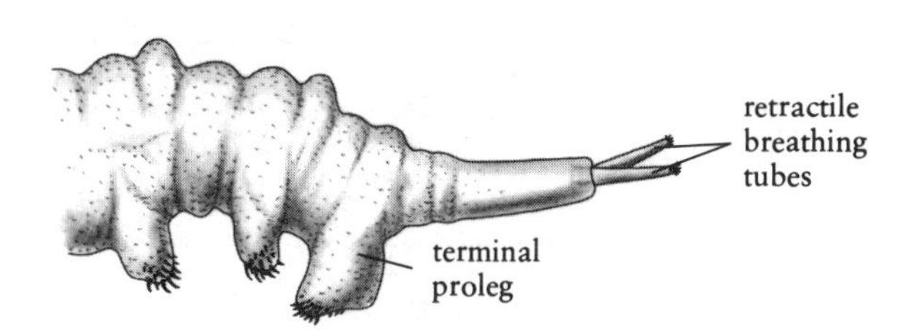

Figure 16.92. *Ephydra* (Ephydrinae) larva, end of abdomen (lateral)

Figure 16.93. *Brachydeutera* puparium

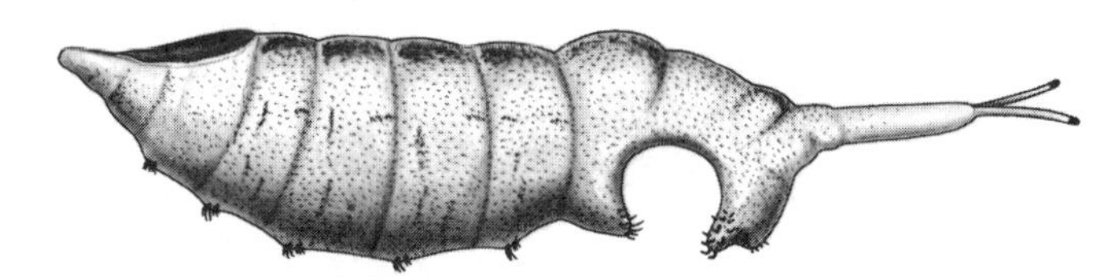

Figure 16.94. *Ephydra* puparium

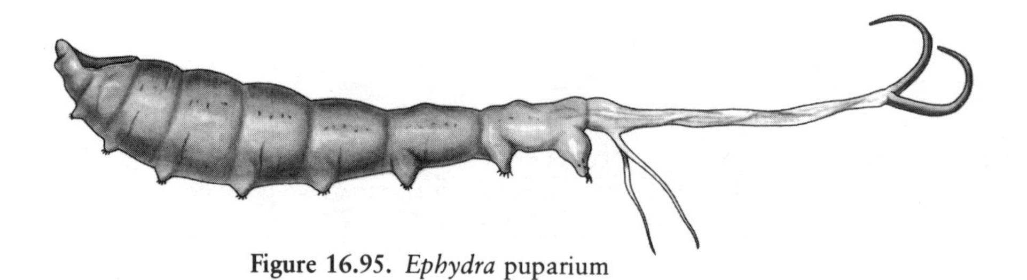

Figure 16.95. *Ephydra* puparium

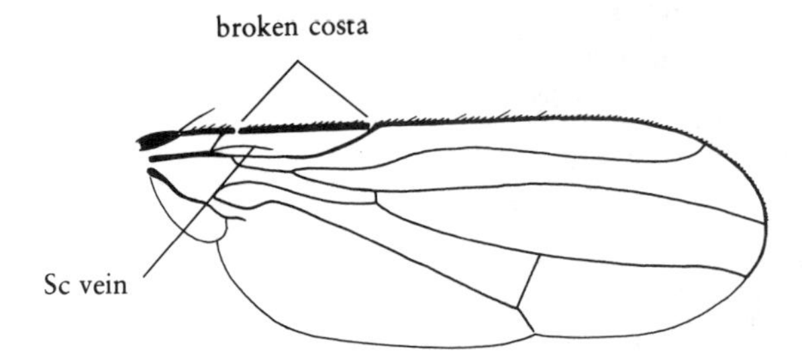

Figure 16.96. Wing

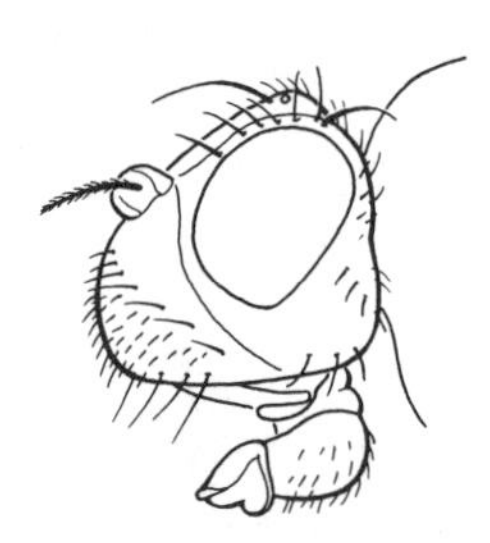

Figure 16.97. Adult head (lateral)

The following diagnoses of subfamilies are based on very few described larvae and must be considered very preliminary.

SUBFAMILY PSILOPINAE: **Larvae** have a pair of very short breathing tubes that are black, hardened at the tip, and as broad as they are long. The group is widespread in North America.

SUBFAMILY NOTIPHILINAE: **Larvae** (Figs. 16.89, 16.90) of some genera have a pair of minute terminal spines and are associated with aquatic vegetation. The group is widespread in North America.

SUBFAMILY PARYDRINAE: **Larvae** (Fig. 16.91) usually have a pair of very short terminal breathing tubes that are longer or as long as they are broad. Body of many is highly wrinkled. The group is widespread in North America.

SUBFAMILY EPHYDRINAE: **Larvae** (Fig. 16.92; Plate XVI, Fig. 121) have well-developed terminal breathing tubes; they may or may not have well-developed abdominal prolegs. The group is widespread in North America.

AQUATIC MUSCIDS
(Families Muscidae and Anthomyiidae)

LARVAL DIAGNOSIS: (Figs. 16.98–16.101; Plate XVI, Fig. 123) Larvae are generally elongate forms, usually measuring 6–14 mm when mature. Abdominal prolegs when present are usually in the form of welts having

AQUATIC MUSCIDS
(Muscidae and Anthomyiidae)

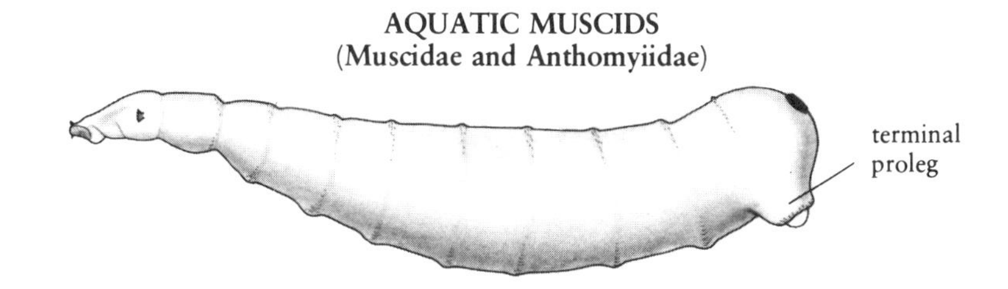

Figure 16.98. Larva (lateral)

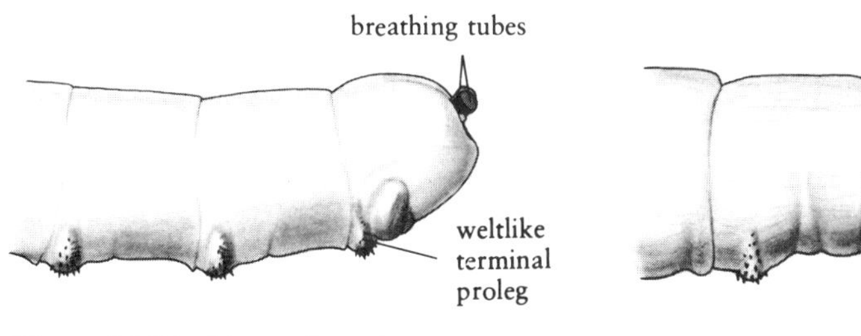

Figure 16.99. *Limnophora* larva, end of abdomen (lateral)

breathing tubes

terminal proleg

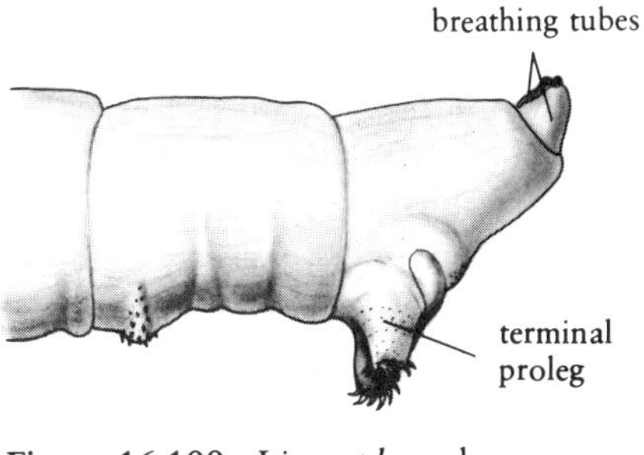

Figure 16.100. *Limnophora* larva, end of abdomen (lateral)

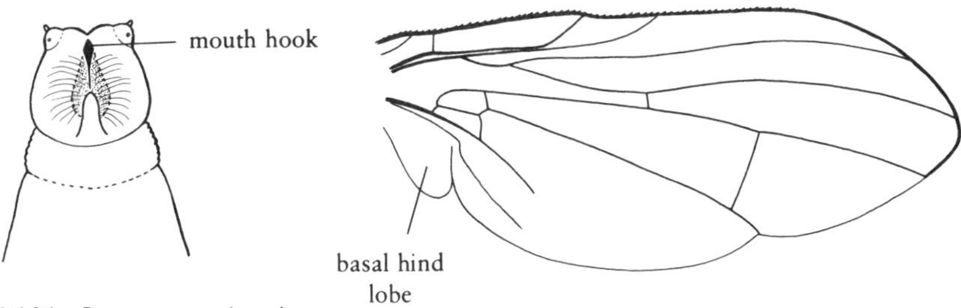

Figure 16.101. Larva, anterior tip of body (ventral schematic)

Figure 16.102. Wing

short, scattered, well-developed spines. Terminal prolegs are usually as long or longer than the pair of short terminal breathing tubes.

ADULT DIAGNOSIS: (Fig. 16.102) These resemble house flies and have a drab or dark, robust body. Basal hind lobes of wings are well developed.

DISCUSSION: Aquatic larvae are poorly known. They may resemble marsh flies or shore flies. Those with a distinct pair of terminal prolegs as long or longer than the terminal breathing tubes can be distinguished from most shore flies. The complete lack of head structure and well-developed prolegs distinguish others from marsh flies. Very close examination of aquatic muscids reveals very smooth, slender mouth hooks.

The families Muscidae and Anthomyiidae both include aquatic muscid flies. Together these families constitute a large and primarily terrestrial group, but about eight genera are presently known to have some species with aquatic larvae. The aquatic and semiaquatic genera *Fucellia* and *Hydrophoria* (Anthomyiidae) are perhaps the best known. Aquatic forms occur throughout North America.

Larvae inhabit ponds and streams or their margins. Some prefer silt bottoms or rocky surfaces, and others prefer filamentous plant growths. Most aquatic larvae are apparently predaceous. Adults of *Fucellia* (Fig. 17.22) are one of the most common flies seen along wet sea beaches with rotting seaweed.

DUNG FLIES
(Family Scatophagidae)

Some members of this primarily terrestrial family develop in the plant tissue of aquatic vegetation; see Chapter 18.

FLESH FLIES
(Family Sarcophagidae)

A few species of this otherwise terrestrial family develop in the water held by pitcher plants; see Chapter 19.

TACHINID FLIES
(Family Tachinidae)

Some species of this family develop as internal parasites of other aquatic or semiaquatic insects. They are therefore only indirectly associated with the aquatic environment.

More Information On Aquatic Diptera

General Coverage

The families and genera of North American Diptera by Curran (1934). Although this book was published some time ago, it remains highly useful for keying adults to genus, at least among families that have not recently undergone extensive taxonomic revision.

Aquatic Diptera by Johannsen (1934, 1935, 1937, 1938). These works have been reprinted into a single volume. Although out-of-date in many respects, it remains one of the best single sources of information on the aquatic stages of North American Diptera.

A catalogue of the Diptera of America north of Mexico by Stone et al. (1965). This large volume is indispensible to the serious student of Diptera, since it contains a list of species along with some biological and geographic information.

Chironomidae

Biology of the larval chironomids by Beck (1976). This is a useful introduction to the midge larvae and includes one of the better keys to genera as well as techniques necessary for the taxonomic study of these larvae.

Environmental requirements and pollution tolerance of common freshwater Chironomidae by Beck (1977). This work is an up-to-date summary of the known habitat ecology of North American freshwater midges.

"Life history of the Chironomidae" by Oliver (1971). This is an excellent introduction to the midges as well as a relatively comprehensive review of general chironomid biology.

References

Because the aquatic Diptera encompasses such a large and diverse group of families, the following references are divided into general works and works dealing with single families. Both categories should be checked for more information on families of interest. As an example, the works by Byers (1978) and Teskey (1969) are very useful for those wishing additional larval identification of Tipulidae or Tabanidae, respectively.

General or Comprehensive

Adamus, P. R., and A. R. Gaufin. 1976. A synopsis of Nearctic taxa found in aquatic drift. *Amer. Midl. Natural.* 95:198–204.

Curran, C. H. 1934. *The families and genera of North American Diptera.* C. H. Curran, New York. 515 pp. (Reprint: 1965. Henry Tripp, Woodhaven, N.Y.).

Heath, B. L., and W. P. McCafferty. 1975. *Aquatic and semi-aquatic Diptera of Indiana.* Purdue Univ. Ag. Exp. Sta. Res. Bull. 930. 17 pp.

Hennig, W. 1948–1952. *Die larvenformen der Dipteran.* Akademie-Verlag, Berlin. Pt. 1 (1948), 185 pp.; Pt. 2 (1950), 458 pp.; Pt. 3 (1952), 628 pp.

Hinton, H. E. 1968. Spiracular gills. *Adv. Ins. Physiol.* 5:65–162.

Johannsen, O. A. 1934–1938. *Aquatic Diptera. Pt. I* (1934) *Nematocera, exclusive of Chironomidae and Ceratopogonidae. Pt. II.* (1935) *Orthorrhapha-Brachycera and Cyclorrhapha. Pt. III.* (1937) *Chironomidae: Subfamilies Tanypodinae, Diamesinae, and Orthocladiinae. Pt. IV.* (1938) *Chironomidae: Subfamily Chironominae.* Mem. Cornell Univ. Ag. Exp. Sta. 164, 71 pp.; 171, 62 pp.; 205, 3–84 pp.; 210, 3–80 pp. (Reprint: 1969. Entomol. Rep. Specialists, Los Angeles).

Kimerle, R. A., and W. R. Enns. 1968. Aquatic insects associated with midwestern waste stabilization lagoons. *J. Wat. Poll. Control Fed.* 40:31–41.

Komnick, H. 1977. Chloride cells and chloride epithelia of aquatic insects. *Inter. Rev. Cyt.* 49:285–329.

McAlpine, J. F. 1979. Diptera, pp. 389–395. In *Canada and its insect fauna* (H. V. Danks, ed.). Mem. Canad. Entomol. Soc. 108.

Revill, D. L., K. W. Stewart, and H. E. Schlichting, Jr. 1967. Dispersal of viable algae and protozoa by horse flies and mosquitoes (Diptera: Tabanidae, Culicidae). *Ann. Entomol. Soc. Amer.* 60:1077–1081.

Revill, D. L., K. W. Stewart, and H. E. Schlichting, Jr. 1967. Passive dispersal of viable algae and protozoa by certain craneflies and midges. *Ecology* 48: 1023–1027.

Schwiebert, E. 1973. *Nymphs.* Winchester Press, New York. 339 pp.

Stone, A., et al. 1965. *A catalog of the Diptera of America north of Mexico.* U.S.D.A. Handbook 276. 1969 pp.

Wirth, W. W., and A. Stone. 1956. Aquatic Diptera, pp. 372–482. In *Aquatic insects of California* (R. L. Usinger, ed.). Univ. Calif. Press, Berkeley.

Athericidae

Webb, D. W. 1977. The Nearctic Athericidae (Insecta: Diptera). *J. Kans. Entomol. Soc.* 50:473–495.

Blephariceridae

Alexander, C. P. 1963. Blephariceridae and Deuterophlebiidae, pp. 39–83. In *Guide to the insects of Connecticut. Part VI. The Diptera or true flies of Connecticut.* Fasc. 8. St. Geol. Nat. Hist. Surv. Conn. Bull. No. 93.

Hogue, C. L. 1978. *The net-winged midges of eastern North America, with notes on new taxonomic characters in the family Blephariceridae (Diptera).* Nat. Hist. Mus. Los Angeles Co. Contr. Sci. 291. 41 pp.

Ceratopogonidae

Battle, F. V., and F. C. Turner, Jr. 1971. *The insects of Virginia: No. 3. A systematic review of the genus Culicoides (Diptera: Ceratopogonidae) of Virginia with a geographic catalog of species occurring in eastern United States north of Florida.* Va. Poly. Inst. St. Univ. Res. Div. Bull. 44. 129 pp.

Blanton, F. S., and W. W. Wirth. 1979. *The sand flies (Culicoides) of Florida (Diptera: Ceratopogonidae).* Fla. Dept. Ag. Cons. Serv., Gainesville. 204 pp.

Dow, M. I., and E. C. Turner, Jr. 1976. *A revision of the Nearctic species of the genus Bezzia (Diptera: Ceratopogonidae).* Va. Poly. Inst. St. Univ. Res. Div. Bull. 103. 162 pp.

Downes, J. A. 1978. *Feeding and mating in the insectivorous Ceratopogonidae (Diptera).* Mem. Entomol. Soc. Canada 104. 62 pp.

Grogan, W. L., Jr. 1977. A revision of the Nearctic species of *Parabezzia* Malloch (Diptera: Ceratopogonidae). *J. Kans. Entomol. Soc.* 50:49–84.

Grogan, W. L., Jr., and W. W. Wirth. 1975. *A revision of the genus Palpomyia Meigen of northeastern North America (Diptera: Ceratopogonidae).* Contr. Maryland Ag. Exp. Sta. 5076. 49 pp.

Linley, J. R. 1968. Studies on the larval biology of *Leptoconops becquaerti* (Kieff) (Diptera: Ceratopogonidae). *Bull. Entomol. Res.* 58:1–24.

Waugh, W. T., and W. W. Wirth. 1976. A revision of the genus *Dasyhelea* Kieffer of the eastern United States north of Florida (Diptera: Ceratopogonidae). *Ann. Entomol. Soc. Amer.* 69:219–247.

Wirth, W. W., N. C. Ratanaworabhan, and D. H. Messersmith. 1977. Natural history of Plummers Island, Maryland. XXII. Biting midges (Diptera: Ceratopogonidae). 1. Introduction and key to genera. *Proc. Biol. Soc. Wash.* 90: 615–647.

Chaoboridae

Cook, E. F. 1956. *The Nearctic Chaoborinae (Diptera: Culicidae).* Univ. Minn. Ag. Exp. Sta. Tech. Bull. 218. 102 pp.

Hitchcock, S. W. 1965. The seasonal fluctuation of limnetic *Chaoborus punctipennis* and its role as a pest in drinking water. *J. Econ. Entomol.* 58:902–904.

Parma, S. 1971. *Chaoborus flavicans (Meigen) (Diptera, Chaoboridae): An autecological study.* Bronder-Offset n.v., Rotterdam. 128 pp.

Chironomidae

Beck, W. M., Jr. 1976. *Biology of the larval chironomids.* St. Fla. Dept. Environ. Reg. Tech. Ser. 2. 58 pp.

Beck, W. M., Jr. 1977. *Environmental requirements and pollution tolerance of common freshwater Chironomidae.* Environ. Monit. Sup. Lab., Off. Res. Dev. U.S.E.P.A., Cincinnati. 417 pp.

Brock, E. M. 1960. Mutualism between the midge *Cricotopus* and the alga *Nostoc.* *Ecology* 41:474–483.

Bryce, D., and A. Hobart. 1972. The biology and identification of the larvae of the Chironomidae (Diptera). *Entomol. Gaz.* 23:175–217.

Danks, H. V. 1971. Overwintering of some north temperate and arctic Chironomidae. II. Chironomid biology. *Canad. Entomol.* 103:1875–1910.

Danks, H. V., and J. W. Jones. 1978. Further observations on winter cocoons in Chironomidae (Diptera). *Canad. Entomol.* 110:667–669.

Dendy, J. S. 1973. Predation on chironomid eggs and larvae by *Nanocladius alternantherae* Dendy and Sublette (Diptera: Chironomidae, Orthocladiinae). *Entomol. News* 84:91–95.

Hamilton, A. L., O. A. Saether, and D. R. Oliver. 1969. *A classification of the Nearctic Chironomidae.* Fish. Res. Brd. Canada Tech. Rep. 124. 42 pp.

Hilsenhoff, W. L. 1966. The biology of *Chironomus plumosus* (Diptera: Chironomidae) in Lake Winnebago, Wisconsin. *Ann. Entomol. Soc. Amer.* 59:465–473.

Hinton, H. E. 1960. A fly larva that tolerates dehydration and temperatures of −270° to +102°C. *Nature* 188:336–337.

Iovino, A. J., and W. H. Bradley. 1969. The role of larva Chironomidae in the production of lacustrine copropel in Mud Lake, Marion County, Florida. *Limnol. Oceonogr.* 14:898–905.

Neumann, D. 1976. Adaptations of chironomids to intertidal environments. *Annu. Rev. Entomol.* 21:387–414.

Oliver, D. R. 1971. Life history of the Chironomidae. *Annu. Rev. Entomol.* 16:211–230.

Roback, S. S. 1971. *The adults of the subfamily Tanypodinae in North America (Diptera: Chironomidae).* Monogr. Acad. Nat. Sci. Phil. 17. 410 pp.

Roback, S. S. 1976. The immature chironomids of the eastern United States. I. Introduction and Tanypodinae—Coelotanypodini. *Proc. Acad. Nat. Sci. Phil.* 127:147–201.

Rosenberg, D. M., A. P. Wiens, and O. A. Saether. 1977. Life histories of *Cricotopus (Cricotopus) bicinctus* and *C. (C.) mackenziensis* (Diptera: Chironomidae) in the Fort Simpson area, Northwest Territories. *J. Fish. Res. Brd. Canada* 34:247–253.

Saether, O. A. 1979. Chironomid communities as water quality indicators. *Hol. Ecol.* 2:65–74.

Soponis, A. R. 1977. *A revision of the Nearctic species of Orthocladius (Orthocladius) van der Wulp (Diptera: Chironomidae).* Mem. Entomol. Soc. Canada. 102. 187 pp.

Steffan, A. W. 1967. Ectosymbiosis in aquatic insects, pp. 207–289. In *Symbiosis,* Vol. II (S. M. Henry, ed.). Academic Press, New York.

Usinger, R. L., and W. R. Kellen. 1955. The role of insects in sewage disposal beds. *Hilgardia* 23:263–321.

Culicidae

Carpenter, S. J., and W. J. LaCasse. 1955. *Mosquitoes of North America.* Univ. Calif. Press, Berkeley. 360 pp.

Deuterophlebiidae

Alexander, C. P. 1963. Blephariceridae and Deuterophlebiidae, pp. 39–83. In *Guide to the insects of Connecticut. Part VI. The Diptera or true flies of Connecticut.* Fasc. 8. St. Geol. Nat. Hist. Surv. Conn. Bull. No. 93.

Kennedy, H. D. 1958. Biology and life history of a new species of mountain midge, *Deuterophlebia nielsoni,* from eastern California (Diptera: Deuterophlebiidae). *Trans. Amer. Micr. Soc.* 89:201–228.

Dixidae

Nowell, W. R. 1963. Dixidae, pp. 85–111. In *Guide to the insects of Connecticut. Part VI. The Diptera or true flies of Connecticut.* Fasc. 8. St. Geol. Nat. Hist. Surv. Conn. Bull. No. 93.

Peters, T. M., and E. F. Cook. 1966. The Nearctic Dixidae (Diptera). *Misc. Publ. Entomol. Soc. Amer.* 5:233–278.

Ephydridae

Deonier, D. L. 1964. Keys to the shore flies of Iowa (Diptera, Ephydridae). *Iowa St. J. Sci.* 39:103–126.

Deonier, D. L. 1971. *A systematic and ecological study of Nearctic Hydrellia (Diptera: Ephydridae).* Smith. Contr. Zool. 68. 147 pp.

Deonier, D. L. (ed.). 1979. *First symposium on the systematics and ecology of Ephydridae (Diptera).* N. A. Benthological Soc. 147 pp.

Scheiring, J. F. 1976. Ecological notes on the brine flies of northwestern Oklahoma (Diptera: Ephydridae). *J. Kans. Entomol. Soc.* 49:450–452.

Simpson, K. W. 1976. Shore flies and brine flies (Diptera: Ephydridae), pp. 465–495. In *Marine insects* (L. Cheng, ed.). North-Holland, Amsterdam.

Sturtevant, A. H., and M. R. Wheeler. 1954. Synopses of Nearctic Ephydridae (Diptera). *Trans. Amer. Entomol. Soc.* 79:151–261.

Thier, R. W., and B. A. Foote. 1980. Biology of mud-shore Ephydridae (Diptera). *Proc. Entomol. Soc. Wash.* 82:517–535.

Wirth, W. W. 1971. The brine flies of the genus *Ephydra* in North America (Diptera: Ephydridae). *Ann. Entomol. Soc. Amer.* 64:357–377.

Nymphomyiidae

Ide, F. P. 1965. A fly of the archaic family Nymphomyiidae (Diptera) from North America. *Canad. Entomol.* 97:496–507.

Kevan, D. K. McE., and F. E. A. Cutten-Ali-Khan. 1975. Canadian Nymphomyiidae (Diptera). *Canad. J. Zool.* 53:853–866.

Psychodidae

Quate, L. W. 1960. *Guide to the insects of Connecticut. Part VI. The Diptera or true flies of Connecticut. Fasc. 7. Psychodidae.* St. Geol. Nat. Hist. Surv. Conn. Bull. 92. 47 pp.

Ptychopteridae

Hodkinson, I. D. 1973. The immature stages of *Ptychoptera lenis lenis* (Diptera: Ptychopteridae) with notes on their biology. *Canad. Entomol.* 105:1091–1099.

Sciomyzidae

Berg, C. O. 1953. Sciomyzid larvae (Diptera) that feed on snails. *J. Parasit.* 39:630–636.

Berg, C. O. 1964. Snail-killing sciomyzid flies: Biology of the aquatic species. *Verh. Int. Verein. Limnol.* 15:926–932.

Berg, C. O., and L. Knutson. 1978. Biology and systematics of the Sciomyzidae. *Annu. Rev. Entomol.* 23:239–258.

Bratt, A. D., L. V. Knutson, B. A. Foote, and C. O. Berg. 1969. *Biology of Pherbellia (Diptera: Sciomyzidae).* Mem. Cornell Univ. Exp. Sta. 404. 247 pp.

Foote, B. A. 1976. Biology and larval feeding habits of three species of *Renocera* (Diptera: Sciomyzidae) that prey on fingernail clams (Mollusca: Sphaeriidae). *Ann. Entomol. Soc. Amer.* 69:121–133.

Knutson, L. V., and C. O. Berg. 1964. Biology and immature stages of snail-killing flies: The genus *Elgiva* (Diptera: Sciomyzidae). *Ann. Entomol. Soc. Amer.* 57:173–192.

Neff, S. E., and C. O. Berg. 1966. Biology and immature stages of malacophagous Diptera of the genus *Sepedon* (Sciomyzidae). *Trans. Amer. Entomol. Soc.* 88:77–93.

Simuliidae

Jamnback, H. 1973. Recent developments in control of blackflies. *Annu. Rev. Entomol.* 18:281–304.

Mulla, M. S., and L. A. Lacey. 1976. Feeding rates of *Simulium* larvae on particulates in natural streams (Diptera: Simuliidae). *Environ. Entomol.* 5:283–287.

Stone, A. 1964. *Guide to the insects of Connecticut. Part VI. The Diptera or true flies of Connecticut. Fasc. 9. Simuliidae and Thaumaleidae.* St. Geol. Nat. Hist. Surv. Conn. Bull. 97. 123 pp.

Tarshis, I. B., and W. Neil. 1970. Mass movement of black fly larvae on silken threads (Diptera: Simuliidae). *Ann. Entomol. Soc. Amer.* 63:608–610.

Stratiomyidae

McFadden, M. W. 1967. *Soldier fly larvae in America north of Mexico.* Proc. U.S. Nat. Mus. 121. 72 pp.

Syrphidae

Lavallee, A. G., and J. B. Wallace. 1974. Immature stages of Milesiinae (Syrphidae). II. *Sphegina keeniana* and *Chrysogaster nitida. J. Ga. Entomol. Soc.* 9:8–15.

Roberts, M. J. 1970. The structure of the mouth parts of syrphid larvae (Diptera) in relation to feeding habits. *Acta Zool.* 51:43–65.

Tabanidae

Goodwin, J. T. 1976. Immature stages of some eastern Nearctic Tabanidae (Diptera). VII. *Haematopota* Meigen and *Whitneyomyia* Bequaert plus other Tabanini. *Fla. Entomol.* 59:369–390.

Teskey, H. J. 1969. *Larvae and pupae of some eastern North American Tabanidae (Diptera).* Mem. Entomol. Soc. Canada 63. 147 pp.

Teskey, H. J., and J. F. Burger. 1976. Further larvae and pupae of eastern North American Tabanidae (Diptera). *Canad. Entomol.* 108:1085–1096.

Tanyderidae

Exner, K., and D. A. Craig. 1976. Larvae of Alberta Tanyderidae (Diptera: Nematocera). *Quaest. Entomol.* 12:219–237.

Thaumaleidae

Stone, A. 1964. *Guide to the insects of Connecticut. Part VI. The Diptera or true flies of Connecticut. Fasc. 9. Simuliidae and Thaumaleidae.* St. Geol. Nat. Hist. Surv. Conn. Bull. 97. 123 pp.

Valliant, F. 1959. The Thaumaleidae (Diptera) of the Appalachian mountains. *J. N.Y. Entomol. Soc.* 67:31–37.

Tipulidae

Alexander, C. P. 1942. Tipulidae, pp. 196–485. In *Guide to the insects of Connecticut. Part VI. The Diptera or true flies of Connecticut.* Fasc. 1. St. Geol. Nat. Hist. Surv. Conn. Bull. No. 64.

Byers, G. W. 1978. Tipulidae, pp. 285–310. In *An introduction to the aquatic insects of North America* (R. W. Merritt and K. W. Cummins, eds.). Kendall/Hunt, Dubuque.

Hinton, H. E. 1968. Spiracular gills. *Adv. Ins. Physiol.* 5:65–162.

COLOR PLATES

Glory be to God for dappled things—
For skies of couple-colour as a brindled cow;
For rose-moles all in stipple upon trout that swim;
Fresh-firecoal chestnut-falls; finches wings;
Landscape plotted and pieced—fold, fallow, and plough;
And all trades, their gear and tackle and trim.

Gerard Manley Hopkins

Mayflies

EPHEMEROPTERA

SMALL MINNOW MAYFLIES Family **Baetidae,** p. 102

1. *Baetis longipalpus,* larva: 4–5 mm excluding tails; commonly found on rocks, driftwood, and gravel in fast-flowing areas of some larger rivers of the Midwest; August emergence.
2. *Baetis tricaudatus,* adult ♂ (the Rusty Spinner): 5–8 mm excluding tails; commonly found near riffles and rapids of streams and rivers throughout most of North America except extreme southeastern states; has been known also as *B. vagans* and by several common names in eastern and central regions, including Bluewinged Olive, which generally applies to *Baetis*; size and color variable depending on local environment and emergence time.
3. *Baetis tricaudatus,* subimago ♀ (the Little Iron Blue Quill): 5–8 mm excluding tails; this dun, known by several fishermen's names, provides some of the earliest major hatches on trout streams; hatches periodically through spring, summer, and fall.
4. *Callibaetis pacificus,* adult ♀ (the Medium Specklewinged Quill): 7–8 mm excluding tails; develops in ponds, lakes, and slower waters in western and westernmost central states, dry fly imitations of this and other *Callibaetis,* sometimes known as Cream Hen Spinners, can be used for fishing lake margins and ponds throughout the season.

BRUSHLEGGED MAYFLIES Family **Oligoneuriidae,** p. 105

5. *Isonychia bicolor,* adult ♂ (the Whitegloved Howdy): 10–12 mm excluding tails; occurs in eastern and central states and eastern Canada; May through August emergence with significant spinner falls on trout streams in July.
6. *Isonychia bicolor,* larva: 11–14 mm excluding tails; found primarily on substrates associated with moderate- to fast-flowing areas of streams and rivers; provides excellent model for nymph fishing, and the duns of this and related species, commonly known as the Leadwing Coachman or Mahogany Dun, have best fishing hatches in June and July.

1
2
3
4
5
6

Mayflies

EPHEMEROPTERA

FLATHEADED MAYFLIES Family **Heptageniidae,** p. 106

7. *Epeorus longimanus,* subimago ♂ (the Quill Gordon or Western Quill Gordon): 9–11 mm excluding tails; develops in cold, clean streams above 5,000 feet elevation throughout much of western North America; *E. albertae* may occur at lower elevations of same streams; late-June to early-September hatches signal excellent fly fishing.

8. *Stenacron interpunctatum,* larva: 8–11 mm excluding tails; commonly found in streams and rivers east of Rocky Mountains, often on the undersides of rocks in pools or slow to moderate curents and also in ponds and lakes in more northern and mountainous regions; a variable and ubiquitous mayfly.

9. *Stenacron interpunctatum,* subimago ♀ (the Light Cahill): 7–12 mm excluding tails; this well-known mayfly provides excellent hatches on the slower reaches of trout and bass waters of the Midwest and East from May through September.

PRONGGILLS Family **Leptophlebiidae,** p. 110

10. *Paraleptophlebia debilis,* larva: 6–8 mm excluding tails; inhabits rocks and gravel in small to medium-sized streams throughout much of North America; most common in northern states and Canada and in the West at elevations above 3,000 feet; mature individuals often migrate to quiet pools prior to hatching.

11. *Paraleptophlebia debilis,* adult ♂ : 6–8 mm excluding tails; typical of many species of *Paraleptophlebia,* which as subimagos are commonly referred to as the Dark Blue Quill; small, distinct swarms common during afternoons, primarily August through October.

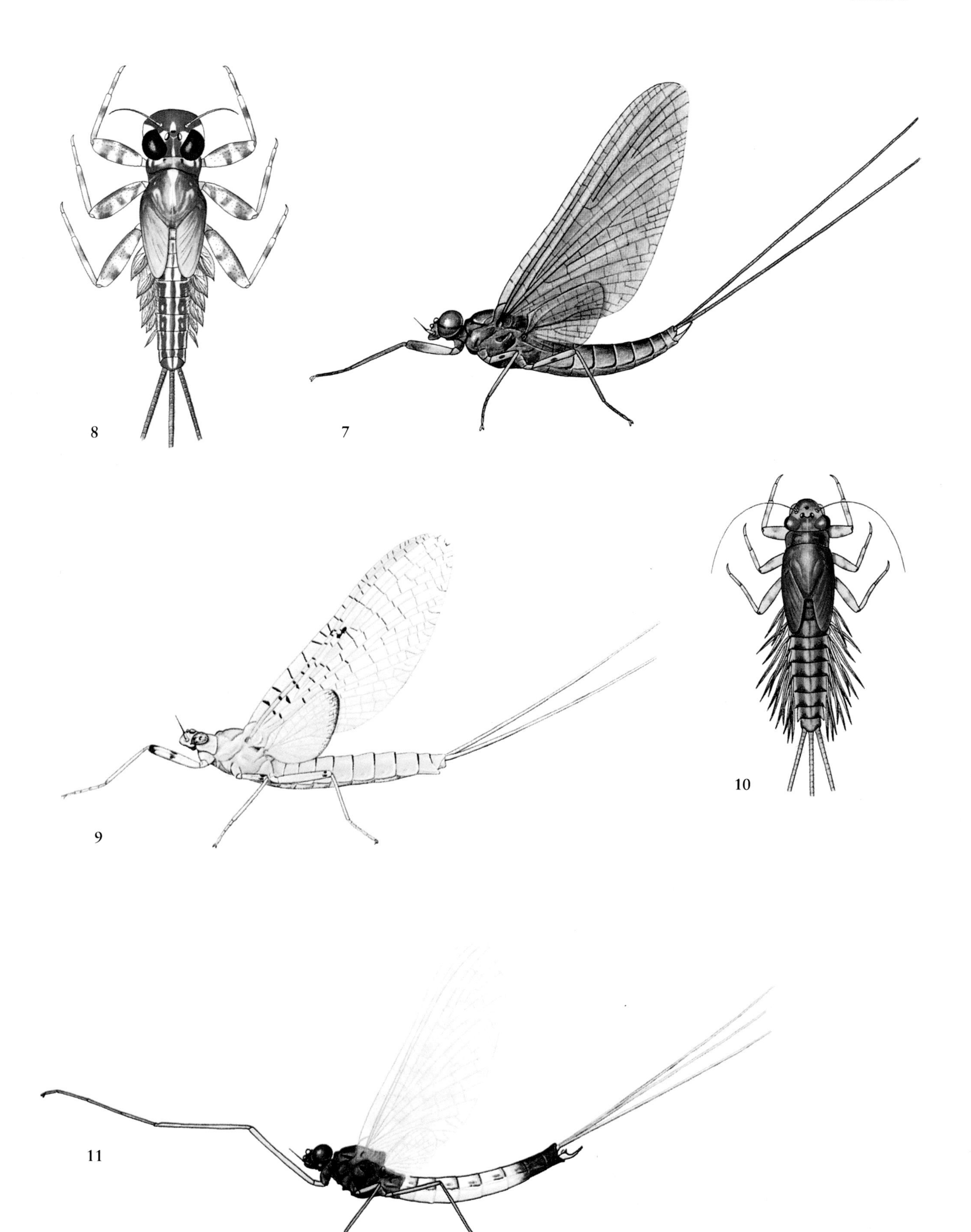
8
7
10
9
11

Mayflies

EPHEMEROPTERA

HACKLEGILLS Family **Potamanthidae,** p. 112

12. *Potamanthus rufous,* larva: 10–12 mm excluding tails; usually found on rocks in moderate currents of larger streams in several areas of eastern and central North America; hatches of this and other *Potamanthus,* commonly known as the Golden or Yellow Drake, begin in late June.

COMMON BURROWERS Family **Ephemeridae,** p. 115

13. *Hexagenia limbata,* larva: 17–33 mm excluding tails; burrows in silt deposits of lakes, ponds, streams, and rivers throughout much of North America; this common burrower provides excellent live bait for a variety of game fishes throughout the year and is an excellent model for nymph fishing.

14. *Hexagenia limbata,* subimago ♂ (the Michigan Caddis): 20–28 mm excluding tails; the well-known duns sometimes hatch in massive numbers periodically throughout the warmer months; fishermen, who also know this species as the Sandfly or Fishfly, can find productive wet and dry fly fishing with imitations during mid-summer hatches on trout streams.

15. *Ephemera guttulata,* subimago ♂ (the Green Drake): 18–21 mm excluding tails; develops in clean trout streams throughout the East; imitations of this large and popular dun are best fished during the first few days of a hatch; primarily late May and June emergence.

SPINY CRAWLERS Family **Ephemerellidae,** p. 118

16. *Drunella grandis,* larva: 12–15 mm excluding tails; occurs throughout most of mountainous western North America in moderately flowing streams and rivers up to 7,000 feet elevation; this distinctive mayfly was previously considered in the genus *Ephemerella.*

17. *Drunella grandis,* subimago ♂ (the Great Leadwinged Olive Drake): 14–18 mm excluding tails; western fly fishermen have referred to this dun as the Western Green Drake because it approaches the Green Drake not only in size but also in the ability of its hatches to stimulate trout feeding; mid-June through mid-August emergence.

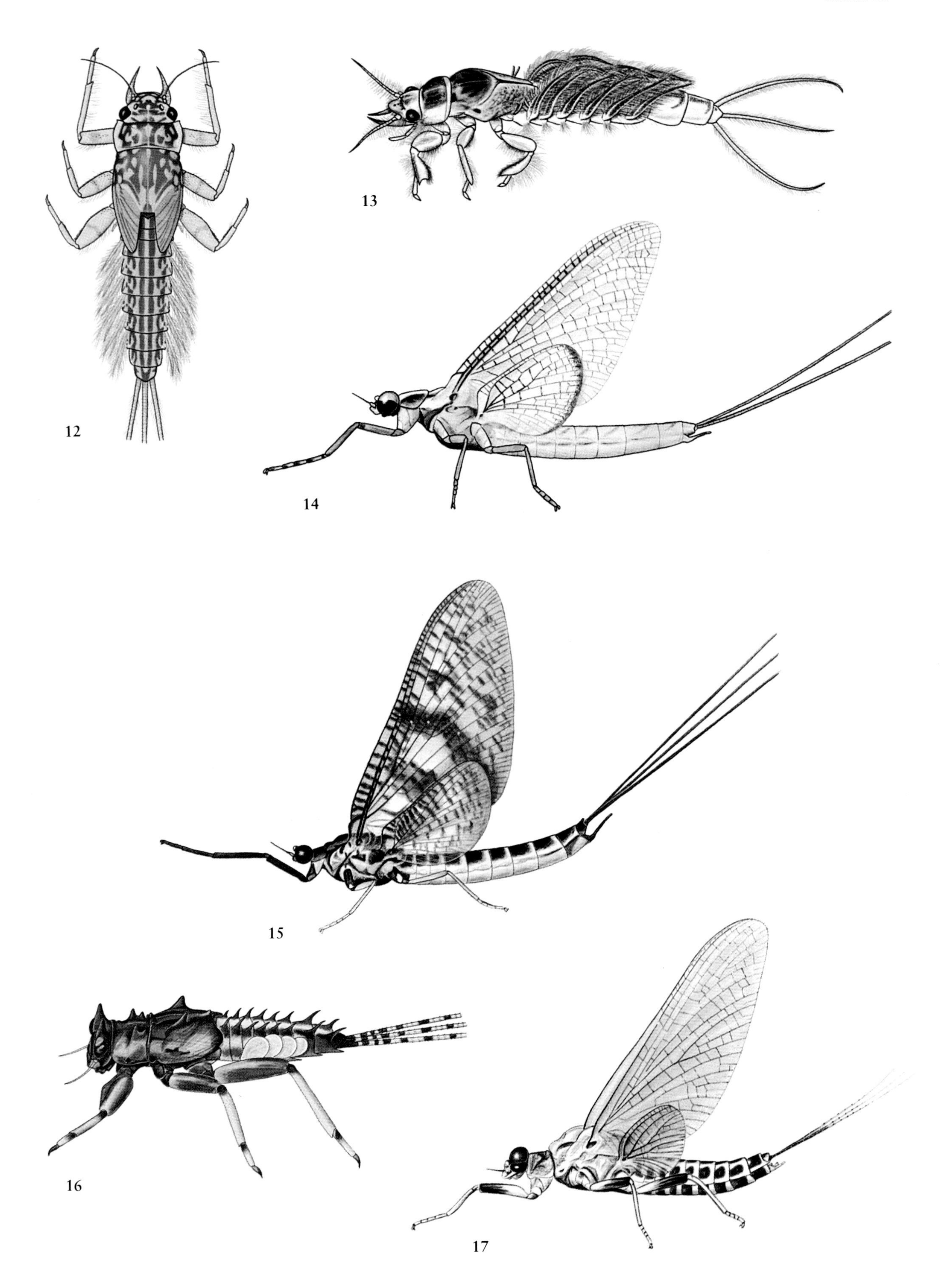
12
13
14
15
16
17

Mayflies

EPHEMEROPTERA

SPINY CRAWLERS Family **Ephemerellidae,** p. 118

18. *Ephemerella subvaria,* larva: 8–10 mm excluding tails; inhabits a variety of streams in northeastern and central states and eastern Canada, often associated with gravel bottoms and vegetation; emerger patterns can be fished in early May, since larvae become quite active during this period prior to hatching; bottom nymph fishing is good in late summer and fall.

19. *Ephemerella subvaria,* subimago ♂ (the Hendrickson): 8–9 mm excluding tails; matching the hatches of this popular dun in May provides some of the best early-season fly fishing on north-central and eastern trout waters.

LITTLE STOUT CRAWLERS Family **Tricorythidae,** p. 119

20. *Tricorythodes atratus,* subimago ♂ (the Pale Olive Dun or Snowflake Mayfly): 3–3.5 mm excluding tails; develops in moderately flowing streams of central states and eastern and central Canada; large hatches of this small dun occur in June and August; adult swarms are often conspicuous for long distances over the stream.

SMALL SQUAREGILLS Family **Caenidae,** p. 121

21. *Caenis simulans,* adult ♂ (the Tiny Whitewinged Sulphur): 3–4 mm excluding tails; known from throughout North America excluding the extreme southeastern states; large numbers are often attracted to lights at night in the vicinity of lakes, ponds, and slower reaches of streams; the extremely short-lived adults emerge from May through August.

22. *Caenis simulans,* larva: 3–4 mm excluding tails; often found in association with detrital sediments or thick algal growths of stillwater habitats.

ARMORED MAYFLIES Family **Baetiscidae,** p. 122

23. *Baetisca laurentina,* larva: 8–10 mm excluding tails; occurs on the substrate near the banks of moderately sized streams of central and northeastern states and eastern Canada; because of the development of the thoracic shield, this genus is one of the most striking of mayflies; May and June emergence.

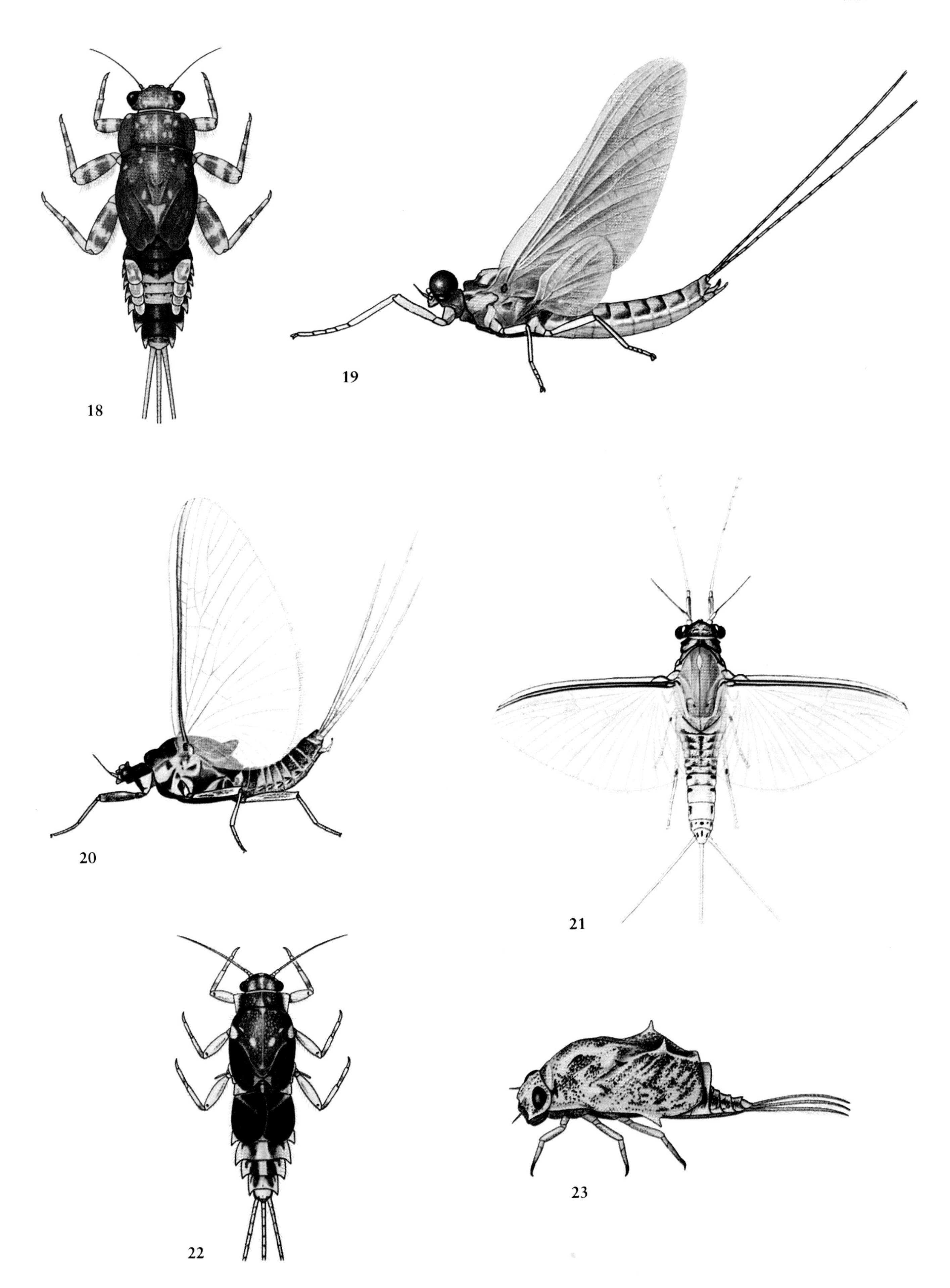
18
19
20
21
22
23

Damselflies and Dragonflies

ODONATA

BROADWINGED DAMSELFLIES Family **Calopterygidae,** p. 132

24. *Hetaerina americana,* larva: 25–30 mm; found climbing among aquatic vegetation, roots, and debris near the banks of rivers and streams in central, southwestern, and eastern states, and eastern Canada.

25. *Hetaerina americana,* adult ♂ (the American Ruby Spot): 40–46 mm; this striking damselfly can be seen flying along the edges of streams from mid-June through October; numbers may congregate along river banks in late afternoon in the Midwest.

SPREADWINGED DAMSELFLIES Family **Lestidae,** p. 133

26. *Lestes congener,* larva: 22–26 mm; found climbing among aquatic vegetation of ponds throughout North America excluding southeastern states.

27. *Lestes congener,* adult ♂ (the Little Spreadwing): 34–36 mm; flight period August through mid-November; generally in tandem during oviposition, which may take place well above water on cattails, sedges, and willows; although wings shown together, they are usually spread somewhat at rest.

NARROWWINGED DAMSELFLIES Family **Coenagrionidae,** p. 134

28. *Argia vivida,* larva: 17 mm; occurs on rocks and vegetation of spring-fed stream and pool bottoms throughout much of western North America.

29. *Argia vivida,* adult ♂ (the Vivid Dancer): 30–34 mm; this common damselfly flies from June through August, usually in the vicinity of spring habitats.

CLUBTAILS Family **Gomphidae,** p. 137

30. *Gomphus vastus,* larva: 27–30 mm; inhabits soft substrate of larger lakes and rivers of central states south to Alabama and Kansas, northeastern states, and eastern and central Canada; most common in lower Great Lakes region.

31. *Gomphus vastus,* adult ♂ (the Desolate Clubtail): 49–52 mm; flight period May through July; mating occurs on adjacent trees and shrubs; oviposition takes place far out over the water.

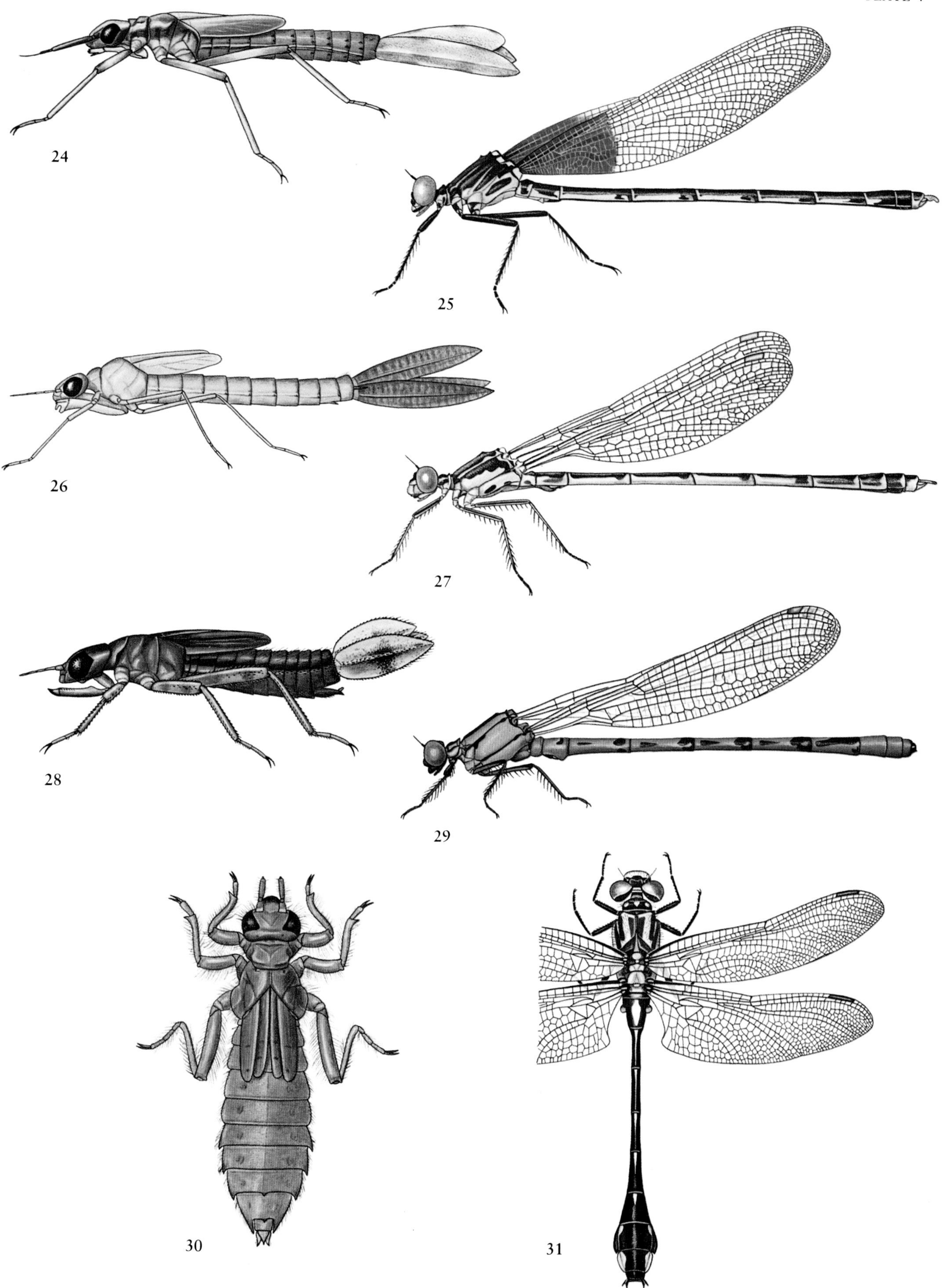
24
25
26
27
28
29
30
31

Dragonflies
ODONATA

DARNERS — Family **Aeshnidae,** p. 139

32. *Anax junius,* adult ♂ (the Common Green Darner): 78–80 mm; this large and beautiful dragonfly is common throughout the United States and southern Canada; in most regions it flies from May through October, but flies in all months of the year in Florida; wild nuptial flights occur soon after emergence, and large numbers of individuals sometimes migrate along river courses or lake margins.

33. *Anax junius,* larva: 43–47 mm; climbs among plants in still waters with emergent vegetation, including ponds, lakes, and slow streams; completes development within a year and undergoes winter dormancy in colder regions.

GREENEYED SKIMMERS — Family **Corduliidae,** p. 143

34. *Epitheca princeps,* adult ♂ (the Royal Skimmer): 59–62 mm; develops in lakes and quiet reaches of rivers throughout most of central and eastern North America; mid-May to mid-September flight period.

COMMON SKIMMERS — Family **Libellulidae,** p. 144

35. *Libellula pulchella,* adult ♂ (the Ten-Spot Skimmer): 49–52 mm; this widespread species develops in ponds, lake margins, and slow streams; primarily May through September flight period; often drives away other dragonflies from its territory, and perches on vegetation between short flights over the water.

36. *Libellula luctuosa,* larva (the Widow Skimmer): 20–22 mm; common inhabitant of ponds, lakes, and marshes throughout most of North America excluding far western regions; wings of the adults differ from other common skimmers by being dark basally, white in the middle, and hyaline at the apex.

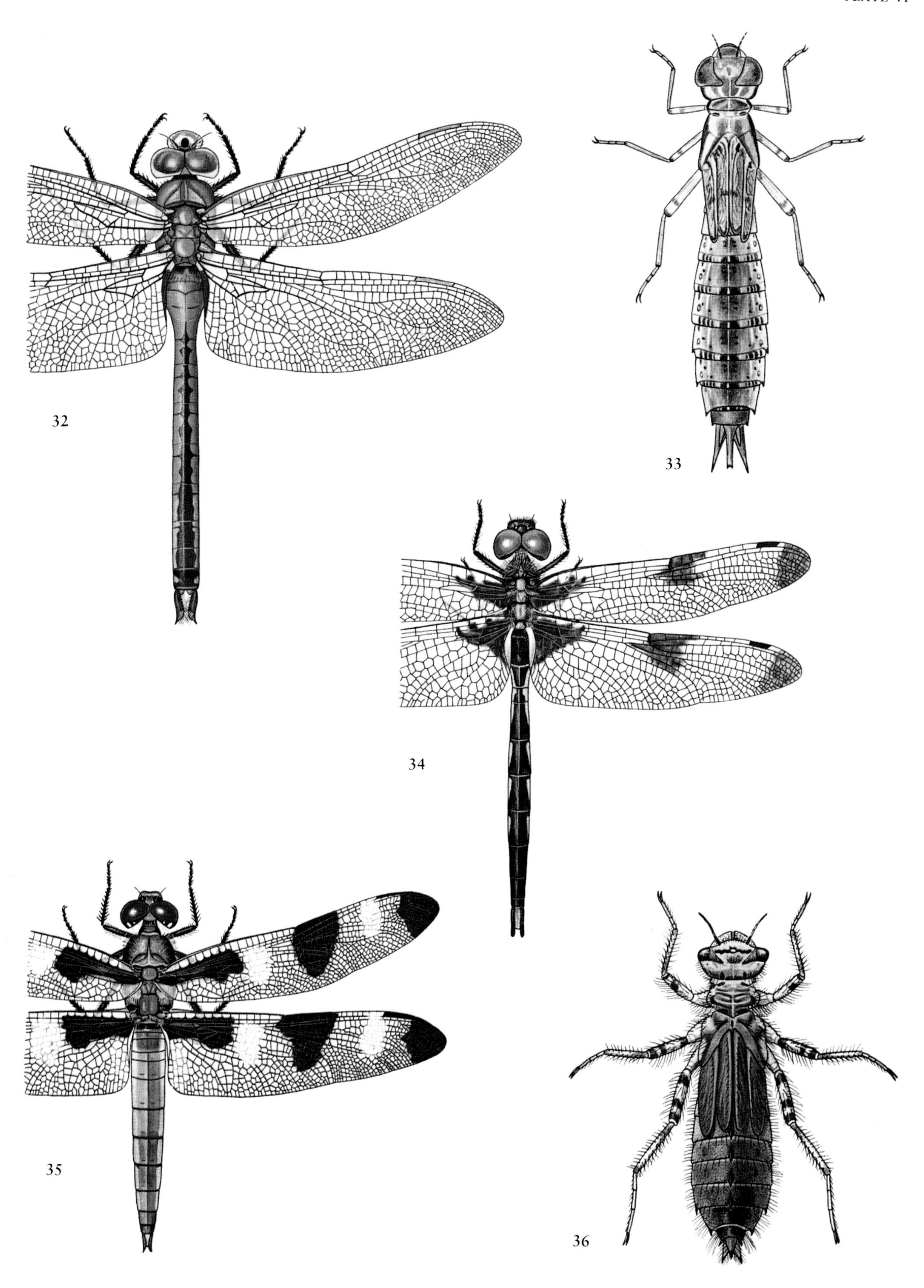
32
33
34
35
36

Stoneflies

PLECOPTERA

NEMOURID BROADBACKS — Family **Nemouridae**, p. 153

37. *Amphinemura nigritta,* larva: 4–6 mm excluding tails; occurs on sandy bottoms and leaf packs in small streams of central and eastern North America; April through mid-June emergence.

SLENDER WINTER STONEFLIES — Family **Capniidae,** p. 156

38. *Capnia vernalis,* larva: 6–8 mm excluding tails; found in streams and rivers of northeastern and north-central states and central and eastern Canada; imitations are used for early-season nymph fishing on northern trout streams.
39. *Capnia vernalis,* adult (the Early Black Stonefly): 5–7 mm excluding tails; primarily late winter and spring emergence.
40. *Allocapnia granulata,* larva: 6–8 mm excluding tails; inhabits medium-sized and large streams and rivers in central and eastern North America; winter and spring emergence.

GIANT STONEFLIES — Family **Pteronarcyidae,** p. 159

41. *Pteronarcys californica,* larva: 38–43 mm excluding tails; occurs under rocks with accumulated debris or in beds of aquatic plants in streams and rivers in mountainous western North America; an excellent live bait, and larval imitations are used throughout the fly fishing season on western trout streams.
42. *Pteronarcys californica,* adult (the Salmonfly): 38–44 mm to wing tips; mid-April through early-August emergence, with largest numbers flying in June; the downwing imitations provide excellent dry fly fishing, especially when females are laying eggs; the flies are floated dead drift.

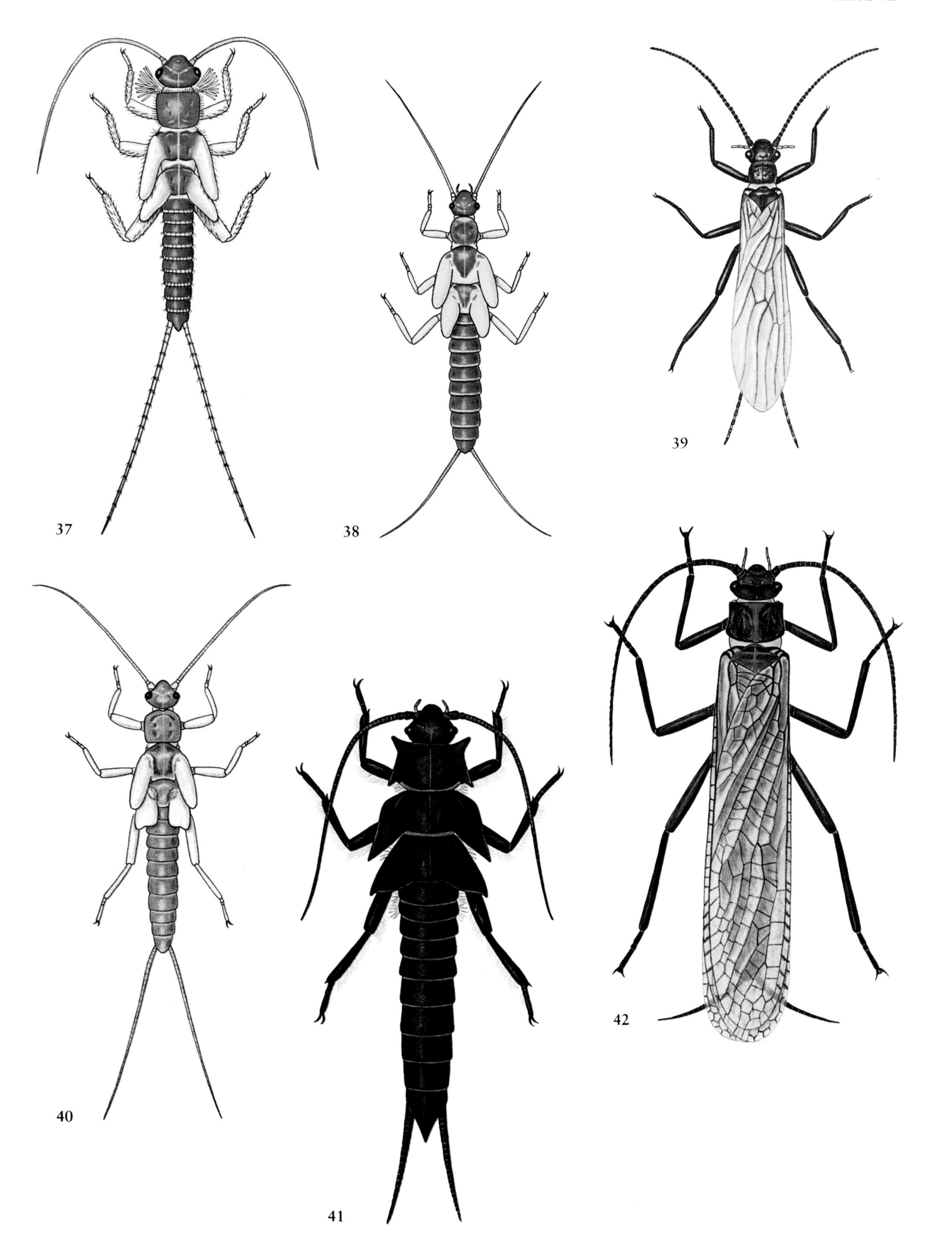
37
38
39
40
41
42

Stoneflies

PLECOPTERA

ROACHLIKE STONEFLIES Family **Peltoperlidae,** p. 160

43. *Peltoperla arcuata,* larva: 9–12 mm excluding tails; found among leaf detritus and debris in spring-fed mountain streams in eastern Canada and northeastern states south to Virginia; primarily June and July emergence.

COMMON STONEFLIES Family **Perlidae,** p. 160

44. *Acroneuria evoluta,* larva: 17–23 mm excluding tails; occurs on rocks and gravelly substrates of streams and rivers of central states south to Louisiana and east to Pennsylvania and Virginia.

45. *Acroneuria evoluta,* adult: 28 mm to wing tips; mid-May through July emergence; large numbers may be attracted to lights at night.

PERLODID STONEFLIES Family **Perlodidae,** p. 163

46. *Isoperla bilineata,* larva: 9–11 mm excluding tails; common inhabitant of medium-sized to large rivers of central and eastern North America.

47. *Isoperla bilineata,* adult (the Little Yellow Stonefly): 13–15 mm to wing tips; this species has been restricted by water pollution; downwing imitations can be fished on larger, clean trout streams of the East and north-central regions in late spring and early summer.

GREEN STONEFLIES Family **Chloroperlidae,** p. 164

48. *Sweltsa coloradensis,* larva (the Little Olive Stonefly): 11 mm excluding tails; commonly found in small streams and rivers in mountainous regions of western North America; this is an excellent model for nymph fishing on western trout streams in early to mid-summer.

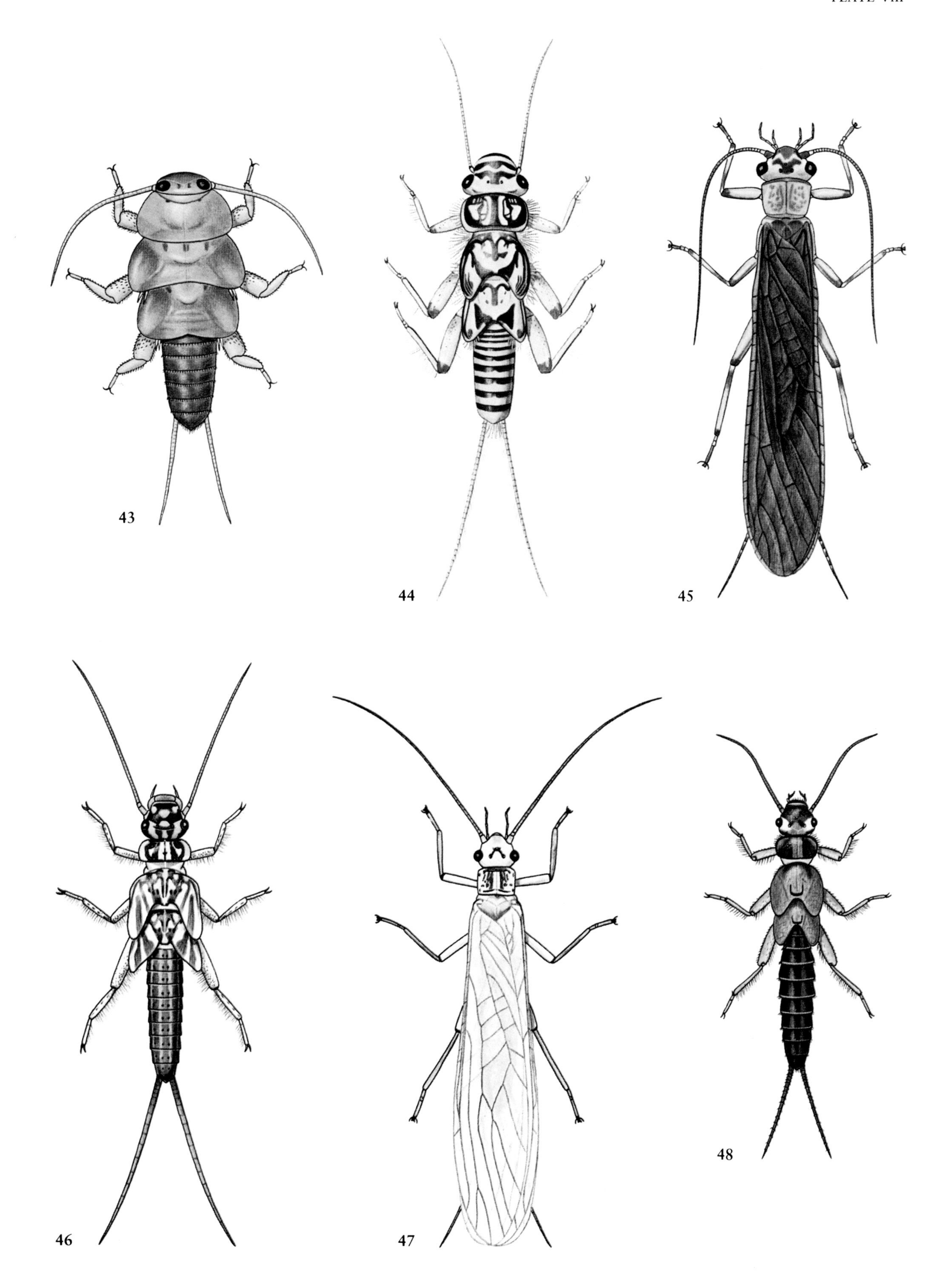
43
44
45
46
47
48

Water Bugs

HEMIPTERA

BACKSWIMMERS — Family **Notonectidae,** p. 173

49. *Notonecta undulata,* adult (the Grousewinged Backswimmer): 11–12 mm; this widespread species primarily inhabits ponds, often resting with its posterior end touching the water surface.

CREEPING WATER BUGS — Family **Naucoridae,** p. 176

50. *Pelocoris femoratus,* adult: 9–12 mm; this vigorous swimmer is found throughout much of eastern and central North America in ponds and swampy pools with abundant emergent vegetation.

WATER BOATMEN — Family **Corixidae,** p. 179

51. *Sigara alternata,* adult (the Corixa Bug): 5–7 mm; this widespread species primarily inhabits ponds and deep pools of streams, frequently surfacing for air.

MARSH TREADERS — Family **Hydrometridae,** p. 184

52. *Hydrometra martini,* adult: 8–11 mm; inhabits vegetation along ponds, streams, and swamps in central and eastern North America.

GIANT WATER BUGS — Family **Belostomatidae,** p. 178

53. *Lethocerus americanus,* adult: 41–62 mm; this large and widespread species occurs in shallow ponds with abundant vegetation.

WATER SCORPIONS — Family **Nepidae,** p. 175

54. *Ranatra nigra,* adult: 29–32 mm; found among debris in ponds and shallow waters of rivers throughout much of central and eastern North America.

SHORTLEGGED STRIDERS — Family **Veliidae,** p. 183

55. *Rhagovelia oriander,* adult: 3–4 mm; this riffle bug is found in large numbers at the surface of the water, most commonly along quiet edges of streams and rivers of the central states.

WATER STRIDERS — Family **Gerridae,** p. 181

56. *Gerris remigis,* adult: 11–16 mm; this large, well-known, and widespread species commonly inhabits the surface of quiet waters of streams and occasionally ponds.

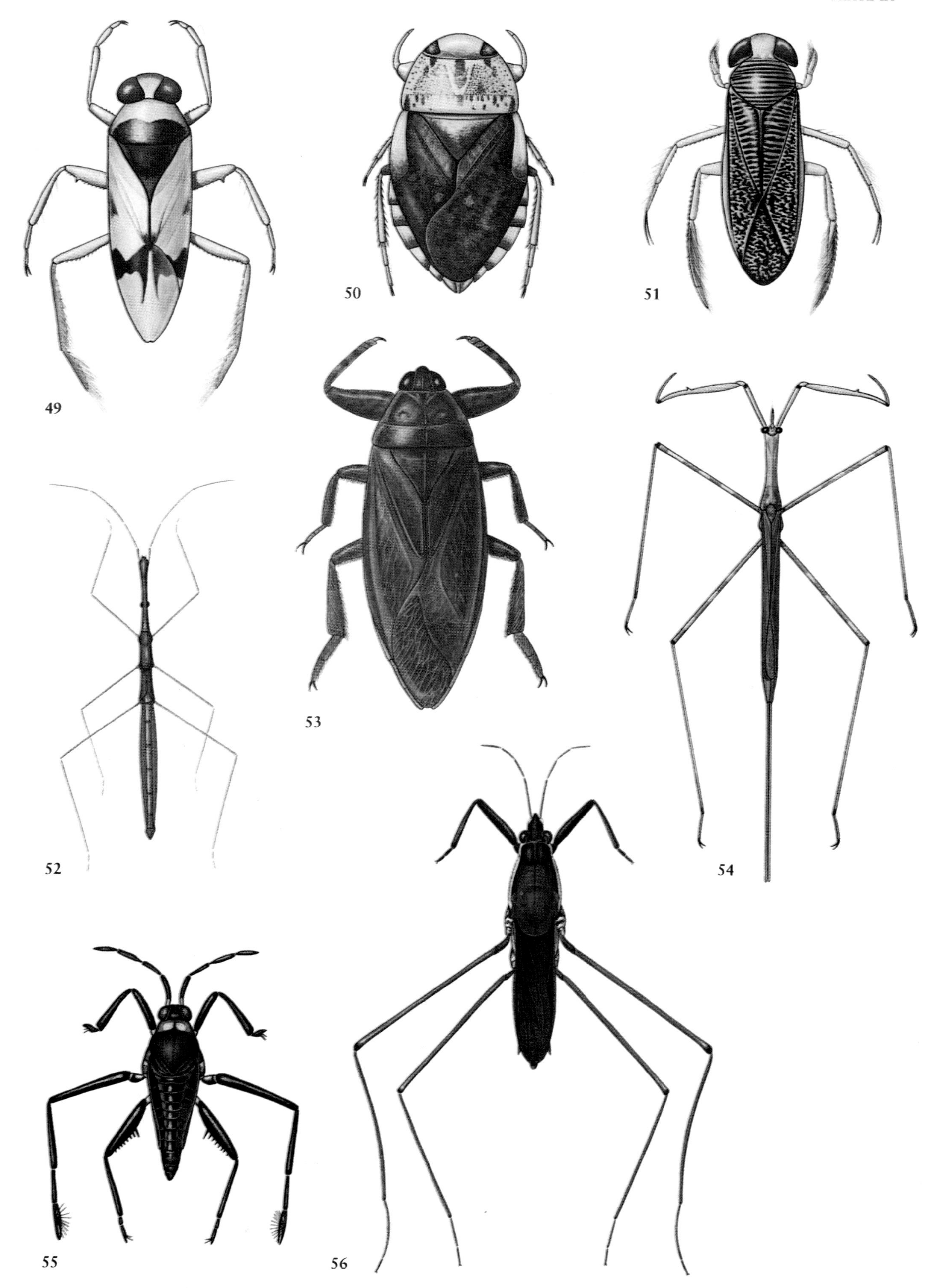
49
50
51
52
53
54
55
56

Water Beetles

COLEOPTERA

CRAWLING WATER BEETLES Family **Haliplidae**, p. 213

57. *Peltodytes lengi,* adult: 3–4 mm; occurs throughout the eastern and central states, primarily in the shallow marginal waters of ponds with abundant filamentous algae.

58. *Peltodytes* sp., larva: 6 mm; genus widely distributed; usually associated with aquatic vegetation and soft bottoms of ponds and pools.

PREDACEOUS DIVING BEETLES Family **Dytiscidae**, p. 214

59. *Coptotomus interrogatus,* adult: 6–8 mm; occurs in woodland pools, ditches, and swampy areas adjacent to ponds, lakes, and streams in central and eastern states.

60. *Dytiscus fasciventris,* adult: 25–28 mm; often found in muddy and leaf-choked margins of pools, ponds, lakes, and streams of eastern North America.

61. *Dytiscus* sp., larva: 48 mm; this widespread genus includes the largest of the North American water tigers.

WHIRLIGIG BEETLES Family **Gyrinidae**, p. 219

62. *Dineutus assimilis,* larva: 18 mm; inhabits ponds and quiet waters of streams throughout most of North America.

63. *Dineutus assimilis,* adult: 10–11 mm; same habitat as larva but at the surface, often found congregated in large schools of individuals.

WATER SCAVENGER BEETLES Family **Hydrophilidae**, p. 221

64. *Berosus striatus,* adult: 6–7 mm; this widespread species occurs primarily in the marginal waters of deep pools, ponds, and lakes.

65. *Tropisternus lateralis,* larva: 13–16 mm; this widespread species is common in almost any shallow-water habitat, preferring standing water but also found in running water with thick marginal vegetation.

66. *Tropisternus lateralis,* adult: 8–10 mm; a ubiquitous and commonly found water scavenger beetle; same habitat as larva; often associated with highly enriched water.

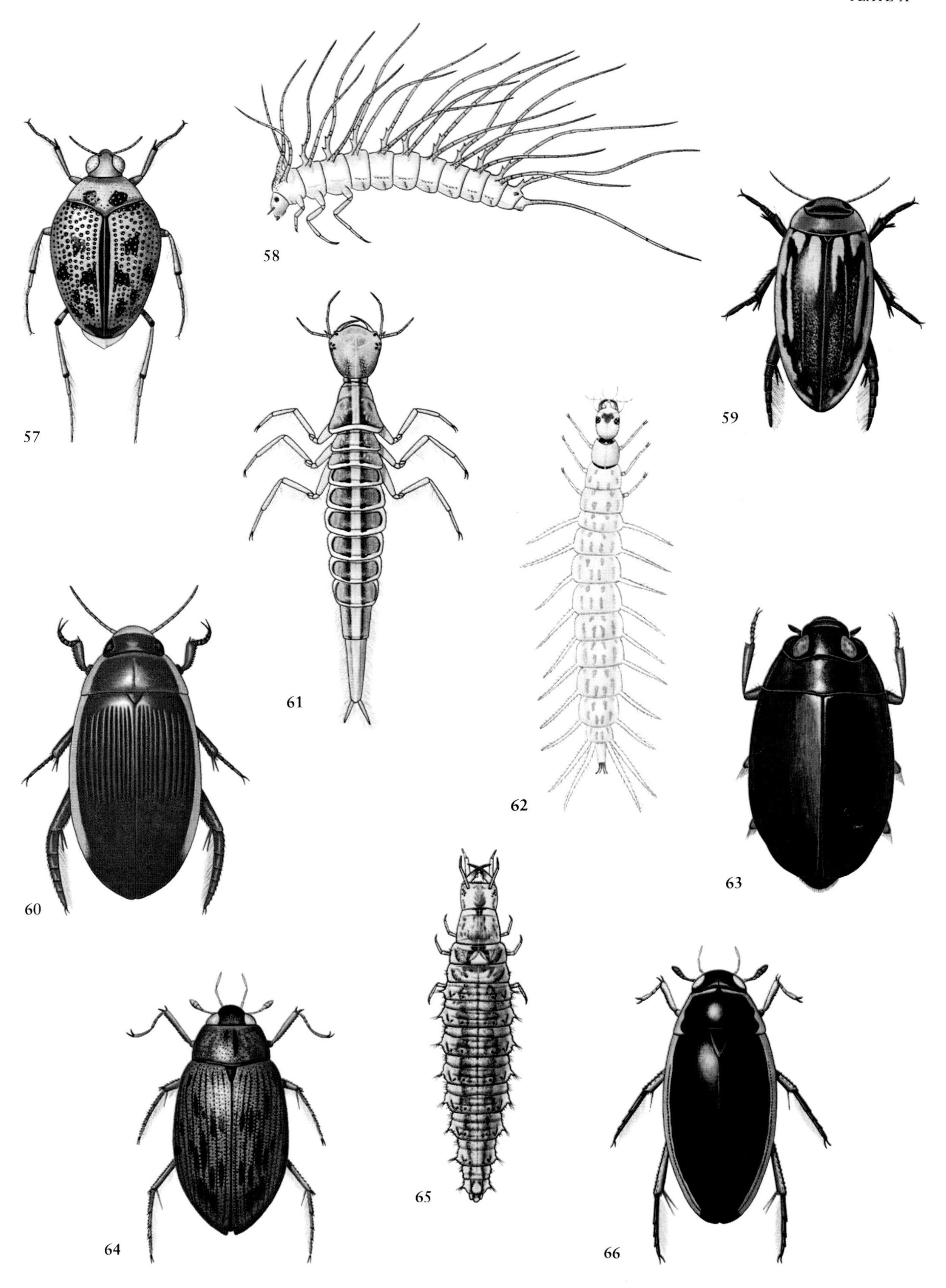
57
58
59
60
61
62
63
64
65
66

Water Beetles, Fishflies and Alderflies, Spongillaflies, and Aquatic Caterpillars

COLEOPTERA, MEGALOPTERA, NEUROPTERA, AND LEPIDOPTERA

LONGTOED WATER BEETLES — Family **Dryopidae**, p. 229

67. *Helichus fastigiatus,* adult: 5–6 mm; occurs under rocks in streams of eastern and central North America.

RIFFLE BEETLES — Family **Elmidae**, p. 231

68. *Stenelmis sexlineata,* adult: 3–4 mm; inhabits substrates of a variety of streams in south-central states, northernmost southeastern states, and Texas.

69. *Stenelmis* sp., larva: 6.5 mm; this widespread genus is associated with coarse substrates in strong to moderate currents of streams.

WATER PENNIES — Family **Psephenidae**, p. 226

70. *Psephenus herricki,* larva: 8 mm; found clinging to rocks in moderate to strong currents of streams throughout much of central and eastern North America excluding Florida.

AQUATIC LEAF BEETLES — Family **Chrysomelidae**, p. 232

71. *Donacia palmata,* adult: 8 mm; occurs on the emergent parts of white and yellow pond lilies of central and eastern North America; larvae associated with underwater parts of host plant.

FISHFLIES — Family **Corydalidae**, p. 193

72. *Chauliodes pectinicornis,* larva: 38–45 mm; inhabits shallows of pools, ponds, and swamps in central and eastern North America, often among debris and detritus.

73. *Chauliodes pectinicornis,* adult: 45–48 mm to wing tips; emergence period from March (southern) to mid-November (northern).

ALDERFLIES — Family **Sialidae**, p. 193

74. *Sialis hamata,* larva: 19–24 mm; occurs primarily in streams of western Canada and northwestern states.

SPONGILLAFLIES — Family **Sisyridae**, p. 197

75. *Climacia areolaris,* larva: 4–6 mm; occurs in association with sponge hosts in streams, rivers, ponds, and lakes in North America west to New Mexico.

AQUATIC PYRALID MOTHS — Family **Pyralidae**, p. 277

76. *Parargyractis* sp., larva: 15 mm; individuals of this widespread genus develop on alga-covered rocks in moderate to strong currents of streams.

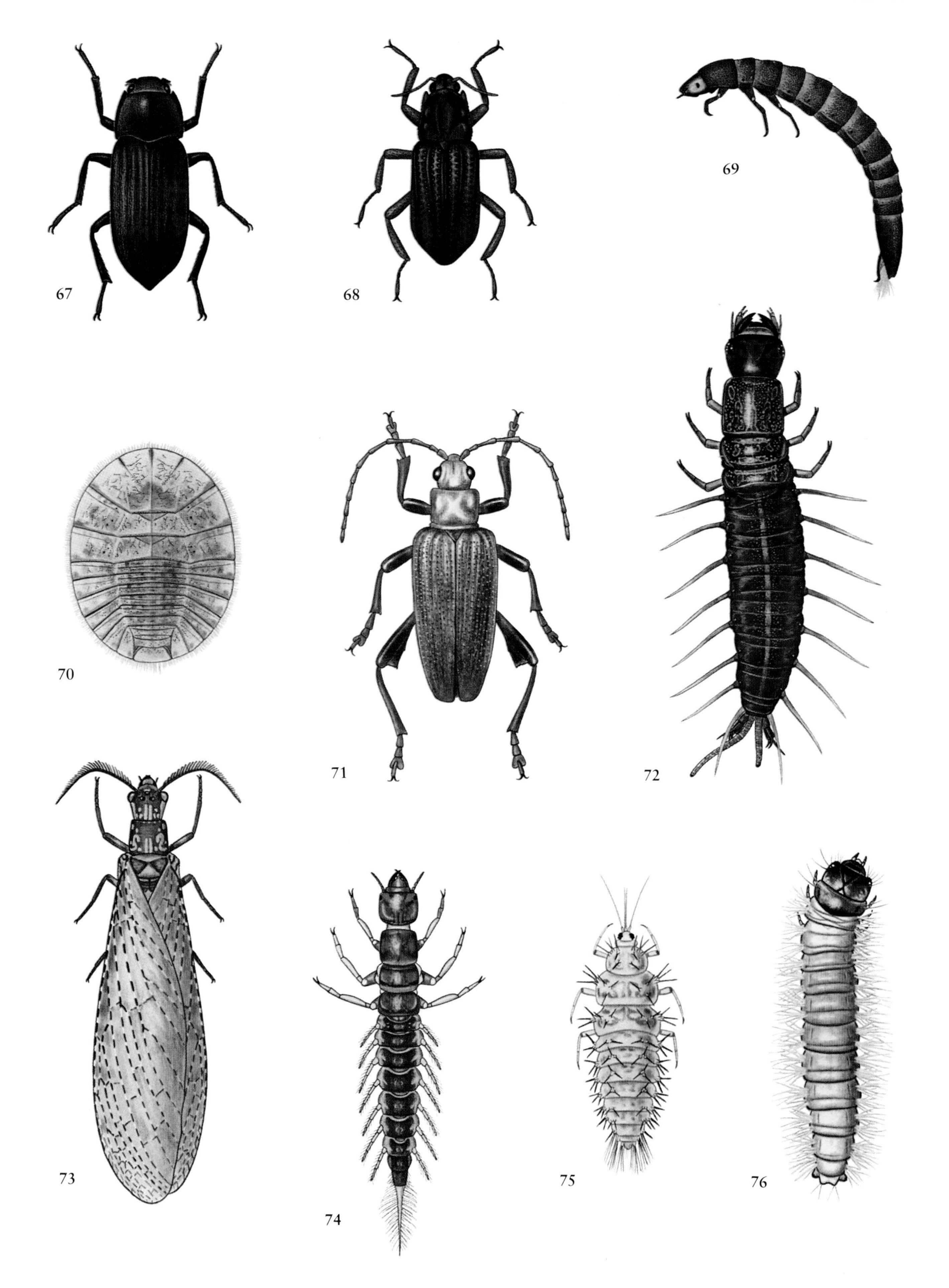
67
68
69
70
71
72
73
74
75
76

Caddisflies

TRICHOPTERA

TRUMPETNET AND TUBEMAKING CADDISFLIES Family **Polycentropodidae,** p. 247

77. *Polycentropus cinereus,* larva: 11–14 mm; occupies net retreats primarily under rocks of clear lakes and large rivers throughout most of North America excluding southwestern states.

COMMON NETSPINNERS Family **Hydropsychidae,** p. 249

78. *Symphitopsyche slossanae,* larva: 15–16 mm; primarily inhabits medium- to large-sized rocks in cold, clear streams of eastern and central Canada and central and northeastern states south through the Appalachian Mountains; has also been considered in the genus *Hydropsyche.*
79. *Symphitopsyche slossanae,* pupa: 11 mm.
80. *Symphitopsyche slossanae,* adult (the Spotted Sedge): 8–11 mm to wing tips; an excellent model for both emerger and wet fly patterns throughout warmer months on eastern and north-central trout streams.
81. *Potamyia flava,* adult: 10–11 mm; one of the most common and abundant species of the central and southeastern states, developing in large streams and rivers; May through September emergence, sometimes massive.

FREELIVING CADDISFLIES Family **Rhyacophilidae,** p. 252

82. *Rhyacophila lobifera,* larva (the Green Caddis): 15 mm; found in small, sometimes temporary, clear streams of central states; April through June emergence; patterns of related species used for bottom nymph fishing in most North American trout streams.

SADDLECASE MAKERS Family **Glossosomatidae,** p. 254

83. *Glossosoma* sp., larva: 8 mm (most smaller); occurs in cold streams throughout North America, but most species represented in western and northeastern mountainous regions; saddlecase not shown.

MICRO CADDISFLIES Family **Hydroptilidae,** p. 255

84. *Hydroptila hamata,* larva: 3 mm; inhabits the bottoms of clear lakes, streams, and rivers throughout hilly and mountainous regions of North America excluding extreme southeastern states.
85. *Hydroptila hamata,* adult: 2.5 mm to wing tips; spring to late summer emergence; tiny orange emerger patterns sometimes fished for this species.

GIANT CASE MAKERS Family **Phryganeidae,** p. 257

86. *Ptilostomis* sp., larva: 44 mm to end of case; this widespread genus occurs in a variety of still and running water habitats throughout North America.

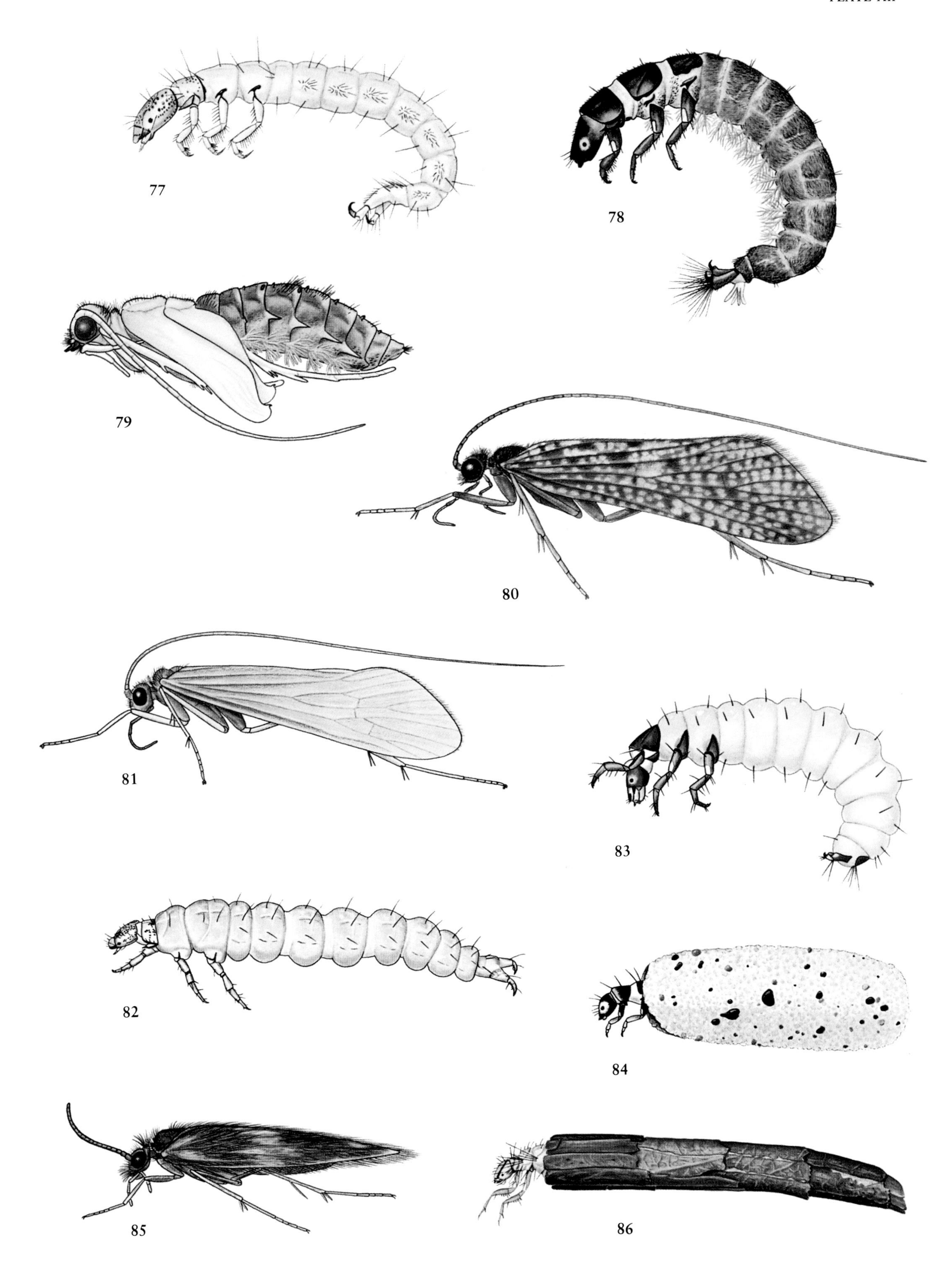
77
78
79
80
81
83
82
84
85
86

Caddisflies

TRICHOPTERA

HUMPLESS CASE MAKERS Family **Brachycentridae,** p. 259

87. *Brachycentrus americanus,* larva: 10–13 mm; occurs in rivers and streams with fast currents in northern and mountainous regions throughout North America; case is cemented to rocks in riffles and rapids.

88. *Brachycentrus americanus,* case: 15 mm; a cylindrical tube of silk lines the chimneylike case.

89. *Brachycentrus americanus,* adult (the American Sedge): 14 mm to wing tips; flies from May to October and provides a favorite pattern for fly fishing northern trout streams.

NORTHERN CASE MAKERS Family **Limnephilidae,** p. 260

90. *Limnephilus* sp., larva: 14 mm (often larger); individuals of this genus, which is found in most of North America, inhabit ponds, lake margins, marshes, streams, and cold springs.

91. *Limnephilus* sp., case: 20 mm; one of many kinds of plant and rock cases for this genus; some range to 50 mm in length.

92. *Limnephilus coloradensis,* adult: 10 mm; the Orange Sedge is a popular tied fly imitation of this small, western species; develops in cold water at higher elevations of the Rocky Mountains.

STRONGCASE MAKERS Family **Odontoceridae,** p. 266

93. *Psilotreta frontalis,* pupa (the Dark Blue Sedge): 11–12 mm; this species develops in trout streams and rivers of eastern Canada and northeastern states west to Wisconsin and south to North Carolina; pupae occupy cases layered on the undersides of rocks.

SNAILCASE MAKERS Family **Helicopsychidae,** p. 267

94. *Helicopsyche borealis,* larva: case width 4–5 mm; this widespread species is found in clear, swift streams and occasionally along wave-washed lake shores; mid-May to mid-September emergence.

LONGHORNED CASE MAKERS Family **Leptoceridae,** p. 269

95. *Oecetis inconspicua,* larva: 9–11 mm to end of case; this widespread species occurs in ponds, lakes, streams, and rivers and is very common and abundant in certain regions; fed on heavily by a variety of fishes; May to October emergence, sometimes massive.

96. *Nectopsyche exquisita,* adult (the White Miller): 14–17 mm to wing tips; develops in lakes and slower reaches of rivers throughout most of central and eastern North America; common name applies to any species of the genus; provides an excellent light-colored pattern for fly fishing in mid-summer.

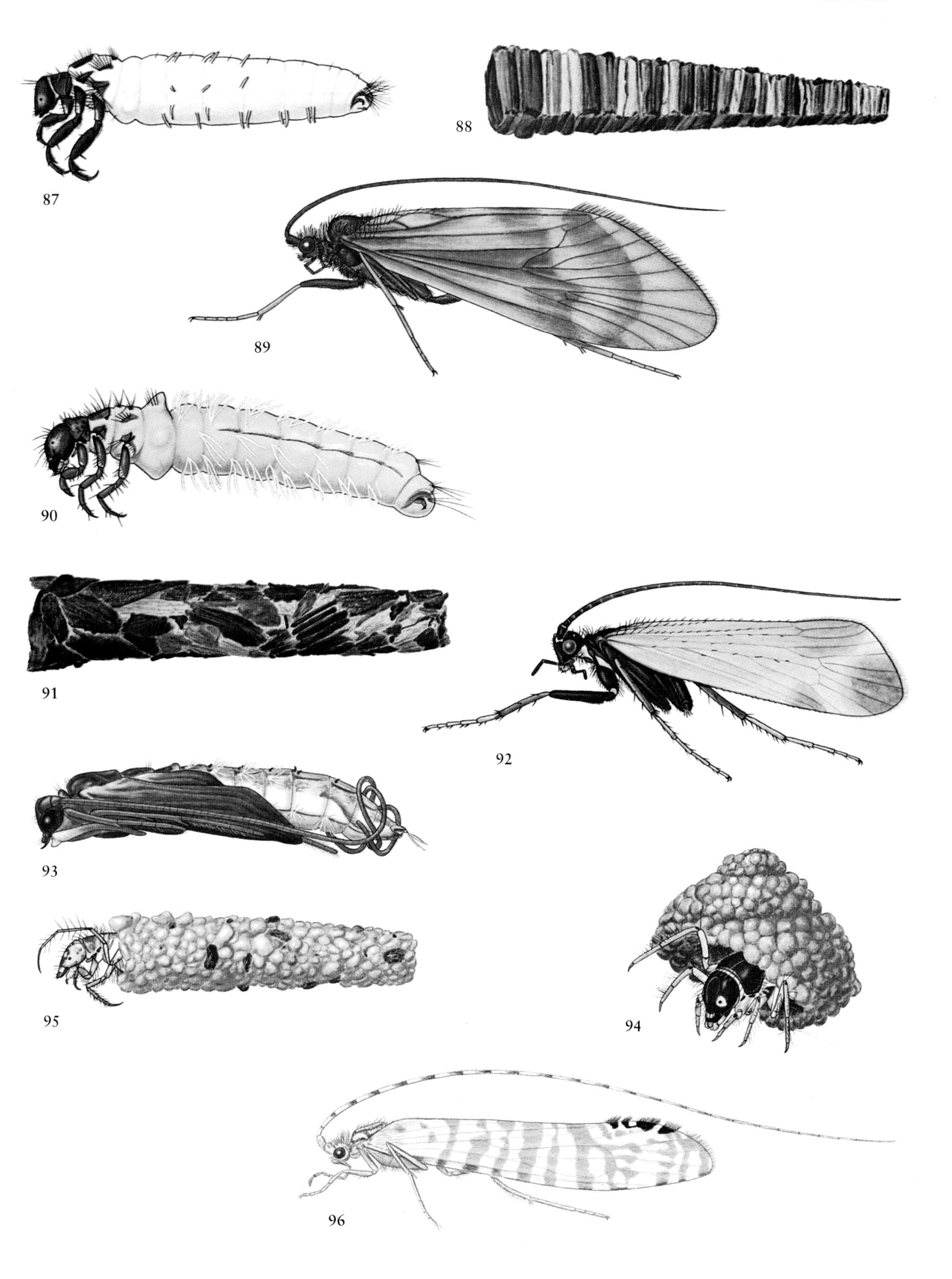
87
88
89
90
91
92
93
94
95
96

Crane Flies and Mosquitoes

DIPTERA

CRANE FLIES — Family **Tipulidae,** p. 294

97. *Tipula abdominalis,* larva: 35–45 mm; found under rocks, in debris, and in muddy bottoms of a variety of streams throughout central and eastern North America; provides a large live bait (spikes) as well as a model for nymph fishing.

98. *Tipula abdominalis,* adult (the Giant Crane Fly): 25–38 mm; this striking species has two generations per year with first emergence from May through July and second emergence in August or September, when adults are usually more abundant.

MOSQUITOES — Family **Culicidae,** p. 305

99. *Culex pipiens,* larva: 6–9 mm; this wriggler is found in artificial containers, ponds, marshy pools, lakes, and slack waters of streams; it is often present in polluted waters and sewage lagoons, cesspools, and catch basins.

100. *Culex pipiens,* pupa: 4 mm; these active pupae, or tumblers, are common wherever the larvae occur.

101. *Culex pipiens,* adult ♀ (the House Mosquito): 5–7 mm to wing tips; this mosquito can be a serious pest and biter of man and animals during warm months of the year; it can transmit heartworm of dogs, bird malaria, avian pox virus, and in central and eastern North America it is an important vector of St. Louis encephalitis, which it can transmit to man.

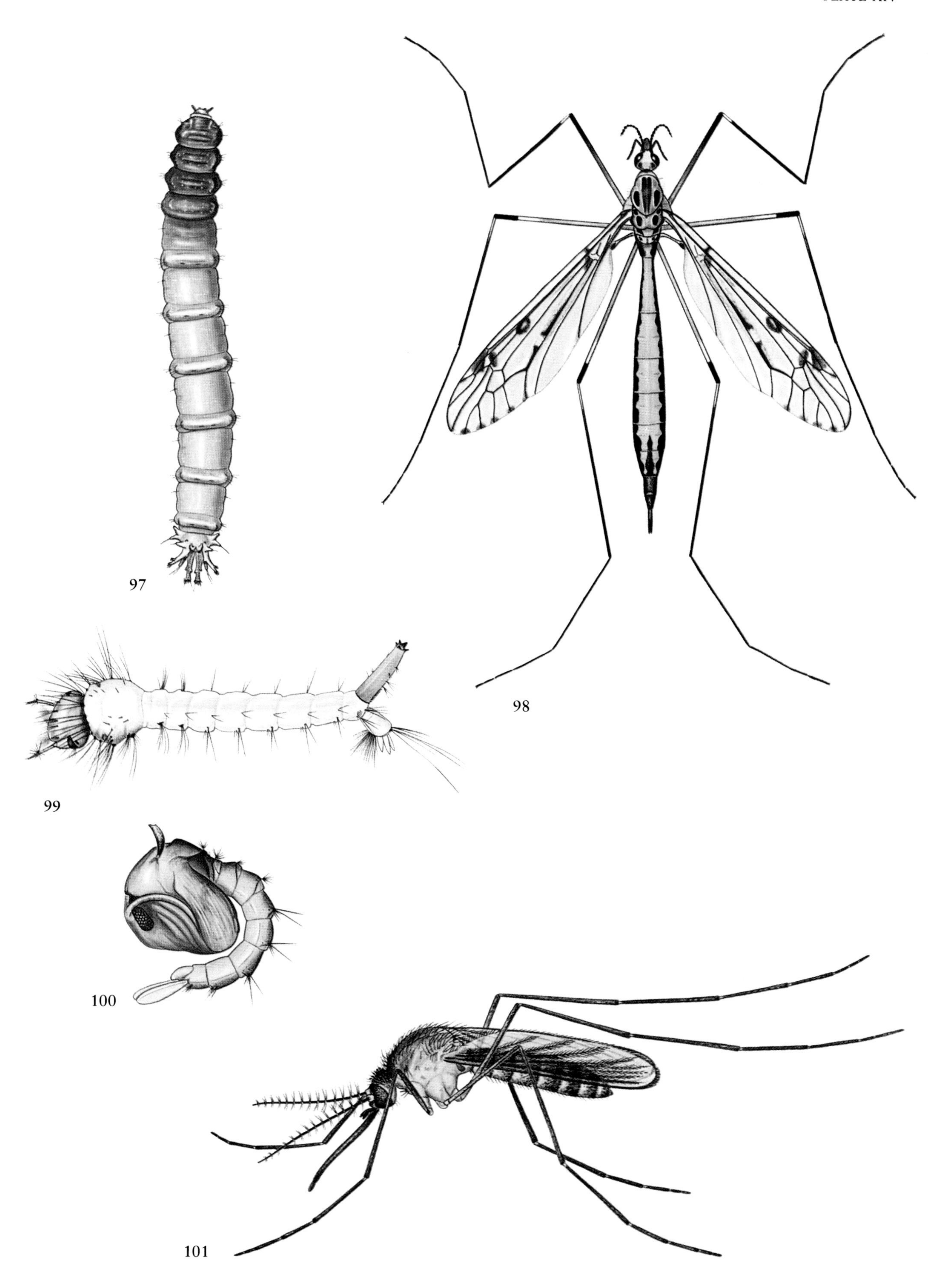
97
98
99
100
101

Midges and Other Aquatic Flies

DIPTERA

MOTH FLIES — Family **Psychodidae,** p. 297

102. *Psychoda* sp., larva: 6 mm; this widespread genus inhabits the soft bottoms, detritus, and sludge of ponds, sewage lagoons, pools of streams, and ocean beaches.

103. *Psychoda* sp., pupa: 3–4 mm; floats at the surface of larval habitat.

104. *Psychoda* sp., adult: 4–5 mm; sometimes emerges in large numbers; common at lights, and often rests on walls near putrid waters.

MIDGES — Family **Chironomidae,** p. 309

105. *Cricotopus* sp., larva: 4 mm; this widespread genus occurs in a variety of habitats, depending on the species, including standing and flowing waters; some build tubes on or in sediments, some live in association with algae, and some mine in vascular plants.

106. *Pseudodiamesa* sp., larva: 10 mm; this genus primarily inhabits colder streams of northern and mountainous regions.

107. *Chironomus attenuatus,* larva: 12–17 mm; this widespread bloodworm is commonly found in shallow ponds, pulluted pools of streams, and sewage lagoons.

108. *Chironomus attenuatus,* adult: 8–9 mm; as many as six generations per year may develop in warmer regions; large numbers may be seen swarming near stillwater habitats.

BLACK FLIES — Family **Simuliidae,** p. 314

109. *Simulium vittatum,* larva: 6 mm; this widespread species is found on gravel, rocks, driftwood, and vegetation in moderately flowing to swift waters of streams and rivers.

110. *Simulium vittatum,* pupa: 3–4 mm to end of case; occupies larval habitat.

111. *Simulium vittatum,* adult: 4 mm; primarily attacks deer, horses, and cattle, but will swarm around humans, often entering eyes, ears, and mouth but rarely biting; April through September emergence with several generations per year.

BITING MIDGES — Family **Ceratopogonidae,** p. 307

112. *Bezzia* sp., larva: 11 mm; individuals of this widespread genus develop primarily in soft margins and bottoms and algal mats of ponds, lakes, and streams.

113. *Culicoides variipennis,* adult: 2.5 mm to wing tips; this widespread sand fly develops in muddy habitats associated with stagnant and flowing waters, either saline or fresh, clean or polluted; flies in all warm months; females readily bite both humans and livestock.

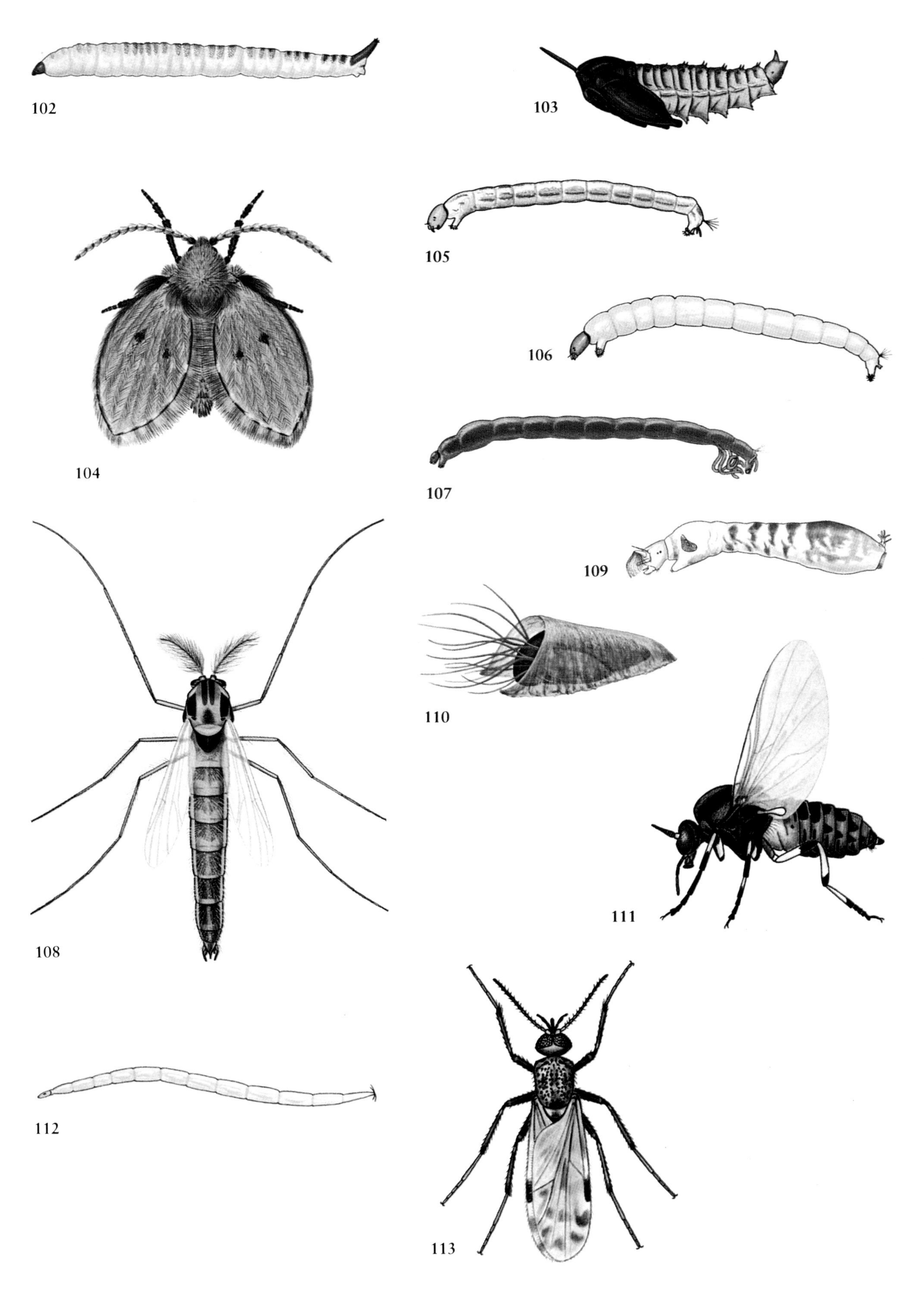
102
103
104
105
106
107
108
109
110
111
112
113

Aquatic Flies

DIPTERA

AQUATIC SOLDIER FLIES — Family **Stratiomyidae,** p. 316

114. *Stratiomys* sp., larva: 28 mm; individuals of this widespread genus develop primarily in the shallows or shallow margins of streams and rivers.

HORSE AND DEER FLIES — Family **Tabanidae,** p. 317

115. *Tabanus atratus,* larva: 50–55 mm; this large, ornate, and widespread species is most common on banks of drainage and irrigation ditches, small streams, and farm ponds; adult is a common black form known as the Mourning Horse Fly.

116. *Chrysops montanus,* adult: 8–9 mm; this typical-appearing deer fly occurs in central and eastern North America, where it develops in muddy banks of streams and rivers.

WATERSNIPE FLIES — Family **Athericidae,** p. 318

117. *Atherix pachypus,* larva: 18–20 mm; occurs on rocky substrates of streams and rivers in western North America.

AQUATIC DANCE FLIES — Family **Empididae,** p. 319

118. *Chelifera* sp., larva: 7 mm; individuals of this widespread genus develop primarily in slower reaches of streams.

RATTAILED MAGGOTS — Family **Syrphidae,** p. 322

119. *Eristalis arbustorum,* adult: 10 mm; this hover fly occurs in central and eastern North America.

120. *Eristalis* sp., larva: 16 mm excluding rattail; this widespread genus inhabits stagnant waters, pools of streams, marshes, and lagoons.

SHORE FLIES — Family **Ephydridae,** p. 324

121. *Ephydra cinerea,* larva: 14–16 mm including tail; occurs in coastal and inland saltwater habitats throughout western North America; one of three species of brine flies occurring in Great Salt Lake.

122. *Ephydra cinerea,* adult: 3–4 mm; sometimes a nuisance because it may emerge in massive numbers in warm months.

AQUATIC MUSCIDS — Family **Muscidae,** p. 327

123. *Lispoides aequifrons,* larva: 14–18 mm; this widespread species inhabits the substrate of moderate to swift currents of streams.

MARSH FLIES — Family **Sciomyzidae,** p. 323

124. *Sepedon fuscipennis,* adult: 11 mm to wing tips; this widespread species develops in ponds and marshes, where its larvae attack snails; adults are often found on emergent vegetation.

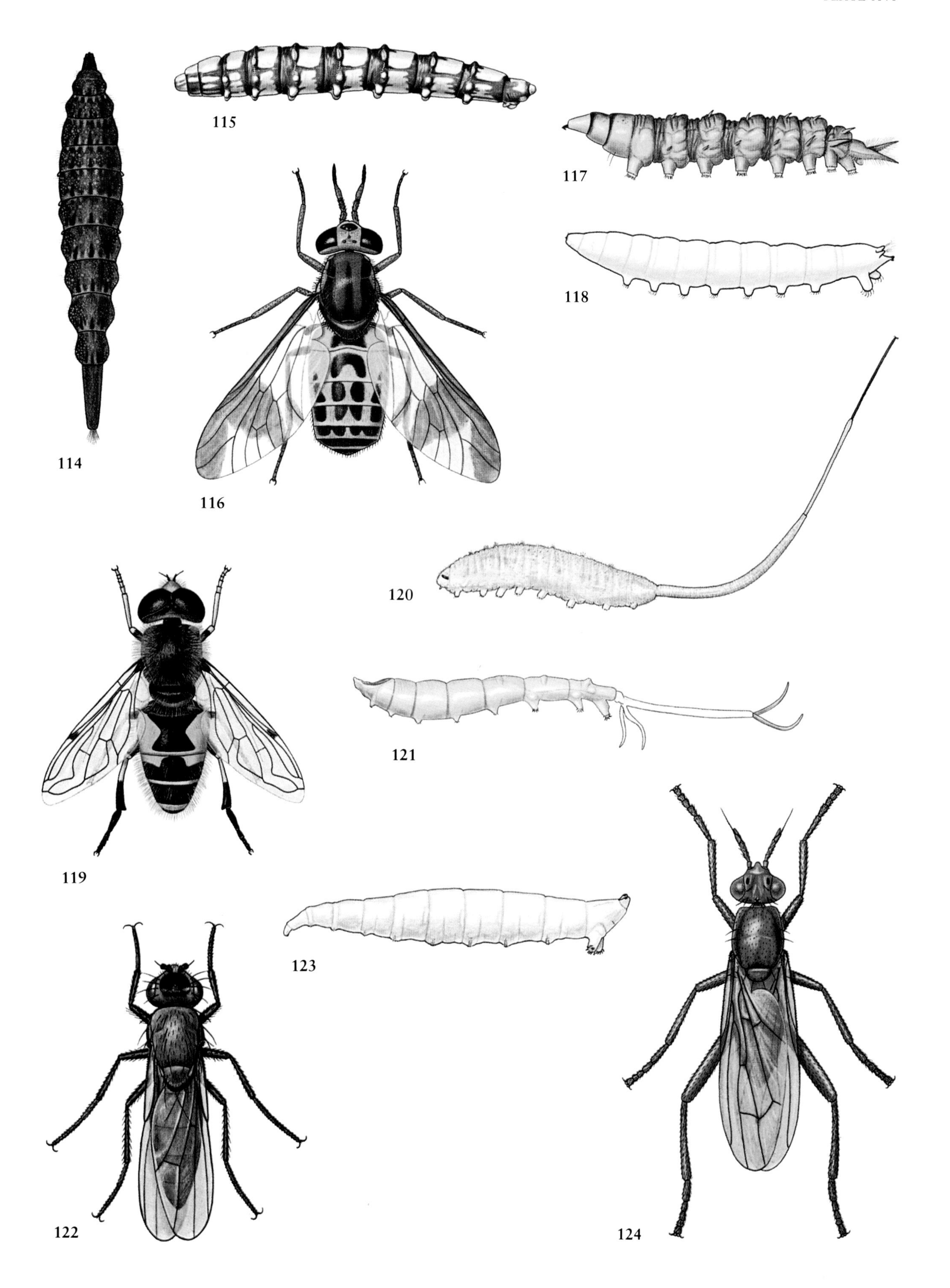
114
115
116
117
118
119
120
121
122
123
124

PART III

OTHER INSECTS ASSOCIATED WITH AQUATIC ENVIRONMENTS

CHAPTER 17

Shore-Dwelling Insects

A LARGE NUMBER of diverse insect species regularly inhabit areas adjacent to bodies of water, and many demonstrate special adaptations for living in this marginal environment. These species, which might be termed semiaquatic, are sometimes taken in water samples and are often studied within the realm of aquatic entomology. It is for these reasons that shore-dwelling insects are a subject deserving at least minimal coverage in this book.

Semiaquatic shore habitats can be typified by moist sand, mud flats, a mixture of rocks and sand with various intermittent pools (e.g., tide pools), and a variety of marginal or riparian vegetation. (Those insects occurring on the exposed parts of floating or emergent vegetation are treated in Chapter 18.) The marginal shore habitat is subject to inundation either by the regular action of waves and tides or by the irregular increase of water level due to fluctuating rainfall or discharge. Another important ecological aspect that influences this marginal area as an environment for sustaining insect life is the presence of such materials as driftwood and various animal and plant detritus, including seaweed stranded on ocean beaches. These materials provide specific food sources, shelter, and sites for breeding and development.

Along with common shore-dwelling insects (of both freshwater and marine shores), the more common insects of tide pools and moist seepage areas are also conveniently treated in this chapter. Many insects that might qualify as shore dwellers are not covered herein because the fine line that

. . . those very many flies, worms, and little living creatures with which the sun and the summer adorn and beautify the river-banks and meadows, both for the recreation and contemplation of us Anglers: pleasures which, I think, myself enjoy more than any other man that is not my profession.

Izaak Walton

separates the semiaquatic shore environment from the strictly terrestrial environment is interpreted differently from one ecologist to the next; thus the emphasis here is on those insects that are important from the standpoint of aquatic entomology. Not included here are such insects as tiger beetles, which often occur on dry sandy shores, and coastal Thysanura, which generally remain high enough so as not to be subject to high tides. The springtails, or Collembola (sometimes considered insects), are common inhabitants of shore environments, both freshwater and marine. The Collembola, however, are treated exclusively in Chapter 21.

Several families of insects (especially of Diptera and Coleoptera) contain some species that are aquatic and some that are shore-dwelling as well as species that are aquatic in one stage and shore-dwelling in another. Any families containing at least some aquatic species are treated in detail in chapters dealing with the aquatic orders, but if they also contain shore-dwelling species they are cross-referenced below.

Order Orthoptera

The Orthoptera is a large and important group of terrestrial insects and includes the jumping insects known as grasshoppers, crickets, and katydids. Although the group is not generally associated with aquatic environments, several species commonly occur in shore environments. A few are specifically adapted to living in soft marginal substrates, and several others are closely associated with marginal vegetation. Many species are most commonly associated with marshy environments, bogs, and low wetlands. Metamorphosis is incomplete, and immature individuals generally resemble adults.

Diagnosis

These are generally elongate insects with variously developed antennae and wings. Mouthparts are adapted for chewing. Head is short and vertically oriented. When wings are developed, the fore wings are narrow and thickened and are used to cover the more expanded and membranous hind wings that are folded at rest. Hind legs are usually elongate and adapted for jumping or at least have enlarged hind femora. Abdomen ends in two variously developed tails.

Similar Orders

Those Orthoptera with enlarged hind legs should not be confused with any other insects. Several bugs (Hemiptera) that occur in shore habitats will readily jump when disturbed but are quite dissimilar to the Orthoptera. Orthoptera that do not have such conspicuously well-developed hind legs have strongly expanded and sculptured digging fore legs (Fig. 17.3) that are quite unlike those of any of the other shore-dwelling insects.

PYGMY MOLE CRICKETS (Family Tridactylidae)

DIAGNOSIS: (Fig. 17.1) These are very small, dark forms, less than 10 mm in length. Antennae are less than half the length of the body. Fore legs are modified for digging. Hind legs are modified for jumping.

DISCUSSION: Two species of this family (genera *Ellipes* and *Neotridactylus*) occur in North America, where they have a widespread distribution and are sometimes found together in the same habitat. Pygmy mole crickets prefer sandy banks of streams, rivers, ponds, and lakes. They are active on

PYGMY MOLE CRICKET
(Tridactylidae)

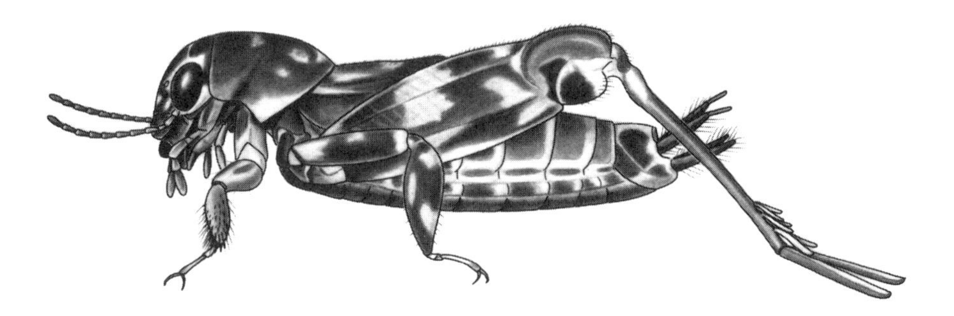

Figure 17.1. *Ellipes* adult

the sand during daylight in summer. They will readily jump when disturbed (often onto adjacent water) and can be difficult to catch. They construct burrows in the sand but evidently leave these burrows to forage. Food consists of minute particles or organisms associated with the sandy substrate.

Pygmy mole crickets possess special swimming plates on the tarsi of the hind legs. These enable individuals to swim and to jump easily onto or from the water surface. If these swimming plates are removed, the pygmy mole cricket quickly flounders in the water.

GROUSE LOCUSTS
(Family Tetrigidae)

DIAGNOSIS: (Fig. 17.2) These are small or medium-sized, dark forms. Antennae are less than half the length of body. Fore legs are not modified for digging. Hind legs are modified for jumping. Pronotum extends posteriorly as a shield over the back of the body.

DISCUSSION: Several species of this widespread North American family commonly occur adjacent to ponds and streams on wet sandy margins, on riparian vegetation, or on emergent parts of aquatic vascular plants. Young larvae are often associated with some of the emergent vegetation of slow or standing water. They do not possess any special adaptations for aquatic or semiaquatic life and should be considered essentially terrestrial.

GROUSE LOCUST
(Tetrigidae)

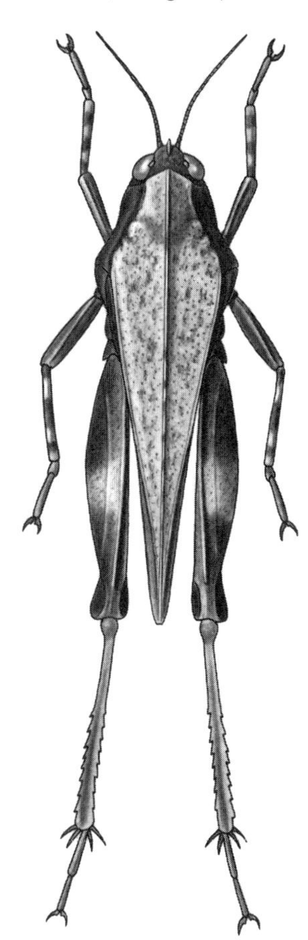

Figure 17.2. *Tettigidea* adult

CRICKETS
(Family Gryllidae)

DIAGNOSIS: These are small or medium-sized, usually dark forms. Antennae are generally longer than the body. Fore legs are not modified for digging. Hind legs are modified for jumping. All tarsi are three-segmented.

DISCUSSION: Several species of crickets are apt to be found in marginal, often grassy areas adjacent to ponds, marshes, and streams. They are essentially a terrestrial group.

MOLE CRICKETS
(Family Gryllotalpidae)

DIAGNOSIS: (Fig. 17.3) These are large and robust insects when mature, and are brownish and quite hairy. Antennae are less than half the length of body. Fore legs are distinctly modified for digging, being enlarged, flattened, and having toothed tibiae. Hind legs are somewhat enlarged but not as in jumping Orthoptera.

MOLE CRICKET
(Gryllotalpidae)

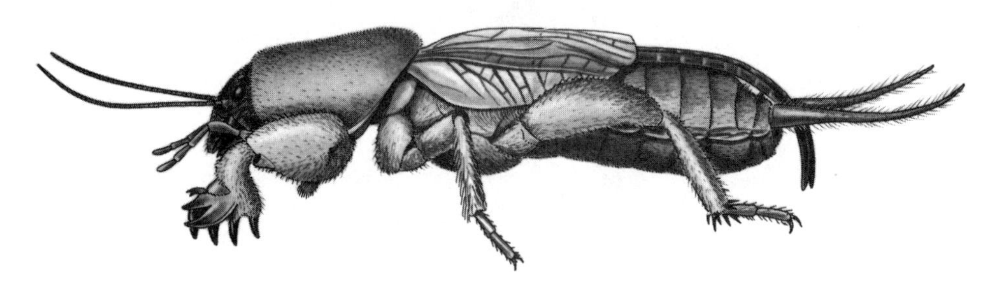

Figure 17.3. *Neocurtilla* adult

DISCUSSION: Two species of the genus *Neocurtilla* commonly occur in freshwater shore habitats of central and eastern North America. These peculiar looking insects sometimes show up in water samples that are taken near the banks of ponds and lakes. They do not live in the water but are easily knocked from burrows formed in the banks.

Order Hemiptera

A complete introduction to this order is given in Chapter 10.

TOAD BUG
(Gelastocoridae)

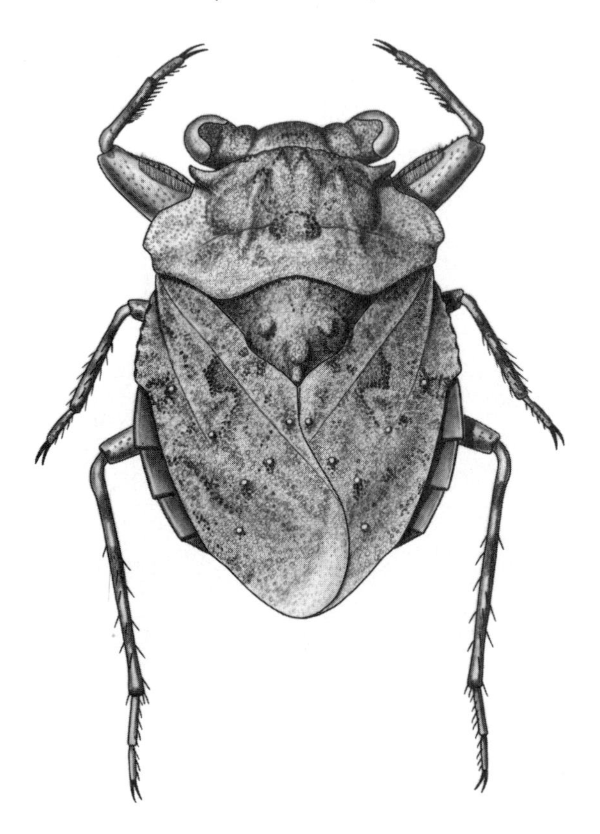

Figure 17.4. *Gelastocoris* adult

TOAD BUGS
(Family Gelastocoridae)

DIAGNOSIS: (Fig. 17.4) These are squat bugs that usually measure 6–9 mm as adults. Head is very short and has protuberant eyes and inconspicuous antennae. Fore legs are modified for grasping prey.

DISCUSSION: As their common name implies, these bugs are reminiscent of miniature toads, having a squat, warty body, bulging eyes, and in some, a fine jumping ability. These characteristics and their small, concealed antennae make them easily identifiable among shore bugs. This is a small family with two genera and few species. *Gelastocoris* is widespread, ranging into southern Canada but not known in Alaska.

Toad bugs primarily inhabit damp sand and mud. They feed on other shore-dwelling insects and suck water directly from the substrate. Toad bugs are generally associated with inland habitats, although one species of *Nerthra* has occasionally been found along the beaches of southern California.

VELVET SHORE BUGS
(Family Ochteridae)

DIAGNOSIS: (Fig. 17.5) These are dark, oval bugs that measure 3–7 mm as adults. Antennae are shorter than the head but usually visible from above.

DISCUSSION: The velvet shore bugs are distinguished from toad bugs by several characters; and the somewhat similar-looking shore bugs (Saldidae) possess long antennae. This small family consists of only about six species of the genus *Ochterus* in North America. They are found along sandy marginal freshwater habitats, often in the sand and carrying sand grains on

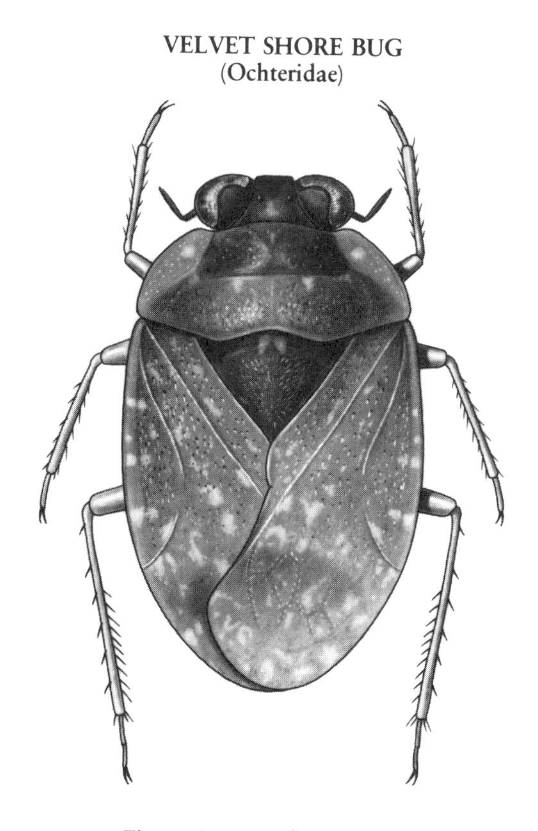

Figure 17.5. *Ochterus* adult

their bodies. They also inhabit sphagnum mats around bog lakes. They are rare and are *not* known from northwestern states or western and central Canada.

SHORTLEGGED STRIDERS
(Family Veliidae)

Occasionally, members of the genera *Paravelia* and *Microvelia* occur along shores, particularly on shore or emergent vegetation. The family is treated in Chapter 10.

VELVET WATER BUGS
(Family Hebridae)

Individuals are occasionally found on freshwater shores or debris accumulated there. The family is treated in detail in Chapter 10.

MACROVELIID SHORE BUGS
(Family Macroveliidae)

DIAGNOSIS: (Fig. 17.6) These are small, slender bugs that measure 4–6 mm at maturity. Antennae are much longer than the head. Legs are moderately long but do not possess black spines on the femora.

DISCUSSION: Macroveliid shore bugs resemble small water striders,

MACROVELIID SHORE BUG
(Macroveliidae)

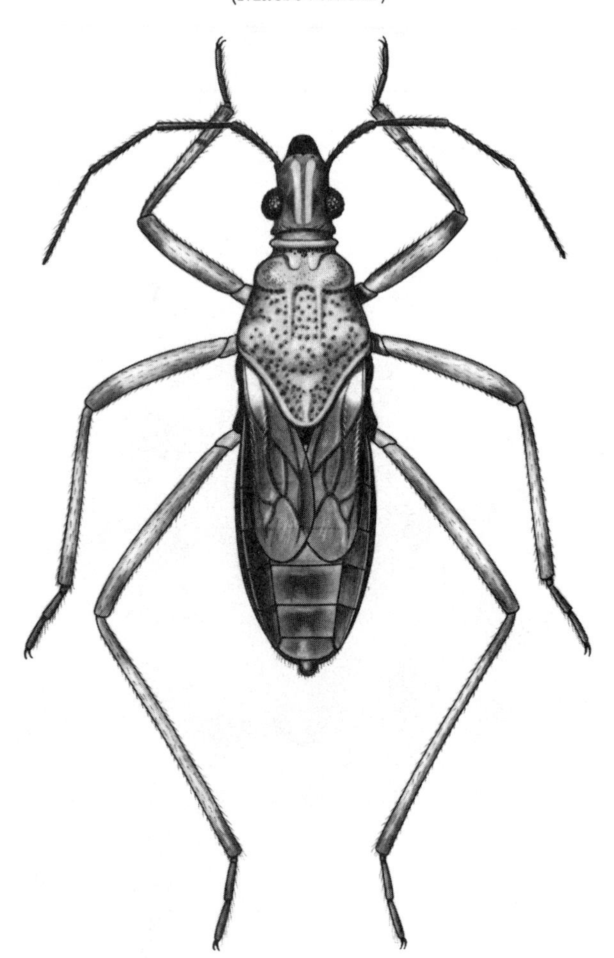

Figure 17.6. *Macrovelia* adult

but their claws are apical. Two rare species in two genera occur primarily in protected areas in the vicinities of permanent or intermittent streams. They are restricted in distribution to the western states and westernmost central states.

SHORE BUGS
(Family Saldidae)

DIAGNOSIS: (Fig. 17.7) These are small, oval bugs that measure 2–8 mm at maturity. Head appears to be set off from the remainder of body due to the laterally protruding eyes. Hemielytra, when developed, have veined membranous areas.

DISCUSSION: Most small oval bugs that occur along the shores or marginal vegetation of aquatic environments and readily take flight or jump when disturbed are apt to be saldids. This is a large and widespread family, with 70 or more species in North America. A large number of species belong to the genus *Saldula*. Shore bugs occur along ocean beaches and salt marshes as well as the shores of every type of freshwater habitat.

SPINY SHORE BUGS
(Family Leptopodidae)

DIAGNOSIS: These are somewhat similar to shore bugs; adults are about 3.5 mm in length. Head, pronotum, hemielytra, and fore legs are very spiny.

DISCUSSION: This group, which is similar in habit to the shore bugs, is represented in North America by one species, *Patapius spinosus*, that has been introduced into California. The group is more terrestrial than semiaquatic.

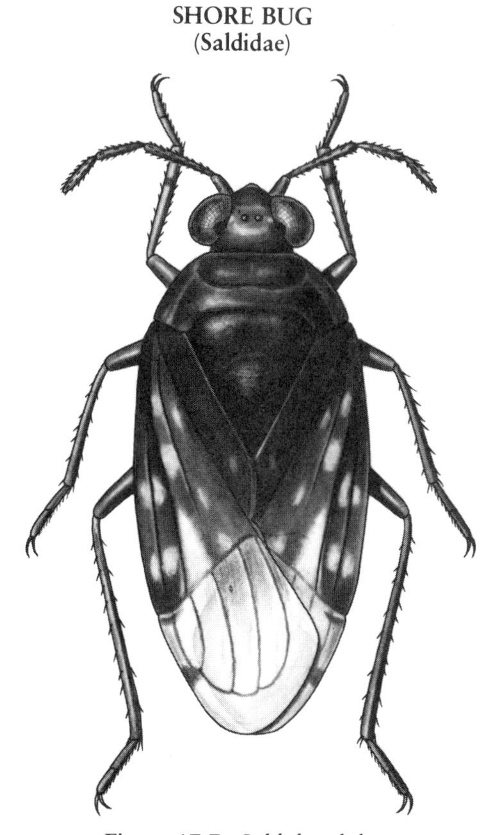

Figure 17.7. *Saldula* adult

JUMPING GROUND BUGS
(Family Dipsocoridae)

DIAGNOSIS: These are minute bugs (1–1.5 mm). They are somewhat similar to shore bugs, but their first and second antennal segments are thickened.

DISCUSSION: Three infrequently taken species of the genus *Cryptostemma* sometimes occur at the edges of streams under rocks. The group is presently known only from Georgia and California.

As the dawn breeze stirs
The milky blanket of mist,
Lifting for a time the nocturnal
Ghost borne of the lake,
Plethoric shores appear—
And the insects are not the least.
McCafferty

Order Coleoptera

A complete introduction to this order is given in Chapter 13.

GROUND BEETLES
(Family Carabidae)

ADULT DIAGNOSIS: (Figs. 17.8, 17.9) These are of variable form and size. Tarsi are all five-segmented. Hind coxae divide the posterior margin of first abdominal sternite medially. Hind coxae do not completely prevent the ventral thorax from contacting the first abdominal sternite.

LARVAL DIAGNOSIS: Legs are six-segmented (including claw as separate segment). Claws are double. Abdomen is nine-segmented.

DISCUSSION: Many genera (e.g., *Hemicarabus* and *Bembidion,* Fig. 17.9) are commonly represented on the sandy or muddy margins of streams, rivers, ponds, and lakes throughout North America. A single species of *Thalassotrechus* occurs in rock crevices along the intertidal area of the California coast.

GROUND BEETLES
(Carabidae)

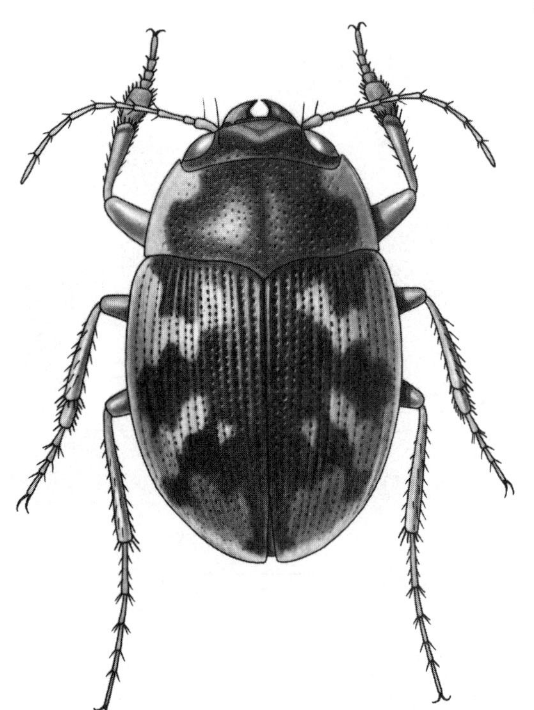

Figure 17.8. *Omophron* adult

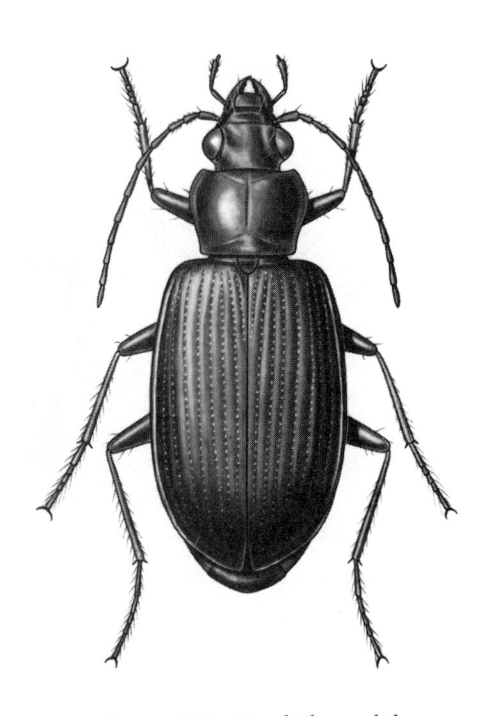

Figure 17.9. *Bembidion* adult

Species of the genus *Omophron* (Fig. 17.8) inhabit wet sandy shores and can be found by turning over driftwood or pouring water on the sand and forcing these small, brownish, oval beetles out of their burrows. Typical of many of the ground beetles, they are nocturnal, predaceous, and gregarious. Some bombardier beetles (carabids) attack the pupae of water scavenger beetles found under rocks along freshwater shore areas of the West.

MINUTE BOG BEETLES
(Family Sphaeridae)

ADULT DIAGNOSIS: These are tiny, oval beetles that are less than 1 mm in length. All tarsi are three-segmented.

LARVAL DIAGNOSIS: These small larvae possess small fingerlike lobes on abdominal segments 1–8. Second segment of antennae possesses a minute lateral filament.

DISCUSSION: These distinctive little beetles are represented by about three species in North America. The adults and presumably the larvae occur in mud, under stones, or in bog habitats in central Canada, western and central states, and Texas.

WATER SCAVENGER BEETLES
(Family Hydrophilidae)

Some species of this primarily aquatic family occur in semiaquatic shore habitats. Legless larvae of the genus *Cercyon* (Fig. 17.10) occur under kelp on ocean beaches or among rotting vegetation along shores of streams and ponds. Larvae are sometimes washed into the water. See Chapter 13 for a detailed treatment of the family.

MINUTE MOSS BEETLES
(Family Hydraenidae)

Individuals of this family are as apt to be found among debris washed up along wet freshwater shores as they are to be found in the water. The family is treated in Chapter 13.

MINUTE MUDLOVING BEETLES
(Family Georyssidae)

ADULT DIAGNOSIS: These are small, oval forms that measure 1.5–3 mm. Head is not visible from above. Antennae are clubbed.

LARVAL DIAGNOSIS: This stage remains unknown in North America, but probably will be found to have reduced legs, a nine-segmented abdomen, and short terminal filaments.

DISCUSSION: Two species are rare and known only from western Canada and western and central states. Adults live along the margins of streams and lakes in sand or mud.

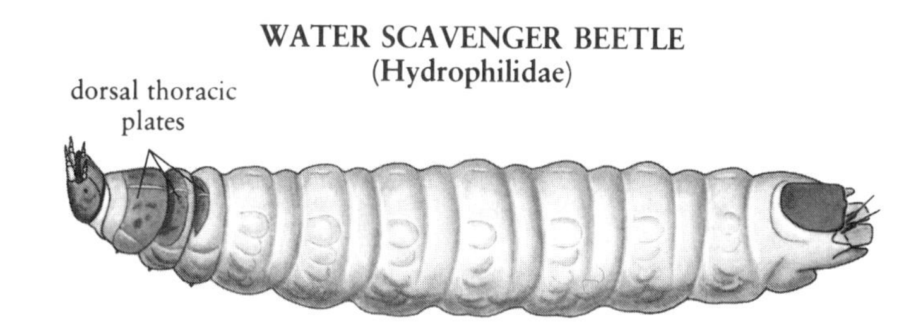

Figure 17.10. *Cercyon* larva (dorsolateral)

ROVE BEETLES
(Staphylinidae)

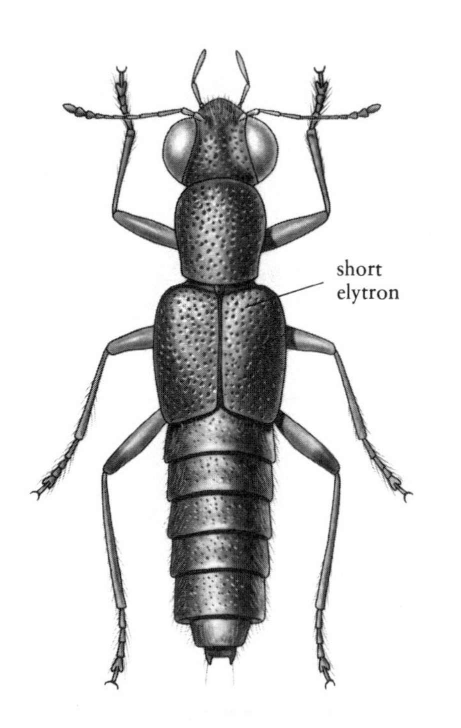

Figure 17.11. *Stenus* adult

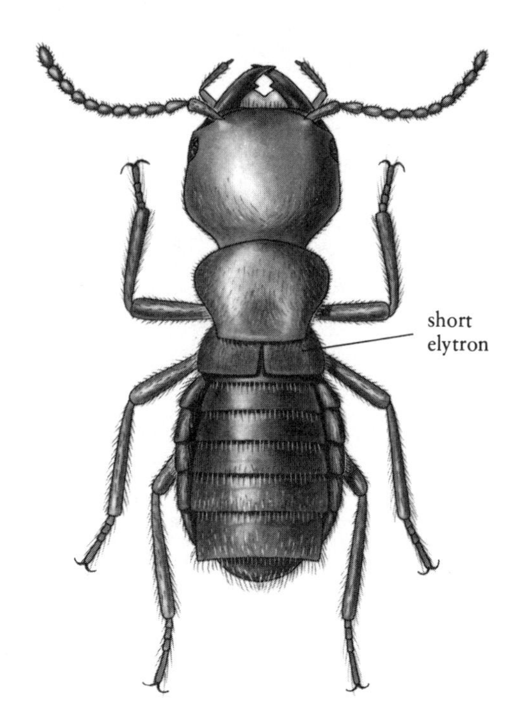

Figure 17.12. *Liparocephalus* adult

ROVE BEETLES
(Family Staphylinidae)

ADULT DIAGNOSIS: (Figs. 17.11, 17.12) These are elongate, usually small forms. Elytra are short and square-shaped at their apex. Much of the abdomen is exposed.

LARVAL DIAGNOSIS: These are generally similar to those of minute moss beetles (Chapter 13), but they usually lack well-developed molar regions of the mandibles.

DISCUSSION: A number of genera of this large and widespread group of primarily terrestrial beetles inhabit wet areas adjacent to aquatic environments. All are active predators. Some genera typically occur along rocky shores or sandy beaches of oceans. A few species of the subfamily Aleocharinae sometimes venture into the water of small tide pools for short periods of time. The genus *Liparocephalus* (Fig. 17.12) is commonly encountered along the coastal beaches of Washington and Oregon. *Diaulota* is another common genus of west coast intertidal areas. Some species of rove beetles may eventually be found to live as larvae within freshwater (this has not been documented in North America).

Some adults of the subfamily Steninae are able to move about on the surface of the water with ease. They secrete a small amount of fluid from the posterior end of their body onto the surface film. This fluid acts to decrease the surface tension of water. The differential between the decreased surface tension behind and the higher surface tension in front facilitates a continuous movement of the beetle across the water.

SHORTWINGED MOLD BEETLES
(Family Pselaphidae)

Several species of this family are especially adapted for living in the saturated environments of sphagnum bogs in North America. It would be tempting to call these species semiaquatic.

MARSH BEETLES
(Family Helodidae)

Most adults of this family are shore dwellers. The family is treated in detail in Chapter 13 because larvae are aquatic.

WATER PENNIES
(Family Psephenidae)

Adults of this family remain in the vicinity of the aquatic habitat of the larvae (streams and rivers); see Chapter 13 for more detail.

PTILODACTYLID BEETLES
(Family Ptilodactylidae)

Adults of this family remain in the vicinity of the aquatic habitat of the larvae (streams and rivers); see Chapter 13 for more detail.

VARIEGATED MUDLOVING BEETLES
(Family Heteroceridae)

ADULT DIAGNOSIS: (Fig. 17.13) These are somewhat elongate forms possessing yellowish bands or spots dorsally. They measure 4–6 mm. Head has large, flattened mandibles projecting forward. Fore legs are adapted for digging (tibiae are expanded and possess long spines laterally).

VARIEGATED MUDLOVING BEETLES
(Heteroceridae)

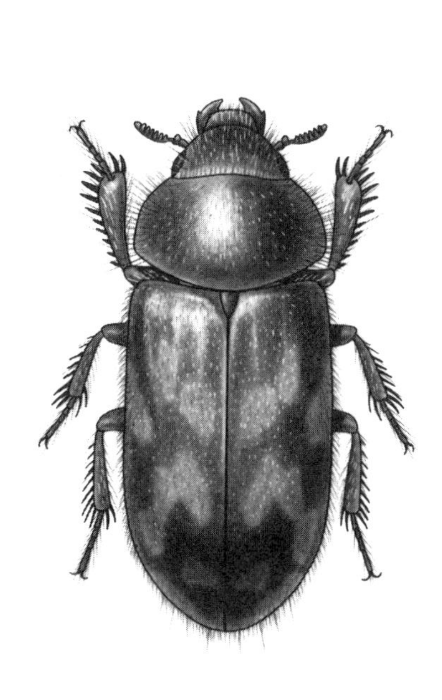

Figure 17.13. *Dampfius* adult

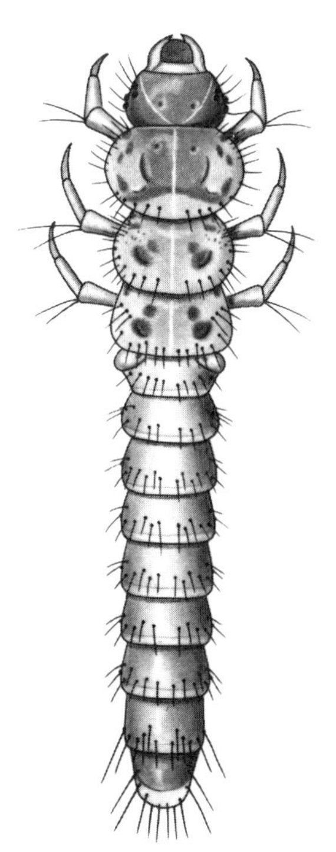

Figure 17.14. Larva

LARVAL DIAGNOSIS: (Fig. 17.14) These are elongate and 5–10 mm in length. Mouthparts are highly developed and directed anteriorly. Legs are large. Abdomen consists of similar segments and terminates in a two-segmented projection.

DISCUSSION: Several genera of these distinctive beetles occur in North America, each with generally few species. They live in galleries in wet sand and mud banks. They are most common along the margins of streams and rivers, but they are also known from margins of lentic habitats and from ocean beaches along the Atlantic coast. Very large numbers of adults are sometimes attracted to lights at night.

MINUTE MARSHLOVING BEETLES
(Family Limnichidae)

Adults are found along shores of streams and rivers throughout much of North America and on intertidal mudflats of Texas and California; see Chapter 13.

RIFFLE BEETLES
(Family Elmidae)

Certain adults of this aquatic family sometimes occur adjacent to streams and rivers; see Chapter 13.

SOFTWINGED FLOWER BEETLES
(Family Melyridae)

ADULT DIAGNOSIS: These are elongate and measure 5–10 mm. Elytra are short and expose at least half of the abdominal segments. Fleshy, yellow or orange vesicles occur laterally on the thorax and abdomen.

DISCUSSION: Several species of the genus *Endeodes* occur in rock crevices and along sandy beaches of the Pacific coast.

ROOTEATING BEETLES
(Family Rhizophagidae)

ADULT DIAGNOSIS: These are elongate, flattened, and measure 1.5–3 mm. Antennae are 10-segmented, including a two-segmented club. Elytra are squared at the ends, exposing the tip of the abdomen.

DISCUSSION: One species of *Phyconomus* occurs under driftwood in the upper intertidal area of the California coast.

DARKLING BEETLES
(Family Tenebrionidae)

ADULT DIAGNOSIS: These are small and elongate. Tarsi of fore and middle legs are five-segmented, and tarsi of hind legs are four-segmented. Hind coxae are widely separated by a broadly rounded abdominal process.

DISCUSSION: One species of *Epantius* is known from the upper intertidal area of the California coast.

NARROWWINGED BARK BEETLES
(Family Salpingidae)

ADULT DIAGNOSIS: (Fig. 7.15) These are somewhat elongate forms, measuring 2–4 mm. Pronotum is constricted basally and rounded laterally. Tarsi of fore and middle legs are five-segmented, and tarsi of hind legs are four-segmented.

DISCUSSION: A few species of the genus *Aegialites* are known from intertidal rock crevices of the Pacific coast.

NARROWWINGED BARK BEETLE
(Salpingidae)

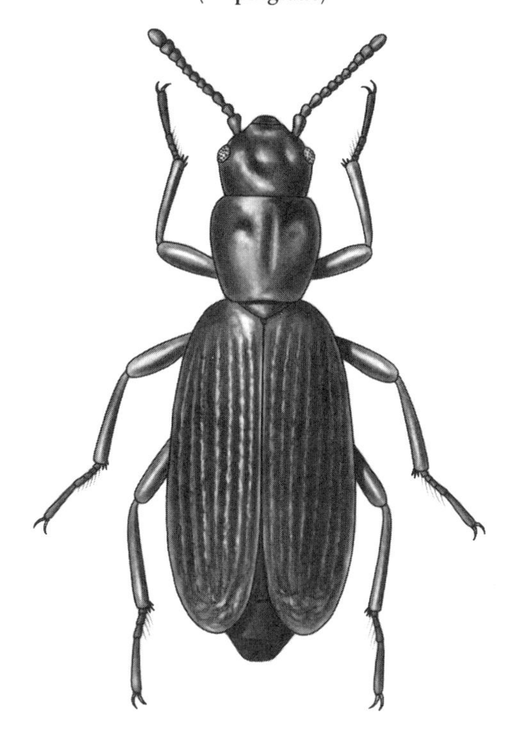

Figure 17.15. *Aegialites* adult

WEEVILS
(Family Curculionidae)

Several species of weevils occur along shore areas of both freshwater and marine environments. For a more detailed discussion of this family, see Water Weevils in Chapter 13.

Order Diptera

A complete introduction to this order is given in Chapter 16.

CRANE FLIES
(Family Tipulidae)

Many species of this family develop as larvae and pupae in wet margins of aquatic environments or moist seepage areas. Some species, such as *Limonia marmorata*, inhabit the intertidal areas of ocean beaches. See Chapter 16 for a more complete treatment of this family.

PRIMITIVE CRANE FLIES
(Family Tanyderidae)

Larvae of this group are sometimes found within the wet margins of streams and rivers; see Chapter 16.

MOTH FLIES
(Family Psychodidae)

Some larvae of this otherwise aquatic family develop in moist sand (both inland and coastal); see Chapter 16.

PHANTOM CRANE FLIES
(Family Ptychopteridae)

Larvae of this group often occur in marginal freshwater environments; see Chapter 16.

MOSQUITOES
(Family Culicidae)

These are not shore-dwelling insects, but some species develop in coastal tide pools; see Chapter 16.

BITING MIDGES
(Family Ceratopogonidae)

Several important pest species of biting midges develop in shore areas of marine and freshwater environments; see Chapter 16.

MIDGES
(Family Chironomidae)

This major aquatic family includes a few species in North America that develop in semiaquatic marginal habitats. There is also a significant marine component of the family (the marine genus *Paraclunio* is shown in Fig. 17.16); see Chapter 16.

SOLDIER FLIES
(Family Stratiomyidae)

Some semiaquatic larvae of this family occur among thick shore vegetation of ponds, and some occur in the marginal mud of streams, ponds, and lakes. For more detail see Aquatic Soldier Flies in Chapter 16.

HORSE AND DEER FLIES
(Family Tabanidae)

Many of these flies commonly develop in moist areas adjacent to many types of freshwater environments. They are also known from salt marshes, seepages, and moist soil. See Chapter 16 for more detail.

MARINE MIDGE
(Chironomidae)

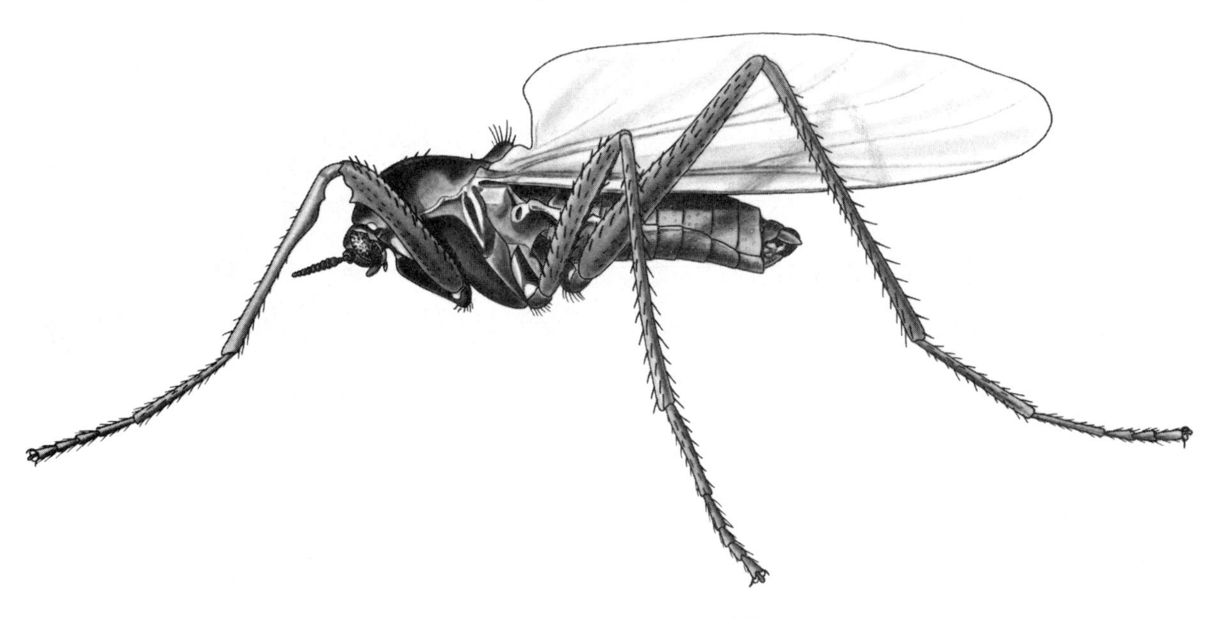

Figure 17.16. *Paraclunio* adult

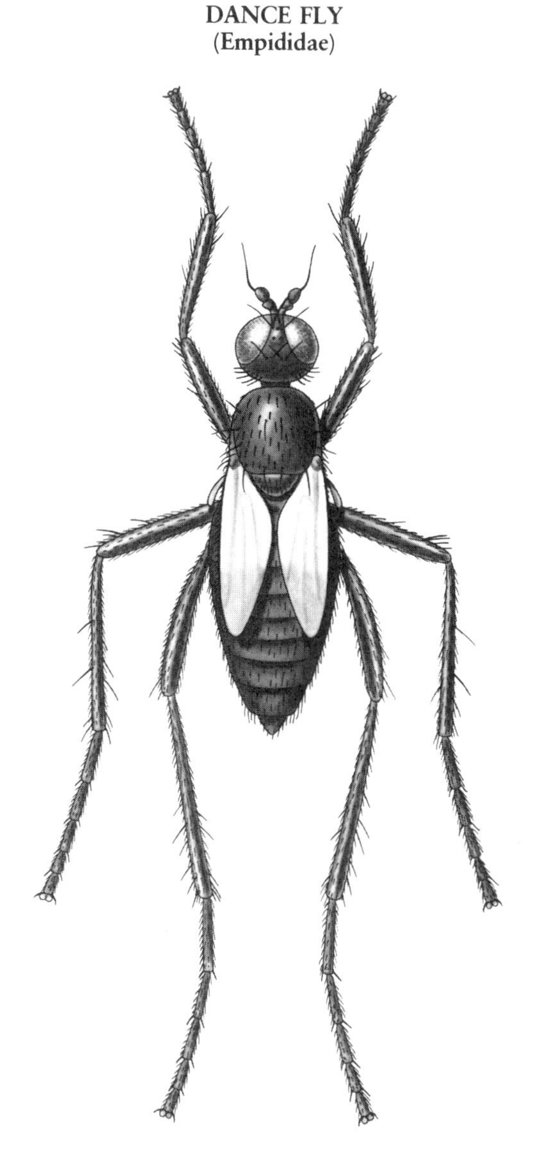

Figure 17.17. *Chersodromia* adult

DANCE FLIES
(Family Empididae)

Several species of this family develop in marginal areas of inland and marine environments (Fig. 17.17); see Aquatic Dance Flies in Chapter 16.

LONGLEGGED FLIES
(Family Dolichopodidae)

Several species of this family develop in marginal areas of inland and marine environments; see Aquatic Longlegged Flies in Chapter 16.

HUMPBACKED FLIES
(Family Phoridae)

Larvae of humpbacked flies occasionally inhabit wet marginal habitats that have decaying vegetation or high organic content. They have also been taken in sewage ponds and sewage filtering systems.

SEABEACH FLY
(Dryomyzidae)

Figure 17.18. *Helcomyza* adult

SEABEACH AND BARNACLE FLIES
(Family Dryomyzidae)

ADULT DIAGNOSIS: (Fig. 17.18) These are medium-sized, blackish flies that are somewhat similar to marsh flies (Chapter 16), but their antennae are usually smaller and not conspicuously projecting, and femoral bristles are not developed. Front margin of wings usually possesses small spines for nearly the entire length.

DISCUSSION: Three species of seabeach flies (subfamily Helcomyzinae) occur in North America, where they are known from coastal beaches of northwestern states and western Canada. The subfamily Dryomyzinae is primarily terrestrial but includes two species of the genus *Oedoparena* (the barnacle flies), which occur along the west coast from Alaska to California. The larvae of barnacle flies, which possess fleshy lateral processes along the body, are predators of intertidal barnacles. The adults remain on or near barnacle beds and oviposit eggs on the operculum of the barnacle prey. Puparia reside in empty barnacle tests, and adults emerge during morning low tide.

SEAWEED FLIES
(Family Coelopidae)

ADULT DIAGNOSIS: (Fig. 17.19) These are small or medium-sized, dark flies. Body and legs have many bristles. Thorax is somewhat flattened on top.

SEAWEED FLY
(Coelopidae)

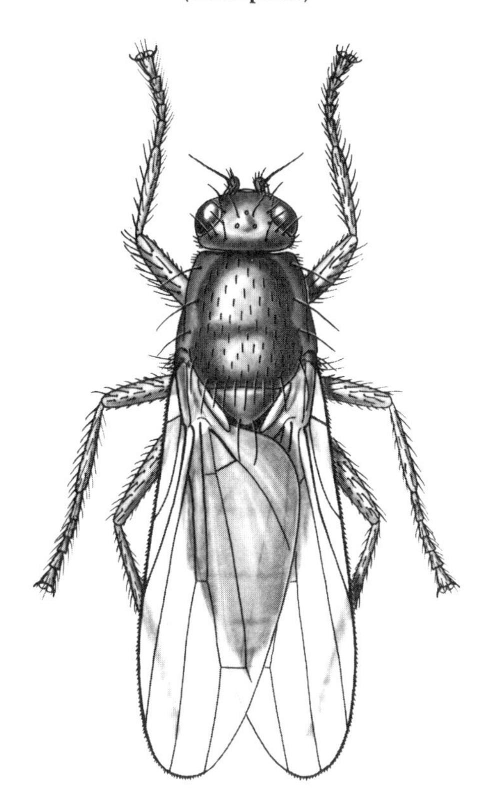

Figure 17.19. Adult

DISCUSSION: Only five species of this family occur in North America. They are known from coastal beaches of western states and western Canada and northeastern states and eastern Canada. Huge numbers of these flies are sometimes associated with rotting kelp. Adults are often numerous enough to be a nuisance.

MARSH FLIES
(Family Sciomyzidae)

Several species of this family develop in semiaquatic shore habitats; see Chapter 16.

SMALL DUNG FLIES
(Family Sphaeroceridae)

ADULT DIAGNOSIS: (Fig. 17.20) These are small, dark flies. Hind tarsi usually have a slightly swollen first segment that is shorter than the second segment.

DISCUSSION: Some species of this widespread group develop in decaying organic matter of marginal aquatic environments. They are sometimes common in rotting seaweed along ocean beaches.

SMALL DUNG FLY
(Sphaeroceridae)

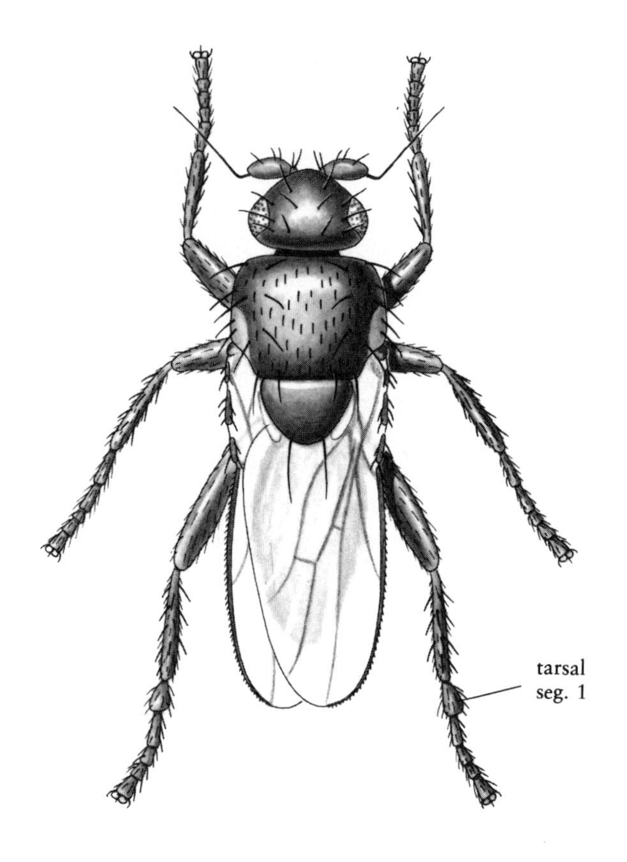

Figure 17.20. *Thoracochaeta* adult

TETHINIDS
(Family Tethinidae)

ADULT DIAGNOSIS: (Fig. 17.21) These are small, gnatlike flies. Costal vein is broken near the end of R_1 vein. All head bristles along the inner margins of the eyes are directed outward. A pair of bristles is present on the venter of head.

DISCUSSION: About six genera of this family occur in North America, mostly along coastal beach areas, but at least a few species are found inland. Some species occur in large numbers and are common around beach grasses or salt marshes. *Neopelomyia* and *Phycomyza* are restricted to the Pacific coast and often swarm in newly stranded seaweed. They are often found together with *Fucellia* (Anthomyiidae) and *Coelopa* (Coelopidae).

MILICHIIDS
(Family Milichiidae)

A species of this family that develops on the wet surface of sewage filters in the central states is easily distinguished in the adult stage by a distinct M-shaped marking on the head.

TETHINID FLY
(Tethinidae)

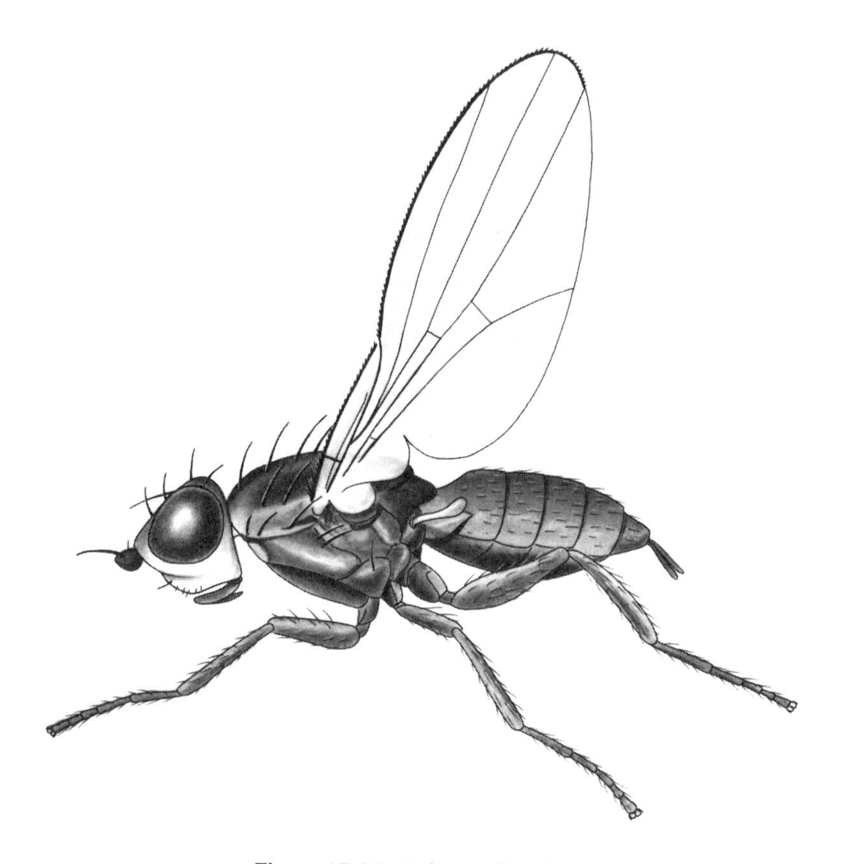

Figure 17.21. *Pelomyiella* adult

BEACH FLIES
(Family Canaceidae)

ADULT DIAGNOSIS: These small flies resemble shore flies (Chapter 16), but the costal vein in the wing is broken only once, and an anal cell is present.

DISCUSSION: Only five rare species are known from North America. The group is known from coastal regions of southwestern states including Texas and from southeastern states including Florida.

SHORE FLIES
(Family Ephydridae)

Some species of this otherwise aquatic family develop in marginal aquatic environments (freshwater, marine, and inland brackish water); see Chapter 16.

MUSCID FLIES
(Families Muscidae and Anthomyiidae)

Several species develop in wet, semiaquatic shore environments. Adults of *Fucellia* (Fig. 17.22) are often common along ocean beaches. See Aquatic Muscids in Chapter 16 for more detail.

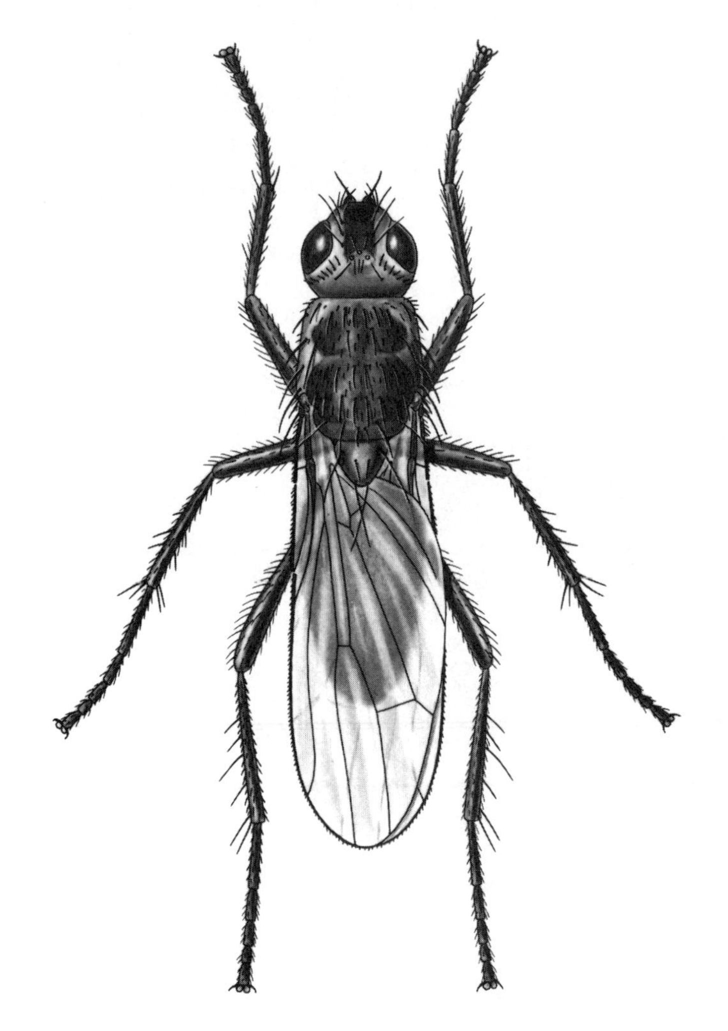

Figure 17.22. *Fucellia* adult

References

(See also references for Chapters 10, 13, and 16.)

Andersen, N. M. 1979. Phylogenetic inference as applied to the study of evolutionary diversification of semiaquatic bugs (Hemiptera: Gerromorpha). *Syst. Zool.* 28:554–578.

Arnett, R. H. 1963. *The beetles of the United States.* Catholic Univ. Amer. Press, Wash., D.C. 1112 pp.

Benedetti, R. 1973. Notes on the biology of *Neomachilis halophila* on a California sandy beach. *Pan-Pac. Entomol.* 49:246–249.

Burger, J. F., J. R. Anderson, and M. F. Knudsen. 1980. The habits and life history of *Oedoparena glauca* (Diptera: Dryomyzidae), a predator of barnacles. *Proc. Entomol. Soc. Wash.* 82:360–377.

Cameron, G. N. 1976. Do tides affect coastal insect communities? *Amer. Midl. Natural.* 95:279–287.

Campbell, J. M. 1979. Coleoptera, pp. 357–363. In *Canada and its insect fauna* (H. V. Danks, ed.). Mem. Entomol. Soc. Canada 108.

Chapman, H. C. 1958. Notes on the identity, habitat and distribution of some semi-aquatic Hemiptera of Florida. *Fla. Entomol.* 41:117–124.

Cheng, L. (ed.). 1976. *Marine insects*. North-Holland, Amsterdam. 581 pp.

Deonier, D. L., S. P. Kincaid, and J. F. Scheiring. 1976. Substrate and moisture preferences in the common toad bug, *Gelastocoris oculatus* (Hemiptera: Gelastocoridae). *Entomol. News* 87:257–264.

Dobson, T. 1976. Seaweed flies (Diptera: Coelopidae, etc.), pp. 447–463. In *Marine insects* (L. Cheng, ed.). North-Holland, Amsterdam.

Doyen, J. T. 1976. Marine beetles (Coleoptera excluding Staphylinidae), pp. 497–519. In *Marine insects* (L. Cheng, ed.). North-Holland, Amsterdam.

Drake, C. J., and H. C. Chapman. 1963. A new genus and species of water-strider from California (Hemiptera: Macroveliidae). *Proc. Biol. Soc. Wash.* 76: 227–234.

Erwin, T. L. 1967. Bombardier beetles (Coleoptera, Carabidae) of North America: Part II. Biology and behavior of *Brachinus pallidus* Erwin in California. *Coleop. Bull.* 21:41–55.

Evans, G. 1975. *The life of beetles.* Allen and Unwin Ltd., London. 232 pp.

Evans, W. G. 1968. Some intertidal insects from western Mexico. *Pan-Pac. Entomol.* 44:236–241.

Holeski, P. M., and R. C. Graves. 1978. An analysis of the shore beetle communities of some channelized streams in northwest Ohio (Coleoptera). *Gr. Lakes Entomol.* 11:23–36.

Mathis, W. N., and G. C. Steyskal. 1980. A revision of the genus *Oedoparena* Curran (Diptera: Dryomyzidae: Dryomyzinae). *Proc. Entomol. Soc. Wash.* 82:349–359.

Melander, A. L. 1951. The North American species of Tethinidae (Diptera). *J. N.Y. Entomol. Soc.* 59:187–212.

Menke, A. S. (ed.). 1979. *The semiaquatic and aquatic Hemiptera of California (Heteroptera: Hemiptera).* Bull. Calif. Ins. Surv. 21. 166 pp.

Moore, I., and E. F. Legner. 1976. Intertidal rove beetles (Coleoptera: Staphylinidae), pp. 521–551. In *Marine insects* (L. Cheng, ed.). North-Holland, Amsterdam.

Pacheco, M. F. 1978. *A catalog of the Coleoptera of Ameria north of Mexico. Family: Heteroceridae.* U.S.D.A. Handbook 529–47. 8 pp.

Polhemus, J. T. 1976. Shore bugs (Hemiptera: Saldidae, etc.), pp. 225–262. In *Marine insects* (L. Cheng, ed.). North-Holland, Amsterdam.

Polhemus, J. T., and H. C. Chapman. 1979. Family Saldidae, pp. 16–33. In *The semiaquatic and aquatic Hemiptera of California (Heteroptera: Hemiptera)* (A. S. Menke, ed.). Bull. Calif. Ins. Surv. 21.

Reichle, D. E. 1967. The temperature and humidity relations of some bog pselaphid beetles. *Ecology* 48:208–215.

Stock, M. W., and J. D. Lattin. 1976. Biology of intertidal *Saldula palustris* (Douglas) on the Oregon coast (Heteroptera: Saldidae). *J. Kans. Entomol. Soc.* 49:313–326.

Usinger, R. L. 1956. Aquatic Hemiptera, pp. 182–228. In *Aquatic insects of California* (R. L. Usinger, ed.). Univ. Calif. Press, Berkeley.

CHAPTER 18

Insects Associated with Emergent Vegetation

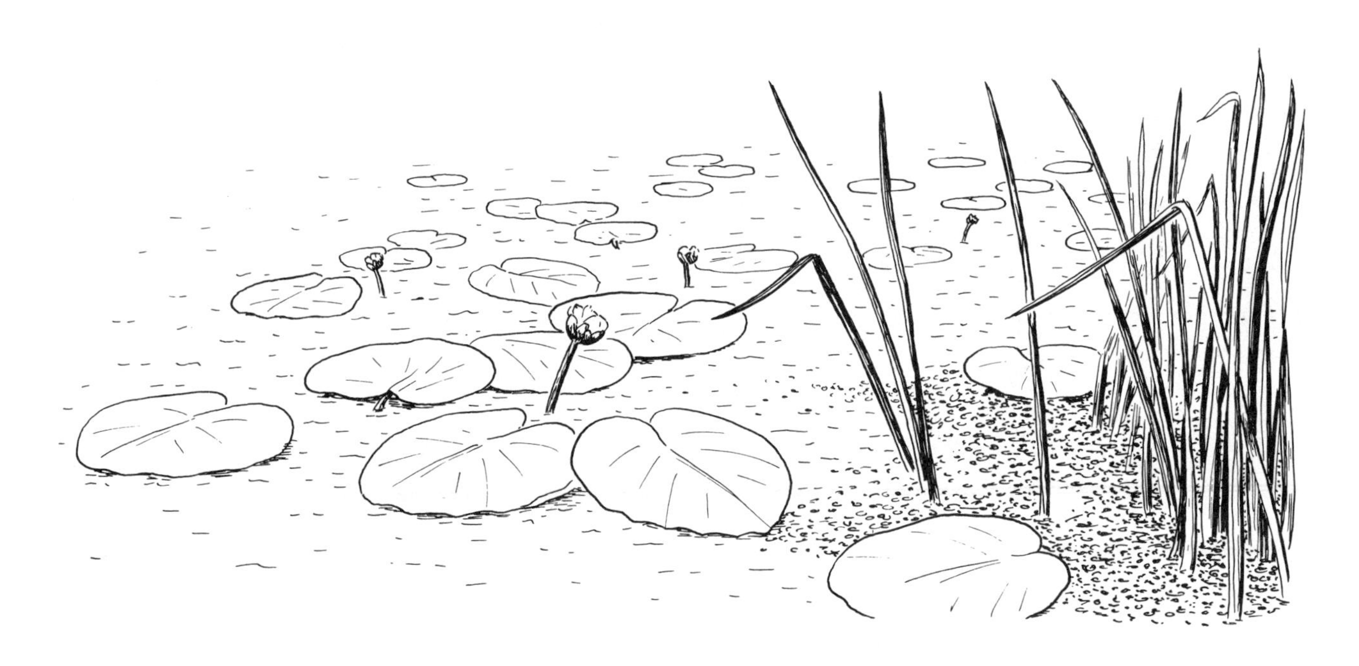

MANY AQUATIC, semiaquatic, and terrestrial insects can frequently be found on exposed parts of aquatic vegetation. The adults of many insects that are aquatic as larvae rest on such vegetation (dragonflies, mayflies, stoneflies, caddisflies, midges, and others). This chapter, however, deals with those insects that are not strictly aquatic but are closely associated with the emergent parts of aquatic vascular plants (those rooted in water or the substrates of aquatic habitats). Most of these insects are dependent on such plants as a food source, but some prey on other insects that they find on the plants.

A number of the insects associated with emergent plants belong to families that also include fully aquatic species; these families are treated in the various ordinal chapters and are cross-referenced below. Those families treated in some detail below are all primarily terrestrial and have only small numbers of so-called semiaquatic species.

Order Orthoptera

The introduction to this order of insects is given in Chapter 17.

GROUSE LOCUSTS
(Family Tetrigidae)

This family was treated in some detail in Chapter 17; however, individuals (especially immatures) of shore-dwelling species are found just as commonly on emergent vegetation close to shore as they are on the shore itself.

GRASSHOPPERS
(Family Acrididae)

DIAGNOSIS: These are variously sized, elongate forms. Antennae are less than half the length of the body. Fore legs are not modified for digging. Hind legs are modified for jumping. All tarsi are three-segmented.

DISCUSSION: A very few species of grasshoppers are commonly associated with emergent vegetation. The semiaquatic species are found primarily in eastern, central, and southwestern areas of North America. They are secretive and occasionally dive into the water when disturbed.

Short-horned (having short antennae) grasshoppers that are terrestrial have been commonly used as fish bait for centuries. Most grasshoppers are large and attractive, remain active, and can be easily fastened to a hook because of their tough body wall. Some imitation flies (e.g., the Yellowbodied Grayback) that are fashioned after grasshoppers are productive lures when large grasshoppers are being blown onto the water. Other flies, such as the Crazy Goof, Humpy, and Goofus Bug, are all good floaters in swift water and may be taken as grasshoppers by fishes.

MEADOW GRASSHOPPERS
(Family Tettigoniidae)

DIAGNOSIS: (Fig. 18.1) These are slender forms. Adults have elongate wings that often extend beyond the abdomen. Antennae are longer than the body. Fore legs are not modified for digging. Hind legs are modified for jumping. All tarsi are four-segmented.

DISCUSSION: Several species of meadow grasshoppers (also known as meadow katydids) can be found along the margins of water as well as on some emergent vegetation. They swim quite efficiently if forced into the water, as do most Orthoptera. The family is widespread in North America.

A fly pattern known as the Meadow Grasshopper is popular on Oregon trout streams during the fall.

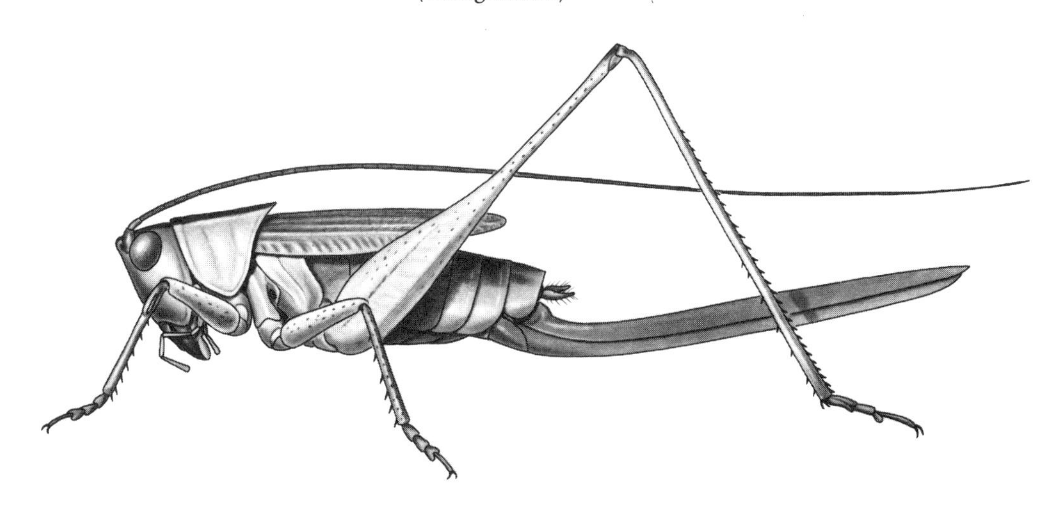

Figure 18.1. *Conocephalus* adult ♀

CRICKETS
(Family Gryllidae)

Crickets are often associated with marsh vegetation. They are treated in detail in Chapter 17.

Order Hemiptera—suborder Heteroptera

This group of insects was introduced in Chapter 10.

PYGMY BACKSWIMMERS
(Family Pleidae)

Pygmy backswimmers commonly rest on floating leaves or emergent vegetation near the waterline. They are treated in some detail in Chapter 10.

The water weeds
Like islands,
And guarding reeds
In their stands,
Enlist the breeds
Nature demands.

McCafferty

VELVET SHORE BUGS
(Family Ochteridae)

This shore-dwelling group (see Chapter 17 for details) may also be collected on aquatic plants near the shore.

SHORTLEGGED STRIDERS
(Family Veliidae)

The genus *Paravelia* is often associated with emergent vegetation. This group is treated in Chapter 10.

MARSH TREADERS
(Family Hydrometridae)

As the common name seems to imply, these surface-dwelling forms occur on marshy vegetation near the waterline as well as on the water itself. They are treated in Chapter 10.

VELVET WATER BUGS
(Family Hebridae)

These insects are often found in association with littoral vegetation of ponds; see Chapter 10.

WATER TREADERS
(Family Mesoveliidae)

Water treaders commonly crawl about emergent vegetation; see Chapter 10.

SHORE BUGS
(Family Saldidae)

These shore-dwelling bugs are also commonly found on vegetation near the shore. They are treated in detail in Chapter 17.

Order Hemiptera—suborder Homoptera

This is a large and important group of terrestrial, herbivorous insects, closely related to the bugs. It includes such groups as cicadas, treehoppers, spittlebugs, leafhoppers, planthoppers, scales, and aphids. A few species utilize emergent vascular plants as their host plants (from which they suck plant juices) and thus are associated to some degree with the aquatic environment (a few are periodically inundated by tides). The more common of these are treated below.

Diagnosis

All members of this group have mouthparts modified into a beak arising from the posterior part of the head. Most semiaquatic forms have two pairs of wings, and the first pair is somewhat leathery and held rooflike over the body. Some members of this group, however, are wingless, and others exhibit wing-size polymorphism. Aphids are wingless or have membranous wings. Hind legs are usually well developed. Scale insects are highly modified, wingless, and often legless forms that are usually scalelike or grublike.

Similar Groups

The Homoptera are quite similar to the Heteroptera. Most are distinguished, however, by the beak, which in the Heteroptera arises from the front part of the head but in the Homoptera arises more from the back of the head. Moreover, the fore wings of most Heteroptera possess a basal leathery portion and distal membranous area; the fore wings of Homoptera are uniformly textured.

LEAFHOPPERS
(Family Cicadellidae)

DIAGNOSIS: (Fig. 18.2) These are small or medium-sized forms. Antennae originate anterior to eyes. All tarsi are three-segmented. Tibia of the hind legs possesses one or more rows of small spines.

DISCUSSION: Several genera of leafhoppers (e.g., *Macrosteles, Draeculacephala, Limotettix, Forcipata,* and *Cicadula*) commonly occur on emergent vegetation but usually well above the waterline. Grasses of intertidal salt-marsh and freshwater-marsh habitats are often occupied by leafhoppers. A few also occur on littoral vegetation of lakes, ponds, streams, and rivers. *Macrosteles fascifrons* can be periodically submerged by tides in arctic marshes.

The Jassid, a tied fly pattern for a leafhopper, is fished dry on trout streams. Black is the most popular color for this fly. It may function primarily as an attractor.

LEAFHOPPER
(Cicadellidae)

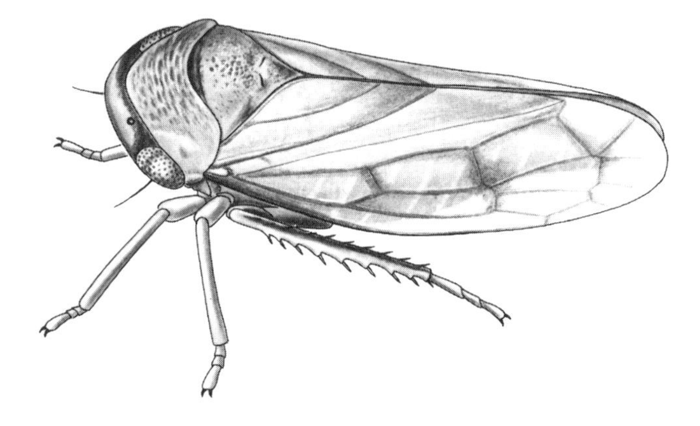

Figure 18.2. *Oncopsis* adult

SPITTLEBUG
(Cercopidae)

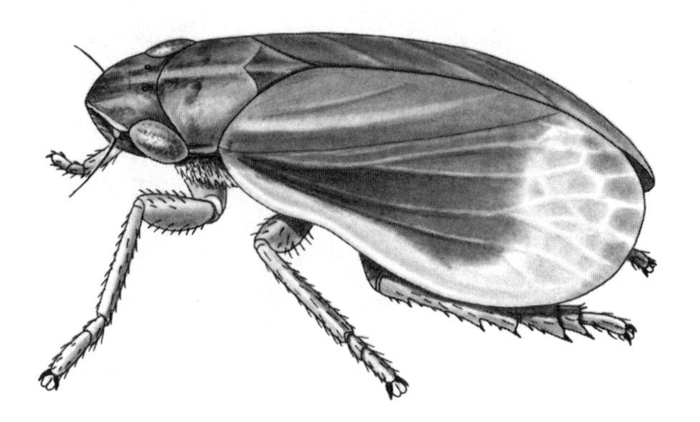

Figure 18.3. *Philaenarcys* adult

SPITTLEBUGS
(Family Cercopidae)

DIAGNOSIS: (Fig. 18.3) These are similar to the leafhoppers, except tibia of hind legs possesses a few stout processes rather than rows of small spines.

DISCUSSION: *Philaenarcys spartina* feeds on alkalie cordgrass along the Atlantic coast. Otherwise, little is known about spittlebugs associated with emergent vegetation in North America.

DELPHACID PLANTHOPPERS
(Family Delphacidae)

DIAGNOSIS: (Fig. 18.4) These differ from the leafhoppers and spittlebugs primarily by the position of the antennae relative to the eyes. Antennae originate below the eyes, on the sides of the head. Adults are either fully winged or short-winged.

DISCUSSION: Of the planthoppers, the Delphacidae are most apt to be encountered in association with saltwater and freshwater plants. However, at least one species each of the planthopper families Issidae and Dic-

DELPHACID PLANTHOPPER
(Delphacidae)

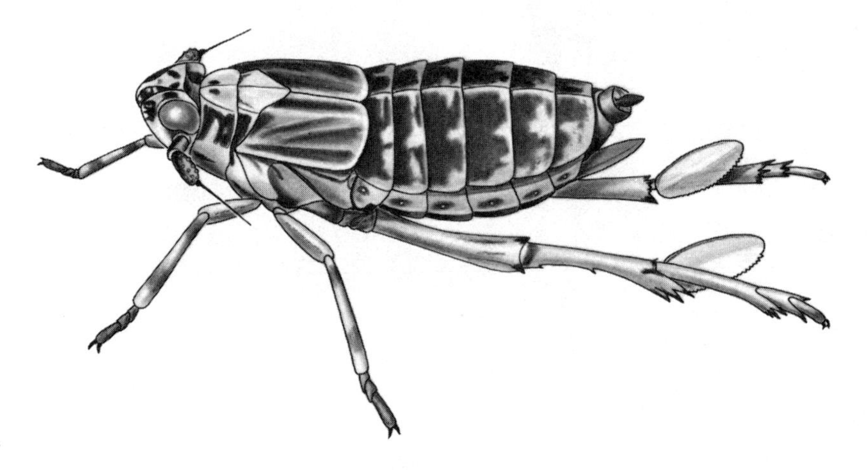

Figure 18.4. *Megamelus* adult

tyopharidae occur in intertidal salt marshes. Delphacids have a broad moveable spur on the apex of the tibia of the hind legs, whereas the other planthopper families do not.

Megamelus davisi (Fig. 18.4) is common on the exposed parts and floating leaves of pond lilies. Many more species occur in marsh environments, and *Prokelesia marginata* is an abundant species on cordgrass along the Atlantic coast. Short-winged forms of species are often most common when host plants are at high densities, whereas fully winged forms (better able to disperse) tend to be prevalent when host plants are at low densities.

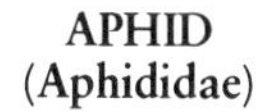

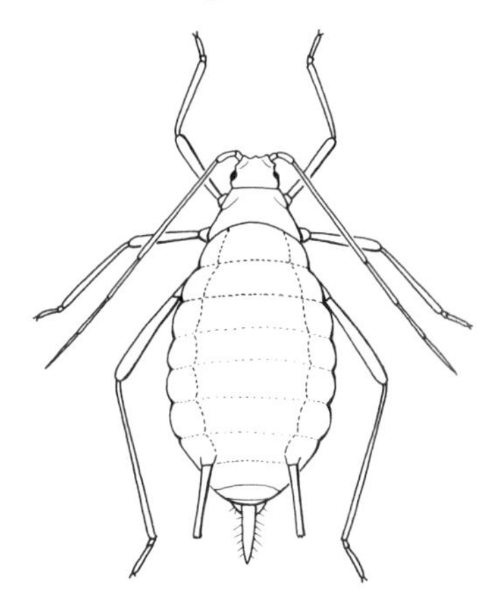

Figure 18.5. Wingless adult

APHIDS
(Family Aphididae)

DIAGNOSIS: (Fig. 18.5) These are small insects that have membranous wings or are wingless. The abdomen is somewhat oval and possesses a pair of short tubelike structures originating anterior to the end of the abdomen.

DISCUSSION: This is a large, terrestrial group of plant-feeding insects that are sometimes found on the exposed parts of emergent vegetation. A common species on pond lilies is *Rhopaosiphum nympheae.*

SCALE INSECTS
(Superfamily Coccoidea)

DIAGNOSIS: Females, which occur on the host plants, are very atypical insects, since they are wingless and often legless forms that may be covered with a scalelike structure or a waxy covering.

DISCUSSION: This is another important terrestrial, plant-feeding group that occurs on all kinds of vascular plants, including those rooted in water.

Order Coleoptera

The order Coleoptera is introduced and treated in detail in Chapter 13.

AQUATIC LEAF BEETLES
(Family Chrysomelidae)

Adult beetles, principally of the genus *Donacia* (Fig. 18.6; Plate XI, Fig. 71), occur commonly on emergent vegetation, especially floating leaves in lentic environments. The larvae of these beetles are aquatic; see Chapter 13 for more detail.

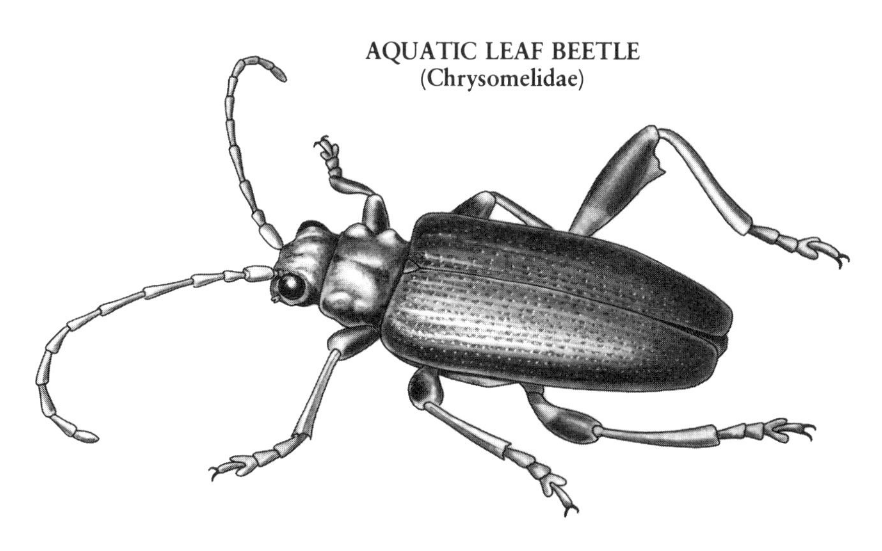

Figure 18.6. *Donacia* adult

WEEVILS
(Family Curculionidae)

Weevils occur commonly on the emergent parts of both freshwater and saltwater plants. Some species are considered to be aquatic; see Chapter 13 for more detail.

Order Lepidoptera

The general ecology, biology, and characters of caterpillars associated with aquatic vegetation were treated in Chapter 15.

PYRALID MOTHS
(Family Pyralidae)

This family contains an array of aquatic and semiaquatic species but is primarily terrestrial. It is treated in considerable detail in Chapter 15.

COSMOPTERYGID MOTHS
(Family Cosmopterygidae)

LARVAL DIAGNOSIS: These are generally less than 12 mm in length. Gills are absent. Thoracic legs are well developed and segmented. Prolegs are present on abdominal segments 3–6 and 10. Crochets occur in circles and possess hooks all of the same size.

ADULT DIAGNOSIS: Mouth siphon is reduced. Labial palps are recurved upwards.

DISCUSSION: Two small genera in North America have species that are miners and stem borers in some sedges and cattails.

SERPENTINE MOTHS
(Family Nepticulidae)

LARVAL DIAGNOSIS: These are small forms that have highly reduced thoracic legs without distinct segments. Prolegs are usually present on abdominal segments 2–7 and lack crochets.

ADULT DIAGNOSIS: These are less than 3 mm in length. Mouth siphon is reduced. Palps extend downward.

DISCUSSION: Some *Nepticula* in the central states are serpentine miners and stem borers in spikerush and bulrush.

NOCTUID MOTHS
(Family Noctuidae)

LARVAL DIAGNOSIS: (Figs. 18.7, 18.8) These generally measure 30–70 mm when mature. Thoracic legs are well developed. Prolegs are present on abdominal segments 5, 6, and 10.

ADULT DIAGNOSIS: (Fig. 18.9) These are generally large, somewhat drab moths. Mouth siphon is well developed.

DISCUSSION: A few species of this otherwise large and important terrestrial family are miners and petiole or stem borers, especially in water lilies. These particular species are sometimes known as "diver moths."

The Yellow Water Lily Borer *(Bellura gortynoides)* mines in leaves as a young caterpillar, then becomes a petiole borer. Within the petiole, larvae are submerged in water and must periodically back out in order to expose specialized posterior spiracles (Fig. 18.8) to the air before submerging for another few minutes. Larvae swim to shore to overwinter in dry protected areas.

NOCTUID MOTHS
(Noctuidae)

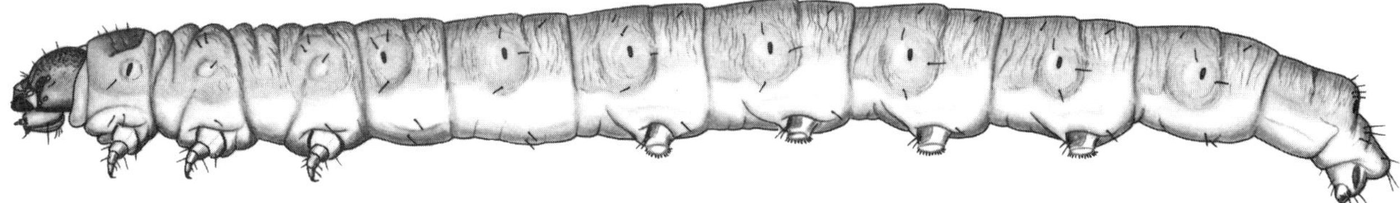

Figure 18.7. *Bellura* larva

Figure 18.8. *Bellura* larva, end of abdomen (posterolateral)

Figure 18.9. *Bellura* adult

Host plants of the semiaquatic noctuids include water lilies, water lotus, water hyacinth, arrowhead, pickerelweed, burreed, rushes, cattail, and bulrush.

Order Diptera

This order was introduced in Chapter 16. Adults of a large variety of aquatic flies rest or forage on the emergent parts of aquatic plants. The larvae of some Tipulidae, Chironomidae, Culicidae, Dolichopodidae, and Ephydridae are closely associated with the underwater parts of emergent vegetation. The following two additional families are noteworthy as plant associates.

GALL GNATS
(Family Cecidomyiidae)

LARVAL DIAGNOSIS: These are small, maggotlike, and often brightly colored forms with poorly developed heads.

DISCUSSION: Larvae of several species develop within the galls formed on leaves or stems of aquatic vegetation (especially sedges and reeds). The galls usually appear as warts, protuberances, or globular masses of plant scar tissue. Although galls are originally formed on the emergent parts of the host plants, they sometimes subsequently become submerged.

DUNG FLY
(Scatophagidae)

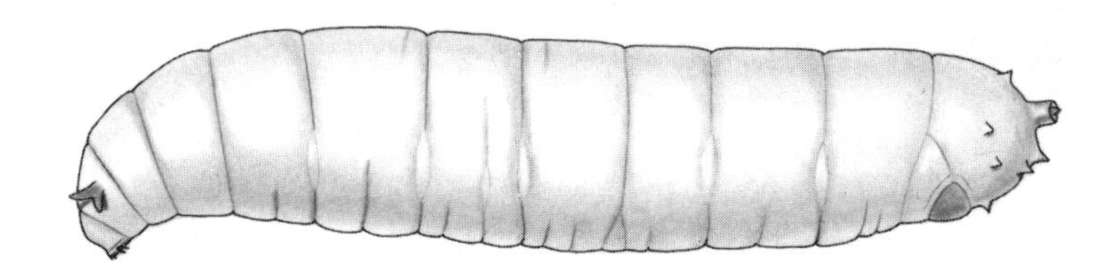

Figure 18.10. *Cordilura* larva

DUNG FLIES
(Family Scatophagidae)

LARVAL DIAGNOSIS: (Fig. 18.10) These are white, maggotlike forms that usually measure 5–13 mm. Head is completely reduced. Body is somewhat blunt at both ends. Abdomen may end in a spiracular disc, with spiracles of some species protruding slightly on very short breathing tubes; abdomen of others ends in a pair of short terminal spines.

ADULT DIAGNOSIS: These are similar to aquatic muscids.

DISCUSSION: The larvae of most species of this family live in dung or rotting vegetation. Species of *Hydromyza* and *Cordilura* sometimes occur as miners in the leaves, petioles, and culms of aquatic vascular plants. Oviposition sometimes occurs below the waterline.

References

Berg, C. O. 1949. Limnological relations of insects to plants of the genus *Potamogeton*. *Trans. Amer. Micr. Soc.* 68:279–291.

Cameron, G. N. 1976. Do tides affect coastal insect communities? *Amer. Midl. Natural.* 95:279–287.

Cantrall, I. J. 1978. Semiaquatic Orthoptera, pp. 99–103. In *An introduction to the aquatic insects of North America* (R. W. Merritt and K. W. Cummins, eds.). Kendall / Hunt, Dubuque.

Claassen, P. W. 1921. *Typha* insects: Their ecological relationships. *Mem. Cornell Ag. Exp. Sta.* 47:457–531.

Davis, L. V., and I. E. Gray. 1966. Zonal and seasonal distribution of insects in North Carolina salt marshes. *Ecol. Monogr.* 36:275–295.

DeLong, D. M. 1965. Ecological aspects of North American leafhoppers and their role in agriculture. *Bull. Entomol. Soc. Amer.* 11:9–26.

Denno, R. F. 1976. Ecological significance of wing polymorphism in Fulgoroidea which inhabit tidal salt marshes. *Ecologic. Entomol.* 1:257–266.

Frohne, W. C. 1938. Contribution to knowledge of the limnological role of the higher plants. *Trans. Amer. Micr. Soc.* 57:256–268.

Frohne, W. C. 1939. Observations on the biology of three semiaquatic lacustrine moths. *Trans. Amer. Micr. Soc.* 58:327–348.

Gerlach, R. 1974. *Creative fly tying and fly fishing.* Winchester Press, New York. 231 pp.

Hamilton, K. G. A. 1979. Synopsis of the North American Philaenini (Rhynchota: Homoptera: Cercopidae) with a new genus and four new species. *Canad. Entomol.* 111:127–141.

Johannsen, O. A. 1934. *Aquatic Diptera. Pt. I. Nematocera, exclusive of Chironomidae and Ceratopogonidae.* Mem. Cornell Univ. Ag. Exp. Sta. 164. 71 pp.

Johannsen, O. A. 1935. *Aquatic Diptera. Pt. II. Orthorrhapha-Brachycera and Cyclorrhapha.* Mem. Cornell Univ. Ag. Exp. Sta. 171. 62 pp.

Lange, W. H. 1978. Aquatic and semiaquatic Lepidoptera, pp. 187–201. In *An introduction to the aquatic insects of North America* (R. W. Merritt and K. W. Cummins, eds.). Kendall / Hunt, Dubuque.

Levine, E., and L. Chandler. 1976. Biology of *Bellura gortynoides* (Lepidoptera: Noctuidae), a yellow water lily borer, in Indiana. *Ann. Entomol. Soc. Amer.* 69:405–414.

McCafferty, W. P., and M. C. Minno. 1979. The aquatic and semiaquatic Lepidoptera of Indiana and adjacent areas. *Gr. Lakes Entomol.* 12:179–187.

McGaha, Y. J. 1952. The limnological relations of insects to certain aquatic flowering plants. *Trans. Amer. Micr. Soc.* 71:355–381.

Petersen, A. 1956. *Fishing with natural insects.* Spahr & Glenn, Columbus. 176 pp.

Wallace, J. B., and S. E. Neff. 1971. Biology and immature stages of the genus *Cordilura* (Diptera: Scatophagidae) in the eastern United States. *Ann. Entomol. Soc. Amer.* 64:1310–1330.

Welch, P. S. 1914. Habits of the larva of *Bellura melanopyga* Grote (Lepidoptera). *Biol. Bull.* 27:97–114.

CHAPTER 19

Tree Hole and Plant Cup Residents

WATER ACCUMULATES in certain small natural depressions formed by living or dead plant parts, and these small reservoirs provide a special aquatic environment inhabited by a number of insect species. Some species, especially of mosquitoes and a few other Diptera, are restricted to such habitats, whereas the majority of inhabitants are more generally adapted to stagnant waters.

Aquatic tree hole habitats are formed when rain collects in bark depressions. They are most commonly located where main branches fork, behind scar tissue, and behind outgrowths of bracket fungi. In addition to water, these holes usually contain rotting vegetation and fallen leaves. The insect larvae that reside in such tree holes feed either on decaying organic matter or other insects.

Pitcher plants, which are found primarily in bog environments of North America, have large tubular leaves in which rainwater and secretions from the plant itself accumulate. These plants (*Sarracenia,* depicted in chapter heading) are adapted specifically for entrapping flying insects. Such insects are digested in the plant cup medium by plant enzymes and thus provide part of the nutritional requirements of the plants. A few mosquitoes, flesh flies, and midges are specialized for living as larvae in the aqueous environment of the pitcher plant cup. Unlike other insects, they are not affected by the digestive enzymes of these carnivorous plants.

Plant cup habitats formed by bracts, stems, or leaves of a variety of plants are especially common in tropical environments. However, in North

America, such habitats are associated primarily with pitcher plants, and the bromeliads of Florida. Mosquitoes are the primary aquatic residents of bromeliad plant cups.

Three approximate categories of insects are associated with the vegetative reservoirs: (1) those that are accidental and found in such habitats very rarely, (2) those that are attracted to a variety of semiaquatic or moist environments and thus are sometimes associated with such reservoirs, and (3) those that are regular inhabitants. This chapter deals primarily with the regular inhabitants.

Order Coleoptera

The beetles were introduced and treated in detail in Chapter 13.

WATER SCAVENGER BEETLES
(Family Hydrophilidae)

Water scavenger beetles rarely occur in tree holes; they are probably accidental residents in such environments. This family is treated in detail in Chapter 13.

MARSH BEETLES
(Family Helodidae)

Larvae of marsh beetles are often found in tree holes and other small reservoirs of water; see Chapter 13.

Order Diptera

This order was introduced and treated in detail in Chapter 16.

CRANE FLIES
(Family Tipulidae)

Larvae of several species of crane flies occur in association with moist environments provided by tree holes and rotting wood in general. Submergent forms, however, are not known from water-filled plant reservoirs. Crane flies are treated in detail in Chapter 16.

MOTH FLIES
(Family Psychodidae)

The genera *Telmatoscopus, Brunettia,* and *Psychoda* contain some species that are able to develop in aquatic or semiaquatic habitats associated with tree holes; see Chapter 16 for a detailed family treatment.

MOSQUITOES
(Family Culicidae)

This important aquatic family was treated in Chapter 16. About 21 species in North America have been found in association with tree holes, and about eight species (e.g., in *Haemagogus, Orthopodomyia,* and *Toxorhynchites*) are restricted to such environments. Some are detritivores and others are carnivorous.

Wyeomyia (Fig. 19.1) develops in the plant cup environment of pitcher plants and in water accumulated in other living or dead plants.

MOSQUITO
(Culicidae)

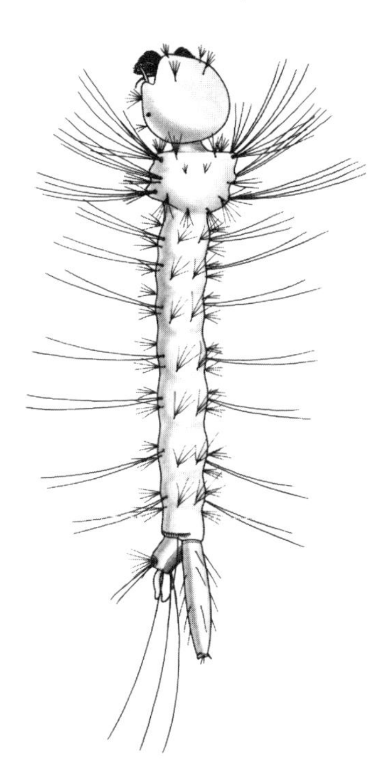

Figure 19.1.
Wyeomyia larva

BITING MIDGES
(Family Ceratopogonidae)

About three genera are known to have larvae that develop either in the water or along the edge of the water of tree holes; see Chapter 16 for details of this family.

MIDGES
(Family Chironomidae)

Larvae of a few species of this large and varied group have been reported from tree holes. One species of *Metriocnemus* is a relatively common resident of pitcher plant cups.

HORSE AND DEER FLIES
(Family Tabanidae)

Larvae (especially of *Leucotabanus*) occasionally occur in association with the tree hole environment, particularly where moist decaying vegetation is available; see Chapter 16 for details of this family.

LONGLEGGED FLIES
(Family Dolichopodidae)

Some longlegged flies (especially of *Systenus*) are associated with moist debris of tree holes; see Chapter 16 for details of the family.

RATTAILED MAGGOTS
(Family Syrphidae)

Tree holes provide adequate habitat for developing larvae of such genera as *Mallota, Eristalis,* and *Meromacrus*. This group is treated in detail in Chapter 16.

FLESH FLIES
(Family Sarcophagidae)

LARVAL DIAGNOSIS: (Fig. 19.2) These measure up to 16 mm in length. They are maggotlike forms with an anteriorly pointed body. Posterior end of the abdomen forms a pronounced, slightly upturned cup within which terminal spiracles are located.

ADULT DIAGNOSIS: (Fig. 19.3) These are very similar to common house flies.

Amidst the misty bogs
And humid forests,
Within bract, stem, and hole,
And vesicles uncountable—
Microcosmic systems pervade.
McCafferty

DISCUSSION: The flesh flies are a primarily terrestrial group of higher (cyclorrhaphan) Diptera. The larvae of most species live as scavengers in rotting animal or vegetable matter (some species are parasitic). A few species of the genus *Blaesoxipha,* however, are associated with pitcher plants, primarily in eastern states and central and eastern Canada.

Larvae associated with pitcher plants can live only in the leaf cups of this carnivorous plant. Their bodies are submerged in the accumulated water of the plant cup except for the posterior end, which floats. If they become totally submerged, they will drown. They are able to withstand the digestive enzymes secreted by the plant, and feed on other insects that fall into the plant cup as victims of the plant. There is generally only one individual flesh fly larva per cup. Larvae crawl to the ground to pupate.

References

Bradshaw, W. E., and L. P. Lounibos. 1972. Photoperiodic control of development in the pitcher-plant mosquito, *Wyeomyia smithii. Canad. J. Zool.* 50:713–719.

Forsyth, A. B., and R. J. Robertson. 1975. K reproductive strategy and larval behavior of the pitcher plant sarcophagid fly, *Blaesoxipha fletcheri. Canad. J. Zool.* 53:174–179.

FLESH FLIES
(Sarcophagidae)

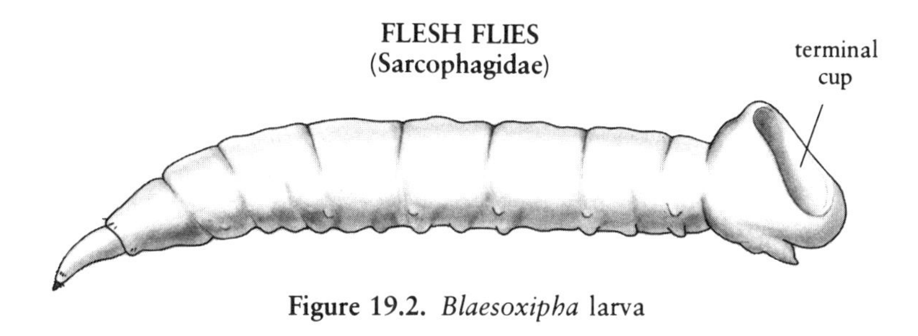

Figure 19.2. *Blaesoxipha* larva

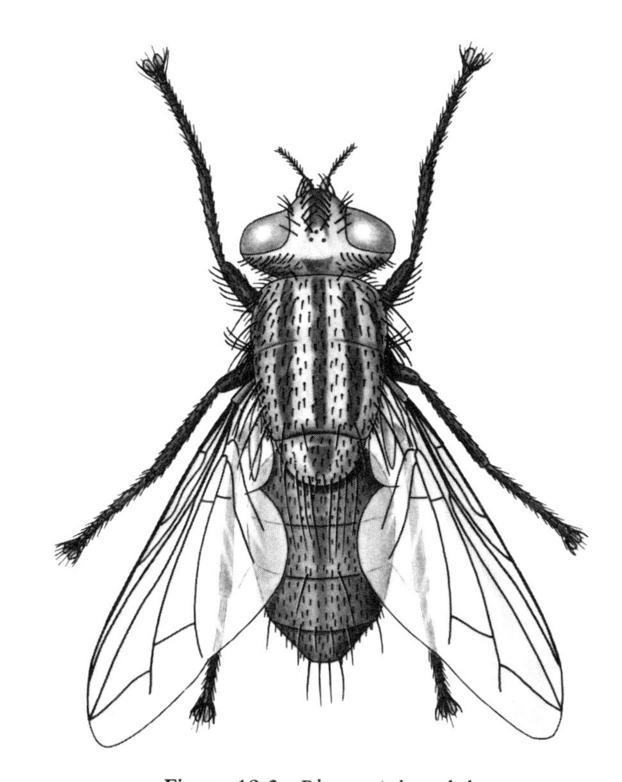

Figure 19.3. *Blaesoxipha* adult

Jenkins, D. W., and S. J. Carpenter. 1946. Ecology of the tree-hole breeding mosquitoes of Nearctic North America. *Ecol. Monogr.* 16:31–47.

Kitching, R. L. 1971. An ecological study of water-filled tree holes and their position in the woodland ecosystem. *J. Anim. Ecol.* 40:281–302.

Leech, H. B., and H. P. Chandler. 1956. Aquatic Coleoptera, pp. 293–371. In *Aquatic insects of California* (R. L. Usinger, ed.). Univ. Calif. Press, Berkeley.

Swales, D. E. 1969. *Sarracenia purpurea* L. as host and carnivore at Lac Carré, Terrebonne Co., Quebec. *Natural. Canad.* 96:759–763.

Teskey, H. J. 1976. *Diptera larvae associated with trees in North America.* Mem. Entomol. Soc. Canada 100. 53 pp.

Wirth, W. W., and A. Stone. 1956. Aquatic Diptera, pp. 372–482. In *Aquatic insects of California* (R. L. Usinger, ed.). Univ. Calif. Press, Berkeley.

CHAPTER 20

Diving Wasps

(ORDER HYMENOPTERA)

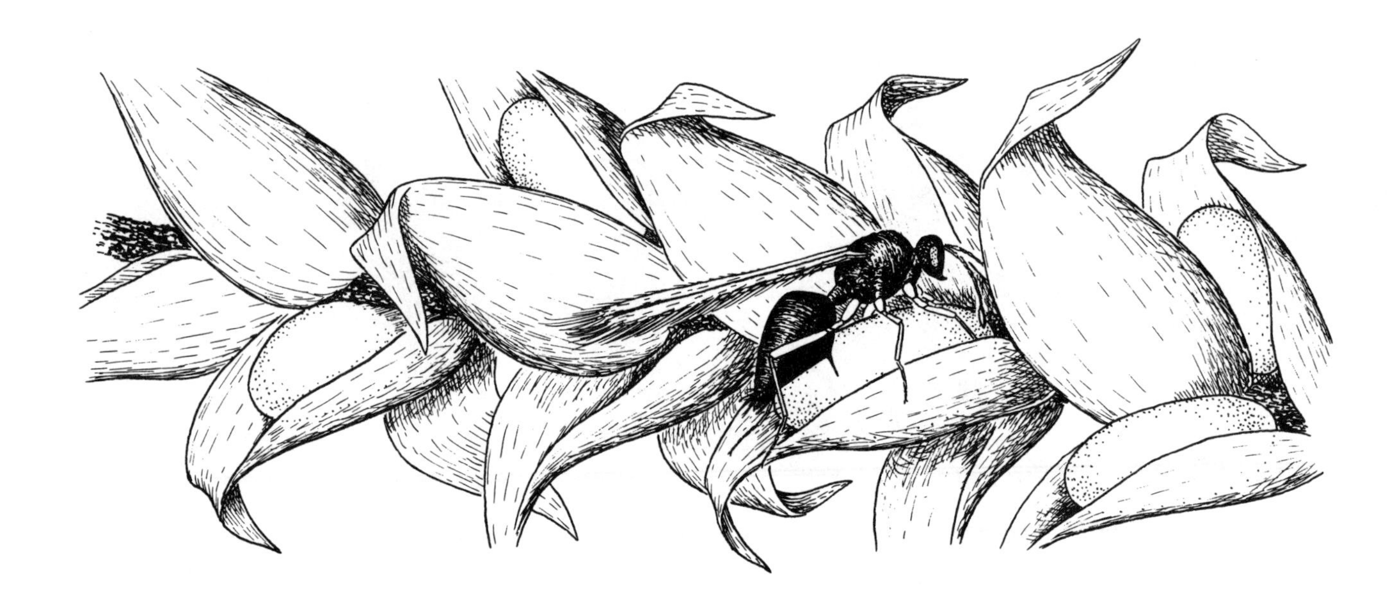

THE ORDER HYMENOPTERA is a large, primarily terrestrial group of insects that includes bees, wasps, ants, sawflies, and other similar forms. The larvae of many species of wasps parasitize other insects. The hosts are mostly terrestrial insects, but the immature stages of some aquatic insects also fall victim to these parasites. Since the adult females of certain wasps enter the water to seek out the aquatic insect host for oviposition, these so-called diving wasps (some crawl, some swim) are appropriate subjects of aquatic entomology.

Diving wasps attack eggs, larvae, or pupae, depending on the species of parasite and the species of the insect host. Nonsubmerged stages of aquatic and semiaquatic insects are also often parasitized by wasps. This type of parasitism, in which the wasp is not required to enter the water, is much more common. This chapter treats only those families of Hymenoptera known to include some species that actually enter the water.

The diving wasps undergo complete metamorphosis; many emerge as adults from their hosts. Larvae of certain species change radically in form from one instar to the next. There may be one or more than one individual parasite per individual host. Respiration of internal parasites is cutaneous. The female adults are generally equipped with a well-developed ovipositor.

Adult Diagnosis

With few exceptions, diving wasps are minute forms. Major body regions are usually well defined, and a strong constriction is usually apparent between the thorax and abdomen. Antennae are slender. Thorax possesses three pairs of legs and usually two pairs of wings. Wings may be broad and membranous, narrow and bordered by a fringe of hairs, or absent.

Similar Orders

And tho at depths,
There is no sanctuary.

McCafferty

Wasps are usually very distinctive. They are distinguished from flies (Diptera) of the same small size class by the presence of four rather than two wings. The constriction between the thorax and abdomen of wasps is much more defined than in most other insects.

BRACONIDS
(Family Braconidae)

ADULT DIAGNOSIS: (Fig. 20.1) These are small wasps whose fore wings have relatively well-developed venation and only one recurrent vein. Antennae possess more than 15 segments.

DISCUSSION: Some species of this large family parasitize aquatic flies (mostly shore flies of the leaf-mining genus *Hydrellia*) and, less frequently, aquatic moths. They develop in eggs, larvae, or pupae of the host insect. Females of some seek out the host in water, especially on aquatic vegetation.

BRACONID WASP
(Braconidae)

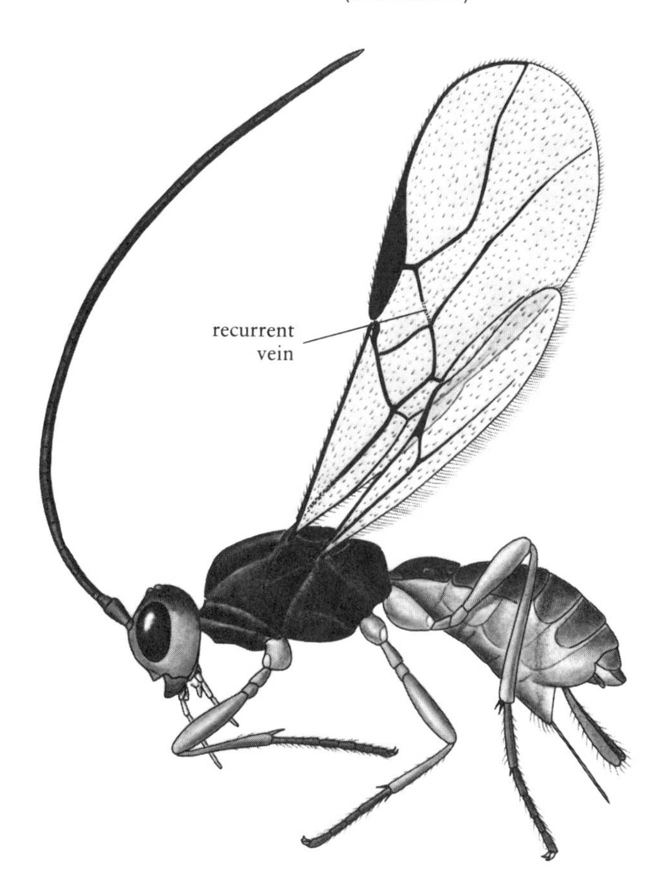

Figure 20.1. Adult ♀

ICHNEUMONID WASP
(Ichneumonidae)

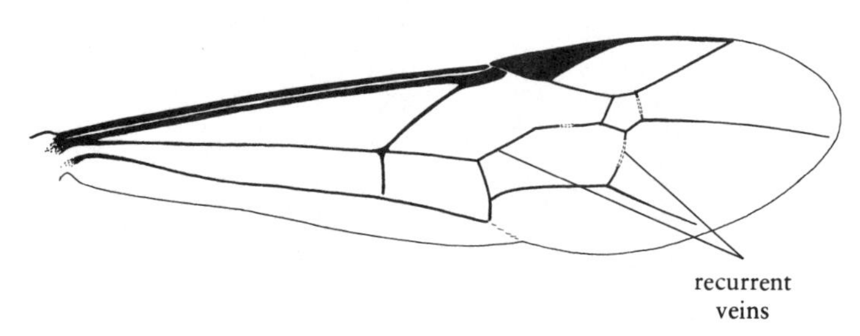

Figure 20.2. Fore wing

ICHNEUMONIDS
(Family Ichneumonidae)

ADULT DIAGNOSIS: (Fig. 20.2) These are usually small wasps whose fore wings have relatively well-developed venation and two recurrent veins. Antennae possess more than 15 segments.

DISCUSSION: Members of this large family of parasitic wasps are sometimes confused with those of the Braconidae. Ichneumonids have two recurrent veins (there appears to be a crossvein in the distal posterior part of the wing), whereas braconids have only one recurrent vein (there appears to be no crossvein in the distal posterior part of the wing). These wasps often parasitize the larvae and pupae of aquatic and semiaquatic moths, and sometimes aquatic soldier flies. At least some species enter the water to find the host.

FAIRYFLIES
(Family Mymaridae)

ADULT DIAGNOSIS: (Fig. 20.3) Fore wings of these small wasps are narrow, have reduced venation, and have a marginal fringe of hairs. Tarsi are four-segmented.

DISCUSSION: About three species of fairyflies are known to parasitize eggs of aquatic insects. Females may swim with their wings and legs or walk

FAIRYFLY
(Mymaridae)

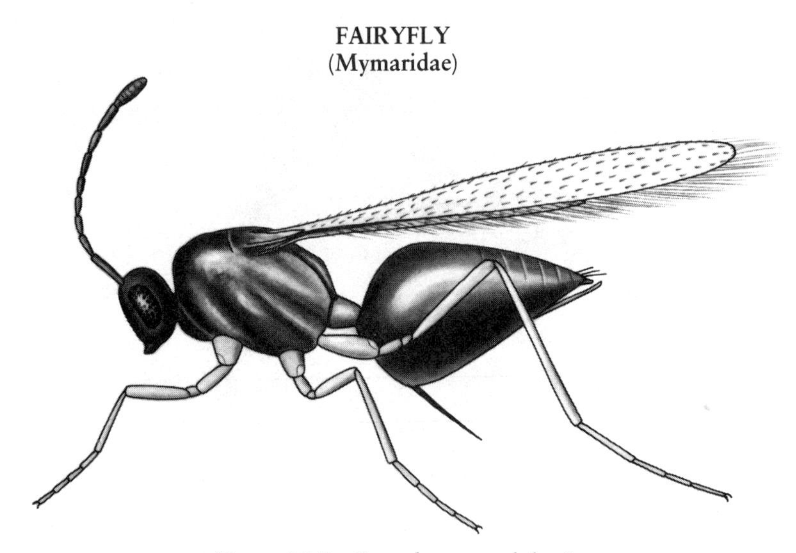

Figure 20.3. *Caraphractus* adult ♀

TRICHOGRAMMATID WASP
(Trichogrammatidae)

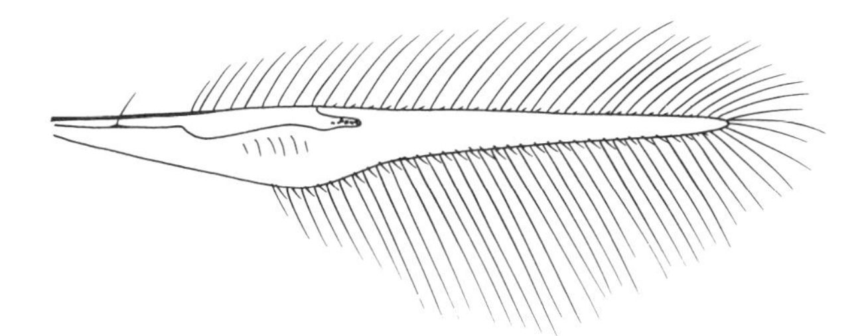

Figure 20.4. Fore wing

about in the water. Hosts include the eggs of backswimmers, water striders, predaceous diving beetles, and broadwinged damselflies.

TRICHOGRAMMATIDS
(Family Trichogrammatidae)

ADULT DIAGNOSIS: (Fig. 20.4) Fore wings of these small wasps are narrow, have reduced venation, and have a marginal fringe of long hairs. Tarsi are three-segmented.

DISCUSSION: About four species of trichogrammatids parasitize the eggs of damselflies, water scorpions, predaceous diving beetles, fishflies, and alderflies. It has yet to be established whether North American trichogrammatids actually enter the water to oviposit; they evidently are able to swim by using their wings.

DIAPRIIDS
(Family Diapriidae)

ADULT DIAGNOSIS: (Fig. 20.5) These small wasps have reduced wing venation and five-segmented tarsi. Antennae originate on a facial shelf.

DISCUSSION: A few species of this family parasitize pupae of shore flies, marsh flies, deer flies, and water pennies.

DIAPRIID WASP
(Diapriidae)

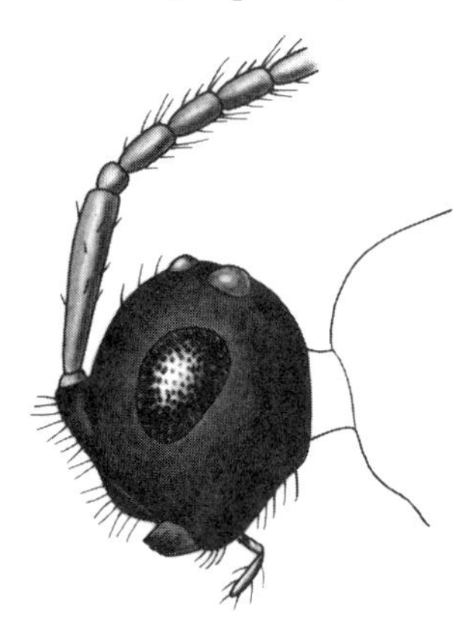

Figure 20.5. *Trichopria* adult, head

SCELIONIDS
(Family Scelionidae)

ADULT DIAGNOSIS: These are small wasps with reduced wing venation and five-segmented tarsi. Antennae do not originate from a facial shelf.

DISCUSSION: About three species of North American scelionids are known to parasitize the eggs of water striders, deer flies, and semiaquatic moths. Adults swim with both wings and legs.

SPIDER WASPS
(Family Pompilidae)

ADULT DIAGNOSIS: (Fig. 20.6) These are large, black wasps (about 15 mm or more) whose fore wings have relatively well-developed venation. Antennae have less than 15 segments and are curled somewhat.

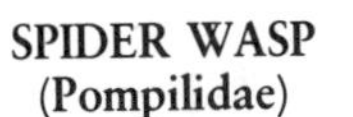

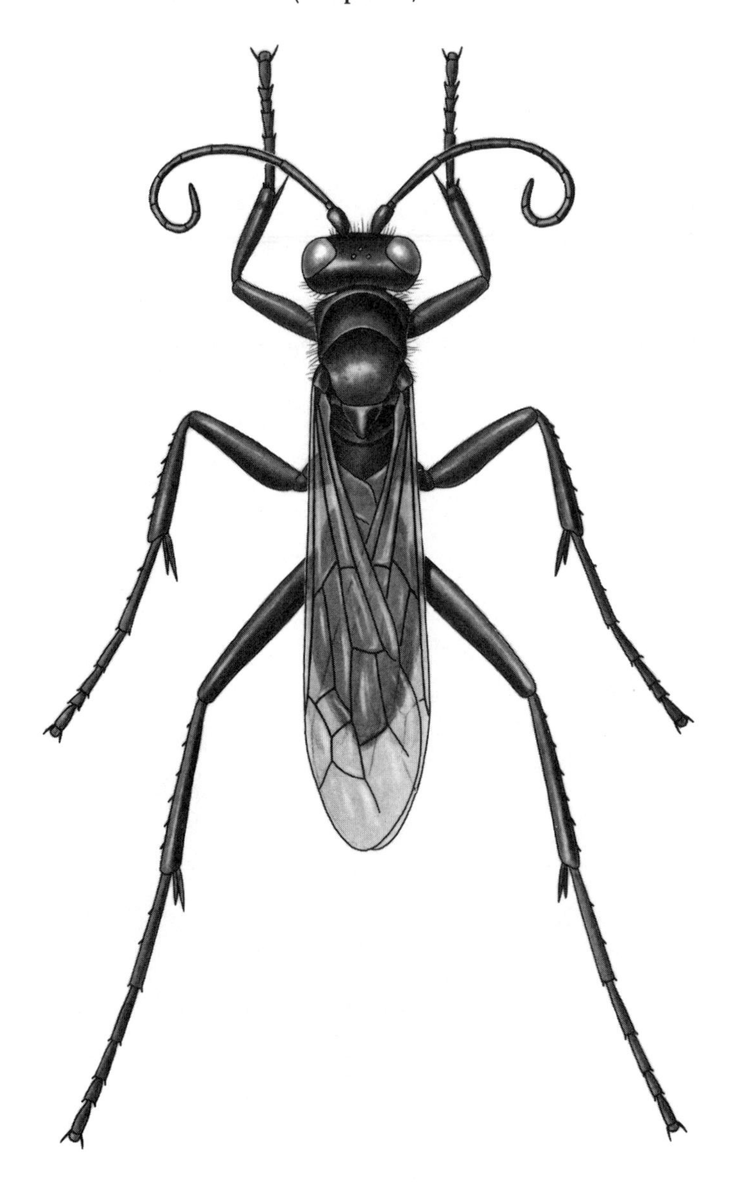

Figure 20.6. *Anoplius* adult

DISCUSSION: The spider wasp, *Anoplius depressipes,* attacks fishing spiders of the genus *Dolomedes* (see Chapter 23). They will enter the water if necessary to sting their prey and then remove it to a nest made in nearby banks. The captured prey is then fed upon by the developing wasp larvae.

References

Berg, C. O. 1949. Limnological relations of insects to plants of the genus *Potamogeton. Trans. Amer. Micr. Soc.* 68:279–291.

Evans, H. E., and C. M. Yoshimoto. 1962. The ecology and nesting behavior of the Pompilidae (Hymenoptera) of the northeastern United States. *Misc. Publ. Entomol. Soc. Amer.* 3:67–119.

Hagen, K. S. 1978. Aquatic Hymenoptera, pp. 233–241. In *An introduction to the aquatic insects of North America* (R. W. Merritt and K. W. Cummins, eds.). Kendall / Hunt, Dubuque.

Jackson, D. J. 1958. Observations on the biology of *Caraphractus cinctus* Walker (Hym: Mymaridae), a parasitoid of the eggs of Dytiscidae. I. Methods of rearing and numbers bred on different host eggs. *Trans. Entomol. Soc. Lond.* 110:533–554.

Jackson, D. J. 1961. Observations on the biology of *Caraphractus cinctus* Walker (Hym: Mymaridae), a parasitoid of the eggs of Dytiscidae (Coleoptera). 2. Immature stages and seasonal history with a review of mymarid larvae. *Parasitology* 51:269–294.

Jackson, D. J. 1966. Observations on the biology of *Caraphractus cinctus* Walker (Hym: Mymaridae), a parasitoid of the eggs of Dytiscidae. III. The adult life and sex ratio. *Trans. Entomol. Soc. Lond.* 118:23–49.

Lange, W. H. 1956. A generic revision of the aquatic moths of North America (Lepidoptera: Pyralidae, Nymphulinae). *Wasmann J. Biol.* 14:59–144.

McFadden, M. W. 1967. *Soldier fly larvae in America north of Mexico.* Proc. U.S. Nat. Mus. 121. 72 pp.

PART IV

OTHER FRESHWATER ARTHROPODS

CHAPTER 21

Freshwater Springtails

(ORDER COLLEMBOLA)

SPRINGTAILS CONSTITUTE a specialized group of small, six-legged arthropods (Hexapoda) that are related to the insects but fundamentally distinct from them. The common name of this group refers to their springlike jumping organ located on the underside of the abdomen. This unique structure allows springtails to jump several centimeters off the ground or water. Over 300 species of springtails occur in North America, and about 50 species have been reported from various freshwater habitats. Only about 10 to 15 species, however, are found regularly associated with the surface of freshwater (families Isotomidae, Poduridae, Smithuridae).

Freshwater springtails are most apt to be found on quiet edgewaters, where they may move about on the surface film or marginal vegetation (most are rarely if ever submerged). They are often overlooked because of their small size, and they are difficult to catch because they quickly jump when disturbed.

Little is known of the biology or ecology of the freshwater species of Collembola. They are found in both lentic and lotic habitats. Food probably consists primarily of microorganisms or small detritus. The eggs of some are laid beneath the water surface, and the hatched young of at least one species are submergent for a short period of time. Several generations can develop during a year. Some species swarm in large numbers on the water surface.

. . . and of the Water-flea: these last are remarkable for skipping up and down upon the water, as if they were at play. . . .

M. DeBuffon

Diagnosis

These organisms are usually less than 3 mm in length, rarely up to 6 mm. They are slender to robust and may or may not appear segmented. Head possesses one pair of antennae and somewhat inconspicuous mouthparts. Eyes may or may not be well developed. Thorax, which may or may not be distinct, possesses three pairs of legs and lacks wings or wing pads. Abdomen possesses an elongate spring (furcula) ventrally that is either tucked underneath the abdomen or extended posteriorly from the abdomen, at which time it may resemble a forked tail. Abdomen possesses other less conspicuous ventral appendages.

ONYCHIURIDS
(Family Onychiuridae)

DIAGNOSIS: Body is elongate and has clearly segmented thorax and abdomen. Eyes are absent. Spring is absent.

DISCUSSION: About five species of this family have occasionally been taken in association with freshwater. They are terrestrial, however, and probably occur in freshwater situations accidentally.

HYPOGASTRURIDS
(Family Hypogastruridae)

DIAGNOSIS: Body is elongate and has clearly segmented thorax and abdomen. Eyes are present. Mouthparts are directed forward on head. First thoracic segment is distinct dorsally and possesses dorsal hairs.

DISCUSSION: Only one species has been taken in association with freshwater, and it was probably accidental. Some hypogastrurids, however, are known from intertidal marine environments.

ISOTOMIDS
(Family Isotomidae)

DIAGNOSIS: (Fig. 21.1) Body is elongate and has clearly segmented thorax and abdomen; body is not covered with scales. Eyes are present or absent. First thoracic segment is usually reduced and devoid of dorsal hairs. Abdominal segment 4 is not longer than segment 3 as seen in dorsal view.

DISCUSSION: About 12 species of isotomids in North America have been found in association with freshwater habitats. Of these, seven are regular inhabitants that exhibit adaptations for aquatic existence. Some are highly tolerant of cold temperatures. Two of the species are known only from stream environments.

ISOTOMID SPRINGTAIL
(Isotomidae)

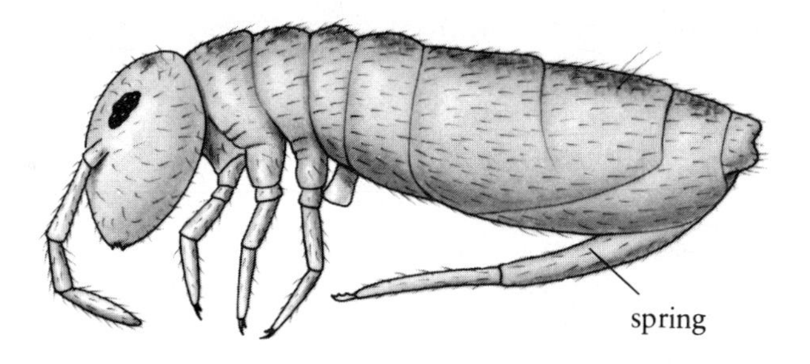

Figure 21.1. *Isotomurus* adult

TOMOCERIDS
(Family Tomoceridae)

DIAGNOSIS: These are similar to the isotomids but possess scales on the body.

DISCUSSION: Two species in North America are transient in freshwater environments and are not considered to be regular freshwater inhabitants.

ENTOMOBRYIDS
(Family Entomobryidae)

DIAGNOSIS: These are similar to isotomids, but abdominal segment 4 is longer than segment 3 as seen in dorsal view.

DISCUSSION: About 11 species of entomobryids in North America occasionally occur in association with freshwater. One additional species occurs in intertidal environments. These do not constitute a primary freshwater group.

PODURIDS
(Family Poduridae)

DIAGNOSIS: (Figs. 21.2, 21.3) Body is somewhat elongate and appears segmented at least dorsally. Mouthparts are directed downwards. First thoracic segment is distinct dorsally. Spring is elongate, and its forks are distinctly convergent at their tips.

DISCUSSION: The widespread species *Podura aquatica* is perhaps the best known of the freshwater Collembola. It is found along edgewaters of ponds, lakes, swamps, streams, and rivers. It is a primary surface-dwelling species (where it is sometimes present in large numbers) but also inhabits emergent vegetation and exposed algal mats.

Eggs develop in the water, and the newly hatched young briefly reside there. Once the young are exposed to the surface, they never again become submerged.

PODURID SPRINGTAILS
(Poduridae)

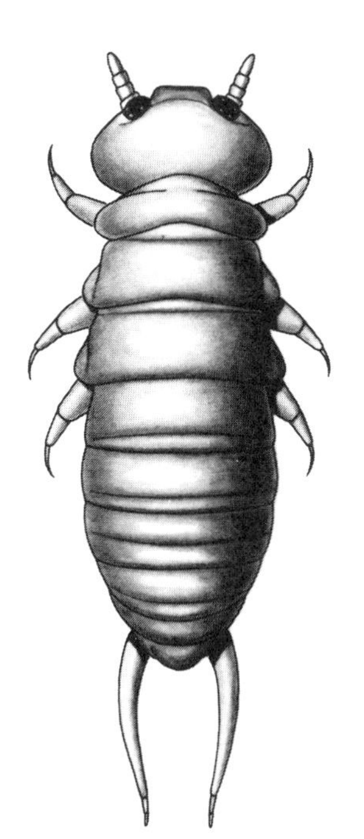

Figure 21.2.
Podura adult (dorsal)

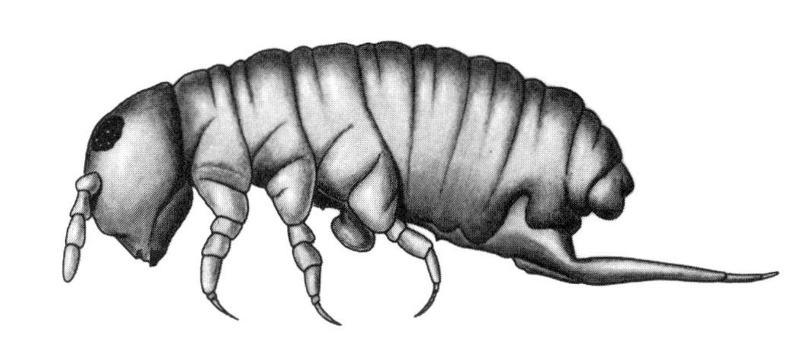

Figure 21.3.
Podura adult (lateral)

SMINTHURID SPRINGTAILS
(Sminthuridae)

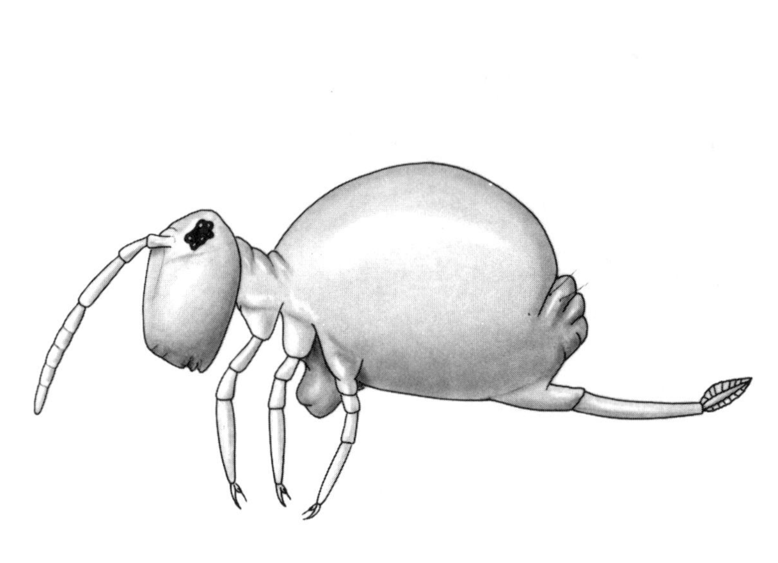

Figure 21.4.
Sminthurides adult (lateral)

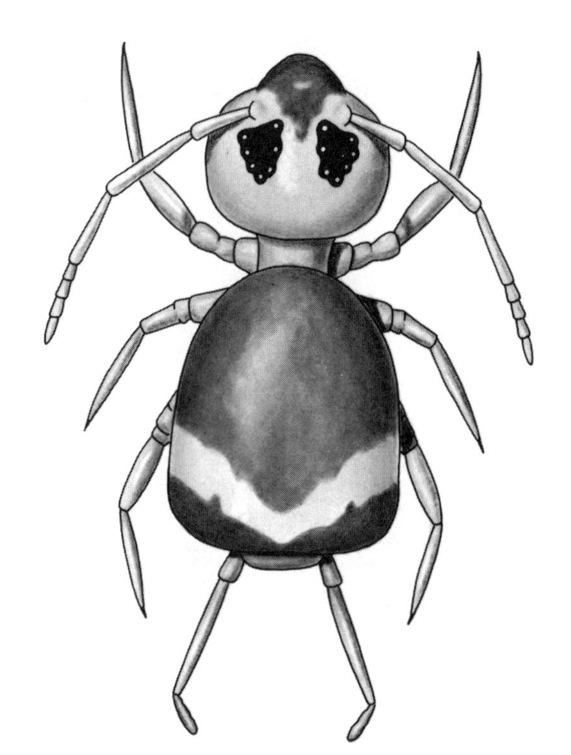

Figure 21.5.
Sminthurides adult (dorsal)

SMINTHURIDS
(Family Sminthuridae)

DIAGNOSIS: (Figs. 21.4, 21.5) Body is robust and lacks apparent segmentation.

DISCUSSION: This distinctive family contains about 12 species in North America that are regularly associated with freshwater environments. *Sminthurides aquaticus* is one of the better known species. It occurs on the surface of quiet waters throughout North America. Males are sometimes carried about on the water by females (being held by antennae). Males can live for four or five weeks, whereas females can live up to four months. Overwintering usually occurs in the egg stage. Food consists of soft decaying or living vegetation, fungal spores, and unicellular algae.

References

Chang, S. L. 1966. Some physiological observations on two aquatic Collembola. *Trans. Amer. Micr. Soc.* 85:359–371.

Joosse, E. N. G. 1976. Littoral apterygotes (Collembola and Thysanura), pp. 151–186. In *Marine insects* (L. Cheng, ed.). North-Holland, Amsterdam.

Mills, H. B. 1934. *A monograph of the Collembola of Iowa.* Collegiate Press, Ames. 143 pp.

Rapaport, E. H., and L. Sanchez. 1963. On the epineuston or the super aquatic fauna. *Oikos* 14:96–109.

Waltz, R. D., and W. P. McCafferty. 1979. *Freshwater springtails (Hexapoda: Collembola) of North America.* Purdue Univ. Exp. Sta. Res. Bull. 960. 32 pp.

CHAPTER 22

Common Freshwater Crustaceans

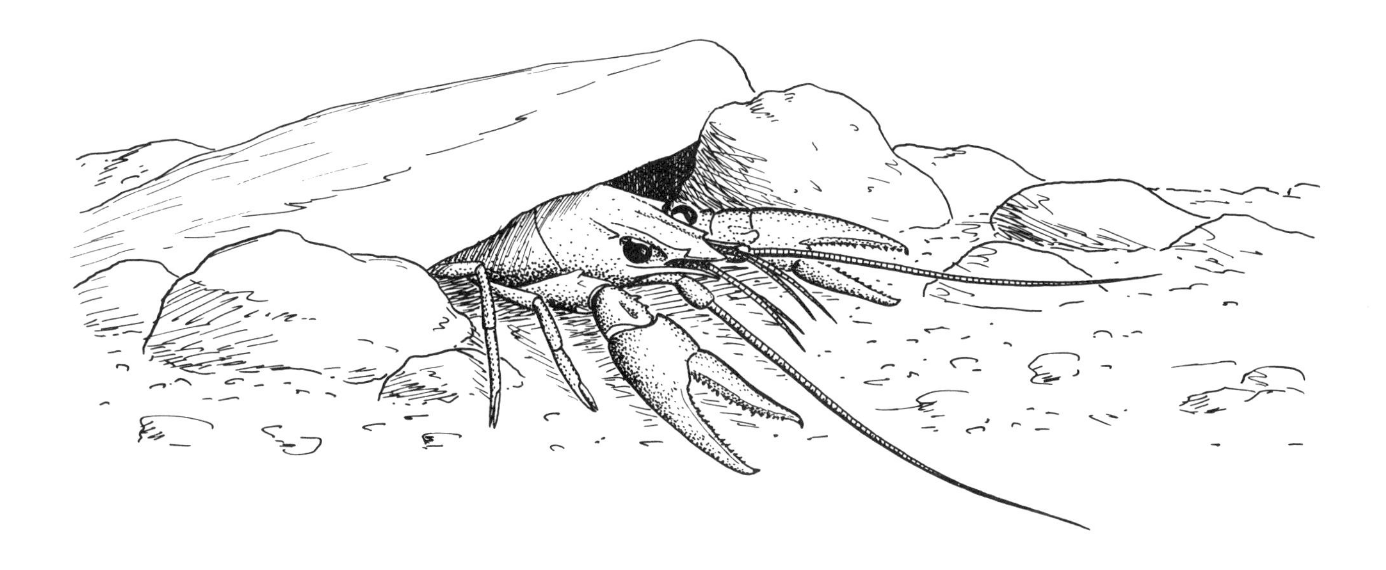

OVER 1,100 FRESHWATER crustacean species are known from North America. They occur in virtually all kinds of freshwater environments, where most are free-living but some are parasitic. Some freshwater crustaceans are the only arthropods to be found in many small temporary pools or ponds and in some subterranean aquatic habitats.

Although not strictly a subject of aquatic entomology, many of the larger and common benthic crustaceans, such as scuds, sowbugs, and crayfishes, are often taken when studying aquatic insects and should not be ignored. Many freshwater crustaceans, including the small cladocerans and copepods, many of which are planktonic, are extremely important in the food chain of aquatic ecosystems.

Diagnosis

These are more-or-less cylindrical or dorsoventrally flattened or laterally flattened forms that range in size from less than 1 mm to as much as 150 mm. Body segmentation is usually apparent. Body may or may not be divided into recognizable regions. Body regions consist either of head and trunk or of head, thorax, and abdomen. Head is often fused with the thorax or some anterior thoracic or trunk segments to form a so-called **cephalothorax.** Part or all of the body is sometimes covered with a shieldlike or valvelike carapace. Head bears two pairs of antennae. Three to 71 pairs of legs are present. Abdomen or trunk usually ends in paired appendages of various types.

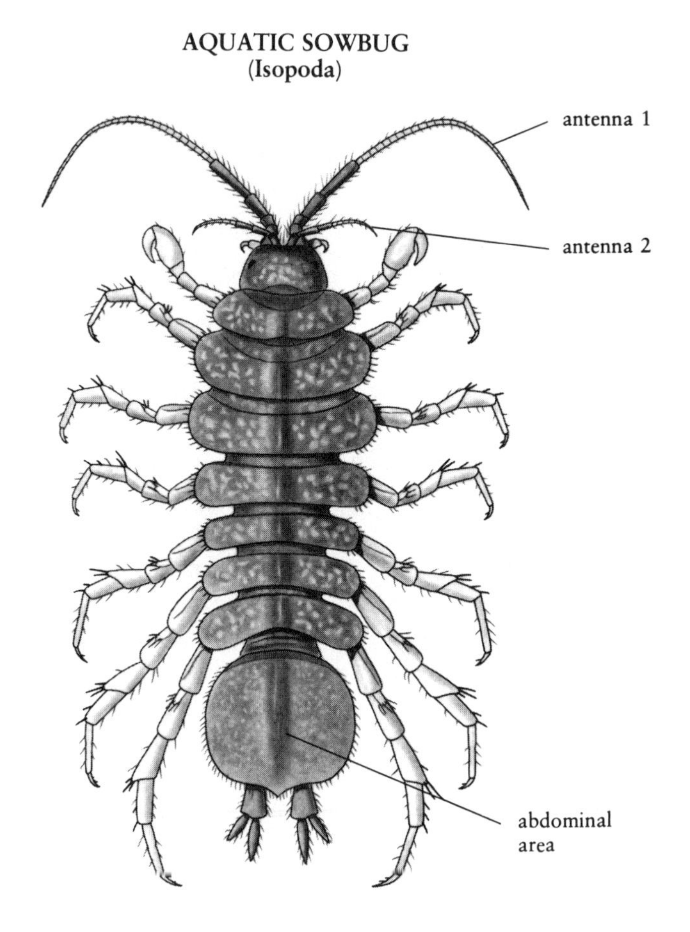

Figure 22.1. *Asellus* adult (dorsal)

AQUATIC SOWBUGS
(Isopoda)

DIAGNOSIS: (Fig. 22.1) These are strongly dorsoventrally flattened forms that usually measure from 5 to 20 mm when mature. Head and first thoracic segment form a cephalothorax. Remainder of thorax possesses seven pairs of well-developed legs, the first pair being modified for grasping. Abdominal segments are fused into a relatively short region.

DISCUSSION: Aquatic sowbugs constitute a significant group of freshwater arthropods occurring generally throughout North America. Of the four families, the Asellidae is by far the most important, being the most widely distributed and containing the vast majority of the roughly 100 species. The family Sphaeromidae contains only a few species known from still fresh and brackish waters of southwestern and southeastern states and Texas and Florida. The family Cirolanidae contains two blind species from caves in Texas and Virginia, and the family Bopyridae contains one highly modified species that is parasitic on river shrimps and freshwater prawns.

Aquatic sowbugs are generally sprawling benthic organisms that inhabit a wide variety of shallow freshwater environments, often among rocks and leaf detritus. Many species are known only from restricted spring habitats. *Asellus* sometimes occurs in large numbers in certain areas of streams. Common species are omnivore-detritivores and utilize a large variety of food; they are rarely predaceous.

At least one species of aquatic sowbugs, *Asellus communis* (the American Sowbug), is used as a fly pattern for nymph fishing in slow-flowing waters with considerable aquatic plant growth.

SCUDS
(Amphipoda)

DIAGNOSIS: (Fig. 22.2) These are laterally flattened, often colorful forms that usually measure 5–20 mm when mature. Head and first thoracic segment form a cephalothorax. The remainder of the thorax possesses seven pairs of legs, the first two pairs being modified for grasping.

DISCUSSION: Three families and approximately 90 species of scuds occur in North America. The family Talitridae contains one widely distributed North American species, *Hyalella azteca,* which is common in springs, streams, lakes, and ponds. The family Haustoriidae also contains only one species in North America, *Pontoporeia hoyi.* Somewhat atypical of scuds, this species is confined to the bottom and open waters of deep, cold lakes. The family Gammaridae is the most important group and is divided into about eight genera.

Scuds occur primarily in shallow waters of all kinds. They are benthic and often rest among vegetation and debris or occasionally slightly within soft substrate. They also swim, however, and are sometimes known as "sideswimmers." They are generally omnivore-detritivores but rarely predaceous. Several species are restricted to particular spring or cave habitats, whereas others are more widespread in larger surface-water habitats and sometimes occur in very large numbers.

Scuds are an important food source for many fishes, and several specific fly patterns for nymph fishing are based on scuds. These include the Yellow Freshwater Scud *(Gammarus minus),* Olive Freshwater Scud *(Gammarus lacustris),* Gray Freshwater Scud *(Gammarus fasciatus),* Tan Freshwater Scud *(Crangonyx gracilus),* and Tiny Olive Scud *(Hyalella azteca).* Other names, such as the Tiny Yellow, Tiny Gray, and Aztecan Scud, also refer to this latter species. In addition, other wet fly patterns, such as the Werner Shrimp, Otter Shrimp, Nyerges Nymph, and Henry's Lake Nymph, are essentially imitations of scuds and fairy shrimps.

SCUD
(Amphipoda)

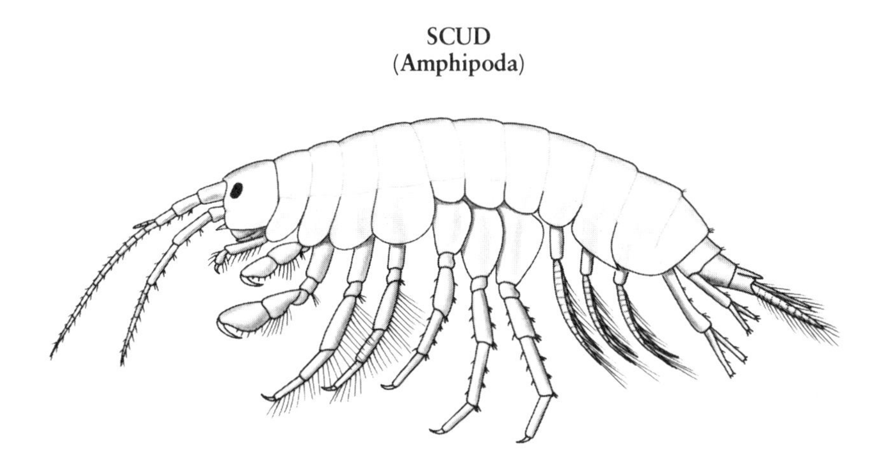

Figure 22.2. *Gammarus* adult (lateral)

CRAYFISH AND SHRIMP
(Decapoda)

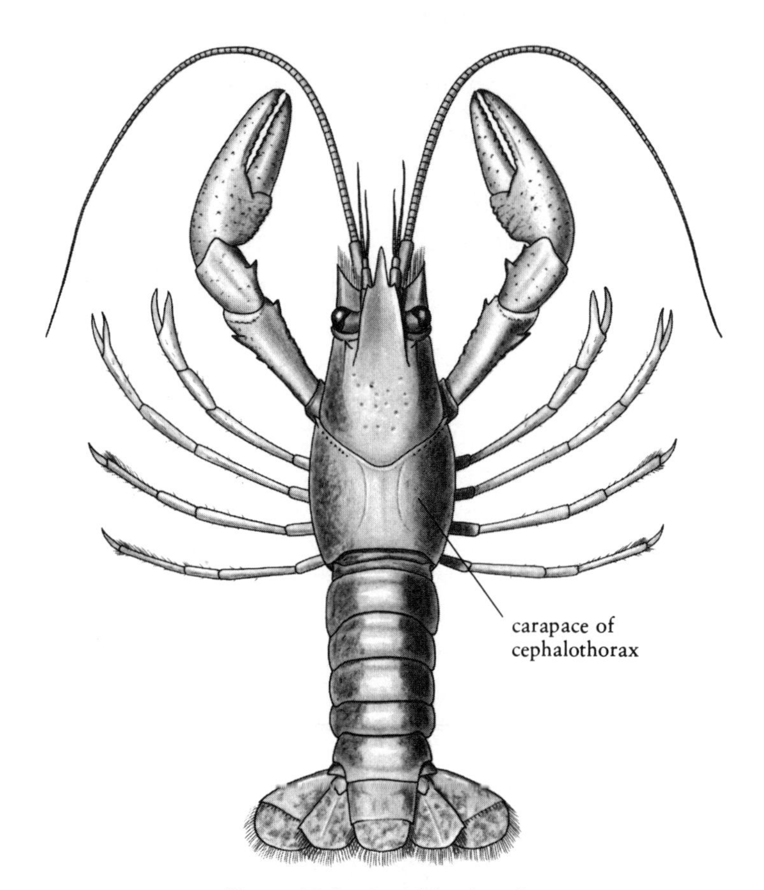

Figure 22.3. Crayfish (dorsal)

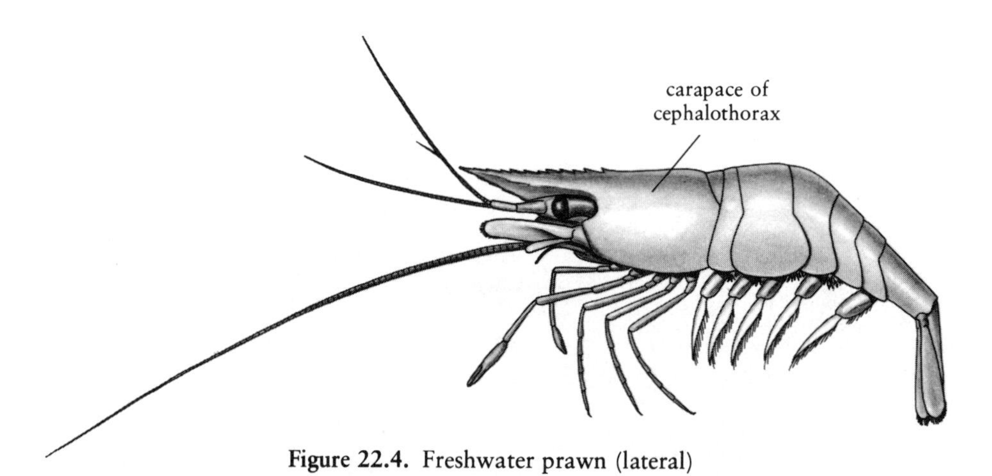

Figure 22.4. Freshwater prawn (lateral)

CRAYFISHES AND SHRIMPS
(Decapoda)

DIAGNOSIS: (Figs. 22.3, 22.4) These are somewhat flattened either dorsoventrally or laterally and range in size from 10 to 150 mm. Head and entire thorax form a large cephalothorax covered by a carapace. Cephalothorax possesses five pairs of legs; first two or three pairs are pincerlike at their ends; and first pair is often very robust.

DISCUSSION: The freshwater Decapoda in North America comprise four species of the family Atyidae, which are restricted to certain caves of the southeastern states and coastal streams of California; about 12 species of the family Palaemonidae, which includes the river shrimps, freshwater prawns, and glass shrimps of certain rivers, streams, and coastal waters; and about 280 species and subspecies of the family Astacidae, or crayfishes (also known as "crawfishes" and "crawdads"). Crayfishes are widely distributed, except that they are not generally found in the Rocky Mountain region.

Crayfishes occur in a wide variety of shallow freshwater habitats, and some live in swamps and wetlands. They are benthic and, at least in daylight hours, usually remain hidden in burrows or under stones and debris. They retreat rapidly backwards when disturbed. Depending on the species, crayfishes may be herbivores, carnivores, detritivores, or omnivores; their very robust first pair of legs (chelae) is used to cut or crush food. These chelae are also used as defensive weapons. Prawns and river shrimps are generally swimmers.

Crayfishes have long been used as fish bait, and edible species are often harvested and in some areas are cultivated for human consumption. The Freshwater Shrimp fly pattern is based on the freshwater prawn *Palaemonetes paludosus*.

OPOSSUM SHRIMPS
(Mysidacea)

DIAGNOSIS: These are 8–30 mm in length. They generally resemble crayfishes, but their five pairs of legs are filamentous rather than robust, and their carapace does not completely cover the thorax.

DISCUSSION: There are three species of opossum shrimps in North America. The first is a plankter of cold, deep lakes of Canada and central and northeastern states and is an important food of lake trout. The second inhabits some lakes and rivers of western coastal states. The third is known only from roadside ditches of Louisiana.

FAIRY SHRIMPS
(Anostraca)

DIAGNOSIS: (Fig. 22.5) These are cylindrical forms that range in size from 7 to 100 mm. Body is divided into head and trunk and does not possess a carapace. Eyes are stalked. Trunk possesses 11 pairs of leaflike swimming legs.

DISCUSSION: About 30 species of fairy shrimps occur in North America. They generally inhabit temporary pools or ponds, except for the Brine Shrimp, *Artemia salina,* which occurs in salt lakes and evaporation basins. Most species swim upside down.

FAIRY SHRIMP
(Anostraca)

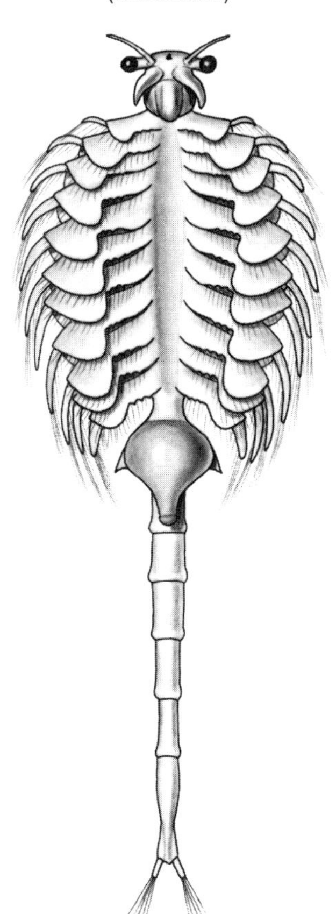

Figure 22.5. *Artemia* adult ♀ (ventral)

TADPOLE SHRIMPS
(Notostraca)

DIAGNOSIS: (Fig. 22.6) These are somewhat dorsoventrally flattened forms that have a large shieldlike carapace covering the head and much of the trunk. They measure 10–58 mm. Trunk possesses 35–71 pairs of leaflike legs and ends in two tails.

DISCUSSION: Six species of tadpole shrimps are known from North America, where they are restricted to pools and ponds that are generally devoid of predators. They swim upright and crawl about the bottom.

TADPOLE SHRIMP
(Notostraca)

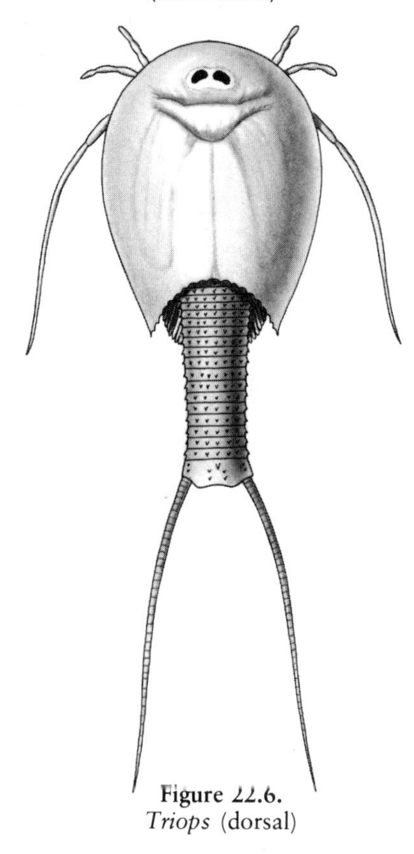

Figure 22.6. *Triops* (dorsal)

CLAM SHRIMPS
(Conchostraca)

DIAGNOSIS: (Fig. 22.7) These are small, laterally flattened forms that measure 2–16 mm. Body is covered by a carapace consisting of two lateral valves reminiscent of those of clams. Valves of most possess concentric growth lines. Trunk possesses 10–32 pairs of legs.

DISCUSSION: About 28 species of clam shrimps are found in North America, and like the fairy shrimps and tadpole shrimps, they occur in temporary pools and ponds, where they swim or crawl about the bottom.

CLAM SHRIMP
(Conchostraca)

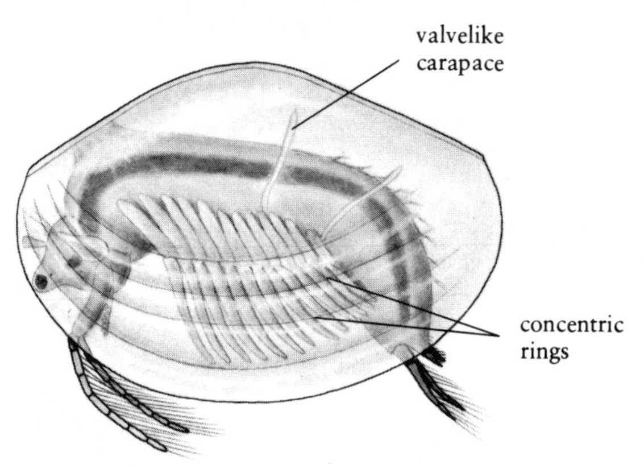

Figure 22.7. *Eulimnadia* (lateral)

WATER FLEA
(Cladocera)

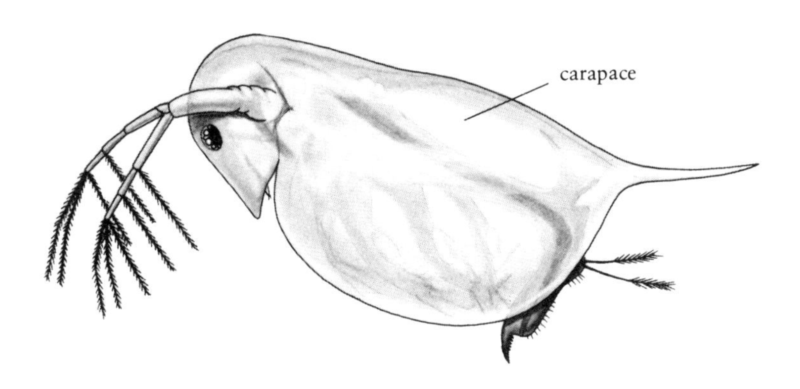

Figure 22.8. *Daphnia* (lateral)

WATER FLEAS
(Cladocera)

DIAGNOSIS: (Fig. 22.8) These are small, laterally flattened forms that usually measure 0.2–3 mm. Body is covered by a carapace, but head and antennae are usually apparent. Body does not appear segmented and possesses five or six pairs of legs. Carapace often ends in a spine.

DISCUSSION: Some 135 species of freshwater water fleas are known from North America, where the group is widespread and can be found in most freshwater environments. Most species occur in open waters, where they swim intermittently. The second pair of antennae is used primarily to propel them. Movement is generally vertical, with the head directed upwards. Many of these open-water forms are also known for their vertical migration, which generally consists of upward movement in the dark and downward migration during daylight hours. Some water fleas are primarily benthic.

Daphnia is commonly maintained in laboratories for assaying toxic substances in water. Water fleas are often of great importance in the diets of fishes, especially young fishes, and predaceous insects, such as many of the Diptera larvae.

SEED SHRIMPS
(Ostracoda)

DIAGNOSIS: (Fig. 22.9) These are small, laterally flattened forms that measure up to 3 mm in length but are usually less than 1 mm. Body segmentation is not apparent (head is not distinct from trunk), and entire body is covered with a bivalve carapace. Carapace does not possess concentric growth lines. Three pairs of highly modified legs are present.

DISCUSSION: About 200 species of freshwater seed shrimps occur in North America, where the group is widespread. They inhabit temporary and permanent running or still waters. Most species are benthic and occur where there is little if any current, often in association with silt bottoms and rooted vegetation. A few are found only on the gills of crayfishes. The first pair of antennae are often modified for digging, climbing, or swimming.

SEED SHRIMP
(Ostracoda)

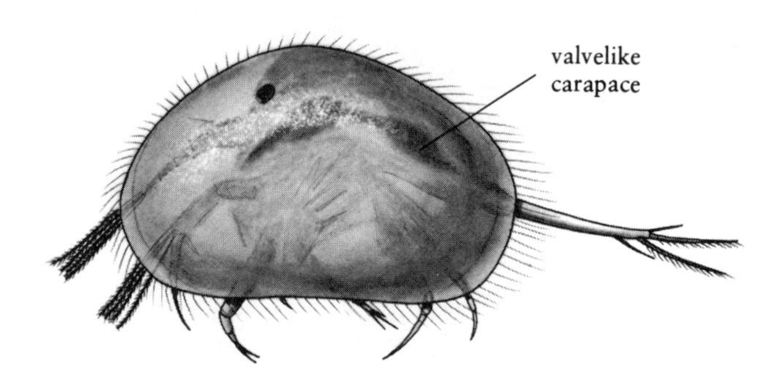

Figure 22.9. Adult (lateral)

COPEPODS
(Copepoda)

DIAGNOSIS: (Fig. 22.10) These are small, more-or-less cylindrical forms that are generally less than 3 mm in length. Body is divided into a cephalothorax, thorax, and abdomen. Cephalothorax is covered by a carapace. Six pairs of legs are usually present, the first of which is modified for feeding and the remaining five pairs for swimming. Body lacks lateral abdominal appendages.

DISCUSSION: About 180 species of copepods occur in North American freshwaters. Two groups of copepods (the Caligoida and Lernaeopodoida) are parasitic on fishes and are highly modified for this type of existence. These groups are sometimes referred to as "fish lice," but this term is often restricted to branchiurans. The vast majority of copepods are free-living and

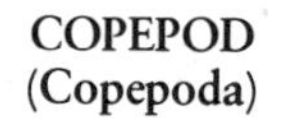

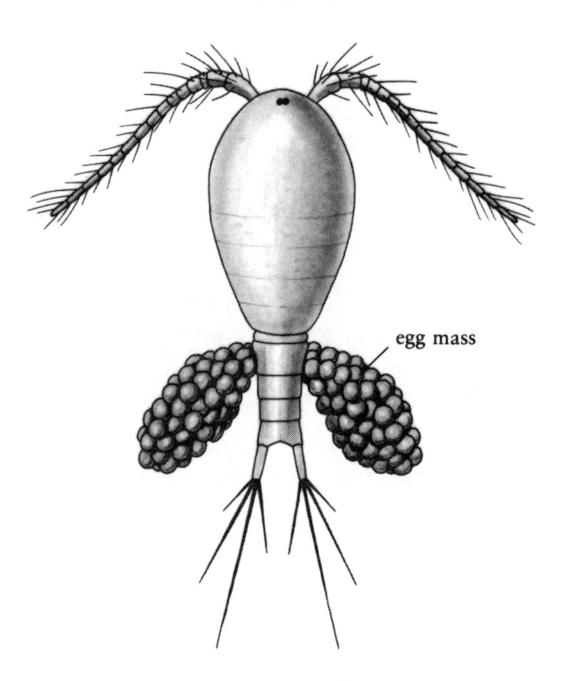

Figure 22.10. *Cyclops* adult ♀ (dorsal)

belong to the Calanoida, Harpacticoida, and Cyclopoida. One genus of Cyclopoida, however, is parasitic on fishes.

Free-living copepods are planktonic or benthic in a wide variety of freshwater environments. Some species of cyclopoid and calanoid copepods occur in extremely high densities. Some of the planktonic copepods have a daily vertical migration in lakes, similar to that of some water fleas. Parasitic copepods can become a serious economic problem in fish hatcheries. Many free-living copepods are important in the food chain of many fishes.

FISH LICE
(Branchiura)

DIAGNOSIS: These are broad, dorsoventrally flattened forms that are 5–25 mm in length. Body consists of a cephalothorax, thorax, and abdomen. Cephalothorax and thorax are covered by a carapace. Four pairs of swimming legs are present.

DISCUSSION: About 23 species of the genus *Argulus* occur in North American freshwaters, where as adults they parasitize fishes. They become attached to the branchial chamber or general body surface of the host and feed on the host's blood. The group is sometimes classified with the Copepoda.

References

Barnes, R. D. 1968. *Invertebrate zoology* (2nd ed.). W. B. Saunders, Philadelphia. 743 pp.

Cressey, R. F. 1972. *The genus Argulus (Crustacea: Branchiura) of the United States.* Biota of Freshwater Ecosystems, U.S.E.P.A. Ident. Man. No. 2. Wash., D.C. 14 pp.

Gerlach, R. 1974. *Creative fly tying and fly fishing.* Winchester Press, New York. 231 pp.

Hobbs, H. H. 1972. *Crayfishes (Astacidae) of North and middle America.* Wat. Poll. Cont. Res. Ser., Cincinnati. 173 pp.

Holsinger, J. R. 1972. *The freshwater amphipod crustaceans (Gammaridae) of North America.* Biota of Freshwater Ecosystems, U.S.E.P.A. Ident. Man. No. 5. Wash., D.C. 89 pp.

Lorman, J. G., and J. J. Magnuson. 1978. The role of crayfishes in aquatic ecosystems. *Fisheries* 3:8–19.

Meglitsch, P. A. 1972. *Invertebrate zoology* (2nd ed.). Oxford Univ. Press, New York. 834 pp.

Pennak, R. W. 1978. *Fresh-water invertebrates of the United States* (2nd ed.). John Wiley & Sons, New York. 803 pp.

Schwiebert, E. 1973. *Nymphs.* Winchester Press, New York. 339 pp.

Williams, W. D. 1972. *Freshwater isopods (Asellidae) of North America.* Biota of Freshwater Ecosystems, U.S.E.P.A. Ident. Man. No. 7. Wash., D.C. 45 pp.

CHAPTER 23

Common Freshwater Arachnids

THE ARACHNIDA is a class of arthropods that includes both terrestrial and aquatic forms. Freshwater arachnids in North America consist of mites (primarily water mites, or the Hydracarina, of which there are nearly 600 species). Spiders are also members of the Arachnida, and although there are no truly aquatic spiders in North America, those that occur commonly in marginal freshwater habitats are treated in this chapter.

The small, globular-shaped, and often brightly colored water mites are commonly encountered by aquatic entomologists. They occupy a variety of freshwater habitats, where they may be found swimming, crawling along or in the substrate, or attached to the bodies of aquatic insects. Much remains to be learned of their habits, relationships, and roles in aquatic ecosystems.

Diagnosis

Freshwater arachnids are generally oval to somewhat elongate forms (rarely flattened) and are usually less than 4 mm in length, although the spiders are considerably longer. Head is fused either with the remainder of the body to form a single body region or with the thorax to form a cephalothorax. Antennae are absent. Body is not segmented. Four pairs of legs are present except among larval mites, which have only three pairs.

WATER MITES
(Hydracarina)

DIAGNOSIS: (Figs. 23.1–23.4) These are usually globular, oval forms that measure 0.4–3 mm in length. They are often strikingly colored with red, green, yellow, or blue, or combinations of these colors. Body is undivided and may be soft or covered with a variable number of thick leathery plates. Modified head (**capitulum**) of the adult possesses five-segmented palps that are short and seldom extend much beyond the anterior end of the body. Larvae possess only three pairs of legs.

DISCUSSION: Water mites occur in a variety of freshwater habitats throughout North America and can be found in all seasons of the year. Swimming forms, which are most common in ponds and littoral regions of lakes, are usually equipped with swimming hairs on their legs. Benthic forms, which include many of the running-water species as well as others, are usually equipped with spines and bristles on their legs. Many species are conspicuous because of their bright coloration. Aquatic respiration is primarily cutaneous.

Although there are some exceptions, the general life history of water mites is as follows: Newly hatched, six-legged larvae, after a brief active period in the water, seek out and parasitize a host (only a few lack this parasitic phase). They attach to external membranous areas of the host's body with their capitulum and thereby derive bodily fluids. Hosts are usually aquatic or semiaquatic insects but not always (e.g., some species are parasitic on mussels, snails, or sponges).

Most larvae of the more primitive water mites, or the so-called red water mites (not all red water mites are red), which include the Hydrovolzioidea, Hydrachnoidea, Eylaoidea, and Hydryphantoidea, leave the water and seek out hosts in the terrestrial habitat and are thus termed **aerial.** Among the red water mites, the Hydrachnoidea and one genus of Hydryphantoidea are exceptions and do not leave the water to find a host. The larvae of higher water mites, which constitute a much larger group, including the Lebertioidea, Hygrobatoidea, and Arrenuroidea, seek out their hosts in the water and are well adapted for swimming. Although larvae of some groups commonly leave the water with their host (e.g., when the insect emerges or leaves the water to disperse), they are termed aquatic if they initially parasitize their host in the water.

Although it is not known whether many species have specific hosts, there are definitely some broad preferences (see Table 23.1). Water bugs, dragonflies, damselflies, stoneflies, and true flies are commonly parasitized. Caddisflies and water beetles are sometimes parasitized, but mayflies and some water beetles are only rarely parasitized. One or several mites may occur on a single host.

After considerable growth as a parasitic larva, the mite transforms to a nymph stage, which becomes free-living and similar to the adult except for sexual maturity. This transformation usually takes place on the host. Those mites that develop on the terrestrial stage of an aquatic insect must re-enter the water by falling off the host either directly into the water or onto adjacent areas so that they can subsequently crawl into the water. The nymph then undergoes a short quiescent period followed by transformation to the sexually mature and free-living adult stage.

TABLE 23.1
INSECT HOSTS OF WATER MITE LARVAE[1]

Water Mites	(Number of Mite Genera on Which Data Are Based)	Insect Hosts
Lower (Red) Water Mites		
Hydrovolzioidea	(1)	Hemiptera
Hydrachnoidea	(1)	Hemiptera
	(1)	Coleoptera
Eylaoidea	(1)	Odonata
	(2)	Hemiptera
	(2)	Coleoptera
Hydryphantoidea	(1)	Odonata
	(3)	Plecoptera
	(1)	Trichoptera
	(8)	Diptera
	(1)	Hymenoptera
Higher Water Mites		
Lebertioidea	(3)	Diptera
Hygrobatoidea	(1)	Ephemeroptera
	(1)	Plecoptera
	(1)	Coleoptera
	(3)	Trichoptera
	(11)	Diptera
Arrenuroidea	(1)	Odonata
	(1)	Coleoptera
	(1)	Diptera

[1]Adapted primarily from Prasad and Cook (1972).

The free-living stages of water mites are generally carnivorous, but several are omnivore-detritivores. Living and dead worms, aquatic insects, and small crustaceans are common food items for the free-living mites.

HYDROVOLZIOIDEA: **Adults** are red water mites that have a thick body wall. Genital acetabula are absent. They live mostly in streams and springs. **Larvae** seek their hosts aerially.

HYDRACHNOIDEA: **Adults** are red water mites that are nearly spherical and are covered with small papillae. First and third segments of chelate palps are uniquely elongate. They inhabit all types of lentic waters and slow reaches of streams and rivers. **Larvae** seek their hosts aquatically and are well adapted for swimming.

EYLAOIDEA: **Adults** (Figs. 23.1, 23.2) are red water mites that often have stalked genital acetabula. Eyes, when present, are located on a common eye plate. They occur in temporary and permanent standing water, swamps, and occasionally in streams. **Larvae** seek their hosts aerially.

HYDRYPHANTOIDEA: **Adults** are red water mites that typically have moveable genital flaps in both sexes. They are known from all types of freshwater habitats. **Larvae** seek their hosts aerially (a few aquatically).

LEBERTIOIDEA: **Adults** are variously colored water mites that have genital flaps and usually two medial, parallel rows of genital acetabula. They inhabit surface and interstitial waters of streams (some from waterfalls); a few inhabit standing waters. **Larvae** seek their hosts aquatically.

WATER MITES
(Hydracarina)

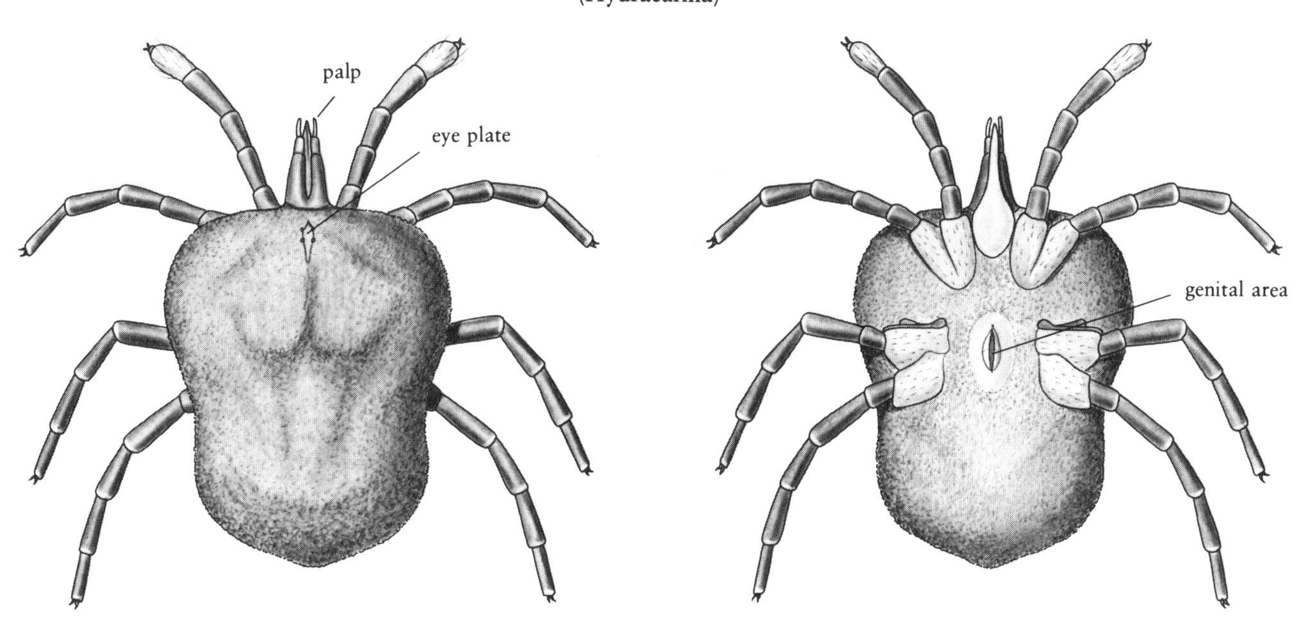

Figure 23.1. Eylaoidea adult (dorsal)

Figure 23.2. Eylaoidea adult (ventral)

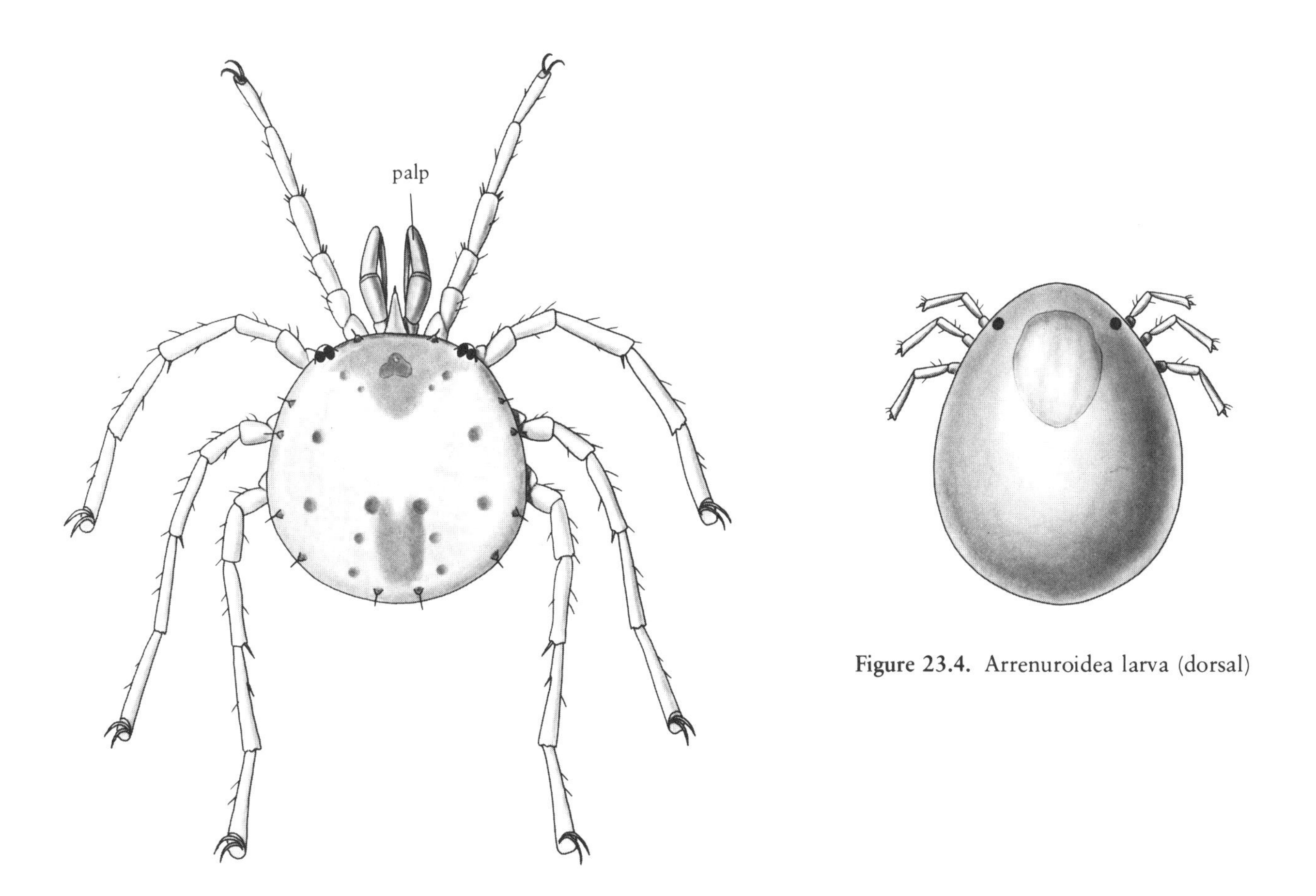

Figure 23.4. Arrenuroidea larva (dorsal)

Figure 23.3. Hygrobatoidea adult (dorsal)

HYGROBATOIDEA: **Adults** (Fig. 23.3) are variously colored water mites that lack genital flaps. They live in a variety of aquatic habitats but mainly in surface and interstitial waters of streams. **Larvae** seek their hosts aquatically.

ARRENUROIDEA: **Adults** are variously colored water mites that are generally heavily sclerotized. They are found in a wide variety of aquatic habitats, and many are known from interstitial waters. **Larvae** (Fig. 23.4) seek their hosts aquatically.

HALACARID WATER MITES
(Halacaridae)

DIAGNOSIS: (Fig. 23.5) These are tiny mites, generally less than 0.5 mm in length. They differ from water mites in that their palps are only three- or four-segmented.

DISCUSSION: Halacarid water mites constitute primarily a marine group, but several genera occur in freshwater environments. These forms are not well known, but they are all thought to be predaceous. They are generally found crawling on vegetation and substrates of streams, lakes, and ponds. They are also known from subterranean and interstitial habitats.

HALACARID WATER MITE
(Halacaridae)

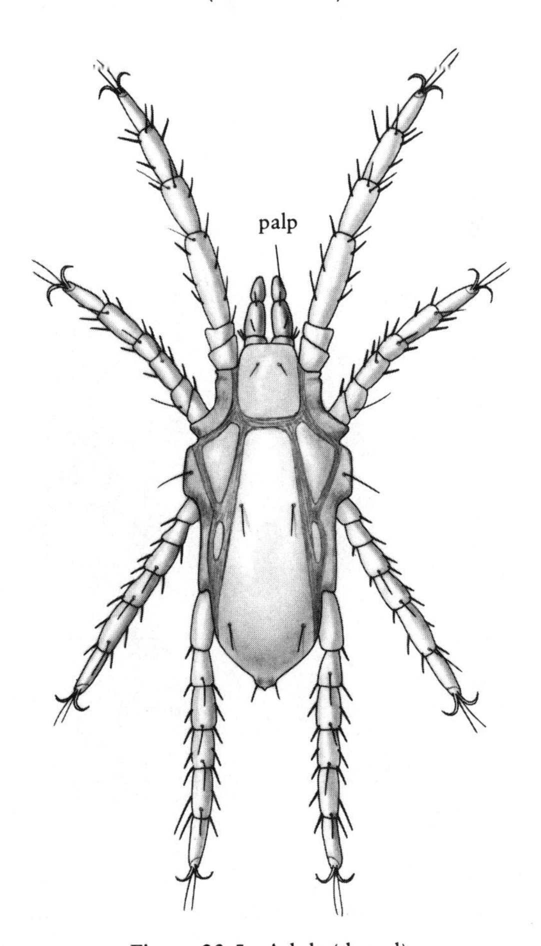

Figure 23.5. Adult (dorsal)

ORIBATID WATER MITE
(Oribatei)

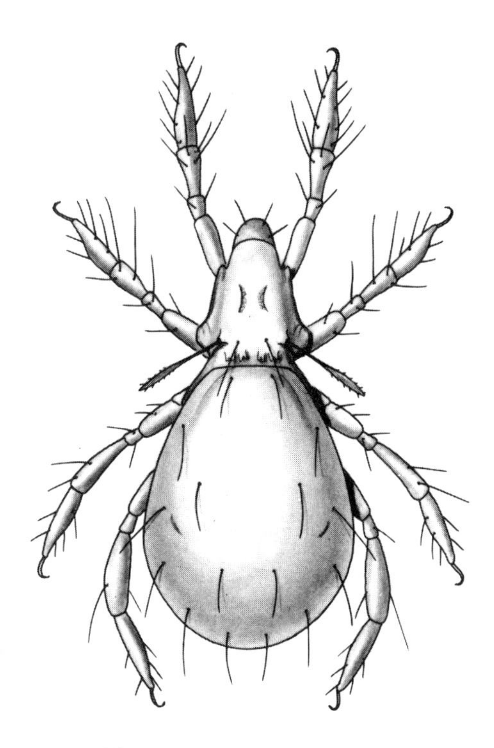

Figure 23.6. Adult (dorsal)

ORIBATID WATER MITES
(Oribatei)

DIAGNOSIS: (Fig. 23.6) These mites differ from the water mites and halacarid water mites by having the anterior part of the body broadly triangular.

DISCUSSION: This is a large group of soil mites in which a few species are adapted for living in aquatic or semiaquatic environments. Some of the more common aquatic species are associated with woody debris in ponds and are primarily detritivores. Some are found in cold springs and snow melt.

FISHING SPIDERS AND OTHERS
(Araneae)

DIAGNOSIS: (Figs. 23.7, 23.8) These spiders measure up to 50 mm or more in length. Body is divided into two primary regions: the cephalothorax and abdomen.

DISCUSSION: Spiders often occur along shores of lakes, ponds, rivers, and streams, among emergent vegetation or on the water surface. They are all predaceous and forage to a great extent on aquatic and semiaquatic organisms. Some are able to move about on the water surface with great agility. Spiders most commonly encountered in association with freshwater environments in North America include the fishing spiders in the genus *Dolomedes* of the family Pisauridae (Fig. 23.7), small wolf spiders in the genus *Pirata* of the family Lycosidae, and a few of the longjawed spiders of the family Tetragnathidae (Fig. 23.8).

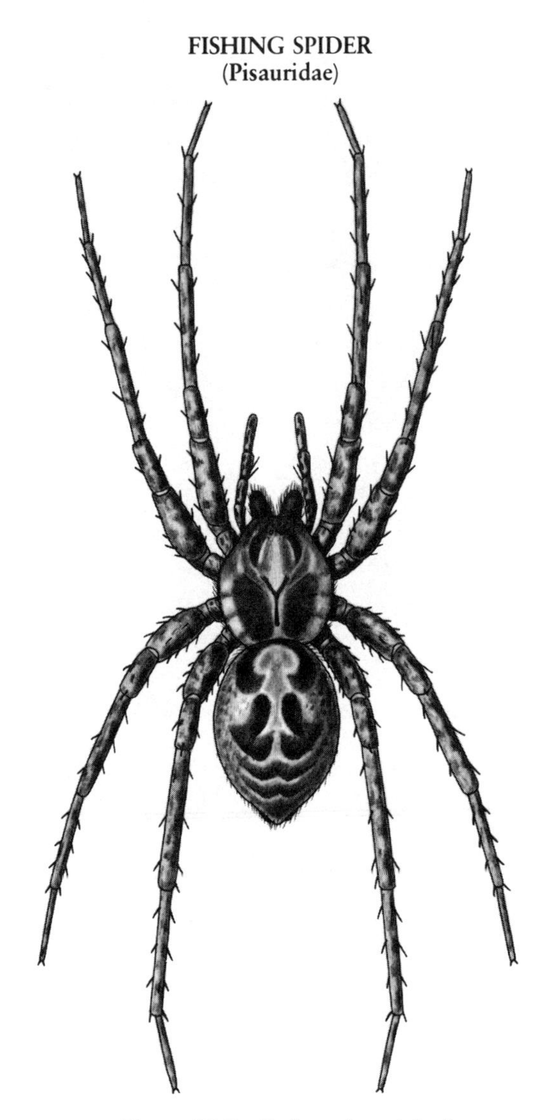

Figure 23.7. *Dolomedes* adult ♀

The longjawed spiders weave orb webs in which to snare prey. Those in the vicinity of water often weave horizontal orbs on overhanging vegetation or man-made structures and capture flying aquatic insect adults. A quick check of these webs by the fly fisherman will often indicate which aquatic insects are flying in the area.

Fishing spiders and semiaquatic wolf spiders do not weave webs to snare prey but are hunters that rely primarily on capturing prey that have fallen onto the water. In a sense, the water surface itself acts as web for these spiders. For instance, fishing spiders often rest on a floating or emergent object but keep their first pair of legs on the water surface. Ripples caused by would-be prey evidently provide the stimulus for the predatory response. Fishing spiders also will dive into the water either to escape predation or to prey upon submergent organisms. They can remain submerged for up to several minutes and apparently use a plastron during this period. Only small fishes have been observed to be preyed upon in this fashion, although other submergent animals may also fall victim to the spiders. *Dolomedes triton* has been known to cause considerable damage in fish hatcheries.

LONGJAWED SPIDER
(Tetragnathidae)

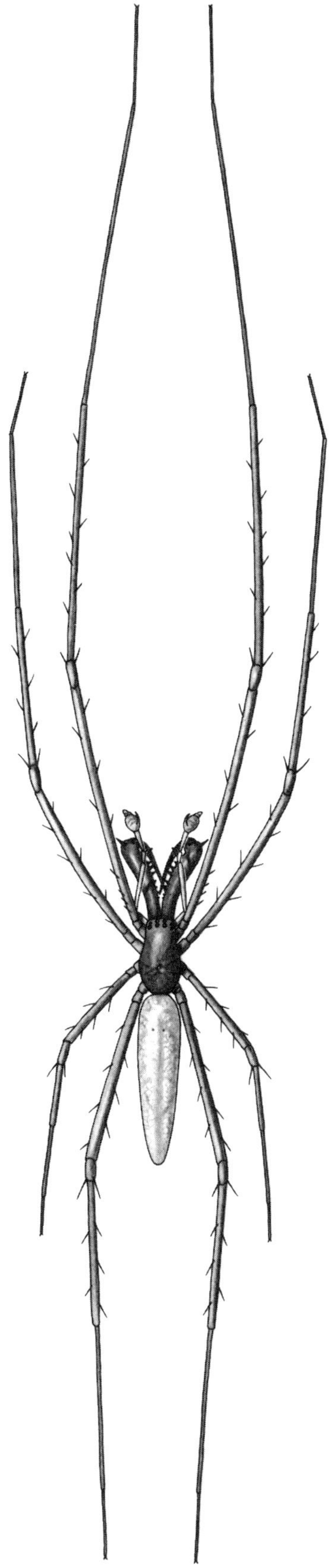

Figure 23.8. *Tetragnatha* adult ♂

References

Baker, E. W., and G. W. Wharton. 1952. *An introduction to acarology.* Macmillan, New York. 465 pp.

Böttger, K. 1970. Feeding of water mites. *Int. Rev. Ges. Hydrobiol.* 55:895–912.

Carico, J. E. 1973. The Nearctic species of the genus *Dolomedes* (Araneae: Pisauridae). *Bull. Mus. Comp. Zool.* 144:435–488.

Cook, D. R. 1974. *Water mite genera and subgenera.* Mem. Amer. Entomol. Inst. 21. 860 pp.

Meehean, O. L. 1934. Spiders that fish. *Natur. Hist.* 34:538–540.

Pennak, R. W. 1978. *Fresh-water invertebrates of the United States* (2nd ed.). John Wiley & Sons, New York. 803 pp.

Prasad, V., and D. R. Cook. 1972. *The taxonomy of water mite larvae.* Mem. Amer. Entomol. Inst. 18. 326 pp.

Wallace, H. K., and H. Exline. 1978. Spiders of the genus *Pirata* in North America, Central America and West Indies (Araneae: Lycosidae). *J. Arachnol.* 5:1–112.

APPENDIX
A Guide to Fishermen's Mayflies in North America

The fisherman has studied them in his own way. He has learned their names and seasons, and their favorite haunts. He has captured them when they hovered near enough, and has used them to bait his line. Finding them too delicate for rough handling, he has invented effigies of them for his hook, crudely fashioned out of feathers and wire and silk. He has given them his own names—soft, tripping, facile names, such as grow readily outside the Nomenclatural Garden.

James G. Needham

THE VAST AND UNRULY array of common names that have been applied to mayfly species in North America makes the intelligent application of such names almost impossible for the angler. Three basic problems are inevitable in such usage: (1) several common names sometimes are applied to a single species, (2) single common names are in some cases applied to several different species, and (3) increasing knowledge of the makeup of mayfly species in North America necessitates a continual revision of the scientific nomenclature, thus resulting in some changes of long-used scientific names. In addition, common names in North America often are different for subimagos and adults of the same species, and some names are applied differently depending on what part of North America is being fished.

To help alleviate the great confusion and inconsistent usage that the fly fisherman is confronted with, the following guide is given. Fishermen's common names for North American mayflies are listed alphabetically with separate lists for (1) eastern and central North America and (2) western North America. For each fishermen's name, the proper scientific name of the species to which it has been applied is given. If in the literature the species has appeared under different (no longer scientifically valid) names, these synonyms are given in parentheses. If the name applies only to the subimago, then the scientific name is followed by (Sub); if the name applies only to the adult stage, then the scientific name is followed by (Ad). If a specific sex is involved, it is also indicated. If (Sub) or (Ad) do not follow the name, then both stages apply. All fishermen's names can be used for larvae, but a few apply only to the larval stage; these names are followed by (L). When the abbreviation spp. follows a generic name, it indicates a reference to all species of the genus.

The family to which each species belongs is also indicated so that the pertinent section of Chapter 7 may be referred to if desired. Names that have been applied exclusively to tie patterns are not included in the list. Finally, fishermen's names that are not specific to at least the genus level are not included.

EASTERN AND CENTRAL NORTH AMERICA

FISHERMEN'S NAMES	SCIENTIFIC NAMES	FAMILY
American Brown	*Stenonema vicarium*	Heptageniidae
American Iron Blue Quill	*Paraleptophlebia mollis*	Leptophlebiidae
American March Brown	*Stenonema vicarium* (Sub)	Heptageniidae
Beaverkill	*Ephemerella subvaria* (Sub)	Ephemerellidae
Big Slate Drake	*Hexagenia atrocaudata* (Sub)	Ephemeridae
Black and White Spinner	*Ephemera guttulata* (♂ Ad)	Ephemeridae
Black Angel	*Leptophlebia cupida*	Leptophlebiidae
Black Hen Spinner	*Tricorythodes* spp. (Ad)	Tricorythidae
Black Quill	*Leptophlebia cupida* (Sub) " *nebulosa* (Sub)	Leptophlebiidae "
Blue Dun	*Paraleptophlebia adoptiva* (Sub) *Pseudocloeon* spp. (Sub)	Leptophlebiidae Baetidae
Blue Quill	*Paraleptophlebia adoptiva* (Sub)	Leptophlebiidae
Blue Quill Spinner	*Leptophlebia johnsoni* (♀ Ad)	Leptophlebiidae
Bluewinged Hendrickson	*Ephemerella subvaria* (Sub)	Ephemerellidae
Bluewinged Olive	*Attenella attenuata* (Sub) (*Ephemerella attenuata*) *Baetis* spp. (Sub) *Pseudocloeon* spp. (Sub)	Ephemerellidae " Baetidae "
Bluewinged Olive Dun	*Attenella attenuata* (Sub) (*Ephemerella attenuata*) *Drunella cornuta* (Sub) (*Ephemerella cornuta*) *Drunella* spp. (Sub) (*Ephemerella* spp.)	Ephemerellidae " " " " "
Bluewinged Yellow Quill	*Eurylophella doris* (*Ephemerella doris*)	Ephemerellidae "
Borcher's	*Ephemerella subvaria* (Sub)	Ephemerellidae
Borcher's Drake	*Leptophlebia cupida* (Sub) " *nebulosa* (Sub)	Leptophlebiidae "
Brown and White Spinner	*Ephemera guttulata* (♀ Ad)	Ephemeridae
Brown Drake	*Ephemera simulans* *Litobrancha recurvata* (Ad) (*Hexagenia recurvata*)	Ephemeridae " "
Brown Hen Spinner	*Ephemerella dorothea* (Ad) " *subvaria* (Ad)	Ephemerellidae "
Burrowing Mayfly	*Hexagenia limbata*	Ephemeridae
Cahill	*Stenacron interpunctatum* (Sub) (*Stenonema interpunctatum*) *Stenonema ithaca* (Sub)	Heptageniidae " "
Chocolate Dun	*Ephemera simulans* (Sub) *Eurylophella bicolor* (Sub) (*Ephemerella bicolor*) *Ephemerella needhami* (Sub)	Ephemeridae Ephemerellidae " "
Chocolate Spinner	*Eurylophella bicolor* (Ad) (*Ephemerella bicolor*) *Ephemerella needhami* (Ad)	Ephemerellidae " "
Coffinfly	*Ephemera guttulata* (Ad)	Ephemeridae
Cream Cahill	*Stenonema* spp. (Sub)	Heptageniidae

EASTERN AND CENTRAL NORTH AMERICA (cont.)

FISHERMEN'S NAMES	SCIENTIFIC NAMES	FAMILY
Cream Cahill Spinner	*Stenonema* spp. (Ad)	Heptageniidae
Cream Dun	*Potamanthus distinctus* (Sub)	Potamanthidae
Cream May	*Stenacron* spp.	Heptageniidae
	(*Stenonema* spp.)	"
Cream Variant	*Ephemera varia* (Sub)	Ephemeridae
	Potamanthus distinctus (Sub)	Potamanthidae
Dark Blue Hendrickson	*Ephemerella invaria* (Sub)	Ephemerellidae
Dark Blue Quill	*Paraleptophlebia adoptiva* (Sub)	Leptophlebiidae
	" *debilis* (Sub)	"
	" *guttata* (Sub)	"
	" *mollis* (Sub)	"
	Leptophlebia johnsoni (Sub)	"
Dark Blue Upright	*Rhithrogena pellucida*	Heptageniidae
Dark Bluewinged Olive	*Attenella attenuata*	Ephemerellidae
	(*Ephemerella attenuata*)	"
	Baetis flavistriga	Baetidae
	(*Baetis cingulatus*)	"
Dark Brown Spinner	*Paraleptophlebia adoptiva* (Ad)	Leptophlebiidae
	" *guttata* (♀ Ad)	"
	" *mollis* (♀ Ad)	"
	Tricorythodes atratus (♂ Ad)	Tricorythidae
	" *stygiatus* (♂ Ad)	"
Dark Cahill	*Stenonema vicarium* (Ad)	Heptageniidae
	" *ithaca* (Ad)	"
Dark Golden Drake	*Potamanthus rufous*	Potamanthidae
Dark Golden Quill	*Epeorus pleuralis*	Heptageniidae
	(*Epeorus fraudator*)	"
Dark Gray Fox	*Stenonema vicarium* (Sub)	Heptageniidae
	(*Stenonema fuscum*)	"
Dark Graywinged Olive	*Baetis flavistriga* (Sub)	Baetidae
	(*Baetis phoebus*)	"
Dark Graywinged Quill	*Epeorus rubidus* (Sub)	Heptageniidae
Dark Green Drake	*Litobrancha recurvata* (Sub)	Ephemeridae
	(*Hexagenia recurvata*)	"
Dark Iron Blue Quill	*Serratella serrata* (Sub)	Ephemerellidae
	(*Ephemerella serrata*)	"
Dark Ironwinged Olive	*Drunella cornuta* (Sub)	Ephemerellidae
	(*Ephemerella cornuta*)	"
Dark Leadwinged Olive	*Serratella deficiens* (Sub)	Ephemerellidae
	(*Ephemerella deficiens*)	"
Dark Olive Spinner	*Attenella attenuata* (Ad)	Ephemerellidae
	(*Ephemerella attenuata*)	"
	Drunella cornuta (Ad)	"
	(*Ephemerella cornuta*)	"
	Drunella spp. (Ad)	"
	(*Ephemerella* spp.)	"
Dark Red Quill	*Ephemerella rotunda* (Sub)	Ephemerellidae
	Rhithrogena impersonata (Sub)	Heptageniidae
Dark Rusty Spinner	*Hexagenia atrocaudata* (Ad)	Ephemeridae
Dark Slatewinged Olive	*Dannella simplex* (Sub)	Ephemerellidae
	(*Ephemerella simplex*)	"
Drakefly	*Litobrancha recurvata* (Sub)	Ephemeridae
	(*Hexagenia recurvata*)	"
Dun Variant	*Isonychia bicolor* (Sub)	Oligoneuriidae
Early Black Quill	*Leptophlebia cupida* (Sub)	Leptophlebiidae

EASTERN AND CENTRAL NORTH AMERICA (cont.)

FISHERMEN'S NAMES	SCIENTIFIC NAMES	FAMILY
Early Blue Quill	*Paraleptophlebia adoptiva* (Sub)	Leptophlebiidae
Early Brown Quill	*Leptophlebia nebulosa* (Sub)	Leptophlebiidae
Early Brown Spinner	*Leptophlebia cupida* (Ad)	Leptophlebiidae
Eastern Brown Quill	*Siphlonurus quebecensis* (Sub)	Siphlonuridae
Evening Dun	*Potamanthus distinctus* (Sub)	Potamanthidae
Fishfly	*Hexagenia limbata* (Sub)	Ephemeridae
Giant Michigan Mayfly	*Hexagenia limbata*	Ephemeridae
Ginger Quill	*Stenonema ithaca* (Sub)	Heptageniidae
	" *vicarium* (Sub)	"
	" spp. (Sub)	"
	Stenacron interpunctatum (Sub)	"
	(*Stenacron canadense*)	"
	(*Stenonema canadense*)	"
Ginger Quill Spinner	*Stenonema vicarium* (Ad)	Heptageniidae
	(*Stenonema fuscum*)	"
Golden Drake	*Potamanthus distinctus* (Sub)	Potamanthidae
Golden Spinner	*Potamanthus distinctus* (Ad)	Potamanthidae
Gordon Quill	*Epeorus pleuralis* (Sub)	Heptageniidae
Gray Drake	*Siphlonurus* spp. (Sub)	Siphlonuridae
Gray Fox	*Stenonema ithaca* (Sub)	Heptageniidae
	" *vicarium* (Sub)	"
	(*Stenonema fuscum*)	"
Graywinged Yellow Quill	*Epeorus vitreus* (Sub)	Heptageniidae
Great Brown Drake	*Hexagenia bilineata*	Ephemeridae
Great Dark Green Drake	*Litobrancha recurvata* (Sub)	Ephemeridae
	(*Hexagenia recurvata*)	"
Great Leadwinged Coachman	*Isonychia bicolor* (Sub)	Oligoneuriidae
Great Leadwinged Drake	*Hexagenia atrocaudata* (Sub)	Ephemeridae
Great Mahogany Drake	*Isonychia sadleri* (Sub)	Oligoneuriidae
Great Olive Drake	*Hexagenia rigida*	Ephemeridae
Great Olivewinged Drake	*Hexagenia limbata* (Sub)	Ephemeridae
Great Pale Yellow Drake	*Hexagenia munda*	Ephemeridae
	(*Hexagenia carolina*)	"
Great Red Spinner	*Stenonema vicarium* (Ad)	Heptageniidae
Great Speckled Lake Olive	*Siphloplecton basale* (Sub)	Metretopodidae
Great Sulphur Drake	*Hexagenia munda* (Sub)	Ephemeridae
	(*Hexagenia carolina*)	"
Green Drake	*Ephemera guttulata* (Sub)	Ephemeridae
Green Egg Spinner	*Siphlonurus quebecensis* (♀ Ad)	Siphlonuridae
Green May	*Ephemera guttulata* (Sub)	Ephemeridae
Hendrickson	*Ephemerella subvaria* (Sub)	Ephemerellidae
Iron Blue Dun	*Leptophlebia johnsoni* (Sub)	Leptophlebiidae
	Paraleptophlebia adoptiva (Sub)	"
Iron Dun	*Epeorus pleuralis* (Sub)	Heptageniidae
Jenny Spinner	*Paraleptophlebia guttata* (♂ Ad)	Leptophlebiidae
	" *mollis* (♂ Ad)	"
	Leptophlebia johnsoni (♂ Ad)	"
Lady Beaverquill	*Ephemerella subvaria* (Sub)	Ephemerellidae
Large Gray Fox Variant	*Ephemera guttulata* (Sub)	Ephemeridae
Large Leadwinged Olive Dun	*Drunella walkeri* (Sub)	Ephemerellidae
	(*Ephemerella walkeri*)	"
Large Mahogany Dun	*Isonychia bicolor* (Sub)	Oligoneuriidae

EASTERN AND CENTRAL NORTH AMERICA (cont.)

FISHERMEN'S NAMES	SCIENTIFIC NAMES	FAMILY
Large Summer Olive	*Siphlonurus mirus*	Siphlonuridae
Leadwinged Coachman	*Isonychia bicolor* (Sub)	Oligoneuriidae
Light Blue Quill	*Epeorus pleuralis* (Sub)	Heptageniidae
	(*Epeorus confusus*)	"
Light Bluewinged Olive	*Drunella lata*	Ephemerellidae
	(*Ephemerella lata*)	"
Light Cahill	*Stenacron interpunctatum* (Sub)	Heptageniidae
	(*Stenacron canadense*)	"
	(" *areion*)	"
	(" *heterotarsale*)	"
	(" *frontale*)	"
	(*Stenonema interpunctatum*)	"
	Stenonema ithaca	"
	" *luteum* (Sub)	"
	" *femoratum* (Sub)	"
	(*Stenonema tripunctatum*)	"
	Stenonema modestum (Sub)	"
	(*Stenonema rubrum*)	"
	Epeorus vitreus (♂ Sub)	"
Light Gray Fox	*Stenonema ithaca*	Heptageniidae
Light Rusty Spinner	*Baetis tricaudatus* (Ad)	Baetidae
	(*Baetis intermedius*)	"
Light Blue Dun	*Baetis tricaudatus* (Sub)	Baetidae
	(*Baetis vagans*)	"
	Baetis spp. (Sub)	"
Little Bluewinged Red Quill	*Ephemerella needhami* (Sub)	Ephemerellidae
Little Bluewinged Rusty Dun	*Baetis brunneicolor* (Sub)	Baetidae
Little Graywinged Brown Quill	*Baetis intercalaris* (Sub)	Baetidae
Little Graywinged Red Quill	*Ephemerella excrucians*	Ephemerellidae
Little Dark Olive	*Siphlonurus rapidus* (Sub)	Siphlonuridae
Little Evening Sulphur	*Leucrocuta minerva* (Sub)	Heptageniidae
	(*Heptagenia minerva*)	"
Little Evening Yellow	*Leucrocuta hebe* (Sub)	Heptageniidae
	(*Heptagenia hebe*)	"
Little Gray Cahill	*Stenonema luteum* (Sub)	Heptageniidae
Little Graywinged Olive	*Heptagenia pulla* (Sub)	Heptageniidae
Little Iron Blue Quill	*Baetis tricaudatus*	Baetidae
	(*Baetis vagans*)	"
Little Marryat	*Epeorus vitreus* (Sub)	Heptageniidae
	(*Epeorus humeralis*)	"
	(*Iron humeralis*)	"
	Epeorus spp. (Sub)	"
	Ephemerella dorothea (Sub)	Ephemerellidae
Little Medium Olive	*Siphlonurus barbarus*	Siphlonuridae
Little Medium Olive Dun	*Baetis pygmaeus* (Sub)	Baetidae
Little Pale Evening Dun	*Leucrocuta juno* (Sub)	Heptageniidae
	(*Heptagenia juno*)	"
Little Pale Fox	*Stenacron pallidum* (Sub)	Heptageniidae
Little Slatewinged Brown Quill	*Baetis brunneicolor*	Baetidae
	(*Baetis hiemalas*)	"
Little Slatewinged Olive Quill	*Baetis propinquus*	Baetidae
	(*Baetis spinosus*)	"
Little Sulphur Cahill	*Stenacron carolina* (Sub)	Heptageniidae
	(*Stenonema carolina*)	"
Little Summer Quill	*Paraleptophlebia praepedita*	Leptophlebiidae

EASTERN AND CENTRAL NORTH AMERICA (cont.)

FISHERMEN'S NAMES	SCIENTIFIC NAMES	FAMILY
Little Yellow Fox	*Stenacron interpunctatum* (Sub)	Heptageniidae
	(*Stenonema proximum*)	"
Little Yellow Quill	*Epeorus dispar* (Sub)	Heptageniidae
Mackerel	*Ephemera* spp.	Ephemeridae
Mahogany Drake	*Isonychia sadleri* (Sub)	Oligoneuriidae
Mahogany Dun	*Isonychia bicolor* (Sub)	Oligoneuriidae
	" *sadleri* (Sub)	"
March Brown	*Stenonema vicarium* (Sub)	Heptageniidae
	Ephemera simulans (Sub)	Ephemeridae
Maroon Drake	*Isonychia sadleri* (Sub)	Oligoneuriidae
Medium Claret Mayfly	*Isonychia harperi*	Oligoneuriidae
Medium Olive Dun	*Drunella allegheniensis* (Sub)	Ephemerellidae
	(*Ephemerella allegheniensis*)	"
Medium Olive Upright	*Rhithrogena jejuna*	Heptageniidae
Medium Slatewinged Olive	*Drunella longicornis*	Ephemerellidae
	(*Ephemerella longicornis*)	"
Medium Speckled Lake Olive	*Siphloplecton basale*	Metretopodidae
	(*Siphloplecton signatum*)	"
Medium Specklewinged Dun	*Callibaetis americanus*	Baetidae
Medium Summer Olive	*Siphlonurus alternatus*	Siphlonuridae
Michigan Caddis	*Hexagenia limbata* (Sub)	Ephemeridae
Michigan Spinner	*Hexagenia limbata* (Ad)	Ephemeridae
Minute Bluewinged Olive	*Baetis flavistriga*	Baetidae
	(*Baetis levitans*)	"
Minute Graywinged Brown	*Pseudocloeon carolina* (Sub)	Baetidae
Minute Graywinged Olive	*Pseudocloeon anoka* (Sub)	Baetidae
Olive Cahill Quill	*Stenonema mediopunctatum* (Sub)	Heptageniidae
	(*Stenonema nepotellum*)	"
Olive Cahill Spinner	*Stenonema luteum* (Ad)	Heptageniidae
Olive Dun	*Drunella lata* (Sub)	Ephemerellidae
	(*Ephemerella lata*)	"
Pale Cahill Quill	*Stenonema integrum* (Sub)	Heptageniidae
Pale Cream Drake	*Potamanthus diaphanus* (Sub)	Potamanthidae
Pale Evening Dun	*Ephemerella dorothea* (Sub)	Ephemerellidae
	" *invaria*	"
	" *rotunda* (Sub)	"
	" *septentrionalis* (Sub)	"
	Leucrocuta aphrodite	Heptageniidae
	(*Heptagenia aphrodite*)	"
	Leucrocuta hebe	"
	(*Heptagenia hebe*)	"
	Epeorus spp. (Sub)	"
Pale Evening Olive	*Epeorus fragilis* (Sub)	Heptageniidae
Pale Evening Quill	*Epeorus punctatus* (Sub)	Heptageniidae
Pale Evening Spinner	*Ephemerella rotunda* (Ad)	Ephemerellidae
Pale Golden Drake	*Potamanthus distinctus* (Sub)	Potamanthidae
Pale Gray Fox	*Stenonema pulchellum* (Sub)	Heptageniidae
Pale Gray Fox Quill	*Stenacron interpunctatum* (Sub)	Heptageniidae
	(*Stenonema heterotarsale*)	"
Pale Olive Cahill	*Stenonema modestum* (Sub)	Heptageniidae
	(*Stenonema rubrum*)	"
Pale Olive Cahill Quill	*Stenacron interpunctatum* (Sub)	Heptageniidae
	(*Stenonema frontale*)	"

EASTERN AND CENTRAL NORTH AMERICA (cont.)

FISHERMEN'S NAMES	SCIENTIFIC NAMES	FAMILY
Pale Olive Dun	*Tricorythodes atratus* (Sub) " *stygiatus* (Sub)	Tricorythidae "
Pale Red Fox	*Stenonema femoratum* (Sub) (*Stenonema tripunctatum*)	Heptageniidae "
Pale Specklewinged Dun	*Callibaetis fluctuans*	Baetidae
Pale Specklewinged Olive	*Callibaetis ferrugineus*	Baetidae
Pale Sulphur Dun	*Ephemerella dorothea* (Sub)	Ephemerellidae
Pale Summer Quill	*Paraleptophlebia strigula* (Sub)	Leptophlebiidae
Pale Watery Dun	*Ephemerella dorothea* (Sub) *Centroptilum album* (Sub)	Ephemerellidae Baetidae
Pale Whitewinged Watery	*Centroptilum walshi* (Sub)	Baetidae
Pink Cahill	*Epeorus vitreus* (♀ Sub)	Heptageniidae
Quill Gordon	*Epeorus pleuralis* (Sub) (*Iron fraudator*)	Heptageniidae "
Red Quill	*Ephemerella subvaria* (Sub) *Leptophlebia cupida* (*Blasturus cupidus*)	Ephemerellidae Leptophlebiidae "
Red Quill Spinner	*Epeorus pleuralis* (Ad)	Heptageniidae
Redgilled Nymph	*Rhithrogena impersonata* (L) (*Rhithrogena sanguinea*)	Heptageniidae "
Reverse Jenny Spinner	*Tricorythodes atratus* (♀ Ad) " *stygiatus* (♀ Ad)	Tricorythidae "
Rusty Spinner	*Baetis tricaudatus* (Ad) " spp. (Ad) *Pseudocloeon* spp. (Ad)	Baetidae " "
Salmon Spinner	*Epeorus vitreus* (Ad)	Heptageniidae
Sand Drake	*Stenonema vicarium* (Sub)	Heptageniidae
Sandfly	*Hexagenia limbata* (Sub)	Ephemeridae
Shaddie	*Ephemera guttulata* (Sub)	Ephemeridae
Shadfly	*Ephemera guttulata*	Ephemeridae
Slate Drake	*Isonychia bicolor* (Sub) " *harperi* (Sub) " *sadleri* (Sub)	Oligoneuriidae " "
Slatewinged Mahogany Dun	Paraleptophlebia adoptiva (Sub)	Leptophlebiidae
Slatewinged Olive	*Drunella flavilinea* (Sub) (*Ephemerella flavilinea*) *Drunella lata* (Sub) (*Ephemerella lata*) *Attenella attenuata* (Sub) (*Ephemerella attenuata*) *Drunella flavilinea* (*Ephemerella flavilinea*)	Ephemerellidae " " " " " " "
Small Cream Variant	*Ephemerella dorothea* (Sub) *Epeorus vitreus* (Sub) (*Iron humeralis*)	Ephemerellidae Heptageniidae "
Sulphur	*Ephemerella dorothea* (Sub) *Epeorus* spp. (Sub)	Ephemerellidae Heptageniidae
Sulphur Dun	*Ephemerella dorothea* (Sub)	Ephemerellidae
Sulphury Dun	*Ephemerella dorothea*	Ephemerellidae
Tiny Bluewinged Olive	*Centroptilum bellum* (Sub) *Pseudocloeon anoka* (Sub)	Baetidae "
Tiny Bluewinged Quill	*Centroptilum simile* (Sub)	Baetidae
Tiny Dun Variant	*Attenella attenuata* (Sub) (*Ephemerella attenuata*)	Ephemerellidae "

EASTERN AND CENTRAL NORTH AMERICA (cont.)

FISHERMEN'S NAMES	SCIENTIFIC NAMES	FAMILY
Tiny Graywinged Blue Dun	*Caenis amica* (Sub)	Caenidae
Tiny Graywinged Red Quill	*Cloeon rubropictum* (Sub)	Baetidae
Tiny Graywinged Rusty Dun	*Caenis latipennis* (Sub)	Caenidae
Tiny Graywinged Sherry Quill	*Cloeon insignificans* (Sub)	Baetidae
Tiny Graywinged Sulphur	*Caenis jocosa* (Sub)	Caenidae
Tiny Graywinged Yellow	*Cloeon alamance* (Sub)	Baetidae
	(*Neocloeon alamance*)	"
Tiny Leadwinged Red Quill	*Centroptilum rufostrigatum* (Sub)	Baetidae
Tiny Whitewinged Black	*Tricorythodes* spp. (Sub)	Tricorythidae
Tiny Whitewinged Black Quill	*Tricorythodes stygiatus* (Sub)	Tricorythidae
Tiny Whitewinged Brown Quill	*Tricorythodes allectus* (Sub)	Tricorythidae
Tiny Whitewinged Curse	*Caenis hilaris* (Sub)	Caenidae
Tiny Whitewinged Olive Quill	*Tricorythodes texanus* (Sub)	Tricorythidae
Tiny Whitewinged Red Quill	*Brachycercus nitidus* (Sub)	Caenidae
Tiny Whitewinged Sepia Quill	*Brachycercus lacustris* (Sub)	Caenidae
Tiny Whitewinged Sulphur	*Caenis simulans* (Sub)	Caenidae
Tiny Whitewinged Trico	*Tricorythodes* spp.	Tricorythidae
Trailer	*Ephoron* spp.	Polymitarcyidae
Whirling Dun	*Leptophlebia cupida* (Sub)	Leptophlebiidae
	" *nebulosa* (Sub)	"
White Dun	*Ephemera varia* (Sub)	Ephemeridae
White Mayfly	*Ephoron album*	Polymitarcyidae
	" *leukon*	"
Whitegloved Howdy	*Isonychia bicolor* (Ad)	Oligoneuriidae
	" *harperi* (Ad)	"
	" *sadleri* (Ad)	"
Yellow Drake	*Ephemera varia*	Ephemeridae
	Potamanthus distinctus (Sub)	Potamanthidae
Yellow Dun	*Ephemera varia* (Sub)	Ephemeridae
Yellow Egg Spinner	*Ephemerella invaria*	Ephemerellidae
Yellow May	*Ephemera varia*	Ephemeridae
	Stenacron spp.	Heptageniidae
	(*Stenonema* spp.)	"

WESTERN NORTH AMERICA

FISHERMEN'S NAMES	SCIENTIFIC NAMES	FAMILY
Beaverkill	*Ephemerella inermis*	Ephemerellidae
Bandwing	*Callibaetis pallidus*	Baetidae
Black Hen Spinner	*Tricorythodes* spp. (Ad)	Tricorythidae
Black Quill	*Leptophlebia cupida* (Sub)	Leptophlebiidae
	" *nebulosa* (Sub)	"
Blue Dun	*Pseudocloeon* spp. (Sub)	Baetidae
Bluewinged Olive	*Baetis* spp. (Sub)	Baetidae
	Pseudocloeon spp. (Sub)	"
Bluewinged Olive Dun	*Drunella flavilinea* (Sub)	Ephemerellidae
	(*Ephemerella flavilinea*)	"
Borcher's Drake	*Leptophlebia cupida* (Sub)	Leptophlebiidae
	" *nebulosa* (Sub)	"
Brown Drake	*Ephemera simulans*	Ephemeridae
Brown Quill Spinner	*Siphlonurus occidentalis* (Ad)	Siphlonuridae
Burrowing Mayfly	*Hexagenia limbata*	Ephemeridae
Chocolate Dun	*Ephemera simulans* (Sub)	Ephemeridae
Colorado Mayfly	*Drunella coloradensis*	Ephemerellidae
	(*Ephemerella coloradensis*)	"
Coquette	*Epeorus longimanus*	Heptageniidae
Cream Hen Spinner	*Callibaetis* spp. (Ad)	Baetidae
Dandy	*Heptagenia elegantula*	Heptageniidae
Dark Blue Quill	*Paraleptophlebia bicornuta* (Sub)	Leptophlebiidae
	" *debilis* (Sub)	"
	" *heteronea* (Sub)	"
	" *memorialis* (Sub)	"
	" *vaciva* (Sub)	"
Dark Bluewinged Red Quill	*Drunella doddsi*	Ephemerellidae
	(*Ephemerella doddsi*)	"
Dark Brown Dun	*Ameletus cooki* (Sub)	Siphlonuridae
	Baetis hageni (Sub)	Baetidae
	(*Baetis parvus*)	"
Dark Brown Spinner	*Paraleptophlebia bicornuta* (Ad)	Leptophlebiidae
	" *debilis* (Ad)	"
	" *heteronea* (Ad)	"
	" *memorialis* (Ad)	"
	Ameletus cooki (Ad)	Siphlonuridae
	Baetis hageni (Ad)	Baetidae
	(*Baetis parvus*)	"
	Drunella coloradensis (Ad)	Ephemerellidae
	(*Ephemerella coloradensis*)	"
	Tricorythodes minutus (♀ Ad)	Tricorythidae
Dark Brownwinged Olive	*Ephemerella maculata*	Ephemerellidae
Dark Morning Olive	*Drunella grandis*	Ephemerellidae
	(*Ephemerella yosemite*)	"
Dark Olive Dun	*Drunella coloradensis* (Sub)	Ephemerellidae
	(*Ephemerella coloradensis*)	"
Dark Olive Spinner	*Drunella flavilinea* (Ad)	Ephemerellidae
	(*Ephemerella flavilinea*)	"
Dark Red Quill	*Cinygmula ramaleyi* (Sub)	Heptageniidae
	Rhithrogena undulata (Ad)	"
Dark Rusty Spinner	*Baetis tricaudatus* (Ad)	Baetidae
	(*Baetis intermedius*)	"
	Cinygmula reticulata (Ad)	Heptageniidae
Dark Slatewinged Olive	*Drunella flavilinea* (Sub)	Ephemerellidae
	(*Ephemerella flavilinea*)	"

WESTERN NORTH AMERICA (cont.)

FISHERMEN'S NAMES	SCIENTIFIC NAMES	FAMILY
Dark Slatewinged Purple Quill	*Ironodes nitidus*	Heptageniidae
	(*Epeorus nitidus*)	"
Dark Specklewinged Quill	*Callibaetis nigritus* (Sub)	Baetidae
Dark Tan Spinner	*Rhithrogena hageni* (Ad)	Heptageniidae
Early Brown Quill	*Leptophlebia nebulosa* (Sub)	Leptophlebiidae
Early Brown Spinner	*Leptophlebia cupida* (Ad)	Leptophlebiidae
Fishfly	*Hexagenia limbata* (Sub)	Ephemeridae
Giant Michigan Mayfly	*Hexagenia limbata*	Ephemeridae
Ginger Quill	*Drunella doddsi*	Ephemerellidae
	(*Ephemerella doddsi*)	"
Ginger Quill Spinner	*Heptagenia solitaria* (Ad)	Heptageniidae
Gray Drake	*Siphlonurus occidentalis* (Sub)	Siphlonuridae
Gray Fox	*Heptagenia solitaria* (Sub)	Heptageniidae
Graywinged Pink Quill	*Epeorus albertae*	Heptageniidae
Great Bluewinged Red Quill	*Timpanoga hecuba* (Sub)	Ephemerellidae
	(*Ephemerella hecuba*)	"
Great Brown Spinner	*Timpanoga hecuba* (Ad)	Ephemerellidae
	(*Ephemerella hecuba*)	"
Great Leadwinged Olive Drake	*Drunella grandis* (Sub)	Ephemerellidae
	(*Ephemerella grandis*)	"
Great Olivewinged Drake	*Hexagenia limbata* (Sub)	Ephemeridae
Great Red Quill	*Timpanoga hecuba* (Sub)	Ephemerellidae
	(*Ephemerella hecuba*)	"
Great Red Spinner	*Drunella doddsi* (Ad)	Ephemerellidae
	(*Ephemerella doddsi*)	"
	Drunella grandis (Ad)	"
	(*Ephemerella grandis*)	"
Great Summer Drake	*Siphlonurus occidentalis*	Siphlonuridae
Great Western Leadwing	*Isonychia velma* (Sub)	Oligoneuriidae
Green Drake	*Drunella grandis* (Sub)	Ephemerellidae
	(*Ephemerella grandis*)	"
Greengill	*Rhithrogena hageni* (L)	Heptageniidae
Humpbacked Nymph	*Baetisca obesa* (L)	Baetiscidae
Light Cahill	*Cinygma dimicki*	Heptageniidae
Light Rusty Spinner	*Baetis bicaudatus* (Ad)	Baetidae
	" *tricaudatus* (Ad)	"
	(*Baetis intermedius*)	"
Light Slatewinged Olive	*Drunella coloradensis* (Sub)	Ephemerellidae
	(*Ephemerella coloradensis*)	"
Little Blue Dun	*Baetis tricaudatus* (Sub)	Baetidae
	(*Baetis intermedius*)	"
	Baetis spp. (Sub)	"
Little Blue Quill	*Epeorus deceptivus*	Heptageniidae
Little Bluewinged Red Quill	*Cinygmula ramaleyi*	Heptageniidae
Little Graywinged Dun	*Nixe criddlei* (Sub)	Heptageniidae
	(*Heptagenia criddlei*)	"
Little Marryat	*Epeorus* spp. (Sub)	Heptageniidae
Little Specklewinged Quill	*Callibaetis coloradensis*	Baetidae
Little Western Red Quill	*Paraleptophlebia heteronea*	Leptophlebiidae
Little Yellow Mayfly	*Epeorus hesperus*	Heptageniidae
March Brown	*Ephemera simulans* (Sub)	Ephemeridae
Margaret's Mayfly	*Attenella margarita*	Ephemerellidae
	(*Ephemerella margarita*)	"

WESTERN NORTH AMERICA (cont.)

FISHERMEN'S NAMES	SCIENTIFIC NAMES	FAMILY
Medium Blue Quill	*Epeorus longimanus* (*Epeorus proprius*)	Heptageniidae "
Medium Specklewinged Dun	*Callibaetis americanus* (Sub)	Baetidae
Medium Specklewinged Quill	*Callibaetis pacificus* (Sub)	Baetidae
Michigan Caddis	*Hexagenia limbata* (Sub)	Ephemeridae
Michigan Spinner	*Hexagenia limbata* (Ad)	Ephemeridae
Minute Graywinged Watery	*Pseudocloeon futile* (Sub)	Baetidae
Pack's Tusker	*Paraleptophlebia packi* (L)	Leptophlebiidae
Pale Brown Dun	*Cinygmula reticulata* (Sub) *Rhithrogena hageni* (Sub)	Heptageniidae "
Pale Evening Dun	*Heptagenia elegantula* (Sub) *Epeorus* spp. (Sub)	Heptageniidae "
Pale Evening Spinner	*Heptagenia elegantula* (Ad)	Heptageniidae
Pale Graywinged Olive	*Baetis adonis* (Sub)	Baetidae
Pale Graywinged Sulphur	*Centroptilum convexum* (Sub)	Baetidae
Pale Morning Dun	*Ephemerella infrequens* (Sub) " *inermis* (Sub) " *lacustris*	Ephemerellidae " "
Pale Morning Olive	*Ephemerella inermis* (Sub)	Ephemerellidae
Pale Olive Dun	*Tricorythodes minutus* (Sub) *Baetis bicaudatus* (Sub)	Tricorythidae Baetidae
Pale Specklewinged Sulphur	*Callibaetis pallidus*	Baetidae
Pale Western Leadwing	*Isonychia sicca* (*Isonychia campestris*)	Oligoneuriidae "
Pale Whitewinged Sulphur	*Centroptilum elsa* (Sub)	Baetidae
Palefoot	*Leptophlebia nebulosa* (L)	Leptophlebiidae
Palewinged Pink Quill	*Epeorus dulciana* (Sub)	Heptageniidae
Pink Lady	*Epeorus albertae* (Sub)	Heptageniidae
Prickleback	*Drunella grandis* (L) (*Ephemerella grandis*)	Ephemerellidae "
Quill Gordon	*Epeorus longimanus* (Sub) *Rhithrogena futilis* (Sub) " *undulata* (Sub)	Heptageniidae " "
Redgill	*Cinygmula mimus* (L)	Heptageniidae
Red Quill	*Rhithrogena undulata* (Ad) *Serratella tibialis* (Sub) (*Ephemerella tibialis*)	Heptageniidae Ephemerellidae "
Red Quill Spinner	*Cinygmula ramaleyi* (Ad) *Epeorus longimanus* (Ad)	Heptageniidae "
Reverse Jenny Spinner	*Tricorythodes minutus* (♂ Ad)	Tricorythidae
Rusty Spinner	*Baetis tricaudatus* (Ad) " spp. (Ad) *Pseudocloeon* spp. (Ad) *Ephemerella infrequens* (Ad)	Baetidae " " Ephemerellidae
Salmon Spinner	*Epeorus albertae* (Ad)	Heptageniidae
Sandfly	*Hexagenia limbata* (Sub)	Ephemeridae
Sawtooth	*Timpanoga hecuba* (L) (*Ephemerella hecuba*)	Ephemerellidae "
Slate Brown Dun	*Epeorus longimanus* (Sub)	Heptageniidae
Slate Cream Dun	*Epeorus albertae* (Sub)	Heptageniidae
Slate Gray Dun	*Heptagenia elegantula* (Sub)	Heptageniidae
Slate Maroon Drake	*Ironodes nitidus* (Sub) (*Epeorus nitidus*)	Heptageniidae "

WESTERN NORTH AMERICA (cont.)

FISHERMEN'S NAMES	SCIENTIFIC NAMES	FAMILY
Snowflake	*Tricorythodes explicatus*	Tricorythidae
Speckled Spinner	*Callibaetis coloradensis* (Ad)	Baetidae
Specklewinged Dun	*Callibaetis coloradensis* (Sub)	Baetidae
Specklewinged Spinner	*Callibaetis coloradensis* (Ad)	Baetidae
Sulphur	*Epeorus* spp. (Sub)	Heptageniidae
Tiny Bluewinged Olive	*Pseudocloeon edmundsi*	Baetidae
Tiny Darkwinged Olive	*Centroptilum asperatum*	Baetidae
Tiny Graywinged Olive Quill	*Cloeon implicatum*	Baetidae
Tiny Graywinged Sepia Quill	*Cloeon simplex*	Baetidae
Tiny Whitewinged Black	*Tricorythodes* spp. (Sub)	Tricorythidae
Tiny Whitewinged Claret Quill	*Tricorythodes minutus*	Tricorythidae
Tiny Whitewinged Sulphur	*Caenis simulans*	Caenidae
Tiny Whitewinged Trico	*Tricorythodes* spp.	Tricorythidae
Western Black Quill	*Leptophlebia gravastella*	Leptophlebiidae
Western Blue Dun	*Paraleptophlebia packi*	Leptophlebiidae
Western Blue Quill	*Paraleptophlebia californica*	Leptophlebiidae
Western Brown Quill	*Siphlonurus phyllis*	Siphlonuridae
Western Gordon Quill	*Epeorus longimanus*	Heptageniidae
Western Gray Fox	*Cinygmula mimus*	Heptageniidae
Western Graywinged Yellow Quill	*Epeorus albertae*	Heptageniidae
	(*Epeorus youngi*)	"
Western Green Drake	*Drunella doddsi* (Sub)	Ephemerellidae
	(*Ephemerella doddsi*)	"
	Drunella grandis (Sub)	"
	(*Ephemerella grandis*)	"
Western Little Marryat	*Epeorus lagunitas*	Heptageniidae
Western Pale Evening Dun	*Heptagenia elegantula*	Heptageniidae
Western Red Quill	*Rhithrogena hageni*	Heptageniidae
	(*Rhithrogena doddsi*)	"
Western Slate Olive Dun	*Drunella flavilinea* (Sub)	Ephemerellidae
	(*Ephemerella flavilinea*)	"
Western Yellow Drake	*Siphlonurus spectabilis*	Siphlonuridae
Whirling Dun	*Leptophlebia cupida* (Sub)	Leptophlebiidae
	" *nebulosa* (Sub)	"
White Mayfly	*Ephoron album*	Polymitarcyidae
Whitegloved Howdy	*Serratella tibialis* (Ad)	Ephemerellidae
	(*Ephemerella tibialis*)	"

GLOSSARY

Abdominal. Pertaining to structures or parts of the abdomen (the third and most posterior major body region).

Acetabulum (pl. **acetabula**). A small cuplike structure associated with the genitalia of mites, located ventrally.

Acute. Pointed or sharply angled.

Aerial. Pertaining to water mite larvae that seek out their hosts in the terrestrial environment.

Aerobic. Living in an environment where free oxygen is used for respiration.

Aeropneustic. Utilizing atmospheric oxygen for respiration.

Aestivation. A type of dormancy during drought or warm periods.

Algae. Unicellular, multicellular, or colonial aquatic plants having chlorophyll and not having true roots, stems, or leaves.

Amorphous. Not having any definite form.

Amphibious. Existing in or on water part of the time and existing on land part of the time.

Anaerobic. Living in environments devoid of oxygen and not requiring free oxygen for respiration.

Anal. Pertaining to the hind or most posterior part of the body.

Anal cell. A part of the posterior region of a wing enclosed by veins (esp. Diptera wings).

Anal loop. A distinctively shaped group of cells located in the hind basal region of the hind wings of dragonflies.

Annulation. A ring or subsegment within a true segment, as in some Diptera antennae.

Annuli. Subsegments within true body segments, as in some Diptera larvae.

Antenna (pl. **antennae**). A variously shaped appendage of the head, occurring in pairs, commonly located between the eyes.

Antenodal. Pertaining to crossveins located between the wing base and nodus in Odonata wings.

Anterior. At or directed toward the head or forward part of the body.

Apical. At the end or tip of a structure.

Appendage. Any part or structure attached to the body or other main structure, usually attached by a joint.

Aquaculture. The practice of raising aquatic organisms for commercial purposes, usually specific fishes.

Aquatic. Living in or on water; or being composed of water.

Arculus. A transverse vein between the radius and cubitus formed by the base of the medius in Odonata wings.

Atrophy. A withering of a structure so that it is no longer fully formed or functional.

Attachment disc. Any of several structures by which an insect attaches to a substrate, usually somewhat circular or spherical.

Attractor. A tied fly pattern that does not specifically imitate any particular aquatic insect species.

Basal. At the base of a structure or part, generally the region at or near the point of attachment to a major or central region.

Benthic. Being on or in any substrate, usually in reference to bottom-dwelling organisms.

Benthos. Collectively, bottom-dwelling or substrate-oriented organisms.

Bilaterally symmetrical. Having two equal or symmetrical sides with reference to a longitudinal axis.

Bioassay. The determination of effects of chemical or physical properties (usually unnatural) on live organisms.

Biomass. The amount of organisms, measured as the total weight per unit of area.

Biomonitoring. The use of organisms to assess or monitor environmental conditions.

Borer. Generally, any insect that burrows into plant tissue, especially stems of plants.

Brackish. Having a salt content greater than freshwater.

Breathing tube. A structure used to contact the air-water interface so as to facilitate the acquisition of air while the body remains submerged.

Brood. One of the generations of a population of species that has more than one generation per year; cohort.

Cannibalistic. Pertaining to animals that will prey on members of their own species.

Capitulum. A complex headlike plate in mites to which mouthparts are attached.

Carapace. A hardened shield covering part of the body dorsally.

Carnivorous. Eating animal tissue (the eater is referred to as a carnivore).

Cast skin. The shed exoskeleton left after molting in arthropods.

Catchnet. A net that is spun and used for capturing food in moving water.

Caudal. Pertaining to the posterior end of the body.

Cephalothorax. A single body region consisting of a head and thorax that are little differentiated from each other.

Cervical. Pertaining to that area immediately behind the head and between the head and thorax.

Character. An identifying trait whose variability among groups can be used to differentiate them.

Chelate. Pincerlike, with opposable grasping or crushing parts.

Chloride cell. A specialized cell that functions in the uptake of ions from the water.
Chloride epithelia. Specialized tissue areas that function in the uptake of ions from water.
Climber. Generally, any aquatic insect that climbs about aquatic vegetation.
Clinger. An aquatic insect that is able to cling to substrates and maintain itself in fast-flowing water.
Clubbed. A type of antenna that is expanded or enlarged distally or laterally.
Cocoon. A case in which some larvae and pupae reside, often spun with silk and incorporating various other material, usually closed.
Complete metamorphosis. A kind of developmental process in the life cycle in which there is a distinctive pupal stage that precedes the adult stage.
Costa. The front or fore margin of a wing; the longitudinal vein forming the anterior margin of the wing.
Courtship. Behavior that precedes or accompanies mating.
Coxa (pl. **coxae**). The most basal major segment of the insect leg.
Crepuscular. Active at twilight, dusk, or dawn, as opposed to being strictly day or night active.
Crochet. A series or circlet of tiny hooks at the end of abdominal prolegs of Lepidoptera larvae.
Crossvein. A short vein extending between longitudinal veins and often connecting them in insect wings.
Culm. The stem of grasses or sedges.
Cutaneous respiration. A type of external respiration in which oxygen is taken up directly through the body wall.
Cyclorrhaphan. The higher Diptera, including the aquatic families Syrphidae, Ephydridae, Anthomyiidae, and Muscidae.

Detritus. Dead and decomposing animal or plant material.
Diatom. A unicellular form of algae with walls impregnated with silica and often of intricate design.
Diel. Pertaining to a 24-hour period or a regular occurrence in every 24-hour period.
Dispersal. The act of moving out of or away from the place of birth in order to extend the range of the species.
Dissolved oxygen. Oxygen that is available in water.
Distal. At or toward the outermost area or end of a structure, as opposed to basal.
Diurnal. Active during daylight hours.
Dormancy. A resting or quiescent condition often occurring during periods when climatic factors are not conducive to normal activity or in order to synchronize certain life history events.
Dorsal. Pertaining to the top or upper surface.
Drift. Benthic organisms during the time when they passively move downstream in a suspended manner; to be temporarily suspended and carried by the current.
Dry fly. A fisherman's imitation lure that is usually patterned after an insect and fished on the surface of the water.
Dun pattern. A fisherman's fly imitation that is patterned after a mayfly subimago.

Eclosion. Hatching of the egg.
Ecology. Generally, the study of the relationships of organisms and their environment; pertaining to the relationships of organisms and their environment.
Ecosystem. All of the component organisms of a community and their environment that, together, form an interacting system.
Ectoparasite. An organism that feeds on another organism and resides on it externally.
Effluent. An outward movement of water, as a stream from a lake or waste water from a treatment plant.

Elytron (pl. **elytra**). The hardened and commonly platelike modified fore wing of beetle adults.

Emergence. Transformation to the adult stage.

Emergent. Pertaining to objects or organisms that are partly in water and partly exposed, such as plants that are rooted in water but whose upper parts are aerial or floating.

Emerger pattern. A fisherman's imitation fly patterned after an insect that would be in the act of leaving the water—for example, a caddisfly pupa—and which is usually fished with a rising motion in the water.

Estuarine. Pertaining to the aquatic environment of a coastal stream or river that is subject to the tide of the body of water into which it flows; an area where freshwater and marine waters mix.

Eutrophic. An aquatic environment that is enriched with nutrients, especially nitrogen and phosphorous, and leads to increased production of organic matter.

Eversible. Capable of being protruded outward or turned inside-out.

Exoskeleton. The external body wall of arthropods.

Eyespot. A single eye or eyelike structure found on the head.

Facet. One of the parts or lenslike divisions of a compound (multifaceted) eye.

Facial. Pertaining to the fore part of the head.

Fauna. Animal life, usually in reference to the animal species found within a certain geographic region.

Femur (pl. **femora**). The third major division or part of a leg, immediately distal to the trochanter and preceding the tibia.

Filamentous. Slender and fingerlike or threadlike, often in reference to antennae or gills.

Filter feeder. Generally, an animal that filters its food by means of hairs, specialized mouthparts, catchnets, or other structures.

Fixative. Any chemical agent that fixes or preserves an organism so that it can be used as a study specimen.

Flotsam. Material floating freely on the open ocean surface.

Forage. The act of searching out and eating food.

Free-living. Generally not living in contact with another organism or not encased in a fixed burrow, retreat, or case.

Fry. Small, young fishes.

Fusiform. Streamlined and tapering at both ends or at least posteriorly.

Generalist. An organism that has a broad spectrum of food preferences or is not specific in its feeding habits; an organism with broad ecological preferences.

Generation. A single complete life cycle—for example, from egg to egg or adult to adult.

Genitalia. External sexual organs or structures.

Gravid. Having fertilized eggs.

Gregarious. Living in the company of other individuals.

Hatch. Those insects coming off or from the water, usually corresponding to the metamorphic transition to subimago or adult; also the birth of a larva from the egg.

Hemielytron (pl. **hemielytra**). The modified fore wing of true bugs (Heteroptera), which is composed of a leathery basal portion and a distal membranous portion.

Hemolymph. Generally, the internal body fluids of arthropods.

Herbivore. A plant feeder, specifically of living plant tissue.

Herbivore-detritivore. An animal that feeds on both living and decomposing plant tissue.

Hook plate. A small sclerotized plate that bears tiny hooks, occurring as pairs on the dorsal abdomen of caddisfly pupae.
Host. A plant or animal on which another animal resides, parasitizes, or feeds.
Hydropneustic. Utilizing oxygen dissolved in water for external respiration.
Hydrostatic. Causing or aiding flotation.
Hypertrophy. Excessive growth of a structure via increase in cell or membrane size.
Hyporheic. Pertaining to an underground or subterranean aquatic environment.

Imitation. A fishing lure patterned after a specific organism, usually an aquatic insect.
Immature. The larva or pupa of an insect; also any subadult postembryonic stage.
Incomplete metamorphosis. A kind of developmental process in the life cycle in which no pupal stage precedes the adult.
Incubation. A period during the egg stage necessary for development and subsequent eclosion.
Insectivorous. Pertaining to an animal or plant that feeds on insects.
Instar. The individual insect between two molting events; also includes the individual between egg eclosion and the first larval molt.
Intercalary. A longitudinal vein of various length situated between major, named longitudinal veins of an insect wing.
Intermittent. Occurring temporarily or periodically rather than continuously throughout the year.
Interstitial. Within the bottom substrate or interspersed deeply among bottom substrate particles, as within a stream or lake bed.
Intertidal. Pertaining to that coastal area between low and high tide.

Kelp. Any of various large, coarse seaweeds.
Killing agent. Any substance used to kill; some are also preservatives.

Labium (pl. **labia**). The lower lip or most posterior whole mouthpart of the insect head.
Labrum. The upper lip or most anterior, unpaired mouthpart of the insect head.
Lamella (pl. **lamellae**). A leaflike structure located at the end of the abdomen of damselflies; there are three such lamellae on each individual.
Larva (pl. **larvae**). The first major mobile life stage or form of insects; a developmental stage following egg eclosion, having various numbers of instars, and preceding either the adult stage, pupal stage, or subimaginal stage, depending on the kind of insect.
Lateral. At or toward the sides.
Lateral lobe. A distal part of the labium of Odonata, paired and often possessing variously developed teeth.
Lentic. Pertaining to aquatic environments with nonflowing waters, such as ponds, lakes, and swamps; also to inhabitants of such environments.
Limnetic. Pertaining to the open-water zone of lakes and deep ponds beyond the area where rooted emergent vegetation grows; the zone is no deeper than the depth of light penetration; also pertaining to inhabitants of such environments.
Lingua. A specialized, sclerotized mouthpart of midge larvae.
Littoral. Pertaining to the near-shore zone of ponds and lakes where rooted and emergent plants grow; also to inhabitants of such environments.
Lotic. Pertaining to aquatic environments with flowing waters, such as streams and rivers; also to inhabitants of such environments.

Macroinvertebrate. An animal lacking a backbone and generally visible to the unaided eye or generally larger than 0.5 mm at its geatest dimension.

Macrovore. An insect whose food items are macroscopic or generally larger than 10 cubic microns, although only part of the macroscopic food item may actually be ingested.

Mandible. One of a pair of mouthparts located directly behind the upper lip, or labrum, of the insect head; mandibles are opposable in the unmodified condition.

Marginal. Pertaining to the transitional environmental area between an aquatic environment and a terrestrial one—for example, a shoreline.

Marine. Of or pertaining to the ocean.

Maxilla (pl. **maxillae**). One of a pair of mouthparts located between the mandibles and lower lip, or labium, of the insect head; each maxilla possesses a lateral appendage, or palp, in the unmodified condition.

Medial. At or toward the midline of the body.

Membrane. The distal parchmentlike region of the hemielytra of true bugs, or Heteroptera; or generally any thin covering.

Mesothorax. The second (middle) thoracic segment.

Metathorax. The third and most posterior thoracic segment.

Microbiota. Microscopic organisms.

Microvore. An insect whose food items are microscopic organisms or particles generally less than 10 cubic microns.

Migration. The movement of a population away from and then back to a point of propagation, sometimes requiring more than one generation to complete the migratory cycle.

Miner. An insect that feeds on leaf tissue by tunneling within the leaf itself.

Molt. To shed the exoskeleton and form a new one.

Morphology. The study of the structure and form of organisms; also used with reference to the structure of any particular organism.

Mouth brush. A specialized structure of the head equipped with a brush of hairs and used for filtering or sweeping food particles from water.

Nematocyst. A stinging cell of Coelenterata.

Nocturnal. Active during the night.

Nodus. A strong crossvein near the middle of the anterior margin of Odonata wings.

Notum. The dorsal surface of a body segment, especially of thoracic segments.

Nymph fishing. The use of larval or pupal imitations under water in order to catch fish.

Ocellus (pl. **ocelli**). A simple eye; sets of two or three are located on the head between the compound eyes of some insects.

Omnivore. An animal that eats both plant and animal matter, specifically living matter.

Omnivore-detritivore. An animal that eats living and decomposing plant and animal matter.

Operculate. Functioning as a covering for other structures—for example, the enlarged platelike gills that cover succeeding gills in certain mayfly larvae.

Operculum. A covering of a chamber—for example, the covering of the gill chamber of riffle beetle larvae.

Orb. A more-or-less spherical web.

Organism. Any living individual plant or animal.

Osmoregulation. The maintenance of optimal and necessary internal osmotic concentrations in an animal by regulation of the rate of ion uptake and ion excretion.

Overwinter. To live through a winter or cold climatic season; usually taken to mean in a state of dormancy, but not necessarily.

Oviposition. The act of laying eggs.

Ovipositor. A specialized structure of female adults of certain insects used to lay or deposit eggs.

Palp. An armlike appendage, especially of the maxilla or labium.

Papilla (pl. **papillae**). A thin-walled structure usually found in the caudal region of the larval abdomen of certain insects and used for ion uptake; it may be eversible or expandable, such as anal papillae of mosquito larvae.

Parasitic. Living off another organism for a period of time prior to killing it (or without ever killing it); commonly used with reference to one animal (parasite) living off another animal (host).

Parthenogenetic. Pertains to reproduction without mating or egg fertilization.

Pathogen. Any disease-producing microorganism.

Periphyton. Generally, any benthic plant growth.

Petiole. The stalk by which a leaf is attached to a stem.

Pharate adult. The newly formed adult that still resides within the pupal skin.

Pheromone. A substance given off by one individual that causes a specific reaction by other individuals of the same species, such as a sex attractant.

Physical gill. A bubble of air that is held by an aquatic insect and which, when the insect is submerged in water, acts to take up dissolved oxygen in the water to various degrees.

Physiology. The study of function; also pertains to the function of some particular structure, organ, or organism.

Plankter. A planktonic or suspended aquatic organism.

Plankton. Organisms that live suspended in water and have little or no power of locomotion.

Plant cup. Any depression formed by part of a terrestrial plant in which rainwater accumulates.

Plastron. A thin film of air held by many tiny unwettable hairs or scales, essentially keeping areas of an insect's body dry while it is submerged.

Polymorphism. The condition of having different forms or characters among members of the same stage and species—for example, being either winged or wingless or being differently colored.

Postembryonic. Pertaining to any life history stages that occur after the egg or after embryonic development.

Posterior. At or toward the hind or tail end of the body.

Preanal brush. A small brush of hairs located at the end of a short stalklike structure that occurs dorsally near the end of the abdomen of midge larvae.

Preanal papilla (pl. **papillae**). A small eversible papilla that can be externally protruded from the anus of black fly larvae; or any papilla occurring subterminally at the end of the abdomen.

Preapical. Occurring near the end of a structure.

Predaceous. Feeding by attacking other animals (prey), which are usually killed quickly (the feeder is referred to as a predator).

Prehensile. Adapted for seizing and grasping, as the modified labia of Odonata larvae.

Prey. The animal or animals attacked and eaten by another animal (predator).

Productivity. The production of organic material.

Profundal. Pertaining to the deep-water zone of lakes, below the depth of light penetration; also to inhabitants of such environments.

Proleg. A fleshy, unsegmented, leglike or lobelike structure; usually occurring in pairs and located on the thorax of some Diptera larvae and on the abdomen (usually ventrally or terminally) in many different insect larvae.

Pronotum. The dorsal surface of the first, most anterior thoracic segment.

Prosternal horn. A slender unpaired structure found between the fore legs of some caddisfly larvae.

Prothorax. The first, most anterior thoracic segment.

Pupa (pl. **pupae**). The transitional life history stage that immediately precedes the adult stage in insects with complete metamorphosis.

Puparium (pl. **puparia**). A case formed by a larval skin in which the pupal stage resides; found among certain Diptera.

Pupation. The act of becoming a pupa.

Quadrate. More-or-less square-shaped.

Quiescent. Generally inactive.

Radially symmetrical. Arranged in similar parts around a median vertical axis.

Raptorial. Modified for capturing and holding prey.

Rasping. Scraping.

Rectal chamber. The internal open area in the posterior region of Odonata larvae, opening at the anus and capable of taking in water from the external environment.

Respiratory horn. A tubelike, hornlike, or rodlike structure occurring in pairs on the dorsal thorax of many Diptera pupae, containing or leading to spiracles and used for aeropneustic respiration.

Rheophilic. Found in flowing waters, usually with strong currents.

Riffle. A fast-flowing area of a stream where shallow water races over stones and gravel.

Riparian. Typically occurring or growing along the banks of rivers or streams.

Scale. A flat unicellular outgrowth of the body wall.

Scavenger. A feeder on dead animals.

Sclerotized. Hardened body wall.

Scutellum. In beetle adults, a small dorsal, triangular area found medially at the base of the elytra; in caddisfly adults, a large dorsal area on the mesothorax behind the pronotum.

Scutum. The medial, posterior, and dorsal area of the mesothorax of caddisfly adults.

Secretive. Prone to hide or be hidden from view.

Semiaquatic. Generally, living in nonaquatic environments that are otherwise closely associated with aquatic environments, or living partially in aquatic environments; also pertaining to a wet but nonsubmerged environment.

Sessile. Attached and incapable of actively moving from place to place.

Seston. All of the suspended solid matter in a column of water.

Simple. Unbranched and without arms or appendages.

Silt. Substrate material composed of very small particles.

Spacing hump. A fleshy protuberance found on the first abdominal segment of many case-making caddisfly larvae.

Sphagnum. Any of a number of related mosses found in bogs.

Spinner. A mayfly adult or adult imitation.

Spiracle. An external opening along the body wall of insects used for air intake.

Spiracular disc. A circular disc of lobes that surround a pair of terminal spiracles.

Spiracular gill. A filamentous and branched outgrowth of a spiracle that facilitates hydropneustic respiration, often located on the thorax of Diptera pupae.

Spate. A condition of sudden flooding.

Sprawler. A benthic aquatic insect that generally crawls about bottom substrate in moderately flowing to still waters.

Stage. Any major life form in the metamorphosis of an insect: egg, larva, subimago, pupa, adult.

Sternite. The ventral plate of an abdominal segment.

Stridulatory. Pertaining to the production of sound or vibrations by rubbing two structures or surfaces together.

Subcosta. The second major vein or wing area proceeding from the anterior margin of the insect wing.
Subimago. The fully winged stage that precedes the adult in mayflies.
Submergent. Within water or under the water.
Substrate. Any more-or-less solid surface or surfaces within the aquatic environment; sometimes more specifically the bed or bottom of an aquatic environment.
Succession. The transitional development of ecological communities in a particular environment.
Suspension feeder. An animal that feeds on food suspended in the water column.
Suture. A line or slight furrow in the body wall.
Swarm. An aggregation of several to many individuals, often for the purposes of mating.
Swarm marker. Any object to which a swarm will orient.
Swimming hairs. A series of long hairs on the legs of certain swimming aquatic insects that aid in swimming.
Symbiotic. Living in close association or contact with another species.
Synonym. One of two or more names that apply to the same species.

Tandem. A situation in which a male and female are connected, as in Odonata adults.
Tarsus (pl. **tarsi**). The last or most distal major leg segment, immediately beyond the tibia, usually with attached distal claws, and often subdivided into two to five tarsal segments.
Taxon (pl. **taxa**). Any named group of organisms.
Taxonomy. The study of the classification of organisms; or the classification of organisms.
Terminal. Pertaining to the posterior end of the body or distal end of a structure.
Terrestrial. Living on land or in the air, as opposed to *aquatic;* also pertaining to the land.
Territorial. Being found in and defending a certain area of space from competing individuals, usually for foraging or mating purposes.
Thoracic. Pertaining to the thorax or second (middle) major body region of an insect.
Tibia (pl. **tibiae**). The fourth major segment of the insect leg, immediately distal to the femur and preceding the tarsus.
Tolerance. The degree to which an organism is able to endure certain environmental factors.
Tolerant. Generally pertaining to an organism that is able to tolerate relative extremes in environmental conditions.
Torrential. Pertaining to rapids or "white water."
Tracheal. Pertaining to internal stemlike structures in insects, appearing as trunks, veins, or branches.
Tree hole. Any depression on or in a tree that collects rainwater.
Trochanter. The second major segment of insect legs, immediately distal to the coxa and preceding the femur; the trochanter is sometimes subdivided.
Trochantin. A separate, small, platelike appendage in caddisfly larvae located anteriorly at the base of the fore leg.
Trophic. The relative position in the food web in terms of securing nutrients.
Truncate. Squared off or blunt at the end.
Trunk. The undifferentiated body region posterior to the head and encompassing the remainder of the body.
Tubecase. A case in which larvae and pupae of certain aquatic insects reside, often portable.
Turbinate. Having a differentiated top part; in mayfly eyes, having two very different parts, the top part having dorsal facets only.

Underwing chamber. The area between the dorsal abdomen and folded wings, used to capture or store air.

Valve. Any part that forms or contributes to a shell.
Vascular plant. Any stemmed or rooted plant.
Vector. A carrier of pathogens.
Veinlet. A short vein in insect wings that leads from a major longitudinal vein to a wing margin, particularly in the hind region.
Venter. The undersurface.
Ventilation. In aquatic insects, the act of moving water over a part of the body.
Ventral. Pertaining to the bottom or underside.
Vernal. Temporary or existing only part of the year.

Wet fly. A fisherman's imitation (usually of an aquatic insect) that is fished under the water.
Wing pad. A developing wing or sheath of a developing wing.

Zooplankton. Floating or suspended aquatic animals, usually very small or microscopic.

INDEX

Page numbers in **bold face** indicate principal references to a subject. Page numbers in *italic* indicate references to a figure.